HANDBOOK OF BIOLOGICAL DYES AND STAINS

HANDBOOK OF BIOLOGICAL DYES AND STAINS

SYNTHESIS AND INDUSTRIAL APPLICATIONS

R. W. SABNIS
Pfizer Inc.
Madison, NJ

WILEY
A JOHN WILEY & SONS, INC., PUBLICATION

Published by John Wiley & Sons, Inc., Hoboken, New Jersey
Published simultaneously in Canada

Library of Congress Cataloging-in-Publication Data

ISBN: 978-0-470-40753-0

10 9 8 7 6 5 4 3 2 1

Dedicated to

My Father

Late Mr. Wasudeo S. Sabnis

&

My Mother

Late Mrs. Suhasini W. Sabnis

Contents

Preface

Color has been a fascination for individuals for a long time. The book is intended as a reference guide for dyes used in biology, chemistry, histology, cytology, medicine, microscopy, and all color- and medical-related allied fields.

Even though the use of biological dyes is widespread, it is growing rapidly, and has exploded in the last decade, there is no book available in the market directly on these dyes that provides information, such as CAS registry numbers, safety/toxicity data, and various applications, in one source. Hence, there was a need to publish a book that provided an immediate incentive for compiling the notes to update the scientific community with the wealth of information on biological dyes and stains. The dyestuff literature, particularly on biological dyes and stains, is largely in patents. This book, as a reference handbook, provides systematic and up-to-date library of information on 200 + biological dyes and stains. The book is compiled as a resource guide for biologists, chemists, histologists, cytologists, medical professionals, and nonchemists in industry as well as in university.

Biological dyes and stains are arranged alphabetically by the most commonly used name. Again, the choice of primary name is somewhat arbitrary, but an effort has been made to strike a balance between names that are easily recognizable and names that are chemically informative. The detailed information on each biological dye or stain is covered in the following order: CAS registry number, chemical structure, CA index name, other names, Merck index number (Merck Index 14th Edition, 2006), chemical/dye class, molecular formula, molecular weight, physical form, solubility, melting point, boiling point, pH range, color change at pH, pK_a, absorption (λ_{max}), emission (λ_{max}), synthesis, staining applications, biological applications, industrial applications, safety/toxicity, certification/approval, and references. Where there are discrepancies between different values, the author has used his judgment on selecting the most likely value.

Numerous recent references have been provided on various synthetic methods, staining applications, biological applications, industrial applications, and safety/toxicity data. Space and format limitations prevent giving all the references for each dye. This is the first ever book that provides safety/toxicity data with reference to acute toxicity, aquatic toxicity, carcinogenicity, cytotoxicity, chronic toxicity, ecotoxicity, genotoxicity, hematotoxicity, hepatotoxicity, immunotoxicity, microbial toxicity, mutagenicity, nephrotoxicity, neurotoxicity, nucleic acid damage, oral toxicity, phototoxicity, phytotoxicity, skin toxicity, reproductive toxicity, and so on. The book also provides Biological Stain Commission (BSC)-certified dyes and Food & Drug Administration (FDA)-approved dyes.

Several appendixes have been provided at the end of the book for scientists to conveniently and easily find a dye as per their need. These appendixes include CAS registry numbers, BSC-certified dyes, FDA-approved dyes, metal indicators, nucleic acid stains, organelle probes, and pH indicators.

Omissions as well as errors of fact and interpretation are inevitable in dealing with so vast a subject as biological dyes. I shall be glad to have my attention drawn to errors and to incorporate suggestions for improvement when a revision becomes possible.

I express my profound respect and appreciation to my guru/mentor/advisor, Prof. D. W. Rangnekar, who brought me to this wonderful world of color science in the Dyestuff Technology Department of Mumbai University Institute of Chemical Technology (MUICT), where I laid the foundation stone for my research career in dye chemistry.

It is my pleasure to make grateful acknowledgement to Dr. Alan Fanta, Dr. Ganapati Shankarling, and Dr. Jeffrey Talkington for their extremely useful discussions, encouragement, and inspiration.

Words are inadequate to express my sincere appreciation for my wife Madhuri and daughter Anika. It would not have been possible to write this book without their encouragement and patience. It is a great pleasure to express my gratitude to John Wiley & Sons, Inc. for giving me an opportunity to write this book.

R. W. Sabnis
E-mail: ramsabnis@yahoo.com

About the Author

Ram W. Sabnis is currently a patent agent at Pfizer Inc. in Madison, NJ. His interests include dyes, pigments, organic chemistry, heterocycles, polymers, synthesis, formulations, coatings, biotechnology, medicinal chemistry, medical devices, and patents. Presently, he focuses on drafting and prosecuting U.S. and international patents. He is a registered patent agent with U.S. Patent & Trademark Office (USPTO) and is also the inventor of more than 50 U.S. and international patents (issued/published). Before entering the legal (patents) field, he was a research chemist for Ascadia, General Electric, Brewer Science, U.S. Textiles, and Invitrogen in the United States. He had also worked as a patent agent at Squire, Sanders & Dempsey L.L.P. in San Francisco, CA.

Dr. Sabnis was born and raised in Mumbai, India. He received his M.Sc. in Organic Chemistry from University of Mumbai and Ph.D. in Organic Chemistry (Dyes) from University Institute of Chemical Technology (UICT), University of Mumbai, India. He received FAIC from American Institute of Chemists, USA. He was awarded CCol FSDC (Chartered Colourists, Fellow of Society of Dyers & Colourists), Society of Dyers & Colourists, UK.

Dr. Sabnis is one of the world's foremost experts in dyes, inventing world's first colored bubbles (nonstaining) and color changing dye system with many applications. He has more than 25 years of industrial and academic research experience in dye chemistry, particularly, dyes for biomedical (fluorescent probes), personal care products, health/beauty products, displays, inks, paints, plastics, textiles, and toys. He has over 150 publications that include books, book chapters, encyclopedia chapters, patents, reviews, papers, and symposium presentations. Dr. Sabnis is the recipient of Perkin Innovation Award by Society of Dyers & Colourists (SDC), UK; Grand Innovation Award by Popular Science, USA; Six Sigma Green Belt & Competitive Spirit Award by GE, USA; and Best Doctoral Thesis Award by University of Mumbai, India. He will continue to focus his activities on fascinating dye chemistry as well as demanding intellectual property in the years to come.

ACID FUCHSIN

CAS Registry Number 3244-88-0

Chemical Structure

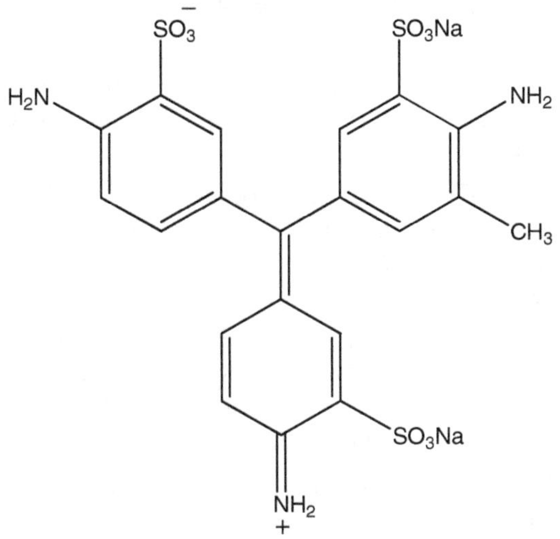

CA Index Name Benzenesulfonic acid, 2-amino-5-[(4-amino-3-sulfophenyl)(4-imino-3-sulfo-2,5-cyclohexadien-1-ylidene)methyl]-3-methyl-, sodium salt

Other Names Benzenesulfonic acid, 2-amino-5-[(4-amino-3-sulfophenyl)(4-imino-3-sulfo-2,5-cyclohexadien-1-ylidene)methyl]-3-methyl-, disodium salt; C.I. Acid Violet 19; C.I. Acid Violet 19, disodium salt; Rubine S; Acid Fuchsine; Acid Fuchsine FB; Acid Fuchsine N; Acid Fuchsine O; Acid Fuchsine S; Acid Leather Magenta A; Acid Magenta; Acid Magenta O; Acid Violet 19; Acid fuchsin sodium salt; Acid rosein; Acid rubin; Acidal Fuchsine; Acidal Magenta; Albion Acid Magenta; Andradeindicator; C.I. 42685; Fuchine Acid Photo Grade; Fuchsin S; Fuchsin acid; Fuchsine acid; Kiton Magenta A; Triacid Magenta; *p*-Fuchsine acid

Merck Index Number 107

Chemical/Dye Class Triphenylmethane

Molecular Formula $C_{20}H_{17}N_3Na_2O_9S_3$

Molecular Weight 585.54

Physical Form Olive to dark olive-green crystals or powder

Solubility Very soluble in water; slightly soluble to insoluble in ethanol; insoluble in xylene

Melting Point >250 °C

pH Range 12.0–14.0

Color Change at pH Red (12.0) to colorless (14.0)

Absorption (λ_{max}) 546 nm

Emission (λ_{max}) 630 nm

Synthesis Synthetic methods[1–3]

Staining Applications Antigen;[4] bacteria;[5] collagen;[6] fungi;[7] fats;[8] neurons;[9,10] paraffin sections;[11] proteins;[8] starch;[8] processed food;[12] tumor cells;[1,13] decayed teeth;[14] lips;[1,15,16] hairs[1,17]

Biological Applications Detecting enzyme activity,[18] proteins,[19] tumor cells[1,13]

Industrial Applications Color filter;[1,20] recording material;[1,21] photographic film;[22] inks;[1,23] highlighters;[1,24] paints;[1,23] explosives;[1,25] corrosion inhibitors;[1,26] leathers;[27] textiles[1,28]

Safety/Toxicity Acute oral toxicity;[1,29] genotoxicity;[1,30] neurotoxicity[1,31,32]

Certification/Approval Certified by Biological Stain Commission (BSC)

REFERENCES

1. Sabnis, R. W. *Handbook of Acid–Base Indicators*; CRC Press: Boca Raton, **2008**; pp 6–7.
2. Kazymov, A. V.; Skryabneva, T. N.; Valeeva, A. M.; Abramenko, P. I.; Voronin, E. K.; Kazymova, M. A.; Khakimov, M. G.; Gil'fanov, M. M. Method of preparing acid fuchsin dye. Russ. RU 2126431, 1999; *Chem. Abstr.* **2000**, *133*, 194631.
3. Fierz, H. E.; Koechlin, H. Study of Heumann's fuchsine synthesis. *Helv. Chim. Acta* **1918**, *1*, 218–226; *Chem. Abstr.* **1919**, *13*, 15880.
4. Mamat, H.; Rummel, C.; Mamat, R.; Beutelspacher, R.; Kroh, G. Preparation of stained bacterial antigen reagents for immunoassays. Ger. (East) DD 252686, 1987; *Chem. Abstr.* **1989**, *110*, 228166.
5. Utsugi, R. Adhesive dressing with indicators to determine bacteria growth. U.S. Pat. Appl. Publ. US 2004044299, 2004; *Chem. Abstr.* **2004**, *140*, 223388.
6. Nielsen, L. F.; Moe, D.; Kirkeby, S.; Garbarsch, C. Sirius red and acid fuchsin staining mechanisms. *Biotech. Histochem.* **1998**, *73*, 71–77.

7. Mehta, P.; Graham, M.; Pomerantz, A. Method for staining biological specimens by combining stable reagents on a microscope slide to make an unstable staining solution. PCT Int. Appl. WO 9963342, 1999; *Chem. Abstr.* **1999**, *132*, 20777.

8. Maeda, T.; Yamada, S.; Taketani, K. Fluorescent image analysis of food dispersion three-dimensional structure. Jpn. Kokai Tokkyo Koho JP 2002323449, 2002; *Chem. Abstr.* **2002**, *137*, 324537.

9. Zsombok, A.; Toth, Z.; Gallyas, F. Basophilia, acidophilia and argyrophilia of "dark" (compacted) neurons during their formation, recovery or death in an otherwise undamaged environment. *J. Neurosci. Methods* **2005**, *142*, 145–152.

10. Victorov, I. V.; Prass, K.; Dirnagl, U. Improved selective, simple, and contrast staining of acidophilic neurons with vanadium acid fuchsin. *Brain Res. Protoc.* **2000**, *5*, 135–139.

11. Kiernan, J. A. Staining paraffin sections without prior removal of the wax. *Biotech. Histochem.* **1996**, *71*, 304–310.

12. Maeda, T. Method of forming tissue structure image of proces-sed food or raw material using fluorescent dye. PCT Int. Appl. WO 2008078752, 2008; *Chem. Abstr.* **2008**, *149*, 78340.

13. Steven, F. Screening technique for detecting tumor cells by fluorescence of isoenzyme marker. PCT Int. Appl. WO 9616334, 1996; *Chem. Abstr.* **1996**, *125*, 108863.

14. Fukunishi, K.; Hino, K. Antibacterial solutions for dental caries staining. Jpn. Kokai Tokkyo Koho JP 10236914, 1998; *Chem. Abstr.* **1998**, *129*, 221228.

15. Blin, X. Cosmetic film products for lips. PCT Int. Appl. WO 2007007294, 2007; *Chem. Abstr.* **2007**, *146*, 148465.

16. Legendre, J. Y.; Filippi, V. Anhydrous film for make-up or lip care. Fr. Demande FR 2865130, 2005; *Chem. Abstr.* **2005**, *143*, 138653.

17. Takahashi, T.; Kurita, N.; Nakamura, T. Integral hair treatment and dyeing compositions. Jpn. Kokai Tokkyo Koho JP 10029919, 1998; *Chem. Abstr.* **1998**, *128*, 208772.

18. Lawrence, D. S.; Sharma, V.; Agnes, R. S. Deeply quenched enzyme sensors and protein–protein binding sensors. PCT Int. Appl. WO 2008070152, 2008; *Chem. Abstr.* **2008**, *149*, 26898.

19. Toyonaga, M.; Watanabe, M. Composition for the detection of protein. Jpn. Kokai Tokkyo Koho JP 61164158, 1986; *Chem. Abstr.* **1987**, *106*, 192294.

20. Galves, J. P.; Gomez, G. A color filter for photosensors. Eur. Pat. Appl. EP 65465, 1982; *Chem. Abstr.* **1983**, *98*, 117132.

21. Ikeda, T. Heat-developable diazo recording material containing anionic dye. Jpn. Kokai Tokkyo Koho JP 2004106340, 2004; *Chem. Abstr.* **2004**, *140*, 312091.

22. McGuckin, H. G.; Cole, R. M. Photographic film. Ger. Offen. DE 2343561, 1974; *Chem. Abstr.* **1974**, *81*, 71104.

23. Miller, R. E. Color-changing compositions for inks and paints. U.S. Patent 5232494, 1993; *Chem. Abstr.* **1994**, *120*, 56822.

24. Davies-Smith, L.; Sum, V. W. Highlightable and highlighted mixtures, marking instruments, eradicator solution and kits, using the mixtures in highlighting. U.S. Pat. Appl. Publ. US 2005120919, 2005; *Chem. Abstr.* **2005**, *143*, 9331.

25. Rao, V. M.; Singh, A.; Srivastava, D.; Seshan, S. Production of slurried or emulsion explosive composition. Indian IN 167782, 1990; *Chem. Abstr.* **1992**, *117*, 236744.

26. Talati, J. D.; Gandhi, D. K. Some dyes as corrosion inhibitors for aluminum-copper alloy in higher concentrations of hydrochloric acid. *Indian J. Technol.* **1985**, *23*, 232–236.

27. Grall, F.; Gardere, H. Lightfastness of dyed leathers. *Technicuir* **1978**, *12*, 14–18; *Chem. Abstr.* **1978**, *89*, 112457.

28. Deeds, S. M. Fabric repair coloring device and using the device to color unwanted bleached out spots on fabric. U.S. Patent 6739779, 2004; *Chem. Abstr.* **2004**, *140*, 424935.

29. Schafer, E. W., Jr.; Bowles, W. A., Jr.; Hurlbut, J. The acute oral toxicity, repellency, and hazard potential of 998 chemicals to one or more species of wild and domestic birds. *Arch. Environ. Contam. Toxicol.* **1983**, *12*, 355–382.

30. Hagiwara, M.; Watanabe, E.; Barrett, J. C.; Tsutsui, T. Assessment of genotoxicity of 14 chemical agents used in dental practice: ability to induce chromosome aberrations in Syrian hamster embryo cells. *Mutat. Res.* **2006**, *603*, 111–120.

31. Lee, J. Y.; Hwang, J. J.; Park, M. H.; Koh, J. Y. Cytosolic labile zinc: a marker for apoptosis in the developing rat brain. *Eur. J. Neurosci.* **2006**, *23*, 435–442.

32. Lee, J.; Kim, D.; Hong, H.; Han, S.; Kim, J. Protective effect of etomidate on kainic acid-induced neurotoxicity in rat hippocampus. *Neurosci. Lett.* **2000**, *286*, 179–182.

ACRIDINE HOMODIMER

CAS Registry Number 57576-49-5

Chemical Structure

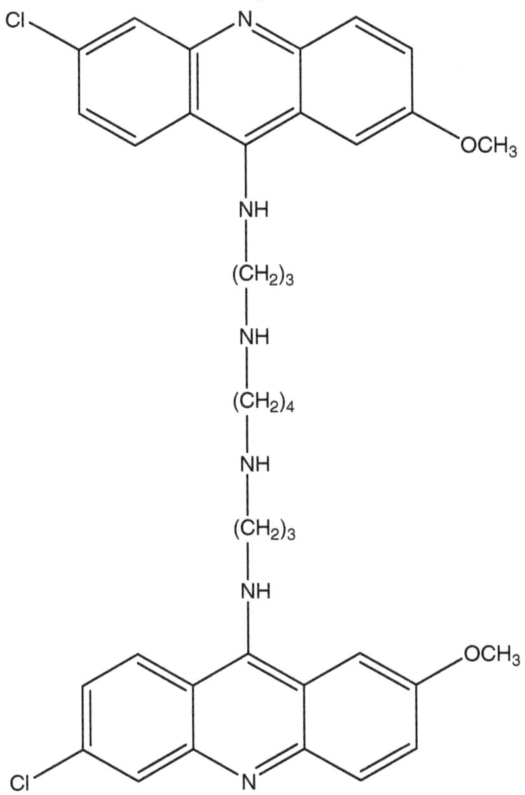

CA Index Name 1,4-Butanediamine, *N,N'*-bis[3-[(6-chloro-2-methoxy-9-acridinyl)amino]propyl]-

Other Names Acridine homodimer; NSC 219743

Merck Index Number Not listed

Chemical/Dye Class Acridine

Molecular Formula $C_{38}H_{42}Cl_2N_6O_2$

Molecular Weight 685.69

Physical Form Orange-brown powder

Solubility Soluble in *N,N*-dimethylformamide, dimethyl sulfoxide

Melting Point 169–170 °C

Boiling Point (Calcd.) 885.4 ± 65.0 °C, pressure: 760 Torr

pK$_a$ (Calcd.) 10.63 ± 0.19, most basic, temperature: 25 °C

Absorption (λ_{max}) 431 nm

Emission (λ_{max}) 498 nm

Synthesis Synthetic methods[1–4]

Staining Applications Nucleic acids;[3–8] chromosomes[9]

Biological Applications Diagnosis of tissue necrosis;[10] treating malformed proteins causing neurodegenerative disease[11]

Industrial Applications Not reported

Safety/Toxicity Neurotoxicity[11]

REFERENCES

1. May, B. C. H.; Fafarman, A. T.; Hong, S. B.; Rogers, M.; Deady, L., W.; Prusiner, S. B.; Cohen, F. E. Potent inhibition of scrapie prion replication in cultured cells by bis-acridines. *Proc. Natl. Acad. Sci. U.S.A.* **2003**, *100*, 3416–3421.
2. Canellakis, E. S.; Shaw, Y. H.; Hanners, W. E.; Schwartz, R. A. Diacridines: bifunctional intercalators. I. Chemistry, physical chemistry and growth inhibitory properties. *Biochim. Biophys. Acta* **1976**, *418*, 277–289.
3. Barbet, J.; Roques, B. P.; Le Pecq, J. B. Compounds from polyintercalating DNA. Synthesis of acridine dimers. *Compt. Rend. Seances Acad. Sci., Ser. D* **1975**, *281*, 851–853.
4. Le Pecq, J. B.; Le Bret, M.; Barbet, J.; Roques, B. DNA polyintercalating drugs. DNA binding of diacridine derivatives. *Proc. Natl. Acad. Sci. U.S.A.* **1975**, *72*, 2915–2919.
5. Markovits, J.; Garbay-Jaureguiberry, C.; Roques, B. P.; Le Pecq, J. B. Acridine dimers: influence of the intercalating ring and of the linking-chain nature on the equilibrium and kinetic DNA-binding parameters. *Eur. J. Biochem.* **1989**, *180*, 359–366.
6. Bottiroli, G.; Giordano, P.; Prosperi, E. Fluorescent probes in nucleic acid research. *Acta Histochem., Suppl.* **1982**, *26*, 189–194.
7. Bottiroli, G.; Giordano, P.; Doglia, S.; Cionini, P. G. Employment of bis-intercalating dyes for the "*in situ*" study of DNA composition. *Basic Appl. Histochem.* **1979**, *23*, 59–63.
8. Le Bret, M.; Le Pecq, J. B.; Barbet, J.; Roques, B. P. A reexamination of the problem of resonance energy transfer between DNA intercalated chromophores using bis-intercalating compounds. *Nucleic Acids Res.* **1977**, *4*, 1361–1379.

9. Van de Sande, J. H.; Lin, C. C.; Deugau, K. V. Clearly differentiated and stable chromosome bands produced by a spermine bis-acridine, a bifunctional intercalating analog of quinacrine. *Exp. Cell Res.* **1979**, *120*, 439–444.

10. Mills, R. L. Pharmaceuticals and apparatus based on Moessbauer isotopic resonant absorption of γ emission (MIRAGE) providing diagnosis and selective tissue necrosis. Can. Pat. Appl. CA 2005039, 1991; *Chem. Abstr.* **1992**, *116*, 37120.

11. Prusiner, S. B.; Korth, C.; May, B. C. H. Cyclic bis-compounds clearing malformed proteins. U.S. Pat. Appl. Publ. US 2004229898, 2004; *Chem. Abstr.* **2004**, *141*, 406119.

ACRIDINE ORANGE

CAS Registry Number 65-61-2

Chemical Structure

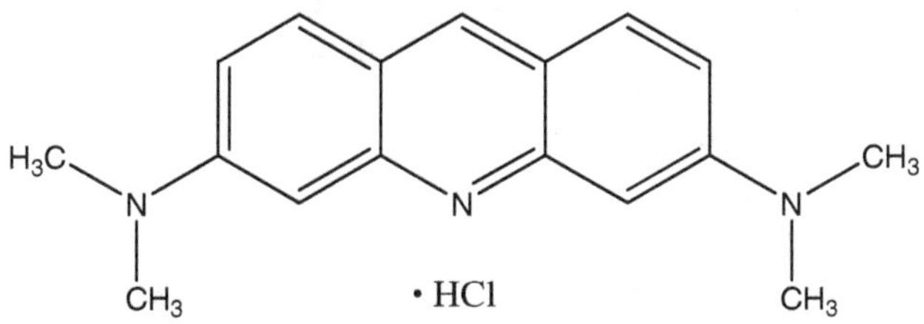

CA Index Name 3,6-Acridinediamine, *N*3,*N*3,*N*6,*N*6-tetramethyl-, hydrochloride (1 : 1)

Other Names 3,6-Acridinediamine, *N*,*N*,*N*′,*N*′-tetramethyl-, monohydrochloride; Acridine Orange R; Acridine, 3,6-bis(dimethylamino)-, hydrochloride; Acridine, 3,6-bis(dimethylamino)-, monohydrochloride; 3,6-Bis (dimethylamino)acridine hydrochloride; Acridine Orange; Acridine Orange N; Acridine Orange NO; Acridine Orange NS; Basic Orange 14; Basic Orange 3RN; C.I. 46005; C.I. Basic Orange 14; Euchrysine 3R; Rhoduline Orange NO; Sumitomo Acridine Orange NO; Sumitomo Acridine Orange RK conc

Merck Index Number Not listed

Chemical/Dye Class Acridine

Molecular Formula $C_{17}H_{20}ClN_3$

Molecular Weight 301.81

Physical Form Burnt orange to brick-red powder

Solubility Soluble in water, ethanol, dimethyl sulfoxide

Melting Point 284–287 °C

pK$_a$ −3.2, 10.5

Absorption (λ_{max}) 500 nm

Emission (λ_{max}) 526 nm

Synthesis Synthetic methods[1–8]

Staining Applications Lysosomes;[9] Golgi apparatus;[9] acidic compartments;[9]secretory granules (SGs);[10] synaptic-like microvesicles (SLMVs);[10] liposomes;[10] acidic organelle membranes;[10] bacteria;[11] bacterial endospores;[12] apoptotic cells;[13] blood smears;[14] nucleic acids;[15–17] cells;[18] malignant musculoskeletal tumors;[19] micronucleus;[20] microorganisms;[21,22] nuclei;[23] peptides;[24] proteins;[24] antibodies;[24] parasites;[25] sperms[26]

Biological Applications Detecting cancer cells,[27,28] spores,[28] human papilloma virus (HPV),[29] stress biomarkers;[30] treating amyloid associated diseases,[31] lupus,[32] pathogen infections;[33] apoptosis assays;[34] cytotoxicity assays[35]

Industrial Applications Thin films;[36] wiring boards[37]

Safety/Toxicity Carcinogenicity;[38,39] cytotoxicity;[40] DNA damage;[41] embryotoxicity;[42] genotoxicity;[43,44] mutagenicity;[45,46] phototoxicity[47]

REFERENCES

1. Glushko, V. N.; Parbuzina, I. L.; Petrova, G. S. Acridine orange hydrochloride. *Khim. Promyshl., Ser.: Reakt. Osobo Chistye Veshch.* **1980**, 3–4; *Chem. Abstr.* **1981**, *94*, 32154.
2. Glushko, V. N.; Parbuzina, I. L.; Petrova, G. S. 3,6-Bis (dimethylamino)acridine hydrochloride (acridine orange). U.S.S.R. SU 694525, 1979; *Chem. Abstr.* **1980**, *92*, 43278.
3. Zanker, V. Methods and results of dye purification: determination of dye content of different technical acridine orange products. *Acta Histochem., Suppl.* **1973**, *13*, 281–289.
4. Saunders, A. M. Absorption and fluorescence spectra of euchrysine GGNX and acridine orange: effects of heparin, sodium chloride, and a cation exchange resin. *Stain Technol.* **1968**, *43*, 207–215.
5. Tomita, G. Molecular complexes of acridine orange and nucleosides. *Biophysik* **1967**, *4*, 118–128.
6. Ratajak, E. J.; Nubel, R. C. Oleandomycin. U.S. Patent 2842481, 1958; *Chem. Abstr.* **1958**, *52*, 107566.
7. Albert, A. Acridine syntheses and reactions. III. Synthesis of aminoacridines from formic acid and amines. *J. Chem. Soc.* **1947**, 244–250.
8. Karr, A. E. Acridine oranges. *Text. Colorist* **1940**, *62*, 604–607, 634, 676–679, 763–767, 836–837, 852.
9. Krolenko, S. A.; Adamyan, S. Ya.; Belyaeva, T. N.; Mozhenok, T. P. Acridine orange bioaccumulation in acid organelles of normal and vacuolated frog skeletal muscle fibres. *Cell Biol. Int.* **2006**, *30*, 933–939.
10. Wang, R.; Hosaka, M.; Han, L.; Yokota-Hashimoto, H.; Suda, M.; Mitsushima, D.; Torii, S.; Takeuchi, T. Molecular probes for sensing the cholesterol composition of subcellular organelle membranes. *Biochim. Biophys. Acta* **2006**, *1761*, 1169–1181.

11. Meseguer, M.; de Rafael, L.; Baquero, M.; Martinez, F. M; Lopez-Brea, M. Acridine orange stain in the early detection of bacteria in blood cultures. *Eur. J. Clin. Microbiol.* **1984**, *3*, 113–115.

12. Schichnes, D.; Nemson, J. A.; Ruzin, S. E. Fluorescent staining method for bacterial endospores. *Microscope* **2006**, *54*, 91–93.

13. Kim, C. S.; Kim, C. S.; Kim, H. S.; Lim, Y. K.; Park, S. Y.; Yang, K. H. Rapid determination method of apoptotic cells in large quantity samples using acridine orange dye staining method. Repub. Korean Kongkae Taeho Kongbo KR 2005005250, 2005; *Chem. Abstr.* **2006**, *145*, 244441.

14. Sciotto, C. G.; Lauer, B. A.; White, W. L.; Istre, G. R. Detection of *Borrelia* in acridine orange-stained blood smears by fluorescence microscopy. *Arch. Pathol. Lab. Med.* **1983**, *107*, 384–386.

15. Exner, M.; Rogers, A. Methods for identifying nucleic acids and determining melting temperature using multiple fluorophores and FRET. U.S. Pat. Appl. Publ. US 2007172836, 2007; *Chem. Abstr.* **2007**, *147*, 182489.

16. Martins, C. F.; Dode, M. N.; Bao, S. N.; Rumpf, R. The use of the acridine orange test and the TUNEL assay to assess the integrity of freeze-dried bovine spermatozoa DNA. *Genet. Mol. Res.* **2007**, *6*, 94–104.

17. Lai, S.; Chang, X.; Tian, L.; Wang, S.; Bai, Y.; Zhai, Y. Fluorometric determination of DNA using nano-SiO_2 particles as an effective dispersant and stabilizer for acridine orange. *Microchim. Acta* **2007**, *156*, 225–230.

18. Lacombe, F.; Belloc, F.; Veriac, S.; Lefevre, D. Method of discrimination of at least two cellular populations by immunofluorescence flow cytometry and application to oncopathology. Fr. Demande FR 2895087, 2007; *Chem. Abstr.* **2007**, *147*, 93934.

19. Matsubara, T.; Kusuzaki, K.; Matsumine, A.; Shintani, K.; Satonaka, H.; Uchida, A. Acridine orange used for photodynamic therapy accumulates in malignant musculoskeletal tumors depending on pH gradient. *Anticancer Res.* **2006**, *26*, 187–193.

20. Nersesyan, A.; Kundi, M.; Atefie, K.; Schulte-Hermann, R.; Knasmueller, S. Effect of staining procedures on the results of micronucleus assays with exfoliated oral mucosa cells. *Cancer Epidemiol., Biomarkers Prev.* **2006**, *15*, 1835–1840.

21. Noda, N.; Mizutani, T. Microorganism-measuring method using multiple staining. Jpn. Kokai Tokkyo Koho JP 2006340684, 2006; *Chem. Abstr.* **2006**, *146*, 77563.

22. Martens-Habbena, W.; Sass, H. Sensitive determination of microbial growth by nucleic acid staining in aqueous suspension. *Appl. Environ. Microbiol.* **2006**, *72*, 87–95.

23. Horobin, R. W.; Stockert, J. C.; Rashid-Doubell, F. Fluorescent cationic probes for nuclei of living cells: why are they selective? A quantitative structure–activity relations analysis. *Histochem. Cell Biol.* **2006**, *126*, 165–175.

24. Hahn, K. M.; Toutchkine, A.; Muthyala, R.; Kraynov, V.; Bark, S. J.; Burton, D. R.; Chamberlain, C. Labeled peptides, proteins and antibodies and processes and intermediates useful for their preparation. U.S. Pat. Appl. Publ. US 2002055133, 2002; *Chem. Abstr.* **2002**, *136*, 366139.

25. Rickman, L. S.; Long, G. W.; Oberst, R.; Cabanban, A.; Sangalang, R.; Smith, J. I.; Chulay, J. D.; Hoffman, S. L. Rapid diagnosis of malaria by acridine orange staining of centrifuged parasites. *Lancet* **1989**, *1*, 68–71.

26. Chohan, K. R.; Griffin, J. T.; Lafromboise, M.; De Jonge, C. J.; Carrell, D. T. Comparison of chromatin assays for DNA fragmentation evaluation in human sperm. *J. Androl.* **2006**, *27*, 53–59.

27. Chang, T. C.; Chang, C. C.; Kang, C. C.; Cheng, J. Y. Cancer cell detecting devices. U.S. Pat. Appl. Publ. US 2007098233, 2007; *Chem. Abstr.* **2007**, *146*, 457991.

28. Levy, M.; Ellington, A. D.; Pai, S. Proximity ligation assays with peptide conjugate 'burrs' and aptamers for the sensitive detection of spores and cancer cells. PCT Int. Appl. WO 2007027796, 2007; *Chem. Abstr.* **2007**, *146*, 312239.

29. Poetter, K.; Gould, T. Human papilloma virus (HPV) detection using nucleic acid probes, microbeads, and fluorescence-activated cell sorter (FACS). PCT Int. Appl. WO 2006060872, 2006; *Chem. Abstr.* **2006**, *145*, 41223.

30. Cullum, M. E.; Duplessis, C. A.; Crepeau, L. J. Method for the detection of stress biomarkers including cortisol by fluorescence polarization. U.S. Pat. Appl. Publ. US 2006105397, 2006; *Chem. Abstr.* **2006**, *144*, 463810.

31. Yang, J.; Inbar, P. Compounds and method for the diagnosis and treatment of amyloid associated diseases. PCT Int. Appl. WO 2007011834, 2007; *Chem. Abstr.* **2007**, *146*, 177223.

32. Kallick, C. A. Specific bacterial inclusions in bone marrow cells indicate systematic lupus erythematosus, and treatment for lupus. U.S. Pat.

Appl. Publ. US 2007031820, 2007; *Chem. Abstr.* **2007**, *146*, 201609.

33. Gazit, E.; Cherny, I. Use of anti-amyloid agents for treating and typing pathogen infections. PCT Int. Appl. WO 2006006172, 2006; *Chem. Abstr.* **2006**, *144*, 121763.
34. Baskic, D.; Popovic, S.; Ristic, P.; Arsenijevic, N. N. Analysis of cycloheximide-induced apoptosis in human leukocytes: fluorescence microscopy using annexin V/propidium iodide versus acridine orange/ethidium bromide. *Cell Biol. Int.* **2006**, *30*, 924–932.
35. Odawara, K. Visible light cytotoxicity expression ability assay method, and its use. Jpn. Kokai Tokkyo Koho JP 2007143465, 2007; *Chem. Abstr.* **2007**, *147*, 3418.
36. Li, F.; Pfeiffer, M.; Werner, A.; Harada, K.; Leo, K.; Hayashi, N.; Seki, K.; Liu, X.; Dang, X. D. Acridine orange base as a dopant for n doping of C60 thin films. *J. Appl. Phys.* **2006**, *100*, 023716/1–023716/9.
37. Mori, K.; Ito, K. Method for pattern formation of metal deposition layers, and manufacture of wiring boards. Jpn. Kokai Tokkyo Koho JP 2007177322, 2007; *Chem. Abstr.* **2007**, *147*, 155374.
38. Kowalski, L. A.; Laitinen, A. M.; Martazavi-Asl, B.; Wee, R. K. H.; Erb, H. E.; Assi, K. P.; Madden, Z. *In vitro* determination of carcinogenicity of sixty-four compounds using a bovine papilloma virus DNA-carrying C3H/10T1/2 cell line. *Environ. Mol. Mutagen.* **2000**, *35*, 300–311.
39. Heil, J.; Reifferscheid, G. Detection of mammalian carcinogens with an immunological DNA synthesis-inhibition test. *Carcinogenesis* **1992**, *13*, 2389–2394.
40. Chang, Y. S.; Wu, C. L.; Tseng, S. H.; Kuo, P. Y.; Tseng, S. Y. Cytotoxicity of triamcinolone acetonide on human retinal pigment epithelial cells. *Invest. Ophthalmol. Vis. Sci.* **2007**, *48*, 2792–2798.
41. McCarroll, N. E.; Piper, C. E.; Keech, B. H. An *E. coli* microsuspension assay for the detection of DNA damage induced by direct-acting agents and promutagens. *Environ. Mutagen.* **1981**, *3*, 429–444.
42. Kohler, M.; Kundig, A.; Reist, H. W.; Michel, C. Modification of *in vitro* mouse embryogenesis by X-rays and fluorochromes. *Radiat. Environ. Biophys.* **1994**, *33*, 341–351.
43. He, L.; Jurs, P. C.; Custer, L. L.; Durham, S. K.; Pearl, G. M. Predicting the genotoxicity of polycyclic aromatic compounds from molecular structure with different classifiers. *Chem. Res. Toxicol.* **2003**, *16*, 1567–1580.
44. Fernandez, M.; Gauthier, L.; Jaylet, A. Use of newt larvae for *in vivo* genotoxicity testing of water: results on 19 compounds evaluated by the micronucleus test. *Mutagenesis* **1989**, *4*, 17–26.
45. Lee, I. E.; Nguyen, V. C.; Hayase, F.; Kato, H. Desmutagenicity of melanoidins against various kinds of mutagens and activated mutagens. *Biosci., Biotechnol., Biochem.* **1994**, *58*, 18–23.
46. Klopman, G.; Frierson, M. R.; Rosenkranz, H. S. The structural basis of the mutagenicity of chemicals in *Salmonella typhimurium*: the Gene-Tox database. *Mutat. Res.* **1990**, *228*, 1–50.
47. Lukasiak-Bachurzewska, B.; Dulczewska-Klopotowska, M. Studies on the phototoxic properties of some coal derivatives. *Przeglad Dermatol.* **1981**, *68*, 33–37; *Chem. Abstr.* **1981**, *95*, 19097.

ALCIAN BLUE 8GX

CAS Registry Number 33864-99-2

Chemical Structure

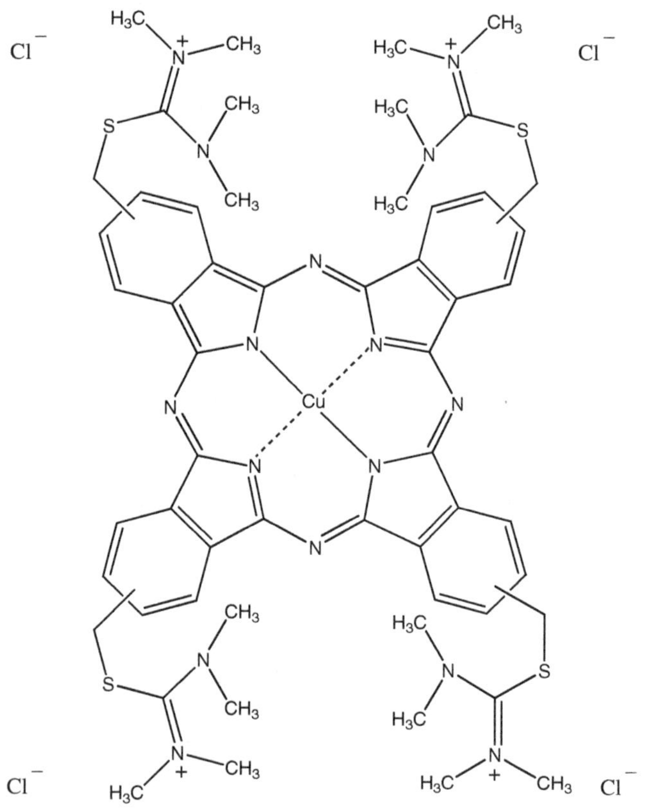

CA Index Name Copper(4 +), [[*N*,*N'*,*N''*,*N'''*-[(29*H*,31*H*-phthalocyanine-*C*,*C*,*C*,*C*-tetrayl-κ*N*29,κ*N*30, κ*N*31,κ*N*32)tetrakis[methylenethio[(dimethylamino)methylidyne]]]tetrakis[*N*-methylmethanaminiumato]](2-)]-, chloride (1 : 4)

Other Names Alcian Blue; Alcian Blue 8G; Alcian Blue 8GN; Alcian Blue 8GS; Alcian Blue 8GX; C.I. 74240; C.I. Ingrain Blue 1; Chloromethylated copper phthalocyanine-thiourea reaction products; Copper (4 +), [[*N*,*N'*,*N''*,*N'''*-[(29*H*,31*H*-phthalocyaninetetrayl-κ *N*29,κ*N*30,κ*N*31,κ*N*32)tetrakis[methylenethio[(dimethylamino)methylidyne]]]tetrakis[*N*-methylmethanaminiumato]](2-)]-, tetrachloride; Copper(4 +), [[*N*,*N'*,*N''*,*N'''*-[29*H*,31*H*-phthalocyaninetetrayltetrakis[methylenethio[(dimethylamino)methylidyne]]]tetrakis[*N*-methylmethanaminiumato]](2-)-*N*29,*N*30,*N*31,*N*32]-, tetrachloride; Methanaminium, *N*,*N'*,*N''*,*N'''*-[29*H*,31*H*-phthalocyaninetetrayltetrakis[methylenethio[(dimethylamino)methylidyne]]]tetrakis[*N*-methyl-, copper complex

Merck Index Number 218

Chemical/Dye Class Copper phthalocyanine

Molecular Formula $C_{56}H_{68}Cl_4CuN_{16}S_4$

Molecular Weight 1298.93

Physical Form Dark blue-violet powder

Solubility Soluble in water, ethanol

Melting Point 148 °C

Absorption (**λ_{max}**) 615 nm

Synthesis Synthetic methods[1–6]

Staining Applications Bonghan threads inside lymphatic vessels;[7] chromatin;[8,9] glycoproteins;[10] glycosaminoglycans;[11] lung goblet cells;[12] neurons;[13] nuclear basic proteins;[14] surface coat of cells;[15] urine sediments;[16]

Biological Applications Glycoconjugates assay;[2] hyaluronic acid assay;[17] hybridization assay;[18] cell agglutionation;[19,20] detecting humic acid,[21,22] detecting pancreatic β-granules;[23] locomotion of amoeba,[24] quantitative determination of glycosaminoglycans,[25,26]

Industrial Applications Inks;[4] leather products;[27] textiles[28]

Safety/Toxicity No data available

Certification/Approval Certified by Biological Stain Commission (BSC)

REFERENCES

1. Yang, Z.; Lu, Z.; Xie, M. Synthesis of alcian blue. *Huaxue Yanjiu Yu Yingyong* 2002, *14*, 222–224; *Chem. Abstr.* **2002**, *138*, 355185.
2. Rees-Milton, K. J.; Anastassiades, T. P. High-throughput quantitation of metabolically labeled anionic glyco-conjugates by scintillation proximity assay utilizing binding to cationic dyes. *Methods Mol. Biol.* **2006**, *347*, 267–275.
3. Carrado, K. A.; Forman, J. E.; Botto, R. E.; Winans, R. E. Incorporation of phthalocyanines by cationic and anionic clays via ion exchange and direct synthesis. *Chem. Mater.* **1993**, *5*, 472–478.
4. Ide, Y.; Hikosaka, M.; Kawamura, T. Manufacture of pigments with high coloring power and gloss for inks and coatings. Jpn. Kokai Tokkyo Koho JP 02102272, 1990; *Chem. Abstr.* **1990**, *113*, 193531.

5. Scott, J. E. Histochemistry of Alcian blue. III. Molecular biological basis of staining by Alcian blue 8GX and analogous phthalocyanins. *Histochemie* **1972**, *32*, 191–212.

6. Herovici, C. Mucopolychrome: demonstration of acidic mucopolysaccharides. *Pathol. Biol., Semaine Hop.* **1960**, *8*, 493–497; *Chem. Abstr.* **1961**, *55*, 54781.

7. Lee, C.; Seol, S. K.; Lee, B. C.; Hong, Y. K.; Je, J. H.; Soh, K. S. Alcian blue staining method to visualize Bonghan threads inside large caliber lymphatic vessels and X-ray micro-tomography to reveal their microchannels. *Lymph. Res. Biol.* **2006**, *4*, 181–190.

8. Stockert, J. C.; Colman, O. D.; Juarranz, A. Chromatin staining reactions by oxidized *p*-phenylenediamine and Alcian blue. *Acta Histochem., Suppl.* **1982**, *25*, 129–136.

9. Juarranz, A.; Stockert, J. C. Monastral fast blue. Cytochemical properties of a reaction product from Alcian blue-stained chromatin. *Acta Histochem.* **1982**, *70*, 130–134.

10. Wardi, A. H.; Michos, G. A. Alcian blue staining of glycoproteins in acrylamide disc electrophoresis. *Anal. Biochem.* **1972**, *49*, 607–609.

11. Tas, J. Polyacrylamide films as a tool for investigating qualitative and quantitative aspects of the staining of glycosaminoglycans with basic dyes. *Histochem. J.* **1977**, *9*, 267–276.

12. Perry, S. F. Alcian blue as an en bloc stain for turtle lung goblet cells, with and without prior periodate oxidation. *Stain Technol.* **1971**, *46*, 191–194.

13. Castejon, O. J.; Castejon, H. V. Tinctorial potentiality of two basic stains in the electron-histochemical study of polyanionic compounds in nerve tissue. I. Synaptic region. *Acta Histochem.* **1972**, *43*, 153–163.

14. Labelle, J. L.; Briere, N. Staining of nuclear basic proteins without deoxyribonucleic acid hydrolysis, using heparin and alcian blue. *Acta Histochem.* **1971**, *41*, 338–348.

15. Shea, S. M. Lanthanum staining of the surface coat of cells. Its enhancement by the use of fixatives containing Alcian blue or cetylpyridinium chloride. *J. Cell Biol.* **1971**, *51*, 611–620.

16. Sternheimer, R. Urinary sediment stain. U.S. Patent 3961039, 1976; *Chem. Abstr.* **1976**, *85*, 43435.

17. Liu, J.; Zhou, X.; Zhang, H.; Zhang, T. A simple spectro-photometric method for hyaluronic acid assay. *Wuxi Qinggongye Xueyuan Xuebao* **1995**, *14*, 43–48; *Chem. Abstr.* **1995**, *123*, 159809.

18. Cubbage, M. L.; Bresser, J.; Blick, M.; Ju, S. C. Use of analogs of reporter groups to lower background in hybridization assays. U.S. Patent 5501952, 1996; *Chem. Abstr.* **1996**, *124*, 309553.

19. Halbhuber, K. J.; Geyer, G. Alcian Blue-induced mixed cell agglutination. *Folia Haematol.* **1977**, *104*, 98–101.

20. Halbhuber, K. J.; Geyer, G.; Feuerstein, H. Agglutination behavior of red blood cells following experimental glycocalix alteration. *Folia Haematol.* **1977**, *104*, 85–97.

21. Kloecking, R.; Eichhorn, U.; Blumoehr, T. Disk electro-phoretic method for detecting humic acid-type substances in blood serum. *Fresenius' Z. Anal. Chem.* **1978**, *292*, 408–411.

22. Kloecking, R. Localization of humic acids on polyacrylamide gels by staining with Alcian Blue after lead(II) acetate fixation. *Z. Chem.* **1974**, *14*, 317–318.

23. Von Denffer, H.; Mertz, M. Sensitivity of various dyes for detecting pancreatic β-granules in white mice during ontogenesis. *Histochemie* **1972**, *29*, 54–64.

24. Oshima, N.; Takeda, F.; Ishii, K. Effects of chemical substances on rate of locomotion in the amoeba *Mayorella penardi*. *J. Protozool.* **1985**, *32*, 509–512.

25. Whiteman, P. Quantitative measurements of Alcian Blue–glycosaminoglycan complexes. *Biochem. J.* **1973**, *131*, 343–350.

26. Whiteman, P. Quantitative determination of glycosam-inoglycans in urine with Alcian Blue 8GX. *Biochem. J.* **1973**, *131*, 351–357.

27. Mozersky, S. M.; Wildermuth, R. J.; Marmer, W. N. Estimation of the sulfated glycosaminoglycan content of bovine skin with Alcian blue. *J. Am. Leather Chem. Assoc.* **2003**, *98*, 337–343.

28. Segal, L.; Timpa, J. D.; Drake, G. L., Jr. Effect of dyes on the flammability of cotton fabric. 1. *Text. Res. J.* **1974**, *44*, 839–846.

ALIZARIN

CAS Registry Number 72-48-0

Chemical Structure

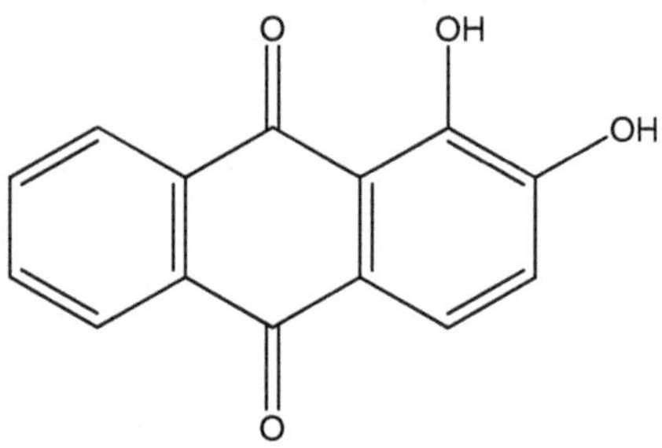

CA Index Name 9,10-Anthracenedione, 1,2-dihydroxy-

Other Names Alizarin B; Anthraquinone, 1,2-dihydroxy-; 1,2-Anthraquinonediol; 1,2-Dihydroxy-9,10-anthracenedione; 1,2-Dihydroxy-9,10-anthraquinone; 1,2-Dihydroxyanthraquinone; 1,2-Dihydroxyanthrachinone; Acid Metachrome Red B; Acid Mordant Red B; Alizarin; Alizarin Red; Alizarina; Alizarine; Alizarine 3B; Alizarine B; Alizarine Indicator; Alizarine L Paste; Alizarine Lake Red 2P; Alizarine Lake Red 3P; Alizarine Lake Red IPX; Alizarine NAC; Alizarine Paste 20 percent Bluish; Alizarine Red; Alizarine Red B; Alizarine Red B2; Alizarine Red IP; Alizarine Red IPP; Alizarine Red L; C Ext. Red 62; C.I. 58000; C.I. Mordant Red 11; Certiqual Alizarine; C.I. Pigment Red 83; D And C Orange Number 15; Deep Crimson Madder 10821; Eljon Madder; Mitsui Alizarine B; Mordant Red 11; NSC 7212; Sanyo Carmine L2B; Turkey Red

Merck Index Number 251

Chemical/Dye Class Anthraquinone

Molecular Formula $C_{14}H_8O_4$

Molecular Weight 240.21

Physical Form Orange needles or powder

Solubility Virtually insoluble in water; moderately soluble in ethanol; soluble in benzene, toluene, xylene, pyridine, acetic acid; freely soluble in hot methanol, ether

Melting Point 290 °C

Boiling Point 430 °C

pH Range 5.5–6.8; 10.1–12.1

Color Change at pH Yellow (5.5) to red (6.8); red (10.1) to purple (12.1)

pK_a 6.77

Absorption (λ_{max}) 567 nm, 609 nm

Synthesis Synthetic methods[1–16]

Staining Applications Bacteria;[17] human adipose-derived stem cells;[18] multipotent adult progenitor cells;[19] parasite;[5] lips;[1,20] skin;[1,20,21] hairs;[1,22,23] keratin fibers[24]

Biological Applications Detecting microorganisms;[25] treating dermatological conditions[26]

Industrial Applications Solar cells;[27,28] plasma display panel;[1,29] antireflective coatings;[1,30] chemical mechanical polishing;[1,31] optical recording materials;[32] inks;[33,34] paints;[1,34,35] adhesives;[34] thermoplastics;[1,36] detergents;[1,37] textiles;[1,38,39] wood[1,40]

Safety/Toxicity Acute oral toxicity;[1,41] carcinogenicity;[1,42] clastogenicity;[43] environmental toxicity;[1,44] estrogenicity;[1,45] genotoxicity;[1,46,47] hypersensitivity;[1,48] mutagenicity;[1,49,50] photoinduced toxicity[1,51]

REFERENCES

1. Sabnis, R. W. *Handbook of Acid–Base Indicators*; CRC Press: Boca Raton, **2008**; pp 11–13.
2. Orban, N.; Boldizsar, I.; Szucs, Z.; Danos, B. Influence of different elicitors on the synthesis of anthraquinone derivatives in *Rubia tinctorum* L. cell suspension cultures. *Dyes Pigm.* **2008**, *77*, 249–257.
3. Ashnagar, A.; Naseri, N. Gharib; Zadeh, A. S. Isolation and identification of 1,2-dihydroxy-9,10-anthraquinone (alizarin) from the roots of maddar plant (*Rubia tinctorum*). *Int. J. Chem. Sci.* **2007**, *5*, 683–688.
4. Nair, M. G.; Dhananjeyan, M. R.; Kron, M. A.; Milev, Y. Process for preparation of anthraquinones from phthalic anhydrides and hydroxyphenols in the presence of molten Lewis acid–alkali metal halide mixtures and their use as parasiticides. PCT Int. Appl. WO 2005113478, 2005; *Chem. Abstr.* **2005**, *144*, 22729.
5. Dhananjeyan, M. R.; Milev, Y. P.; Kron, M. A.; Nair, M. G. Synthesis and activity of substituted anthraquinones against a human filarial parasite, *Brugia malayi*. *J. Med. Chem.* **2005**, *48*, 2822–2830.
6. Darvekar, M.; Ghorpade, B.; Vankar, P. S. Microwave assisted improved syntheses of indigoid and anthraquinoid dyes. *Asian J. Chem.* **2004**, *16*, 965–970.
7. Derksen, C. C. H.; Van Beek, T. A.; De Groot, A.; Capelle, A. Dye preparation based on madder root

and its production. Eur. Pat. Appl. EP 1191070, 2002; *Chem. Abstr.* **2002**, *136*, 249015.

8. Pozzo, J.; Clavier, G. M.; Colomes, M.; Bouas-Laurent, H. Different synthetic routes towards efficient organogelators: 2,3-substituted anthracenes. *Tetrahedron* **1997**, *53*, 6377–6390.
9. Sartori, G.; Casnati, G.; Bigi, F.; Foglio, F. A new methodological approach to anthraquinone and anthracy-clidone synthesis. *Gazz. Chim. Ital.* **1990**, *120*, 13–19.
10. Bigi, F.; Casiraghi, G.; Casnati, G.; Sartori, G. Process for the preparation of *p*-quinones. Belg. BE 904981, 1986; *Chem. Abstr.* **1987**, *107*, 58672.
11. Broadbent, A. D.; Stewart, J. M. Photochemical hydroxylation of anthracene-9,10-dione in sulfuric acid solution. *J. Chem. Soc., Chem. Commun.* **1980**, 676–677.
12. Bell, K. A.; Flatman, I. J.; Goldborn, P.; Pachl, A.; Scheinmann, F. Elimination of an allyloxy-group during the Claisen rearrangement of 1,2-diallyloxyanthraquinone. *J. Chem. Soc., Chem. Commun.* **1978**, 900–901.
13. Duclaux, J.; Cohn, C. Formation of color lakes. *Bull. Soc. Chim. Fr.* **1964**, 1600–1603. *Chem. Abstr.* **1964**, *61*, 65595.
14. Scherf, K.; Nienburg, H. Hydroxyanthraquinones. DE 1160124, 1963; *Chem. Abstr.* **1964**, *61*, 4143.
15. Scherf, K. Hydroxyanthraquinones. DE 1165180, 1964; *Chem. Abstr.* **1964**, *60*, 90648.
16. Voyatzakis, E.; Jannakoudakis, D.; Dorfmuller, T.; Sipitanos, C. Photochemical conversion of anthracene into alizarin. *Compt. Rend.* **1960**, *250*, 112–114.
17. Louvet, O.; Thuault, D.; Vaillant, R. Shelf life indicator based on acidifying bacteria. PCT Int. Appl. WO 2005026383, 2005; *Chem. Abstr.* **2005**, *142*, 335317.
18. Wolbank, S.; Peterbauer, A.; Wassermann, E.; Hennerbichler, S.; Voglauer, R.; van Griensven, M.; Duba, H. C.; Gabriel, C.; Redl, H. Labelling of human adipose-derived stem cells for non-invasive *in vivo* cell tracking. *Cell Tissue Banking* **2007**, *8*, 163–177.
19. Price, E. M.; Prather, R. S.; Foley, C. Multipotent adult progenitor cell lines originating from the peripheral blood of green fluorescent protein transgenic swine. *Stem Cells Dev.* **2006**, *15*, 507–522.
20. Dumousseaux, C. Cosmetic composition for application to the skin and lips containing pigments. Fr. Demande FR 2845277, 2004; *Chem. Abstr.* **2004**, *140*, 326628.
21. Lin, C. B. Vanillin polymers for use in darkening the skin. U.S. Pat. Appl. Publ. US 2005129633, 2005; *Chem. Abstr.* **2005**, *143*, 65136.
22. De Boni, M. Use of natural dyes for dyeing human hair. PCT Int. Appl. WO 2008047055, 2008; *Chem. Abstr.* **2008**, *148*, 479061.
23. Plos, G. Hair dye composition containing an oxidase alcohol and a quinonoid direct dye for dyeing keratin fibers. Fr. Demande FR 2865397, 2005; *Chem. Abstr.* **2005**, *143*, 179095.
24. Kainz, S.; Kruegermann, I.; Hoeffkes, H. Oxidation colorant for coloring keratin-containing fibers. Ger. Offen. DE 102005062830, 2007; *Chem. Abstr.* **2007**, *146*, 106792.
25. Martin, S. M.; MacDonald, J. G.; Lye, J.; Sayre, C.; Thompson, K. Array for rapid detection of a microorganism based on dyes. PCT Int. Appl. WO 2008026104, 2008; *Chem. Abstr.* **2008**, *148*, 302867.
26. Pellet, A. Exfoliant composition containing a cellulosic polymer and its applications in cosmetic or topical dermatological treatments. Fr. Demande FR 2902322, 2007; *Chem. Abstr.* **2007**, *148*, 61286.
27. Harima, H.; Oyama, Y. Oxide semiconductor electrodes of dye-sensitized solar cells with high photoelectric efficiency and their manufacture. Jpn. Kokai Tokkyo Koho JP 2007287455, 2007; *Chem. Abstr.* **2007**, *147*, 472238.
28. Harima, Y.; Kawabuchi, K.; Kajihara, S.; Ishii, A.; Ooyama, Y.; Takeda, K. Improvement of photovoltages in organic dye-sensitized solar cells by Li intercalation in particulate TiO_2 electrodes. *Appl. Phys. Lett.* **2007**, *90*, 103517/1–103517/3.
29. Yamaguchi, N.; Suzuki, M. Plasma display panel, glass substrates with dielectric layer containing color material, and its manufacture. Jpn. Kokai Tokkyo Koho JP 2005327542, 2005; *Chem. Abstr.* **2005**, *143*, 485917.
30. Kennedy, J.; Baldwin, T.; Hacker, N. P.; Spear, R. Spin-on-glass anti-reflective coating for photolithography. U.S. Patent 6506497, 2003; *Chem. Abstr.* **2003**, *138*, 98200.
31. Carter, P. W.; Zhang, J.; Grumbine, S. K.; Rege Thesauro, F. D. Compositions and methods for tantalum CMP. U.S. Pat. Appl. Publ. US 2006030158, 2006; *Chem. Abstr.* **2006**, *144*, 203428.
32. Takano, S. Optical recording material using anthraquinone compound metal complex. Jpn. Kokai Tokkyo Koho JP 11301116, 1999; *Chem. Abstr.* **1999**, *131*, 305208.

33. Takemasa, K.; Kakino, R. Fabrication of presensitized lithographic plates and clogging-resistant photocurable inks therefor. Jpn. Kokai Tokkyo Koho JP 2008075056, 2008; *Chem. Abstr.* **2008**, *148*, 390923.

34. Wenning, A.; Loest, D. Coating compositions based on universal pigment preparations for use as paints, adhesives and printing inks. PCT Int. Appl. WO 2007141115, 2007; *Chem. Abstr.* **2007**, *148*, 56686.

35. Machii, Y.; Hirano, M. Paint composition. Jpn. Kokai Tokkyo Koho JP 2006090971, 2006; *Chem. Abstr.* **2006**, *144*, 362266.

36. Van Den Oever, M. J. A.; Boeriu, C. G.; Blaauw, R.; Van Haveren, J. Colorants based on renewable resources and food-grade colorants for application in thermoplastics. *J. Appl. Polym. Sci.* **2004**, *92*, 2961–2969.

37. Macdonald, J. G.; Huang, Y.; Yang, K.; Kim, J.; Wei, N. Single phase color change agents for liquid detergents for cleaning surfaces. U.S. Pat. Appl. Publ. US 2005049157, 2005; *Chem. Abstr.* **2005**, *142*, 263578.

38. Vankar, P. S.; Shanker, R.; Mahanta, D.; Tiwari, S. C. Ecofriendly sonicator dyeing of cotton with *Rubia cordifolia* Linn. using biomordant. *Dyes Pigm.* **2007**, *76*, 207–212.

39. Park, J. H.; Gatewood, B. M.; Ramaswamy, G. N. Naturally occurring quinones and flavonoid dyes for wool: insect feeding deterrents. *J. Appl. Polym. Sci.* **2005**, *98*, 322–328.

40. Kerle, T. Modern, low-cost and rapid method for pictorial coloring of wood surfaces. PCT Int. Appl. WO 2004101176, 2004; *Chem. Abstr.* **2004**, *141*, 425484.

41. Schafer, E. W., Jr.; Bowles, W. A., Jr.; Hurlbut, J. The acute oral toxicity, repellency, and hazard potential of 998 chemicals to one or more species of wild and domestic birds. *Arch. Environ. Contam. Toxicol.* **1983**, *12*, 355–382.

42. Wolfle, D.; Schmutte, C.; Westendorf, J.; Marquardt, H. Hydroxyanthraquinones as tumor promoters: enhancement of malignant transformation of C3H mouse fibroblasts and growth stimulation of primary rat hepatocytes. *Cancer Res.* **1990**, *50*, 6540–6544.

43. Simi, S.; Morelli, S.; Gervasi, P. G.; Rainaldi, G. Clastogenicity of anthraquinones in V79 and in three derived cell lines expressing P450 enzymes. *Mutat. Res.* **1995**, *347*, 151–156.

44. Mallakin, A.; McConkey, B. J.; Miao, G.; McKibben, B.; Snieckus, V.; Dixon, D. G.; Greenberg, B. M. Impacts of structural photomodification on the toxicity of environ-mental contaminants: anthracene photooxidation products. *Ecotoxicol. Environ. Saf.* **1999**, *43*, 204–212.

45. Kurihara, R.; Shiraishi, F.; Tanaka, N.; Hashimoto, S. Presence and estrogenicity of anthracene derivatives in coastal Japanese waters. *Environ. Toxicol. Chem.* **2005**, *24*, 1984–1993.

46. Ohno, K.; Tanaka-Azuma, Y.; Yoneda, Y.; Yamada, T. Genotoxicity test system based on p53R2 gene expression in human cells: examination with 80 chemicals. *Mutat. Res.* **2005**, *588*, 47–57.

47. Westendorf, J.; Marquardt, H.; Poginsky, B.; Dominiak, M.; Schmidt, J.; Marquardt, H. Genotoxicity of naturally occurring hydroxyanthraquinones. *Mutat. Res.* **1990**, *240*, 1–12.

48. Fujii, S. Evaluation of hypersensitivity to anthraquinone-related substances. *Toxicology* **2003**, *193*, 261–267.

49. Jaeger, I.; Hafner, C.; Welsch, C.; Schneider, K.; Iznaguen, H.; Westendorf, J. The mutagenic potential of madder root in dyeing processes in the textile industry. *Mutat. Res.* **2006**, *605*, 22–29.

50. Krivobok, S.; Seigle-Murandi, F.; Steiman, R.; Marzin, D. R.; Betina, V. Mutagenicity of substituted anthraquinones in the Ames/*Salmonella* microsome system. *Mutat. Res.* **1992**, *279*, 1–8.

51. Lampi, M. A.; Gurska, J.; McDonald, K. I. C.; Xie, F.; Huang, X. D.; Dixon, D. G.; Greenberg, B. M. Photoinduced toxicity of polycyclic aromatic hydrocarbons to *Daphnia magna*: ultraviolet-mediated effects and the toxicity of polycyclic aromatic hydrocarbon photoproducts. *Environ. Toxicol. Chem.* **2006**, *25*, 1079–1087.

ALIZARIN RED S

CAS Registry Number 130-22-3

Chemical Structure

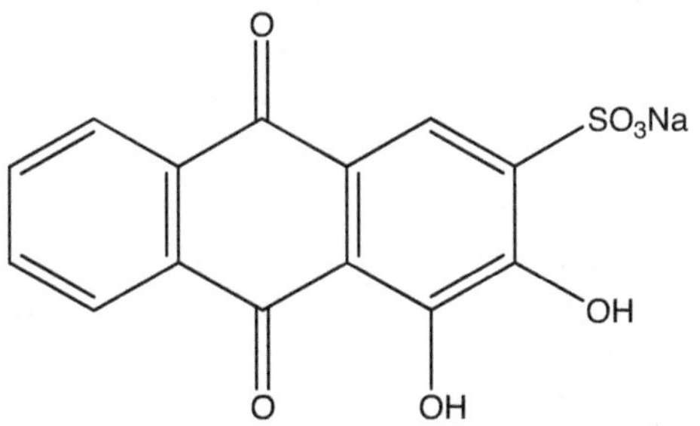

CA Index Name 2-Anthracenesulfonic acid, 9,10-dihydro-3,4-dihydroxy-9,10-dioxo-, monosodium salt

Other Names 2-Anthraquinonesulfonic acid, 3,4-dihydroxy-, sodium salt; Acid Mordant Red SW; Acid Red Alizarine; Ahcoquinone Red S; Alizarin Carmine; Alizarin Red S; Alizarin S; Alizarine Carmine; Alizarine Carmine Indicator; Alizarine Red A; Alizarine Red AS; Alizarine Red Indicator; Alizarine Red S; Alizarine Red S sodium salt; Alizarine Red SW; Alizarine Red SZ; Alizarine Red W; Alizarine Red WA; Alizarine Red WS; Alizarine Red for Wool; Alizarine S; Alizarine S Extra Conc. A Export; Alizarine S Extra Pure A; Alizarinsulfonate; C.I. 58005; C.I. Mordant Red 3; Calcochrome Alizarine Red SC; Carnelio Rubine Lake; Chrome Red Alizarine; Diamond Red W; Ext D and C Red No. 7; Fenakrom Red W; Mitsui Alizarine Red S; Mordant Red 3; Oxanal Fast Red SW; Sodium 3,4-dihydroxyanthraquinone-2-sulfonate; Sodium alizarin-3-sulfonate; Sodium alizarinesulfonate; Sodium alizarinsulfonate

Merck Index Number 8573

Chemical/Dye Class Anthraquinone

Molecular Formula $C_{14}H_7NaO_7S$

Molecular Weight 342.26

Physical Form Orange-yellow powder

Solubility Freely soluble in water; soluble in ethanol; insoluble in ether

Melting Point 287–289 °C

pH Range 3.5–6.5; 9.4–12.0

Color Change at pH Yellow (3.5) to red (6.5); orange (9.4) to violet (12.0)

pK_a 4.5, 11

Absorption (λ_{max}) 556 nm, 596 nm, 423 nm, 546 nm

Synthesis Synthetic methods[1–6]

Staining Applications Bacteria;[7] bones;[8,9] cartilage;[8] dental plaques;[10] fetal skeletons;[11] fish;[12] myocardium;[13] protein particles;[14] calcified tissues;[15] hairs;[1,16] keratin fibers[17,18]

Biological Applications Detecting *Candida*,[19] lactic acid bacteria,[7] microorganisms;[20] treating gastropathy,[21] viral diseases[22]

Industrial Applications Chemical mechanical polishing;[1,23] inks;[24] adhesives;[25] ozone sensor;[26] chemical warfare agent;[27] detergents;[28] coloring fabrics[1,29]

Safety/Toxicity Carcinogenicity;[1,30] cytotoxicity;[1,31] genotoxicity;[1,31] fish toxicity;[1,32] environmental pollutants;[1,33] viral toxicity[31]

Certification/Approval Certified by Biological Stain Commission (BSC)

REFERENCES

1. Sabnis, R. W. *Handbook of Acid-Base Indicators*; CRC Press: Boca Raton, **2008**; pp 14–15.
2. Welcher, F. J. *Organic Analytical Reagents*; D. Van Nostrand Company: New York, **1948**; Vol. *4*, pp 423–436.
3. Ookuma, T.; Ito, N. Manufacture of sodium 3-alizarinsulfonate. Jpn. Kokai Tokkyo Koho JP 08027385, 1996; *Chem. Abstr.* **1996**, *124*, 292291.
4. Geckeler, K. E.; Zhou, R. Functionalized materials based on amino and amide polymers by an easy one-pot preparation. *Naturwissenschaften* **1993**, *80*, 270–271; *Chem. Abstr.* **1993**, *119*, 50060.
5. Broadbent, A. D.; Newton, R. P. Photolysis of aqueous solutions of 9,10-anthraquinone-2-sulfonate. I. Dependence of product yields on reaction conditions. *Can. J. Chem.* **1972**, *50*, 381–387.
6. Hosoda, Y. Anthraquinone mordant dyes. VII. Alizarin Red S. *J. Soc. Org. Synth. Chem.* **1951**, *9*, 187–188; *Chem. Abstr.* **1953**, *47*, 7925.
7. Horikoshi, N.; Okada, Y.; Takeshita, K.; Samejima, T. Semi-solid medium for detecting lactic acid bacteria.

Jpn. Kokai Tokkyo Koho JP 2006136272, 2006; *Chem. Abstr.* **2006**, *144*, 484206.

8. Walker, M. B.; Kimmel, C. B. A two-color acid-free cartilage and bone stain for zebrafish larvae. *Biotech. Histochem.* **2007**, *82*, 23–28.
9. Hong, Y. C.; Yen, P. K.; Shaw, J. H. Microscopic evaluation of the effects of some vital staining agents on growing bone in rabbits. *Calcif. Tissue Res.* **1968**, *2*, 286–295.
10. Matsumoto, Y. Material for evaluating dental caries activity. U.S. Pat. Appl. Publ. US 2003113266, 2003; *Chem. Abstr.* **2003**, *139*, 26734.
11. Young, A. D.; Phipps, D. E.; Astroff, A. B. Large-scale double-staining of rat fetal skeletons using Alizarin Red S and Alcian Blue. *Teratology* **2000**, *61*, 273–276.
12. Moen, V.; Nordtug, T.; Zahlsen, K. Mixture and procedure for marking fish. Norw. NO 317169, 2004; *Chem. Abstr.* **2007**, *146*, 141497.
13. Thornell, L. E.; Holmbom, B.; Eriksson, A.; Reiz, S.; Marklund, S.; Naeslund, U. Enzyme and immunohistochemical assessment of myocardial damage after ischemia and reperfusion in a closed-chest pig model. *Histochemistry* **1992**, *98*, 341–353.
14. Wunderlich, G.; Dreyer, R.; Fischer, S. Radiochemical labeling of protein particles with astatine. Ger. (East) DD 234870, 1986; *Chem. Abstr.* **1986**, *105*, 178470.
15. Kariyama, M.; Akai, M.; Nishijima, S. Three-color fluorescent labeling method for calcified tissues in a reptile, *Caiman crocodilus*. *Arch. Oral Biol.* **1969**, *14*, 1349–1350.
16. Grollier, J. F.; Rosenbaum, G.; Cotteret, J. Hair dye containing hydroxyanthraquinones and methods for dyeing human hair. Ger. Offen. DE 3207036, 1982; *Chem. Abstr.* **1982**, *97*, 203097.
17. Plos, G. Hair dye composition containing an oxidase alcohol and a quinonoid direct dye for dyeing keratin fibers. Fr. Demande FR 2865397, 2005; *Chem. Abstr.* **2005**, *143*, 179095.
18. Grollier, J. F. Dyes for keratin fibers. Ger. Offen. DE 3423349, 1985; *Chem. Abstr.* **1985**, *102*, 137582.
19. Martin, S. M.; Long, A. M.; MacDonald, J. G.; Lye, J. Method for detecting *Candida* on skin using dyes. PCT Int. Appl. WO 2008026105, 2008; *Chem. Abstr.* **2008**, *148*, 302866.
20. Martin, S. M.; MacDonald, J. G.; Lye, J.; Sayre, C.; Thompson, K. Array for rapid detection of a microorganism based on dyes. PCT Int. Appl. WO 2008026104, 2008; *Chem. Abstr.* **2008**, *148*, 302867.
21. Gao, Q. New oral formulation of traditional Chinese medicine mixture for treating gastropathy. Faming Zhuanli Shenqing Gongkai Shuomingshu CN 1879711, 2006; *Chem. Abstr.* **2006**, *146*, 128554.
22. May, G.; Leonhadt, K.; Ott, H. Use of anthraquinone derivatives in the prophylaxis and treatment of viral diseases. Eur. Pat. Appl. EP 454058, 1991; *Chem. Abstr.* **1992**, *116*, 51547.
23. Carter, P. W.; Zhang, J.; Grumbine, S. K.; Rege Thesauro, F. D. Compositions and methods for tantalum CMP. U.S. Pat. Appl. Publ. US 2006030158, 2006; *Chem. Abstr.* **2006**, *144*, 203428.
24. Ryu, S. M.; Jung, Y. K. Inkjet ink composition and multi-color inkjet ink set having the same. U.S. Pat. Appl. Publ. US 2007081061, 2007; *Chem. Abstr.* **2007**, *146*, 403849.
25. Kataoka, K. Optical fiber arrays allowing visual confirmation of fiber position and adhesives therefor. Jpn. Kokai Tokkyo Koho JP 2005232255, 2005; *Chem. Abstr.* **2005**, *143*, 275285.
26. 26. Maruo, Y.; Kunioka, T.; Miwa, T.; Nakamura, J. Compact portable system for ozone exposure assessment. Jpn. Kokai Tokkyo Koho JP 2008107138, 2008; *Chem. Abstr.* **2008**, *148*, 501859.
27. Stevens, A. E.; Ehret, A.; Stuhl, L. S. Chemical warfare agent and toxic industrial chemical detection by degradation combined with acid–base indicator dyes deployed as coatings. PCT Int. Appl. WO 2008018853, 2008; *Chem. Abstr.* **2008**, *148*, 207994.
28. Macdonald, J. G.; Huang, Y.; Yang, K.; Kim, J.; Wei, N. Single phase color change agents for liquid detergents for cleaning surfaces. U.S. Pat. Appl. Publ. US 2005049157, 2005; *Chem. Abstr.* **2005**, *142*, 263578.
29. Deeds, S. M. Fabric repair coloring device and using the device to color unwanted bleached out spots on fabric. U.S. Patent 6739779, 2004; *Chem. Abstr.* **2004**, *140*, 424935.
30. Yang, W.; Jin, J. Spectrophotometric study on copper (II)-*o*-phenanthroline–alizarine Red S mixed-ligand complex—a tentative model for metal carcinogenesis. *Huanjing Huaxue* **1982**, *1*, 116–122; *Chem. Abstr.* **1982**, *97*, 176679.
31. Barnard, D. L.; Fairbairn, D. W.; O'Neill, K. L.; Gage, T. L.; Sidwell, R. W. Anti-human cytomegalovirus activity and toxicity of sulfonated

anthraquinones and anthraquinone derivatives. *Antiviral Res.* **1995**, *28*, 317–329.

32. Tonogai, Y.; Ito, Y. Toxicity of organic nitrogen compounds to fish. Syncopic effect of aniline derivatives on fish. *Seitai Kagaku* **1984**, *7*, 17–26; *Chem. Abstr.* **1984**, *101*, 164999.

33. Tonogai, Y.; Ogawa, S.; Ito, Y.; Iwaida, M. Actual survey on TLm (median tolerance limit) values of environmental pollutants, especially on amines, nitriles, aromatic nitrogen compounds and artificial dyes. *J. Toxicol. Sci.* **1982**, *7*, 193–203.

ALLURA RED

CAS Registry Number 25956-17-6

Chemical Structure

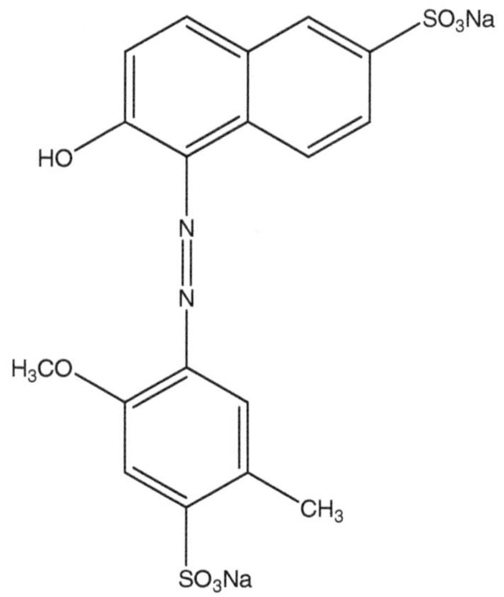

CA Index Name 2-Naphthalenesulfonic acid, 6-hydroxy-5-[2-(2-methoxy-5-methyl-4-sulfophenyl)diazenyl]-, sodium salt (1 : 2)

Other Names 2-Naphthalenesulfonic acid, 6-hydroxy-5-[(2-methoxy-5-methyl-4-sulfophenyl)azo]-, disodium salt; 2-Naphthalenesulfonic acid, 6-hydroxy-5-[(6-methoxy-4-sulfo-*m*-tolyl)azo]-, disodium salt; Allura Red 40; Allura Red AC; Allura red; C.I. 16035; C.I. Food Red 17; E 129; FD and C Red No. 40; FD&C Red No. 40; FDC Red 40; FDC Red 40 dye; Food Red 17; Food Red No. 17; Food Red No. 40; Japan Food Red No. 40; Japan Red 40; Red 40; Red No. 40

Merck Index Number 284

Chemical/Dye Class Azo

Molecular Formula $C_{18}H_{14}N_2Na_2O_8S_2$

Molecular Weight 496.42

Physical Form Dark red powder

Solubility Soluble in water, ethanol

Melting Point >300 °C

Absorption (λ_{max}) 504 nm

Synthesis Synthetic methods[1–6]

Staining Applications Alcohol;[7] beverage;[8,9] drinks;[7,10,11] syrup;[9] candies;[11] cotton candy;[12] baked food;[13] frozen food;[14] aerated fruits;[15] seafood;[16] sweetener;[17] capsules;[18] pharmaceutical dosage;[19] tablets;[20] toothpaste;[21] suncreeen;[22] dental bleaching gel;[23] eyebrows;[24] lips;[25–27] skin;[26,27] body surfaces;[27] tattoos;[28] hairs;[29] keratin fibers;[29,30] cocoa bath;[31] waters[32]

Biological Applications Antifungal formulation;[33] treating bone metabolic diseases,[34] coughing,[35] sneezing,[35] rhinorrhea,[35] nasal obstruction,[35] periodontal disease,[36] rhinitis[37]

Industrial Applications Antireflective coatings;[38] inks;[39] detergents;[40] colored bubbles;[41] fabric softeners;[42] textiles;[43] entertainment products;[44] toys[45]

Safety/Toxicity Carcinogenicity;[46,47] chromosomal aberration;[48] developmental toxicity;[49] DNA damage;[50] genotoxicity;[51] hyperactive behavior in children;[52] mutagenicity;[53–55] neurotoxicity;[56] psychotoxicity,[49] reproductive toxicity[56]

Certification/Approval Approved by Food & Drugs Administration (FDA)

REFERENCES

1. Rogers, R.; Horwitz, E. P.; Bond, A. H. Process for separating and recovering an anionic dye from an aqueous solution. U.S. Patent 5707525, 1998; *Chem. Abstr.* **1998**, *128*, 142665.
2. Waltonen, J. R.; Schutz, R. W. Preparation of a flexible piezoelectric membrane. U.S. Patent 5230921, 1993; *Chem. Abstr.* **1993**, *119*, 260915.
3. Rebhahn, R. W. J.; Cook, W. L. Ultrafiltration purification of food dyes. U.S. Patent 4560746, 1985; *Chem. Abstr.* **1986**, *105*, 116539.
4. Rast, G. E.; Steiner, R. I. Red phenyl-azo-naphthol dyestuffs for edible compositions. U.S. Patent 3519617, 1970.
5. Allied Chemical Corporation, Red monoazo food dyes. Brit. GB 1164249, 1969; *Chem. Abstr.* **1970**, *72*, 68195.
6. Richfield-Fratz, N.; Bailey, J. E., Jr. Determination of *p*-cresidine in FD&C red no. 40 by the diazotization and coupling procedure followed by reversed-phase high-performance liquid chromatography. *J. chromatogr.* **1987**, *405*, 283–294.
7. Donhowe, E. T. Product and process of making an alcohol containing sport drink. U.S. Pat. Appl. Publ. US 2003134007, 2003; *Chem. Abstr.* **2003**, *139*, 100270.

8. Smith, L. C. Carbonated beverage national school lunch meal. U.S. Pat. Appl. Publ. US 2007281059, 2007; *Chem. Abstr.* **2007**, *148*, 10213.
9. Chang, P. K. Color-stable syrup and beverage compositions fortified with vitamin C. U.S. Patent 5336510, 1994; *Chem. Abstr.* **1994**, *121*, 254387.
10. Sherwood, S.; Jenkins, D. High energy carbonated protein drink and method of making. U.S. Pat. Appl. Publ. US 2006147602, 2006; *Chem. Abstr.* **2006**, *145*, 102723.
11. Yoshioka, N.; Ichihashi, K. Determination of 40 synthetic food colors in drinks and candies by high-performance liquid chromatography using a short column with photodiode array detection. *Talanta* **2008**, *74*, 1408–1413.
12. Litvin, A. Sugar-substitute cotton candy and procedure for its production. Ger. Offen. DE 102008037185, 2009; *Chem. Abstr.* **2009**, *150*, 258878.
13. Stewart, D. R. Method of making and using an edible cellulose-based polymeric film for decorating foods, especially baked goods. U.S. Patent 6616958, 2003; *Chem. Abstr.* **2003**, *139*, 229693.
14. Stanley, S. A.; Pachivas, J. A.; Annis, J. R. Frozen product and method of oral delivery of active ingredients. U.S. Patent 6258384, 2001; *Chem. Abstr.* **2001**, *135*, 97458.
15. Kepplinger, J.; Bath, D. E. Aerated fruit-containing food additive. PCT Int. Appl. WO 2002071872, 2002; *Chem. Abstr.* **2002**, *137*, 231733.
16. Altemueller, A. G. Simulated seafood compositions comprising structured plant protein products and fatty acids. U.S. Pat. Appl. Publ. US 2008069927, 2008; *Chem. Abstr.* **2008**, *148*, 330362.
17. Catani, S. J.; Liao, S.; Loades, M. Self-mixing tabletop sweetener. U.S. Pat. Appl. Publ. US 2007059418, 2007; *Chem. Abstr.* **2007**, *146*, 315560.
18. Yu, S.; Liu, M. Method for manufacturing soft capsule. Faming Zhuanli Shenqing Gongkai Shuomingshu CN 101069678, 2007; *Chem. Abstr.* **2007**, *148*, 39571.
19. Darmuzey, O.; Macleod, G.; Cengic, D.; Stokes, K. M. Solid pharmaceutical dosage form. PCT Int. Appl. WO 2008140461, 2008; *Chem. Abstr.* **2008**, *150*, 10869.
20. Moroni, A.; Drefko, W. J. Pharmaceutical tablet coating composition. U.S. Pat. Appl. Publ. US 2004001884, 2004; *Chem. Abstr.* **2004**, *140*, 65216.
21. Wright, H. W. Color-changing toothpaste containing dyes. U.S. Patent 6419902, 2002; *Chem. Abstr.* **2002**, *137*, 98684.
22. Tanner, P. R. Multi-phase personal care composition compris-ing a polymeric thickener and a sunscreen. PCT Int. Appl. WO 2007148292, 2007; *Chem. Abstr.* **2007**, *148*, 85150.
23. Banerjee, A.; Friedman, J. Dental bleaching gel composition and activator. U.S. Pat. Appl. Publ. US 2003118519, 2003; *Chem. Abstr.* **2003**, *139*, 57671.
24. Park, M. S.; Yun, P.; Park, H. K. Composition of eyebrow pencil. Repub. Korea KR 135388, 1998; *Chem. Abstr.* **2004**, *140*, 47021.
25. Gross, P. Lip coloring agent. Ger. Offen. DE 19627931, 1998; *Chem. Abstr.* **1998**, *128*, 132267.
26. Blin, X. Range of compositions for application to the skin, lips, and/or superficial body growths. Fr. Demande FR 2868697, 2005; *Chem. Abstr.* **2005**, *143*, 392523.
27. Dumousseaux, C. Cosmetic composition for application to the skin, lips, and/or body surface. Fr. Demande FR 2868696, 2005; *Chem. Abstr.* **2005**, *143*, 392522.
28. Kundakovic, L.; Klitzman, B.; Koger, K. E.; Mathiowitz, E.; Drill, C. A. Cellular or organelle-entrapped nanoparticles containing dyes for tattoos. PCT Int. Appl. WO 2007024429, 2007; *Chem. Abstr.* **2007**, *146*, 280376.
29. Sallwey, A.; Schmitt, M.; Dieckow, A. Dyestuff for hair and fibers. Eur. Pat. Appl. EP 976385, 2000; *Chem. Abstr.* **2000**, *132*, 127457.
30. Javet, M.; Mueller, C.; Weber, I. Bleaching keratin fibers with compositions containing an agent having a basic pH and sulfamate bleach booster. Eur. Pat. Appl. EP 1759684, 2007; *Chem. Abstr.* **2007**, *146*, 301740.
31. Smith, J. W. Whipped cocoa bath. U.S. Pat. Appl. Publ. US 2003109392, 2003; *Chem. Abstr.* **2003**, *139*, 26329.
32. Cave, A. Process and composition for coloring bath, bathing and/or irrigation waters. Fr. Demande FR 2918068, 2009; *Chem. Abstr.* **2009**, *150*, 105384.
33. Oppenheim, F. G.; Cheng, D. J.; Helmerhorst, E. J. Antifungal formulation and method of preparation. PCT Int. Appl. WO 2009005798, 2009; *Chem. Abstr.* **2009**, *150*, 106182.
34. Diaz, L. E. Liquid composition comprising alendronic acid for prevention and/or treatment of different bone metabolic diseases, uses thereof, and preparation process therefore. Eur. Pat. Appl. EP 1894557, 2008; *Chem. Abstr.* **2008**, *148*, 292170.

35. Giordano, J. A.; Brown, K. M. Compositions and methods for treatment of coughing, sneezing, rhinorrhea, and/or nasal obstruction. U.S. Pat. Appl. Publ. US 2007160689, 2007; *Chem. Abstr.* **2007**, *147*, 150809.
36. Andersen, S. N.; Wilson, J. B. Dye composition and laser energy for treating periodontal disease. U.S. Pat. Appl. Publ. US 2003059379, 2003; *Chem. Abstr.* **2003**, *138*, 276252.
37. Hall, M.; Roberts, A.; Heil, M. Kits comprising immunonutritional agents that contain pharmaceutically acceptable form of zinc, for prevention and treatment of rhinitis. PCT Int. Appl. WO 2008033155, 2008; *Chem. Abstr.* **2008**, *148*, 339026.
38. Dammel, R. R.; Norwood, R. A. Light-absorbing antireflective layers with improved performance due to refractive index optimization. U.S. Patent 6274295, 2001; *Chem. Abstr.* **2001**, *135*, 160160.
39. Barreto, M. A.; Garcia, J. L. Consumable inks containing colorants and water-soluble organic solvents with improved image performance. U.S. Pat. Appl. Publ. US 2007132823, 2007; *Chem. Abstr.* **2007**, *147*, 54283.
40. Janke, H.; Harer, J.; Holderbaum, T. Colored detergent for dishwashing. PCT Int. Appl. WO 9960088, 1999; *Chem. Abstr.* **1999**, *132*, 4281.
41. Sabnis, R. W.; Kehoe, T. D. Composition and method for producing colored bubbles. U.S. Pat. Appl. Publ. US 2006004110, 2006; *Chem. Abstr.* **2006**, *144*, 110164.
42. Wahl, E. H.; Vimba, A. I. Dyes for imparting pink color to aqueous fabric softener compositions. Eur. Pat. Appl. EP 304410, 1989; *Chem. Abstr.* **1989**, *111*, 176797.
43. Lupia, J. A.; Suhadolnik, J.; Wood, M. G.; Martin, D. W. H. Stabilized body care products, household products, textiles and fabrics. PCT Int. Appl. WO 2006122889, 2006; *Chem. Abstr.* **2006**, *146*, 12598.
44. Sundaram, M.; Ihatsu, P.; Bourhis, Y. Dry ice products for the entertainment industry. PCT Int. Appl. WO 2007088442, 2007; *Chem. Abstr.* **2007**, *147*, 191690.
45. Ono, Y. Water-erasable markers for toys and figure toy sets therewith. Jpn. Kokai Tokkyo Koho JP 2002129085, 2002; *Chem. Abstr.* **2002**, *136*, 371209.
46. Rosenkranz, H. S.; Klopman, G. Structural basis of carcinogenicity in rodents of genotoxicants and non-genotoxicants. *Mutat. Res.* **1990**, *228*, 105–124.
47. Borzelleca, J. F.; Olson, J. W.; Reno, F. E. Lifetime toxicity/carcinogenicity studies of FD and C Red No. 40 (allura red) in mice. *Food Chem. Toxicol.* **1991**, *29*, 313–319.
48. Serra, J. R.; Thompson, E. D.; Jurs, P. C. Development of binary classification of structural chromosome aberrations for a diverse set of organic compounds from molecular structure. *Chem. Res. Toxicol.* **2003**, *16*, 153–163.
49. Vorhees, C. V.; Butcher, R. E.; Brunner, R. L.; Wootten, V.; Sobotka, T. J. Developmental toxicity and psychotoxicity of FD and C red dye No. 40 (Allura Red AC) in rats. *Toxicology* **1983**, *28*, 207–217.
50. Tsuda, S.; Murakami, M.; Matsusaka, N.; Kano, K.; Taniguchi, K.; Sasaki, Y. F. DNA damage induced by red food dyes orally administered to pregnant and male mice. *Toxicol. Sci.* **2001**, *61*, 92–99.
51. Haveland-Smith, R. B.; Combes, R. D. Screening of food dyes for genotoxic activity. *Food Cosmet. Toxicol.* **1980**, *18*, 215–221.
52. McCann, D.; Barrett, A.; Cooper, A.; Crumpler, D.; Dalen, L.; Grimshaw, K.; Kitchin, E.; Lok, K.; Porteous, L.; Prince, E.; Sonuga-Barke, E.; Warner, J. O.; Stevenson, J. Food additives and hyperactive behaviour in 3-year-old and 8/9-year-old children in the community: a randomized, double-blinded placebo-controlled trial. *Lancet* **2007**, *370*, 1560–1567.
53. Zeiger, E.; Margolin, B. H. The proportions of mutagens among chemicals in commerce. *Regul. Toxicol. Pharmacol.* **2000**, *32*, 219–225.
54. Klopman, G.; Frierson, M. R.; Rosenkranz, H. S. The structural basis of the mutagenicity of chemicals in *Salmonella typhimurium*: the Gene-Tox data base. *Mutat. Res.* **1990**, *228*, 1–50.
55. Chung, K. T.; Fulk, G. E.; Andrews, A. W. Mutagenicity testing of some commonly used dyes. *Appl. Environ. Microbiol.* **1981**, *42*, 641–648.
56. Tanaka, T. Reproductive and neurobehavioral effects of Allura Red AC administered to mice in the diet. *Toxicology* **1994**, *92*, 169–177.

AMARANTH

CAS Registry Number 915-67-3

Chemical Structure

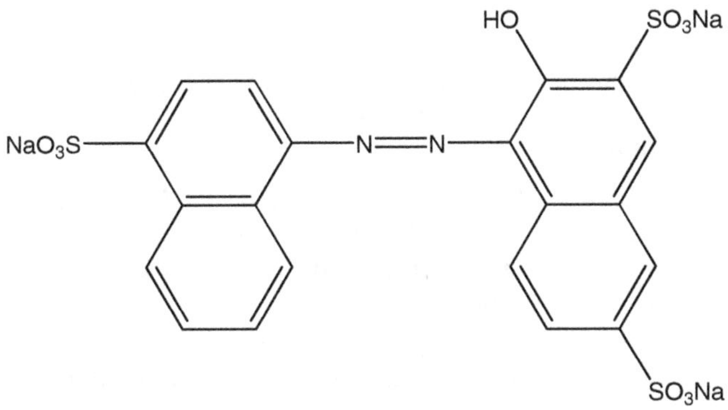

CA Index Name 2,7-Naphthalenedisulfonic acid, 3-hydroxy-4-[2-(4-sulfo-1-naphthalenyl)diazenyl]-, sodium salt (1 : 3)

Other Names 2,7-Naphthalenedisulfonic acid, 3-hydroxy-4-[(4-sulfo-1-naphthalenyl)azo]-, trisodium salt; Bordeaux S; C.I. Acid Red 27; C.I. Acid Red 27, trisodium salt; 1-(4-Sulfo-1-naphthylazo)-2-naphthol-3,6-disulfonic acid trisodium salt; 1302 Red; 1508 Red; 2-Hydroxy-1,1′-azonaphthalene-3,6,4′-trisulfonic acid trisodium salt; Acid Amaranth; Acid Amaranth I; Acid Amaranth J; Acid Amaranth N; Acid Leather Red I 2BW; Acid Leather Rubine S; Acid Red 27; Aizen Amaranth; Amacid Amaranth; Amaranth; Amaranth (dye); Amaranth 307018; Amaranth 36010; Amaranth 85; Amaranth A; Amaranth B; Amaranth BPC; Amaranth Extra; Amaranth Lake; Amaranth R; Amaranth S; Amaranth S Specially Pure; Amaranth USP; Amaranth WD; Amaranth red; Amaranthe; Azo Red R; Azo Rubine S-FQ; Azo Rubine SF; Azo Ruby S; Azorubin S; Basovit Red 470E; Bordeaux S Extra Conc. A Export; Bordeaux S Extra Pure A; Borunil Red A-B; C.I. 16185; C.I. Food Red 9; Canacert Amaranth; Certicol Amaranth S; Cogilor Red 318.11; Cranberry Red; D & C Red 2; D and C Red No. 2; D&C Red No. 2; Daishiki Amaranth; Dolkwal Amaranth; Dye Red Raspberry; E 123; Edicol Amaranth; Edicol Supra Amaranth A; Edicol Supra Amaranth AS; Euorcert Allura 311801; Eurocert Amaranth; FD and C Red No. 2; FD&C Red 2; Fast Red; Food Red 2; Food Red 9; Food Red No. 2; Fruit Red A Geigy; HD Amaranth B; HD Amaranth Supra; Hexacert Red No. 2; Hexacol Amaranth B Extra; Hidacid Amaranth; Hidacid Amaranth 21-6008; Hispacid Red AM; Japan Food Red No. 2; Japan Red 2; Japan Red No. 2; Java Amaranth; KCA Foodcol Amaranth A; Kayaku Amaranth; Kayaku Food Colour Red No. 2; Kiton Rubine S; L Red 3; L Red Z 3050; Lissamine Amaranth AC; Maple Amaranth; Naphthol Red LZS; Naphthol Red S; Naphthol Red S Conc. Specially Pure; Naphthol Red S Specially Pure; Naphthol Red SI; Necol Amaranth; Neklacid Red A; PuriColor Red ARE 27; Rakuto Amaranth; Red No. 2; Red dye no. 2; Ritacid Red G; Rouge Solid E; S-Azo Rubine; San-ei Amaranth; Sandal Amaranth I; Shikiso Amaranth; Solar Red O; Takaoka Amaranth; Tertracid Red A; Toyo Amaranth; Triacid Amaranth A; Trisodium salt of 1-(4-sulfo-1-naphthylazo)-2-naphthol-3,6-disulfonic acid; Usacert Red No. 2; Victoria Rubine O; Victory Scarlet; Water Red 176574; Whortleberry Red; Wool Bordeaux 6RK; Wool Red 40F

Merck Index Number 375

Chemical/Dye Class Azo

Molecular Formula $C_{20}H_{11}N_2Na_3O_{10}S_3$

Molecular Weight 604.47

Physical Form Dark reddish-brown powder

Solubility Soluble in water; slightly soluble in ethanol

Melting Point >300 °C

Absorption (λ_{max}) 520 nm

Synthesis Synthetic methods[1–13]

Staining Applications Candies;[14,15] cereals;[14] chewing gums;[14] chocolate confetti;[14] snacks;[14] drinks;[15] micro-organisms;[16] oral liquids;[17] soft drinks;[18] sweetener;[19] drugs;[20] dietary supplements;[20] tablets;[20] granules;[20] drug crystals;[20] pellets;[20] capsules;[20] skin;[21] hairs;[22,23] keratin fibers[24,25]

Biological Applications Detecting membrane-potential change;[26] treating acquired resistance to GABAergic (ARG) agents[27]

Industrial Applications Color filters;[28] inks;[29,30] thermoplastics;[31] laundry detergent;[32] textiles[33,34]

Safety/Toxicity Acute toxicity;[35] carcinogenicity;[36] cytotoxicity;[37] DNA damage;[38] embryotoxicity;[39] genotoxicity;[40–42] hepatotoxicity;[43] mutagenicity;[44–48] neurotoxicity;[49] teratogenicity[50,51]

REFERENCES

1. Lin, X. Method for extracting edible amaranth red pigment from Amaranthus. Faming Zhuanli Shenqing Gongkai Shuomingshu CN 101133825, 2008; *Chem. Abstr.* **2008**, *148*, 330343.

2. Yu, S.; Takeda, S.; Tari, I.; Nishijima, S.; Nakahira, A. Development of a recovery process for organic dyes using high-gradient magnetic separation. *Teion Kogaku* **2003**, *38*, 77–82; *Chem. Abstr.* **2003**, *139*, 198787.
3. Yamada, M.; Kato, Y.; Nakamura, M.; Yamada, T.; Maitani, T.; Goda, Y. Structural determination of unknown subsidiary colors in commercial food red Nos. 2 and 102. *Chem. Pharm. Bull.* **1998**, *46*, 494–499.
4. Mogi, H.; Takahashi, M.; Inoe, H.; Sakashita, M. Industrial manufacture of Bordeaux solution. Jpn. Kokai Tokkyo Koho JP 08073307, 1996; *Chem. Abstr.* **1996**, *124*, 335657.
5. Liu, P.; Xu, M.; Wang, D. Study on extraction of bordeaux for colorizing food. *Huaxue Shijie* **1995**, *36*, 359–362; *Chem. Abstr.* **1996**, *124*, 258777.
6. Kabasawa, Y.; Tanimura, T.; Nakazawa, H.; Shinomiya, K. Application of counter alternative current chromatography to purification of food mono-azo dyes. *Anal. Sci.* **1992**, *8*, 351–353.
7. Imai, T.; Iwano, K. Transparent solid cosmetics containing solidified oils and water-soluble dyes. Jpn. Kokai Tokkyo Koho JP 03294212, 1991; *Chem. Abstr.* **1992**, *116*, 158600.
8. Rebhahn, R. W. J.; Cook, W. L. Ultrafiltration purification of food dyes. U.S. Patent 4560746, 1985; *Chem. Abstr.* **1986**, *105*, 116539.
9. Smith, A. M.; Perletz, P. Rapidly soluble food colors. U.S. Patent 3447933, 1969; *Chem. Abstr.* **1969**, *71*, 37606.
10. Pla-Delfina, J. M. Systematic identification of food, drug, and cosmetic azo dyes. *J. Soc. Cosmet. Chem.* **1962**, *13*, 214–244.
11. McKeown, G. G. Separation of amaranth and tartrazine. *J. Assoc. Off. Agric. Chem.* **1954**, *37*, 527–529.
12. Jablonski, C. F. Report on (the determination of) coloring matters in foods. *J. Assoc. Off. Agric. Chem.* **1929**, *12*, 354–360.
13. Jablonski, C. F. Report on (the determination of) coloring matters in foods: (Quantitative separation of amaranth from tartrazine). *J. Assoc. Off. Agric. Chem.* **1928**, *11*, 434–438.
14. Prado, M. A.; Godoy, H. T. Contents of synthetic dyes in foods determined by high performance liquid chromatography. *Quim. Nova* **2007**, *30*, 268–273; *Chem. Abstr.* **2007**, *146*, 480797.
15. Yoshioka, N.; Ichihashi, K. Determination of 40 synthetic food colors in drinks and candies by high-performance liquid chromatography using a short column with photodiode array detection. *Talanta* **2008**, *74*, 1408–1413.
16. Noda, N.; Mizutani, T. Microorganism-measuring method using multiple staining. Jpn. Kokai Tokkyo Koho JP 2006340684, 2006; *Chem. Abstr.* **2006**, *146*, 77563.
17. Kamiyasu, H. Odorless oral liquid preparations containing methionine and vitamin B2. Jpn. Kokai Tokkyo Koho JP 2006111566, 2006; *Chem. Abstr.* **2006**, *144*, 419709.
18. Gianotti, V.; Angioi, S.; Gosetti, F.; Marengo, E.; Gennaro, M. C. Chemometrically assisted development of IP-RP-HPLC and spectrophotometric methods for the identification and determination of synthetic dyes in commercial soft drinks. *J. Liq. Chromatogr. Relat. Technol.* **2005**, *28*, 923–937.
19. Mori, K. Sweetener composition. U.S. Pat. Appl. Publ. US 2006257550, 2006; *Chem. Abstr.* **2006**, *145*, 488169.
20. Petereit, H. U.; Meier, C.; Roth, E. Method for spray-coating drugs and dietary supplements with pigment-containing polymers. Ger. Offen. DE 10260920, 2004; *Chem. Abstr.* **2004**, *141*, 59757.
21. Lee, H.; Soistier, N.; Simon, J. C. Cosmetic makeup foundations for darkened skins. Fr. Demande FR 2848822, 2004; *Chem. Abstr.* **2004**, *141*, 76373.
22. Hoeffkes, H.; Gross, W. Hair dyes containing silylcarbonates and/or silylcarbamates for simultaneous dying and bleaching of hair. Ger. Offen. DE 102004052060, 2006; *Chem. Abstr.* **2006**, *144*, 419031.
23. Plos, G. Hair dye composition containing an oxidase alcohol and a direct dye having an acid function or its salt for dyeing keratin fibers. Fr. Demande FR 2865396, 2005; *Chem. Abstr.* **2005**, *143*, 179094.
24. Kravtchenko, S. A composition for simultaneously bleaching and coloring of keratin fibers comprising an anionic or nonionic dye and a polymer. Eur. Pat. Appl. EP 1738800, 2007; *Chem. Abstr.* **2007**, *146*, 128003.
25. Kravtchenko, S. Agent for simultaneously bleaching and coloring of keratin fibers comprising an anionic or nonionic dye and an inert organic liquid. Eur. Pat. Appl. EP 1738801, 2007; *Chem. Abstr.* **2007**, *146*, 128004.
26. Kataoka, T.; Matsubara, M.; Hisada, M. Membrane-potential change detecting method, drug screening method, and well plate manufacture. Jpn. Kokai Tokkyo Koho JP 2006126073, 2006; *Chem. Abstr.* **2006**, *144*, 447614.

27. Johnson, J. Methods and compositions for treating ARG. PCT Int. Appl. WO 2006124609, **2006**; *Chem. Abstr.* **2006**, *145*, 500136.

28. Deno, T.; Eichenberger, T.; Voigt, J. Use of pyrimido [5,4-*g*]pteridines as shading component in color filter colorant compositions. PCT Int. Appl. WO 2006037728, 2006; *Chem. Abstr.* **2006**, *144*, 400791.

29. Mukai, H. Polyoxyalkylene surfactant-containing water-thinned jet-printing inks with good breakage prevention of heads, their manufacture, and containers, heads, and printers containing them. Jpn. Kokai Tokkyo Koho JP 2006016532, 2006; *Chem. Abstr.* **2006**, *144*, 130597.

30. Yamazaki, H.; Tsuda, M. Aqueous jet-printing inks containing acid and basic dyes without precipitating nor clogging. Jpn. Kokai Tokkyo Koho JP 2004263151, 2004; *Chem. Abstr.* **2004**, *141*, 262228.

31. Van Den Oever, M. J. A.; Boeriu, C. G.; Blaauw, R.; Van Haveren, J. Colorants based on renewable resources and food-grade colorants for application in thermoplastics. *J. Appl. Polym. Sci.* **2004**, *92*, 2961–2969.

32. Kabuto, S.; Tamura, N.; Nagayasu, K.; Nagata, S. Laundry detergent compositions containing coated particles contain-ing pigments and/or fluorescent agents. Jpn. Kokai Tokkyo Koho JP 2005179530, 2005; *Chem. Abstr.* **2005**, *143*, 117172.

33. Grier, W. Time and temperature additive scheduling during fabric sublimation. PCT Int. Appl. WO 2007112037, 2007; *Chem. Abstr.* **2007**, *147*, 428814.

34. Jocic, D.; Vilchez, S.; Topalovic, T.; Molina, R.; Navarro, A.; Jovancic, P.; Julia, M. R.; Erra, P. Effect of low-temperature plasma and chitosan treatment on wool dyeing with acid red 27. *J. Appl. Polym. Sci.* **2005**, *97*, 2204–2214.

35. Tonogai, Y.; Iwaida, M.; Tati, M.; Ose, Y.; Sato, T. Biochemical decomposition of coal-tar dyes. II. Acute toxicity of coal-tar dyes and their decomposed products. *J. Toxicol. Sci.* **1978**, *3*, 205–214.

36. Johnson, F. M. Carcinogenic chemical-response "fingerprint" for male F344 rats exposed to a series of 195 chemicals: implications for predicting carcinogens with transgenic models. *Environ. Mol. Mutagen.* **1999**, *34*, 234–245.

37. Yamashoji, S.; Isshiki, K. Rapid detection of cytotoxicity of food additives and contaminants by a novel cytotoxicity test, menadione-catalyzed H_2O_2 production assay. *Cytotechnology* **2002**, *37*, 171–178.

38. Tsuda, S.; Murakami, M.; Matsusaka, N.; Kano, K.; Taniguchi, K.; Sasaki, Y. F. DNA damage induced by red food dyes orally administered to pregnant and male mice. *Toxicol. Sci.* **2001**, *61*, 92–99.

39. Piersma, A. H.; Attenon, P.; Bechter, R.; Govers, M. J. A. P.; Krafft, N.; Schmid, B. P.; Stadler, J.; Verhoff, A.; Verseil, C. Interlaboratory evaluation of embryotoxicity in the postimplantation rat embryo culture. *Reprod. Toxicol.* **1995**, *9*, 275–280.

40. Das, A.; Mukherjee, A. Genotoxicity testing of the food colours amaranth and tartrazine. *Int. J. Hum. Genet.* **2004**, *4*, 277–280.

41. Tripathy, N. K.; Nabi, Md. J.; Sahu, G. P.; Kumar, A. A. Genotoxicity testing of two red dyes in the somatic and germ line cells of *Drosophila*. *Food Chem. Toxicol.* **1995**, *33*, 923–927.

42. Haveland-Smith, R. B.; Combes, R. D. Screening of food dyes for genotoxic activity. *Food Cosmet. Toxicol.* **1980**, *18*, 215–221.

43. Yoshimoto, M.; Yamaguchi, M.; Hatano, S.; Watanabe, T. Configurational changes in rat liver nuclear chromatin caused by azo dyes. *Food Chem. Toxicol.* **1984**, *22*, 337–344.

44. Varella, S. D.; Pozetti, G. L.; Vilegas, W.; Varanda, E. A. Mutagenic activity of sweepings and pigments from a household wax factory assayed with *Salmonella typhimurium*. *Food Chem. Toxicol.* **2004**, *42*, 2029–2035.

45. Rosenkranz, H. S.; Klopman, G. Structural basis of the mutagenicity of 1-amino-2-naphthol-based azo dyes. *Mutagenesis* **1990**, *5*, 137–146.

46. Ishidate, M., Jr.; Sofuni, T.; Yoshikawa, K.; Hayashi, M.; Nohmi, T.; Sawada, M.; Matsuoka, A. Primary mutagenicity screening of food additives currently used in Japan. *Food Chem. Toxicol.* **1984**, *22*, 623–636.

47. Al-Mossawi, M. A. J. The mutagenic effect of amaranth (FD and C Red No. 2) in bacteria and yeast. *Environ. Int.* **1983**, *9*, 145–148.

48. Chung, K. T.; Fulk, G. E.; Andrews, A. W. Mutagenicity testing of some commonly used dyes. *Appl. Environ. Microbiol.* **1981**, *42*, 641–648.

49. Khera, K. S.; Whalen, C. Detection of neuroteratogens with an *in vitro* cytotoxicity assay using primary monolayers cultured from dissociated fetal rat brains. *Toxicol. In Vitro* **1988**, *2*, 257–273.

50. Cicurel, L.; Schmid, B. P. Post-implantation embryo culture: validation with selected compounds for teratogenicity testing. *Xenobiotica* **1988**, *18*, 617–624.

51. Khera, K. S.; Roberts, G.; Trivett, G.; Terry, Geraldine; Whalen, C. A teratogenicity study with amaranth in cats. *Toxicol. Appl. Pharmacol.* **1976**, *38*, 389–398.

9-AMINO-6-CHLORO-2-METHOXYACRIDINE (ACMA)

CAS Registry Number 3548-09-2

Chemical Structure

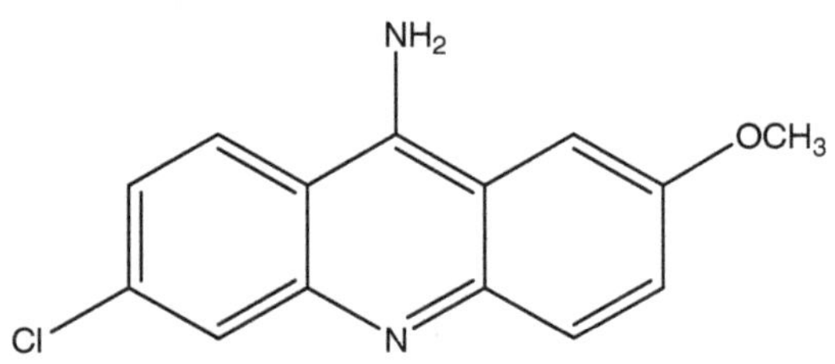

CA Index Name 9-Acridinamine, 6-chloro-2-methoxy-

Other Names Acridine, 9-amino-6-chloro-2-methoxy-; 2-Methoxy-6-chloro-9-aminoacridine; 3-Chloro-7-methoxy-9-aminoacridine; 6-Chloro-9-amino-2-methoxyacridine; 9-Amino-3-chloro-7-methoxyacridine; 9-Amino-6-chloro-2-methoxyacridine; G 185; NSC 15300

Merck Index Number Not listed

Chemical/Dye Class Acridine

Molecular Formula $C_{14}H_{11}ClN_2O$

Molecular Weight 258.70

Physical Form Yellow crystals

Solubility Soluble in methanol, *N,N*-dimethylformamide, dimethylsulfoxide

Melting Point 272–274 °C

Boiling Point (Calcd.) 475.1 ± 35.0 °C, pressure: 760 Torr

pK_a 8.6

Absorption (λ_{max}) 412 nm

Emission (λ_{max}) 471 nm

Synthesis Synthetic methods[1–10]

Staining Applications Nucleic acids;[11–15] cells;[16] chromosomes;[17] microorganisms[18]

Biological Applications Antimalarial;[19,20] bactericidal;[21,22] detection of cancer cells,[23] nucleic acids;[11–15] treating malformed proteins causing neurodegenerative disease[24]

Industrial Applications Not reported

Safety/Toxicity Genotoxicity;[25] mutagenicity[26,27]

REFERENCES

1. Mansour, M.; Thaller, S.; Parlar, H.; Korte, F. Photoinduced reaction of Atebrine. *Z. Naturforsch. B* **1984**, *39B*, 1626–1628; *Chem. Abstr.* **1985**, *102*, 184582.
2. Shibnev, V. A.; Finogenova, M. P.; Gazumyan, A. K.; Poletaev, A. I.; Mar'yash, L. I. 2-Methoxy-6,9-dichloroacridine in peptide synthesis as a fluorescent label. *Bioorg. Khim.* **1984**, *10*, 610–617; *Chem. Abstr.* **1984**, *101*, 171702.
3. Albert, A. Acridine syntheses and reactions. VI. A new dehalogenation of 9-chloroacridine and its derivatives. Further acridine ionization constants and ultraviolet spectra. *J. Chem. Soc.* **1965**, 4653–4657.
4. Kitani, K. The syntheses of 9-substituted acridines. III. Reaction of 9-amino- and 9-(alkylamino) acridines with amines. *Nippon Kagaku Kaishi* **1954**, *75*, 477–480; *Chem. Abstr.* **1955**, *49*, 53549.
5. Barber, H. J. Amino-substituted acridines. U.S. Patent 2450367, 1948; *Chem. Abstr.* **1949**, *43*, 2882.
6. Barber, H. J.; Wilkinson, J. H.; Edwards, W. G. H. The reaction of 9-alkoxyacridines with bases and their salts. *J. Soc. Chem. Ind.* **1947**, *66*, 411–415.
7. Guha, P. C.; Mukherjee, S. P. Synthesis of new antimalarial drugs related to atebrin. II. *J. Indian Inst. Sci.* **1946**, *28A*, 70–74.
8. Barber, H. J. 9-Aminoacridines. Brit. GB 581695, 1946; *Chem. Abstr.* **1947**, *41*, 17272.
9. Albert, A.; Goldacre, R.; Heymann, E. Amino acridines: some partition and surface phenomena. *J. Chem. Soc.* **1943**, 651–654.
10. Gerchuk, M. P.; Arbuzova, P. G.; Kel'manskaya, I. A. Synthesis of new chemotherapeutic pyroplasmocidic compounds. II. Synthesis of acridyl-substituted ureas. *Zh. Obshch. Khim.* **1941**, *11*, 948–953; *Chem. Abstr.* **1943**, *37*, 2075.
11. Park, H. O.; Kim, H. B.; Chi, S. M. Detection method of DNA amplification using probes labeled with intercalating dyes. PCT Int. Appl. WO 2006004267, 2006; *Chem. Abstr.* **2006**, *144*, 122707.
12. MacFarlane, D. E. 9-Aminoacridine and 9-aminoquinoline compounds inhibiting immunostimulatory DNA-associated responses. U.S. Pat. Appl. Publ. US 2003232856, 2003; *Chem. Abstr.* **2004**, *140*, 386030.
13. McNally, A. J.; Wu, R. S.; Li, Z. Immunoassay based on DNA replication using labeled primer. U.S. Pat.

Appl. Publ. US 2002072053, 2002; *Chem. Abstr.* **2002**, *137*, 30238.

14. Chin, A. M. A library of modified primers for nucleic acid sequencing, and method of use thereof. PCT Int. Appl. WO 2000028087, 2000; *Chem. Abstr.* **2000**, *133*, 1455.
15. Zeleznick, L. D.; Crim, J. A.; Gray, G. D. Immunosuppression by compounds which complex with deoxyribonucleic acid. *Biochem. Pharmacol.* **1969**, *18*, 1823–1827.
16. Teuber, M.; Rogner, M.; Berry, S. Fluorescent probes for non-invasive bioenergetic studies of whole cyanobacterial cells. *Biochim. Biophys. Acta, Bioenerg.* **2001**, *1506*, 31–46.
17. Tsou, K. C.; Giles, B.; Kohn, G. Chemical basis of chromosome banding patterns. *Stain Technol.* **1975**, *50*, 293–295.
18. Noda, N.; Mizutani, T. Microorganism-measuring method using multiple staining. Jpn. Kokai Tokkyo Koho JP 2006340684, 2006; *Chem. Abstr.* **2006**, *146*, 77563.
19. Winter, R. W.; Kelly, J. X.; Smilkstein, M. J.; Dodean, R.; Bagby, G. C.; Rathbun, R. K.; Levin, J. I.; Hinrichs, D.; Riscoe, M. K. Evaluation and lead optimization of antimalarial acridones. *Exp. Parasitol.* **2006**, *114*, 47–56.
20. Girault, S.; Delarue, S.; Grellier, P.; Berecibar, A.; Maes, L.; Quirijnen, L.; Lemiere, P.; Debreu-Fontaine, M. A.; Sergheraert, C. Antimalarial *in-vivo* activity of bis(9-amino-6-chloro-2-methoxyacridines). *J. Pharm. Pharmacol.* **2001**, *53*, 935–938.
21. Wainwright, M.; Phoenix, D. A.; Marland, J.; Wareing, D. R. A.; Bolton, F. J. A comparison of the bactericidal and photobactericidal activities of aminoacridines and bis(aminoacridines). *Lett. Appl. Microbiol.* **1998**, *26*, 404–406.
22. Wainwright, M.; Phoenix, D. A.; Marland, J.; Wareing, D. R. A.; Bolton, F. J. *In vitro* photobactericidal activity of amino-acridines. *J. Antimicrob. Chemother.* **1997**, *40*, 587–589.
23. Schwarz, G.; Wittekind, D. Selected aminoacridines as fluorescent probes in cytochemistry in general and in the detection of cancer cells in particular. *Anal. Quant. Cytol.* **1982**, *4*, 44–54.
24. Prusiner, S. B.; Korth, C.; May, B. C. H. Cyclic bis-compounds clearing malformed proteins. U.S. Pat. Appl. Publ. US 2004229898, 2004; *Chem. Abstr.* **2004**, *141*, 406119.
25. He, L.; Jurs, P. C.; Custer, L. L.; Durham, S. K.; Pearl, G. M. Predicting the genotoxicity of polycyclic aromatic compounds from molecular structure with different classifiers. *Chem. Res. Toxicol.* **2003**, *16*, 1567–1580.
26. Henry, D. R.; Lavine, B. K.; Jurs, P. C. Electronic factors and acridine frameshift mutagenicity—a pattern recognition study. *Mutat. Res.* **1987**, *179*, 115–121.
27. Brown, B. R.; Firth, W. J., III; Yielding, L. W. Acridine structure correlated with mutagenic activity in *Salmonella*. *Mutat. Res.* **1980**, *72*, 373–388.

ANILINE BLUE (METHYL BLUE)

CAS Registry Number 28983-56-4

Chemical Structure

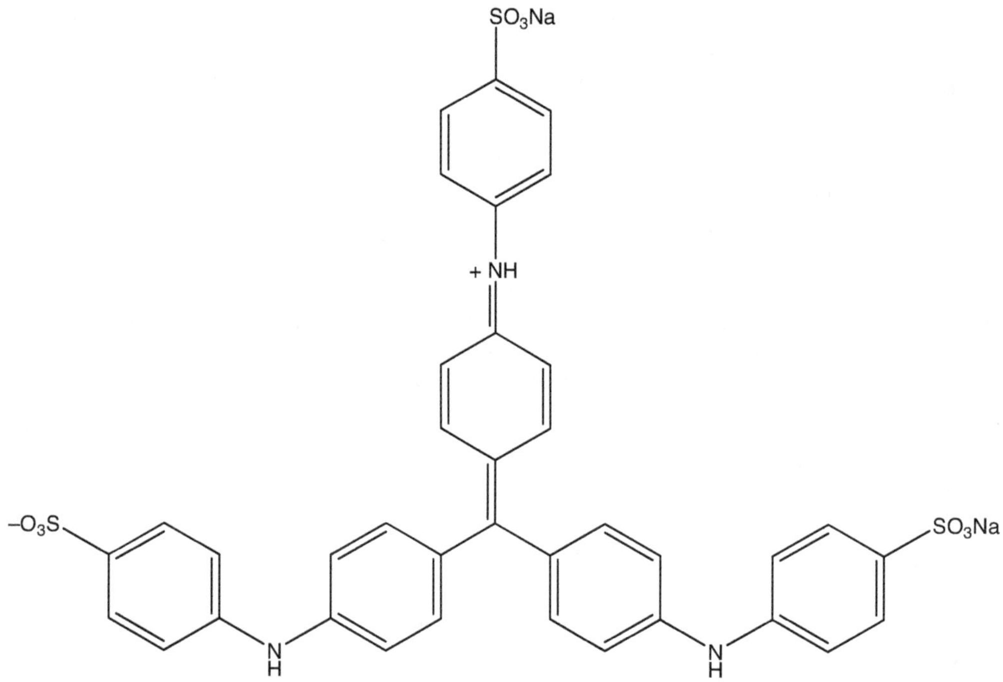

Merck Index Number 6028

Chemical/Dye Class Triphenylmethane

Molecular Formula $C_{37}H_{27}N_3Na_2O_9S_3$

Molecular Weight 799.80

CA Index Name Benzenesulfonic acid, [[4-[bis[4-[(sulfophenyl)amino]phenyl]methylene]-2,5-cyclohexadien-1-ylidene]amino]-, sodium salt (1 : 2)

Other Names Benzenesulfonic acid, [[4-[bis[4-[(sulfophenyl)amino]phenyl]methylene]-2,5-cyclohexadien-1- ylidene]amino]-, disodium salt; C.I. Acid Blue 93; C.I. Acid Blue 93, disodium salt; Methyl blue; Acid Blue 93; Acid Blue FG; Acid Ink Blue G; Acid Leather Blue HER; Aniline Blue; Aniline Blue water soluble; Brilliant Lake Blue G; C.I. 42780; Conacid Blue NC; Cotton blue; Dycosacid Ink Blue G; Helvetia Blue; Helvetia Blue I; Helvetia Blue Pure I; Ink Blue; Ink Blue BA; Ink Blue BJT; Ink Blue BJTBN 80; Ink Blue BJTBNA 80; Ink Blue BJTN; Ink Blue G; Ink Blue M; Ink Blue N; Ink Blue Special; Ink Blue WGS; Ink Blue WRS; Methyl blue (biological stain); Orient Soluble Blue OBB; Orient Soluble Blue OBC; Orient Soluble Blue OBX; Poirriers Blue; Pure Soluble Blue I; Silk Blue H; Sky Blue G; Solar Soluble Blue BN; Soluble Blue; Soluble Blue 8B; Soluble Blue OBB; Soluble Blue OBX; Special Soluble Blue HT; Vicoacid Ink Blue; Water Blue; Water Blue 6B; Water Blue B; Water Blue I Old Type 1240a; Water Blue IN

Physical Form Dark blue or dark brown powder

Solubility Soluble in water, ethanol; insoluble in xylene

Melting Point >250 °C

Absorption (λ_{max}) 600 nm

Synthesis Synthetic methods[1–9]

Staining Applications Cartilage;[10] chitin;[11] hyaline droplets;[12] louse eggs;[13] mitochondria;[14] proteins;[15] sphagnum moss;[16] urine sediment;[17]

Biological Applications Assay for nucleic acids,[18] protein determination,[19] identification of *Candida albicans*;[20] method for counting leukocytes,[21] controlled drug release;[22] radiotherapy;[23] treating apolipoprotein E-related diseases;[24] oral care products[25]

Industrial Applications Color filter;[26,27] magnetic toners;[28] inks;[29] highlighters;[30] colored pencils;[31] packaging materials;[32] paints;[33] textiles;[34] asbestos content measurement;[35] paper products[36]

Safety/Toxicity Mutagenicity[37]

Certification/Approval Certified by Biological Stain Commission (BSC)

REFERENCES

1. Luo, G.; Lu, Y.; Zhu, S.; Dai, Y. Recovery of dyestuffs from dilute solution with two-phase electrophoresis. *Chin. J. Chem. Eng.* **2000**, *8*, 80–84.
2. Phillips, N. J.; Porter, D. Organically accelerated bleaches: their role in holographic image formation. *J. Phys. E* **1977**, *10*, 96–98.
3. Furlotti, A. Ball point pen ink. U.S. Patent 3468679, 1969; *Chem. Abstr.* **1969**, *71*, 126166.
4. Safarikova, M.; Ptackova, L.; Kibrikova, I.; Safarik, I. Biosorption of water-soluble dyes on magnetically modified *Saccharomyces cerevisiae* subsp. uvarum cells. *Chemosphere* **2005**, *59*, 831–835.
5. Yucesoy, M.; Esen, N.; Yulug, N. Use of chromogenic tube and methyl blue-sabouraud agar for the identification of *Candida albicans* strains. *Kobe J. Med. Sci.* **2001**, *47*, 161–167.
6. Li, C.; Zhang, C. Cobalt-free blue color-changeable indicator and its preparation. Faming Zhuanli Shenqing Gongkai Shuomingshu CN 1740781, 2006; *Chem. Abstr.* **2006**, *145*, 64452.
7. Kern, A. Manufacture of soluble methyl-blue from rosaniline. U.S. Patent 00333861, 1886; *Chem. Abstr.* **1906**, *0*, 138362.
8. Kern, A. Methyl-blue coloring-matter. U.S. Patent 00297415, 1884; *Chem. Abstr.* **1906**, *0*, 136719.
9. Kern, A. Methyl-blue coloring-matter. U.S. Patent 00297414, 1884; *Chem. Abstr.* **1906**, *0*, 136718.
10. Yamada, T. Selective staining methods for cartilage of rat fetal specimens previously treated with alizarin red S. *Teratology* **1991**, *43*, 615–619.
11. Salthouse, T. N. E. I. Histochemistry and staining of chitin. *J. Histochem. Cytochem.* **1962**, *10*, 109.
12. De Rijk, E. P. C. T.; Ravesloot, W. T. M.; Wijnands, Y.; Van Esch, E. A fast histochemical staining method to identify hyaline droplets in the rat kidney. *Toxicol. Pathol.* **2003**, *31*, 462–464.
13. Reid, L. F.; Kross, R. D. Staining method for removing louse nits from hair. U.S. Patent 5972987, 1999; *Chem. Abstr.* **1999**, *131*, 291360.
14. Rashid, F.; Horobin, R. W. Accumulation of fluorescent non-cationic probes in mitochondria of cultured cells: observations, a proposed mechanism, and some implications. *J. Microsc.* **1991**, *163*, 233–241.
15. Becker, K. F.; Schott, C.; Becker, I.; Hoefler, H. Guided protein extraction from formalin-fixed tissues for quantitative multiplex analysis avoids detrimental effects of histological stains. *Proteomics* **2008**, *2*, 737–743.
16. Levesque, Y.; Coude, J.; Pelletier, N. Colored sphagnum moss stained with biological stains. Can. Pat. Appl. CA 2256295, 2000; *Chem. Abstr.* **2005**, *143*, 263616.
17. Matsuoka, M.; Fukada, S.; Shimada, M.; Furuhata, T.; Shimidzu, M.; Itabashi, A.; Okada, S. Investigation of a new urinary staining as an alternative to Sternheimer staining. *Igaku Kensa* **2003**, *52*, 187–193; *Chem. Abstr.* **2003**, *138*, 381508.
18. Cai, C.; Gong, H.; Chen, X. Simple and sensitive assay for nucleic acids by use of the resonance light-scattering technique with the anionic dye methyl blue in the presence of cety-ltrimethylammonium bromide. *Microchim. Acta* **2007**, *157*, 165–171.
19. Dong, L.; Li, Y.; Zhang, Y.; Chen, X.; Hu, Z. A flow injection sampling resonance light scattering system for total protein determination in human serum. *Spectrochim. Acta* **2007**, *66A*, 1317–1322.
20. Goldschmidt, M. C.; Fung, D. Y. C.; Grant, R.; White, J.; Brown, T. New aniline blue dye medium for rapid identification and isolation of *Candida albicans*. *J. Clin. Microbiol.* **1991**, *29*, 1095–1099.
21. Takarada, K.; Kouzuki, C.; Hyousa, Y.; Sakata, T.; Akai, Y. A method for classifying and counting leukocytes. Eur. Pat. Appl. EP 695936, 1996; *Chem. Abstr.* **1996**, *124*, 283704.
22. Fischer, A.; Adde, C.; Hersloef, B. Lipid carriers for controlled drug release. PCT Int. Appl. WO 2001066086, 2001; *Chem. Abstr.* **2001**, *135*, 216022.
23. Lawaczeck, R.; Platzek, J.; Raduchel, B. Metal macrocycles for two-step forms of radiotherapy. PCT Int. Appl. WO 2000045857, 2000; *Chem. Abstr.* **2000**, *133*, 168385.
24. Crutcher, K. A.; Harmony, J. A. K. Methods for the treatment of apolipoprotein E-related diseases. PCT Int. Appl. WO 2000050042, 2000; *Chem. Abstr.* **2000**, *133*, 203003.
25. Ling, J.; Ye, H.; Buckley, C. D. Composition and process for indicating the presence of soluble fluoride ion in oral care compositions and method of making the same. U.S. Pat. Appl. Publ. US 2003031633, 2003; *Chem. Abstr.* **2003**, *138*, 158578.
26. De Keyzer, G.; Lamatsch, B.; Muehlebach, A.; Rime, F.; Schmitt, G. Surface-modified nanoparticles comprising a cationic colorant for use in color

filters. PCT Int. Appl. WO 2008107304, 2008; *Chem. Abstr.* **2008**, *149*, 357452.

27. Shirota, K.; Myazaki, T.; Nakazawa, K.; Yamada, S. Ink-jet printing ink for manufacture of liquid crystal display color filter. Jpn. Kokai Tokkyo Koho JP 08327812, 1996; *Chem. Abstr.* **1997**, *126*, 164272.
28. Suzuki, A. Color magnetic toners. Jpn. Kokai Tokkyo Koho JP 60173553, 1985; *Chem. Abstr.* **1986**, *104*, 99470.
29. Ogasawara, H. Aqueous inks with efficient permeation through cores for fountain pens. Jpn. Kokai Tokkyo Koho JP 2007297519, 2007; *Chem. Abstr.* **2007**, *147*, 523978.
30. Kwan, W. S. V.; Duong, C. Highlighting marking compositions, highlighting kits, highlighted complexes, and application of eradicator. U.S. Pat. Appl. Publ. US 2007017413, 2007; *Chem. Abstr.* **2007**, *146*, 164696.
31. Lugert, G. Leads for colored pencils. Eur. Pat. Appl. EP 924273, 1999; *Chem. Abstr.* **1999**, *131*, 60132.
32. Kato, T. Packaging material containing oxygen indicator. Jpn. Kokai Tokkyo Koho JP 2007147297, 2007; *Chem. Abstr.* **2007**, *147*, 62690.
33. Miller, R. E.; Dereamus, R. C. Color changing systems using pan paint compositions and markers. U.S. Patent 5498282, 1996; *Chem. Abstr.* **1996**, *124*, 346248.
34. De La Torre, M. Method of treating textile carrying a stain. PCT Int. Appl. WO 2008122752, 2008; *Chem. Abstr.* **2008**, *149*, 473420.
35. Oke, Y.; Hashida, T.; Maeda, N. Asbestos color distinguishing method and asbestos content measurement. Jpn. Kokai Tokkyo Koho JP 2008209392, 2008; *Chem. Abstr.* **2008**, *149*, 369257.
36. Allison, K. J.; Li, J.; Rowan, D.; Kressler, A. Use of reducing agents in ruled paper products. U.S. Pat. Appl. Publ. US 2008213483, 2008; *Chem. Abstr.* **2008**, *149*, 309689.
37. Ferguson, L. R.; Baguley, B. C. Verapamil as a co-mutagen in the *Salmonella*/mammalian microsome mutagenicity test. *Mutat. Res. Lett.* **1988**, *209*, 57–62.

AURAMINE O

CAS Registry Number 2465-27-2

Chemical Structure

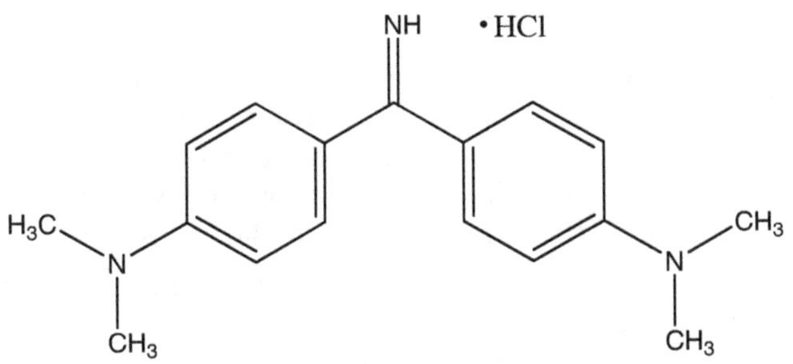

CA Index Name Benzenamine, 4,4′-carbonimidoylbis[*N*,*N*-dimethyl-, hydrochloride (1 : 1)

Other Names Benzenamine, 4,4′-carbonimidoylbis[*N*,*N*-dimethyl-, monohydrochloride; C.I. Basic Yellow 2; C.I. Basic Yellow 2, monohydrochloride; ADC Auramine O; Aizen Auramine; Aizen Auramine OH; Aizen Auramine OW 100; Aizen Auramine conc. SFA; Arazine Yellow; Auramin; Auramine; Auramine 0-100; Auramine A1; Auramine Extra; Auramine Extra 0-100; Auramine Extra 0-125; Auramine Extra Conc. A; Auramine FA; Auramine FWA; Auramine II; Auramine Lake Yellow O; Auramine N; Auramine O; Auramine O Extra Conc. A Export; Auramine ON; Auramine OO; Auramine OOO; Auramine OS; Auramine Pure; Auramine SP; Auramine Yellow; Auramine chloride; Auramine hydrochloride; Basic Flavine Yellow O; Basic Light Yellow; Basic Light Yellow O; Basic Yellow 2; Basonyl Yellow 120; C.I. 41000; Calcozine Yellow OX; Flexo Yellow 110; Mitsui Auramine O

Merck Index Number Not listed

Chemical/Dye Class Diphenylmethane

Molecular Formula $C_{17}H_{22}ClN_3$

Molecular Weight 303.84

Physical Form Yellow powder

Solubility Soluble in water, ethanol; very slightly soluble in xylene

Melting Point >250 °C (decompose)

pK$_a$ 9.8, 10.7

Absorption (λ_{max}) 370 nm, 432 nm

Emission (λ_{max}) 550 nm

Synthesis Synthetic method[1–18]

Staining Applications Acid fast bacilli;[19] cells;[20,21] nuclei;[22] smear;[23] sputum;[24] hairs[25]

Biological Applications Detecting bacteria;[26,27] treatment of cancers,[28] cardiovascular disease,[29] diabetes-associated pain,[30] mechanical allodynia,[31] metabolic syndrome,[32] protozoan infections in fish;[33] targeted drug delivery;[34] as fungal inhibitors[35]

Industrial Applications Color filter;[36] plasma displays;[37] semiconductor devices;[38] inks;[39] highlighters;[40] image-recording materials;[41] printing plates;[42] toners;[43] adhesives;[44] fuels;[45] pesticides;[46] textiles;[47,48] wood[49]

Safety/Toxicity Acute oral toxicity;[50] carcinogenicity;[51–53] cytotoxicity;[54] DNA damage;[55] genotoxicity;[56,57] mutagenicity[58,59]

Certification/Approval Certified by Biological Stain Commission (BSC)

REFERENCES

1. Wang, G.; Lu, C. Synthesis of auramine by iminoation. *Huagong Shikan* **1998**, *12*, 36–37; *Chem. Abstr.* **1998**, *129*, 202707.
2. Rogers, R.; Horwitz, E. P.; Bond, A. H. Process for separating and recovering an anionic dye from an aqueous solution. U.S. Patent 5707525, 1998; *Chem. Abstr.* **1998**, *128*, 142665.
3. Schmeidl, K.; Eisert, M. Auramine chloride. Ger. Offen. DE 2261064, 1974; *Chem. Abstr.* **1974**, *81*, 137595.
4. Grosklos, R. L. Solvent process for preparing auramine and ethyl auramine. U.S. Patent 3668252, 1972; *Chem. Abstr.* **1972**, *77*, 103332.
5. Aida, Y.; Ishii, T.; Kubota, H.; Nakamura, A. Auramine. Jpn. Tokkyo Koho JP 45032545, 1970; *Chem. Abstr.* **1971**, *74*, 65583.
6. Grosklos, R. L. Auramine and its tetraethyl analogue. Ger. Offen. DE 1958330, 1970; *Chem. Abstr.* **1970**, *73*, 121557.
7. Korsunovskii, G. A. Photochemical transformations of an auramine base in carbon tetrachloride. *Zh. Fiz. Khim.* **1969**, *43*, 2537–2542; *Chem. Abstr.* **1970**, *72*, 61308.
8. Schoenberg, A.; Junghans, K. Preparation of imines by the action of liquid ammonia on geminal dihalides. *Chem. Ber.* **1966**, *99*, 1015–1018; *Chem. Abstr.* **1966**, *64*, 84474.
9. Nicolaus, B. J. R.; Testa, E. *O*-(β-Acyloxyethyl)-*N*,*N*-dialky-lhydroxylamines. U.S. Patent 3184500, 1965; *Chem. Abstr.* **1965**, *63*, 62713.
10. Grosklos, R. L. Auramine. BE 649474, 1964; *Chem. Abstr.* **1965**, *63*, 62712.

11. Weyker, R. G.; Yarrington, R. M.; Gerber, S. M. Auramines. BE 615885, 1962; *Chem. Abstr.* **1963**, *58*, 66278.
12. Konz, W.; Sehring, R. Diaryloxyalkanes and α,α-diaryloxy fatty acids. DE 891549, 1953; *Chem. Abstr.* **1958**, *52*, 77477.
13. Endo, K.; Aoki, T.; Mimori, H.; Ishikawa, T. Diarylketimines. JP 32008086, 1957; *Chem. Abstr.* **1958**, *52*, 77476.
14. Matsuo, M.; Watanabe, M. Diarylketimine. JP 26000111, 1951; *Chem. Abstr.* **1953**, *47*, 28849.
15. Simakov, D.; Konovalenko, M. Production of auramine. *Promysh. Organ. Khim.* **1936**, *2*, 206–209; *Chem. Abstr.* **1937**, *31*, 7814.
16. Lukin, A. M. Preparation of auramine by Sandmeyer's method. *Zh. Khim. Promysh.* **1928**, *5*, 34–37; *Chem. Abstr.* **1928**, *22*, 21995.
17. Sansone, R. The present and the future in the manufacture of auramines. *Color Trade J.* **1922**, *10*, 53–55, 119–121.
18. Guyot, A. Synthesis of auramines by means of diethyl oxalate. *Compt. Rend.* **1907**, *144*, 1219–1220; *Chem. Abstr.* **1907**, *1*, 10803.
19. Cserni, G. Elimination of phenol from the auramine fluorescence staining method for acid fast bacilli. *J. Histotechnol.* **1998**, *21*, 241–242.
20. Garini, Y.; Mcnamara, G.; Soenksen, D. G.; Cabib, D.; Buckwald, R. A. *In situ* method of analyzing cells by staining with multiple stains and using a spectral data collection device. PCT Int. Appl. WO 2000031534, 2000; *Chem. Abstr.* **2000**, *133*, 14300.
21. Sakata, T.; Matsumoto, H. A staining reagent and method for detecting malaria infected cells using Auramine analog and condensed benzene derivative. Eur. Pat. Appl. EP 613003, 1994; *Chem. Abstr.* **1994**, *121*, 225851.
22. Horobin, R. W.; Stockert, J. C.; Rashid-Doubell, F. Fluorescent cationic probes for nuclei of living cells: why are they selective? A quantitative structure–activity relations analysis. *Histochem. Cell Biol.* **2006**, *126*, 165–175.
23. Affolabi, D.; Bankole, H.; Ablordey, A.; Hounnouga, J.; Koutchakpo, P.; Sopoh, G.; Aguiar, J.; Dossou, A.; Johnson, R. C.; Anagonou, S.; Portaels, F. Effects of grinding surgical tissue specimens and smear staining methods on Buruli ulcer microscopic diagnosis. *Trop. Med. Int. Health* **2008**, *13*, 187–190.
24. Strumpf, I. J.; Tsang, A. Y.; Sayre, J. W. Re-evaluation of sputum staining for the diagnosis of pulmonary tuberculosis. *Am. Rev. Respir. Dis.* **1979**, *119*, 599–602.
25. Brun, G.; Gourlaouen, L. Hair-dyeing composition containing a reactive silicone, a fluorescent dye or optical blueing agent, and a process of coloring using the composition. Fr. Demande FR 2910298, 2008; *Chem. Abstr.* **2008**, *149*, 111757.
26. Jeyanathan, M.; Alexander, D. C.; Turenne, C. Y.; Girard, C.; Behr, M. A. Evaluation of *in situ* methods used to detect *Mycobacterium avium* subsp. paratuberculosis in samples from patients with Crohn's disease. *J. Clin. Microbiol.* **2006**, *44*, 2942–2950.
27. Oka, A. Dyeability discriminant method for identifying Gram negative bacteria. Jpn. Kokai Tokkyo Koho JP 2004208526, 2004; *Chem. Abstr.* **2004**, *141*, 119787.
28. Wang, H. C. R. Antibody compositions specific to p33QIK and p63krs1 polypeptides for diagnosis, prognosis and treatment of cell proliferation diseases or cancers. U.S. Pat. Appl. Publ. US 2003170252, 2003; *Chem. Abstr.* **2003**, *139*, 244712.
29. Foulkes, J. G.; Liechtfried, F. E.; Pieler, C.; Stephenson, J. R.; Case, C. C. Reporter gene methods for identification of compounds that modulate transcription of genes associated with cardiovascular disease. U.S. Patent 5580722, 1996; *Chem. Abstr.* **1996**, *126*, 102570.
30. Buschman, H. H. Use of compounds binding to the sigma receptor for the treatment of diabetes-associated pain. Eur. Pat. Appl. EP 1787679, 2007; *Chem. Abstr.* **2007**, *146*, 514793.
31. Baeyens Cabrera, J. M. Use of compounds active on the sigma receptor for the treatment of mechanical allodynia. PCT Int. Appl. WO 2006010587, 2006; *Chem. Abstr.* **2006**, *144*, 184692.
32. Buschmann, H. H.; Vela Hernandez, J. M. Use of compounds binding to the sigma receptor for the treatment of metabolic syndrome. Eur. Pat. Appl. EP 1829534, 2007; *Chem. Abstr.* **2007**, *147*, 315088.
33. Blair, B. G. Method of treatment of protozoan infections in fish. U.S. Pat. Appl. Publ. US 2002037921, 2002; *Chem. Abstr.* **2002**, *136*, 260249.
34. Safavy, A. Single-drug multi-ligand conjugates for targeted drug delivery. PCT Int. Appl. WO 2006060664, 2006; *Chem. Abstr.* **2006**, *145*, 50997.
35. Bragulat, M. R.; Abarca, M. L.; Castella, G.; Cabanes, F. J. Dyes as fungal inhibitors: effect on colony enumeration. *J. Appl. Bacteriol.* **1995**, *79*, 578–582.
36. Kudo, T.; Fukuzawa, J. Photosensitive resin composition for color filters. Jpn. Kokai Tokkyo

Koho JP 07325392, 1995; *Chem. Abstr.* **1996**, *124*, 274634.

37. Yamaguchi, N.; Suzuki, M. Plasma display panel, glass substrates with dielectric layer containing color material, and its manufacture. Jpn. Kokai Tokkyo Koho JP 2005327542, 2005; *Chem. Abstr.* **2005**, *143*, 485917.
38. Goto, T. Process control of etchant solutions in manufacture of semiconductor devices and planar display devices. Jpn. Kokai Tokkyo Koho JP 2002344111, 2002; *Chem. Abstr.* **2002**, *137*, 392085.
39. Wakae, T. Aqueous color ink compositions with good decoloration resistance for writing implements and writing implement sets. Jpn. Kokai Tokkyo Koho JP 2004083767, 2004; *Chem. Abstr.* **2004**, *140*, 255111.
40. Davies-Smith, L.; Sum, V. W. Highlightable and highlighted mixtures, marking instruments, eradicator solution and kits, using the mixtures in highlighting. U.S. Pat. Appl. Publ. US 2005120919, 2005; *Chem. Abstr.* **2005**, *143*, 9331.
41. Murota, Y. Photopolymerizable image-recording material containing 350–450-nm light absorbing dye. Jpn. Kokai Tokkyo Koho JP 2003107718, 2003; *Chem. Abstr.* **2003**, *138*, 311593.
42. Baumann, H. Organic photoconductive composition useful for the preparation of printing plates. PCT Int. Appl. WO 2002012963, 2002; *Chem. Abstr.* **2002**, *136*, 175513.
43. Orihara, M.; Tosaka, H.; Tomita, M.; Tanaka, K.; Hagiwara, T. Electrostatographic toner and developing method. Jpn. Kokai Tokkyo Koho JP 63318569, 1988; *Chem. Abstr.* **1989**, *110*, 202865.
44. Hoffmann, G.; Richter, P. Adhesives for relief printing plates. Eur. Pat. Appl. EP 59385, 1982; *Chem. Abstr.* **1983**, *98*, 225306.
45. Thrasher, D. A.; Greiner, L.; Cooper, G. Methanol fuel mixture. U.S. Patent 4932979, 1990; *Chem. Abstr.* **1990**, *113*, 155654.
46. Wang, H. Colored smoke compositions, especially for fireworks and pesticide formulations. Faming Zhuanli Shenqing Gongkai Shuomingshu CN 101070259, 2007; *Chem. Abstr.* **2007**, *148*, 36050.
47. Gamblin, R. L. Surfactant-enhanced dyeing of textiles. U.S. Patent 5593459, 1997; *Chem. Abstr.* **1997**, *126*, 187320.
48. Sugimura, M.; Ichikawa, M. Basic dye compositions for dyeing of textiles. Jpn. Kokai Tokkyo Koho JP 07033999, 1995; *Chem. Abstr.* **1995**, *123*, 172520.
49. Leach, R. M.; Zhang, J. Composition and process for coloring and preserving wood. PCT Int. Appl. WO 2005056257, 2005; *Chem. Abstr.* **2005**, *143*, 61604.
50. Schafer, E. W., Jr.; Bowles, W. A., Jr. Acute oral toxicity and repellency of 933 chemicals to house and deer mice. *Arch. Environ. Contam. Toxicol.* **1985**, *14*, 111–129.
51. Cunningham, A. R.; Moss, S. T.; Iype, S. A.; Qian, G.; Qamar, S.; Cunningham, S. L. Structure–activity relationship analysis of rat mammary carcinogens. *Chem. Res. Toxicol.* **2008**, *21*, 1970–1982.
52. Gold, L. S.; Slone, T. H.; Stern, B. R.; Bernstein, L. Comparison of target organs of carcinogenicity for mutagenic and non-mutagenic chemicals. *Mutat. Res.* **1993**, *286*, 75–100.
53. Heil, J.; Reifferscheid, G. Detection of mammalian carcinogens with an immunological DNA synthesis-inhibition test. *Carcinogenesis* **1992**, *13*, 2389–2394.
54. Huang, R.; Southall, N.; Cho, M. H.; Xia, M.; Inglese, J.; Austin, C. P. Characterization of diversity in toxicity mechanism using *in vitro* cytotoxicity assays in quantitative high throughput screening. *Chem. Res. Toxicol.* **2008**, *21*, 659–667.
55. Parodi, S.; Santi, L.; Russo, P.; Albini, A.; Vecchio, D.; Pala, M.; Ottaggio, L.; Carbone, A. DNA damage induced by auramine O in liver, kidney, and bone marrow of rats and mice, and in a human cell line (alkaline elution assay and SCE induction). *J. Toxicol. Environ. Health* **1982**, *9*, 941–952.
56. Ohno, K.; Tanaka-Azuma, Y.; Yoneda, Y.; Yamada, T. Genotoxicity test system based on p53R2 gene expression in human cells: examination with 80 chemicals. *Mutat. Res.* **2005**, *588*, 47–57.
57. Martelli, A.; Brambilla-Campart, G.; Canonero, R.; Carrozzino, R.; Mattioli, F.; Robbiano, L.; Cavanna, M. Evaluation of auramine genotoxicity in primary rat and human hepatocytes and in the intact rat. *Mutat. Res.* **1998**, *414*, 37–47.
58. Varella, S. D.; Pozetti, G. L.; Vilegas, W.; Varanda, E. A. Mutagenic activity of sweepings and pigments from a household-wax factory assayed with *Salmonella typhimurium*. *Food Chem. Toxicol.* **2004**, *42*, 2029–2035.
59. Zeiger, E.; Anderson, B.; Haworth, S.; Lawlor, T.; Mortelmans, K. *Salmonella* mutagenicity tests. V. Results from the testing of 311 chemicals. *Environ. Mol. Mutagen.* **1992**, *19*, 2–141.

AZURE A

CAS Registry Number 531-53-3

Chemical Structure

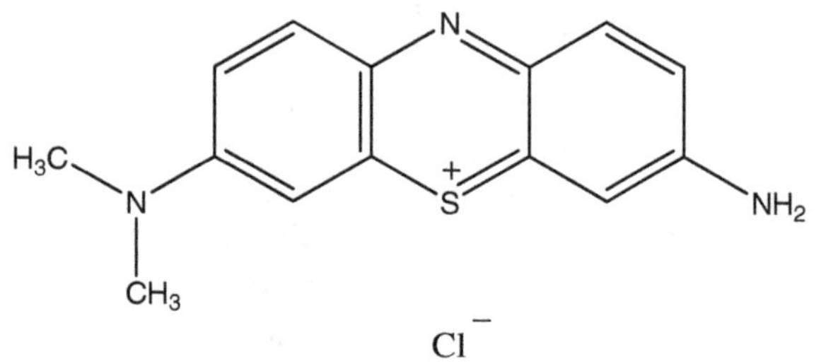

CA Index Name Phenothiazin-5-ium, 3-amino-7-(dimethylamino)-, chloride (1 : 1)

Other Names Azure A; Phenothiazin-5-ium, 3-amino-7-(dimethylamino), chloride; 3-Amino-7-(dimethylamino)phenazathionium chloride; 5-Chloro-3-dimethylamino-7-amino-5*H*-phenothiazine; Azur A; Azurea dye; C.I. 52005; *N,N*-Dimethylthionine

Merck Index Number 927

Chemical/Dye Class Phenothiazine

Molecular Formula $C_{14}H_{14}ClN_3S$

Molecular Weight 291.80

Physical Form Dark green powder or crystals

Solubility Soluble in water, methanol, glycerol; sparingly soluble in ethanol

Melting Point 290 °C (decompose)

Absorption (λ_{max}) 620–634 nm

Synthesis Synthetic method[1–9]

Staining Applications Bacteria;[10] fungi;[10] blood products;[11,12] cells;[13] nucleic acids;[14–16] proteins;[16] viruses;[16] nuclei;[17] plant virus inclusions;[18] tissues;[19] hairs;[20] keratin fibers;[21] malaria-infected cells[22]

Biological Applications Antimalarial;[22,23] medical devices;[24] diagnosis of amyloid accumulation related diseases;[25] reducing the extent of cardiac arrhythmias;[26] treating angiogenic diseases,[27] avian influenza virus,[28] oral cavity infection,[29] neurodegenerative diseases,[30] pathological tau–tau association in Alzheimer disease,[31,32] protozoan infections,[33] viral diseases[34]

Industrial Applications Electrochromic devices;[35] optical sensors;[36] photovoltaic cells;[37] solar cells[37]

Safety/Toxicity Phototoxicity[38]

Certification/Approval Certified by Biological Stain Commission (BSC)

REFERENCES

1. Sorunmu, Y. E.; Nguyen, M.; Sapp, J. B.; Gorski, W.; Phan, T. D.; Wei, X. Study of factors affecting molecular behaviors in phenothiazine-mediated biosensing by electrochemical and spectroscopic methods. *Electroanalysis* **2006**, *18*, 2375–2380.
2. Gigi, S.; Paucescu, V.; Rotaru, M. Total synthesis of phenothiazine blue dyes A + B. Rom. RO 63231, 1978; *Chem. Abstr.* **1979**, *91*, 194625.
3. Nikolaev, L. A.; Novozhilov, Yu. A.; Murzina, L. P. Interconversions of thiazine dyes. *Zh. Fiz. Khim.* **1976**, *50*, 2106–2108; *Chem. Abstr.* **1976**, *85*, 178955.
4. Marshall, P. N. The composition of stains produced by the oxidation of methylene blue. *Histochem. J.* **1976**, *8*, 431–442.
5. Gabriel, E.; Jahn, H. Methylthionine eosinates. Ger. Offen. DE 2334277, 1974; *Chem. Abstr.* **1974**, *81*, 107366.
6. Toepfer, K. Preparation of dimethylthionine (azure A) for histochemical reactions. *Acta Histochem.* **1970**, *35*, 195–197.
7. Bonneau, R.; Faure, J.; Joussot-Dubien, J. Purification of thiazine dyes, azure A, azure B, and azure C, by a partition method. *Talanta* **1967**, *14*, 121–122.
8. Kodousek, R. A new dithionate-leucofuchsin reagent and other similar reagents. *Acta Histochem.* **1965**, *21*, 150–153.
9. Taylor, K. B. Chromatographic separation and isolation of metachromatic thiazine dyes. *J. Histochem. Cytochem.* **1960**, *8*, 248–257.
10. Webster, R. Simple staining of bacteria and fungi in hide, skin and leather. *Stain Technol.* **1983**, *58*, 315–318.
11. Dittmer, D. P.; Floyd, R. A. RNA virus inactivated by phenothiazine dye and visible light for use as vaccines against *Flaviviridae* infection in mammal. PCT Int. Appl. WO 2007094854, 2007; *Chem. Abstr.* **2007**, *147*, 299319.
12. Mohr, H. Photodynamic treatment and UV-B irradiation of thrombocyte suspension blood products. PCT Int. Appl. WO 2002002152, 2002; *Chem. Abstr.* **2002**, *136*, 90903.
13. Gelvan, D. J.; Goltsman, L.; Chausovsky, A. Staining methods and compositions for identifying a cell

phenotype. PCT Int. Appl. WO 2007102146, 2007; *Chem. Abstr.* **2007**, *147*, 339010.

14. Gupta, A.; Will, S. G. Light emission modifiers and their uses in nucleic acid detection, amplification and analysis. Eur. Pat. Appl. EP 1739190, 2007; *Chem. Abstr.* **2007**, *146*, 115979.
15. Garner, D. M.; Todorovic, C.; Lee, W. E. Cytological stain composition for cytological analysis of cellular DNA. U.S. Pat. Appl. Publ. US 2006199243, 2006; *Chem. Abstr.* **2006**, *145*, 288084.
16. Althaus, J. S.; Kyonghoon, L.; Namasivayam, V.; Brahmasandra, S. N.; Handique, K. Microfluidic devices and method for electrochemical detection of nucleic acids, proteins or viruses. PCT Int. Appl. WO 2004094986, 2004; *Chem. Abstr.* **2004**, *141*, 362747.
17. Horobin, R. W.; Stockert, J. C.; Rashid-Doubell, F. Fluorescent cationic probes for nuclei of living cells: why are they selective? A quantitative structure–activity relations analysis. *Histochem. Cell Biol.* **2006**, *126*, 165–175.
18. Hoefert, L. L.; McCreight, J. D.; Christie, R. D. Microwave enhanced staining for plant virus inclusions. *Biotech. Histochem.* **1992**, *67*, 40–44.
19. Chieco, P.; Pagnoni, M.; Romagnoli, E.; Melchiorri, C. A rapid and simple staining method, using toluidine blue, for analysing mitotic figures in tissue sections. *Histochem. J.* **1993**, *25*, 569–577.
20. Moeller, H.; Hoeffkes, H.; Oberkobusch, D. Oxidative hair dyes composed of primary amino group-containing chromophores and reactive carbonyl compounds. Ger. Offen. DE 10260881, 2004; *Chem. Abstr.* **2004**, *141*, 93976.
21. Meinigke, B.; Moeller, H.; Rose, D. Use of alkylated phenothiazines for dyeing keratin fibers. Ger. Offen. DE 19750303, 1999; *Chem. Abstr.* **1999**, *130*, 356890.
22. Howarth, J.; Lloyd, D. G. Redox systems as conduits for antimalarial compounds. *J. Antimicrob. Chemother.* **2001**, *47*, 122–124.
23. Vennerstrom, J. L.; Makler, M. T.; Angerhofer, C. K.; Williams, J. A. Antimalarial dyes revisited: xanthenes, azines, oxazines, and thiazines. *Antimicrob. Agents Chemother.* **1995**, *39*, 2671–2677.
24. Ash, S. R.; Steczko, J. Antimicrobial medical devices. U.S. Pat. Appl. Publ. US 2008279907, 2008; *Chem. Abstr.* **2008**, *149*, 541798.
25. Kudo, K.; Suemoto, T.; Suzuki, M.; Tojo, H.; Shimazu, H. Azure A analogs as imaging agents and probes for diagnosis of diseases related to amyloid accumulation. Jpn. Kokai Tokkyo Koho JP 2000344685, 2000; *Chem. Abstr.* **2000**, *134*, 27007.
26. Weiss, S. M. Reducing myocardial damage and the incidence of arrhythmia arising from loss, reduction or interruption in coronary blood flow. PCT Int. Appl. WO 2007022568, 2007; *Chem. Abstr.* **2007**, *146*, 280994.
27. Weisz, P. B. Anti-angiogenic protein isolation from cartilage using anionic oligosaccharides, and use of the protein– oligosaccharide complex for treatment of angiogenic and related diseases, including inhibition of tumor growth. PCT Int. Appl. WO 9711090, 1997; *Chem. Abstr.* **1997**, *126*, 272367.
28. Wood, C.; Sterling, R. C. Methylene blue therapy of avian influenza. U.S. Pat. Appl. Publ. US 2007123520, 2007; *Chem. Abstr.* **2007**, *146*, 514711.
29. Kishen, A.; George, S.; Neo, C. L. J.; Song, K. P. Photosensitizing compositions for oral cavity infection treatment. PCT Int. Appl. WO 2006135344, 2006; *Chem. Abstr.* **2006**, *146*, 68740.
30. Wischik, C. M.; Horsley, D.; Rickard, J. E.; Harrington, C. R. Drug screening for effectors of tau protein proteolytic processing and expression systems of controlled aggre-gation in treatment of neurodegenerative disease. PCT Int. Appl. WO 2002055720, 2002; *Chem. Abstr.* **2002**, *137*, 88474.
31. Wischik, C. M.; Edwards, P. C.; Harrington, C. R.; Roth, M.; Klug, A. Method for screening for agents inhibiting tau–tau-association and use of phenothiazine compounds for treatment of pathological tau–tau association. PCT Int. Appl. WO 9630766, 1996; *Chem. Abstr.* **1996**, *125*, 322333.
32. Wischik, C. M.; Edwards, P. C.; Lai, R. Y.; Roth, M.; Harrington, C. R. Selective inhibition of Alzheimer disease-like tau aggregation by phenothiazines. *Proc. Natl. Acad. Sci. U.S.A.* **1996**, *93*, 11213–11218.
33. Blair, B. G. Method of treatment of protozoan infections in fish. U.S. Pat. Appl. Publ. US 2002037921, 2002; *Chem. Abstr.* **2002**, *136*, 260249.
34. Wood, C.; Habib, N. Methylene blue therapy of viral disease. U.S. Pat. Appl. Publ. US 2006264423, 2006; *Chem. Abstr.* **2006**, *146*, 703.
35. Varaprasad, D. V.; Agrawal, A.; Zhao, M.; Allemand, P. M.; Dornan, C. A.; Lynam, N. R. Electrochromic polymeric solid films, manufacturing electrochromic devices using such solid films, and processing for making such solid films and devices. Eur. Pat. Appl. EP 612826, 1994; *Chem. Abstr.* **1995**, *122*, 326083.

36. Caputo, G.; Della, C. L. General method for the covalent immobilisation of indicators in, or onto a support useful for the fabrication of optical sensors. Eur. Pat. Appl. EP 1213583, 2002; *Chem. Abstr.* **2002**, *137*, 14966.

37. Chane, C. J. Y.; Corma, A.; Garcia, H. Multivalent metal cation-doped cerium oxide materials for photovoltaic cells and solar cells. Fr. Demande FR 2857783, 2005; *Chem. Abstr.* **2005**, *142*, 159487.

38. Sayed, Z.; Harris, F.; Phoenix, D. A. A study on the bacterial phototoxicity of phenothiazinium based photosensitisers. *FEMS Immunol. Med. Microbiol.* **2005**, *43*, 367–372.

AZURE B

CAS Registry Number 531-55-5

Chemical Structure

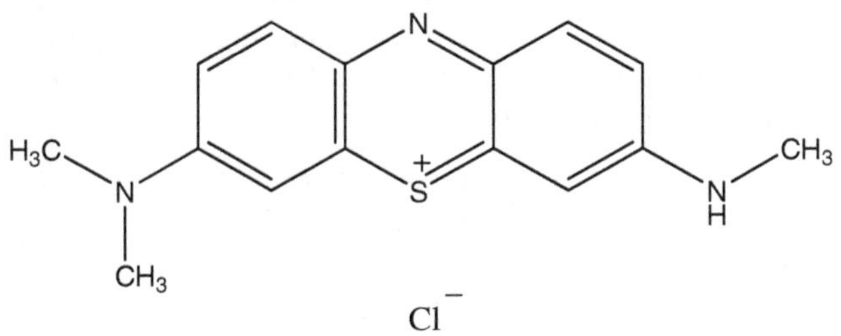

CA Index Name Phenothiazin-5-ium, 3-(dimethylamino)-7-(methylamino)-, chloride (1 : 1)

Other Names Azure B; Phenothiazin-5-ium, 3-(dimethylamino)-7-(methylamino)-, chloride; Trimethylthionine; 3-Methylamino-7-dimethylaminophenazathionium chloride; Azur I; Azure I; C.I. 52010; Methylene Azure; Trimethylthionine chloride

Merck Index Number 928

Chemical/Dye Class Phenothiazine

Molecular Formula $C_{15}H_{16}ClN_3S$

Molecular Weight 305.83

Physical Form Dark green powder or crystals

Solubility Soluble in water; sparingly soluble in ethanol

Melting Point 205–210 °C (decompose)

Absorption (λ_{max}) 648–655 nm

Synthesis Synthetic method[1–14]

Staining Applications Blood products;[15–17] cells;[18] epithelial tissues;[19] nucleic acids;[20–24] proteins;[23,24] viruses;[23] nuclei;[25] nucleolus;[26] lignins;[27] melanin;[28,29] reticulocytes;[30] keratin fibers;[31] malaria-infected cells[32]

Biological Applications Antimalarial;[32,33] biofuel cell;[34] medical devices;[35] diagnosis of amyloid accumulation related diseases,[36] diabetes,[37] malignant melanoma;[28,29] detecting oral cancer,[38] cells,[18] nucleic acids;[20–24] treating avian influenza virus,[39] nail infection,[40] oral cavity infection,[41,42] neurodegenerative diseases,[43] pathological tau–tau association in Alzheimer disease,[44,45] protozoan infections,[46] viral diseases[47]

Industrial Applications Inks;[48] electrochromic devices;[49] solar energy[50]

Safety/Toxicity Carcinogenicity;[51] phototoxicity[52,53]

Certification/Approval Certified by Biological Stain Commission (BSC)

REFERENCES

1. Storey, J.; Mervyn, D.; Sinclair, J. P.; Marshall, C.; Tan, H. W.; Wischik, C. M. Synthesis and purification of diamino-phenothiazinium compounds including methylthioninium chloride (MTC) via oxidative coupling and ring closure. PCT Int. Appl. WO 2006032879, 2006; *Chem. Abstr.* **2006**, *144*, 350699.
2. Sorunmu, Y. E.; Nguyen, M.; Sapp, J. B.; Gorski, W.; Phan, T. D.; Wei, X. Study of factors affecting molecular behaviors in phenothiazine-mediated biosensing by electrochemical and spectroscopic methods. *Electroanalysis* **2006**, *18*, 2375–2380.
3. Couto, S. R.; Dominguez, A.; Sanroman, A. Photocatalytic degradation of dyes in aqueous solution operating in a fluidised bed reactor. *Chemosphere* **2002**, *46*, 83–86.
4. Moura, J. C. V.; Oliveria-Campos, A. M. F.; Griffiths, J. Synthesis and evaluation of phenothiazine singlet oxygen sensitizing dyes for application in cancer phototherapy. *Phosphorus, Sulfur Silicon Relat. Elem.* **1997**, *120–121*, 459–460.
5. Zipfel, E.; Grezes, J. R.; Seiffert, W.; Zimmermann, H. W. Romanowsky dyes and Romanowsky-Giemsa effect. 1. Azure B, purity and content of dye samples, association. *Histochemistry* **1981**, *72*, 279–290.
6. Gigi, S.; Paucescu, V.; Rotaru, M. Total synthesis of phenothiazine blue dyes A + B. Rom. RO 63231, 1978; *Chem. Abstr.* **1979**, *91*, 194625.
7. Shigehara, K.; Matsunaga, H.; Tsuchida, E. Synthesis and properties of polymeric azine and thiazine dyes. *J. Polym. Sci., Polym. Chem. Ed.* **1978**, *16*, 1853–1863.
8. Nikolaev, L. A.; Novozhilov, Yu. A.; Murzina, L. P. Interconversions of thiazine dyes. *Zh. Fiz. Khim.* **1976**, *50*, 2106–2108; *Chem. Abstr.* **1976**, *85*, 178955.
9. Marshall, P. N. The composition of stains produced by the oxidation of methylene blue. *Histochem. J.* **1976**, *8*, 431–442.
10. Marshall, P. N.; Bentley, S. A.; Lewis, S. M. Standardized Romanowsky stain prepared from purified dyes. *J. Clin. Pathol.* **1975**, *28*, 920–923.
11. Loehr, W.; Grubhofer, N.; Sohmer, I.; Wittekind, D. Azure dyes. Purification and physicochemical properties. II. Purification of azure B. *Stain Technol.* **1975**, *50*, 149–156.

12. Bonneau, R.; Faure, J.; Joussot-Dubien, J. Purification of thiazine dyes, azure A, azure B, and azure C, by a partition method. *Talanta* **1967**, *14*, 121–122.

13. Taylor, K. B. Chromatographic separation and isolation of metachromatic thiazine dyes. *J. Histochem. Cytochem.* **1960**, *8*, 248–257.

14. Wilhelmi, G. Influence of the phenothiazine series on regeneration in planarians and axolotls. *Helv. Physiol. Pharmacol. Acta* **1955**, *13*, C40–C42.

15. Dittmer, D. P.; Floyd, R. A. RNA virus inactivated by phenothiazine dye and visible light for use as vaccines against *Flaviviridae* infection in mammal. PCT Int. Appl. WO 2007094854, 2007; *Chem. Abstr.* **2007**, *147*, 299319.

16. Stokes, B. O.; Nelson, P. A.; Bradshaw, G. D. Method and staining reagent for staining hematology sample in an automated staining apparatus. U.S. Pat. Appl. Publ. US 2004043495, 2004; *Chem. Abstr.* **2004**, *140*, 195855.

17. Mohr, H. Photodynamic treatment and UV-B irradiation of thrombocyte suspension blood products. PCT Int. Appl. WO 2002002152, 2002; *Chem. Abstr.* **2002**, *136*, 90903.

18. Clampitt, R.; Barron, R. Stain and capillary slide to detect animal and plant cells. PCT Int. Appl. WO 9709616, 1997; *Chem. Abstr.* **1997**, *126*, 248565.

19. Burkett, D. D. Improved diagnostic method for detecting dysplastic epithelial tissue. PCT Int. Appl. WO 2002007693, 2002; *Chem. Abstr.* **2002**, *136*, 131224.

20. Gupta, A.; Will, S. G. Light emission modifiers and their uses in nucleic acid detection, amplification and analysis. Eur. Pat. Appl. EP 1739190, 2007; *Chem. Abstr.* **2007**, *146*, 115979.

21. Garner, D. M.; Todorovic, C.; Lee, W. E. Cytological stain composition for cytological analysis of cellular DNA. U.S. Pat. Appl. Publ. US 2006199243, 2006; *Chem. Abstr.* **2006**, *145*, 288084.

22. Li, Y. F.; Huang, C. Z.; Li, M. Study of the interaction of Azur B with DNA and the determination of DNA based on resonance light scattering measurements. *Anal. Chim. Acta* **2002**, *452*, 285–294.

23. Althaus, J. S.; Kyonghoon, L.; Namasivayam, V.; Brahmasandra, S. N.; Handique, K. Microfluidic devices and method for electrochemical detection of nucleic acids, proteins or viruses. PCT Int. Appl. WO 2004094986, 2004; *Chem. Abstr.* **2004**, *141*, 362747.

24. Colurso, G. J.; Kan, R. K.; Anthony, A. Microdensitometric measures of cytoplasmic RNA and total protein in pyramidal neurons of the insular cortex and midfrontal gyrus in patients with Alzheimer's disease. *Cell Biochem. Funct.* **1995**, *13*, 287–292.

25. Horobin, R. W.; Stockert, J. C.; Rashid-Doubell, F. Fluorescent cationic probes for nuclei of living cells: why are they selective? A quantitative structure–activity relations analysis. *Histochem. Cell Biol.* **2006**, *126*, 165–175.

26. Weng, B.; Yang, Z. Vital nucleolus dyeing method. Faming Zhuanli Shenqing Gongkai Shuomingshu CN 1256317, 2000; *Chem. Abstr.* **2001**, *134*, 53484.

27. Fineran, B. A. Cyto- and histochemical demonstration of lignins in plant cell walls: an evaluation of the chlorine water/ethanolamine-silver nitrate method of Coppick and Fowler. *Protoplasma* **1997**, *198*, 186–201.

28. Kligora, C. J.; Fair, K. P.; Clem, M. S.; Patterson, J. W. A comparison of melanin bleaching and azure blue counterstaining in the immunohistochemical diagnosis of malignant melanoma. *Mod. Pathol.* **1999**, *12*, 1143–1147.

29. Kamino, H.; Tam, S. T. Immunoperoxidase technique modified by counterstain with azure B as a diagnostic aid in evaluating heavily pigmented melanocytic neoplasms. *J. Cutan. Pathol.* **1991**, *18*, 436–439.

30. Studholme, R. M.; Marshall, P. N.; Embleton, A. M.; Glazier, J. G.; Van Hove, L. Reagent system and method for the differentiation and identification of reticulocytes. PCT Int. Appl. WO 9719356, 1997; *Chem. Abstr.* **1997**, *127*, 62858.

31. Meinigke, B.; Moeller, H.; Rose, D. Use of alkylated phenothiazines for dyeing keratin fibers. Ger. Offen. DE 19750303, 1999; *Chem. Abstr.* **1999**, *130*, 356890.

32. Howarth, J.; Lloyd, D. G. Redox systems as conduits for antimalarial compounds. *J. Antimicrob. Chemother.* **2001**, *47*, 122–124.

33. Vennerstrom, J. L.; Makler, M. T.; Angerhofer, C. K.; Williams, J. A. Antimalarial dyes revisited: xanthenes, azines, oxazines, and thiazines. *Antimicrob. Agents Chemother.* **1995**, *39*, 2671–2677.

34. Minteer, S. D.; Martin, R. S.; Moore, C. M. Microfluidic biofuel cell. PCT Int. Appl. WO 2005096430, 2005; *Chem. Abstr.* **2005**, *143*, 389794.

35. Ash, S. R.; Steczko, J. Antimicrobial medical devices. U.S. Pat. Appl. Publ. US 2008279907, 2008; *Chem. Abstr.* **2008**, *149*, 541798.

36. Kudo, K.; Suemoto, T.; Suzuki, M.; Tojo, H.; Shimazu, H. Azure A analogs as imaging agents and probes for diagnosis of diseases related to

amyloid accumulation. Jpn. Kokai Tokkyo Koho JP 2000344685, 2000; *Chem. Abstr.* **2000**, *134*, 27007.

37. Umegae, Y.; Machida, R.; Takagi, H.; Irie, Y.; Yokoyama, T.; Tanabe, T. Test kits comprising redox enzyme, stabilizer and electrochemical electrode sensor chip for 1,5-anhydroglucitol determination in blood and diabetes diagnosis. PCT Int. Appl. WO 2008072702, 2008; *Chem. Abstr.* **2008**, *149*, 76607.
38. Pomerantz, E. Methods and compositions for *in vivo* detection of oral cancers and precancerous conditions. PCT Int. Appl. WO 9726018, 1997; *Chem. Abstr.* **1997**, *127*, 158773.
39. Wood, C.; Sterling, R. C. Methylene blue therapy of avian influenza. U.S. Pat. Appl. Publ. US 2007123520, 2007; *Chem. Abstr.* **2007**, *146*, 514711.
40. Loebel, N. G.; Mckenzie, F. R.; Street, C.; Andersen, R.; Scott, R. Composition containing antifungal agents and photosensitizers, therapy and device for treatment of nail infections. PCT Int. Appl. WO 2008109424, 2008; *Chem. Abstr.* **2008**, *149*, 362854.
41. Kishen, A.; George, S.; Neo, C. L. J.; Song, K. P. Photosensitizing compositions for oral cavity infection treatment. PCT Int. Appl. WO 2006135344, 2006; *Chem. Abstr.* **2006**, *146*, 68740.
42. Kennedy, J. Treatment of microorganisms in the oral cavity by photodynamic therapy. Can. Pat. Appl. CA 2486475, 2006; *Chem. Abstr.* **2006**, *144*, 440202.
43. Wischik, C. M.; Horsley, D.; Rickard, J. E.; Harrington, C. R. Drug screening for effectors of tau protein proteolytic processing and expression systems of controlled aggregation in treatment of neurodegenerative disease. PCT Int. Appl. WO 2002055720, 2002; *Chem. Abstr.* **2002**, *137*, 88474.
44. Wischik, C. M.; Edwards, P. C.; Harrington, C. R.; Roth, M.; Klug, A. Method for screening for agents inhibiting tau–tau-association and use of phenothiazine compounds for treatment of pathological tau–tau association. PCT Int. Appl. WO 9630766, 1996; *Chem. Abstr.* **1996**, *125*, 322333.
45. Wischik, C. M.; Edwards, P. C.; Lai, R. Y.; Roth, M.; Harrington, C. R. Selective inhibition of Alzheimer disease-like tau aggregation by phenothiazines. *Proc. Natl. Acad. Sci. U.S.A.* **1996**, *93*, 11213–11218.
46. Blair, B. G. Method of treatment of protozoan infections in fish. U.S. Pat. Appl. Publ. US 2002037921, 2002; *Chem. Abstr.* **2002**, *136*, 260249.
47. Wood, C.; Habib, N. Methylene blue therapy of viral disease. U.S. Pat. Appl. Publ. US 2006264423, 2006; *Chem. Abstr.* **2006**, *146*, 703.
48. Lee, E. R.; Perl, M. L. Ink set and ink composition used for document security and printing markings on security documents. PCT Int. Appl. WO 2000077104, 2000; *Chem. Abstr.* **2000**, *134*, 58042.
49. Varaprasad, D. V.; Agrawal, A.; Zhao, M.; Allemand, P. M.; Dornan, C. A.; Lynam, N. R. Electrochromic polymeric solid films, manufacturing electrochromic devices using such solid films, and processing for making such solid films and devices. Eur. Pat. Appl. EP 612826, 1994; *Chem. Abstr.* **1995**, *122*, 326083.
50. Gangotri, K. M.; Genwa, K. R. Comparative studies in anionic, cationic and nonionic surfactants in photogalvanic cells from solar energy conversion and storage point of view: NTA-Azure B system. *J. Indian Chem. Soc.* **2004**, *81*, 592–594.
51. Bogdanovic, Z.; Culo, F.; Marusic, M. Effect of three biological response modifiers on chemical carcinogenesis in mice. *Neoplasma* **1993**, *40*, 229–233.
52. Hussain, S.; Harris, F.; Phoenix, D. A. The phototoxicity of phenothiazinium-based photosensitizers to bacterial membranes. *FEMS Immunol. Med. Microbiol.* **2006**, *46*, 124–130.
53. Sayed, Z.; Harris, F.; Phoenix, D. A. A study on the bacterial phototoxicity of phenothiazinium based photosensitisers. *FEMS Immunol. Med. Microbiol.* **2005**, *43*, 367–372.

BAPTA

CAS Registry Number 85233-19-8

Chemical Structure

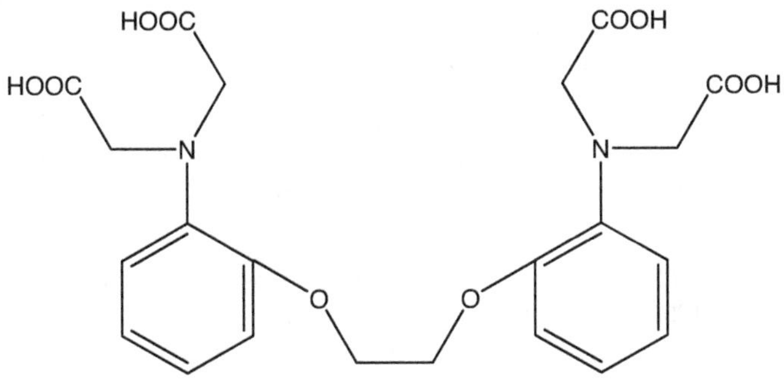

CA Index Name Glycine, *N,N'*-[1,2-ethanediylbis(oxy-2,1-phenylene)]bis[*N*-(carboxymethyl)-

Other Names 1,2-Bis(2-aminophenoxy)ethane-*N,N,N',N'*-tetraacetic acid; 1,2-Bis(*o*-aminophenoxy)ethane-*N,N,N',N'*-tetraacetic acid; Ethylenedioxybis(*o*-phenylenenitrilo)tetraacetic acid; 2,2'-(Ethylenedioxy)dianiline-*N,N,N',N'*-tetraacetic acid; BAPTA

Merck Index Number 957

Chemical/Dye Class Aromatic

Molecular Formula $C_{22}H_{24}N_2O_{10}$

Molecular Weight 476.43

Physical Form Beige powder

Solubility Soluble in dimethyl sulfoxide

Melting Point 149–159°C

Boiling Point (Calcd.) 766.6 ± 60.0°C, pressure: 760 Torr

pKa (Calcd.) 1.40 ± 0.10, most acidic, temperature: 25°C; 5.96 ± 0.38, most basic, temperature 25°C

Absorption (λ_{max}) 209 nm, 254 nm, 203 nm, 274 nm

Emission (λ_{max}) 363 nm

Synthesis Synthetic methods[1–3]

Staining Applications Calcium ions;[1–3,6–22] zinc ions;[23,24] bone;[4] proteins[5]

Biological Applications Calcium indicators;[1–3,6–22] zinc indicators;[23,24] treating acute cell death,[25] cerebral infarction,[25] myocardial infarction,[25] hepatonecrosis,[25] kidney ischemic necrosis,[25] necrotizing pancreatitis,[25] amyloidosis,[26] atherosclerosis,[26] diseases characterized by calcification and/or plaque formation,[27] osteoporosis,[28] Paget's disease,[28] heterotropic ossification,[28] hypercalcemia,[28] cancer,[29] inflammation,[29] diabetes mellitus,[30] epilepsy,[31] epithelial disorders,[32] glaucoma,[33] HIV-associated conditions,[34] respiratory disorders,[35] streptococcal infection,[36] viral diseases[37]

Industrial Applications Inks[38]

Safety/Toxicity Carcinogenicity;[39] cytotoxicity;[40] excitotoxicity;[41,42] hydrogen peroxide toxicity;[43] neurotoxicity[44–48]

REFERENCES

1. Diwu, Z.; He, J.; Liao, J. Fluorescent ion indicators and their applications to detn. of metal cations. U.S. Pat. Appl. Publ. US 2008254498, 2008; *Chem. Abstr.* **2008**, *149*, 443327.
2. Tsien, R. Y.; Minta, A. Preparation and properties of calcium-specific, long-wavelength indicator dyes. Eur. Pat. Appl. EP 314480, 1989; *Chem. Abstr.* **1990**, *112*, 135620.
3. Tsien, R. Y. New calcium indicators and buffers with high selectivity against magnesium and protons: design, synthesis, and properties of prototype structures. *Biochemistry* **1980**, *19*, 2396–2404.
4. Pautke, C.; Vogt, S.; Tischer, T.; Wexel, G.; Deppe, H.; Milz, S.; Schieker, M.; Kolk, A. Polychrome labeling of bone with seven different fluorochromes: enhancing fluorochrome discrimination by spectral image analysis. *Bone* **2005**, *37*, 441–445.
5. Czerney, P. T.; Desai, S.; Lehmann, F. G.; Murtaza, Z. S.; Schweder, B. G.; Wenzel, M. S.; Wolf, B. D. Protein probe compounds, compositions, and methods. U.S. Pat. Appl. Publ. US 2008026478, 2008; *Chem. Abstr.* **2008**, *148*, 209647.
6. Nagasawa, K.; Tarui, T.; Yoshida, S.; Sekiguchi, F.; Matsunami, M.; Ohi, A.; Fukami, K.; Ichida, S.; Nishikawa, H.; Kawabata, A. Hydrogen sulfide evokes neurite outgrowth and expression of high-voltage-activated Ca^{2+} currents in NG108-15 cells:

Handbook of Biological Dyes and Stains By R. W. Sabnis

involvement of T-type Ca^{2+} channels. *J. Neurochem.* **2009**, *108*, 676–684.

7. Flagg, T. P.; Cazorla, O.; Remedi, M. S.; Haim, T. E.; Tones, M. A.; Bahinski, A.; Numann, R. E.; Kovacs, A.; Schaffer, J. E.; Nichols, C. G.; Nerbonne, J. M. Ca^{2+}-independent alterations in diastolic sarcomere length and relaxation kinetics in a mouse model of lipotoxic diabetic cardiomyopathy. *Circ. Res.* **2009**, *104*, 95–103.
8. Yoshida, Y.; Tsuchiya, R.; Matsumoto, N.; Morita, M.; Miyakawa, H.; Kudo, Y. Ca^{2+}-dependent induction of intracellular Ca^{2+} oscillation in hippocampal astrocytes during metabotropic glutamate receptor activation. *J. Pharmacol. Sci.* **2005**, *97*, 212–218.
9. Shang, J.; Lehrman, M. A. Inhibition of mammalian RNA synthesis by the cytoplasmic Ca^{2+} buffer BAPTA: analyses of [^{3}H]uridine incorporation and stress-dependent transcription. *Biochemistry* **2004**, *43*, 9576–9582.
10. Nelson, G.; Kozlova-Zwinderman, O.; Collis, A. J.; Knight, M. R.; Fincham, J. R. S.; Stanger, C. P.; Renwick, A.; Hessing, J. G. M.; Punt, P. J.; Van Den Hondel, C. A. M. J. J.; Read, N. D. Calcium measurement in living filamentous fungi expressing codon-optimized aequorin. *Mol. Microbiol.* **2004**, *52*, 1437–1450.
11. Rubart, M.; Wang, E.; Dunn, K. W.; Field, L. J. Two-photon molecular excitation imaging of Ca^{2+} transients in Langendorff-perfused mouse hearts. *Am. J. Physiol.* **2003**, *284*, C1654–C1668.
12. Rintoul, G. L.; Baimbridge, K. G. Effects of calcium buffers and calbindin-D28k upon histamine-induced calcium oscillations and calcium waves in HeLa cells. *Cell Calcium* **2003**, *34*, 131–144.
13. Lemonnier, L.; Vitko, Y.; Shuba, Y. M.; Vanden Abeele, F.; Prevarskaya, N.; Skryma, R. Direct modulation of volume-regulated anion channels by Ca^{2+} chelating agents. *FEBS Lett.* **2002**, *521*, 152–156.
14. Lomax, R. B.; Camello, C.; Van Coppenolle, F.; Petersen, O. H.; Tepikin, A. V. Basal and physiological Ca(2+) leak from the endoplasmic reticulum of pancreatic acinar cells: second messenger-activated channels and translocons. *J. Biol. Chem.* **2002**, *277*, 26479–26485.
15. Branchaw, J. L.; Banks, M. I.; Jackson, M. B. Ca^{2+} and voltage-dependent inactivation of Ca^{2+} channels in nerve terminals of the neurohypophysis. *J. Neurosci.* **1997**, *17*, 5772–5781.
16. Wang, L.; Bhatnagar, A.; Ansari, N. H.; Dhir, P.; Srivastava, S. K. Mechanism of calcium-induced disintegrative globulization of rat lens fiber cells. *Invest. Ophthalmol. Vis. Sci.* **1996**, *37*, 915–922.
17. Spigelman, I.; Tymianski, M.; Wallace, C. M.; Carlen, P. L.; Velumian, A. A. Modulation of hippocampal synaptic transmission by low concentrations of cell-permeant Ca^{2+} chelators: effects of Ca^{2+} affinity, chelator structure and binding kinetics. *Neuroscience* **1996**, *75*, 559–572.
18. Skutella, M.; Rueegg, U. T. Increase of empty pool-activated Ca^{2+} influx using an intracellular Ca^{2+} chelating agent. *Biochem. Biophys. Res. Commun.* **1996**, *218*, 837–841.
19. Li, L.; Tucker, R. W.; Hennings, H.; Yuspa, S. H. Chelation of intracellular Ca^{2+} inhibits murine keratinocyte differentiation *in vitro*. *J. Cell. Physiol.* **1995**, *163*, 105–114.
20. Packham, D. E.; Jiang, L.; Conigrave, A. D. Arachidonate and other fatty acids mobilize Ca^{2+} ions and stimulate beta-glucuronidase release in a Ca (2+)-dependent fashion from undifferentiated HL-60 cells. *Cell Calcium* **1995**, *17*, 399–408.
21. Subbaiah, C. C.; Zhang, J.; Sachs, M. M. Involvement of intracellular calcium in anaerobic gene expression and survival of maize seedlings. *Plant Physiol.* **1994**, *105*, 369–376.
22. Kiang, J. G. Effect of intracellular pH on cytosolic free [Ca^{2+}] in human epidermoid A-431 cells. *Eur. J. Pharmacol.* **1991**, *207*, 287–296.
23. Aballay, A.; Sarrouf, M. N.; Colombo, M. I.; Stahl, P. D.; Mayorga, L. S. Zn^{2+} depletion blocks endosome fusion. *Biochem. J.* **1995**, *312*, 919–923.
24. Frederickson, C. J.; Costello, L. C.; Franklin, R. B. Zinc-based screening test and kit for early diagnosis of prostate cancer. U.S. Pat. Appl. Publ. US 2007292900, 2007; *Chem. Abstr.* **2007**, *148*, 49163.
25. Liu, J. Use of bapta derivative for the manufacture of clinical medicament. PCT Int. Appl. WO 2005053668, 2005; *Chem. Abstr.* **2005**, *143*, 13427.
26. Rosenberg, G. I.; Friedman, J. E.; Angel, I.; Kozak, A. Use of lipophilic diesters of chelating agent for the treatment of amyloidosis and atherosclerosis. PCT Int. Appl. WO 2005016229, 2005; *Chem. Abstr.* **2005**, *142*, 254646.
27. Kajander, E. O.; Aho, K.; Ciftcioglu, N.; Millican, H. B.; Maniscalco, B. Methods and compositions for the treatment of diseases characterized by calcification and/or plaque formation. U.S. Pat. Appl. Publ. US 2006083727, 2006; *Chem. Abstr.* **2006**, *144*, 398343.
28. Athayde, A. M. Preparation and use of a pharmaceutical composition containing bisphosphonate for the treatment of diseases related to calcium and

phosphate metabolism. Braz. Pedido PI BR 2001006601, 2003; *Chem. Abstr.* **2004**, *141*, 400886.

29. Pitson, S. M.; Vadas, M. A.; Xia, P.; Moretti, P. A.; Leclercq, T.; Sutherland, C. Methods of modulating sphingosine kinase signaling or localization, sphingosine kinase sequence variants, and therapeutic uses against cancer and inflammation. PCT Int. Appl. WO 2005123115, 2005; *Chem. Abstr.* **2005**, *144*, 83059.
30. Engelmayer, J.; Varadhachary, A. Compositions comprising recombinant lactoferrin and its variants in the treatment of diabetes mellitus. PCT Int. Appl. WO 2004103285, 2004; *Chem. Abstr.* **2004**, *141*, 420455.
31. Nedergaard, M.; Tian, G. F. Method for treatment and prevention of epilepsy. PCT Int. Appl. WO 2006062683, 2006; *Chem. Abstr.* **2006**, *145*, 55958.
32. Maclay, P. B., Jr.; Wang, G.; Davidson, B.; Bottner, M.; Herman, S. M.; Jolly, D. J. Methods and compositions for increasing infectivity of retrovirus vectors to epithelial tissues and treatment of epithelial disorders. Jpn. Kokai Tokkyo Koho JP 2000143548, 2000; *Chem. Abstr.* **2000**, *132*, 343324.
33. Garst, M. E.; Adorante, J. S. Bis(aminophenoxy) ethanetetraacetic acid calcium chelators for treatment of glaucoma. PCT Int. Appl. WO 2000059546, 2000; *Chem. Abstr.* **2000**, *133*, 276377.
34. Halstead, B. Methods for the treatment of HIV-associated conditions. U.S. Pat. Appl. Publ. US 2002182272, 2002; *Chem. Abstr.* **2002**, *138*, 11389.
35. Glynn, P.; Varadhachary, A. Oral lactoferrin in the treatment of respiratory disorders. PCT Int. Appl. WO 2003099207, 2003; *Chem. Abstr.* **2003**, *140*, 752.
36. Bjoerck, L.; Herwald, H.; Moergelin, M.; Russell, W.; Norrby-Teglund, A.; Lindbom, L.; Sollenberg, U.; Cramer, H.; Flodgaard, H. Method for identifying anti-streptococcal agent and treatment of streptococcal infection. PCT Int. Appl. WO 2004094468, 2004; *Chem. Abstr.* **2004**, *141*, 388628.
37. Johansen, L. M.; Owens, C. M.; Mawhinney, C.; Chappell, T. W.; Brown, A. T.; Frank, M. G.; Altmeyer, R. Compositions and methods for treatment of viral diseases. PCT Int. Appl. WO 2008033466, 2008; *Chem. Abstr.* **2008**, *148*, 394354.
38. Uozumi, S.; Yamamoto, Y. Non-aqueous ink-jet ink composition containing a chelating agent. U.S. Pat. Appl. Publ. US 2005223939, 2005; *Chem. Abstr.* **2005**, *143*, 368898.
39. Lan, M.; Shi, Y.; Sun, L.; Liu, L.; Guo, X.; Lu, Y.; Wang, J.; Liang, J.; Fan, D. KCl depolarization increases HIF-1 transcriptional activity via the calcium-independent pathway in SGC7901 gastric cancer cells. *Tumor Biol.* **2007**, *28*, 173–180.
40. Lees, G. J.; Leong, W. Neuronal cytotoxicity of inositol hexakisphosphate (phytate) in the rat hippocampus. *Brain Res.* **1996**, *741*, 134–141.
41. Shuttleworth, C. W.; Connor, J. A. Strain-dependent differences in calcium signaling predict excitotoxicity in murine hippocampal neurons. *J. Neurosci.* **2001**, *21*, 4225–4236.
42. Clementi, E.; Racchetti, G.; Melino, G.; Meldolesi, J. Cytosolic Ca^{2+} buffering, a cell property that in some neurons markedly decreases during aging, has a protective effect against NMDA/nitric oxide-induced excitotoxicity. *Life Sci.* **1996**, *59*, 389–397.
43. Lomonosova, E. E.; Kirsch, M.; de Groot, H. Calcium vs. iron-mediated processes in hydrogen peroxide toxicity to L929 cells: effects of glucose. *Free Radical Biol. Med.* **1998**, *25*, 493–503.
44. Ndountse, L. T.; Chan, H. M. Methylmercury increases *N*-methyl-D-aspartate receptors on human SH-SY 5Y neuroblastoma cells leading to neurotoxicity. *Toxicology* **2008**, *249*, 251–255.
45. Kelly, B. L.; Ferreira, A. Beta-amyloid-induced dynamin 1 degradation is mediated by *N*-methyl-D-aspartate receptors in hippocampal neurons. *J. Biol. Chem.* **2006**, *281*, 28079–28089.
46. Grolleau, F.; Gamelin, L.; Boisdron-Celle, M.; Lapied, B.; Pelhate, M.; Gamelin, E. A possible explanation for a neurotoxic effect of the anticancer agent oxaliplatin on neuronal voltage-gated sodium channels. *J. Neurophysiol.* **2001**, *85*, 2293–2297.
47. Yang, C. W.; Borowitz, J. L.; Gunasekar, P. G.; Isom, G. E. Cyanide-stimulated inositol 1,4,5-trisphosphate formation: an intracellular neurotoxic signaling cascade. *J. Biochem. Toxicol.* **1996**, *11*, 251–256.
48. Singh, A. K. Neurotoxicity in rats chronically exposed to lead ingestion: measurement of intracellular concentrations of free calcium and lead ions in resting or depolarized brain slices. *Neurotoxicology* **1995**, *16*, 133–138.

BAPTA AM

CAS Registry Number 126150-97-8

Chemical Structure

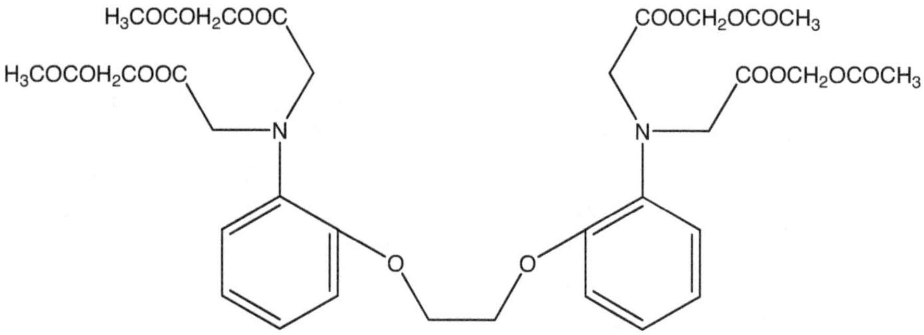

CA Index Name Glycine, *N*,*N*′-[1,2-ethanediylbis(oxy-2,1-phenylene)]bis[*N*-[2-[(acetyloxy)methoxy]-2-oxoethyl]-, 1,1′-bis[(acetyloxy)methyl] ester

Other Names 1,2-Bis(2-aminophenoxy)ethane-*N*,*N*,*N*′,*N*′-tetraacetic acid tetrakis(acetoxymethyl ester); Glycine, *N*,*N*′-[1,2-ethanediylbis(oxy-2,1-phenylene)]bis[*N*-[2-[(acetyloxy)methoxy]-2-oxoethyl]-, bis[(acetyloxy)methyl] ester; BAPTA-AM

Merck Index Number Not listed

Chemical/Dye Class Aromatic

Molecular Formula $C_{34}H_{40}N_2O_{18}$

Molecular Weight 764.68

Physical Form White powder

Solubility Soluble in chloroform, dimethyl sulfoxide

Melting Point >200°C

Boiling Point (Calcd.) 796.1 ± 60.0°C, pressure: 760 Torr

pK_a (Calcd.) 1.69 ± 0.50, most basic, temperature: 25°C

Absorption (λ_{max}) 287 nm

Synthesis Synthetic methods[1,2]

Staining Applications Calcium ions;[1–18] fungi;[19] motor neurons[20]

Biological Applications Calcium indicators;[1–18] treating cancer,[21] chronic bacterial infection,[22] glaucoma,[23] ocular hypertension,[23] HIV-associated conditions,[24] infectious diseases,[25] neurodegenerative disorders,[26] neurological conditions,[27] psychiatric conditions[27]

Industrial Applications Not reported

Safety/Toxicity Cytotoxicity;[28] excitotoxicity;[29] neurotoxicity;[20,30,31] renal toxicity[32]

REFERENCES

1. Tsien, R. Y.; Minta, A. Preparation and properties of calcium-specific, long-wavelength indicator dyes. Eur. Pat. Appl. EP 314480 **1989**; *Chem. Abstr.* **1990**, *112*, 135620.
2. Tsien, R. Y. New calcium indicators and buffers with high selectivity against magnesium and protons: design, synthesis, and properties of prototype structures. *Biochemistry* **1980**, *19*, 2396–2404.
3. Kim, J. H.; Kim, J. H.; Ahn, B. J.; Park, J. H.; Shon, H. K.; Yu, Y. S.; Moon, D. W.; Lee, T. G.; Kim, K. W. Label-free calcium imaging in ischemic retinal tissue by TOF-SIMS. *Biophys. J.* **2008**, *94*, 4095–4102.
4. Okuda, T.; Kadotsuji, K.; Takayama, C.; Hanada, K.; Mukaizawa, F.; Ogawara, K. I.; Higaki, K.; Kimura, T. Involvement of intracellular Ca^{2+} dynamics in cytoprotective action by amino acids and cytotoxicity by sodium laurate, an absorption enhancer. *J. Pharm. Sci.* **2006**, *95*, 2256–2265.
5. Lee, D. H.; Han, Y. S.; Han, E. S.; Bang, H.; Lee, C. S. Differential involvement of intracellular Ca^{2+} in 1-methyl-4-phenylpyridinium- or 6-hydroxydopamine-induced cell viability loss in PC12 cells. *Neurochem. Res.* **2006**, *31*, 851–860.
6. Lanteri, M. L.; Pagnussat, G. C.; Lamattina, L. Calcium and calcium-dependent protein kinases are involved in nitric oxide- and auxin-induced adventitious root formation in cucumber. *J. Exp. Bot.* **2006**, *57*, 1341–1351.

7. Benesic, A.; Schwerdt, G.; Mildenberger, S.; Freudinger, R.; Gordjani, N.; Gekle, M. Disturbed Ca^{2+}-signaling by chloroacetaldehyde: a possible cause for chronic ifosfamide nephrotoxicity. *Kidney Int.* **2005**, *68*, 2029–2041.
8. Satoh, E.; Murakami, K.; Nishimura, M. Propylene glycol increases cytosolic free calcium in rat cerebrocortical synaptosomes. *Int. J. Neurosci.* **2004**, *114*, 587–596.
9. Azenabor, A. A.; Chaudhry, A. U.; Yang, S. Macrophage L-type Ca^{2+} channel antagonists alter *Chlamydia pneumoniae* MOMP and HSP-60 mRNA gene expression, and improve antibiotic susceptibility. *Immunobiology* **2003**, *207*, 237–245.
10. Xu, N.; Luo, K. Q.; Chang, D. C. Ca^{2+} signal blockers can inhibit M/A transition in mammalian cells by interfering with the spindle checkpoint. *Biochem. Biophys. Res. Commun.* **2003**, *306*, 737–745.
11. Cousson, A. Pharmacological evidence for a positive influence of the cyclic GMP-independent transduction on the cyclic GMP-mediated Ca^{2+}-dependent pathway within *Arabidopsis* stomatal opening in response to auxin. *Plant Sci.* **2003**, *164*, 759–767.
12. Upadhya, G. A.; Topp, S. A.; Hotchkiss, R. S.; Anagli, J.; Strasberg, S. M. Effect of cold preservation on intracellular calcium concentration and calpain activity in rat sinusoidal endothelial cells. *Hepatology* **2003**, *37*, 313–323.
13. Yi, F. X.; Zhang, A. Y.; Campbell, W. B.; Zou, A. P.; Van Breemen, C.; Li, P. L. Simultaneous *in situ* monitoring of intracellular Ca^{2+} and NO in endothelium of coronary arteries. *Am. J. Physiol.* **2002**, *283*, H2725–H2732.
14. Ahluwalia, J. P.; Topp, J. D.; Weirather, K.; Zimmerman, M.; Stamnes, M. A role for calcium in stabilizing transport vesicle coats. *J. Biol. Chem.* **2001**, *276*, 34148–34155.
15. Strayer, D. S.; Hoek, J. B.; Thomas, A. P.; White, M. K. Cellular activation by Ca^{2+} release from stores in the endoplasmic reticulum but not by increased free Ca^{2+} in the cytosol. *Biochem. J.* **1999**, *344*, 39–46.
16. Tiffert, T.; Lew, V. L. Apparent Ca^{2+} dissociation constant of Ca^{2+} chelators incorporated non-disruptively into intact human red cells. *J. Physiol.* **1997**, *505*, 403–410.
17. Collatz, M. B.; Ruedel, R.; Brinkmeier, H. Intracellular calcium chelator BAPTA protects cells against toxic calcium overload but also alters physiological calcium responses. *Cell Calcium* **1997**, *21*, 453–459.
18. Tymianski, M.; Wallace, M. C.; Spigelman, I.; Uno, M.; Carlen, P. L.; Tator, C. H.; Charlton, M. P. Cell-permeant calcium chelators reduce early excitotoxic and ischemic neuronal injury *in vitro* and *in vivo*. *Neuron* **1993**, *11*, 221–235.
19. Gressel, J.; Michaeli, D.; Kampel, V.; Amsellem, Z.; Warshawsky, A. Ultralow calcium requirements of fungi facilitate use of calcium regulating agents to suppress host calcium-dependent defenses, synergizing infection by a mycoherbicide. *J. Agric. Food Chem.* **2002**, *50*, 6353–6360.
20. Li, B.; Guo, Y. S.; Sun, M. M.; Dong, H.; Wu, S. Y.; Wu, D. X.; Li, C. Y. The NADPH oxidase is involved in lipopolysaccharide-mediated motor neuron injury. *Brain Res.* **2008**, *1226*, 199–208.
21. D'Andrea, A. D.; Taniguchi, T. Compositions and methods for the treatment of cancer. U.S. Pat. Appl. Publ. US 2007105130, 2007; *Chem. Abstr.* **2007**, *146*, 475663.
22. Azenabor, A. A. Treatment of chronic bacterial infection and related pathologies. PCT Int. Appl. WO 2005007082, 2005; *Chem. Abstr.* **2005**, *142*, 127559.
23. Garst, M. E.; Adorante, J. S. Pharmaceutical compositions for treating glaucoma and/or ocular hypertension containing calcium chelators. U.S. Patent 6350780, 2002; *Chem. Abstr.* **2002**, *136*, 205407.
24. Halstead, B. Methods for the treatment of HIV-associated conditions. U.S. Pat. Appl. Publ. US 2002182272, 2002; *Chem. Abstr.* **2002**, *138*, 11389.
25. Brando, L. V.; Posner, G. H.; D'Angelo, J. G.; Yolken, R. H.; Hencken, C. P.; Woodard, L. Preparation of artemisinin derivatives for use in preventing, controlling or treating infectious diseases. PCT Int. Appl. WO 2008127381, 2008; *Chem. Abstr.* **2008**, *149*, 493823.
26. Jin, X.; Staunton, J.; Macdonald, D.; Dong, H.; Kifle, L. Methods and compositions for the treatment of neurodegenerative disorders. PCT Int. Appl. WO 2008133884, 2008; *Chem. Abstr.* **2008**, *149*, 525459.
27. Haydon, P. G.; Halassa, M. M.; Fellin, T.; Ding, S.; Zhu, Y. Methods for treating neurological and psychiatric conditions, and test compound screening methods. PCT Int. Appl. WO 2007002285, 2007; *Chem. Abstr.* **2007**, *146*, 93591.
28. Bai, P.; Bakondi, E.; Szabo, E.; Gergely, P.; Szabo, C.; Virag, L. Partial protection by poly(ADP-ribose) polymerase inhibitors from nitroxyl-induced cytotoxicity in thymocytes. *Free Radical Biol. Med.* **2001**, *31*, 1616–1623.

29. Dubinsky, J. M. Effects of calcium chelators on intracellular calcium and excitotoxicity. *Neurosci. Lett.* **1993**, *150*, 129–132.

30. Ndountse, L. T.; Chan, H. M. Methylmercury increases *N*-methyl-D-aspartate receptors on human SH-SY 5Y neuroblastoma cells leading to neurotoxicity. *Toxicology* **2008**, *249*, 251–255.

31. Paschen, W.; Hotop, S.; Aufenberg, C. Loading neurons with BAPTA-AM activates xbp1 processing indicative of induction of endoplasmic reticulum stress. *Cell Calcium* **2003**, *33*, 83–89.

32. Schnellmann, R. G. Intracellular calcium chelators and oxidant-induced renal proximal tubule cell death. *J. Biochem. Toxicol.* **1991**, *6*, 299–303.

BASIC FUCHSIN

CAS Registry Number 632-99-5

Chemical Structure

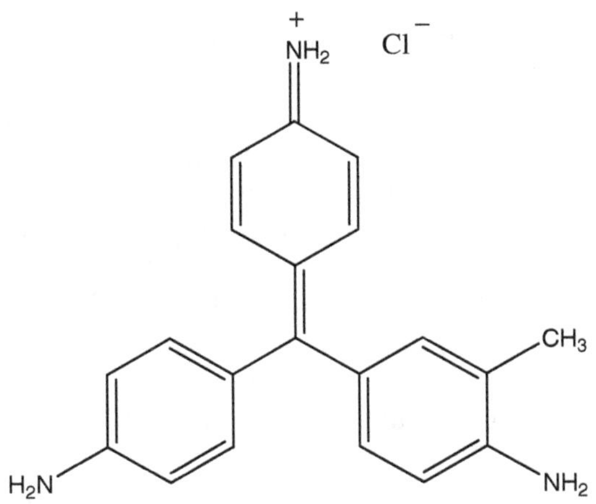

CA Index Name Benzenamine, 4-[(4-aminophenyl)(4-imino-2,5-cyclohexadien-1-ylidene)methyl]-2-methyl-, hydrochloride (1:1)

Other Names Benzenamine, 4-[(4-aminophenyl)(4-imino-2,5-cyclohexadien-1-ylidene)methyl]-2-methyl-, monohydrochloride; C.I. Basic Violet 14, monohydrochloride; Rosaniline; 12418 Red; Aizen Magenta; Astra Fuchsine B; Basic Fuchsine; Basic Magenta; Basic Magenta E 200; Basic Violet 14; Basic fuchsin; C-WR Violet 8; C.I. 42510; C.I. Basic Violet 14; Calcozine Fuchsine HO; Calcozine Magenta RTN; Calcozine Magenta XX; Cerise B; Diabasic Magenta; Diamond Fuchsine; Fuchsin; Fuchsin basic; Fuchsine; Fuchsine A; Fuchsine CS; Fuchsine G; Fuchsine HO; Fuchsine N; Fuchsine RTN; Fuchsine SBP; Fuchsine Y; Lowacryl Violet 14; Magenta; Magenta DP; Magenta E; Magenta G; Magenta I; Magenta PN; Magenta Powder N; Magenta S; Magenta superfine; Orient Basic Magenta; RGB 20; RGN 10; RGN 10 (dye); Rosaniline chloride; Rosaniline hydrochloride; Rosanilinium chloride

Merck Index Number 5652

Chemical/Dye Class Triphenylmethane

Molecular Formula $C_{20}H_{20}ClN_3$

Molecular Weight 337.85

Physical Form Dark green crystals

Solubility Soluble in water, ethanol; insoluble in ether

Melting Point 200°C (decompose)

Absorption (λ_{max}) 543 nm

Synthesis Synthetic method[1–7]

Staining Applications Bacteria;[8,9] cells;[10] cytoplasm;[11] nuclei;[11] collagen;[11] mucus;[11] elastin;[11] fats;[11] lipids;[11] fish;[12] intestinal mucosa;[13] fruit juices;[14] fungi;[15] nucleic acids;[16] pathogens;[17] white blood cells;[18] nails;[19] skin;[20] lips;[20] hairs;[21–23] keratin fibers;[24] shampoos[25]

Biological Applications Detecting cancer;[26] treating Jock itch,[27] oral fungal infection,[28] piles;[29] periodontal disease[30]

Industrial Applications Inks;[31,32] highlighters;[33] textiles[34]

Safety/Toxicity Acute toxicity;[35] carcinogenicity;[36] genotoxicity;[37,38] mutagenicity[39]

Certification/Approval Certified by Biological Stain Commission (BSC)

REFERENCES

1. Potsch, W. R. Fuchsine—the basis of the dyestuff industry. *Melliand Textilber.* **2002**, *83*, 556.
2. Tokarev, V. I. Purification of basic fuchsine. U.S.S.R. SU 188504, 1966; *Chem. Abstr.* **1967**, *67*, 65430.
3. Loiseleur, J.; Petit, M. The direct radiosynthesis of dyes. *Compt. Rend.* **1960**, *250*, 2573–2575.
4. Fry, B. A. Basic triphenylmethane dyes and the inhibition of glutamine synthesis by *Staphylococcus aureus* (*Micrococcus pyogenes* var. *aureus*). *J. Gen. Microbiol.* **1957**, *16*, 341–349.
5. Scanlan, J. T.; Melin, C. G. Production of basic fuchsin suitable for the Feulgen technic. *Stain Technol.* **1936**, *12*, 1–8.
6. Ignatyev, C. N.; Vasin, I. I. Fuchsin. RU 50998, 1929; *Chem. Abstr.* **1931**, *25*, 23166.
7. Suais, E. Process of preparing fuchsins and analogous colors. *Bull. Soc. Ind. Mulhouse* **1911**, *81*, 180–182. *Chem. Abstr.* **1911**, *5*, 21766.
8. Kataoka, N. Quick and simple Gram's staining method, and apparatus. Jpn. Kokai Tokkyo Koho. JP 2006071447, 2006; *Chem. Abstr.* **2006**, *144*, 307911.
9. Kono, N. Test liquid and reagent kit for Gram staining, and Gram staining method. Jpn. Kokai Tokkyo Koho JP 2003169694, 2003; *Chem. Abstr.* **2003**, *139*, 3217.
10. Gelvan, D. J.; Goltsman, L.; Chausovsky, A. Staining methods and compositions for identifying a cell

phenotype. PCT Int. Appl. WO 2007102146, 2007; *Chem. Abstr.* **2007**, *147*, 339010.

11. D'Amico, F. A polychromatic staining method for epoxy embedded tissue: a new combination of methylene blue and basic fuchsine for light microscopy. *Biotech. Histochem.* **2005**, *80*, 207–210.
12. Moen, V.; Nordtug, T.; Zahlsen, K. Mixture and procedure for marking fish. Norw. NO 317169, 2004; *Chem. Abstr.* **2007**, *146*, 141497.
13. Alieva, E. I.; Filin, V. A.; Shcherbakov, I. T.; Gracheva, N. M.; Leont'eva, N. I.; Partin, O. S.; Novikova, A. V. Method for diagnosis of cryptosporidiosis involving histological exami-nation of intestinal mucosa bioptates fixed with methanol, stained with carbolic fuchsine, and dechlorinated with sulfuric acid. Russ. RU 2229128, 2004; *Chem. Abstr.* **2004**, *141*, 172330.
14. Wu, M. C.; Jiang, C. M.; Ho, Y. Y.; Shen, S. C.; Chang, H. M. Convenient quantification of methanol in juices by methanol oxidase in combination with basic fuchsin. *Food Chem.* **2007**, *100*, 412–418.
15. Yu, Z. D.; Li, Y.; Li, D. W. Histological stain technology study of two plant parasitic fungi with their hosts. *Wuhan Zhiwuxue Yanjiu* **2005**, *23*, 588–591. *Chem. Abstr.* **2006**, *145*, 183301.
16. Garner, D. M.; Todorovic, C.; Lee, W. E. Cytological stain composition for cytological analysis of cellular DNA. U.S. Pat. Appl. Publ. US 2006199243, 2006; *Chem. Abstr.* **2006**, *145*, 288084.
17. Potts, M. H. Method for the simultaneous demonstration of various opportunistic pathogens. U.S. Patent 6436663, 2002; *Chem. Abstr.* **2002**, *137*, 152013.
18. Lindberg, S.; Svensson, J. Enumeration of white blood cells by staining and imaging. Eur. Pat. Appl. EP 1701150, 2006; *Chem. Abstr.* **2006**, *145*, 288168.
19. Borsakian, B.; Faraci, J. Color changing nail polish. PCT Int. Appl. WO 2003007675, 2003; *Chem. Abstr.* **2003**, *138*, 142188.
20. Jager-Lezer, N.; Simon, J. C. Colored transparent or translucent cosmetic composition. PCT Int. Appl. WO 2002047627, 2002; *Chem. Abstr.* **2002**, *137*, 52018.
21. Kleen, A.; Terrier, J. Hair dye composition with plant dyes. Eur. Pat. Appl. EP 1800652, 2007; *Chem. Abstr.* **2007**, *147*, 101201.
22. Pollack, G. Hair coloring composition and method. U.S. Patent 6506374, 2003; *Chem. Abstr.* **2003**, *138*, 95211.
23. Grit, M. Hair dye composition comprising direct dyes, quaternary ammonium compounds, and cationic polymers. Eur. Pat. Appl. EP 1224927, 2002; *Chem. Abstr.* **2002**, *137*, 98675.
24. Knuebel, G.; Hoeffkes, H.; Giesa, H. Cationic azo compounds as direct dyes for the coloring of keratin fibers. PCT Int. Appl. WO 2006013036, 2006; *Chem. Abstr.* **2006**, *144*, 198108.
25. Grit, M. Sprayable coloring shampoos containing direct dyes and surfactants. Ger. Offen. DE 10114979, 2002; *Chem. Abstr.* **2002**, *137*, 268150.
26. Carte, C.; Evelegh, M.; Zawydiwski, R. Liquid-phase galactose oxidase-Schiff's assay for carbohydrate markers in samples for detecting cancer or precancerous condition. U.S. Pat. Appl. Publ. US 2007065893, 2007; *Chem. Abstr.* **2007**, *146*, 333652.
27. Narang, U.; Nicholson, W. S. C.; Sherbondy, A.; Szabo, G. N. Adhesive treatment for tinea cruris. U.S. Pat. Appl. Publ. US 2003007946, 2003; *Chem. Abstr.* **2003**, *138*, 78490.
28. Narang, U. Adhesive treatment for oral fungal infection. U.S. Pat. Appl. Publ. US 2003007947, 2003; *Chem. Abstr.* **2003**, *138*, 95591.
29. Sheng, X. Chinese patent medicine for treating pile. Faming Zhuanli Shenqing Gongkai Shuomingshu CN 1457860, 2003; *Chem. Abstr.* **2005**, *142*, 322704.
30. Andersen, S. N.; Wilson, J. B. Dye composition and laser energy for treating periodontal disease. U.S. Pat. Appl. Publ. US 2003059379, 2003; *Chem. Abstr.* **2003**, *138*, 276252.
31. House, G. L.; Wyand, A. T. Polymeric black pigment dispersions and ink jet ink compositions. U.S. Pat. Appl. Publ. US 2007043146, 2007; *Chem. Abstr.* **2007**, *146*, 276243.
32. Aoyama, M.; Tsuda, M.; Sago, H.; Okada, M.; Yamazaki, H.; Higashiyama, S. Fluorescent water inks for ink jet recording. Jpn. Kokai Tokkyo Koho JP 2005264072, 2005; *Chem. Abstr.* **2005**, *143*, 348861.
33. Davies-Smith, L.; Sum, V. W. Highlightable and highlighted mixtures, marking instruments, eradicator solution and kits, using the mixtures in highlighting. U.S. Pat. Appl. Publ. US 2005120919, 2005; *Chem. Abstr.* **2005**, *143*, 9331.
34. Sheikh, M. R. K.; Farouqui, F. I.; Modak, P. R.; Hoque, Md. A.; Yasmin, Z. Dyeing of Rajshahi silk with basic dyes: effect of modification on dyeing properties. *J. Textile Inst.* **2006**, *97*, 295–300.
35. Srivastava, S. J.; Singh, N. D.; Sinha, R.; Srivastava, A. K. Acute toxicity of fuchsin basic (magenta) and its effects on blood parameters of a freshwater catfish, *Heteropneustes fossilis* (Bloch). *J. Adv. Zool.* **1996**, *17*, 38–43.

36. Vachalkova, A.; Novotny, L.; Blesova, M. Polarographic reduction of some triphenylmethane dyes and their potential carcinogenic activity. *Neoplasma* **1996**, *43*, 113–117.

37. Hagiwara, M.; Watanabe, E.; Barrett, J. C.; Tsutsui, T. Assessment of genotoxicity of 14 chemical agents used in dental practice: ability to induce chromosome aberrations in Syrian hamster embryo cells. *Mutat. Res.* **2006**, *603*, 111–120.

38. De Flora, S.; Zanacchi, P.; Camoirano, A.; Bennicelli, C.; Badolati, G. S. Genotoxic activity and potency of 135 compounds in the Ames reversion test and in a bacterial DNA-repair test. *Mutat. Res.* **1984**, *133*, 161–198.

39. Mortelmans, K.; Haworth, S.; Lawlor, T.; Speck, W.; Tainer, B.; Zeiger, E. *Salmonella* mutagenicity tests. II. Results from the testing of 270 chemicals. *Environ. Mutagen.* **1986**, *8*, 1–119.

BISMARK BROWN Y

CAS Registry Number 10114-58-6

Chemical Structure

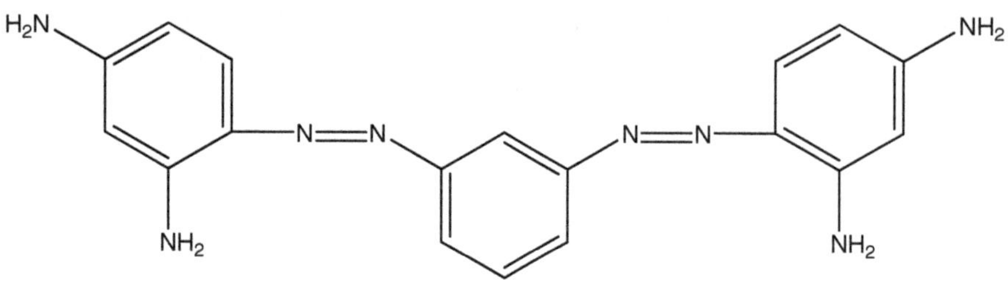

• 2HCl

Solubility Soluble in water, methyl cellosolve, ethylene glycol; slightly soluble in ethanol; insoluble in acetone, benzene, carbon tetrachloride, xylene

Melting Point >200°C

CA Index Name 1,3-Benzenediamine, 4,4′-[1,3-phenylenebis(2,1-diazenediyl)]bis-, hydrochloride (1:2)

Other Names 1,3-Benzenediamine, 4,4′-[1,3-phenylenebis(azo)]bis-, dihydrochloride; C.I. 21000; C.I. Basic Brown 1, dihydrochloride; Basic Brown G; Basic Brown GX; Basic Brown GXP; Excelsior Brown; Leather Brown; Manchester Brown; Phenylene Brown; Vesuvin; Vesuvine

Merck Index Number 1253

Chemical/Dye Class Azo

Molecular Formula $C_{18}H_{20}Cl_2N_8$

Molecular Weight 419.31

Physical Form Blackish-brown or red-brown powder

pK_a 5.0

Absorption (λ_{max}) 457 nm

Synthesis Synthetic methods[1–3]

Staining Applications Brain;[3] spinal cord;[3] sperm head;[4] tissue culture;[5] cytoplasmic vacuoles;[6] nucleic acids;[7] keratin fibers[8]

Biological Applications Differential inhibition of brain specific binding[9]

Industrial Applications Inks;[10] toners;[11] electrophotography;[12] paints;[13] wood[14]

Safety/Toxicity No data available

Certification/Approval Certified by Biological Stain Commission (BSC)

REFERENCES

1. Welcher, F. J. *Organic Analytical Reagents*; D. Van Nostrand Company: New York, **1948**; Vol. *4*, pp 339–340.
2. Green, F. J. *The Sigma–Aldrich Handbook of Stains, Dyes and Indicators*; Aldrich Chemical Company Inc.: Milwaukee, **1991**; pp 144–145.
3. Augulis, V.; Sigg, E. B. Supravital staining and fixation of brain and spinal cord by intravascular perfusion. *Stain Technol.* **1971**, *46*, 183–190.
4. Clark, G. Displacement. *Stain Technol.* **1979**, *54*, 111–119.
5. Everett, M. M.; Miller, W. A. Differential dichrome staining of tissue culture monolayers: alternate dyes and possible mechanism. *Stain Technol.* **1978**, *53*, 315–320.
6. Bader, J. P.; Brown, N. R. Cytoplasmic vacuoles of Rous virus transformed cells are organelles involved in cation uptake. *J. Gen. Virol.* **1978**, *38*, 471–481.
7. Iogannsen, M. G.; Dolgopyatova, M. A. Interaction of nucleic acids isolated from tumor cells with dyes. *Biokhimiya* **1972**, *37*, 1082–1087; *Chem. Abstr.* **1973**, *78*, 54308.
8. Grollier, J. F.; Fourcadier, C. Dyeing of keratin fibers in two steps by modification of pH values. Ger. Offen. DE 2913755, 1979; *Chem. Abstr.* **1980**, *92*, 82244.
9. Smith, T. M.; Squires, R. F. Differential inhibition of brain specific [^{3}H]flunitrazepam binding by several types of dyes. *Neurochem. Res.* **1983**, *8*, 1177–1183.

10. Sasage, D.; Iizuka, S. Ink compositions. Jpn. Kokai Tokkyo Koho JP 52146304, 1977; *Chem. Abstr.* **1978**, *88*, 154534.

11. Kosel, G. E. Hybrid liquid pigmentation charges or toners. Fr. Demande FR 2369599, 1978; *Chem. Abstr.* **1979**, *90*, 213201.

12. Ikeda, T.; Nagahashi, K.; Kamezawa, Y.; Aizawa, T. Liquid developer for electrophotography. Jpn. Tokkyo Koho JP 49038180, 1974; *Chem. Abstr.* **1975**, *83*, 50731.

13. Kaiser, W. D.; Natzschka, H.; Glausch, R.; Baessler, G. Corrosion-resistant paints for corroded iron articles. Ger. (East) DD 210703, 1984; *Chem. Abstr.* **1985**, *102*, 63687.

14. Leach, R. M.; Zhang, J. Composition and process for coloring and preserving wood. PCT Int. Appl. WO 2005056257, 2005; *Chem. Abstr.* **2005**, *143*, 61604.

BLUE TETRAZOLIUM (BT)

CAS Registry Number 1871-22-3

Chemical Structure

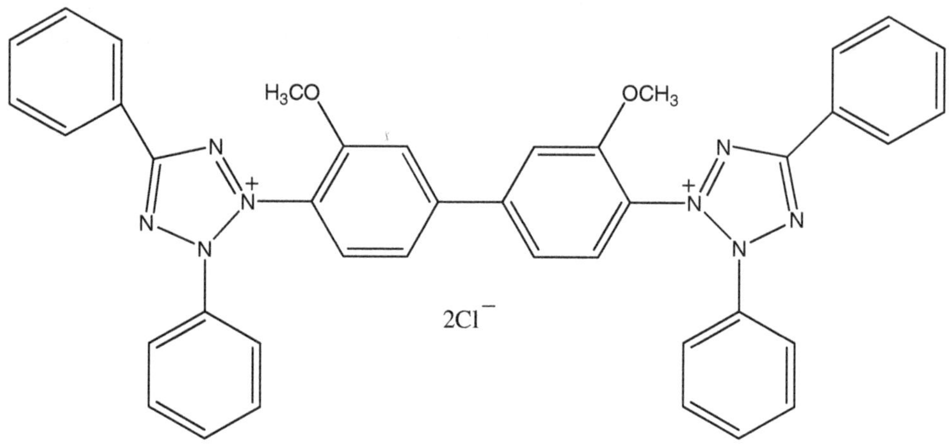

Molecular Weight 727.64

Physical Form Lemon yellow crystals or powder

Solubility Soluble in water; freely soluble ethanol, methanol, chloroform; insoluble in acetone, ether, ethyl acetate

CA Index Name 2*H*-Tetraolium, 2,2′-(3,3′-dimethoxy [1,1′-biphenyl]-4,4′-diyl)bis[3,5-diphenyl-, chloride (1:2)

Other Names 2*H*-Tetrazolium, 2,2′-(3,3′-dimethoxy [1,1′-biphenyl]-4,4′-diyl)bis[3,5-diphenyl-, dichloride; 2*H*-Tetrazolium,3,3′-(3,3′-dimethoxy-4,4′-biphenylylene) bis[2,5-diphenyl-, dichloride; 3,3′-(3,3′-Dimethoxy-4,4′-biphenylylene)bis[2,5-diphenyl-2*H*-tetrazolium chloride]; 2,2′,5,5′-Tetraphenyl-3,3′-(3,3′-dimethoxy-4,4′-diphenylene)ditetrazolium chloride; 2,2′-(*m,m*′-Dimethoxy-*p,p*′-biphenylene)bis(3,5-diphenyltetrazolium chloride); 3,3′-(3,3′-Dimethoxy-4,4′-biphenylene)bis[2,5-diphenyl-2*H*-tetrazolium chloride]; 3,3′-Dianisolebis[4,4′-(3,5-diphenyl)tetrazolium chloride]; BT; BT (dye); Blue tetrazolium; Blue tetrazolium chloride; Ditetrazolium chloride; NSC 27623; Tetrazolium blue; Tetrazolium blue chloride

Merck Index Number 9244

Chemical/Dye Class Tetrazolium salt

Molecular Formula $C_{40}H_{32}Cl_2N_8O_2$

Melting Point 242–245°C (decompose)

Absorption (λ_{max}) 253 nm

Synthesis Synthetic methods[1–8]

Staining Applications Fungi;[9] motoneurons[10]

Biological Applications Cellular response evaluation assays;[11] microbial growth assays;[12,13] tannins assays;[14] anti-cancer agents;[15] diagnostic test strips;[16] detecting lactate dehydrogenase (LDH) isoenzymes,[17] gamma-hydroxybutyric acid (GHB);[18] measuring ATP,[19] number of soil microorganisms,[20] niacin;[21] treating cancer[22]

Industrial Applications Electrochromic displays;[23] recording materials;[24,25] toner;[26] photochromic devices;[27] thin-film sensor;[28] photographic materials[29]

Safety/Toxicity Cytotoxicity;[30–32] dental toxicity;[33,34] hepatotoxicity;[35] *in vitro* toxicity[36,37]

REFERENCES

1. Cui, X.; Vlahakis, J. Z.; Crandall, I. E.; Szarek, W. A. Anti-*Plasmodium* activity of tetrazolium salts. *Bioorg. Med. Chem.* **2008**, *16*, 1927–1947.
2. James, P. R. Tetrazolium salts as steam sterilization process indicators. PCT Int. Appl. WO 9837157, 1998; *Chem. Abstr.* **1998**, *129*, 218237.
3. Li, N.; Yang, Y.; Zho, J.; Yuan, Z.; Ho, P. Preparation and properties of an anti-DNA antibody immunoadsorbent. *Zhongguo Shengwu Yixue Gongcheng Xuebao* **1985**, *4*, 181–186; *Chem. Abstr.* **1986**, *104*, 146650.
4. Vecerek, B. Tetrazolium salts. Czech. CS 147890, 1973; *Chem. Abstr.* **1973**, *79*, 53332.
5. Glantz, M. D.; Fried, F. Microseparation of tetrazolium salts by paper electrophoresis. *Microchem. J.* **1959**, *3*, 214–216.

6. Day, E. A.; Patton, S. Paper chromatography of 2,4-dinitrophenyl sulfide derivatives of mercaptans and mercapto-acids. *Microchem. J.* **1959**, *3*, 137–141.
7. Pannone, L. J.; Rust, J. B. Tetrazolium salts. U. S. Patent 2713581, 1955; *Chem. Abstr.* **1956**, *50*, 64811.
8. Rutenburg, A. M.; Gofstein, R.; Seligman, A. M. Preparation of a new tetrazolium salt which yields a blue pigment on reduction and its use in the demonstration of enzymes in normal and neoplastic tissues. *Cancer Res.* **1950**, *10*, 113–121.
9. Komatsu, S. Coloration culture medium for identifying *Candida*. Jpn. Kokai Tokkyo Koho. JP 2005080574, 2005; *Chem. Abstr.* **2005**, *142*, 351732.
10. De Graaf, F.; Van, R. W.; Van Asselt, E.; Diegenbach, P. C. Histochemical profiles of motoneurons innervating muscle fibres with different activity patterns in the zebrafish, *Brachydanio rerio*. *Histochem. J.* **1991**, *23*, 273–280.
11. Jaffe, R. M. Assay for evaluation of cellular response to allergens. U.S. Pat. Appl. Publ. US 2002187516, 2002; *Chem. Abstr.* **2002**, *138*, 3656.
12. Ushiyama, M. Tetrazolium salts and polyhydric alcohols for microbial growth and detection. Jpn. Kokai Tokkyo Koho. JP 2005287452, 2005; *Chem. Abstr.* **2005**, *143*, 363250.
13. Storey, D.; Field, T.; Ramage, G. Microbial growth assay and assay device. PCT Int. Appl. WO 2005098023, 2005; *Chem. Abstr.* **2005**, *143*, 402159.
14. Amin, A. S. Utilization of tetrazolium blue for the colorimetric assay of tannins in tea. *Mikrochim. Acta* **1997**, *126*, 105–108.
15. Kraemer, A. Centrosomal cluster-inhibitor screening for use as anti-cancer agents. Eur. Pat. Appl. EP 1884773, 2008; *Chem. Abstr.* **2008**, *148*, 206600.
16. Bell, D. E.; Chu, A. H. Preparation of diagnostic test strips containing tetrazolium salt indicators. U.S. Patent. 5360595, 1994; *Chem. Abstr.* **1995**, *122*, 4937.
17. Kuskov, M. V.; Semke, V. Ya.; Ivanova, S. A.; Terovskii, S. S.; Fedorenko, O. Yu.; Epanchintseva, E. M. Determination of dynamics of latent activity of lactate dehydrogenase (LDH) isoenzymes in leukocyte homogenate. Russ. RU 2293332, 2007; *Chem. Abstr.* **2007**, *146*, 223470.
18. Parsons, S. M.; Harris, D. O.; Bravo, D. T. Methods, compositions and apparatuses for detection of gamma-hydroxybutyric acid (GHB). U.S. Pat. Appl. Publ. US 2003175846, 2003; *Chem. Abstr.* **2003**, *139*, 256601.
19. Iwata, K.; Suzuki, T. ATP-measuring method allowing visual evaluation, and its reagent. PCT Int. Appl. WO 2003044222, 2003; *Chem. Abstr.* **2003**, *138*, 398374.
20. Iwata, K.; Suzuki, T.; Yamada, M.; Sakamoto, I. Method and reagent for measuring number of soil microorganisms by enzymic analysis. Jpn. Kokai Tokkyo Koho. JP 2003225098, 2003; *Chem. Abstr.* **2003**, *139*, 146186.
21. Jacobson, E. L.; Jacobson, M. K. Biochemical method to measure niacin status in a biological sample. U.S. Patent. 6287796, 2001; *Chem. Abstr.* **2001**, *135*, 223772.
22. Kong, Q. Method and composition for treating cancer using cellular organelle crystallizing agents. U.S. Pat. Appl. Publ. US 2006286170, 2006; *Chem. Abstr.* **2006**, *146*, 55482.
23. Inami, Y.; Wada, T. Electrochromic display devices. Jpn. Tokkyo Koho. JP 60044666, 1985; *Chem. Abstr.* **1986**, *104*, 159674.
24. Abe, Y.; Tsuchida, T.; Omura, H. Thermal recording materials containing tetrazolium leuco dyes, reducing agents, and salts of basic organic compounds. Jpn. Kokai Tokkyo Koho. JP 04270684, 1992; *Chem. Abstr.* **1993**, *118*, 158046.
25. Takatori, Y.; Haruta, M.; Shimozawa, A.; Nishide, K. Image reproduction or recording material. Ger. Offen. DE 2429729, 1975; *Chem. Abstr.* **1975**, *83*, 155788.
26. Yamamoto, A.; Takagi, S.; Fukumoto, H.; Tanaka, K. Toner for development of electrostatic images. Jpn. Kokai Tokkyo Koho. JP 61137162, 1986; *Chem. Abstr.* **1987**, *106*, 58907.
27. Iuchi, K.; Ohko, Y.; Tatsuma, T.; Fujishima, A. Cathode-separated TiO_2 photocatalysts applicable to a photochromic device responsive to backside illumination. *Chem. Mater.* **2004**, *16*, 1165–1167.
28. Hsu, C. K.; Al-Sheikhly, M.; Mclaughlin, W. L.; Christou, A. Radiochromic thin-film sensor using blue tetrazolium in polyvinyl alcohol. *Mater. Res. Soc. Symp. Proc.* **1997**, *441*, 743–748.
29. Nishina, Y.; Matsuo, T.; Hase, T.; Takahashi, K.; Hanyu, T. Lith-type silver halide photographic materials and their processing. Jpn. Kokai Tokkyo Koho. JP 53093015, 1978; *Chem. Abstr.* **1979**, *90*, 64397.
30. Figenschau, Y.; Yousef, M. I.; Sveinbjornsson, B.; Bertheussen, K. A sensitive serum-free colorimetric assay for the detection of cytotoxic effects of pesticides. *J. Environ. Sci. Health, Part B* **1997**, *32*, 177–194.
31. Viale, M.; Cafaggi, S.; Parodi, B.; Esposito, M. Cytotoxicity and cellular accumulation of a new cis-diammineplatinum (II) complex containing

procaine in murine L1210 cells sensitive and resistant to cis-diamminedichloroplatinum (II). *Cancer Chemother. Pharmacol.* **1995**, *35*, 371–376.

32. Dacasto, M.; Nebbia, C.; Bollo, E. Triphenyltin acetate (TPTA)-induced cytotoxicity to mouse thymocytes. *Pharmacol. Res.* **1994**, *29*, 179–186.
33. Aranha, A. M. F.; Giro, E. M. A.; Souza, P. P. C.; Hebling, J.; de Souza, C. C. A. Effect of curing regime on the cytotoxicity of resin-modified glass-ionomer lining cements applied to an odontoblast-cell line. *Dent. Mater.* **2006**, *22*, 864–869.
34. Chang, Y. C.; Chou, M. Y. Cytotoxicity of halothane on human gingival fibroblast cultures *in vitro*. *J. Endod.* **2001**, *27*, 82–84.
35. Hwang, J. M.; Tseng, T. H.; Hsieh, Y. S.; Chou, F. P.; Wang, C. J.; Chu, C. Y. Inhibitory effect of atractylon on *tert*-butyl hydroperoxide induced DNA damage and hepatic toxicity in rat hepatocytes. *Arch. Toxicol.* **1996**, *70*, 640–644.
36. Jelinek, A.; Klocking, H. P. *In vitro* toxicity of surfactants in U937 cells: cell membrane integrity and mitochondrial function. *Exp. Toxicol. Pathol.* **1998**, *50*, 472–476.
37. Santa, M. A.; Lopez, A.; Diaz, M. M.; Munoz-Mingarro, D.; Pozuelo, J. M. Evaluation of the toxicity of guarana with *in vitro* bioassays. *Ecotoxicol. Environ. Saf.* **1998**, *39*, 164–167.

BOBO 1

CAS Registry Number 169454-13-1

Chemical Structure

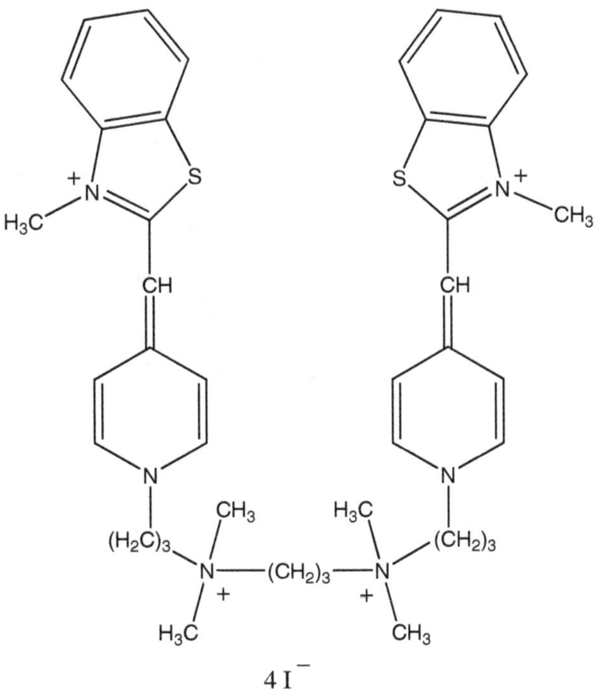

CA Index Name Benzothiazolium, 2,2′-[1,3-propanediylbis[(dimethyliminio)-3,1-propanediyl-1(4*H*)-pyridinyl-4-ylidenemethylidyne]]bis[3-methyl-, iodide (1:4)

Other Names Benzothiazolium, 2,2′-[1,3-propanediylbis[(dimethyliminio)-3,1-propanediyl-1(4*H*)-pyridinyl-4-ylidenemethylidyne]]bis[3-methyl-, tetraiodide; BOBO 1; BOBO 1 iodide

Merck Index Number Not listed

Chemical/Dye Class Cyanine

Molecular Formula $C_{41}H_{54}I_4N_6S_2$

Molecular Weight 1202.66

Physical Form Yellow-brown powder

Solubility Soluble in dimethyl sulfoxide

Melting Point >250°C

Absorption (λ_{max}) 462 nm

Emission (λ_{max}) 481 nm

Synthesis Synthetic method[1]

Staining Applications Nucleic acids;[1–4] cells;[5,6] chromatin;[7] leukocytes;[15] nuclei;[15] micronuclei;[7] megakaryocyte;[8] microorganisms;[9] sperms;[10] hairs[11]

Biological Applications Nucleic acid hybridization;[12,13] detecting nucleic acids,[1–4] cells,[5,6] pathogens;[14] counting embryoblasts;[15] characterization of DNA/lipid complexes[16]

Industrial Applications Not reported

Safety/Toxicity No data available

REFERENCES

1. Yue, S. T.; Haugland, R. P. Dimers of unsymmetrical cyanine dyes containing pyridinium moieties. U.S. Patent 5410030, 1995; *Chem. Abstr.* **1995**, *123*, 259753.
2. Exner, M.; Rogers, A. Methods for identifying nucleic acids and determining melting temperature using multiple fluorophores and FRET. U.S. Pat. Appl. Publ. US 2007172836, 2007; *Chem. Abstr.* **2007**, *147*, 182489.
3. Wittwer, C. T.; Dujols, V. E.; Reed, G.; Zhou, L. Genotyping by thermal denaturation analysis using loss of fluorescence of double-stranded DNA-specific dyes as a reporter to monitor denaturation. PCT Int. Appl. WO 2004038038, 2004; *Chem. Abstr.* **2004**, *140*, 386988.
4. Erikson, G. H.; Daksis, J. I.; Kandic, I.; Picard, P. Conditions for formation of three- and four-stranded nucleic acid complexes and their detection and analytical use. PCT Int. Appl. WO 2002103051, 2002; *Chem. Abstr.* **2002**, *138*, 50812.
5. Anderson, A. L.; Knutson, C. R.; Mueth, D.; Plewa, J.; Tanner, E. Methods for staining cells for identification and sorting. U.S. Pat. Appl. Publ. US 2006172315, 2006; *Chem. Abstr.* **2006**, *145*, 183714.
6. Millard, P. J.; Roth, B. L.; Yue, S. T.; Haugland, R. P. Fluorescent viability assay using cyclic-substituted unsymmetrical cyanine dyes. U.S. Patent 5534416, 1996; *Chem. Abstr.* **1996**, *125*, 162751.
7. Dertinger, S. D.; Cairns, S. E.; Avlasevich, S. L.; Torous, D. K. Method for enumerating and differentiating between mammalian cell micronuclei by differentially staining micronuclei and the chromatin of dead, dying and viable cells. PCT Int. Appl. WO 2006007479, 2006; *Chem. Abstr.* **2006**, *144*, 124521.
8. Minakami, T.; Mori, Y.; Tsuji, T.; Ikeuchi, Y. Megakaryocyte classification/counting method by double fluorescent staining and flow cytometry. Jpn. Kokai Tokkyo Koho JP 2006275985, 2006; *Chem. Abstr.* **2006**, *145*, 413679.

9. Noda, N.; Mizutani, T. Microorganism-measuring method using multiple staining. Jpn. Kokai Tokkyo Koho JP 2006340684, 2006; *Chem. Abstr.* **2006**, *146*, 77563.

10. Matsumoto, T.; Okada, H.; Hamaguchi, Y. Method and reagent for counting sperm by flow cytometry. Jpn. Kokai Tokkyo Koho JP 2001242168, 2001; *Chem. Abstr.* **2001**, *135*, 207869.

11. Lagrange, A. Hair dye compositions containing a polycationic direct dye. Fr. Demande FR 2848840, 2004; *Chem. Abstr.* **2004**, *141*, 76344.

12. Erikson, G. H.; Daksis, J. I. Improving the signal/noise ratio of nucleic acid hybridization assays by preincubation of primer and target with nucleic acid binding agents. U.S. Pat. Appl. Publ. US 2004180345, 2004; *Chem. Abstr.* **2004**, *141*, 255469.

13. Erikson, G. H. Method for modifying transcription and/or translation in an organism by heteropolymeric probes and duplex, triplex or quadruplex hybridization for therapeutic, prophylactic and/or analytic uses. U.S. Pat. Appl. Publ. US 2003181412, 2003; *Chem. Abstr.* **2003**, *139*, 272000.

14. Vannier, E. Methods for detection of pathogens in red blood cells. PCT Int. Appl. WO 2006031544, 2006; *Chem. Abstr.* **2006**, *144*, 307966.

15. Heuven, B.; Wong, F.; Tsuji, T.; Sakata, T.; Hamaguchi, I. Method for classifying and counting erythroblasts by flow cytometry. Jpn. Kokai Tokkyo Koho JP 11326323, 1999; *Chem. Abstr.* **1999**, *132*, 1804.

16. Madeira, C.; Loura, L. M. S.; Aires-barros, M. R.; Fedorov, A.; Prieto, M. Characterization of DNA/lipid complexes by fluorescence resonance energy transfer. *Biophys. J.* **2003**, *85*, 3106–3119.

BOBO 3

CAS Registry Number 169454-17-5

Chemical Structure

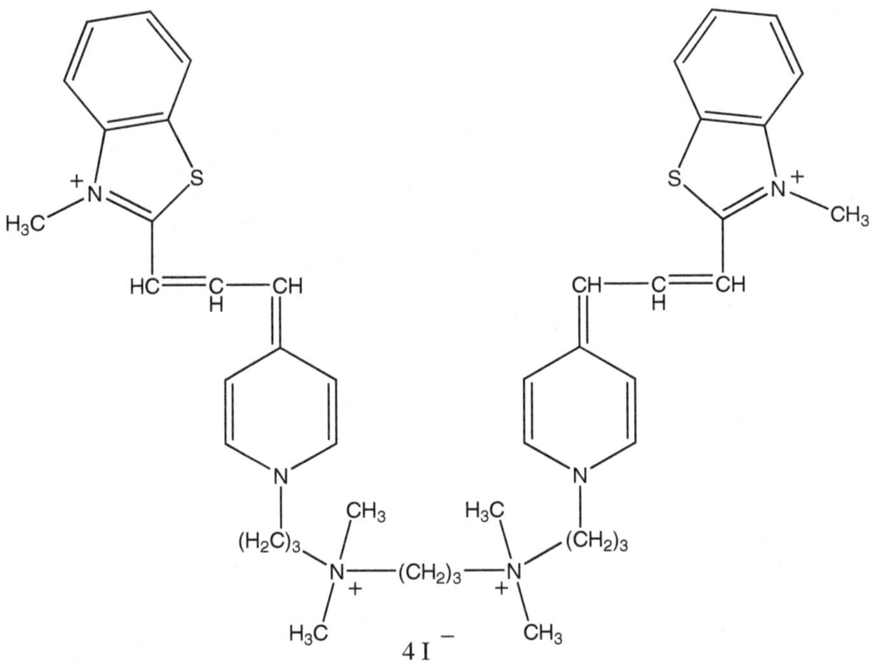

CA Index Name Benzothiazolium, 2,2′-[1,3-propanediylbis[(dimethyliminio)-3,1-propanediyl-1(4*H*)-pyridinyl-4-ylidene-1-propen-1-yl-3-ylidene]]bis[3-methyl-, iodide (1:4)

Other Names Benzothiazolium, 2,2′-[1,3-propanediylbis[(dimethyliminio)-3,1-propanediyl-1(4*H*)-pyridinyl-4-ylidene-1-propen-1-yl-3-ylidene]]bis[3-methyl, tetraiodide; BOBO 3, BIBO 3 iodide

Merck Index Number Not listed

Chemical/Dye Class Cyanine

Molecular Formula $C_{45}H_{58}I_4N_6S_2$

Molecular Weight 1254.73

Physical Form Yellow-brown powder

Solubility Soluble in dimethyl sulfoxide

Melting Point >250°C

Absorption (λ_{max}) 570 nm

Emission (λ_{max}) 602 nm

Synthesis Synthetic method[1]

Staining Applications Nucleic acids;[1–5] bacteria;[6] cells;[7,8] leukocytes;[16] nuclei;[16] megakaryocytes;[9] sperms;[10] hairs[11]

Biological Applications Nucleic acid hybridization;[12] DNA fingerprinting;[13] DNA sequencing;[14] detecting nucleic acids,[1–5] cells,[7,8] pathogens;[15] counting embryoblasts[16]

Industrial Applications Not reported

Safety/Toxicity No data available

REFERENCES

1. Yue, S. T.; Haugland, R. P. Dimers of unsymmetrical cyanine dyes containing pyridinium moieties. U.S. Patent 5410030, 1995; *Chem. Abstr.* **1995**, *123*, 259753.
2. Exner, M.; Rogers, A. Methods for identifying nucleic acids and determining melting temperature using multiple fluorophores and FRET. U.S. Pat. Appl. Publ. US 2007172836, 2007; *Chem. Abstr.* **2007**, *147*, 182489.
3. Wittwer, C. T.; Dujols, V. E.; Reed, G.; Zhou, L. Genotyping by thermal denaturation analysis using loss of fluorescence of double-stranded DNA-specific

dyes as a reporter to monitor denaturation. PCT Int. Appl. WO 2004038038, 2004; *Chem. Abstr.* **2004**, *140*, 386988.

4. Erikson, G. H.; Daksis, J. I.; Kandic, I.; Picard, P. Conditions for formation of three- and four-stranded nucleic acid complexes and their detection and analytical use. PCT Int. Appl. WO 2002103051, 2002; *Chem. Abstr.* **2002**, *138*, 50812.
5. Kricka, L. J. Stains, labels and detection strategies for nucleic acids assays. *Ann. Clin. Biochem.* **2002**, *39*, 114–129.
6. Kitaguchi, A.; Yamaguchi, N.; Nasu, M. Simultaneous enumeration of viable *Enterobacteriaceae* and *Pseudomonas* spp. within three hours by multicolor fluorescence *in situ* hybridization with vital staining. *J. Microbiol. Methods* **2006**, *65*, 623–627.
7. Anderson, A. L.; Knutson, C. R.; Mueth, D.; Plewa, J.; Tanner, E. Methods for staining cells for identification and sorting. U.S. Pat. Appl. Publ. US 2006172315, 2006; *Chem. Abstr.* **2006**, *145*, 183714.
8. Millard, P. J.; Roth, B. L.; Yue, S. T.; Haugland, R. P. Fluorescent viability assay using cyclic-substituted unsymmetrical cyanine dyes. U.S. Patent 5534416, 1996; *Chem. Abstr.* **1996**, *125*, 162751.
9. Minakami, T.; Mori, Y.; Tsuji, T.; Ikeuchi, Y. Megakaryocyte classification/counting method by double fluorescent staining and flow cytometry. Jpn. Kokai Tokkyo Koho JP 2006275985, 2006; *Chem. Abstr.* **2006**, *145*, 413679.
10. Matsumoto, T.; Okada, H.; Hamaguchi, Y. Method and reagent for counting sperm by flow cytometry. Jpn. Kokai Tokkyo Koho JP 2001242168, 2001; *Chem. Abstr.* **2001**, *135*, 207869.
11. Lagrange, A. Hair dye compositions containing a polycationic direct dye. Fr. Demande FR 2848840, 2004; *Chem. Abstr.* **2004**, *141*, 76344.
12. Erikson, G. H.; Daksis, J. I. Improving the signal/noise ratio of nucleic acid hybridization assays by preincubation of primer and target with nucleic acid binding agents. U.S. Pat. Appl. Publ. US 2004180345, 2004; *Chem. Abstr.* **2004**, *141*, 255469.
13. Pena, S. D. J. Multiplex PCR analysis of human microsatellite DNA and use in DNA fingerprinting. Braz. Pedido PI BR 2099004359, 2001; *Chem. Abstr.* **2002**, *136*, 242894.
14. Williams, J. G. K.; Anderson, J. P. Field-switch single molecule DNA sequencing in a two-electrode chamber. PCT Int. Appl. WO 2005111240, 2005; *Chem. Abstr.* **2005**, *143*, 474548.
15. Vannier, E. Methods for detection of pathogens in red blood cells. PCT Int. Appl. WO 2006031544, 2006; *Chem. Abstr.* **2006**, *144*, 307966.
16. Heuven, B.; Wong, F.; Tsuji, T.; Sakata, T.; Hamaguchi, I. Method for classifying and counting erythroblasts by flow cytometry. Jpn. Kokai Tokkyo Koho JP 11326323, 1999; *Chem. Abstr.* **1999**, *132*, 1804.

BODIPY FL C_5-CERAMIDE

CAS Registry Number 133867-53-5

Chemical Structure

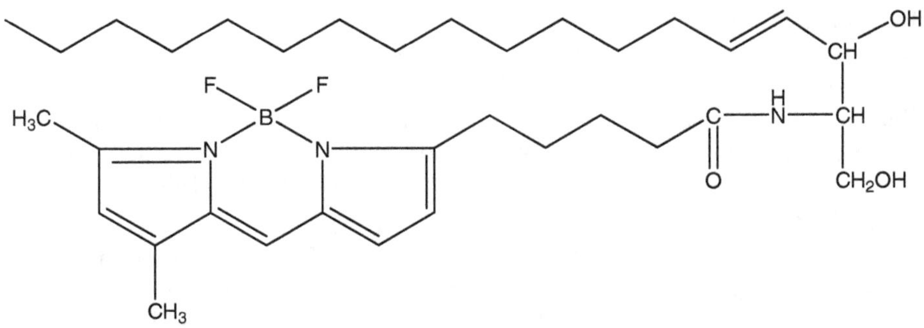

CA Index Name Boron, [5-[(3,5-dimethyl-2*H*-pyrrol-2-ylidene-κ*N*)methyl]-*N*-[(1*S*,2*R*,3*E*)-2-hydroxy-1-(hydroxymethyl)-3-heptadecen-1-yl]-1*H*-pyrrole-2-pentanamidato-κ*N*1]difluoro-, (T-4)-

Other Names Boron, [5-[(3,5-dimethyl-2*H*-pyrrol-2-ylidene)methyl]-*N*-[2-hydroxy-1-(hydroxymethyl)-3-heptadecenyl]-1*H*-pyrrole-2-pentanamidato-*N*1,*N*5]difluoro-, [T-4-[*R*-[*R**,*S**-(*E*)]]]-; Boron, [5-[(3,5-dimethyl-2*H*-pyrrol-2-ylidene-κ*N*)methyl]-*N*-[(1*S*,2*R*,3*E*)-2-hydroxy-1-(hydroxymethyl)-3-heptadecenyl]-1*H*-pyrrole-2-pentanamidato-κ*N*1]difluoro-, (T-4)-; 1*H*-Pyrrole-2-pentanamide, 5-[(3,5-dimethyl-2*H*-pyrrol-2-ylidene)methyl]-*N*-[2-hydroxy-1-(hydroxymethyl)-3-heptadecenyl]-, boron complex, [*R*-[*R**,*S**-(*E*)]]-; BODIPY FL-C_5-ceramide; C_5-DMB-ceramide

Merck Index Number Not listed

Chemical/Dye Class Organometallic, boron based

Molecular Formula $C_{34}H_{54}BF_2N_3O_3$

Molecular Weight 601.63

Physical Form Solid

Solubility Soluble in chloroform, methanol, dimethyl sulfoxide

Melting Point >200°C

Absorption (λ_{max}) 505 nm

Emission (λ_{max}) 511 nm, 620 nm

Synthesis Synthetic method[1]

Staining Applications Golgi apparatus;[1–4] endoplasmic reticulum;[4] bacteria;[5] exosomes;[6] glycoproteins;[7] Madin–Darby canine kidney (MDCK) cells;[8] lipids;[9–12] lipid bilayers;[13] lipoproteins;[14] myelin;[15] sphingolipid;[16–19] transfer RNA[20]

Biological Applications Measuring activity of a hydrolase;[21] studying lipid traffic,[18] membrane traffic;[19] inositol phosphorylceramide synthase activity assay[22]

Industrial Applications Not reported

Safety/Toxicity No data available

REFERENCES

1. Pagano, R. E.; Martin, O. C.; Kang, H. C.; Haugland, R. P. A novel fluorescent ceramide analog for studying membrane traffic in animal cells: accumulation at the Golgi apparatus results in altered spectral properties of the sphingolipid precursor. *J. Cell Biol.* **1991**, *113*, 1267–1279.
2. Moreno, R. D.; Schatten, G.; Ramalho-Santos, J. Golgi apparatus dynamics during mouse oocyte *in vitro* maturation: effect of the membrane trafficking inhibitor brefeldin A. *Biol. Reprod.* **2002**, *66*, 1259–1266.
3. Roth, M. G. Inheriting the Golgi. *Cell* **1999**, *99*, 559–562.
4. Teiten, M. H.; Bezdetnaya, L.; Morlière, P.; Santus, R.; Guillemin, F. Endoplasmic reticulum and Golgi apparatus are the preferential sites of Foscan localisation in cultured tumour cells. *Br. J. Cancer* **2003**, *88*, 146–152.
5. Boleti, H.; Ojcius, D. M.; Dautry-Varsat, A. Fluorescent labelling of intracellular bacteria in living host cells. *J. Microbiol. Methods* **2000**, *40*, 265–274.
6. Laulagnier, K.; Vincent-Schneider, H.; Hamdi, S.; Subra, C.; Lankar, D.; Record, M. Characterization of exosome subpopulations from RBL-2H3 cells

using fluorescent lipids. *Blood Cells, Mol. Dis.* **2005**, *35*, 116–121.

7. Rosenwald, A. G.; Pagano, R. E. Inhibition of glycoprotein traffic through the secretory pathway by ceramide. *J. Biol. Chem.* **1993**, *268*, 4577–4579.
8. Iida-Tanaka, N.; Namekata, I.; Tamura, M.; Kawamata, Y.; Kawanishi, T.; Tanaka, H. Membrane-labeled MDCK cells and confocal microscopy for the analyses of cellular volume and morphology. *Biol. Pharm. Bull.* **2008**, *31*, 731–734.
9. Kuerschner, L.; Ejsing, C. S.; Ekroos, K.; Shevchenko, A.; Anderson, K. I.; Thiele, C. Polyene-lipids: a new tool to image lipids. *Nat. Methods* **2005**, *2*, 39–45.
10. Allan, D. Lipid metabolic changes caused by short-chain ceramides and the connection with apoptosis. *Biochem. J.* **2000**, *345*, 603–610.
11. Bai, J.; Pagano, R. E. Measurement of spontaneous transfer and transbilayer movement of BODIPY-labeled lipids in lipid vesicles. *Biochemistry* **1997**, *36*, 8840–8848.
12. Redman, C. A.; Kusel, J. R. Distribution and biophysical properties of fluorescent lipids on the surface of adult *Schistosoma mansoni*. *Parasitology* **1996**, *113*, 137–143.
13. Johnson, M. E.; Berk, D. A.; Blankschtein, D.; Golan, D. E.; Jain, R. K.; Langer, R. S. Lateral diffusion of small compounds in human stratum corneum and model lipid bilayer systems. *Biophys. J.* **1996**, *71*, 2656–2668.
14. Ping, G.; Zhu, B.; Jabasini, M.; Xu, F.; Oka, H.; Sugihara, H.; Baba, Y. Analysis of lipoproteins by microchip electrophoresis with high speed and high reproducibility. *Anal. Chem.* **2005**, *77*, 7282–7287.
15. Bilderback, T. R.; Chan, J. R.; Harvey, J. J.; Glaser, M. Measurement of the rate of myelination using a fluorescent analog of ceramide. *J. Neurosci. Res.* **1997**, *49*, 497–507.
16. Marks, D. L.; Bittman, R.; Pagano, R. E. Use of Bodipy-labeled sphingolipid and cholesterol analogs to examine membrane microdomains in cells. *Histochem. Cell Biol.* **2008**, *130*, 819–832.
17. Marks, D. L.; Singh, R. D.; Choudhury, A.; Wheatley, C. L.; Pagano, R. E. Use of fluorescent sphingolipid analogs to study lipid transport along the endocytic pathway. *Methods* **2005**, *36*, 186–195.
18. Pagano, R. E.; Watanabe, R.; Wheatley, C.; Dominguez, M. Applications of BODIPY-sphingolipid analogs to study lipid traffic and metabolism in cells. *Methods Enzymol.* **2000**, *312*, 523–534.
19. Pagano, R. E.; Chen, C. S. Use of BODIPY-labeled sphingolipids to study membrane traffic along the endocytic pathway. *Ann. N.Y. Acad. Sci.* **1998**, *845*, 152–160.
20. Cooperman, B. S.; Smilansky, Z.; Goldman, Y. E.; Pan, D. Fluorescent labeling of transfer RNA and study of protein synthesis. PCT Int. Appl. WO 2009002866, 2008; *Chem. Abstr.* **2009**, *150*, 93172.
21. Karuso, P. H.; Choi, H. Y. Determination of the activity of a hydrolase or other hydrolytic agent and real-time monitoring of biomolecule digestion using a fluorescent reporter dye. PCT Int. Appl. WO 2007051257, 2007; *Chem. Abstr.* **2007**, *146*, 477766.
22. Elhammer, A. Novel assays for inositol phosphorylceramide synthase activity and for inhibitors using fluorescence resonance energy transfer. U.S. Pat. Appl. Publ. US 2007269844, 2007; *Chem. Abstr.* **2007**, *148*, 4443.

BRILLIANT BLUE FCF

CAS Registry Number 3844-45-9

Chemical Structure

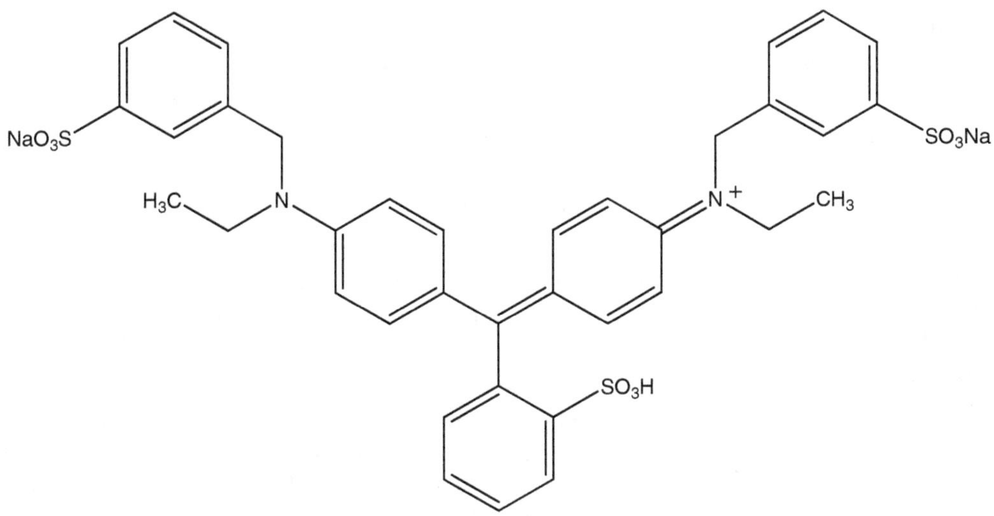

Merck Index Number 1373

Chemical/Dye Class Triphenylmethane

Molecular Formula $C_{37}H_{34}N_2Na_2O_9S_3$

Molecular Weight 792.85

CA Index Name Benzenemethanaminium, *N*-ethyl-*N*-[4-[[4-[ethyl[(3-sulfophenyl)methyl]amino]phenyl](2-sulfophenyl)methylene]-2,5-cyclohexadien-1-ylidene]-3-sulfo-, inner salt, sodium salt (1:2)

Other Names Alphazurine FG; Benzenemethanaminium, *N*-ethyl-*N*-[4-[[4-[ethyl[(3-sulfophenyl)methyl] amino]phenyl](2-sulfophenyl)methylene]-2,5-cyclohexadien-1-ylidene]-3-sulfo-, hydroxide, inner salt, disodium salt; Benzenemethanaminium, *N*-ethyl-*N*-[4-[[4-[ethyl [(3-sulfophenyl)methyl]amino]phenyl](2-sulfophenyl) methylene]-2,5-cyclohexadien-1-ylidene]-3-sulfo-,inner salt, disodium salt; Brilliant Blue FCF; C.I. 42090; C.I. Acid Blue 9, disodium salt; Blue 1; Blue 1206; Blue FCF; Blue No. 1; Brilliant Blue E 133; Brilliant Blue FCF Supra; C.I. Food Blue 2; Canacert Brilliant Blue FCF; D and C Blue No. 1; D&C Blue No. 1; Disodium erioglaucine; Dolkwal Brilliant Blue; Duasyn Acid Blue AE 02; E 133; E 133 (dye); Erioglaucine; Erioglaucine disodium salt; FD and C Blue 1; FD and C Blue No. 1; FD&C Blue No. 1; FDC Blue 1; Food Blue 1; Food Blue 2; Food Blue Dye No. 1; Food Blue No. 1; Food Blue No. 2; Hexacol Brilliant Blue A; Japan Blue 1; Japan Blue No. 1; Japan Food Blue No. 1; Puricolor Blue ABL 9; Sicovit Brilliant Blue 85E133; Usacert Blue No. 1; Usacert FD and C Blue No. 1

Physical Form Reddish-violet or purple powder

Solubility Soluble in water, ethanol, methyl cellosolve

Melting Point 283°C (decompose)

Absorption (λ_{max}) 406 nm, 625 nm

Synthesis Synthetic methods[1–11]

Staining Applications Stain for endoscopy;[12] cells;[13] microorganisms;[14] alcohol;[15] bakery products;[16] beverages;[17,18] candies;[19] cotton candy;[20] canned food;[21] cucumbers;[21] drinks;[15,19] meat products;[22] sweeteners;[23] capsules;[24] tablets;[25] toothpastes;[26] eyelashes;[27] eye shadow;[28] skin;[29] hairs;[30] tattoos;[31] teeth[32]

Biological Applications Treating coughing,[33] sneezing,[33] rhinorrhea,[33] nasal obstruction,[33] rhinitis;[34] medical devices[35]

Industrial Applications Electrochromic display devices;[36] inks;[37,38] paints;[39] photographic materials;[40] detergents;[41] textiles;[42] wood;[43] entertainment products[44]

Safety/Toxicity Acute toxicity;[45] carcinogenicity;[46–48] environmental toxicity;[49] genotoxicity;[50,51] mutagenicity[52,53]

Certification/Approval Approved by Food & Drugs Administration (FDA)

REFERENCES

1. Holtzinger, G.; Marie, G. Pigments based on clay-bound water-soluble dyes. Fr. Demande FR 2879097, 2006; *Chem. Abstr.* **2006**, *145*, 69332.
2. Kubo, Y.; Nakayama, J.; Doi, I.; Higashimura, Y. Pigment compositions with good leaching resistance and manufacture thereof. Jpn. Kokai Tokkyo Koho JP 08302230, 1996; *Chem. Abstr.* **1997**, *126*, 105494.
3. Gessner, T.; Mayer, U. Preparation of di- or triarylmethane dyes by oxidation. Ger. Offen. DE 4211783, 1993; *Chem. Abstr.* **1994**, *120*, 220390.
4. Imai, T.; Iwano, K. Transparent solid cosmetics containing solidified oils and water-soluble dyes. Jpn. Kokai Tokkyo Koho JP 03294212, 1991; *Chem. Abstr.* **1992**, *116*, 158600.
5. Clark, H. W.; Chapman, T. E. Insoluble pigments. Eur. Pat. Appl. EP 206800, 1986; *Chem. Abstr.* **1987**, *107*, 178133.
6. Clark, H. W.; Chapman, T. E.; Yates, R. L. Insoluble pigments. Eur. Pat. Appl. EP 206796, 1986; *Chem. Abstr.* **1987**, *107*, 8971.
7. Clark, H. W.; Chapman, T. E. Insoluble pigments and their preparation. Eur. Pat. Appl. EP 206797, 1986; *Chem. Abstr.* **1987**, *106*, 197899.
8. Guillon, M.; Mondet, J.; Papantoniou, C.; Vandenbossche, C. Polymer salts with dyes for cosmetic preparations. Ger. Offen. DE 3238166, 1983; *Chem. Abstr.* **1983**, *99*, 43322.
9. Kase, T. Electrolytic oxidation of triphenylmethane derivatives. Jpn. Kokai Tokkyo Koho JP 54130481, 1979; *Chem. Abstr.* **1980**, *92*, 163717.
10. Jones, J. H.; Dolinsky, M.; Harrow, L. S.; Heine, K. S., Jr.; Staves, M. C. The triphenylmethane colors derived from ethylbenzylanilinesulfonic acid. *J. Assoc. Off. Agric. Chem.* **1955**, *38*, 977–1010.
11. Salisbury, R.; Guth, E. P. A new preparative cell for high-voltage electrophoresis. *J. Am. Pharm. Assoc.* **1956**, *45*, 456–460.
12. Yamamoto, A.; Iimori, Y.; Saze, M.; Ishiguro, M.; Kurosawa, Y.; Sasaki, H. Histofluorescent stain for endoscopy. Brit. UK Pat. Appl. GB 2441892, 2008; *Chem. Abstr.* **2008**, *148*, 350550.
13. Yamazaki, K.; Fujiwara, N.; Seto, T.; Yang, P. Horny cell-staining liquid, and horny cell-staining method. PCT Int. Appl. WO 2007029437, 2007; *Chem. Abstr.* **2007**, *146*, 312248.
14. Noda, N.; Mizutani, T. Microorganism-measuring method using multiple staining. Jpn. Kokai Tokkyo Koho JP 2006340684, 2006; *Chem. Abstr.* **2006**, *146*, 77563.
15. Donhowe, E. T. Product and process of making an alcohol containing sport drink. U.S. Pat. Appl. Publ. US 2003134007, 2003; *Chem. Abstr.* **2003**, *139*, 100270.
16. Waldock, C. S. Method and apparatus for marking bakery products. PCT Int. Appl. WO 2005013698, 2005; *Chem. Abstr.* **2005**, *142*, 217857.
17. Smith, L. C. Carbonated beverage national school lunch meal. U.S. Pat. Appl. Publ. US 2007281059, 2007; *Chem. Abstr.* **2007**, *148*, 10213.
18. Huang, H. Y.; Shih, Y. C.; Chen, Y. C. Determining eight colorants in milk beverages by capillary electrophoresis. *J. Chromatogr. A* **2002**, *959*, 317–325.
19. Yoshioka, N.; Ichihashi, K. Determination of 40 synthetic food colors in drinks and candies by high-performance liquid chromatography using a short column with photodiode array detection. *Talanta* **2008**, *74*, 1408–1413.
20. Litvin, A. Sugar-substitute cotton candy and procedure for its production. Ger. Offen. DE 102008037185, 2009; *Chem. Abstr.* **2009**, *150*, 258878.
21. Pankova, A. V.; Kvasenkov, O. I. Method for canned food production from semifinished preserved cucumbers. Russ. RU 2289274, 2006; *Chem. Abstr.* **2006**, *146*, 44552.
22. Newman, P. B. Marking composition and method. PCT Int. Appl. WO 2007141557, 2007; *Chem. Abstr.* **2007**, *148*, 56722.
23. Mori, K. Sweetener composition. U.S. Pat. Appl. Publ. US 2006257550, 2006; *Chem. Abstr.* **2006**, *145*, 488169.
24. Fujii, T.; Noami, M.; Kida, A. Colored capsules in which discoloration is prevented. Jpn. Kokai Tokkyo Koho JP 2007091670, 2007; *Chem. Abstr.* **2007**, *146*, 387140.
25. Flanagan, J.; Smith, T. L.; Barkley, A.; Nicholson, R. E.; Callahan, T. P. Gellan gum-based tablet coatings. U.S. Patent 6485747, 2002; *Chem. Abstr.* **2002**, *137*, 389149.
26. Wright, H. W. Color-changing toothpaste containing dyes. U.S. Patent 6419902, 2002; *Chem. Abstr.* **2002**, *137*, 98684.
27. Okuyama, M. Eyelash cosmetic composition containing light interference color fibers. Jpn. Kokai Tokkyo Koho JP 2005314395, 2005; *Chem. Abstr.* **2005**, *143*, 446245.

28. Bouchard, F. Eye shadow comprising least 50% colorants. Eur. Pat. Appl. EP 1757262, 2007; *Chem. Abstr.* **2007**, *146*, 280374.

29. Arnaud, P.; Feltin, C.; Bazelaire, E. Process for coloring of dark skin using a direct dye. Fr. Demande FR 2901475, 2007; *Chem. Abstr.* **2007**, *148*, 39160.

30. Sallwey, A.; Schmitt, M.; Dieckow, A. Dyestuff for hair and fibers. Eur. Pat. Appl. EP 976385, 2000; *Chem. Abstr.* **2000**, *132*, 127457.

31. Kundakovic, L.; Klitzman, B.; Koger, K. E.; Mathiowitz, E.; Drill, C. A. Cellular or organelle-entrapped nanoparticles containing dyes for tattoos. PCT Int. Appl. WO 2007024429, 2007; *Chem. Abstr.* **2007**, *146*, 280376.

32. Brennan, J. V.; Soo, P. P.; Kalgutkar, R. S.; Suhrwardy, R. A. Methods of identifying orthodontic adhesives on teeth by selective staining with pigments and/or food dyes. U.S. Pat. Appl. Publ. US 2007148609, 2007; *Chem. Abstr.* **2007**, *147*, 102288.

33. Giordano, J.; Brown, K. Compositions for treatment of coughing, sneezing, rhinorrhea, and/or nasal obstruction, comprising guaifenesin and tannate salts of phenylephrine and dextromethorphan, and methods thereof. PCT Int. Appl. WO 2007084331, 2007; *Chem. Abstr.* **2007**, *147*, 197346.

34. Hall, M.; Roberts, A.; Heil, M. Kits comprising immunonutritional agents that contain pharmaceutically acceptable form of zinc, for prevention and treatment of rhinitis. PCT Int. Appl. WO 2008033155, 2008; *Chem. Abstr.* **2008**, *148*, 339026.

35. Raad, I. Coating medical device surfaces with antimicrobial agents. PCT Int. Appl. WO 2007062306, 2007; *Chem. Abstr.* **2007**, *147*, 39321.

36. Sawada, S.; Tsuboi, H. Electrochromic display device using triphenylmethane dye and display method. Jpn. Kokai Tokkyo Koho JP 2007052171, 2007; *Chem. Abstr.* **2007**, *146*, 262155.

37. Shibata, Y. Writing instruments containing aqueous inks with good storage stability and discoloration prevention. Jpn. Kokai Tokkyo Koho JP 2007119530, 2007; *Chem. Abstr.* **2007**, *146*, 523561.

38. Iida, Y. Porous silica-containing printing inks for foods and food packaging materials with excellent water resistance. Jpn. Kokai Tokkyo Koho JP 2007056191, 2007; *Chem. Abstr.* **2007**, *146*, 276256.

39. Sabnis, R. W.; Kehoe, T. D.; Balchunis, R. J. Paint compositions comprising color changing phthalein acid–base indicator. PCT Int. Appl. WO 2006105191, 2006; *Chem. Abstr.* **2006**, *145*, 403489.

40. Hayashi, K.; Okada, M. Silver photographic image reduction solution. Jpn. Kokai Tokkyo Koho JP 03007936, 1991; *Chem. Abstr.* **1992**, *116*, 13200.

41. Sumiya, M. Solid detergents for toilet bowls. Jpn. Kokai Tokkyo Koho JP 04253798, 1992; *Chem. Abstr.* **1993**, *118*, 24158.

42. Lupia, J. A.; Suhadolnik, J.; Wood, M. G.; Martin, D. W. H. Stabilized body care products, household products, textiles and fabrics. PCT Int. Appl. WO 2006122889, 2006; *Chem. Abstr.* **2006**, *146*, 12598.

43. Ikeda, M. Wood preservative emulsions containing colorants and their use for treatment of wood. Jpn. Kokai Tokkyo Koho JP 2005035288, 2005; *Chem. Abstr.* **2005**, *142*, 221441.

44. Sundaram, M.; Ihatsu, P.; Bourhis, Y. Dry ice products for the entertainment industry. PCT Int. Appl. WO 2007088442, 2007; *Chem. Abstr.* **2007**, *147*, 191690.

45. Tonogai, Y.; Iwaida, M.; Tati, M.; Ose, Y.; Sato, T. Biochemical decomposition of coal-tar dyes. II. Acute toxicity of coal-tar dyes and their decomposed products. *J. Toxicol. Sci.* **1978**, *3*, 205–214.

46. Helguera, A. M.; Cabrera Perez, M. A.; Gonzalez, M. P.; Ruiz, R. M.; Gonzalez Diaz, H. A topological substructural approach applied to the computational prediction of rodent carcinogenicity. *Bioorg. Med. Chem.* **2005**, *13*, 2477–2488.

47. Tanaka, Y.; Konishi, Y.; Nishimune, T.; Takagaki, Y. Effects of synthetic food colors on [^{3}H]serotonin release from rat basophilic leukemia cells (RBL-2H3). *Jpn. J. Toxicol. Environ. Health* **1995**, *41*, 206–211.

48. Haseman, J. K.; Winbush, J. S.; O'Donnell, M. W., Jr. Use of dual control groups to estimate false positive rates in laboratory animal carcinogenicity studies. *Fundam. Appl. Toxicol.* **1986**, *7*, 573–584.

49. Flury, M.; Fluehler, H. Brilliant Blue FCF as a dye tracer for solute transport studies—a toxicological overview. *J. Environ. Qual.* **1994**, *23*, 1108–1112.

50. Ohno, K.; Tanaka-Azuma, Y.; Yoneda, Y.; Yamada, T. Genotoxicity test system based on p53R2 gene expression in human cells: examination with 80 chemicals. *Mutat. Res.* **2005**, *588*, 47–57.

51. Haveland-Smith, R. B.; Combes, R. D. Screening of food dyes for genotoxic activity. *Food Cosmet. Toxicol.* **1980**, *18*, 215–221.

52. Pereira, L.; Ali, B.; Mohite, K.; Arora, P.; Rao, C. V. A study of mutagenic effects of spurious lipsticks chemicals in *Drosophila melanogaster*. *J. Environ. Biol.* **2000**, *21*, 309–315.

53. Ozaki, A.; Kitano, M.; Itoh, N.; Kuroda, K.; Furusawa, N.; Masuda, T.; Yamaguchi, H. Mutagenicity and DNA-damaging activity of decomposed products of food colors under UV irradiation. *Food Chem. Toxicol.* **1998**, *36*, 811–817.

BRILLIANT CRESYL BLUE

CAS Registry Number 81029-05-2

Chemical Structure

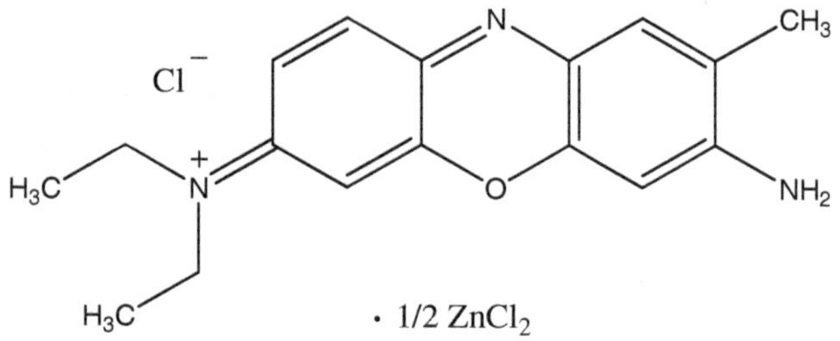

CA Index Name Phenoxazin-5-ium, 3-amino-7-(diethylamino)-2-methyl-, chlorozincate (2:1)

Other Names 3-Amino-7-(diethylamino)-2-methylphenoxazin-5-ium chlorozincate; BCB; Brilliant Cresyl Blue; Brilliant Cresyl Blue ALD; Brilliant Cresyl Blue BB; Brilliant Blue C; C.I. 51010

Merck Index Number Not listed

Chemical/Dye Class Phenoxazine

Molecular Formula $C_{17}H_{20}ClN_3O \cdot 0.5\ ZnCl_2$

Molecular Weight 385.96

Physical Form Green crystalline powder

Solubility Soluble in water, ethanol

Melting Point 233–236°C

pK$_a$ 6.0, 11.0

Absorption (λ_{max}) 622 nm

Synthesis Synthetic methods[1–3]

Staining Applications Brain tissue;[4] nuclei;[5] plant chromosomes;[6] reticulocytes;[7] platelets;[8] reticulated red cells[8]

Biological Applications Detection of biochemical molecules;[9] enzyme assay[9]

Industrial Applications Optical data storage[10]

Safety/Toxicity No data available

Certification/Approval Certified by Biological Stain Commission (BSC)

REFERENCES

1. Green, F. J. *The Sigma–Aldrich Handbook of Stains, Dyes and Indicators*; Aldrich Chemical Company Inc.: Milwaukee, **1991**; pp 155–157.
2. Gurr, E. *Encyclopedia of Microscopic Stains*; Leonard Hill Books Ltd.: London, **1960**; pp 86–87.
3. *Colour Index*, 1st ed.; Society of Dyers and Colourists: Bradford, 1924.
4. Smith, T. M.; Squires, R. F. Differential inhibition of brain specific [^{3}H]flunitrazepam binding by several types of dyes. *Neurochem. Res.* **1983**, *8*, 1177–1183.
5. Horobin, R. W.; Stockert, J. C.; Rashid-Doubell, F. Fluorescent cationic probes for nuclei of living cells: why are they selective? A quantitative structure–activity relations analysis. *Histochem. Cell Biol.* **2006**, *126*, 165–175.
6. Stewart, W. N.; Schertiger, A. M. Brilliant cresyl blue as a stain for plant chromosomes. *Stain Technol.* **1949**, *24*, 39–45.
7. Robertson, O. H. The effects of experimental plethora on blood production. *J. Exp. Med.* **1917**, *26*, 221–237.
8. Cunningham, R. S. A method for permanent staining of reticulated red cells. *Arch. Int. Med.* **1920**, *26*, 405–409.
9. Palmer, J. L.; Johnson, J. B.; Timmerman, M. W. Method, system and devices for the oxygen-independent enzymic assay and detection of biochemical molecules. Eur. Pat. Appl. EP 330517, 1989; *Chem. Abstr.* **1990**, *113*, 111981.
10. Vo-Dinh, T.; Stokes, D. L. SERODS: a new medium for high-density optical data storage. *Proc. SPIE-Int. Soc. Opt. Eng.* **1998**, *3401*, 284–290.

BRILLIANT GREEN

CAS Registry Number 633-03-4

Chemical Structure

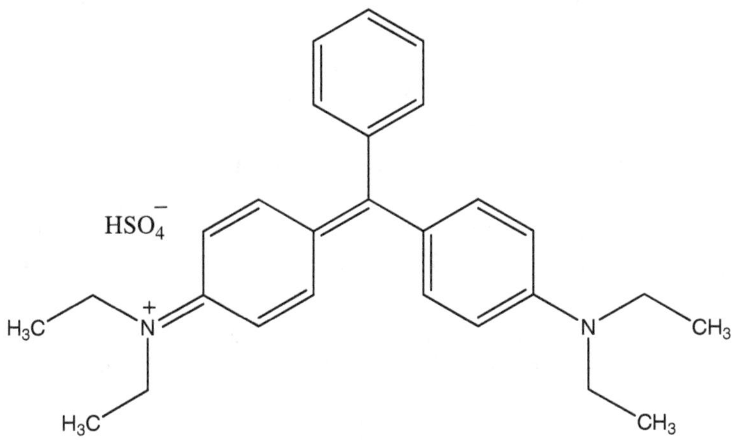

Merck Index Number 1374

Chemical/Dye Class Triphenylmethane

Molecular Formula $C_{27}H_{34}N_2O_4S$

Molecular Weight 482.63

CA Index Name Ethanaminium, *N*-[4-[[4-(diethylamino)phenyl]phenylmethylene]-2,5-cyclohexadien-1-ylidene]-*N*-ethyl-, sulfate

Other Names ADC Brilliant Green crystals; Astradiamant Green GX; C.I. Basic Green 1; 12415 Green; ADC Brilliant Green Crystals; Aizen Diamond Green GH; Aizen Malachite Green GH; Astra Diamond Green GX; Astrazon Green D; Avon Green A 4379; Basic Bright Green; Basic Bright Green Sulfate; Basic Brilliant Green; Basic Green 1; Basic Green V; Brilliant Green; Brilliant Green B; Brilliant Green BP; Brilliant Green BP Crystals; Brilliant Green BPC; Brilliant Green Crystals; Brilliant Green Crystals H; Brilliant Green DSC; Brilliant Green G; Brilliant Green GX; Brilliant Green Lake; Brilliant Green P; Brilliant Green Special; Brilliant Green Sulfate; Brilliant Green WP Crystals; Brilliant Green Y; Brilliant Green YN; Brilliant Green YNS; Brilliant Lake Green Y; Brilliant green aseptic; C.I. 42040; Calcozine Brilliant Green G; Deorlene Green JJO; Diamond Green G; Diamond Green GH; *N*-[4-[[4-(Diethylamino)phenyl]phenylmethylene]-2,5-cyclohexadien-1-ylidene]-*N*-ethylethanaminium sulfate (1:1); Emerald Green; Ethyl green; Fast Green J; Fast Green JJO; Green EN; Hidaco Brilliant Green; Malachite Green G; Mitsui Brilliant Green GX; NSC 5011; Resplendency Green; Solid Green; Solid Green JO; Tertrophene Brilliant Green G; Tokyo Aniline Brilliant Green; Zelen brilantni; Zelen malachitova G; Zelen smaragdova; Zelen zasadita 1

Physical Form Glistering green to green-gold crystals

Solubility Soluble in water, ethanol

Melting Point 210°C (decompose)

pH Range 0.0–2.6

Color Change at pH Yellow (0.0) to green (2.6)

Absorption (λ_{max}) 625 nm, 428 nm

Synthesis Synthetic methods[1–12]

Staining Applications Bone cement;[13] cytoplasm;[14] horny layer cell;[15] large intestine;[14] nucleic acids;[16] skin;[15,17] stratum corneum cell;[18] tissues;[14] hairs[1,19,20]

Biological Applications Antimalarial agent;[1,21] detecting carious tissue;[1,22] treating anococcygeal pain syndrome;[23] wound dressing materials;[1,17,24] catheter;[25] mouthwash[26]

Industrial Applications Color filters;[1,27] black matrix;[1,27] liquid crystal displays;[1,27] photoresist;[1,28] conducting polymer films;[29] optical fiber pH sensor;[30] printed circuit boards;[31] inks;[1,32,33] textiles[34]

Safety/Toxicity Bacterial toxicity;[35] carcinogenicity;[1,36] fish toxicity;[1,37] microbial toxicity;[38] skin toxicity[39]

Certification/Approval Certified by Biological Stain Commission (BSC)

REFERENCES

1. Sabnis, R. W. *Handbook of Acid-Base Indicators*; CRC Press: Boca Raton, **2008**; pp 36–37.
2. Welcher, F. J. *Organic Analytical Reagents*; D. Van Nostrand Company: New York, **1948**; Vol. *4*, p 341.
3. Berneth, H.; Raue, R. Oxidative manufacture of sulfo group-free triphenylmethane dyes. Ger. Offen. DE 3842014, 1990; *Chem. Abstr.* **1991**, *114*, 64262.
4. Aizawa, T. Manufacture of triphenylmethane dye by electrochemical oxidation. Jpn. Kokai Tokkyo Koho JP 02194188, 1990; *Chem. Abstr.* **1991**, *114*, 52054.
5. Leont'eva, K. A.; Kulygina, M. A. Commercial-type Basic Brilliant Green dye. U.S.S.R. SU 724543, 1980; *Chem. Abstr.* **1980**, *93*, 48544.
6. Kast, H.; Mayer, U. Basic dyes. Ger. Offen. DE 2736679, 1979; *Chem. Abstr.* **1979**, *90*, 170166.
7. Feldman, M. L.; Gordon, J. E. Recovery and purification of basic triarylmethane dyes. U.S. Patent 3679713, 1972; *Chem. Abstr.* **1972**, *77*, 141488.
8. Feldman, M. L.; Gordon, J. E. Isolation of triphenylmethane dyes. Ger. Offen. DE 2102404, 1971; *Chem. Abstr.* **1971**, *75*, 89326.
9. Kerr, G. O.; Gregory, G. R. E. C. Simple method for the purification of brilliant green, reagent for antimony. *Analyst* **1969**, *94*, 1036–1037.
10. Yoshino, M. Synthesis of triphenylmethane dyes. *Rep. Tokyo Imp. Ind. Res. Inst. Lab.* **1942**, *37*, 95–189; *Chem. Abstr.* **1948**, *42*, 27489.
11. Laptev, N. G. Brilliant green. *Anilinokrasochnaya Prom.* **1933**, *3*, 462–468; *Chem. Abstr.* **1934**, *28*, 30113.
12. Browning, C. H.; Gulbransen, R.; Kennaway, E. L.; Thornton, L. H. D. Flavine and brilliant green. *Br. Med. J.* **1917**, 73–78.
13. Kuehn, K. D.; Vogt, S. Colored polymethacrylate bone cement and method for preparation. Ger. DE 102005032110, 2006; *Chem. Abstr.* **2006**, *145*, 235931.
14. Yamamoto, A.; Iimori, Y.; Saze, M.; Ishiguro, M.; Kurosawa, Y.; Sasaki, H. Histofluorescent stain for endoscopy. Brit. UK Pat. Appl. GB 2441892, 2008; *Chem. Abstr.* **2008**, *148*, 350550.
15. Yamazaki, K.; Fujiwara, N. Horny layer cell staining solution/method for skin status evaluation and cosmetic selection. Jpn. Kokai Tokkyo Koho JP 2006053117, 2006; *Chem. Abstr.* **2006**, *144*, 249970.
16. Li, L.; Song, G.; Fang, G. Determination of nucleic acid with brilliant green by resonance light scattering. *Am. Biotechnol. Lab.* **2007**, *25*, 34–36.
17. Smelov, V. V.; Smelov, E. V. Remedy for removing benign skin neoplasms. Russ. RU 2259834, 2005; *Chem. Abstr.* **2005**, *143*, 272574.
18. Yamazaki, K.; Sakazaki, Y. Color control of magnified images of stained stratum corneum cell preparations by computer graphic technique. Jpn. Kokai Tokkyo Koho JP 2007114076, 2007; *Chem. Abstr.* **2007**, *146*, 478277.
19. Hoeffkes, H.; Bossmann, B. Use of coated alkali agents for oxidative hair bleaching and dyeing compositions to prevent foaming. Eur. Pat. Appl. EP 1752191, 2007; *Chem. Abstr.* **2007**, *146*, 212227.
20. Hoeffkes, H.; Gross, W. Hair dyes containing silylcarbonates and/or sylilcarbamates for simultaneous dyeing and bleaching of hair. Ger. Offen. DE 102004052060, 2006; *Chem. Abstr.* **2006**, *144*, 419031.
21. Chong, C. R.; Chen, X.; Shi, L.; Liu, J. O.; Sullivan, D. J., Jr. A clinical drug library screen identifies astemizole as an antimalarial agent. *Nat. Chem. Biol.* **2006**, *2*, 415–416.
22. Guggenberger, R.; Luchterhandt, T.; Haeberlein, I.; Kappler, O.; Hansen, M. Dental composition for detection of carious tissue, detection method. PCT Int. Appl. WO 2006125650, 2006; *Chem. Abstr.* **2006**, *146*, 33158.
23. Kazakov, M. V.; Khidiyatov, I. I.; Gainutdinov, F. M. Method for the treatment of anococcygeal pain syndrome by "carpet" blockades. Russ. RU 2289397, 2006; *Chem. Abstr.* **2006**, *146*, 55545.
24. Addison, D.; Greenhalgh, D.; Cullen, B. M. Antioxidant and antimicrobial wound dressing materials. PCT Int. Appl. WO 2005049101, 2005; *Chem. Abstr.* **2005**, *143*, 13454.
25. Rosenblatt, J.; Vitullo, J.; Bouchard, M. Catheter with polymeric coating. U.S. Pat. Appl. Publ. US 2007129690, 2007; *Chem. Abstr.* **2007**, *147*, 39251.
26. Bahna, P.; Hanna, H. A.; Dvorak, T.; Vaporciyan, A.; Chambers, M.; Raad, I. Antiseptic effect of a novel alcohol-free mouthwash: a convenient prophylactic alternative for high-risk patients. *Oral Oncol.* **2007**, *43*, 159–164.
27. Kurita, E.; Igarashi, T. Photocurable resin compositions for forming color filters or black matrixes of liquid crystal displays. Jpn. Kokai Tokkyo Koho JP 2005208480, 2005; *Chem. Abstr.* **2005**, *143*, 183241.

28. Abramov, V. N.; Aref'ev, N. M.; Kochukov, A. V.; Yakovlev, V. B. Film photoresist for stenciling and method for its manufacture. Russ. RU 2321037, 2008; *Chem. Abstr.* **2008**, *148*, 414240.

29. Kato, H.; Takemura, S.; Nishikawa, O.; Taniguchi, M. X-ray photoemission spectroscopy and Fourier transform infrared studies of dye molecule doped conducting polymer films. *J. Vac. Sci. Technol., A* **2005**, *23*, 869–874.

30. Beltran-Perez, G.; Lopez-Huerta, F.; Munoz-Aguirre, S.; Castillo-Mixcoatl, J.; Palomino-Merino, R.; Lozada-Morales, R.; Portillo-Moreno, O. Fabrication and characterization of an optical fiber pH sensor using sol-gel deposited TiO_2 film doped with organic dyes. *Sens. Actuators, B* **2006**, *B120*, 74–78.

31. Murao, K. Fabrication of printed circuit boards provided with blind via-holes plugged with copper by electrodeposition. Jpn. Kokai Tokkyo Koho JP 2006009079, 2006; *Chem. Abstr.* **2006**, *144*, 116371.

32. Kimura, F. Color-changing black ink compositions and marking pen-type color-changing writing tools and writing tool sets containing them. Jpn. Kokai Tokkyo Koho JP 2007197486, 2007; *Chem. Abstr.* **2007**, *147*, 236905.

33. Masushige, N. Color-changing ink compositions and color-changing writing tools and writing tool sets containing them. Jpn. Kokai Tokkyo Koho JP 2007119676, 2007; *Chem. Abstr.* **2007**, *146*, 502658.

34. Sheikh, M. R. K.; Farouqui, F. I.; Modak, P. R.; Hoque, Md. A.; Yasmin, Z. Dyeing of Rajshahi silk with basic dyes: effect of modification on dyeing properties. *J. Text. Inst.* **2006**, *97*, 295–300.

35. Khadzhai, Y. I.; Skubko, T. P.; Kobzar, A. I.; Nikolaeva, A. V.; Pavlova, V. V.; Ershova, N. V. Activity and toxicity of bactericidal adhesive plasters. *Farmatsiya* **1988**, *37*, 56–59; *Chem. Abstr.* **1988**, *108*, 142902.

36. Vachalkova, A.; Novotny, L.; Blesova, M. Polarographic reduction of some triphenylmethane dyes and their potential carcinogenic activity. *Neoplasma* **1996**, *43*, 113–117.

37. Ghisotti, F.; Colombo, G. Toxicological analysis of water-soluble synthetic organic dyes in aquatic animals. *Ecologia* **1972**, *2*, 32–34; *Chem. Abstr.* **1972**, *77*, 122700.

38. Oros, G.; Cserhati, T.; Forgacs, E. Separation of the strength and selectivity of the microbiological effect of synthetic dyes by spectral mapping technique. *Chemosphere* **2003**, *52*, 185–193.

39. Shehade, S. A.; Foulds, I. S. Allergic contact dermatitis to Brilliant Green. *Contact Dermatitis* **1986**, *14*, 186–187.

BSPT

CAS Registry Number 38116-89-1

Chemical Structure

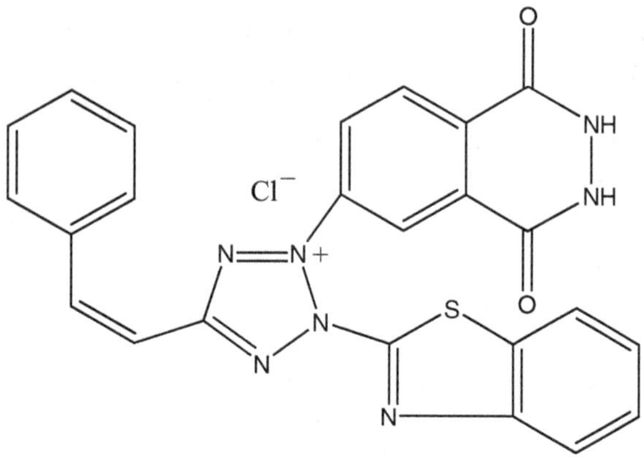

CA Index Name 2*H*-Tetrazolium, 2-(2-benzothiazolyl)-3-(3,4-dihydro-1-hydroxy-4-oxo-6-phthalazinyl)-5-(2-phenylethenyl)-, chloride (1:1)

Other Names 2*H*-Tetrazolium, 2-(2-benzothiazolyl)-5-(2-phenylethenyl)-3-(1,2,3,4-tetrahydro-1,4-dioxo-6-phthalazinyl)-, chloride; 2-(2′-Benzothiazolyl)-5-styryl-3-(4′-phthalhydrazidyl)tetrazolium chloride; 2-(2-Benzothiazolyl)-3-(4-phthalhydrazidyl)-5-styryltetrazolium chloride; 2-(2-Benzothiazolyl)-5-styryl-3-(4-phthalhydrazidyl)tetrazolium chloride; BPST; BPST chloride; BSPT; BSPT (dye); Thiazolyl blue; Thiazolyl blue (Chinese)

Merck Index Number Not listed

Chemical/Dye Class Tetrazolium salt

Molecular Formula $C_{24}H_{16}ClN_7O_2S$

Molecular Weight 501.95

Physical Form Orange crystals or powder

Solubility Soluble in water, ethanol, *N,N*-dimethylformamide

Melting Point 254°C

Absorption (λ_{max}) 300 nm

Synthesis Synthetic methods[1–3]

Staining Applications Enzymes;[4–11] tissues[12,13]

Biological Applications Albumin assays;[14] detecting gamma-hydroxybutyric acid (GHB);[15] treating cancer,[16] cerebrovascular disorder,[17] brain degenerative disease,[17] demyelinating disease,[17] muscle degenerative disease[18]

Industrial Applications Not reported

Safety/Toxicity Skin toxicity[19]

REFERENCES

1. James, P. R. Tetrazolium salts as steam sterilization process indicators. PCT Int. Appl. WO 9837157, 1998; *Chem. Abstr.* **1998**, *129*, 218237.
2. Kong, Q. Anticancer tetrazolium compounds and their application. Faming Zhuanli Shenqing Gongkai Shuomingshu CN 1338256, 2002; *Chem. Abstr.* **2002**, *137*, 363040.
3. Kalina, M.; Plapinger, R. E.; Hoshino, Y.; Seligman, A. M. Nonosmiophilic tetrazolium salts that yield osmiophilic, lipophobic formazans for ultrastructural localization of dehydrogenase activity. *J. Histochem. Cytochem.* **1972**, *20*, 685–695.
4. Goosen, N. K.; Broers, C. A. M.; Hombergen, E. J.; Stumm, C. K.; Vogels, G. D. Effect of fixation on activity and cytochemistry of hydrogenosomal enzymes in *Trichomonas vaginalis. J. Gen. Microbiol.* **1990**, *136*, 2189–2193.
5. Van Noorden, C. J. F.; Butcher, R. G. Histochemical localization of NADP-dependent dehydrogenase activity with four different tetrazolium salts. *J. Histochem. Cytochem.* **1984**, *32*, 998–1004.
6. Shannon, W. A., Jr. Use of BSPT in dehydrogenase histochemistry. *Histochem. J.* **1982**, *14*, 166–168.
7. Van Noorden, C. J. F.; Tas, J.; Sanders, J. A. H. Quantitative aspects of the cytochemical demonstration of glucose-6-phosphate dehydrogenase with tetrazolium salts studied in a model system of polyacrylamide films. *Acta Histochem., Suppl.* **1981**, *24*, 231–236.
8. Altman, F. P.; Hoeyer, P. E.; Andersen, H. Dehydrogenase histochemistry of lipid-rich tissues: a tetrazolium-metal chelation technique to improve localization. *Histochem. J.* **1979**, *11*, 485–488.
9. Gossrau, R. Tetrazolium methods for the histochemical investigation of hydrolases. *Histochemistry* **1978**, *58*, 203–218.
10. Muller, J.; Da Lage, C. Ultracytochemical demonstration of monoamine oxidase activity in nervous and non-nervous tissues of the rat. *J. Histochem. Cytochem.* **1977**, *25*, 337–348.
11. Shannon, W. A., Jr.; Wasserkrug, H. L.; Seligman, A. M. Ultrastructural localization of monoamine

oxidase (MAO) with tryptamine and a new tetrazolium salt, 2-(2′-benzothiazolyl)-5-styryl-3-(4′-phthalhydrazidyl) tetrazolium chloride (BSPT). *J. Histochem. Cytochem.* **1974**, *22*, 170–182.

12. Altman, F. P. The quantification of formazans in tissue sections by microdensitometry. III. The effect of objective power and scanning spot size. *Histochem. J.* **1976**, *8*, 507–511.
13. Altman, F. P. The quantification of formazans in tissue sections by microdensitometry. II. The use of BPST, a new tetrazolium salt. *Histochem. J.* **1976**, *8*, 501–506.
14. Atkinson, A.; Hammond, P.; Hinton, R. J.; Miller, J.; Price, C. P. Assays of albumin and reducing agents and drugs based on albumin–tetrazolium interaction. PCT Int. Appl. WO 9012318, 1990; *Chem. Abstr.* **1991**, *115*, 25540.
15. Parsons, S. M.; Harris, D. O.; Bravo, D. T. Methods, compositions and apparatuses for detection of gamma-hydroxybutyric acid (GHB). U.S. Pat. Appl. Publ. US 2003175846, 2003; *Chem. Abstr.* **2003**, *139*, 256601.
16. Kong, Q. Method and composition for treating cancer using cellular organelle crystallizing agents. U.S. Pat. Appl. Publ. US 2006286170, 2006; *Chem. Abstr.* **2006**, *146*, 55482.
17. Urade, Y.; Eguchi, N.; Aritake, K.; Sato, Y.; Kadoyama, K.; Taniike, M. Drugs for improving the prognosis of brain injury and a method of screening the same. PCT Int. Appl. WO 2004006956, 2004; *Chem. Abstr.* **2004**, *140*, 117405.
18. Urade, Y.; Eguchi, N.; Aritake, K.; Sato, A.; Taniike, M.; Mori, I.; Miyano, M. Remedies for muscle degenerative disease containing hematopoietic prostaglandin D synthase inhibitors, and method for drug screening for muscle degenerative disease. Jpn. Kokai Tokkyo Koho JP 2005119984, 2005; *Chem. Abstr.* **2005**, *142*, 423877.
19. He, G.; Huang, Q.; Huang, J.; Zheng, S.; Huang, J. Determination of chemical-induced skin allergy by thiazolyl blue colorimetry. *Zhongguo Weisheng Jianyan Zazhi* **2006**, *16*, 1007–1008; *Chem. Abstr.* **2007**, *147*, 251816.

BTC

CAS Registry Number 154324-80-8

Chemical Structure

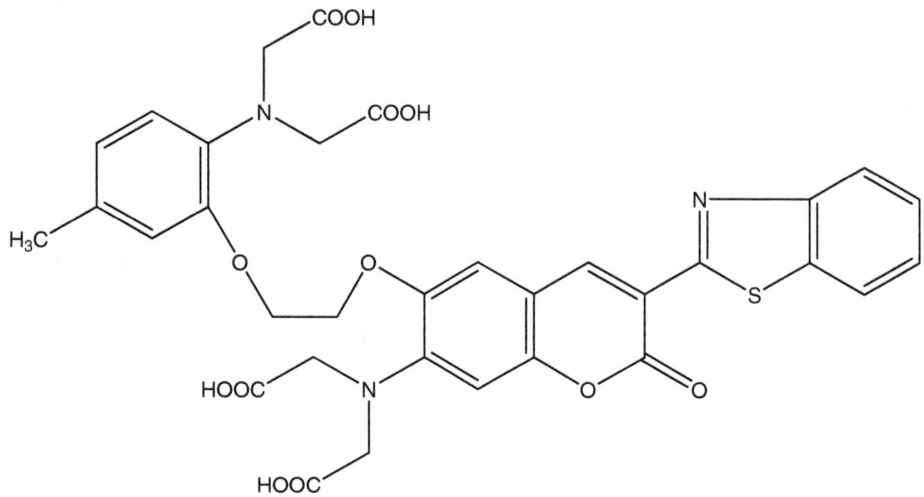

CA Index Name Glycine, *N*-[3-(2-benzothiazolyl)-6-[2-[2-[bis(carboxymethyl)amino]-5-methylphenoxy]ethoxy]-2-oxo-2*H*-1-benzopyran-7-yl]-*N*-(carboxymethyl)-

Other Names BTC

Merck Index Number Not listed

Chemical/Dye Class Benzothiazolylcoumarin

Molecular Formula $C_{33}H_{29}N_3O_{12}S$

Molecular Weight 691.66

Physical Form Solid

Solubility Soluble in dimethyl sulfoxide

Melting Point >250°C

Boiling Point (Calcd.) 1027.8 ± 75.0°C, pressure: 760 Torr

pK$_a$ (Calcd.) 1.70 ± 0.10, most acidic, temperature: 25°C; 6.17 ± 0.50, most basic, temperature: 25°C

Absorption (λ_{max}) 464 nm

Emission (λ_{max}) 533 nm

Synthesis Synthetic methods[1,2]

Staining Applications Calcium ions[1–10]

Biological Applications Calcium indicator;[1–10] identifying ion channels modulators[11]

Industrial Applications Not reported

Safety/Toxicity No data available

REFERENCES

1. Katerinopoulos, H. E.; Iatridou, H.; Foukaraki, E.; Malekzadeh, M. N.; Kuhn, M. A.; Haugland, R. P. Benzazolylcoumarin-based ion indicators. U.S. Patent 5501980, 1996; *Chem. Abstr.* **1996**, *124*, 359466.
2. Iatridou, H.; Foukaraki, E.; Kuhn, M. A.; Marcus, E. M.; Haugland, R. P.; Katerinopoulos, H. E. The development of a new family of intracellular calcium probes. *Cell Calcium* **1994**, *15*, 190–198.
3. Liepouri, F.; Foukaraki, E.; Deligeorgiev, T. G.; Katerinopoulos, H. E. Iminocoumarin-based low affinity fluorescent Ca^{2+} indicators excited with visible light. *Cell Calcium* **2001**, *30*, 331–335.
4. Fukano, T.; Shimozono, S.; Miyawaki, A. Simultaneous dual-excitation ratiometry using orthogonal linear polarized lights. *Biochem. Biophys. Res. Commun.* **2004**, *317*, 77–83.
5. Ohnuma, K.; Whim, M. D.; Fetter, R. D.; Kaczmarek, L. K.; Zucker, R. S. Presynaptic target of Ca^{2+} action on neuropeptide and acetylcholine release in *Aplysia californica*. *J. Physiol.* **2001**, *535*, 647–662.
6. Hyrc, K. L.; Bownik, J. M.; Goldberg, M. P. Ionic selectivity of low-affinity ratiometric calcium indicators: mag-Fura-2, Fura-2FF and BTC. *Cell Calcium* **2000**, *27*, 75–86.

7. Lakowicz, J. R.; Gryczynski, I. Low frequency modulation sensors using nanosecond fluorophores. PCT Int. Appl. WO 2000014515, 2000; *Chem. Abstr.* **2000**, *132*, 205106.

8. Hyrc, K. L.; Bownik, J. M.; Goldberg, M. P. Neuronal free calcium measurement using BTC/AM, a low affinity calcium indicator. *Cell Calcium* **1998**, *24*, 165–175.

9. Zhao, M.; Hollingworth, S.; Baylor, S. M. Properties of tri- and tetracarboxylate Ca^{2+} indicators in frog skeletal muscle fibers. *Biophys. J.* **1996**, *70*, 896–916.

10. Regehr, W. G.; Atluri, P. P. Calcium transients in cerebellar granule cell presynaptic terminals. *Biophys. J.* **1995**, *68*, 2156–2170.

11. Weaver, C. D. Methods for detecting modulators of ion channels using thallium (i) sensitive assays. PCT Int. Appl. WO 2002031508, 2002; *Chem. Abstr.* **2002**, *136*, 321701.

BTC AM

CAS Registry Number 176767-94-5

Chemical Structure

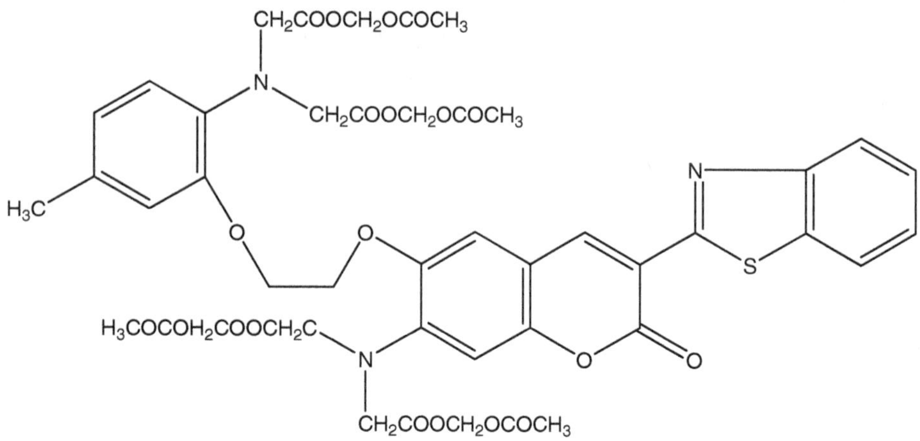

CA Index Name Glycine, *N*-[2-[(acetyloxy)methoxy]-2-oxoethyl]-*N*-[3-(2-benzothiazolyl)-6-[2-[2-[bis[2-[(acetyloxy)methoxy]-2-oxoethyl]amino]-5-methylphenoxy]ethoxy]-2-oxo-2*H*-1-benzopyran-7-yl]-, (acetyloxy)methyl ester

Other Names BTC acetoxymethyl ester, BTC AM

Merck Index Number Not listed

Chemical/Dye Class Benzothiazolylcoumarin

Molecular Formula $C_{45}H_{45}N_3O_{20}S$

Molecular Weight 979.91

Physical Form Solid

Solubility Soluble in methanol, dimethyl sulfoxide

Melting Point >250°C

Boiling Point (Calcd.) 992.9 ± 75.0°C, pressure: 760 Torr

pK_a (Calcd.) 1.91 ± 0.50, most basic, temperature: 25°C

Absorption (λ_{max}) 433 nm

Emission (λ_{max}) 504 nm

Synthesis Synthetic method[1]

Staining Applications Calcium ions;[1,3–11] leukocytes[2]

Biological Applications Calcium indicator;[1,3–11] detecting leukocytes tumor cells[2]

Industrial Applications Not reported

Safety/Toxicity No data available

REFERENCES

1. Katerinopoulos, H. E.; Iatridou, H.; Foukaraki, E.; Malekzadeh, M. N.; Kuhn, M. A.; Haugland, R. P. Benzazolylcoumarin-based ion indicators. U.S. Patent 5501980, 1996; *Chem. Abstr.* **1996**, *124*, 359466.
2. Minakami, T.; Tsuji, T.; Oguni, S.; Hamaguchi, Y.; Tsuruta, K.; Kamihiro, T. Method for detecting leukocyte tumor cells. Jpn. Kokai Tokkyo Koho JP 2002207036, 2002; *Chem. Abstr.* **2002**, *137*, 106038.
3. Fukano, T.; Shimozono, S.; Miyawaki, A. Simultaneous dual-excitation ratiometry using orthogonal linear polarized lights. *Biochem. Biophys. Res. Commun.* **2004**, *317*, 77–83.
4. Ohnuma, K.; Whim, M. D.; Fetter, R. D.; Kaczmarek, L. K.; Zucker, R. S. Presynaptic target of Ca^{2+} action on neuropeptide and acetylcholine release in *Aplysia californica*. *J. Physiol.* **2001**, *535*, 647–662.
5. Hyrc, K. L.; Bownik, J. M.; Goldberg, M. P. Ionic selectivity of low-affinity ratiometric calcium indicators: mag-Fura-2, Fura-2FF and BTC. *Cell Calcium* **2000**, *27*, 75–86.
6. Williams, D. A.; Bowser, D. N.; Petrou, S. Confocal Ca^{2+} imaging of organelles, cells, tissues, and organs. *Methods Enzymol.* **1999**, *307*, 441–469.

7. Ito, K.; Miyashita, Y.; Kasai, H. Kinetic control of multiple forms of Ca(2+) spikes by inositol trisphosphate in pancreatic acinar cells. *J. Cell Biol.* **1999**, *146*, 405–413.

8. Hyrc, K. L.; Bownik, J. M.; Goldberg, M. P. Neuronal free calcium measurement using BTC/AM, a low affinity calcium indicator. *Cell Calcium* **1998**, *24*, 165–175.

9. Ito, K.; Miyashita, Y.; Kasai, H. Micromolar and submicromolar Ca^{2+} spikes regulating distinct cellular functions in pancreatic acinar cells. *EMBO J.* **1997**, *16*, 242–251.

10. Zhao, M.; Hollingworth, S.; Baylor, S. M. AM-loading of fluorescent Ca^{2+} indicators into intact single fibers of frog muscle. *Biophys. J.* **1997**, *72*, 2736–2747.

11. Regehr, W. G.; Atluri, P. P. Calcium transients in cerebellar granule cell presynaptic terminals. *Biophys. J.* **1995**, *68*, 2156–2170.

CALCEIN

CAS Registry Number 1461-15-0

Chemical Structure

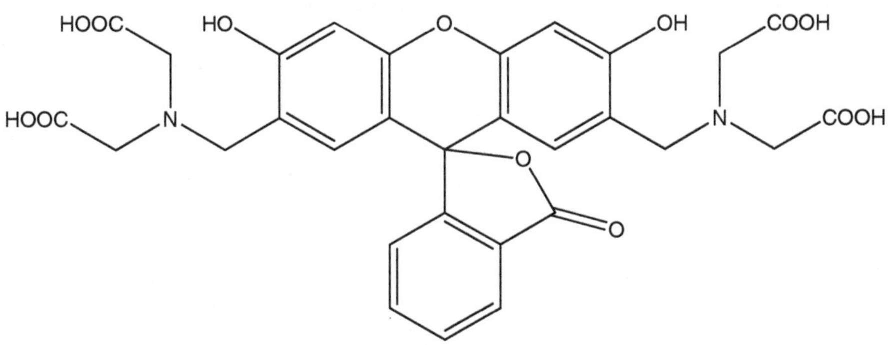

CA Index Name Glycine, *N*,*N*′-[(3′,6′-dihydroxy-3-oxospiro[isobenzofuran-1(3*H*),9′-[9*H*]xanthene]-2′,7′-diyl) bis(methylene)]bis[*N*-(carboxymethyl)-

Other Names Bis[*N*, *N*-bis(carboxymethyl)aminomethyl] fluorescein; Fluorescein, 2′,7′-bis[[bis(carboxymethyl) amino]methyl]-; Fluorescein-bis(methyliminodiacetic acid); Spiro[isobenzofuran-1(3*H*),9′-[9*H*]xanthene], glycine derivative; 2,7-Bis[*N*,*N*-bis(carboxymethyl)aminomethylene]fluorescein; Acetic acid, [(3′,6′-dihydroxy-2′,7′-fluorandiyl)bis(methylenenitrilo)]tetra-; Calcein; Fluorescein complexon; Fluorexon; NSC 298193; Oftasceine

Merck Index Number Not listed

Chemical/Dye Class Xanthene

Molecular Formula $C_{30}H_{26}N_2O_{13}$

Molecular Weight 622.53

Physical Form Yellow-orange to brown powder

Solubility Soluble in water, ethanol, *N*,*N*-dimethyl formamide, dimethyl sulfoxide

Melting Point 200°C

Boiling Point (Calcd.) 952.7 ± 65.0°C, pressure: 760 Torr

pH Range 6.0–7.2

Color Change at pH Weak green fluorescence (6.0) to strong green fluorescence (7.2)

pK$_a$ 6.67

Absorption (λ_{max}) 494 nm

Emission (λ_{max}) 517 nm

Synthesis Synthetic methods[1–7]

Staining Applications Calcium ions;[8,9] fluoride ions;[10] iron ions;[11–13] mercury ions;[14] peptides;[2] proteins;[2] antibodies;[2] atherosclerotic plaque;[1,15] bone;[16,17] cells;[18–20] fish;[21] liposomes;[22,23] neurons;[8] skin;[24] tumor cells;[1,25] inflammations;[1] lymphokines;[1] hepatocytes[3,4]

Biological Applications Calcium indicator;[8,9] fluoride indicator;[10] iron indicator;[11–13] mercury indicator;[14] detecting nucleic acids,[1,26] proteins;[27] treating osteoporosis;[28] drug delivery systems[1,29]

Industrial Applications Chemical mechanical polishing;[1,30] coatings;[31] photoresists[32]

Safety/Toxicity Acute toxicity;[33] cytotoxicity;[1,34] mutagenicity;[35] neurotoxicity;[36] ophthalmotoxicity[37]

REFERENCES

1. Sabnis, R. W. *Handbook of Acid-Base Indicators*; CRC Press: Boca Raton, **2008**; pp 59–60.
2. Hahn, K. M.; Toutchkine, A.; Muthyala, R.; Kraynov, V.; Bark, S. J.; Burton, D. R.; Chamberlain, C. Labeled peptides, proteins and antibodies and processes and intermediates useful for their preparation. U.S. Pat. Appl. Publ. US 2002055133, 2002; *Chem. Abstr.* **2002**, *136*, 366139.

Handbook of Biological Dyes and Stains By R. W. Sabnis

3. Horiuchi, K.; Saji, H.; Arano, Y.; Yokoyama, A. Ligandin binding phthalein complexone complex of technetium for hepatic function studies. *Eur. J. Nucl. Med.* **1990**, *16*, 137–142.
4. Saji, H.; Yokoyama, A.; Arano, Y.; Tanaka, H.; Odori, T.; Morita, R.; Torizuka, K. Phthalein and fluorescein derivatives with an effective technetium-99m labeled state for a hepatobiliary transport: 99mTc-PC, a new hepatobi-liary radiopharmaceutical. *J. Labelled Compd. Radiopharm.* **1979**, *16*, 188–189.
5. Urinovich, E. M. Fluorexon. SU 132233, 1960; *Chem. Abstr.* **1961**, *55*, 59407.
6. Korbl, J.; Vydra, F. Metallochromic indicators. IV. Preparation and properties of "Calcein". *Chem. Listy* **1957**, *51*, 1457–1461; *Chem. Abstr.* **1957**, *51*, 89773.
7. Korbl, J.; Pribil, R. New metallochromic indicators of the complexon type. *Chem. Ind.* **1957**, 233–234.
8. Dravid, S. M.; Murray, T. F. Fluorescent detection of Ca^{2+}-permeable AMPA/kainate receptor activation in murine neocortical neurons. *Neurosci. Lett.* **2003**, *351*, 145–148.
9. Sugita, M.; Hirono, C.; Tanaka, S.; Nakahari, T.; Imai, Y.; Kanno, Y.; Shiba, Y. Visualization of the secretory process involved in Ca^{2+}-activated fluid secretion from rat submandibular glands using the fluorescent dye, calcein. *Eur. J. Cell Biol.* **2000**, *79*, 182–191.
10. Li, H. B.; Chen, F. A highly sensitive fluorometric method for the determination of fluoride in biological material with Al^{3+}–calcein complex. *Fresenius' J. Anal. Chem.* **2000**, *368*, 501–504.
11. Hasinoff, B. B. The intracellular iron sensor calcein is cataly-tically oxidatively degraded by iron(II) in a hydrogen peroxide-dependent reaction. *J. Inorg. Biochem.* **2003**, *95*, 157–164.
12. Ali, A.; Zhang, Q.; Dai, J.; Huang, X. Calcein as a fluorescent iron chemosensor for the determination of low molecular weight iron in biological fluids. *BioMetals* **2003**, *16*, 285–293.
13. Thomas, F.; Serratrice, G.; Beguin, C.; Saint A., Eric; Pierre, J. L.; Fontecave, M.; Laulhere, J. P. Calcein as a fluorescent probe for ferric iron: application to iron nutrition in plant cells. *J. Biol. Chem.* **1999**, *274*, 13375–13383.
14. Lista, A. G.; Palomeque, M. E.; Fernandez Band, B. S. Flow-injection fluorometric determination of mercury(II) with calcein. *Talanta* **1999**, *50*, 881–885.
15. Fisher, E. A.; Fayad, Z. A.; Randolph, G.; Feig, J.; Trogan, E. Methods and compositions for imaging atherosclerotic plaque. U.S. Pat. Appl. Publ. US 2006099148, 2006; *Chem. Abstr.* **2006**, *144*, 456552.
16. Pautke, C.; Vogt, S.; Tischer, T.; Wexel, G.; Deppe, H.; Milz, S.; Schieker, M.; Kolk, A. Polychrome labeling of bone with seven different fluorochromes: enhancing fluorochrome discrimination by spectral image analysis. *Bone* **2005**, *37*, 441–445.
17. Du, S. J.; Frenkel, V.; Kindschi, G.; Zohar, Y. Visualizing normal and defective bone development in zebrafish embryos using the fluorescent chromophore calcein. *Dev. Biol.* **2001**, *238*, 239–246.
18. Lacombe, F.; Belloc, F.; Veriac, S.; Lefevre, D. Method of discrimination of at least two cellular populations by immunofluorescence flow cytometry and application to oncopathology. Fr. Demande FR 2895087, 2007; *Chem. Abstr.* **2007**, *147*, 93934.
19. Ransom, J.; Okun, I.; Okun, A. Cell flow apparatus and method for real-time of cellular responses. U.S. Patent 6280967, 2001; *Chem. Abstr.* **2001**, *135*, 164439.
20. Garini, Y.; Mcnamara, G.; Soenksen, D. G.; Cabib, D.; Buckwald, R. A. *In situ* method of analyzing cells by staining with multiple stains and using a spectral data collection device. PCT Int. Appl. WO 2000031534, 2000; *Chem. Abstr.* **2000**, *133*, 14300.
21. Moen, V.; Nordtug, T.; Zahlsen, K. Mixture and procedure for marking fish. Norw. NO 317169, 2004; *Chem. Abstr.* **2007**, *146*, 141497.
22. Chen, T.; Mcintosh, D.; He, Y.; Kim, J.; Tirrell, D. A.; Scherrer, P.; Fenske, D. B.; Sandhu, A. P.; Cullis, P. R. Alkylated derivatives of poly(ethylacrylic acid) can be inserted into preformed liposomes and trigger pH-dependent intracellular delivery of liposomal contents. *Mol. Membr. Biol.* **2004**, *21*, 385–393.
23. Katsu, T. Application of calcein-loaded liposomes for the determination of membrane channel size. *Biol. Pharm. Bull.* **1999**, *22*, 978–980.
24. Park, J. H.; Lee, J. W.; Kim, Y. C.; Prausnitz, M. R. The effect of heat on skin permeability. *Int. J. Pharm.* **2008**, *359*, 94–103.
25. Kovar, J.; Chen, J.; Draney, D. R.; Olive, D. M.; Volcheck, W. M.; Xu, X.; Lugade, A. G.; Narayanan, N. Optical fluorescent imaging. U.S. Pat. Appl. Publ. US 2006280688, 2006; *Chem. Abstr.* **2006**, *146*, 41053.
26. Miller, B. L.; Krauss, T. D.; Du, H.; Crnkovich, N.; Strohsahl, C. M. Use of sensor arrays containing hairpin probes for detecting nucleic acids of pathogens. PCT Int. Appl. WO 2004061127, 2004; *Chem. Abstr.* **2004**, *141*, 118285.
27. Cheng, D.; Feng, Y.; Yan, J. Dimer of calcein used as fluorescence probe in the determination of proteins. *Microchim. Acta* **2007**, *157*, 173–179.

28. Warman, M. L.; Gong, Y.; Olsen, B. R.; Rawadi, G.; Roman-Roman, S. A regulatory gene for bone strength and mineralization and its use in the diagnosis and treatment of osteoporosis. PCT Int. Appl. WO 2002016553, 2002; *Chem. Abstr.* **2002**, *136*, 197599.

29. Saito, Y.; Tokutome, Y.; Sato, F.; Hinotani, T.; Goto, K. Ceramide-containing liposome for drug delivery systems and composition including the liposome. Jpn. Kokai Tokkyo Koho JP 2006335651, 2006; *Chem. Abstr.* **2006**, *146*, 68676.

30. Gray, C.; Apone, D.; Rogers, C.; Manno, Vi. P.; Barns, C.; Moinpour, M.; Anjur, S.; Philipossian, A. Viewing asperity behavior under the wafer during CMP. *Electrochem. Solid-State Lett.* **2005**, *8*, G109–G111.

31. Garcia, M. A.; Paje, S. E.; Villegas, M. A.; Llopis, J. Preparation and characterization of calcein-doped thin coatings. *Appl. Phys. A* **2002**, *74*, 83–88.

32. Takahashi, O.; Kawabe, Y.; Yasunami, S.; Yamanaka, T. Radiation-sensitive positive-working chemically amplified resist composition. Jpn. Kokai Tokkyo Koho JP 2002214771, 2002; *Chem. Abstr.* **2002**, *137*, 132109.

33. Bumguardner, B. W.; King, T. L. Toxicity of oxytetracycline and calcein to juvenile striped bass. *Trans. Am. Fish. Soc.* **1996**, *125*, 143–145.

34. Barnabe, N.; Zastre, J. A.; Venkataram, S.; Hasinoff, B. B. Deferiprone protects against doxorubicin-induced myocyte cytotoxicity. *Free Radical Biol. Med.* **2002**, *33*, 266–275.

35. Xie, S.; Gu, Z.; Zhou, D. Sensitivity of photobacterial dark mutant for detecting chemical mutagenicity. *Huanjing Kexue Xuebao* **1999**, *19*, 313–318; *Chem. Abstr.* **1999**, *131*, 224722.

36. De Lago, E.; Gustafsson, S. B.; Fernandez-Ruiz, J.; Nilsson, J.; Jacobsson, S. O. P.; Fowler, C. J. Acyl-based anandamide uptake inhibitors cause rapid toxicity to C6 glioma cells at pharmacologically relevant concentrations. *J. Neurochem.* **2006**, *99*, 677–688.

37. Oncel, M.; Khoobehi, B.; Peyman, G. A. Calcein angiography: a preliminary report on an experimental dye. *Int. Ophthalmol.* **1990**, *14*, 245–250.

CALCEIN AM

CAS Registry Number 148504-34-1

Chemical Structure

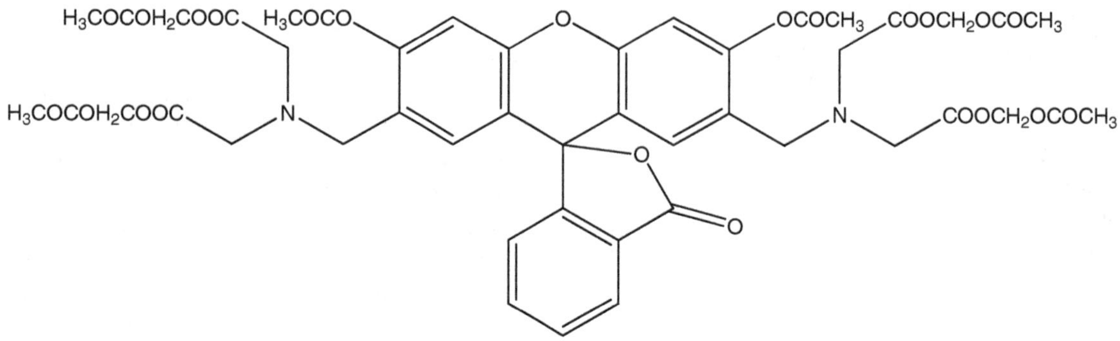

CA Index Name Glycine, *N,N′*-[[3′,6′-bis(acetyloxy)-3-oxospiro[isobenzofuran-1(3*H*),9′-[9*H*]xanthene]-2′,7′-diyl]bis(methylene)]bis[*N*-[2-[(acetyloxy)methoxy]-2-oxoethyl]-, 1,1′-bis[(acetyloxy)methyl] ester

Other Names Calcein *O,O′*-diacetate tetrakis(acetoxymethyl)ester; Glycine, *N,N′*-[[3′,6′-bis(acetyloxy)-3-oxospiro[isobenzofuran-1(3*H*),9′-[9*H*]xanthene]-2′,7′-diyl]bis(methylene)]bis[*N*-[2-[(acetyloxy)methoxy]-2-oxoethyl]-, bis[(acetyloxy)methyl] ester; Spiro[isobenzofuran-1(3*H*),9′-[9*H*]xanthene], glycine derivative; Calcein AM; Calcein acetoxymethyl ester; NSC 689290

Merck Index Number Not listed

Chemical/Dye Class Xanthene

Molecular Formula $C_{46}H_{46}N_2O_{23}$

Molecular Weight 994.86

Physical Form Colorless powder

Solubility Soluble in dimethyl sulfoxide

Melting Point >250°C

Boiling Point (Calcd.) 982.7 ± 65.0°C, pressure: 760 Torr

pK_a (Calcd.) 2.66 ± 0.50, most basic, temperature: 25°C

Absorption (λ_{max}) < 300 nm

Synthesis Synthetic methods[1–4]

Staining Applications Calcium ions;[5–8] zinc ions;[9] bacteria;[10–12] fungi;[11] yeast;[12,13] cells;[14–21] cornea;[22,23] erythrocytes;[24] lymphocytes;[25] microorganisms;[26,27] neurons;[28,29] Schwann cells;[30] sperms;[31–33] tissues[34]

Biological Applications Calcium indicators;[5–8] zinc indicators;[9] cytotoxicity assays;[35–37] apoptosis assays;[38,39] viability assays;[3,4,9,13,14,16,21–24,28,32,33] labile iron pool assays;[40] chemotaxis probes;[41,42] cell adhesion probes;[25,43–46] mitochondrial probes;[46,47] P-glycoprotein probes;[48] multi-drug resistance probes;[49–51] treating atherosclerosis,[52] cancer;[53] ischemic disease[54]

Industrial Applications Not reported

Safety/Toxicity Carcinogenicity;[55] cytotoxicity;[56–61] fish toxicity;[62] nanoparticle toxicity;[63] neurotoxicity;[64–66] phototoxicity;[67] ophthalmotoxicity;[68,69] reproductive toxicity[46]

REFERENCES

1. Brechbiel, M. W.; Xu, H.; Baidoo, K. Preparation of imaging agents containing an amino acid-based trifunctional signaling agent for monoclonal antibody tumor-targeted imaging. PCT Int. Appl. WO 2009012288, 2009; *Chem. Abstr.* **2009**, *150*, 168767.
2. Sinn, H.; Muelbaier, M. Preparation of albumin-fluorescent dye conjugate for diagnostics during surgery. Ger. Offen. DE 102005019104, 2006; *Chem. Abstr.* **2006**, *145*, 426066.
3. Millard, P. J.; Roth, B. L.; Yue, S. T.; Haugland, R. P. Fluorescent viability assay using cyclic-substituted unsymmetrical cyanine dyes. U.S. Patent 5534416, 1996; *Chem. Abstr.* **1996**, *125*, 162751.
4. Haugland, R. P.; MacCoubrey, I. C.; Moore, P. L. Dual-fluorescence cell viability assay using ethidium homodimer and calcein AM. U.S. Patent 5314805, 1994; *Chem. Abstr.* **1994**, *121*, 53487.

5. Kansui, Y.; Garland, C. J.; Dora, K. A. Enhanced spontaneous Ca^{2+} events in endothelial cells reflect signalling through myoendothelial gap junctions in pressurized mesenteric arteries. *Cell Calcium* **2008**, *44*, 135–146.
6. López, J. J.; Camello-Almaraz, C.; Pariente, J. A.; Salido, G. M.; Rosado, J. A. Ca^{2+} accumulation into acidic organelles mediated by Ca^{2+}- and vacuolar H^{+}-ATPases in human platelets. *Biochem. J.* **2005**, *390*, 243–252.
7. Knight, M. M.; Roberts, S. R.; Lee, D. A.; Bader, D. L. Live cell imaging using confocal microscopy induces intracellular calcium transients and cell death. *Am. J. Physiol.* **2003**, *284*, C1083–C1089.
8. Chacon, E.; Ohata, H.; Harper, I. S.; Trollinger, D. R.; Herman, B.; Lemasters, J. J. Mitochondrial free calcium transients during excitation–contraction coupling in rabbit cardiac myocytes. *FEBS Lett.* **1996**, *382*, 31–36.
9. Stork, C. J.; Li, Y. V. Measuring cell viability with membrane impermeable zinc fluorescent indicator. *J. Neurosci. Methods* **2006**, *155*, 180–186.
10. Fuller, M. E.; Streger, S. H.; Rothmel, R. K.; Mailloux, B. J.; Hall, J. A.; Onstott, T. C.; Fredrickson, J. K.; Balkwill, D. L.; DeFlaun, M. F. Development of a vital fluorescent staining method for monitoring bacterial transport in subsurface environments. *Appl. Environ. Microbiol.* **2000**, *66*, 4486–4496.
11. Tokuda, Y.; Ishiyama, Y. Method and apparatus for instantly counting number of bacteria or fungi cells by fluorescent staining. Jpn. Kokai Tokkyo Koho JP 2000316596, 2000; *Chem. Abstr.* **2000**, *133*, 360571.
12. Malacrino, P.; Zapparoli, G.; Torriani, S.; Dellaglio, F. Rapid detection of viable yeasts and bacteria in wine by flow cytometry. *J. Microbiol. Methods* **2001**, *45*, 127–134.
13. Garcia, M. T.; Sanz, R.; Galceran, M. T.; Puignou, L. Use of fluorescent probes for determination of yeast cell viability by gravitational field-flow fractionation. *Biotechnol. Prog.* **2006**, *22*, 847–852.
14. Andersson, K. Method for measurement of cell viability by attaching to cells a status marker such as fluorescent or chemiluminescent mol. or an emitter of radioactive radiation. PCT Int. Appl. WO 2009029039, 2009; *Chem. Abstr.* **2009**, *150*, 278173.
15. Abbott, N.; Murphy, C.; Israel, B.; Sotos, J.; Hansmann, D.; Herber, R.; Burkholder, J.; Hulkower, K.; Bonds, M. Improved devices for cell assays containing cell exclusion zones and cell seeding zones. PCT Int. Appl. WO 2009026359, 2009; *Chem. Abstr.* **2009**, *150*, 254954.
16. Itle, L. J.; Pishko, M. V. Multiphenotypic whole-cell sensor for viability screening. *Anal. Chem.* **2005**, *77*, 7887–7893.
17. McClain, M. A.; Culbertson, C. T.; Jacobson, S. C.; Allbritton, N. L.; Sims, C. E.; Ramsey, J. M. Microfluidic devices for the high-throughput chemical analysis of cells. *Anal. Chem.* **2003**, *75*, 5646–5655.
18. Yoshimura, Y.; Kawasaki, Y.; Tsuji, A.; Kurane, R. Fluorometirc method for detecting living cells. Jpn. Kokai Tokkyo Koho JP 2002034595, 2002; *Chem. Abstr.* **2002**, *136*, 147464.
19. Oberhardt, B. J. Cell analysis methods and apparatus. U.S. Patent 6251615, 2001; *Chem. Abstr.* **2001**, *135*, 58132.
20. Garini, Y.; Mcnamara, G.; Soenksen, D. G.; Cabib, D.; Buckwald, R. A. *In situ* method of analyzing cells by staining with multiple stains and using a spectral data collection device. PCT Int. Appl. WO 2000031534, 2000; *Chem. Abstr.* **2000**, *133*, 14300.
21. Thompson, T. A. Viability assays for cells *in vitro*: the ethidium/calcein assay and the immunofluorescence combination assay. *Methods Mol. Med.* **1999**, *22*, 145–155.
22. Chen, C. C.; Chang, J. H.; Lee, J. B.; Javier, J.; Azar, D. T. Human corneal epithelial cell viability and morphology after dilute alcohol exposure. *Invest. Ophthalmol. Vis. Sci.* **2002**, *43*, 2593–2602.
23. Imbert, D.; Cullander, C. Assessment of cornea viability by confocal laser scanning microscopy and MTT assay. *Cornea* **1997**, *16*, 666–674.
24. Bratosin, D.; Mitrofan, L.; Palii, C.; Estaquier, J.; Montreuil, J. Novel fluorescence assay using calcein-AM for the determination of human erythrocyte viability and aging. *Cytometry* **2005**, *66A*, 78–84.
25. Braut-Boucher, F.; Pichon, J.; Rat, P.; Adolphe, M.; Aubery, M.; Font, J. A non-isotopic, highly sensitive, fluorimetric, cell–cell adhesion microplate assay using calcein AM-labeled lymphocytes. *J. Immunol. Methods* **1995**, *178*, 41–51.
26. Noda, N.; Mizutani, T. Microorganism-measuring method using multiple staining. Jpn. Kokai Tokkyo Koho JP 2006340684, 2006; *Chem. Abstr.* **2006**, *146*, 77563.
27. Tokuda, Y.; Ishiyama, Y. Instant method for evaluating live and dead microorganisms by fluorescent staining. Jpn. Kokai Tokkyo Koho JP 2000232897, 2000; *Chem. Abstr.* **2000**, *133*, 190192.

28. Whitfield, J.; Neame, S. J.; Ham, J. Methods for culturing primary sympathetic neurons and for determining neuronal viability. *Methods Mol. Biol.* **2004**, *282*, 157–168.

29. Bozyczko-Coyne, D.; McKenna, B. W.; Connors, T. J.; Neff, N. T. A rapid fluorometric assay to measure neuronal survival *in vitro*. *J. Neurosci. Methods* **1993**, *50*, 205–216.

30. Decherchi, P.; Cochard, P.; Gauthier, P. Dual staining assessment of Schwann cell viability within whole peripheral nerves using calcein-AM and ethidium homodimer. *J. Neurosci. Methods* **1997**, *71*, 205–213.

31. Kato, M.; Makino, S.; Kimura, H.; Ota, T.; Furuhashi, T.; Nagamura, Y.; Hirano, K. *In vitro* evaluation of acrosomal status and motility in rat epididymal spermatozoa treated with α-chlorohydrin for predicting their fertilizing capacity. *J. Reprod. Dev.* **2002**, *48*, 461–468.

32. Collins, A. M.; Donoghue, A. M. Viability assessment of honey bee, *Apis mellifera*, sperm using dual fluorescent staining. *Theriogenology* **1999**, *51*, 1513–1523.

33. Kimura, H.; Kato, M. Determination of rat sperm viability with fluorescence. Jpn. Kokai Tokkyo Koho JP 08332098, 1996; *Chem. Abstr.* **1997**, *126*, 154802.

34. Connors, K. M.; Monosov, A. Native-state method and system for determining viability and proliferative capacity of tissues *in vitro*. U.S. Patent 5726009, 1998; *Chem. Abstr.* **1998**, *128*, 226223.

35. Cholujova, D.; Jakubikova, J.; Kubes, M.; Arendacka, B.; Sapak, M.; Robert, I.; Sedlak, J. Comparative study of four fluorescent probes for evaluation of natural killer cell cytotoxicity assays. *Immunobiology* **2008**, *213*, 629–640.

36. Guzman, E.; McCrae, M. A. A rapid and accurate assay for assessing the cytotoxicity of viral proteins. *J. Virol. Methods* **2005**, *127*, 119–125.

37. Neri, S.; Mariani, E.; Meneghetti, A.; Cattini, L.; Facchini, A. Calcein-acetoxymethyl cytotoxicity assay: standardization of a method allowing additional analyses on recovered effector cells and supernatants. *Clin. Diagn. Lab. Immunol.* **2001**, *8*, 1131–1135.

38. Gatti, R.; Belletti, S.; Orlandini, G.; Bussolati, O.; Dall'Asta, V.; Gazzola, G. C. Comparison of annexin V and calcein-AM as early vital markers of apoptosis in adherent cells by confocal laser microscopy. *J. Histochem. Cytochem.* **1998**, *46*, 895–900.

39. Palma, P. F. R.; Baggio, G. L.; Spada, C.; Silva, R. D A.; Ferreira, S. I. A C P.; Treitinger, A. Evaluation of annexin V and Calcein-AM as markers of mononuclear cell apoptosis during human immunodeficiency virus infection. *Braz. J. Infect. Dis.* **2008**, *12*, 108–114.

40. Tenopoulou, M.; Kurz, T.; Doulias, P. T.; Galaris, D.; Brunk, U. T. Does the calcein-AM method assay the total cellular 'labile iron pool' or only a fraction of it? *Biochem. J.* **2007**, *403*, 261–266.

41. DeGendt, C. M.; DeClerck, L. S.; Bridts, C. H.; Stevens, W. J. The use of calceinacetomethylester (AM)-labeled poly-morphonuclear cells in a polycarbonate filter chemotaxis assay. *Clin. Chim. Acta* **1996**, *249*, 189–195.

42. Schratzberger, P.; Kaehler, C. M.; Wiedermann, C. J. Use of fluorochromes in the determination of chemotaxis and haptotaxis of granulocytes by micropore filter assays. *Ann. Hematol.* **1996**, *72*, 23–27.

43. Aparicio, C. L.; Strong, L. H.; Yarmush, M. L.; Berthiaume, F. Correction for label leakage in fluorimetric assays of cell adhesion. *BioTechniques* **1997**, *23*, 1056–1060.

44. Petersen, T. K.; Vaever Bysted, B.; Jensen, A. L. Determination of the adhesive capability of canine polymorphonuclear neutrophil granulocytes using a fluorometric microtiter plate cellular adhesion assay. *Vet. Immunol. Immunopathol.* **1999**, *68*, 283–291.

45. Stoltenborg, J. K.; Tsao, P. W.; George, H. J.; Bouchard, P. J.; Wexler, E. J.; Hausner, E. A. A fluorescent cellular adhesion assay using insect cell produced human VCAM1. *J. Immunol. Methods* **1994**, *175*, 59–68.

46. Kato, M.; Makino, S.; Kimura, H.; Ota, T.; Furuhashi, T.; Nagamura, Y. Evaluation of mitochondrial function and membrane integrity by dual fluorescent staining for assessment of sperm status in rats. *J. Toxicol. Sci.* **2002**, *27*, 11–18.

47. Woollacott, A. J.; Simpson, P. B. High throughput fluorescence assays for the measurement of mitochondrial activity in intact human neuroblastoma cells. *J. Biomol. Screen.* **2001**, *6*, 413–420.

48. Liminga, G.; Nygren, P.; Larsson, R. Microfluorometric evaluation of calcein acetoxymethyl ester as a probe for P-glycoprotein-mediated resistance: effects of cyclosporin A and its nonimmunosuppressive analog SDZ PSC 833. *Exp. Cell Res.* **1994**, *212*, 291–296.

49. Schwab, R.; Petak, I.; Sarkadi, B.; Kopper, L.; Keri, G.; Pap, A.; Szakacs, G.; Homolya, L.; Jakab, F.; Moldvay, J.; Feher, A. Method and reagent kit for evaluation of multidrug resistance activity, and

method of screening for MDR transport inhibitors. PCT Int. Appl. WO 2005116648, 2005; *Chem. Abstr.* **2005**, *144*, 17129.

50. Sarkadi, B.; Homolya, L.; Hollo, Z. Assay and reagent kit for evaluation of multi-drug resistance in cells. U.S. Patent 6277655, 2001; *Chem. Abstr.* **2001**, *135*, 175357.
51. Karaszi, E.; Jakab, K.; Homolya, L.; Szakacs, G.; Hollo, Z.; Telek, B.; Kiss, A.; Rejto, L.; Nahajevszky, S.; Sarkadi, B.; Kappelmayer, J. Calcein assay for multidrug resistance reliably predicts therapy response and survival rate in acute myeloid leukemia. *Br. J. Haematol.* **2001**, *112*, 308–314.
52. Tabas, I. Phagocyte enhancement therapy for atherosclerosis. U.S. Pat. Appl. Publ. US 2008267909, 2008; *Chem. Abstr.* **2008**, *149*, 506154.
53. Zerangue, N. GLUT3 transporters expressed in cancer cells, and use in screening for cytotoxic and imaging agents for the treatment and diagnosis of cancer. PCT Int. Appl. WO 2006001986, 2006; *Chem. Abstr.* **2006**, *144*, 81159.
54. Wakita, H.; Igarashi, K.; Oie, K. Diagnosis and treatment of ischemic disease. PCT Int. Appl. WO 2009022756, 2009; *Chem. Abstr.* **2009**, *150*, 245325.
55. Han, Y.; Riwanto, M.; Go, M. L.; Rachel, E. P. L. Modulation of breast cancer resistance protein (BCRP/ABCG2) by non-basic chalcone analogues. *Eur. J. Pharm. Sci.* **2008**, *35*, 30–41.
56. Al-Abd, A. M.; Lee, J. H.; Kim, S. Y.; Kun, N.; Kuh, H. J. Novel application of multicellular layers culture for *in situ* evaluation of cytotoxicity and penetration of paclitaxel. *Cancer Sci.* **2008**, *99*, 423–431.
57. Ho, J. D.; Tsai, R. J. F.; Chen, S. N.; Chen, H. C. Cytotoxicity of indocyanine green on retinal pigment epithelium: implications for macular hole surgery. *Arch. Ophthalmol.* **2003**, *121*, 1423–1429.
58. Dias, N.; Lima, N. A comparative study using a fluorescence-based and a direct-count assay to determine cytotoxicity in *Tetrahymena pyriformis*. *Res. Microbiol.* **2002**, *153*, 313–322.
59. Adler, M.; Shafer, H.; Hamilton, T.; Petrali, J. P. Cytotoxic actions of the heavy metal chelator TPEN on NG108-15 neuroblastoma-glioma cells. *Neurotoxicology* **1999**, *20*, 571–582.
60. Jonsson, B.; Liminga, G.; Csoka, K.; Fridborg, H.; Dhar, S.; Nygren, P.; Larsson, R. Cytotoxic activity of calcein acetoxymethyl ester (calcein/AM) on primary cultures of human hematological and solid tumors. *Eur. J. Cancer* **1996**, *32A*, 883–887.
61. Liminga, G.; Nygren, P.; Dhar, S.; Nilsson, K.; Larsson, R. Cytotoxic effect of calcein acetoxymethyl ester on human tumor cell lines: drug delivery by intracellular trapping. *Anti-Cancer Drugs* **1995**, *6*, 578–585.
62. Caminada, D.; Zaja, R.; Smital, T.; Fent, K. Human phar-maceuticals modulate P-gp1 (ABCB1) transport activity in the fish cell line PLHC-1. *Aquat. Toxicol.* **2008**, *90*, 214–222.
63. Monteiro-Riviere, N. A.; Inman, A. O.; Zhang, L. W. Limitations and relative utility of screening assays to assess engineered nanoparticle toxicity in a human cell line. *Toxicol. Appl. Pharmacol.* **2009**, *234*, 222–235.
64. Yamamoto, R.; Yoneda, S.; Hara, H. Neuroprotective effects of beta-secretase inhibitors against rat retinal ganglion cell death. *Neurosci. Lett.* **2004**, *370*, 61–64.
65. Green, P. S.; Perez, E. J.; Calloway, T.; Simpkins, J. W. Estradiol attenuation of beta-amyloid-induced toxicity: a comparison o. *J. Neurocytol.* **2000**, *29*, 419–423.
66. Otori, Y.; Wei, J. Y.; Barnstable, C. J. Neurotoxic effects of low doses of glutamate on purified rat retinal ganglion cells. *Invest. Ophthalmol. Vis. Sci.* **1998**, *39*, 972–981.
67. Thorpe, W. P.; Toner, M.; Ezzell, R. M.; Tompkins, R. G.; Yarmush, M. L. Dynamics of photoinduced cell plasma membrane injury. *Biophys. J.* **1995**, *68*, 2198–2206.
68. Ho, J. D.; Chen, H. C.; Chen, S. N.; Tsai, R. J. F. Reduction of indocyanine green-associated photosensitizing toxicity in retinal pigment epithelium by sodium elimination. *Arch. Ophthalmol.* **2004**, *122*, 871–878.
69. Geerling, G.; Daniels, J. T.; Dart, J. K.; Cree, I. A.; Khaw, P. T. Toxicity of natural tear substitutes in a fully defined culture model of human corneal epithelial cells. *Invest. Ophthalmol. Vis. Sci.* **2001**, *42*, 948–956.

CALCEIN BLUE

CAS Registry Number 54375-47-2
Chemical Structure

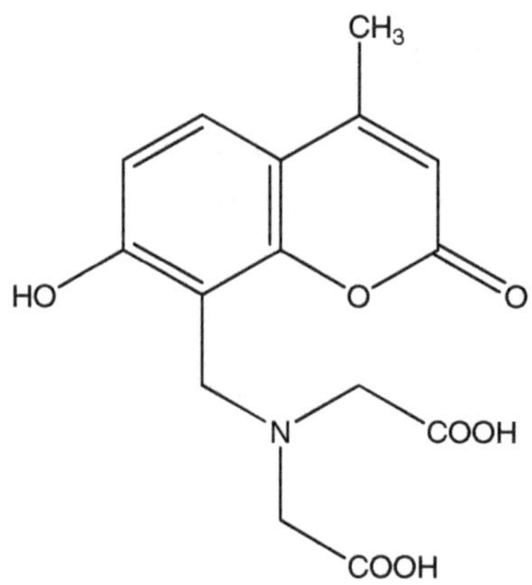

CA Index Name Glycine, *N*-(carboxymethyl)-*N*-[(7-hydroxy-4-methyl-2-oxo-2*H*-1-benzopyran-8-yl)methyl]-
Other Names Calcein blue; 4-Methylumbelliferone-8-methyliminodiacetic acid
Merck Index Number Not listed
Chemical/Dye Class Coumarin
Molecular Formula $C_{15}H_{15}NO_7$
Molecular Weight 321.28
Physical Form White to pale yellow powder
Solubility Soluble in water, methanol, dimethyl sulfoxide
Melting Point >200°C
Boiling Point (calcd.) 636.2 ± 55.0°C, pressure: 760 Torr
pK_a (Calcd.) 1.74 ± 0.10, most acidic, temperature: 25°C
Absorption (λ_{max}) 360 nm
Emission (λ_{max}) 449 nm
Synthesis Synthetic methods[1–4]
Staining Applications Alkaline phosphatase;[5] amino acids;[6] bone;[7] calcification front;[8] fish;[9] mineralized nodule;[10] mineralized tissues;[11] tumor tissues;[12] copper ions;[13,14] calcium ions;[15] cadmium ions;[16] iron (II) ions;[17] iron (III) ions;[3] fluoride ions;[18,19] lanthanides ions;[20] magnesium ions;[21] silver ions;[22] sulfate ions;[23] zinc ions[24]
Biological Applications Copper indicator;[13,14] calcium indicator;[15] cadmium indicator;[16] iron (II) indicator;[17] iron (III) indicator;[3] fluoride indicator;[18,19] lanthanides indicator;[20] magnesium indicator;[21] silver indicator;[22] sulfate indicator;[23] zinc indicator;[24] treating atherosclerosis,[25] restenosis,[25] hypertension[25]
Industrial Applications Laser materials;[26] semiconductor wafer-cleansing solutions;[27] detergent[28]
Safety/Toxicity No data available

REFERENCES

1. Teller, J.; Gruttner, C.; Rudershausen, S.; Westphal, F. Preparation of homogeneous colored or fluorescent silicic acid particles by controlled hydrolysis and polycondensation. Eur. Pat. Appl. EP 1036763, 2000; *Chem. Abstr.* **2000**, *133*, 225158.
2. Ponten, E.; Glad, B.; Stigbrand, M.; Sjoegren, A.; Irgum, K. Non-porous spherical polymer particles for luminescent solid phase detection reactors. *Anal. Chim. Acta* **1996**, *320*, 87–97.
3. Yoshida, H.; Ozawa, T.; Jitsukawa, K.; Einaga, H. Kinetics and mechanism of complex formation reaction of iron(III) with 4-methyl-8-(carboxymethyl)aminomethyl umbelliferrone and related ligands: role of substituent groups of nitrogen donor atoms on the reaction. *Polyhedron* **1993**, *12*, 1319–1328.
4. Wilkins, D. H. Calcein Blue—a new metal fluorochromic indicator for chelatometric titrations. *Talanta* **1960**, *4*, 182–184.
5. Murray, G. I.; Ewen, S. W. B. A new fluorescence method for alkaline phosphatase histochemistry. *J. Histochem. Cytochem.* **1992**, *40*, 1971–1974.
6. Buryak, A.; Severin, K. A chemosensor array for the colorimetric identification of 20 natural amino acids. *J. Am. Chem. Soc.* **2005**, *127*, 3700–3701.
7. Pautke, C.; Vogt, S.; Tischer, T.; Wexel, G.; Deppe, H.; Milz, S.; Schieker, M.; Kolk, A. Polychrome labeling of bone with seven different fluorochromes: enhancing fluorochrome discrimination by spectral image analysis. *Bone* **2005**, *37*, 441–445.
8. Aaron, J. E.; Makins, N. B.; Francis, R. M.; Peacock, M. Staining of the calcification front in human bone using contrasting fluorochromes *in vitro*. *J. Histochem. Cytochem.* **1984**, *32*, 1251–1261.
9. Moen, V.; Nordtug, T.; Zahlsen, K. Mixture and procedure for marking fish. Norw. NO 317169, 2004; *Chem. Abstr.* **2007**, *146*, 141497.

10. Wang, Y. H.; Liu, Y.; Maye, P.; Rowe, D. W. Examination of mineralized nodule formation in living osteoblastic cultures using fluorescent dyes. *Biotechnol. Prog.* **2006**, *22*, 1697–1701.

11. Goto, T.; Kajiwara, H.; Yoshinari, M.; Fukuhara, E.; Kobayashi, S.; Tanaka, T. *In vitro* assay of mineralized-tissue formation on titanium using fluorescent staining with calcein blue. *Biomaterials* **2003**, *24*, 3885–3892.

12. Connors, K. M.; Monosov, A. Native-state method and system for determining viability and proliferative capacity of tissues *in vitro*. U.S. Patent 5726009, 1998; *Chem. Abstr.* **1998**, *128*, 226223.

13. Gholivand, M. B.; Sohrabi, A.; Abbasi, S. Determination of copper by adsorptive stripping voltammetry in the presence of calcein blue. *Electroanalysis* **2007**, *19*, 1609–1615.

14. Royzen, M.; Dai, Z.; Canary, J. W. Ratiometric displacement approach to Cu(II) sensing by fluorescence. *J. Am. Chem. Soc.* **2005**, *127*, 1612–1613.

15. Matsoukas, K. A.; Demertzis, M. A. Fluorometric deter-mination of calcium in serum with calcein blue. *Analyst* **1988**, *113*, 251–253.

16. Yokoi, K.; Mizumachi, M.; Koide, T. Determination of cadmium by adsorptive stripping voltammetry of a cadmium–calcein blue complex. *Anal. Sci.* **1995**, *11*, 257–260.

17. Noire, M. H.; Dureault, B. A ferrous ion optical sensor based on fluorescence quenching. *Sens. Actuators B* **1995**, *B29*, 386–391.

18. Yuchi, A.; Murase, H.; Wada, H. Structural features of organic reagents suitable for spectrophotometric or fluorometric determination of fluoride based on mixed-ligand complex formation. *Anal. Sci.* **1995**, *11*, 221–226.

19. Russell, D. A.; Narayanasawamy, R. An optical-fiber sensor for fluoride. *Anal. Chim. Acta* **1989**, *220*, 75–81.

20. Brittain, H. G. Submicrogram determination of lanthanides through quenching of calcein blue fluorescence. *Anal. Chem.* **1987**, *59*, 1122–1125.

21. Matsoukas, K. A.; Evmiridis, N. P.; Demertzis, M. A. Spectrofluorometric determination of magnesium in blood serum with calcein blue. *Anal. Chim. Acta* **1989**, *227*, 211–218.

22. Escarrilla, A. M. Calcein blue as a fluorescent adsorption indicator in the direct titration of silver ions. *Anal. Chim. Acta* **1968**, *43*, 353–356.

23. Tan, L. H.; West, T. S. Spectrofluorimetric method for the determination of small amounts of sulfate ion. *Analyst* **1971**, *96*, 281–285.

24. Zou, S.; Aodeng, G.; Sai, Y. A new spectrophotometric fluorescence method for the determination of trace zinc (II) in foodstuff using calcein blue. *Fenxi Shiyanshi* **1995**, *14*, 45–48; *Chem. Abstr.* **1996**, *124*, 305850.

25. Foulkes, J. G.; Liechtfried, F. E.; Pieler, C.; Stephenson, J. R.; Case, C. C. Reporter gene methods for identification of compounds that modulate transcription of genes associated with cardiovascular disease. U.S. Patent 5580722, 1996; *Chem. Abstr.* **1996**, *126*, 102570.

26. Ishibashi, N.; Imasaka, T.; Ogawa, T.; Maeda, M.; Miyazoe, Y. Coordination compounds of calcein blue for new laser material. *Chem. Lett.* **1974**, *1315*-1316.

27. Hiratsuka, H.; Myashita, M.; Kato, R. Semiconductor wafer-cleansing solutions. Jpn. Kokai Tokkyo Koho JP 06041773, 1994; *Chem. Abstr.* **1994**, *121*, 219390.

28. Sgargetta, G.; Hecht, S. E.; Shi, J.; Sivik, M. R.; Wilke, D. L. Detergent composition with a metal-complexing dye and surfactant. U.S. Pat. Appl. Publ. US 2006073999, 2006; *Chem. Abstr.* **2006**, *144*, 352774.

CALCEIN BLUE AM

CAS Registry Number 168482-84-6

Chemical Structure

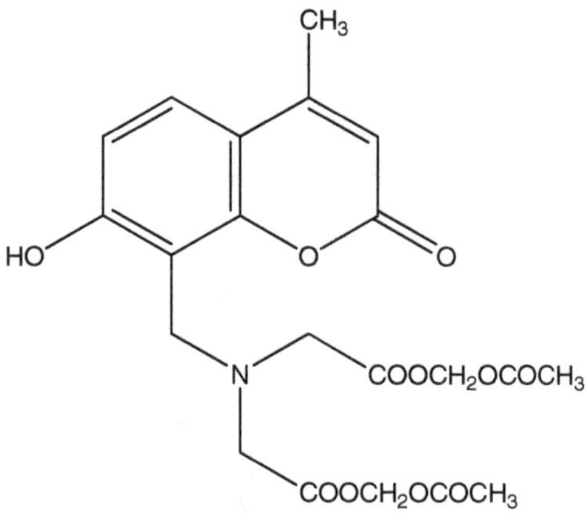

CA Index Name Glycine, *N*-[2-[(acetyloxy)methoxy]-2-oxoethyl]-*N*-[(7-hydroxy-4-methyl-2-oxo-2*H*-1-benzopyran-8-yl)methyl]-, (acetyloxy)methyl ester

Other Names Calcein Blue AM; Calcein Blue acetoxymethyl ester

Merck Index Number Not listed

Chemical/Dye Class Coumarin

Molecular Formula $C_{21}H_{23}NO_{11}$

Molecular Weight 465.41

Physical Form Solid

Solubility Soluble in dimethyl sulfoxide, methanol

Melting Point >200°C

Boiling Point (Calcd.) 617.0 ± 55.0°C, pressure: 760 Torr

pK_a (Calcd.) 8.88 ± 0.40, most acidic, temperature: 25°C; 2.66 ± 0.50, most basic, temperature: 25°C

Absorption (λ_{max}) 322 nm

Emission (λ_{max}) 437 nm

Synthesis Synthetic method[1]

Staining Applications Bacteria,[2] cells[1]

Biological Applications Drug delivery;[3] fluorescent viability assay;[1] monitoring bacterial transport[2]

Industrial Applications Not reported

Safety/Toxicity Cytotoxicity[3]

REFERENCES

1. Millard, P. J.; Roth, B. L.; Yue, S. T.; Haugland, R. P. Fluorescent viability assay using cyclic-substituted unsymmetrical cyanine dyes. U.S. Patent 5534416, 1996; *Chem. Abstr.* **1996**, *125*, 162751.
2. Fuller, M. E.; Streger, S. H.; Rothmel, R. K.; Mailloux, B. J.; Hall, J. A.; Onstott, T. C.; Fredrickson, J. K.; Balkwill, D. L.; DeFlaun, M. F. Development of a vital fluorescent staining method for monitoring bacterial transport in subsurface environments. *Appl. Environ. Microbiol.* **2000**, *66*, 4486–4496.
3. Liminga, G.; Nygren, P.; Dhar, S.; Nilsson, K.; Larsson, R. Cytotoxic effect of calcein acetoxymethyl ester on human tumor cell lines: drug delivery by intracellular trapping. *Anti-Cancer Drugs* **1995**, *6*, 578–585.

CALCIUM GREEN 1

CAS Number 154719-40-1

Chemical Structure

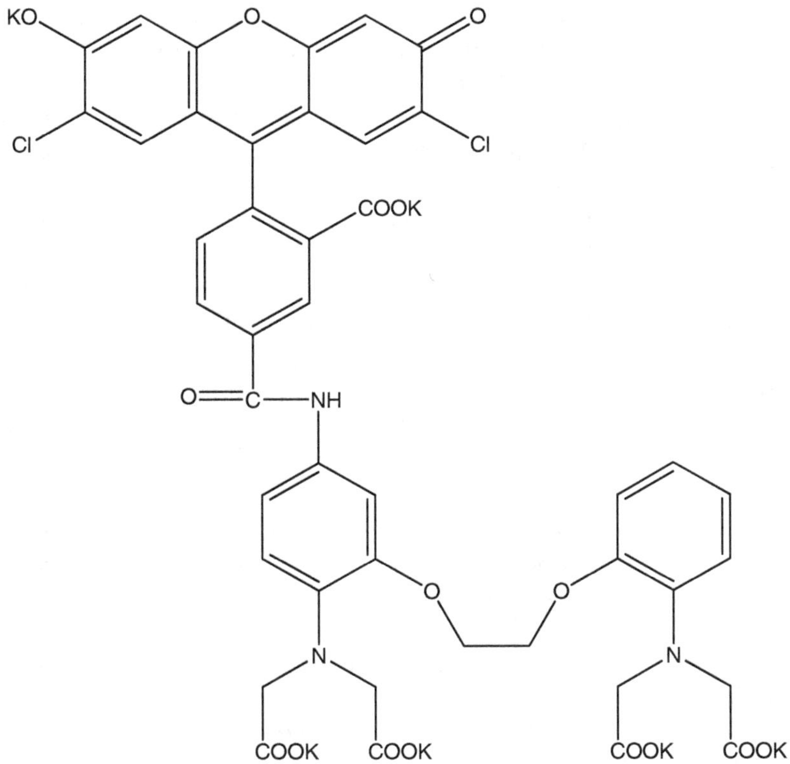

Chemical/Dye Class Xanthene

Molecular Formula $C_{43}H_{27}Cl_2K_6N_3O_{16}$

Molecular Weight 1147.19

Physical Form Solid

CA Index Name Glycine, *N*-[2-[2-[2-[bis(carboxymethyl)amino]-5-[[(2′,7′-dichloro-3′,6′-dihydroxy-3-oxospiro[isobenzofuran-1(3*H*),9′-[9*H*]xanthen]-5-yl)carbonyl]amino]phenoxy]ethoxy]phenyl]-*N*-(carboxymethyl)-, potassium salt (1:6)

Other Names Glycine, *N*-[2-[2-[2-[bis(carboxymethyl)amino]-5-[[(2′,7′-dichloro-3′,6′-dihydroxy-3-oxospiro[isobenzofuran-1(3*H*),9′-[9*H*]xanthen]-5-yl)carbonyl]amino]phenoxy]ethoxy]phenyl]-*N*-(carboxymethyl)-, hexapotassium salt; Spiro[isobenzofuran-1(3*H*),9′-[9*H*]xanthene], glycine deriv.; Calcium Green 1; Calcium Green I

Merck Index Number Not listed

Solubility Soluble in water

Melting Point >250°C

pK_a 6.7

Absorption (λ_{max}) 506 nm

Emission (λ_{max}) 531 nm

Synthesis Synthetic methods[1,2]

Staining Applications Calcium ions;[4–27] cells,[3] neurons[7,8,10,11,13,21]

Biological Applications Calcium indicator[4–27]

Industrial Applications Not reported

Safety/Toxicity Phloem mobility of xenobiotics[28]

REFERENCES

1. Kuhn, M. A.; Haugland, R. P. Reactive derivatives of BAPTA used to make ion-selective chelators. U.S. Patent 5453517, 1995; *Chem. Abstr.* **1995**, *124*, 86585.

2. Kuhn, M. A.; Hoyland, B.; Carter, S.; Zhang, C.; Haugland, R. P. Fluorescent ion indicators for detecting heavy metals. *Proc. SPIE-Int. Soc. Opt. Eng.* **1995**, *2388*, 238–244.

3. Lichtman, J. W.; Wong, W. T.; Wong, R.; Gan, W. B.; Grutzendler, J. Method for labeling individual cells. PCT Int. Appl. WO 2001069244, 2001; *Chem. Abstr.* **2001**, *135*, 238975.

4. Dineley, K. E. On the use of fluorescent probes to distinguish Ca^{2+} from Zn^{2+} in models of excitotoxicity. *Cell Calcium* **2007**, *42*, 341–342.

5. Serulle, Y.; Sugimori, M.; Llinas, R. R. Imaging synap-tosomal calcium concentration microdomains and vesicle fusion by using total internal reflection fluorescent microscopy. *Proc. Natl. Acad. Sci. U.S.A.* **2007**, *104*, 1697–1702.

6. Dong, C. J.; Hare, W. A. Methods of determining activity of ryanodine receptor modulators. U.S. Pat. Appl. Publ. US 2007196856, 2007; *Chem. Abstr.* **2007**, *147*, 292267.

7. Nagayama, S.; Zeng, S.; Xiong, W.; Fletcher, M. L.; Masurkar, A. V.; Davis, Douglas J.; Pieribone, V. A.; Chen, W. R. *In vivo* simultaneous tracing and Ca^{2+} imaging of local neuronal circuits. *Neuron* **2007**, *53*, 789–803.

8. Boyer, S.; Ruel, J.; Puel, J. L.; Chabbert, C. A procedure to label inner ear afferent nerve endings for calcium imaging. *Brain Res. Protoc.* **2004**, *13*, 91–98.

9. Xia, M.; Connolly, T. M.; Bennett, P. B., Jr.; Cohen, C. J. Assay methods for state-dependent calcium channel agonists/antagonists. PCT Int. Appl. WO 2004033647, 2004; *Chem. Abstr.* **2004**, *140*, 335276.

10. Brustein, E.; Marandi, N.; Kovalchuk, Y.; Drapeau, P.; Konnerth, A. *In vivo* monitoring of neuronal network activity in zebrafish by two-photon Ca^{2+} imaging. *Pfluegers Arch.* **2003**, *446*, 766–773.

11. Kettunen, P.; Demas, J.; Lohmann, C.; Kasthuri, N.; Gong, Y.; Wong, R. O. L.; Gan, W. B. Imaging calcium dynamics in the nervous system by means of ballistic delivery of indicators. *J. Neurosci. Methods* **2002**, *119*, 37–43.

12. Kuchitsu, K.; Ward, J. M.; Allen, G. J.; Schelle, I.; Schroeder, J. I. Loading acetoxymethyl ester fluorescent dyes into the cytoplasm of *Arabidopsis* and *Commelina* guard cells. *New Phytol.* **2002**, *153*, 527–533.

13. Yoshida, R.; Iwamoto, A.; Nagahama, T. Calcium imaging for detection and estimation of spike activities in *Aplysia* neurons. *Zool. Sci.* **2001**, *18*, 631–643.

14. Rothe, U.; Glanz, D.; Wolter, A. Method and tool for the determination of ion concentration in the basolateral part of adhered cells by fluorescence microscopy. Ger. Offen. DE 10003673, 2001; *Chem. Abstr.* **2001**, *135*, 119244.

15. Thomas, D.; Tovey, S. C.; Collins, T. J.; Bootman, M. D.; Berridge, M. J.; Lipp, P. A comparison of fluorescent Ca^{2+} indicator properties and their use in measuring elementary and global Ca^{2+} signals. *Cell Calcium* **2000**, *28*, 213–223.

16. Denda, M.; Hosoi, J.; Asida, Y. Visual imaging of ion distribution in human epidermis. *Biochem. Biophys. Res. Commun.* **2000**, *272*, 134–137.

17. Koester, H. J.; Baur, D.; Uhl, R.; Hell, S. W. Ca^{2+} fluorescence imaging with pico- and femtosecond two-photon excitation: signal and photodamage. *Biophys. J.* **1999**, *77*, 2226–2236.

18. Lee, S. K.; Lee, J. Y.; Lee, M. Y.; Chung, S. M.; Chung, J. H. Advantages of Calcium Green-1 over Other fluorescent dyes in measuring cytosolic calcium in platelets. *Anal. Biochem.* **1999**, *273*, 186–191.

19. Lin, K.; Sadee, W.; Quillan, J. M. Rapid measurements of intracellular calcium using a fluorescence plate reader. *BioTechniques* **1999**, *26*, 318–326.

20. Sullivan, E.; Tucker, E. M.; Dale, I. L. Measurement of [Ca^{2+}] using the fluorometric imaging plate reader (FLIPR). *Methods Mol. Biol.* **1999**, *114*, 125–133.

21. Dudkin, E. A.; Myers, P. Z.; Ramirez-Latorre, J. A.; Gruberg, E. R. Calcium signals monitored from leopard frog optic tectum after the optic nerve has been selectively loaded with calcium sensitive dye. *Neurosci. Lett.* **1998**, *258*, 124–126.

22. Oheim, M.; Naraghi, M.; Muller, T. H.; Neher, E. Two dye two wavelength excitation calcium imaging: results from bovine adrenal chromaffin cells. *Cell Calcium* **1998**, *24*, 71–84.

23. Bullen, A.; Saggau, P. Indicators and optical configuration for simultaneous high-resolution recording of membrane potential and intracellular calcium using laser scanning microscopy. *Pfluegers Arch.* **1998**, *436*, 788–796.

24. Xin, Q.; Wightman, R. M. Simultaneous detection of catecholamine exocytosis and Ca^{2+} release from single bovine chromaffin cells using a dual microsensor. *Anal. Chem.* **1998**, *70*, 1677–1681.

25. Prajer, M.; Fleury, A.; Laurent, M. Dynamics of calcium regulation in Paramecium and possible morphogenetic implication. *J. Cell Sci.* **1997**, *110*, 529–535.

26. Belan, P. V.; Gerasimenko, O. V.; Berry, D.; Saftenku, E.; Petersen, O. H.; Tepikin, A. V. A new technique for assessing the microscopic distribution of cellular calcium exit sites. *Pfluegers Arch.* **1996**, *433*, 200–208.

27. Shalom, S.; Strinkovski, A.; Peleg, G.; Druckmann, S.; Krauss, A.; Lewis, A.; Linial, M.; Ottolenghi, M. An optical submicrometer calcium sensor with conductance sensing capability. *Anal. Biochem.* **1997**, *244*, 256–259.

28. Wright, K. M.; Horobin, R. W.; Oparka, K. J. Phloem mobility of fluorescent xenobiotics in *Arabidopsis* in relation to their physicochemical properties. *J. Exp. Bot.* **1996**, *47*, 1779–1787.

CALCIUM GREEN 2

CAS Registry Number 170516-40-2
Chemical Structure

Molecular Weight 1665.58
Physical Form Solid
Solubility Soluble in water
Melting Point >250°C

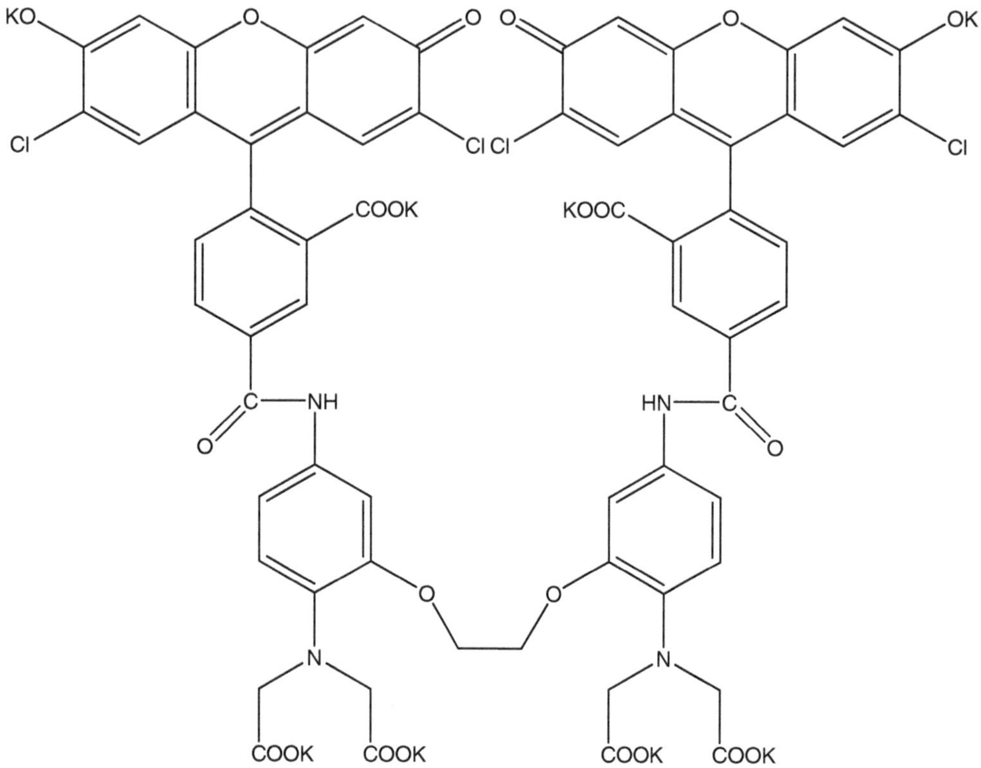

CA Index Name Glycine, *N,N′*-[1,2-ethanediylbis[oxy[4-[[(2′,7′-dichloro-3′,6′-dihydroxy-3-oxospiro[isobenzofuran-1(3*H*),9′-[9*H*]xanthen]-5-yl)carbonyl]amino]-2,1-phenylene]]]bis[*N*-(carboxymethyl)-, potassium salt (1:8)

Other Names Glycine, *N,N′*-[1,2-ethanediylbis[oxy[4-[[(2′,7′-dichloro-3′,6′-dihydroxy-3-oxospiro[isobenzofuran-1(3*H*),9′-[9*H*]xanthen]-5-yl)carbonyl]amino]-2,1-phenylene]]]bis[*N*-(carboxymethyl)-, octapotassium salt; CG 2; Calcium Green 2

Merck Index Number Not listed

Chemical/Dye Class Xanthene

Molecular Formula $C_{64}H_{34}Cl_4K_8N_4O_{22}$

Absorption (λ_{max}) 503 nm

Emission (λ_{max}) 536 nm

Synthesis Synthetic methods[1,2]

Staining Applications Calcium ions[1–6]

Biological Applications Calcium indicator;[1–6] identifying ryanodine receptors modulators,[3] taste receptors modulators[6]

Industrial Applications Not reported

Safety/Toxicity No data available

REFERENCES

1. Kuhn, M. A.; Haugland, R. P. Reactive derivatives of BAPTA used to make ion-selective chelators. U.S. Patent 5453517, 1995; *Chem. Abstr.* **1995**, *124*, 86585.
2. Kuhn, M. A.; Hoyland, B.; Carter, S.; Zhang, C.; Haugland, R. P. Fluorescent ion indicators for detecting heavy metals. *Proc. SPIE-Int. Soc. Opt. Eng.* **1995**, *2388*, 238–244.

3. Dong, C. J.; Hare, W. A. Methods of determining activity of ryanodine receptor modulators. U.S. Pat. Appl. Publ. US 2007196856, 2007; *Chem. Abstr.* **2007**, *147*, 292267.
4. Hsu, M. Over-expressing calcium channel CRAC (calcium release activated channel) gene for the assay of G-protein coupled receptors and their ligands. PCT Int. Appl. WO 2006050214, 2006; *Chem. Abstr.* **2006**, *144*, 463795.
5. Yoshiki, K.; Azuma, H.; Yoshioka, K.; Hashimoto, M.; Araki, T. Finding of optimal calcium ion probes for fluorescence lifetime measurement. *Opt. Rev.* **2005**, *12*, 415–419.
6. Servant, G.; Kamdar, P.; Rivadeneyra, A. Novel cell-based assays for identifying enhancers or inhibitors of T1R taste receptors (T1R2/T1R3 sweet) and umami (T1R1/T1R3 umami) taste receptors. U.S. Pat. Appl. Publ. US 2008305500, 2008; *Chem. Abstr.* **2008**, *150*, 31118.

CALCIUM GREEN 5N

CAS Registry Number 153130-66-6

Chemical Structure

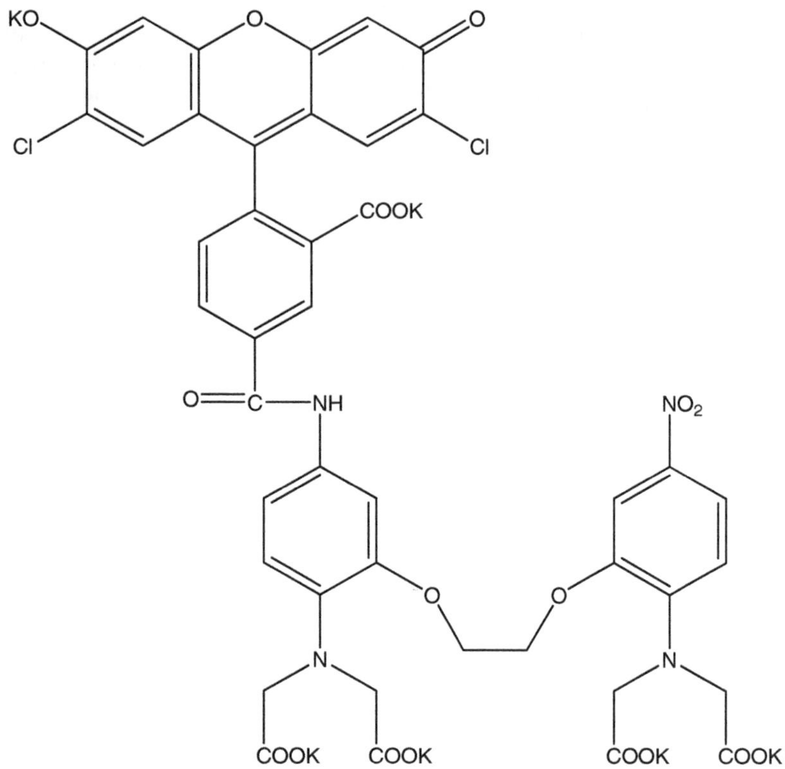

Molecular Weight 1192.19

Physical Form Solid

Solubility Soluble in water

CA Index Name Glycine, *N*-[2-[2-[2-[bis(carboxymethyl)amino]-5-[[(2′,7′-dichloro-3′,6′-dihydroxy-3-oxospiro[isobenzofuran-1(3*H*),9′-[9*H*]xanthen]-5-yl)carbonyl]amino]phenoxy]ethoxy]-4-nitrophenyl]-*N*-(carboxymethyl)-, hexapotassium salt

Other Names Calcium green 5N

Merck Index Number Not listed

Chemical/Dye Class Xanthene

Molecular Formula $C_{43}H_{26}Cl_2K_6N_4O_{18}$

Melting Point >250°C

Absorption (λ_{max}) 506 nm

Emission (λ_{max}) 532 nm

Synthesis Synthetic methods[1,2]

Staining Applications Calcium ions[1–12]

Biological Applications Calcium indicator[1–12]

Industrial Applications Not reported

Safety/Toxicity No data available

REFERENCES

1. Kuhn, M. A.; Hoyland, B.; Carter, S.; Zhang, C.; Haugland, R. P. Fluorescent ion indicators for detecting heavy metals. *Proc. SPIE-Int. Soc. Opt. Eng.* **1995**, *2388*, 238–244.
2. Rajdev, S.; Reynolds, I. J. Calcium green-5N, a novel fluore-scent probe for monitoring high intracellular free calcium concentrations associated with glutamate excitotoxicity in cultured rat brain neurons. *Neurosci. Lett.* **1993**, *162*, 149–152.
3. Yoshiki, K.; Azuma, H.; Yoshioka, K.; Hashimoto, M.; Araki, T. Finding of optimal calcium ion probes for fluorescence lifetime measurement. *Opt. Rev.* **2005**, *12*, 415–419.

4. Murphy, A. N.; Stout, A. K. Screening assays based on intramitochondrial calcium. PCT Int. Appl. WO 2001051923, 2001; *Chem. Abstr.* **2001**, *135*, 117202.

5. Lakowicz, J. R.; Gryczynski, I. Low frequency modulation sensors using nanosecond fluorophores. PCT Int. Appl. WO 2000014515, 2000; *Chem. Abstr.* **2000**, *132*, 205106.

6. Morris, S. A.; Correa, V.; Cardy, T. J. A.; O'Beirne, G.; Taylor, C. W. Interactions between inositol trisphosphate receptors and fluorescent Ca^{2+} indicators. *Cell Calcium* **1999**, *25*, 137–142.

7. Clark, H. A.; Kopelman, R.; Tjalkens, R.; Philbert, M. A. Optical nanosensors for chemical analysis inside single living cells. 2. Sensors for pH and calcium and the intracel-lular application of PEBBLE sensors. *Anal. Chem.* **1999**, *71*, 4837–4843.

8. Stout, A. K.; Reynolds, I. J. High-affinity calcium indicators underestimate increases in intracellular calcium concentrations associated with excitotoxic glutamate stimulations. *Neuroscience* **1999**, *89*, 91–100.

9. Naraghi, M. T-jump study of calcium binding kinetics of calcium chelators. *Cell Calcium* **1997**, *22*, 255–268.

10. Zhao, M.; Hollingworth, S.; Baylor, S. M. Properties of tri- and tetracarboxylate Ca^{2+} indicators in frog skeletal muscle fibers. *Biophys. J.* **1996**, *70*, 896–916.

11. Tucker, T.; Fettiplace, R. Confocal imaging of calcium microdomains and calcium extrusion in turtle hair cells. *Neuron* **1995**, *15*, 1323–1335.

12. Tepikin, A. V.; Llopis, J.; Snitsarev, V. A.; Gallacher, D. V.; Petersen, O. H. The droplet technique: measurement of calcium extrusion from single isolated mammalian cells. *Pfluegers Arch.* **1994**, *428*, 664–670.

CARMINE

CAS Registry Number 1390-65-4

Chemical Structure

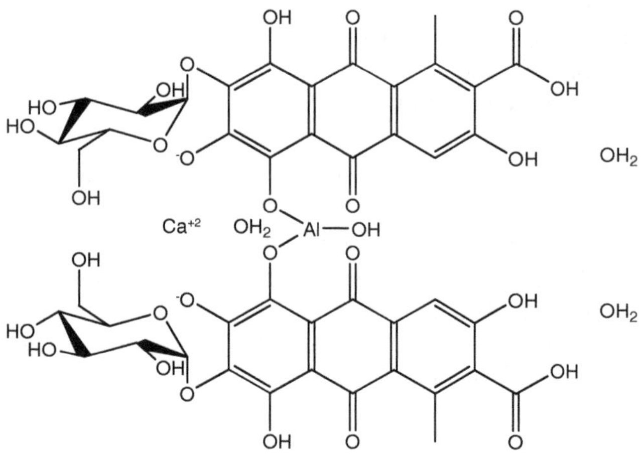

CA Index Name Carmine

Other Names Alum carmine; Alum lake; Alum lake of carminic acid; B Rose Liquid; C.I. 75470; Carmine; Carmine alum lake; Carmine red; Cochineal; Cochineal extract; Nacarat; Natural Red 4

Merck Index Number Not listed

Chemical/Dye Class Anthraquinone

Molecular Formula $C_{44}H_{43}AlCaO_{32}$

Molecular Weight 1150.86

Physical Form Red powder

Solubility Slightly soluble in water; soluble in alkali solution; insoluble in cold water

Melting Point 138–140°C (darkens)

Absorption (λ_{max}) 531 nm, 563 nm

Synthesis Synthetic methods[1–11]

StainingApplications Beverages;[12] candies;[13] cheese;[14] chewing gum;[15] noodle seasoning;[16] sweeteners;[17] confectionery products;[18,19] fish;[20] meat;[20] food casings;[21] fruits;[22] vegetables;[22] bacteria;[23] cells;[24] chromatin;[25] glycogen;[26] mucus;[26] nuclei;[26] microorganisms;[27] nucleic acids;[28] proteins;[29] capsules;[30] cough syrups;[31] tablets;[13,19,32] lips;[35,36,40,41] lip gloss;[33] lipstick;[34] cheek color;[34] nails;[36,37] eyelashes;[35] eyebrows;[36] sunscreen;[38] skin;[35,39–41] hairs[35,42,43]

Biological Applications Drug delivery products;[44] preventing coronary artery disease;[45] cancer chemopreventive activity;[46] lipid metabolism;[47] treating Alzheimer's disease;[48] soaps;[49,50] whitening teeth;[51] in food products;[52] pharmaceuticals;[52] cosmetics;[52] medical devices[53]

Industrial Applications Adhesives;[54] paints;[54] inks;[54] toners;[55] coloring cigarette papers,[56] textiles[57]

Safety/Toxicity Carcinogenicity;[58] chronic toxicity;[58] clastogenic effects;[59] DNA repair;[61] embryotoxicity;[60] food allergy;[62] hepatotoxicity;[61] occupational asthma;[62] reproductive toxicity;[63] teratogenicity[60]

Certification/Approval Certified by Biological Stain Commission (BSC)

REFERENCES

1. Tamura, I.; Hirai, T.; Aryama, H. Deodoring of carmine color by extraction with supercritical carbon dioxide. Jpn. Kokai Tokkyo Koho JP 08283600, 1996; *Chem. Abstr.* **1997**, *126*, 76135.
2. Galfayan, K. K. Experiments of Saak Tsakhkarar on obtaining a dye from the *Ararat cochineal. Istoriya Estestvozn. Tekhn. Armenii, Erevan* **1979**, 116–129; *Chem. Abstr.* **1980**, *92*, 109263.
3. Gibaja Oviedo, S.; Aldana Gibaja, L. D. Preparation of carmine. I. Evaluation of four processes for preparing carmine. *Bol. Soc. Quim. Peru* **1977**, *43*, 133–138; *Chem. Abstr.* **1978**, *89*, 91132.
4. Harms, H. Chemistry of carmine and of its solutions used in microscopy. *Naturwissenschaften* **1957**, *44*, 327; *Chem. Abstr.* **1958**, *52*, 117084.
5. Fester, G. A.; Retamar, J. A. Dyes of ancient Peru. *Rev. Fac. Ing. Quim.* **1953**, *22*, 133–143; *Chem. Abstr.* **1954**, *48*, 37481.
6. Fester, A. Red dyes used by the natives of South America. *Rev. Fac. Ing. Quim.* **1951**, *20*, 105–116; *Chem. Abstr.* **1954**, *48*, 37480.
7. Moore, R. A. Gelatin carmine injections. *Stain Technol.* **1929**, *5*, 35.
8. Moore, R. A. Gelatin carmine injections. *Med. Museums* **1929**, *12*, 55–58; *Chem. Abstr.* **1930**, *24*, 17101.
9. De Puyster, B. Use of organic dyestuffs for lake manufacture. II. Natural dyes of animal origin. *Color Trade J.* **1918**, *3*, 339–342.
10. Yuge, T.; Nishiyama, H.; Koda, T. Red colorants containing carmine, and their manufacture. Jpn.

Kokai Tokkyo Koho JP 2004149643, 2004; *Chem. Abstr.* **2004**, *140*, 422809.

11. Schul, J. Carmine. *IFT Basic Symp. Ser.* **2000**, *14*, 1–10.
12. Samsonov, O. V. Composition for manufacturing alcohol-free beverage. Russ. RU 2335992, 2008; *Chem. Abstr.* **2008**, *149*, 447086.
13. Cottrell, C. E. Aqueous low viscosity dispersions containing high pigment content for coating pharmaceutical tablets and candy. U.S. Patent 4750938, 1988; *Chem. Abstr.* **1989**, *110*, 141557.
14. Zong, X.; Zhang, S.; Zhang, P.; Xing, H.; Zhang, Y.; Liu, W.; Jia, S.; Fu, Z.; Fan, L.; Yang, J. Method for manufacturing natural hard colorful cheese. Faming Zhuanli Shenqing Gongkai Shuomingshu CN 101297664, 2008; *Chem. Abstr.* **2008**, *149*, 554469.
15. Zou, J.; Liu, H.; Zou, J. New chewing gum formulation of tadalafil for treating erecting disturbance. Faming Zhuanli Shenqing Gongkai Shuomingshu CN 101143145, 2008; *Chem. Abstr.* **2008**, *148*, 387243.
16. Ji, A. Method for manufacturing edible bag for packing instant noodle seasoning. Faming Zhuanli Shenqing Gongkai Shuomingshu CN 1699117, 2005; *Chem. Abstr.* **2006**, *145*, 270591.
17. Catani, S. J.; Liao, S.; Loades, M. Self-mixing tabletop sweetener. U.S. Pat. Appl. Publ. US 2007059418, 2007; *Chem. Abstr.* **2007**, *146*, 315560.
18. Royo, A. P.; Yang, M. Coated chewy confectionery product and method. U.S. Pat. Appl. Publ. US 2007148292, 2007; *Chem. Abstr.* **2007**, *147*, 52032.
19. Grattan, T. J. Process for manufacture of coated, intagliated, tableted dosage forms and/or sugar confectionery. Brit. UK Pat. Appl. GB 2191117, 1987; *Chem. Abstr.* **1988**, *108*, 173616.
20. Holzchuh, P.; Buch, G.; Weiland, J. J. Smoking and coloration of food in discrete steps. Fr. Demande FR 2850248, 2004; *Chem. Abstr.* **2004**, *141*, 139400.
21. Ducharme, P. E., Jr.; Nicholson, M. D. Red colorant composition for food casings. Eur. Pat. Appl. EP 1250853, 2002; *Chem. Abstr.* **2002**, *137*, 309922.
22. Nafisi-Novaghar, K.; Keefer, D. K.; Strasser, J. H.; Dravenstadt, L. V.; Lem, R. A. Coloring of fruits and vegetables with anthraquinone metal complexes. U.S. Patent 5164212, 1992; *Chem. Abstr.* **1993**, *118*, 21390.
23. Vinderola, C. G.; Costa, G. A.; Regenhardt, S.; Reinheimer, J. A. Influence of compounds associated with fermented dairy products on the growth of lactic acid starter and probiotic bacteria. *Int. Dairy J.* **2002**, *12*, 579–589.
24. Garcia-Gasca, T.; Paz-Gonzalez, V.; Moncada-Alvarez, M. C.; Blanco-Labra, A.; Salazar-Olivo, L. A. Colorimetric quanti-tation of *in vitro* cell density using carmine, a chromosome-specific stain. *Toxicol. In Vitro* **2002**, *16*, 573–579.
25. Stockert, J. C.; Llorente, A. R.; Del Castillo, P.; Gomez, A. Chromatin fluorescence after carmine staining. *Stain Technol.* **1990**, *65*, 299–302.
26. Dapson, R. W.; Frank, M.; Penney, D. P.; Kiernan, J. A. Revised procedures for the certification of carmine (C.I. 75470, Natural red 4) as a biological stain. *Biotech. Histochem.* **2007**, *82*, 13–15.
27. Schuer, J. P. Symbiotic regenerative compositions containing microorganisms. Eur. Pat. Appl. EP 1228769, 2002; *Chem. Abstr.* **2002**, *137*, 154384.
28. Dutt, M. K. Use of carmine in the specific staining of nucleic acids. *Curr. Sci.* **1973**, *42*, 751–752.
29. Mcmindes, M. K.; Valle, V. Seafood compositions comprising structured protein products. U.S. Pat. Appl. Publ. US 2008254167, 2008; *Chem. Abstr.* **2008**, *149*, 424520.
30. Fantuzzi, M. Chewable soft gelatin capsules. PCT Int. Appl. WO 2007143652, 2007; *Chem. Abstr.* **2007**, *148*, 62014.
31. Neelwarne, B.; Rudrappa, T.; Narayan, M. S.; Chidam-baramurthy, K. N.; Ravishankar, G. A. An improved process for the preparation of natural color stabilized cough syrup mixture. Indian Pat. Appl. IN 2003DE00395, 2007; *Chem. Abstr.* **2007**, *147*, 455620.
32. Mogenet, L.; Gardin, Y.; Bernard, S. Effervescent tablet for tracing animals being treated, method for producing the same, and method for treating an aqueous solution. PCT Int. Appl. WO 2001026622, 2001; *Chem. Abstr.* **2001**, *134*, 316104.
33. Jacobson-Alti, J. Multi-sensory pleasant lip gloss comprising a fatty acid ester. U.S. Pat. Appl. Publ. US 2004101499, 2004; *Chem. Abstr.* **2004**, *140*, 428701.
34. Nishiyama, H.; Yuge, T.; Koda, T. Cosmetic compositions colored by carmine solid fine particles. Jpn. Kokai Tokkyo Koho JP 2004210708, 2004; *Chem. Abstr.* **2004**, *141*, 128487.
35. Zheng, T.; Sandstrom, G. A. Long wear topical composition having improved glossy appearance. U.S. Pat. Appl. Publ. US 2006134035, 2006; *Chem. Abstr.* **2006**, *145*, 89391.

36. Grollier, J. F. Process for preparing a composition for permanent or semipermanent dyeing of skin, eyebrows, lips or nails from a fluid under pressure, and dyestuffs. Eur. Pat. Appl. EP 1559392, 2005; *Chem. Abstr.* **2005**, *143*, 179096.
37. Collins, C.; Bealer, L. Method of promoting a water washable nail composition. U.S. Pat. Appl. Publ. US 2004191195, 2004; *Chem. Abstr.* **2004**, *141*, 301037.
38. Gers-Barlag, H.; Mueller, A.; Grotelueschen, B.; Grundt, W. Visualization of sunscreen agents on the skin by using dyes. Ger. Offen. DE 10035512, 2002; *Chem. Abstr.* **2002**, *136*, 123418.
39. Teboul, K. Method for coloring dark skins. PCT Int. Appl. WO 2008087308, 2008; *Chem. Abstr.* **2008**, *149*, 160219.
40. Blin, X. Range of compositions for application to the skin, lips, and/or superficial body growths. Fr. Demande FR 2868697, 2005; *Chem. Abstr.* **2005**, *143*, 392523.
41. Dumousseaux, C. Cosmetic composition for application to the skin, lips, and/or body surface. Fr. Demande FR 2868696, 2005; *Chem. Abstr.* **2005**, *143*, 392522.
42. De Boni, M. Use of natural dyes for dyeing human hair. PCT Int. Appl. WO 2008047055, 2008; *Chem. Abstr.* **2008**, *148*, 479061.
43. Massoni, J. T. Oxidative hair dye composition with color indication. Eur. Pat. Appl. EP 223572, 1987; *Chem. Abstr.* **1988**, *108*, 156251.
44. Paulsen, N. E.; Johnson, R.; Coffee, M. Process for manufacturing chewable dosage forms for drug delivery and products thereof. U.S. Pat. Appl. Publ. US 2007128251, 2007; *Chem. Abstr.* **2007**, *147*, 39161.
45. Bonfiglio, R. P. Antibacterial oral rinse formulation comprising bioflavonoid polyphenol for preventing coronary artery disease. U.S. Pat. Appl. Publ. US 2007154414, 2007; *Chem. Abstr.* **2007**, *147*, 124648.
46. Kapadia, G. J.; Tokuda, H.; Sridhar, R.; Balasubramanian, V.; Takayasu, J.; Bu, P.; Enjo, F.; Takasaki, M.; Konoshima, T.; Nishino, H. Cancer chemopreventive activity of synthetic colorants used in foods, pharmaceuticals and cosmetic preparations. *Cancer Lett.* **1998**, *129*, 87–95.
47. Nagem, T. J.; Pereira, W. L.; Oliveira, T. T.; Pinto, A. S.; Oliveira, M. G. A.; Stringheta, P. C. Naringin and the natural pigments anthocyanin and carmine and their effects on lipid metabolism. *Rev. Bras. Farm.* **1999**, *80*, 25–28; *Chem. Abstr.* **2000**, *133*, 12706.
48. Yan, Y. Manufacture of memantine hydrochloride effervescent tablet for treating Alzheimer's disease. Faming Zhuanli Shenqing Gongkai Shuomingshu CN 1742711, 2006; *Chem. Abstr.* **2006**, *144*, 398290.
49. Meyer, J.; Moenikes, M. Kneadable soap. Ger. Offen. DE 10339807, 2005; *Chem. Abstr.* **2005**, *142*, 394166.
50. Kim, S. O. Soaps containing amino acid derivatives and coated mica powder. Jpn. Kokai Tokkyo Koho JP 2002121589, 2002; *Chem. Abstr.* **2002**, *136*, 345480.
51. Tarver, J. G.; Card, B. D.; Ladow, M. R.; Livingstone, D. H. Tooth whitening compositions and methods for using the same. U.S. Pat. Appl. Publ. US 2006104922, 2006; *Chem. Abstr.* **2006**, *144*, 474410.
52. Klingenberg, A. Method for manufacturing a colour mixture for use in food products, pharmaceuticals and cosmetics, and colour mixture obtained according to this method. U.S. Pat. Appl. Publ. US 2005084462, 2005; *Chem. Abstr.* **2005**, *142*, 379428.
53. Konrad, F. *In-vitro* diagnostic medical devices and solutions for determining saliva volume and for collecting saliva. U.S. Pat. Appl. Publ. US 2006073538, 2006; *Chem. Abstr.* **2006**, *144*, 365907.
54. Wenning, A.; Loest, D. Coating compositions based on universal pigment preparations for use as paints, adhesives and printing inks. PCT Int. Appl. WO 2007141115, 2007; *Chem. Abstr.* **2007**, *148*, 56686.
55. Kadota, T.; Ishiwatari, T.; Takano, H. Electrophotographic toners with excellent grindability, fixation to recording media, high-temperature offset resistance, and transparency and their manufacture. Jpn. Kokai Tokkyo Koho JP 2004279520, 2004; *Chem. Abstr.* **2004**, *141*, 322535.
56. St. Charles, F. K.; Lauterbach, J. H.; Chao, L. C.; Tang, J. Y.; Chakraborty, B. B.; MacMurtrie, A. Dyes for coloration of cigarette paper. Ger. Offen. DE 4040346, 1991; *Chem. Abstr.* **1991**, *115*, 94714.
57. Tsatsaroni, E.; Liakopoulou-Kyriakides, M. Effect of enzymic treatment on the dyeing of cotton and wool fibers with natural dyes. *Dyes Pigm.* **1995**, *29*, 203–209.
58. Ford, G. P.; Gopal, T.; Grant, D.; Gaunt, I. F.; Evans, J. G.; Butler, W. H. Chronic toxicity/carcinogenicity study of carmine of cochineal in the rat. *Food Chem. Toxicol.* **1987**, *25*, 897–902.
59. Au, W.; Hsu, T. C. Studies on the clastogenic effects of biologic stains and dyes. *Environ. Mutagen.* **1979**, *1*, 27–35.
60. Grant, D.; Gaunt, I. F.; Carpanini, F. M. B. Teratogenicity and embryotoxicity study of

carmine of cochineal in the rat. *Food Chem. Toxicol.* **1987**, *25*, 913–917.

61. Kornbrust, D.; Barfknecht, T. Testing of 24 food, drug, cosmetic, and fabric dyes in the *in vitro* and the *in vivo/in vitro* rat hepatocyte primary culture/DNA repair assays. *Environ. Mutagen.* **1985**, *7*, 101–120.

62. Acero, S.; Tabar, A. I.; Alvarez, M. J.; Garcia, B. E.; Olaguibel, J. M.; Moneo, I. Occupational asthma and food allergy due to carmine. *Allergy* **1998**, *53*, 897–901.

63. Grant, D.; Gaunt, I. F. Three-generation reproduction study on carmine of cochineal in the rat. *Food Chem. Toxicol.* **1987**, *25*, 903–912.

COELENTERAZINE

CAS Registry Number 55779-48-1

Chemical Structure

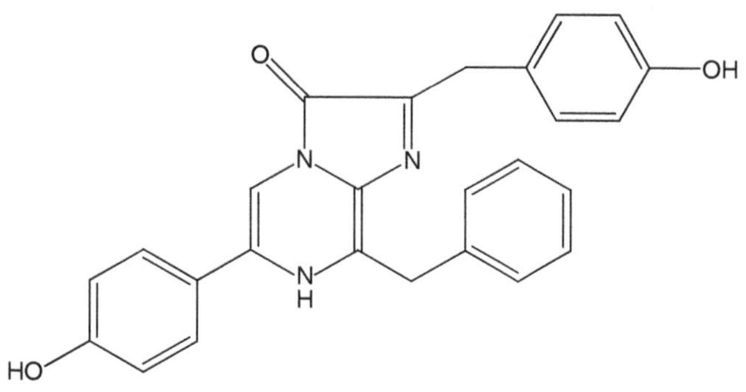

CA Index Name Imidazo[1,2-*a*]pyrazin-3(7*H*)-one, 6-(4-hydroxyphenyl)-2-[(4-hydroxyphenyl)methyl]-8-(phenylmethyl)-

Other Names 8-Benzyl-2-(4-hydroxybenzyl)-6-(4-hydroxyphenyl)imidazo[1,2-a]pyrazin-3(7*H*)-one; CLZN; Coelenterazin; Coelenterazine; Coelenterazine native; 3,2-Dihydro-2-(*p*-hydroxybenzyl)-6-(*p*-hydroxyphenyl)-8-benzylimidazolo[1,2-*a*]pyrazin-3-one; Luciferin; Luciferin (Oplophorus); NanoFuel; Preluciferin; Preluciferin (*Watasenia*); Watasenia preluciferin

Merck Index Number Not listed

Chemical/Dye Class Heterocycle; Imidazo[1,2-*a*]pyrazin-3(7*H*)-one

Molecular Formula $C_{26}H_{21}N_3O_3$

Molecular Weight 423.46

Physical Form Orange-yellow crystals or powder

Solubility Soluble in ethanol, methanol

Melting Point 176–181°C (decompose)

Boiling Point (Calcd.) 641.4 ± 65.0°C, pressure: 760 Torr

pK$_a$ (Calcd.) 9.91 ± 0.15, most acidic, temperature: 25°C; 6.63 ± 0.60, most basic, temperature: 25°C

Absorption (λ_{max}) 429 nm

Emission (λ_{max}) 466 nm; 514 nm

Synthesis Synthetic methods[1–15]

Staining Applications Calcium ions;[12,13,17–31,51] cells,[16] erythrocytes[62]

Biological Applications Calcium indicator;[12,13,17–31,51] detecting luciferase,[34] protease, [35] gene expression,[36] nucleic acids,[37–40] proteins,[41,42] stem cells,[43] quantum dot conjugates,[44–46] superoxide;[47–50] tracing protein dynamics;[51,52] as a substrate for luciferase,[53–56] alkaline phosphatase,[57] galactosidase;[58] treating epilepsy,[59] pain syndromes,[59] hepatitis C virus (HCV) infections,[60] herpes virus infection,[61] *Plasmodium falciparum* infection,[61] human immunodeficiency virus infection[61]

Industrial Applications Not reported

Safety/Toxicity Fish toxicity;[62] vascular toxicity[63]

REFERENCES

1. Adamczyk, M.; Akireddy, S. R.; Johnson, D. D.; Mattingly, P. G.; Pan, Y.; Reddy, R. E. Synthesis of 3,7-dihydroimidazo[1,2-*a*]pyrazine-3-ones and their chemiluminescent properties. *Tetrahedron* **2003**, *59*, 8129–8142.
2. Kakoi, H. Synthesis of 2-amino-3-benzyl-5-(*p*-hydroxyphenyl) pyrazine. *Chem. Pharm. Bull.* **2002**, *50*, 301–302.
3. Adamczyk, M.; Johnson, D. D.; Mattingly, P. G.; Pan, Y.; Reddy, R. E. Synthesis of coelenterazine. *Org. Prep. Proced. Int.* **2001**, *33*, 477–485.
4. Kakoi, H.; Inoue, S. Improved synthesis of *Watasenia preluciferin. Heterocycles* **1998**, *48*, 1669–1672.
5. Hirano, T.; Ohmiya, Y.; Maki, S.; Niwa, H.; Ohashi, M. Bioluminescent properties of fluorinated semi-synthetic aequorins. *Tetrahedron Lett.* **1998**, *39*, 5541–5544.
6. Hibbert, F.; Jones, K.; Keenan, M. Preparation of arylpyrazines. PCT Int. Appl. WO 9743267, 1997; *Chem. Abstr.* **1997**, *128*, 22924.
7. Keenan, M.; Jones, K.; Hibbert, F. Highly efficient and flexible total synthesis of coelenterazine. *Chem. Commun.* **1997**, 323–324.
8. Gonzalez-Trueba, G.; Paradisi, C.; Zoratti, M. Synthesis of coelenterazine. *Anal. Biochem.* **1996**, *240*, 308–310.
9. Jones, K.; Keenan, M.; Hibbert, F. A Suzuki coupling approach to pyrazines related to coelenterazine. *Synlett* **1996**, 509–510.
10. Qi, C. F.; Gomi, Y.; Hirano, T.; Ohashi, M.; Ohmiya, Y.; Tsuji, F. I. Chemi- and bio-luminescence of coelenterazine analogs with phenyl homologs at the C-2 position. *J. Chem. Soc., Perkin Trans. 1* **1992**, 1607–1611.

11. Qi, C. F.; Gomi, Y.; Ohashi, M.; Ohmiya, Y.; Tsuji, F. I. Chemi- and bioluminescence of coelenterazine analogs: effect of substituents at the C-2 position. *J. Chem. Soc., Chem. Commun.* **1991**, 1307–1309.
12. Shimomura, O.; Musicki, B.; Kishi, Y. Semi-synthetic aequorins with improved sensitivity to calcium ions. *Biochem. J.* **1989**, *261*, 913–920.
13. Shimomura, O.; Musicki, B.; Kishi, Y. Semi-synthetic aequorin. An improved tool for the measurement of calcium ion concentration. *Biochem. J.* **1988**, *251*, 405–410.
14. Inoue, S.; Sugiura, S.; Kakoi, H.; Hasizume, K.; Goto, T.; Iio, H. Squid bioluminescence. II. Isolation from *Watasenia scintillans* and synthesis of 2-(*p*-hydroxybenzyl)-6-(*p*-hydroxyphenyl)-3,7-dihydroimidazo[1,2-*a*]pyrazin-3-one. *Chem. Lett.* **1975**, 141–144.
15. Dubois, R. Synthesis of luciferin. *Compt. Rend.* **1918**, *166*, 578–580; *Chem. Abstr.* **1918**, *12*, 11062.
16. Garini, Y.; Mcnamara, G.; Soenksen, D. G.; Cabib, D.; Buckwald, R. A. *In situ* method of analyzing cells by staining with multiple stains and using a spectral data collection device. PCT Int. Appl. WO 2000031534, 2000; *Chem. Abstr.* **2000**, *133*, 14300.
17. Stepanyuk, G. A.; Liu, Z. J.; Vysotski, E. S.; Lee, J.; Rose, J. P.; Wang, B. C. Structure based mechanism of the Ca(2 +)-induced release of coelenterazine from the Renilla binding protein. *Proteins* **2009**, *74*, 583–593.
18. Brini, M. Calcium-sensitive photoproteins. *Methods* **2008**, *46*, 160–166.
19. Inoue, S.; Sasaki, T.; Sahara, Y.; Sato, J. The protein G fusion protein with apoaequorin as a sensitive IgG Fc detection reagent generating calcium ion-dependent fluorescent signals. Jpn. Kokai Tokkyo Koho JP 2008283944, 2008; *Chem. Abstr.* **2008**, *149*, 551271.
20. Brulet, P.; Rogers, K. L.; Curie, T. Non-invasive real-time *in vivo* bioluminescence imaging of local Ca^{2+} dynamics in living organisms. PCT Int. Appl. WO 2008104830, 2008; *Chem. Abstr.* **2008**, *149*, 302509.
21. Rowe, L.; Rothert, A.; Logue, C.; Ensor, C. M.; Deo, S. K.; Daunert, S. Spectral tuning of photoproteins by partnering site-directed mutagenesis strategies with the incorporation of chromophore analogs. *Protein Eng., Design Select.* **2008**, *21*, 73–81.
22. Bovolenta, S.; Foti, M.; Lohmer, S.; Corazza, S. Development of a Ca(2 +)-activated photoprotein, Photina, and its appli-cation to high-throughput screening. *J. Biomol. Screening* **2007**, *12*, 694–704.
23. Mazzotta, J. B.; Nguyen, D.; Ebens, A. J.; Winslow, J. Recombinant cell and method for identifying insect cyclic nucleotide-gated calcium channel agonists for use as insecti-cides. Ger. Offen. DE 102004062273, 2006; *Chem. Abstr.* **2006**, *145*, 120011.
24. Mitchell, K. J.; Pinton, P.; Varadi, A.; Tacchetti, C.; Ainscow, E. K.; Pozzan, T.; Rizzuto, R.; Rutter, G. A. Dense core secretory vesicles revealed as a dynamic Ca(2 +) store in neuroendocrine cells with a vesicle-associated membrane protein aequorin chimaera. *J. Cell Biol.* **2001**, *155*, 41–51.
25. Ungrin, M. D.; Singh, L. M. R.; Stocco, R.; Sas, D. E.; Abramovitz, M. An automated aequorin luminescence-based functional calcium assay for G-protein-coupled receptors. *Anal. Biochem.* **1999**, *272*, 34–42.
26. Creton, R.; Steele, M. E.; Jaffe, L. F. Expression of apo-aequorin during embryonic development; how much is needed for calcium imaging?. *Cell Calcium* **1997**, *22*, 439–446.
27. Kendall, J. M.; Badminton, M. N.; Sala-Newby, G. B.; Campbell, A. K.; Rembold, C. M. Recombinant apoaequorin acting as a pseudo-luciferase reports micromolar changes in the endoplasmic reticulum free Ca^{2+} of intact cells. *Biochem. J.* **1996**, *318*, 383–387.
28. Button, D.; Brownstein, M. Aequorin-expressing mammalian cell lines used to report Ca^{2+} mobilization. *Cell Calcium* **1993**, *14*, 663–671.
29. Knight, M. R.; Read, N. D.; Campbell, A. K.; Trewavas, A. J. Imaging calcium dynamics in living plants using semi-synthetic recombinant aequorins. *J. Cell Biol.* **1993**, *121*, 83–90.
30. Nakajima-Shimada, J.; Iida, H.; Tsuji, F. I.; Anraku, Y. Monitoring of intracellular calcium in *Saccharomyces cerevisiae* with an apoaequorin cDNA expression system. *Proc. Natl. Acad. Sci. U.S.A.* **1991**, *88*, 6878–6882.
31. Knight, M. R.; Campbell, A. K.; Smith, S. M.; Trewavas, A. J. Recombinant aequorin as a probe for cytosolic free calcium in *Escherichia coli. FEBS Lett.* **1991**, *282*, 405–408.
32. Shimomura, O.; Inouye, S.; Musicki, B.; Kishi, Y. Recombinant aequorin and recombinant semi-synthetic aequorins: cellular calcium ion indicators. *Biochem. J.* **1990**, *270*, 309–312.
33. Campbell, A. K.; Patel, A.; Houston, W. A.; Scolding, N. J.; Frith, S.; Morgan, B. P.; Compston, D. A. Photoproteins as indicators of intracellular free Ca^{2+}. *J. Biolumin. Chemilum.* **1989**, *4*, 463–474.

34. Leu, M. P.; Daly, J. M. Luciferase luminescent signal enhancing compositions and reagent composition for determination of luciferase. PCT Int. Appl. WO 2008049160, 2008; *Chem. Abstr.* **2008**, *148*, 489577.

35. Leng, J. Protease-cleavable luciferases and their use in protease determination and screening for protease inhibitors. PCT Int. Appl. WO 2002006458, 2002; *Chem. Abstr.* **2002**, *136*, 130761.

36. Bhaumik, S.; Lewis, X. Z.; Gambhir, S. S. Optical imaging of *Renilla luciferase*, synthetic *Renilla luciferase*, and firefly luciferase reporter gene expression in living mice. *J. Biomed. Opt.* **2004**, *9*, 578–586.

37. Cissell, K. A.; Rahimi, Y.; Shrestha, S.; Deo, S. K. Reassembly of a bioluminescent protein *Renilla luciferase* directed through DNA hybridization. *Bioconjugate Chem.* **2009**, *20*, 15–19.

38. Cissell, K. A.; Campbell, S.; Deo, S. K. Rapid, single-step nucleic acid detection. *Anal. Bioanal. Chem.* **2008**, *391*, 2577–2581.

39. Cissell, K. A.; Rahimi, Y.; Shrestha, S.; Hunt, E. A.; Deo, S. K. Bioluminescence-based detection of microRNA, miR21 in breast cancer cells. *Anal. Chem.* **2008**, *80*, 2319–2325.

40. Walls, Z. F.; Gambhir, S. S. BRET-based method for detection of specific RNA species. *Bioconjugate Chem.* **2008**, *19*, 178–184.

41. Carriba, P.; Navarro, G.; Ciruela, F.; Ferre, S.; Casado, V.; Agnati, L.; Cortes, A.; Mallol, J.; Fuxe, K.; Canela, E. I.; Lluis, C.; Franco, R. Detection of heteromerization of more than two proteins by sequential BRET-FRET. *Nat. Methods* **2008**, *5*, 727–733.

42. Suzuki, T.; Usuda, S.; Ichinose, H.; Inouye, S. Real-time bioluminescence imaging of a protein secretory pathway in living mammalian cells using *Gaussia luciferase*. *FEBS Lett.* **2007**, *581*, 4551–4556.

43. Walia, R. Genetic engineering of siScreen vector encoding luciferase for use in bioluminescent imaging of transfected human stem cells. PCT Int. Appl. WO 2008144052, 2008; *Chem. Abstr.* **2008**, *150*, 16452.

44. Xing, Y.; So, M. K.; Koh, A. L.; Sinclair, R.; Rao, J. Improved QD-BRET conjugates for detection and imaging. *Biochem. Biophys. Res. Commun.* **2008**, *372*, 388–394.

45. So, M. K.; Loening, A. M.; Gambhir, S. S.; Rao, J. Creating self-illuminating quantum dot conjugates. *Nat. Protoc.* **2006**, *1*, 1160–1164.

46. So, M. K.; Xu, C.; Loening, A. M.; Gambhir, S. S.; Rao, J. Self-illuminating quantum dot conjugates for *in vivo* imaging. *Nat. Biotechnol.* **2006**, *24*, 339–343.

47. Daiber, A.; Oelze, M.; August, M.; Wendt, M.; Sydow, K.; Wieboldt, H.; Kleschyov, A. L.; Munzel, T. Detection of superoxide and peroxynitrite in model systems and mito-chondria by the luminol analogue L-012. *Free Radical Res.* **2004**, *38*, 259–269.

48. Tarpey, M. M.; White, C. R.; Suarez, E.; Richardson, G.; Radi, R.; Freeman, B. A; Chemiluminescent detection of oxidants in vascular tissue: lucigenin but not coelente-razine enhances superoxide formation. *Circ. Res.* **1999**, *84*, 1203–1211.

49. Teranishi, K.; Shimomura, O. Coelenterazine analogs as a chemiluminescent probe for superoxide anion. *Anal. Biochem.* **1997**, *249*, 37–43.

50. Lucas, M.; Solano, F. Coelenterazine is a superoxide anion-sensitive chemiluminescent probe: its usefulness in the assay of respiratory burst in neutrophils. *Anal. Biochem.* **1992**, *206*, 273–277.

51. Kim, S. B.; Sato, M.; Tao, H. Split *Gaussia luciferase*-based bioluminescence template for tracing protein dynamics in living cells. *Anal. Chem.* **2009**, *81*, 67–74.

52. Kim, S. B.; Sato, M.; Tao, H. Circularly permutated bioluminescent probes for illuminating ligand-activated protein dynamics. *Bioconjugate Chem.* **2008**, *19*, 2480–2486.

53. Titushin, M. S.; Markova, S. V.; Frank, L. A.; Malikova, N. P.; Stepanyuk, G. A.; Lee, J.; Vysotski, E. S. Coelenterazine-binding protein of *Renilla muelleri*: cDNA cloning, overexpression, and characterization as a substrate of luciferase. *Photochem. Photobiol. Sci.* **2008**, *7*, 189–196.

54. Leu, M. P.; Daly, J. M. Bioluminescent assays utilising secreted luciferases by measuring luminescent signal generated by one or more luciferases in a sample. PCT Int. Appl. WO 2008074100, 2008; *Chem. Abstr.* **2008**, *149*, 72395.

55. Ohmiya, Y.; Wu, C. Stabilization of coelenterazine solution in the presence of antioxidant for high-throughput measurement of luciferase activity. U.S. Pat. Appl. Publ. US 2008020384, 2008; *Chem. Abstr.* **2008**, *148*, 162265.

56. Inouye, S.; Shimomura, O. The use of *Renilla luciferase*, *Oplophorus luciferase*, and apoaequorin as bioluminescent reporter protein in the presence of coelenterazine analogs as substrate. *Biochem. Biophys. Res. Commun.* **1997**, *233*, 349–353.

57. Sobek, D.; Rao, J. Enzyme detection system with caged substrates and BRET-luminescent nanocrystal conjugates. PCT Int. Appl. WO 2008049036, 2008; *Chem. Abstr.* **2008**, *148*, 489576.

58. Graham, R. J.; Sekar, M.; Barbisin, M. *In vivo* assays for enzyme activity using liposome encapsulating chromogenic substrate to facilitate intracellular delivery. PCT Int. Appl. WO 2005059163, 2005; *Chem. Abstr.* **2005**, *143*, 93614.

59. Ng, G. Methods of identifying GABAB receptor subtype-specific agonists. PCT Int. Appl. WO 2001098779, 2001; *Chem. Abstr.* **2001**, *136*, 64146.

60. Herget, T.; Cotten, M.; Obert, S.; Klebl, B. Gastrointestinal glutathione peroxidase as therapeutic target for treatment of HCV infection, methods of treating HCV infection, and compounds useful therefore. U.S. Pat. Appl. Publ. US 2004152073, 2004; *Chem. Abstr.* **2004**, *141*, 167729.

61. Lagunoff, M.; Van Voorhis, W. C.; Jenwitheesuk, E.; Samudrala, R. Compositions and methods for predicting inhibitors of protein targets. PCT Int. Appl. WO 2008006085, 2008; *Chem. Abstr.* **2008**, *148*, 135973.

62. Janssens, B. J.; Marchand-Brynaert, J.; Rees, J. F. Protection of peroxide-treated fish erythrocytes by coelenterazine and coelenteramine. *Free Radical Res.* **2002**, *36*, 967–974.

63. Souza, H. P.; Souza, L. C.; Anastacio, V. M.; Pereira, A. C.; Junqueira, M. L.; Krieger, J. E.; da Luz, P. L.; Augusto, O.; Laurindo, F. R. Vascular oxidant stress early after balloon injury: evidence for increased NAD(P)H oxidoreductase activity. *Free Radical Biol. Med.* **2000**, *28*, 1232–1242.

COELENTERAZINE cp

CAS Registry Number 123437-25-2

Chemical Structure

CA Index Name Imidazo[1,2-*a*]pyrazin-3(7*H*)-one, 8-(cyclopentylmethyl)-6-(4-hydroxyphenyl)-2-[(4-hydroxyphenyl)methyl]-

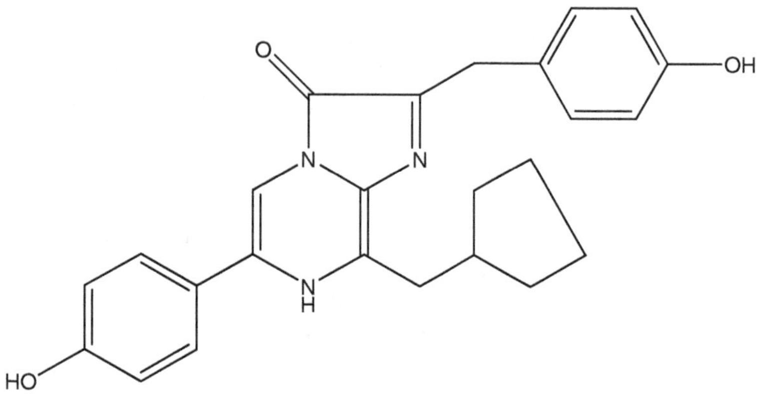

Other Names CLZN-cp; Coelenterazine cp

Merck Index Number Not listed

Chemical/Dye Class Heterocycle; Imidazo[1,2-*a*]pyrazin-3(7*H*)-one

Molecular Formula $C_{25}H_{25}N_3O_3$

Molecular Weight 415.48

Physical Form Yellow powder

Solubility Soluble in ethanol, methanol

Melting Point >200°C

Boiling Point (Calcd.) 615.1 ± 65.0°C, pressure: 760 Torr

pK_a (Calcd.) 9.91 ± 0.15, most acidic, temperature: 25°C; 6.65 ± 0.60, most basic, temperature: 25°C

Absorption (λ_{max}) 430 nm

Emission (λ_{max}) 442 nm

Synthesis Synthetic method[1]

Staining Applications Calcium ions[1–6]

Biological Applications Calcium indicator;[1–6] measuring luciferase activity;[7–9] as a substrate for luciferase; screening HIV-1 protease inhibitors[10]

Industrial Applications Not reported

Safety/Toxicity No data available

REFERENCES

1. Shimomura, O.; Musicki, B.; Kishi, Y. Semi-synthetic aequorins with improved sensitivity to calcium ions. *Biochem. J.* **1989**, *261*, 913–920.
2. Rowe, L.; Rothert, A.; Logue, C.; Ensor, C. M.; Deo, S. K.; Daunert, S. Spectral tuning of photoproteins by partnering site-directed mutagenesis strategies with the incorporation of chromophore analogs. *Protein Eng. Design Select.* **2008**, *21*, 73–81.
3. Inouye, S.; Sasaki, S. Imidazole-assisted catalysis of luminescence reaction in blue fluorescent protein from the photoprotein aequorin. *Biochem. Biophys. Res. Commun.* **2007**, *354*, 650–655.
4. Toma, S.; Chong, K. T.; Nakagawa, A.; Teranishi, K.; Inouye, S.; Shimomura, O. The crystal structures of semi-synthetic aequorins. *Protein Sci.* **2005**, *14*, 409–416.
5. Shimomura, O. Cause of spectral variation in the luminescence of semisynthetic aequorins. *Biochem. J.* **1995**, *306*, 537–543.
6. Shimomura, O.; Kishi, Y.; Inouye, S. The relative rate of aequorin regeneration from apoaequorin and coelenterazine analogs. *Biochem. J.* **1993**, *296*, 549–551.
7. Zhao, H.; Doyle, T. C.; Wong, R. J.; Cao, Y.; Stevenson, D. K.; Piwnica-Worms, D.; Contag, C. H. Characterization of coelenterazine analogs for measurements of *Renilla luciferase* activity in live cells and living animals. *Mol. Imaging* **2004**, *3*, 43–54.
8. Inoue, S.; Sasaki, S. Fluorescent activity-possessing lucife-rase activity enhancement method. Jpn. Kokai Tokkyo Koho JP 2006271327, 2006; *Chem. Abstr.* **2006**, *145*, 413077.

9. Inouye, S.; Shimomura, O. The use of *Renilla luciferase*, *Oplophorus luciferase*, and apoaequorin as bioluminescent reporter protein in the presence of coelenterazine analogs as substrate. *Biochem. Biophys. Res. Commun.* **1997**, *233*, 349–353.

10. Daunert, S.; Deo, S. K.; Dikici, E.; Rowe, L. Aequorin and obelin mutants with differing wavelengths and bio-luminescence and their uses in screening inhibitors of HIV-1 protease. U.S. Pat. Appl. Publ. US 2005214776, 2005; *Chem. Abstr.* **2005**, *143*, 341493.

COELENTERAZINE f

CAS Registry Number 123437-16-1

Chemical Structure

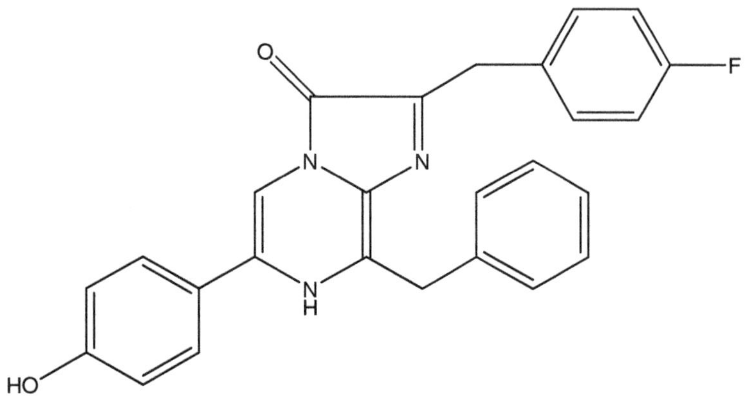

Melting Point >200°C

Boiling Point (Calcd.) 596.7 ± 60.0°C, pressure: 760 Torr

pK$_a$ (Calcd.) 10.48 ± 0.30, most acidic, temperature: 25°C; 6.49 ± 0.60, most basic, temperature: 25°C

CA Index Name Imidazo[1,2-*a*]pyrazin-3(7*H*)-one, 2-[(4-fluorophenyl)methyl]-6-(4-hydroxyphenyl)-8-(phenylmethyl)-

Other Names CLZN-f; Coelenterazine f

Merck Index Number Not listed

Chemical/Dye Class Heterocycle; Imidazo[1,2-*a*]pyrazin-3(7*H*)-one

Molecular Formula $C_{26}H_{20}FN_3O_2$

Molecular Weight 425.45

Physical Form Orange powder

Solubility Soluble in ethanol, methanol

Absorption (λ_{max}) 437 nm

Emission (λ_{max}) 472 nm

Synthesis Synthetic methods[1,2]

Staining Applications Calcium ions[2–9]

Biological Applications Calcium indicator;[2–9] assaying luminescent enzyme;[10] measuring luciferase activity;[11] as a substrate for luciferase;[12,13] screening HIV-1 protease inhibitors[14]

Industrial Applications Not reported

Safety/Toxicity No data available

REFERENCES

1. Isobe, M.; Fujii, T.; Kuse, M.; Miyamoto, K.; Koga, K. ^{19}F-Dehydrocoelenterazine as probe to investigate the active site of symplectin. *Tetrahedron* **2002**, *58*, 2117–2126.
2. Shimomura, O.; Musicki, B.; Kishi, Y. Semi-synthetic aequorins with improved sensitivity to calcium ions. *Biochem. J.* **1989**, *261*, 913–920.
3. Rowe, L.; Rothert, A.; Logue, C.; Ensor, C. M.; Deo, S. K.; Daunert, S. Spectral tuning of photoproteins by partnering site-directed mutagenesis strategies with the incorporation of chromophore analogs. *Protein Eng. Design Select.* **2008**, *21*, 73–81.
4. Inouye, S.; Sahara, Y. Expression, purification and characterization of a photoprotein, clytin, from *Clytia gregarium. Protein Expression Purif.* **2007**, *53*, 384–389.
5. Inouye, S.; Sasaki, S. Imidazole-assisted catalysis of lumine-scence reaction in blue fluorescent protein from the photoprotein aequorin. *Biochem. Biophys. Res. Commun.* **2007**, *354*, 650–655.
6. Shimomura, O. Cause of spectral variation in the luminescence of semisynthetic aequorins. *Biochem. J.* **1995**, *306*, 537–543.
7. Shimomura, O.; Kishi, Y.; Inouye, S. The relative rate of aequorin regeneration from apoaequorin and coelenterazine analogs. *Biochem. J.* **1993**, *296*, 549–551.
8. Shimomura, O.; Musicki, B.; Kishi, Y.; Inouye, S. Light-emitting properties of recombinant semi-synthetic aequorins and recombinant fluorescein-conjugated aequorin for measuring cellular calcium. *Cell Calcium* **1993**, *14*, 373–378.

9. Shimomura, O. Preparation and handling of aequorin solutions for the measurement of cellular Ca^{2+}. *Cell Calcium* **1991**, *12*, 635–643.
10. Inoue, S.; Sasaki, T. Method for assaying luminescent enzyme in presence of organic solvent. Jpn. Kokai Tokkyo Koho JP 2008000073, 2008; *Chem. Abstr.* **2008**, *148*, 115514.
11. Zhao, H.; Doyle, T. C.; Wong, R. J.; Cao, Y.; Stevenson, D. K.; Piwnica-Worms, D.; Contag, C. H. Characterization of coelen-terazine analogs for measurements of *Renilla luciferase* activity in live cells and living animals. *Mol. Imaging* **2004**, *3*, 43–54.
12. Inoue, S.; Sasaki, S. Fluorescent activity-possessing luciferase activity enhancement method. Jpn. Kokai Tokkyo Koho JP 2006271327, 2006; *Chem. Abstr.* **2006**, *145*, 413077.
13. Inouye, S.; Shimomura, O. The use of *Renilla luciferase*, *Oplophorus luciferase*, and apoaequorin as bioluminescent reporter protein in the presence of coelenterazine analogs as substrate. *Biochem. Biophys. Res. Commun.* **1997**, *233*, 349–353.
14. Daunert, S.; Deo, S. K.; Dikici, E.; Rowe, L. Aequorin and obelin mutants with differing wavelengths and bio-luminescence and their uses in screening inhibitors of HIV-1 protease. U.S. Pat. Appl. Publ. US 2005214776, 2005; *Chem. Abstr.* **2005**, *143*, 341493.

COELENTERAZINE h

CAS Registry Number 50909-86-9

Chemical Structure

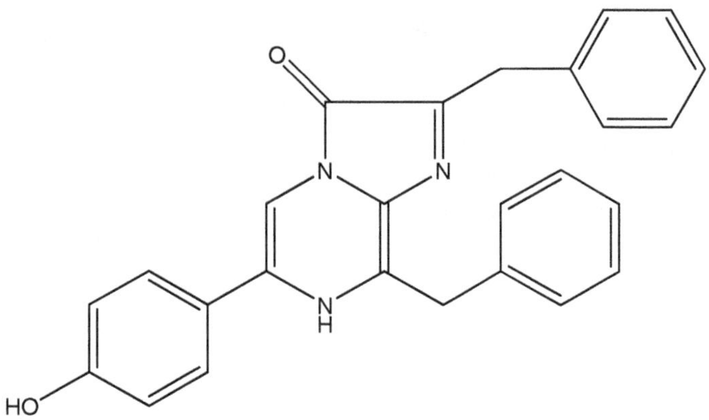

CA Index Name Imidazo[1,2-*a*]pyrazin-3(7*H*)-one, 6-(4-hydroxyphenyl)-2,8-bis(phenylmethyl)-

Other Names CLZN-h; 2-Deoxycoelenterazine; Coelenterazine h; Luciferin; Luciferin (Renilla); Renilla luciferin; h-Coelenterazine

Merck Index Number Not listed

Chemical/Dye Class Heterocycle; Imidazo[1,2-*a*]pyrazin-3(7*H*)-one

Molecular Formula $C_{26}H_{21}N_3O_2$

Molecular Weight 407.46

Physical Form Yellow powder

Solubility Soluble in methanol, ethanol

Melting Point >200°C

Boiling Point (Calcd.) 593.5 ± 60.0°C, pressure: 760 Torr

pK$_a$ (Calcd.) 10.48 ± 0.30, most acidic, temperature: 25°C; 6.15 ± 0.60, most basic, temperature: 25°C

Absorption (λ_{max}) 437 nm

Emission (λ_{max}) 466 nm

Synthesis Synthetic methods[1–4]

Staining Applications Calcium ions[3,5–14]

Biological Applications Calcium indicator;[3,5–14] assaying luminescent enzyme;[15] measuring luciferase activity;[16,17] as a substrate for luciferase;[18] screening HIV-1 protease inhibitors;[19] bioluminescence resonance energy transfer (BRET) detection system[20]

Industrial Applications Not reported

Safety/Toxicity No data available

REFERENCES

1. Wood, K.; Hawkins, E.; Scurria, M.; Klaubert, D. Compositions, methods and kits pertaining to luminescent compounds. PCT Int. Appl. WO 2003040100, 2003; *Chem. Abstr.* **2003**, *138*, 365147.
2. Qi, C. F.; Gomi, Y.; Hirano, T.; Ohashi, M.; Ohmiya, Y.; Tsuji, F. I. Chemi- and bio-luminescence of coelenterazine analogs with phenyl homologs at the C-2 position. *J. Chem. Soc., Perkin Trans. 1* **1992**, 1607–1611.
3. Shimomura, O.; Musicki, B.; Kishi, Y. Semi-synthetic aequorins with improved sensitivity to calcium ions. *Biochem. J.* **1989**, *261*, 913–920.
4. Charbonneau, H.; Cormier, M. J. Calcium ion-induced bioluminescence in *Renilla reniformis*: purification and characterization of a calcium-triggered luciferin-binding protein. *J. Biol. Chem.* **1979**, *254*, 769–780.
5. Rowe, L.; Rothert, A.; Logue, C.; Ensor, C. M.; Deo, S. K.; Daunert, S. Spectral tuning of photoproteins by partnering site-directed mutagenesis strategies with the incorporation of chromophore analogs. *Protein Eng., Design Select.* **2008**, *21*, 73–81.
6. Inouye, S.; Sahara, Y. Expression, purification and characterization of a photoprotein, clytin, from *Clytia gregarium. Protein Expression Purif.* **2007**, *53*, 384–389.
7. Inouye, S.; Sasaki, S. Imidazole-assisted catalysis of lumine-scence reaction in blue fluorescent protein from the photo-protein aequorin. *Biochem. Biophys. Res. Commun.* **2007**, *354*, 650–655.
8. Inouye, S.; Sasaki, S. Blue fluorescent protein from the calcium-sensitive photoprotein aequorin: catalytic properties for the oxidation of coelenterazine as an oxygenase. *FEBS Lett.* **2006**, *580*, 1977–1982.
9. Inouye, S. Fluorescent proteins for ligand marker and calcium determination. PCT Int. Appl. WO 2005014633, 2005; *Chem. Abstr.* **2005**, *142*, 236033.
10. Chemicon International Inc., USA. Engineering aequorin variants with improved bioluminescence

for high-throughput screening of calcium ion flux in cell. PCT Int. Appl. WO 2003082904, 2003; *Chem. Abstr.* **2003**, *139*, 287962.

11. Shimomura, O. Cause of spectral variation in the luminescence of semisynthetic aequorins. *Biochem. J.* **1995**, *306*, 537–543.
12. Shimomura, O.; Kishi, Y.; Inouye, S. The relative rate of aequorin regeneration from apoaequorin and coelenterazine analogs. *Biochem. J.* **1993**, *296*, 549–551.
13. Knight, M. R.; Read, N. D.; Campbell, A. K.; Trewavas, A. J. Imaging calcium dynamics in living plants using semi-synthetic recombinant aequorins. *J. Cell Biol.* **1993**, *121*, 83–90.
14. Kumar, S.; Harrylock, M.; Walsh, K. A.; Cormier, M. J.; Charbonneau, H. Amino acid sequence of the calcium-triggered luciferin binding protein of *Renilla reniformis*. *FEBS Lett.* **1990**, *268*, 287–290.
15. Inoue, S.; Sasaki, T. Method for assaying luminescent enzyme in presence of organic solvent. Jpn. Kokai Tokkyo Koho JP 2008000073, 2008; *Chem. Abstr.* **2008**, *148*, 115514.
16. Zhao, H.; Doyle, T. C.; Wong, R. J.; Cao, Y.; Stevenson, D. K.; Piwnica-Worms, D.; Contag, C. H. Characterization of coelenterazine analogs for measurements of *Renilla luciferase* activity in live cells and living animals. *Mol. Imaging* **2004**, *3*, 43–54.
17. Inoue, S.; Sasaki, S. Fluorescent activity-possessing luciferase activity enhancement method. Jpn. Kokai Tokkyo Koho JP 2006271327, 2006; *Chem. Abstr.* **2006**, *145*, 413077.
18. Inouye, S.; Shimomura, O. The use of *Renilla luciferase*, *Oplophorus luciferase*, and apoaequorin as bioluminescent reporter protein in the presence of coelenterazine analogs as substrate. *Biochem. Biophys. Res. Commun.* **1997**, *233*, 349–353.
19. Daunert, S.; Deo, S. K.; Dikici, E.; Rowe, L. Aequorin and obelin mutants with differing wavelengths and biolumine-scence and their uses in screening inhibitors of HIV-1 protease. U.S. Pat. Appl. Publ. US 2005214776, 2005; *Chem. Abstr.* **2005**, *143*, 341493.
20. Joly, E. A bioluminescence resonance energy transfer (BRET) system with broad spectral resolution between donor and acceptor emission wavelengths and its use. PCT Int. Appl. WO 2001046691, 2001; *Chem. Abstr.* **2001**, *135*, 89517.

COELENTERAZINE hcp

CAS Registry Number 123437-32-1

Chemical Structure

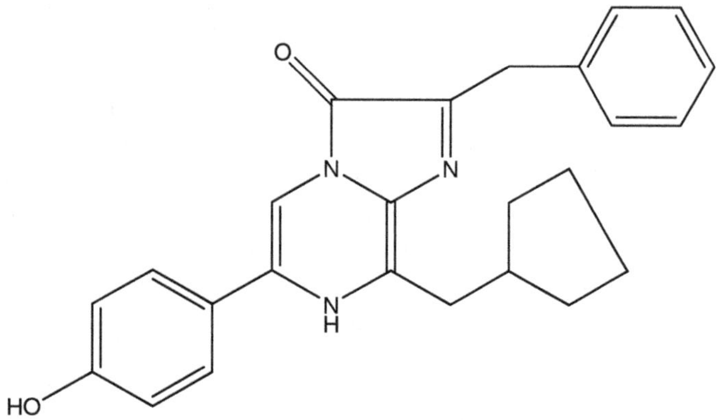

CA Index Name Imidazo[1,2-*a*]pyrazin-3(7*H*)-one, 8-(cyclopentylmethyl)-6-(4-hydroxyphenyl)-2-(phenylmethyl)-

Other Names CLZN-hcp; Coelenterazine hcp

Merck Index Number Not listed

Chemical/Dye Class Heterocycle; Imidazo[1,2-a]pyrazin-3(7*H*)-one

Molecular Formula $C_{25}H_{25}N_3O_2$

Molecular Weight 399.48

Physical Form Yellow powder

Solubility Soluble in ethanol, methanol

Melting Point >200°C

Boiling Point (Calcd.) 568.2 ± 60.0°C, pressure: 760 Torr

pKa (Calcd.) 10.48 ± 0.30, most acidic, temperature: 25°C; 6.17 ± 0.60, most basic, temperature: 25°C

Absorption (λ_{max}) 433 nm

Emission (λ_{max}) 445 nm

Synthesis Synthetic method[1]

Staining Applications Calcium ions[1–7]

Biological Applications Calcium indicator;[1–7] assaying luminescent enzyme;[8] measuring luciferase activity;[9] as a substrate for luciferase;[10] screening HIV-1 protease inhibitors;[11] bioluminescence resonance energy transfer (BRET) detection system[12]

Industrial Applications Not reported

Safety/Toxicity No data available

REFERENCES

1. Shimomura, O.; Musicki, B.; Kishi, Y. Semi-synthetic aequorins with improved sensitivity to calcium ions. *Biochem. J.* **1989**, *261*, 913–920.
2. Rowe, L.; Rothert, A.; Logue, C.; Ensor, C. M.; Deo, S. K.; Daunert, S. Spectral tuning of photoproteins by partnering site-directed mutagenesis strategies with the incorporation of chro-mophore analogs. *Protein Eng. Design Select.* **2008**, *21*, 73–81.
3. Inouye, S.; Sahara, Y. Expression, purification and char-acterization of a photoprotein, clytin, from *Clytia gregarium. Protein Expression Purif.* **2007**, *53*, 384–389.
4. Inouye, S.; Sasaki, S. Imidazole-assisted catalysis of lumine-scence reaction in blue fluorescent protein from the photo-protein aequorin. *Biochem. Biophys. Res. Commun.* **2007**, *354*, 650–655.
5. Shimomura, O. Cause of spectral variation in the luminescence of semisynthetic aequorins. *Biochem. J.* **1995**, *306*, 537–543.
6. Shimomura, O.; Kishi, Y.; Inouye, S. The relative rate of aequorin regeneration from apoaequorin and coelenterazine analogs. *Biochem. J.* **1993**, *296*, 549–551.
7. Shimomura, O.; Inouye, S.; Musicki, B.; Kishi, Y. Recom-binant aequorin and recombinant semisynthetic aequorins: cellular calcium ion indicators. *Biochem. J.* **1990**, *270*, 309–312.
8. Inoue, S.; Sasaki, T. Method for assaying luminescent enzyme in presence of organic solvent. Jpn. Kokai Tokkyo Koho JP 2008000073, 2008; *Chem. Abstr.* **2008**, *148*, 115514.
9. Zhao, H.; Doyle, T. C.; Wong, R. J.; Cao, Y.; Stevenson, D. K.; Piwnica-Worms, D.; Contag, C. H. Characterization of coelenterazine analogs for measurements of *Renilla luciferase* activity in live cells and living animals. *Mol. Imaging* **2004**, *3*, 43–54.
10. Inoue, S.; Sasaki, S. Fluorescent activity-possessing luciferase activity enhancement method. Jpn. Kokai

Tokkyo Koho JP 2006271327, 2006; *Chem. Abstr.* **2006**, *145*, 413077.

11. Daunert, S.; Deo, S. K.; Dikici, E.; Rowe, L. Aequorin and obelin mutants with differing wavelengths and biolumine-scence and their uses in screening inhibitors of HIV-1 protease. U.S. Pat. Appl. Publ. US 2005214776, 2005; *Chem. Abstr.* **2005**, *143*, 341493.
12. Joly, E. A bioluminescence resonance energy transfer (BRET) system with broad spectral resolution between donor and acceptor emission wavelengths and its use. PCT Int. Appl. WO 2001046691, 2001; *Chem. Abstr.* **2001**, *135*, 89517.

COELENTERAZINE n

CAS Registry Number 123437-22-9

Chemical Structure

Melting Point >200°C

Boiling Point (Calcd.) 663.4 ± 65.0°C, pressure: 760 Torr

pK_a (Calcd.) 10.48 ± 0.30, most acidic, temperature: 25°C; 6.52 ± 0.60, most basic, temperature: 25°C

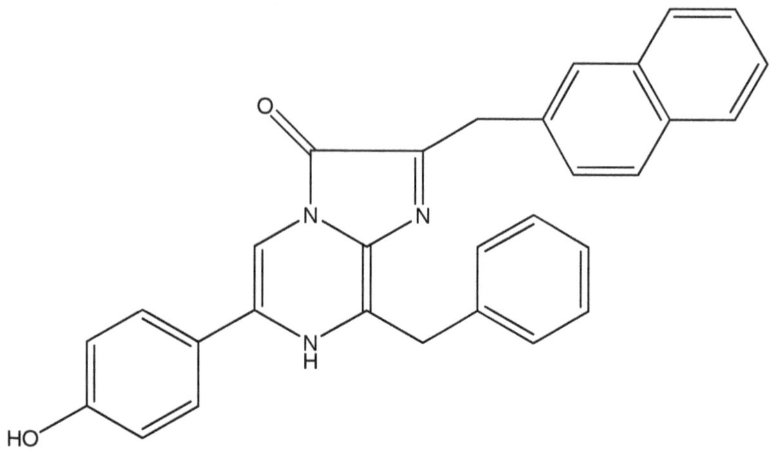

CA Index Name Imidazo[1,2-*a*]pyrazin-3(7*H*)-one, 6-(4-hydroxyphenyl)-2-(2-naphthalenylmethyl)-8-(phenylmethyl)-

Other Names CLZN-n; Coelenterazine n

Merck Index Number Not listed

Chemical/Dye Class Heterocycle; Imidazo[1,2-*a*]pyrazin-3(7*H*)-one

Molecular Formula $C_{30}H_{23}N_3O_2$

Molecular Weight 457.52

Physical Form Yellow powder

Solubility Soluble in ethanol, methanol

Absorption (λ_{max}) 431 nm

Emission (λ_{max}) 468 nm

Synthesis Synthetic method[1]

Staining Applications Calcium ions[1–7]

Biological Applications Calcium indicator;[1–7] assaying luminescent enzyme;[8] measuring luciferase activity;[9] as a substrate for luciferase;[10] screening HIV-1 protease inhibitors[11]

Industrial Applications Not reported

Safety/Toxicity No data available

REFERENCES

1. Shimomura, O.; Musicki, B.; Kishi, Y. Semi-synthetic aequorins with improved sensitivity to calcium ions. *Biochem. J.* **1989**, *261*, 913–920.
2. Rowe, L.; Rothert, A.; Logue, C.; Ensor, C. M.; Deo, S. K.; Daunert, S. Spectral tuning of photoproteins by partnering site-directed mutagenesis strategies with the incorporation of chromophore analogs. *Protein Eng. Design Select.* **2008**, *21*, 73–81.
3. Inouye, S.; Sahara, Y. Expression, purification and characterization of a photoprotein, clytin, from *Clytia gregarium*. *Protein Expression Purif.* **2007**, *53*, 384–389.
4. Inouye, S.; Sasaki, S. Imidazole-assisted catalysis of luminescence reaction in blue fluorescent protein from the photoprotein aequorin. *Biochem. Biophys. Res. Commun.* **2007**, *354*, 650–655.
5. Toma, S.; Chong, K. T.; Nakagawa, A.; Teranishi, K.; Inouye, S.; Shimomura, O. The crystal structures of semi-synthetic aequorins. *Protein Sci.* **2005**, *14*, 409–416.
6. Shimomura, O.; Kishi, Y.; Inouye, S. The relative rate of aequorin regeneration from apoaequorin and coelenterazine analogs. *Biochem. J.* **1993**, *296*, 549–551.

7. Shimomura, O.; Inouye, S.; Musicki, B.; Kishi, Y. Recombinant aequorin and recombinant semisynthetic aequorins: cellular calcium ion indicators. *Biochem. J.* **1990**, *270*, 309–312.

8. Inoue, S.; Sasaki, T. Method for assaying luminescent enzyme in presence of organic solvent. Jpn. Kokai Tokkyo Koho. JP 2008000073, 2008; *Chem. Abstr.* **2008**, *148*, 115514.

9. Zhao, H.; Doyle, T. C.; Wong, R. J.; Cao, Y.; Stevenson, D. K.; Piwnica-Worms, D.; Contag, C. H. Characterization of coelenterazine analogs for measurements of *Renilla luciferase* activity in live cells and living animals. *Mol. Imaging* **2004**, *3*, 43–54.

10. Inouye, S.; Shimomura, O. The use of *Renilla luciferase*, *Oplophorus luciferase*, and apoaequorin as bioluminescent reporter protein in the presence of coelenterazine analogs as substrate. *Biochem. Biophys. Res. Commun.* **1997**, *233*, 349–353.

11. Daunert, S.; Deo, S. K.; Dikici, E.; Rowe, L. Aequorin and obelin mutants with differing wavelengths and biolumine-scence and their uses in screening inhibitors of HIV-1 protease. U.S. Pat. Appl. Publ. US 2005214776, 2005; *Chem. Abstr.* **2005**, *143*, 341493.

CONGO RED

CAS Registry Number 573-58-0

Chemical Structure

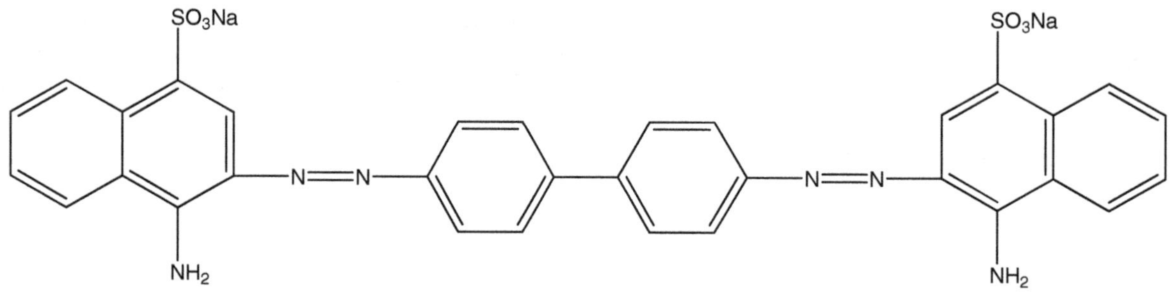

Molecular Weight 696.66

Physical Form Brownish-red powder

Solubility Soluble in water, ethanol; very slightly soluble in acetone; practically insoluble in ether, xylene

CA Index Name 1-Naphthalenesulfonic acid, 3,3′-[[1,1′-biphenyl]-4,4′-diylbis(2,1-diazenediyl)]bis[4-amino-, sodium salt (1 : 2)

Other Names 3,3′-[[1,1′-Biphenyl]-4,4′-diylbis-(azo)] bis[4-amino-1-naphthalenesulfonic acid] disodium salt; 1-Naphthalenesulfonic acid, 3,3′-[[1,1′-biphenyl]-4,4′-diylbis(azo)]bis[4-amino-, disodium salt; C.I. Direct Red 28, disodium salt; Atlantic Congo Red; Atul Congo Red; Azocard Red Congo; Benzo Congo Red; Brasilamina Congo 4B; Cerven Kongo; Cerven Prima 28; C.I. 22120; C.I. Direct Red 28; Congo Red 4B; Congo Red 4BX; Congo Red CR; Congo Red H; Congo Red K; Congo Red L; Congo Red M; Congo Red N; Congo Red R; Congo Red RS; Congo Red TS; Congo Red W; Congo Red WS; Congo Red sodium salt; Congo red; Cotton Red 4BC; Cotton Red 5B; Cotton Red L; Congorot; Diacotton Congo Red; Direct Congo Red; Direct Red 28; Direct Red C; Direct Red DC-CF; Direct Red K; Erie Congo 4B; Haemomedical; Haemonorm; Hemorrhagyl; Hispamin Congo 4B; Kayaku Congo Red; Kongorot; Mitsui Congo Red; Peeramine Congo Red; Red K; Sodium diphenyl-diazo-bis-α-naphthylaminesulfonate; Solucongo; Sugai Congo Red; Tertrodirect Red C; Trisulfon Congo Red; Vondacel Red CL

Merck Index Number 2498

Chemical/Dye Class Azo

Molecular Formula $C_{32}H_{22}N_6Na_2O_6S_2$

Melting Point >360°C

pH Range 3.0–5.0

Color Change at pH Blue (3.0) to red (5.0)

pK$_a$ 4.1

Absorption (λ_{max}) 497 nm, 488 nm

Synthesis Synthetic methods[1–9]

Staining Applications Amyloid-β (Aβ) protein;[1,10–18] bacteria;[19] carious tissue;[20] collagen;[21] fungi;[22] fungal cell wall mutants;[23] liposome;[24] polyglutamine protein oligomers;[25] prion;[26] skin;[27] α-synuclein;[28] tissue section[29]

Biological Applications Detecting bacteria,[30] protein folding disorders;[31] treating dermatological disorders,[32] neurodegenerative diseases,[1,33] Alzheimer's disease[1,34]

Industrial Applications Display devices;[1,35] optical films;[35] alignment layers;[36] fiber-optic sensors;[1,37] optical waveguides;[1,38] highlighters;[39] textiles[40]

Safety/Toxicity Acute toxicity;[1,41] algal toxicity;[42] bacterial toxicity;[42] protozoan toxicity;[42] carcinogenicity;[1,43] cutaneous toxicity;[1,44] cytotoxicity;[1,45,46] environmental toxicity;[47] genotoxicity;[1,48] hematotoxicity;[49] microbial toxicity;[50] mutagenicity;[1,51–54] neurotoxicity;[1,55] yeast toxicity[56]

Certification/Approval Certified by Biological Stain Commission (BSC)

REFERENCES

1. Sabnis, R. W. *Handbook of Acid-Base Indicators*; CRC Press: Boca Raton, **2008**; pp 91–92.
2. Rudyk, H.; Knaggs, M. H.; Vasiljevic, S.; Hope, J.; Birkett, C.; Gilbert, I. H. Synthesis and evaluation of analogues of congo red as potential compounds against transmissible spongiform encephalopathies. *Eur. J. Med. Chem.* **2003**, *38*, 567–579.
3. Welcher, F. J. *Organic Analytical Reagents*; D. Van Nostrand Company: New York, **1948**; Vol. *4*, pp 481–482.
4. Martin, C. F.; Bang, H. Synthesis of tetraiodinated Congo red. *J. Am. Pharm. Assoc., Sci. Ed.* **1948**, *37*, 102.
5. Parks, L. R.; Keller, M. P. The purification and properties of Congo red. *Am. Dyestuff Rep.* **1934**, *23*, 445–448.
6. Whitehead, C. Azo dyestuff technology. VII. Manufacture of Congo red. *Chem. Trade J. Chem. Eng.* **1925**, *77*, 386; *Chem. Abstr.* **1926**, *20*, 17010.
7. Vaubel, W. Commercial Congo colors. *Chem.-Ztg.* **1924**, *48*, 165; *Chem. Abstr.* **1924**, *18*, 25674.
8. Shaposhnikov, V. G.; Bogoyavlenskii, T. Congo Red. *Z. Farben-Ind.* **1913**, *11*, 301–305, 321–323, 337–340, 353–357; *Chem. Abstr.* **1913**, *7*, 8788.
9. Salm, E. Z. A Study of indicators. *Physik. Chem.* **1906**, *57*, 471–501; *Chem. Abstr.* **1907**, *1*, 2788.
10. Howie, A. J.; Brewer, D. B.; Howell, D.; Jones, A. P. Physical basis of colors seen in congo red-stained amyloid in polarized light. *Lab. Invest.* **2008**, *88*, 232–242.
11. Furumoto, S.; Okamura, N.; Iwata, R.; Yanai, K.; Arai, H.; Kudo, Y. *Curr. Top. Med. Chem.* **2007**, *7*, 1773–1789.
12. Maezawa, I.; Hong, H. S.; Liu, R.; Wu, C. Y.; Cheng, R. H.; Kung, M. P.; Kung, H. F.; Lam, K. S.; Oddo, S.; LaFerla, F. M.; Jin, L. W. Congo red and thioflavin-T analogs detect Aβ oligomers. *J. Neurochem.* **2008**, *104*, 457–468.
13. Bartolini, M.; Bertucci, C.; Bolognesi, M. L.; Cavalli, A.; Melchiorre, C.; Andrisano, V. Insight into the kinetic of amyloid β (1-42) peptide self-aggregation: elucidation of inhibitors' mechanism of action. *ChemBioChem* **2007**, *8*, 2152–2161.
14. Chander, H.; Chauhan, A.; Chauhan, V. Binding of proteases to fibrillar amyloid-β protein and its inhibition by congo red. *J. Alzheimer's Dis.* **2007**, *12*, 261–269.
15. Frid, P.; Anisimov, S. V.; Popovic, N. Congo red and protein aggregation in neurodegenerative diseases. *Brain Res. Rev.* **2007**, *53*, 135–160.
16. Hirakura, Y.; Lin, M. C.; Kagan, B. L. Alzheimer amyloid abeta1-42 channels: effects of solvent, pH, and Congo Red. *J. Neurosci. Res.* **1999**, *57*, 458–466.
17. Li, Y. P.; Bushnell, A. F.; Lee, C. M.; Perlmutter, L. S.; Wong, S. K. Beta-amyloid induces apoptosis in human-derived neurotypic SH-SY5Y cells. *Brain Res.* **1996**, *738*, 196–204.
18. Clark, G.; Dodds, H. M. Alkaline Congo red staining for amyloid. *J. Histotechnol.* **1982**, *5*, 167–168.
19. Azizbekyan, R. R.; Smirnova, T. A.; Grigor'eva, T. M.; Kuznetsova, N. I.; Nikolaenko, M. A. Use of Congo red dye for selecting Gram-positive bacteria that produce regular S-layers. *Biotekhnologiya* **2007**, 5–11; *Chem. Abstr.* **2008**, *149*, 98467.
20. Guggenberger, R.; Luchterhandt, T.; Haeberlein, I.; Kappler, O.; Hansen, M. Dental composition for detection of carious tissue, detection method. PCT Int. Appl. WO 2006125650, 2006; *Chem. Abstr.* **2006**, *146*, 33158.
21. Levenson, R.; Hoyt, C. C. Sample imaging and classification using collagen staining for diagnosis. U.S. Pat. Appl. Publ. US 2008074644, 2008; *Chem. Abstr.* **2008**, *148*, 373653.
22. Onkarappa, R.; Seema, Patel, J.; Shobha K. S. Biosorption of the textile dye congo red by fungi. *J. Ecotoxicol. Environ. Monit.* **2006**, *16*, 515–519.
23. Ram, A. F. J.; Klis, F. M. Identification of fungal cell wall mutants using susceptibility assays based on Calcofluor white and Congo red. *Nat. Protoc.* **2006**, *1*, 2253–2256.
24. Steinberg, Y.; Schroeder, A.; Talmon, Y.; Schmidt, J.; Khalfin, R. L.; Cohen, Y.; Devoisselle, J. M.; Begu, S.; Avnir, D. Triggered release of aqueous content from liposome-derived sol–gel nanocapsules. *Langmuir* **2007**, *23*, 12024–12031.
25. Takahashi, Y.; Okamoto, Y.; Popiel, H. A.; Fujikake, N.; Toda, T.; Kinjo, M.; Nagai, Y. Detection of polyglutamine protein oligomers in cells by fluorescence correlation spectroscopy. *J. Biol. Chem.* **2007**, *282*, 24039–24048.
26. Sigurdson, C. J.; Nilsson, K. P. R.; Hornemann, S.; Manco, G.; Polymenidou, M.; Schwarz, P.; Leclerc, M.; Hammarstroem, P.; Wuethrich, K.; Aguzzi, A. Prion strain discrimination using luminescent conjugated polymers. *Nat. Methods* **2007**, *4*, 1023–1030.

27. Fujiwara, N. Skin evaluation method using Congo red. Jpn. Kokai Tokkyo Koho JP 2007263655, 2007; *Chem. Abstr.* **2007**, *147*, 422443.

28. Rao, J. N.; Dua, V.; Ulmer, T. S. Characterization of α-synuclein interactions with selected aggregation-inhibiting small molecules. *Biochemistry* **2008**, *47*, 4651–4656.

29. Farrell, M.; Bieniarz, C.; Reinhardt, K.; Ward, G.; Kosmeder, J. W.; Ghusson, A.; Walk, E. E.; Manriquez, G. H.; Grogan, T. M. Biological sample processing composition and method for staining the sample and contacting the sample with a lipid compound composition. PCT Int. Appl. WO 2007084429, 2007; *Chem. Abstr.* **2007**, *147*, 184152.

30. Andreev, V. M.; Gonikberg, E. M.; Kuznetsova, N. V. Application of the complex of DNA with the congo red anionic diazo dye for detection of nuclease-producing colonies of marine bacteria. *Microbiology* **2007**, *76*, 585–589.

31. Estrada, L.; Soto, C. Methods and compositions for the detection of protein folding disorders. PCT Int. Appl. WO 2008030973, 2008; *Chem. Abstr.* **2008**, *148*, 302851.

32. Pellet, A. Exfoliant composition containing a cellulosic polymer and its applications in cosmetic or topical dermatological treatments. Fr. Demande FR 2902322, 2007; *Chem. Abstr.* **2007**, *148*, 61286.

33. Stockwell, B. R.; Hoffstrom, B.; Varma, H. Agents for treating neurodegenerative diseases. U.S. Pat. Appl. Publ. US 2007149543, 2007; *Chem. Abstr.* **2007**, *147*, 110249.

34. Yang, J.; Inbar, P. Compounds and method for the diagnosis and treatment of amyloid associated diseases. PCT Int. Appl. WO 2007011834, 2007; *Chem. Abstr.* **2007**, *146*, 177223.

35. Iwakawa, Y.; Nakano, S. Polarizer, its manufacture by gelation, optical film, and display device. Jpn. Kokai Tokkyo Koho JP 2007093701, 2007; *Chem. Abstr.* **2007**, *146*, 390948.

36. Nishiyama, I.; Kawamura, S.; Kuwana, Y.; Hajime, S. K. Compositions for preparation of optical alignment layers and optical anisotropic materials. Jpn. Kokai Tokkyo Koho JP 2008076839, 2008; *Chem. Abstr.* **2008**, *148*, 413937.

37. Ganesh, A. B.; Radhakrishnan, T. K. Fiber-optic sensors for the estimation of pH within natural biofilms on metals. *Sens. Actuators B* **2007**, *B123* 1107–1112.

38. Pirasteh, P.; Charrier, J.; Dumeige, Y.; Chaillou, A.; Guendouz, M.; Haji, L. Study of porous silicon optical waveguides impregnated with organic dyes. *Appl. Surf. Sci.* **2007**, *253*, 3440–3443.

39. Kwan, W. S. V.; Duong, C. Highlighting marking compositions, highlighting kits, highlighted complexes, and application of eradicator. U.S. Pat. Appl. Publ. US 2007017413, 2007; *Chem. Abstr.* **2007**, *146*, 164696.

40. El-Gendy, E. H.; Ali, N. M.; Marie, M. M.; El-Shanshoury, I. A. Dyeing kinetics of radiation-grafted polyester fabric using different dyes. *Indian J. Fibre Text. Res.* **2007**, *32*, 232–240.

41. Douglas, M. T.; Chanter, D. O.; Pell, I. B.; Burney, G. M. A proposal for the reduction of animal numbers required for the acute toxicity to fish test (LC_{50} determination). *Aquat. Toxicol.* **1986**, *8*, 243–249.

42. Novotny, C.; Dias, N.; Kapanen, A.; Malachova, K.; Vandrovcova, M.; Itaevaara, M.; Lima, N. Comparative use of bacterial, algal and protozoan tests to study toxicity of azo and anthraquinone dyes. *Chemosphere* **2006**, *63*, 1436–1442.

43. Kauppinen, T.; Pukkala, E.; Saalo, A.; Sasco, A. J. Exposure to chemical carcinogens and risk of cancer among Finnish laboratory workers. *Am. J. Ind. Med.* **2003**, *44*, 343–350.

44. Baynes, R. E.; Monteiro-Riviere, N. A.; Qiao, G. L.; Riviere, J. E. Cutaneous toxicity of the benzidine dye direct red 28 applied as mechanistically-defined chemical mixtures (MDCM) in perfused porcine skin. *Toxicol. Lett.* **1997**, *93*, 159–169.

45. Giunta, S.; Galeazzi, R.; Marcellini, M.; Corder, E. H.; Galeazzi, L. The inflammation-sensitive protein alpha 1-anti-chymotrypsin neutralizes fibrillar aggregation and cytotoxicity of the beta-amyloid peptide more effectively than alpha 1-antitrypsin. *Clin. Biochem.* **2007**, *40*, 887–892.

46. Giunta, S.; Galeazzi, R.; Valli, M. B.; Corder, E. H.; Galeazzi, L. Transferrin neutralization of amyloid beta 25-35 cytotoxicity. *Clin. Chim. Acta* **2004**, *350*, 129–136.

47. Tonogai, Y.; Ogawa, S.; Ito, Y.; Iwaida, M. Actual survey on TLm (median tolerance limit) values of environmental pollutants, especially on amines, nitriles, aromatic nitrogen compounds and artificial dyes. *J. Toxicol. Sci.* **1982**, *7*, 193–203.

48. Choudhary, E.; Capalash, N.; Sharma, P. Genotoxicity of degradation products of textile dyes evaluated with rec-assay after photo Fenton and ligninase treatment. *J. Environ. Pathol. Toxicol. Oncol.* **2004**, *23*, 279–285.

49. Sharma, M. L.; Goel, K. A.; Awasthi, A. K.; Tyagi, S. K. Haematological and biochemical characteristics of *Heter-opneustes fossilis* under the stress of Congo Red (diphenyl disazo binaphthionic acid). *Toxicol. Lett.* **1982**, *14*, 237–241.

50. Oros, G.; Cserhati, T.; Forgacs, E. Separation of the strength and selectivity of the microbiological effect of synthetic dyes by spectral mapping technique. *Chemosphere* **2003**, *52*, 185–193.

51. Kaur, A.; Sandhu, R. S.; Grover, I. S. Screening of azo dyes for mutagenicity with Ames/*Salmonella* assay. *Environ. Mol. Mutagen.* **1993**, *22*, 188–190.

52. De France, B. F.; Carter, M. H.; Josephy, P. D. Comparative metabolism and mutagenicity of azo and hydrazone dyes in the Ames test. *Food Chem. Toxicol.* **1986**, *24*, 165–169.

53. Brown, J. P.; Dietrich, P. S. Mutagenicity of selected sulfonated azo dyes in the *Salmonella*/microsome assay: use of aerobic and anaerobic activation procedures. *Mutat. Res.* **1983**, *116*, 305–315.

54. Robertson, J. A.; Harris, W. J.; McGregor, D. B. Mutagenicity of azo dyes in the *Salmonella*/activation test. *Carcinogenesis* **1982**, *3*, 21–25.

55. Lorenzo, A.; Yankner, B. A. Beta-amyloid neurotoxicity requires fibril formation and is inhibited by congo red. *Proc. Natl. Acad. Sci. U.S.A.* **1994**, *91*, 12243–12247.

56. Kojima, H.; Hashimoto, H.; Yoda, K. Interaction among the subunits of Golgi membrane mannosyltransferase complexes of the yeast *Saccharomyces cerevisiae*. *Biosci., Biotechnol., Biochem.* **1999**, *63*, 1970–1976.

CoroNa GREEN

CAS Registry Number 690993-66-9

Chemical Structure

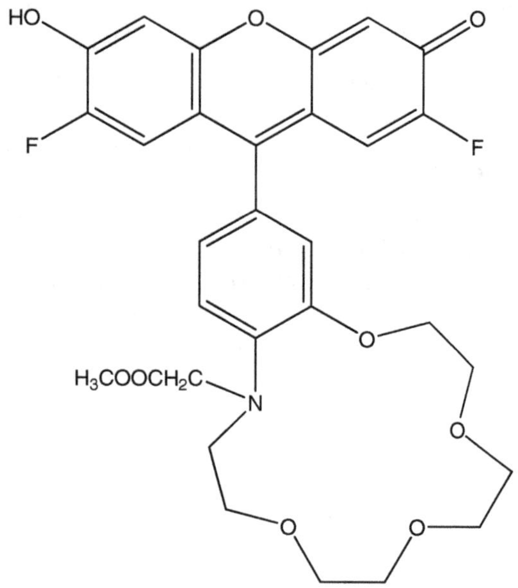

CA Index Name 13*H*-1,4,7,10,13-Benzotetraoxaazacyclopentadecine-13-acetic acid, 16-(2,7-difluoro-6-hydroxy-3-oxo-3*H*-xanthen-9-yl)-2,3,5,6,8,9,11,12-octahydro-, methyl ester

Other Names CoroNA Green; CoroNa Green Sodium Indicator

Merck Index Number Not listed

Chemical/Dye Class Xanthene

Molecular Formula $C_{30}H_{29}F_2NO_9$

Molecular Weight 585.55

Physical Form Orange powder

Solubility Soluble in water

Melting Point >200°C

Boiling Point (Calcd.) 777.7 ± 60.0 °C, pressure: 760 Torr

pK_a (Calcd.) 7.89 ± 0.60, most acidic, temperature: 25°C; 2.92 ± 0.20, most basic, temperature: 25°C

Absorption (λ_{max}) 492 nm

Emission (λ_{max}) 516 nm

Synthesis Synthetic methods[1–3]

Staining Applications Sodium ions;[1–8] neurons;[4] dendrites[4]

Biological Applications Sodium indicator[1–8]

Industrial Applications Not reported

Safety/Toxicity No data available

REFERENCES

1. Martin, V. V.; Rothe, A.; Gee, K. R. Fluorescent metal ion indicators based on benzoannelated crown systems: a green fluorescent indicator for intracellular sodium ions. *Bioorg. Med. Chem. Lett.* **2005**, *15*, 1851–1855.
2. Gee, K.; Martin, V. Preparation of crown ether derivatives as metal chelating agents. PCT Int. Appl. WO 2005016874, 2005; *Chem. Abstr.* **2005**, *142*, 261567.
3. Martin, V. V.; Gee, K. Metal-complexing crown ether fluorescent indicators and their use with biological systems. U.S. Pat. Appl. Publ. US 2004096978, 2004; *Chem. Abstr.* **2004**, *140*, 424969.
4. Meier, S. D.; Kovalchuk, Y.; Rose, C. R. Properties of the new fluorescent Na^+ indicator CoroNa Green: comparison with SBFI and confocal Na^+ imaging. *J. Neurosci. Methods* **2006**, *155*, 251–259.
5. Piper, J. D.; Li, C.; Lo, C. J.; Berry, R.; Korchev, Y.; Ying, L.; Klenerman, D. Characterization and application of con-trollable local chemical changes produced by reagent delivery from a nanopipette. *J. Am. Chem. Soc.* **2008**, *130*, 10386–10393.
6. Zeilinger, C. Process and apparatus for ion channel characterization. PCT Int. Appl. WO 2008022658, 2008; *Chem. Abstr.* **2008**, *148*, 302860.
7. Piper, J. D.; Clarke, R. W.; Korchev, Y. E.; Ying, L.; Klenerman, D. A Renewable nanosensor based on a glass nanopipette. *J. Am. Chem. Soc.* **2006**, *128*, 16462–16463.
8. Ebneth, A.; Netzer, R.; Hahn, U. Method for examining the activity of ion channels. PCT Int. Appl. WO 2005069008, 2005; *Chem. Abstr.* **2005**, *143*, 149507.

CoroNa GREEN AM

CAS Registry Number 690993-67-0

Chemical Structure

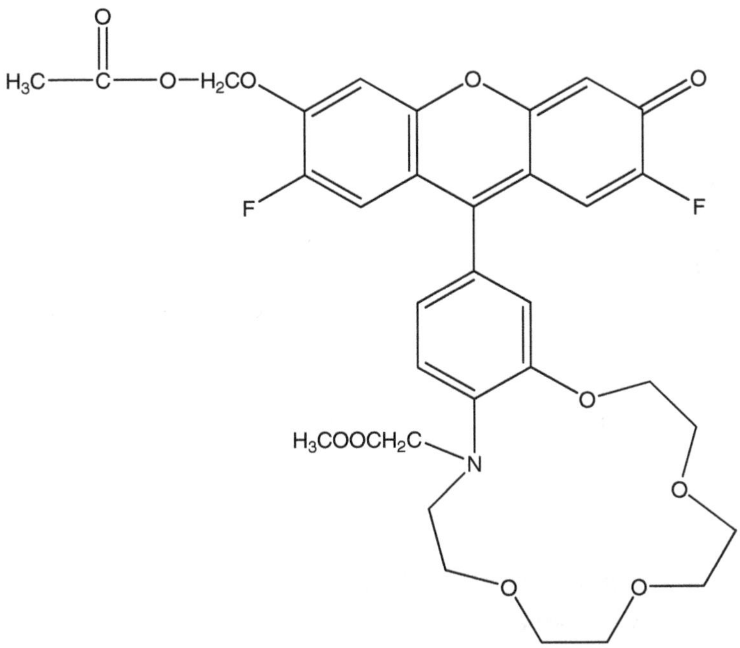

CA Index Name 13*H*-1,4,7,10,13-Benzotetraoxaaza-cyclopentadecine-13-acetic acid, 16-[6-[(acetyloxy)methoxy]-2,7-difluoro-3-oxo-3*H*-xanthen-9-yl]-2,3,5,6,8,9,11,12-octahydro-, methyl ester

Other Names CoroNA Green AM; CoroNa Green acetoxymethyl ester

Merck Index Number Not listed

Chemical/Dye Class Xanthene

Molecular Formula $C_{33}H_{33}F_2NO_{11}$

Molecular Weight 657.61

Physical Form Orange powder

Solubility Soluble in dimethyl sulfoxide

Melting Point >200°C

Boiling Point (Calcd.) 821.4 ± 65.0°C, pressure: 760 Torr

pK_a (Calcd.) 2.85 ± 0.20 most basic, temperature: 25°C

Absorption (λ_{max}) 454 nm

Emission (λ_{max}) 516 nm

Synthesis Synthetic methods[1–3]

Staining Applications Sodium ions[1–3]

Biological Applications Sodium indicators[1–3]

Industrial Applications Not reported

Safety/Toxicity No data available

REFERENCES

1. Martin, V. V.; Rothe, A.; Gee, K. R. Fluorescent metal ion indicators based on benzoannelated crown systems: a green fluorescent indicator for intracellular sodium ions. *Bioorg. Med. Chem. Lett.* **2005**, *15*, 1851–1855.
2. Gee, K.; Martin, V. Preparation of crown ether derivatives as metal chelating agents. PCT Int. Appl. WO 2005016874, 2005; *Chem. Abstr.* **2005**, *142*, 261567.
3. Martin, V.V.; Gee, K. Metal-complexing crown ether fluore-scent indicators and their use with biological systems. U.S. Pat. Appl. Publ. US 2004096978, 2004; *Chem. Abstr.* **2004**, *140*, 424969.

CoroNa RED

CAS Registry Number 481667-01-0
Chemical Structure

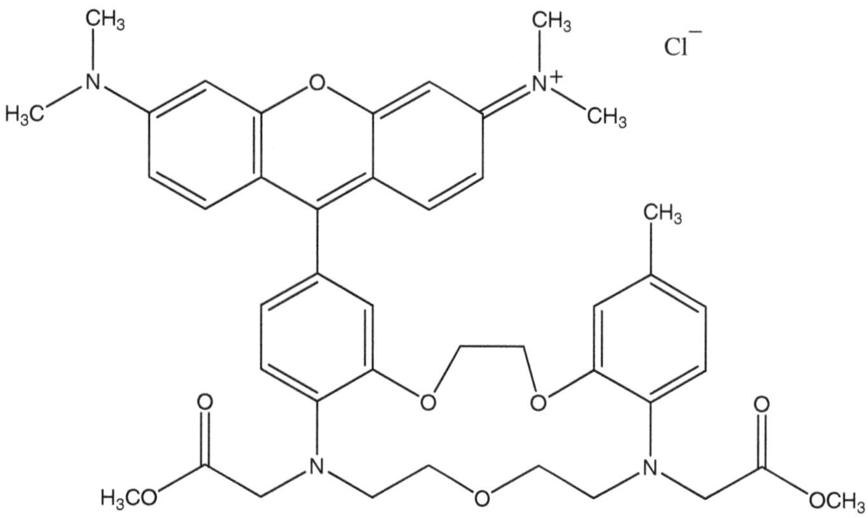

Molecular Formula $C_{42}H_{49}ClN_4O_8$
Molecular Weight 773.32
Physical Form Crimson powder
Solubility Soluble in water, dimethyl sulfoxide

CA Index Name Xanthylium, 3,6-bis(dimethylamino)-9-[6,7,10,11,17,18-hexahydro-5,11-bis(2-methoxy-2-oxoethyl)-14-methyl-5*H*,9*H*-dibenzo[*e*,*n*][1,4,10,7,13]trioxadiazacyclopentadecin-2-yl]-, chloride (1:1)

Other Names Xanthylium, 3,6-bis(dimethylamino)-9-[6,7,10,11,17,18-hexahydro-5,11-bis(2-methoxy-2-oxoethyl)-14-methyl-5*H*,9*H*-dibenzo[*e*,*n*][1,4,10,7,13]trioxadiazacyclopentadecin-2-yl]-, chloride; CoroNa Red; CoroNa Red Sodium Indicator; CoroNa Red chloride

Merck Index Number Not listed

Chemical/Dye Class Xanthene

Melting Point >200°C

Absorption (λ_{max}) 547 nm

Emission (λ_{max}) 570 nm

Synthesis Synthetic methods[1–4]

Staining Applications Sodium ions[1–9]

Biological Applications Sodium indicator;[1–9] detecting prostate cancer[10]

Industrial Applications Not reported

Safety/Toxicity No data available

REFERENCES

1. Gee, K.; Martin, V. Preparation of crown ether derivatives as metal chelating agents. PCT Int. Appl. WO 2005016874, 2005; *Chem. Abstr.* **2005**, *142*, 261567.
2. Martin, V. V.; Rothe, A.; Diwu, Z.; Gee, K. R. Fluorescent sodium ion indicators based on the 1,7-diaza-15-crown-5 system. *Bioorg. Med. Chem. Lett.* **2004**, *14*, 5313–5316.
3. Martin, V. V.; Gee, K. Metal-complexing crown ether fluorescent indicators and their use with biological systems. U.S. Pat. Appl. Publ. US 2004096978, 2004; *Chem. Abstr.* **2004**, *140*, 424969.
4. Martin, V. V.; Gee, K. R.; Haugland, R. P.; Diwu, Z. Metal-chelating crown ether derivatives with an attached dye, reactive group or conjugated substance, their production and their use. Brit. UK Pat. Appl. GB 2372749, 2002; *Chem. Abstr.* **2002**, *138*, 74702.
5. Zeilinger, C. Process and apparatus for ion channel characterization. PCT Int. Appl. WO 2008022658, 2008; *Chem. Abstr.* **2008**, *148*, 302860.
6. Ebneth, A.; Netzer, R.; Hahn, U. Method for examining the activity of ion channels. PCT Int.

Appl. WO 2005069008, 2005; *Chem. Abstr.* **2005**, *143*, 149507.

7. Baron, S.; Caplanusi, A.; van de Ven, M.; Radu, M.; Despa, S.; Lambrichts, I.; Ameloot, M.; Steels, P.; Smets, I. Role of mitochondrial Na^+ concentration, measured by CoroNa red, in the protection of metabolically inhibited MDCK cells. *J. Am. Soc. Nephrol.* **2005**, *16*, 3490–3497.
8. Van Oers, M.; Tareilus, E. W.; Vlak, J. M. Transgenic cells expressing genes for α, β, and γ subunits of human epithelial sodium channels and their use in screening for effectors and flavor enhancers. Brit. UK Pat. Appl. GB 2396414, 2004; *Chem. Abstr.* **2004**, *141*, 68883.
9. Bos, J. W.; Gouka, R. J.; Tareilus, E. W. Expression system for the subunits of the epithelial sodium channel, for screening modulator of salty taste. PCT Int. Appl. WO 2008009565, 2008; *Chem. Abstr.* **2008**, *148*, 184668.
10. Dickman, D. Methods of detecting prostate cancer. PCT Int. Appl. WO 2006054296, 2006; *Chem. Abstr.* **2006**, *145*, 3346.

CRESYL VIOLET ACETATE

CAS Registry Number 10510-54-0

Chemical Structure

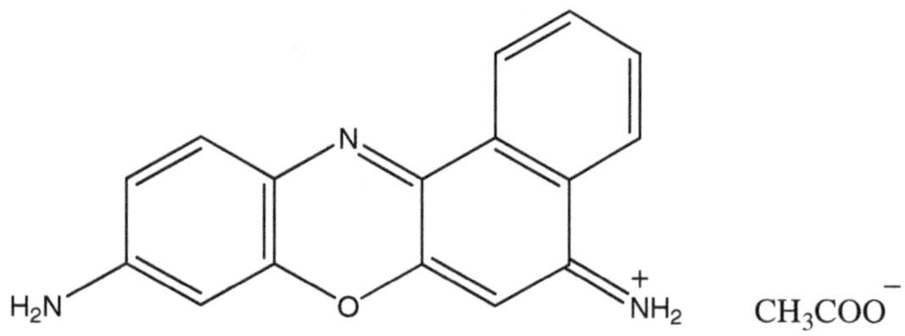

CA Index Name Benzo[*a*]phenoxazin-7-ium, 5,9-diamino-, acetate (1:1)

Other Names 9-Amino-5-imino-5*H*-benzo(*a*)phenoxazine acetate salt; 5*H*-Benzo[*a*]phenoxazin-9-amine, 5-imino-, monoacetate; 5*H*-Benzo[*a*]phenoxazine, 9-amino-5-imino-, acetate; Benzo[*a*]phenoxazin-7-ium, 5,9-diamino-, acetate; 5,9-Diaminobenzo[*a*]phenoxazin-7-ium acetate; Cresyl violet acetate

Merck Index Number Not listed

Chemical/Dye Class Phenoxazine

Molecular Formula $C_{18}H_{15}N_3O_3$

Molecular Weight 321.33

Physical Form Dark green powder

Solubility Soluble in water, ethanol

Melting Point 140–143°C

Absorption (λ_{max}) 596 nm, 601 nm

Emission (λ_{max}) 630 nm

Synthesis Synthetic method[1,2]

Staining Applications Brain;[3–6] spinal cord;[6] sperms;[7] tissues;[8,9] white blood cells[10]

Biological Applications Biochemical and clinical analysis;[11] detecting cellular proteins;[12] treating cardiovascular diseases[13]

Industrial Applications Dye lasers;[14,15] optical recording materials[16]

Safety/Toxicity No data available

Certification/Approval Certified by Biological Stain Commission (BSC)

REFERENCES

1. Fong, D. W.; Hoots, J. E. Fluorescent group-tagged acrylic polymers and their synthesis by post-polymerization (trans)amidation. Eur. Pat. Appl. EP 475602, 1992; *Chem. Abstr.* **1992**, *117*, 27540.
2. Green, F. J. Polychromasia of Cresyl Violet dyes; a chemical explanation of variations in pre-World War II products. *Stain Technol.* **1966**, *41*, 223–228.
3. Churukian, C. J.; Rubio, A.; Lapham, L. W. A simple colloidal silver method (autometallographic technique) for demons-trating inorganic mercury in brain sections. *J. Histotechnol.* **2000**, *23*, 337–339.
4. Alvarez-Buylla, A.; Ling, C. Y.; Kirn, J. R. Cresyl violet: a red fluorescent Nissl stain. *J. Neurosci. Methods* **1990**, *33*, 129–133.
5. Barden, H. Acid fast staining of oxidized neuromelanin and lipofuscin in the human brain. *J. Neuropathol. Exp. Neurol.* **1979**, *38*, 453–462.
6. Augulis, V.; Sigg, E. B. Supravital staining and fixation of brain and spinal cord by intravascular perfusion. *Stain Technol.* **1971**, *46*, 183–190.
7. Carstensen, C. A. Differential-diagnostic sperm determination. Ger. Offen. DE 2651060, 1978; *Chem. Abstr.* **1978**, *89*, 56045.
8. Goldrick, M.; Gonzales, J. C. Methods and compositions for preparing tissue samples for RNA extraction using aqueous alcoholic solutions. U.S. Pat. Appl. Publ. US 2004253661, 2004; *Chem. Abstr.* **2004**, *142*, 34853.
9. Schmitz-Moormann, P. Tissue staining by basic dyes. I. Influence of the pH of the staining solution and of the dye-affinity on the adsorption of the dye. *Histochemie* **1968**, *16*, 23–35.
10. Lindberg, S.; Svensson, J. Enumeration of white blood cells by staining and imaging. Eur. Pat. Appl. EP 1701150, 2006; *Chem. Abstr.* **2006**, *145*, 288168.
11. Theodoropulos, S. Preparation of novel oxazine ureas and thiazine urea chromophors as fluorescent labels for bio-chemical and clinical analysis. U.S. Patent 4714763, 1987; *Chem. Abstr.* **1988**, *108*, 112473.
12. Nithipatikom, K.; McCoy, M. J.; Hawi, S. R.; Nakamoto, K.; Adar, F.; Campbell, W. B. Characterization and application of Raman labels for confocal Raman microspectroscopic detection of cellular proteins in single cells. *Anal. Biochem.* **2003**, *322*, 198–207.
13. Foulkes, J. G.; Liechtfried, F. E.; Pieler, C.; Stephenson, J. R.; Case, C. C. Reporter gene methods for identification of compounds that

modulate transcription of genes associated with cardiovascular disease. U.S. Patent 5580722, 1996; *Chem. Abstr.* **1996**, *126*, 102570.

14. Stokes, E. D.; Dunning, F. B.; Stebbings, R. F.; Walters, G. K.; Rundel, R. D. High efficiency dye laser tunable from the UV to the IR. *Opt. Commun.* **1972**, *5*, 267–270.

15. Runge, P. K. Mode-locking of helium–neon lasers with saturable organic dyes. *Opt. Commun.* **1971**, *3*, 434–436.

16. Simmons, H. E. III; Optical recording material. Braz. Pedido PI BR 8500852, 1985; *Chem. Abstr.* **1986**, *105*, 105851.

CRYSTAL VIOLET

CAS Registry Number 548-62-9

Chemical Structure

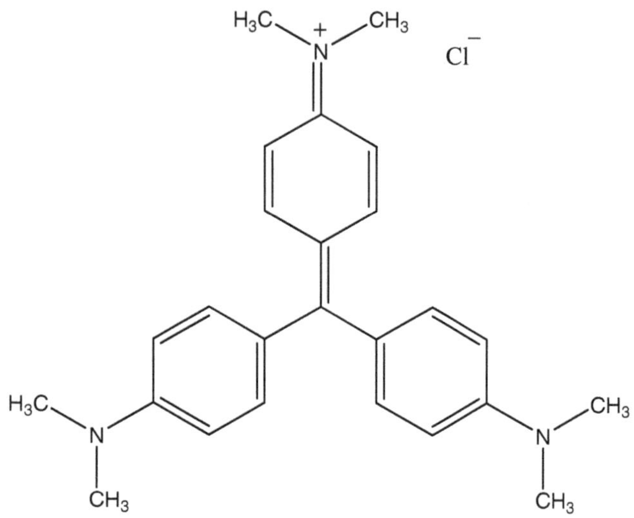

CA Index Name Methanaminium, *N*-[4-[bis[4-(dimethylamino)phenyl]methylene]-2,5-cyclohexadien-1-ylidene]-*N*-methyl-, chloride (1:1)

Other Names C.I. Basic Violet 3; 12416 Violet; Adergon; Aizen Crystal Violet; Aizen Crystal Violet Extra Pure; Aniline violet; Aniline violet pyoktanine; Atmonil; Avermin; Axuris; Badil; Basic Violet 3; Basic Violet BN; Basonyl Violet 610; C.I. 42555; Calcozine Violet 6BN; Calcozine Violet C; Crystal Violet 10B; Crystal Violet 5BO; Crystal Violet 6B; Crystal Violet 6BO; Crystal Violet AO; Crystal Violet AON; Crystal Violet BP; Crystal Violet BPC; Crystal Violet Extra Pure; Crystal Violet Extra Pure APN; Crystal Violet Extra Pure APNX; Crystal Violet FN; Crystal Violet HL 2; Crystal Violet O; Crystal Violet Pure DSC; Crystal Violet Pure DSC Brilliant; Crystal Violet SS; Crystal Violet Technical; Crystal Violet USP; Crystal Violet chloride; Crystal violet; Gentersal; Gentian Violet B; Gentian violet; Gentiaverm; Genticid; Gentioletten; Hecto Violet R; Hectograph Violet SR; Hexamethyl violet; Hexamethyl-*p*-rosaniline chloride; Hexamethylpararosaniline chloride; Hidaco Brilliant Crystal Violet; Kristall-violett; Meroxyl; Meroxyl-Wander; Meroxylan; Meroxylan-Wander; Methyl Violet 10B; Methyl Violet 10BD; Methyl Violet 10BK; Methyl Violet 10BN; Methyl Violet 10BNS; Methyl Violet 10BO; Methyl Violet 5BNO; Methyl Violet 5BO; Methyl Violet 6B; Methyl Violet 6B (biological stain); Methyl Violett; Methylrosaniline chloride; Methylrosanilinium chlorid; Methylrosanilinium chloride; Mitsui Crystal Violet; Oxiuran; Oxycolor; Oxyozyl; Paper Blue R; Pararosaniline, *N,N,N′,N′,N″,N″*-hexamethyl-, chloride; Plastoresin Violet 5BO; Pyoktanin; Sanyo Fanal Violet R; Vermicid; Vianin; Viocid; Violet 5BO; Violet 6BN; Violet CP; Violet XXIII

Merck Index Number 4395

Chemical/Dye Class Triphenylmethane

Molecular Formula $C_{25}H_{30}ClN_3$

Molecular Weight 407.98

Physical Form Dark Green powder

Solubility Soluble in water, acetone, chloroform; very soluble in ethanol; practically insoluble in ether; insoluble in xylene

Melting Point 205–215°C (decompose)

pH Range 0.0–2.0

Color Change at pH Yellow (0.0) to blue-violet (2.0)

Absorption (λ_{max}) 590 nm

Synthesis Synthetic methods[1–13]

Staining Applications β-Amyloid plaques;[14] α-synuclein; bacteria;[15] cells;[16] liposomes;[17] nucleic acids;[18] proteins;[19] skin;[20,21] thrombocytes;[22] vaginal smears;[23] hairs[1,24,25]

Biological Applications Detecting microorganisms;[26] treating atopic dermatitis,[27] dermatological diseases,[28,29] skin wounds,[30] lesions,[30] hemorrhoids,[1,31] multiple myeloma,[32] Non-Hodgkin's lymphoma,[32] breast cancer,[32] neurodegenerative diseases,[33] onychomycosis;[34] wound dressing;[35] drug delivery system;[1,36] dosage form;[37] antimicrobial agent;[1,38] antifungal agent;[1,39] antimalarial agent[1,40]

Industrial Applications Thin film transistors;[41,42] lithographic printing plates;[1,43] batteries;[44] photoresists;[45] inks;[46] printed circuit boards;[47] detergents;[48]

Safety/Toxicity Acute oral toxicity;[1,49] carcinogenicity;[1,50–52] chronic toxicity;[1,51,52] cytotoxicity;[1,53] chromosome damage;[54] DNA damage;[55] genotoxicity;[1,56–58] mutagenicity;[1,59–62] ototoxicity;[1,63] percutaneous toxicity;[1,64] phototoxicity[1,65]

Certification/Approval Certified by Biological Stain Commission (BSC)

REFERENCES

1. Sabnis, R. W. *Handbook of Acid-Base Indicators*; CRC Press: Boca Raton, **2008**; pp 108–110.
2. Sanguinet, L.; Williams, J. C.; Twieg, R. J.; Mao, G.; Wiggers, G.; Petschek, R. G.; Singer, K. D. Synthesis and HRS NLO characterization of new triarylmethyl cations. *Nonlinear Opt., Quant. Opt.* **2005**, *34*, 41–44.
3. Taber, D. F.; Meagley, R. P.; Supplee, D. A colorful Grignard reaction: preparation of the triarylmethane dyes from 4-bromo-*N,N*-dimethylaniline. *J. Chem. Educ.* **1996**, *73*, 259–260.
4. Aizawa, T. Manufacture of triphenylmethane dye by electrochemical oxidation. Jpn. Kokai Tokkyo Koho JP 02194188, 1990; *Chem. Abstr.* **1991**, *114*, 52054.
5. Awad, W. I.; El-Sawi, E. A.; El-Sayed, W. A. Preparation of modified basic dyes and their direct application to cotton fabrics. *Tinctoria* **1985**, *82*, 357–360.
6. Gilbert, M. E. A.; Gilbert, B. Crystal violet and related dyes. Braz. Pedido PI BR 8103521, 1983; *Chem. Abstr.* **1983**, *99*, 6871.
7. Zheryakova, G. I.; Kochkanyan, R. O.; Baranov, S. N. Crystal violet dye. U.S.S.R. SU 806717, 1981; *Chem. Abstr.* **1981**, *94*, 193713.
8. Aida, Y.; Nakamura, A.; Nagata, M. Triarylmethane dyes. Jpn. Tokkyo Koho JP 48020603, 1973; *Chem Abstr.* **1974**, *80*, 28492.
9. Lohmann, G. Y., Jr. Crystal violet. U.S. Patent 3689495, 1972; *Chem. Abstr.* **1973**, *78*, 17632.
10. Kast, H.; Baumann, H.; Mayer, U.; Oberlinner, A. Basic dyes by catalytic oxidation. Ger. Offen. DE 2152703, 1973; *Chem. Abstr.* **1973**, *79*, 6792.
11. Liebusch, W.; Fuerst, H.; Kaeppler, E. Triphenylmethane dye. Ger. (East) DD 62379, 1968; *Chem. Abstr.* **1969**, *70*, 88819.
12. Morton, A. A.; Stevens, J. R. Dyes of the triphenylmethane series. U.S. Patent 2029830, 1936; *Chem. Abstr.* **1936**, *30*, 15313.
13. Gulinov, V. G. Preparation of crystal violet from chloropicrin and dimethylaniline. *Ukr. Khim. Zh.* **1932**, *7*, 75–77; *Chem. Abstr.* **1933**, *27*, 42636.
14. Honson, N. S.; Johnson, R. L.; Huang, W.; Inglese, J.; Austin, C. P.; Kuret, J. Differentiating Alzheimer disease-associated aggregates with small molecules. *Neurobiol. Dis.* **2007**, *28*, 251–260.
15. Kataoka, N. Quick and simple Gram's staining method, and apparatus. Jpn. Kokai Tokkyo Koho JP 2006071447, 2006; *Chem. Abstr.* **2006**, *144*, 307911.
16. Vujanovic, N. L. Improvement of microcytoxicity assay by pre-labeling target cells. PCT Int. Appl. WO 2007123830, 2007; *Chem. Abstr.* **2007**, *147*, 496291.
17. Liu, X; Huan, S.; Bu, Y.; Shen, G.; Yu, R. Liposome-mediated enhancement of the sensitivity in immunoassay based on surface-enhanced Raman scattering at gold nanosphere array substrate. *Talanta* **2008**, *75*, 797–803.
18. Sun, W.; You, J.; Zhao, N.; Jiao, K. Application of linear-sweep voltammetry to the determination of nucleic acids using crystal violet as an electrochemical probe. *J. Anal. Chem.* **2008**, *63*, 265–270.
19. Szent-Gyorgyi, C.; Schmidt, B. A.; Creeger, Y.; Fisher, G. W.; Zakel, K. L.; Adler, S.; Fitzpatrick, J. A. J.; Woolford, C. A.; Yan, Q.; Vasilev, K. V.; Berget, P. B.; Bruchez, M. P.; Jarvik, J. W.; Waggoner, A. Fluorogen-activating single-chain antibodies for imaging cell surface proteins. *Nat. Biotechnol.* **2008**, *26*, 235–240.
20. Arndt, D. C. A ready to use combined skin marking stain compound and applicator requiring only the addition of water. PCT Int. Appl. WO 2008013676, 2008; *Chem. Abstr.* **2008**, *148*, 184985.
21. Magallon, A. G.; Boone, K. W.; Bardwell, J. R. Aqueous antiseptic solution and compatible dye for staining skin. U.S. Pat. Appl. Publ. US 2007253909, 2007; *Chem. Abstr.* **2007**, *147*, 491676.
22. Lindberg, S. Enumeration of thrombocytes. PCT Int. Appl. WO 2008010760, 2008; *Chem. Abstr.* **2008**, *148*, 163110.
23. Liu, X. P.; Fan, S. R. Methyrosaniline chloride stained vaginal smears for the diagnosis of vulvovaginal candidiasis. *Int. J. Gynecol. Obstet.* **2007**, *99*, 83–86.
24. Trigg, D. L.; Jones, J. L. Multistep hair color revitalizing kit comprising hair dye and glossing agent. U.S. Pat. Appl. Publ. US 2008189876, 2008; *Chem. Abstr.* **2008**, *149*, 251967.
25. Hoeffkes, H.; Gross, W. Hair dyes containing silylcarbonates and/or sylilcarbamates for simultaneous dying and bleaching of hair. Ger. Offen. DE 102004052060, 2006; *Chem. Abstr.* **2006**, *144*, 419031.
26. Martin, S. M.; MacDonald, J. G.; Lye, J.; Sayre, C.; Thompson, K. Array for rapid detection of a

microorganism based on dyes. PCT Int. Appl. WO 2008026104, 2008; *Chem. Abstr.* **2008**, *148*, 302867.

27. Rosenberg, E. W.; Skinner, R. B. Use of gentian violet in treatment of atopic dermatitis. U.S. Pat. Appl. Publ. US 2008146674, 2008; *Chem. Abstr.* **2008**, *149*, 62783.
28. Stumpp, O. Bleaching of contrast enhancing agent applied to skin for use with a dermatological treatment system. U.S. Pat. Appl. Publ. US 2008161745, 2008; *Chem. Abstr.* **2008**, *149*, 135505.
29. Pellet, A. Exfoliant composition containing a cellulosic polymer and its applications in cosmetic or topical dermato-logical treatments. Fr. Demande FR 2902322, 2007; *Chem. Abstr.* **2007**, *148*, 61286.
30. Kuhns, J. F. Topical antibacterial compositions used on aquatic animals for the treatment of skin wounds and lesions. U.S. Pat. Appl. Publ. US 2007254018, 2007; *Chem. Abstr.* **2007**, *147*, 508475.
31. Carroll, L. G. Composition comprising gentian violet for treating hemorrhoids. U.S. Pat. Appl. Publ. US 2006008481, 2006; *Chem. Abstr.* **2006**, *144*, 114497.
32. Tiedemann, R. E.; Stewart, A. K. Methods for inhibiting cyclin D by administering kinetin riboside and glucocorticosteroid to cells in treatment of multiple myeloma, Non Hodgkin's lymphoma and breast cancer. PCT Int. Appl. WO 2008045955, 2008; *Chem. Abstr.* **2008**, *148*, 417899.
33. Stockwell, B. R.; Hoffstrom, B.; Varma, H. Agents for treating neurodegenerative diseases. U.S. Pat. Appl. Publ. US 2007149543, 2007; *Chem. Abstr.* **2007**, *147*, 110249.
34. Rolf, D. Antifungal composition, method and kit for topically treating onychomycosis. U.S. Pat. Appl. Publ. US 2006165747, 2006; *Chem. Abstr.* **2006**, *145*, 195664.
35. Svetlik, H. E. Wound treatment-dressing and method of manufacture. U.S. Pat. Appl. Publ. US 2007161936, 2007; *Chem. Abstr.* **2007**, *147*, 150999.
36. Nadkarni, S. S.; Vaya, N.; Karan, R. S.; Gupta, V. K. Novel drug delivery system. Indian Pat. Appl. IN 2005MU01012, 2007; *Chem. Abstr.* **2007**, *148*, 503081.
37. Nadkarni, S. S.; Vaya, N.; Karan, R. S.; Gupta, V. K. Novel dosage form. Indian Pat. Appl. IN 2005MU01013, 2007; *Chem. Abstr.* **2007**, *148*, 387155.
38. Wilson, M.; Parkin, I. P.; Nair, S.; Gil-Tomas, J. Charge-stabilized metallic nanoparticle mixture with photosensitizer as antimicrobial. PCT Int. Appl. WO 2008015453, 2008; *Chem. Abstr.* **2008**, *148*, 221996.
39. Carr, A. D.; Neuss, J. C.; Orchard, M. G.; Porter, D. W. Preparation of substituted aniline derivatives as antifungal agents. PCT Int. Appl. WO 2008012524, 2008; *Chem. Abstr.* **2008**, *148*, 191838.
40. Chong, C. R.; Chen, X.; Shi, L.; Liu, J. O.; Sullivan, D. J. Jr. A clinical drug library screen identifies astemizole as an antimalarial agent. *Nat. Chem. Biol.* **2006**, *2*, 415–416.
41. Takemura, C.; Hirai, K. Organic thin film transistors and fabrication of organic thin film transistors thereof. Jpn. Kokai Tokkyo Koho JP 2008147346, 2008; *Chem. Abstr.* **2008**, *149*, 93154.
42. Takemura, C.; Okubo, Y.; Sugisaki, R.; Hirai, K. Manufacture of organic thin-film transistors. Jpn. Kokai Tokkyo Koho JP 2008060115, 2008; *Chem. Abstr.* **2008**, *148*, 320569.
43. Lingier, S. Method for making a lithographic printing plate. Eur. Pat. Appl. EP 1854627, 2007; *Chem. Abstr.* **2007**, *147*, 551244.
44. Fitter, J. C. System and method for preparing a flowable mixture for a battery. U.S. Pat. Appl. Publ. US 2007240782, 2007; *Chem. Abstr.* **2007**, *147*, 452345.
45. Toshimitsu, E. Negative-working photoresist compositions for blue-violet rays, and photoresist films. Jpn. Kokai Tokkyo Koho JP 2006058597, 2006; *Chem. Abstr.* **2006**, *144*, 243414.
46. Aoyama, M.; Tsuda, M.; Sago, H.; Okada, M.; Yamazaki, H.; Higashiyama, S. Fluorescent water inks for ink jet recording. Jpn. Kokai Tokkyo Koho JP 2005264072, 2005; *Chem. Abstr.* **2005**, *143*, 348861.
47. Murao, K. Fabrication of printed circuit boards provided with blind via-holes plugged with copper by electrodeposition. Jpn. Kokai Tokkyo Koho JP 2006009079, 2006; *Chem. Abstr.* **2006**, *144*, 116371.
48. Sadlowski, E. S.; Cummings, M. D. Laundry detergent compositions with efficient hueing dye. U.S. Pat. Appl. Publ. US 2006183658, 2006; *Chem. Abstr.* **2006**, *145*, 232298.
49. Hodge, H. C.; Indra, J.; Drobeck, H. P.; Duprey, L. P.; Tainter, M. L. Acute oral toxicity of methylrosaniline chloride. *Toxicol. Appl. Pharmacol.* **1972**, *22*, 1–5.
50. Gold, L. S.; Slone, T. H.; Stern, B. R.; Bernstein, L. Comparison of target organs of carcinogenicity for mutagenic and non-mutagenic chemicals. *Mutat. Res.* **1993**, *286*, 75–100.
51. Littlefield, N. A.; Gaylor, D. W.; Blackwell, B. N.; Allen, R. R. Chronic toxicity/carcinogenicity studies

of gentian violet in Fischer 344 rats: two-generation exposure. *Food Chem. Toxicol.* **1989**, *27*, 239–247.

52. Littlefield, N. A.; Blackwell, B. N.; Hewitt, C.; Gaylor, D. W. Chronic toxicity and carcinogenicity studies of gentian violet in mice. *Fundam. Appl. Toxicol.* **1985**, *5*, 902–912.
53. Huang, R.; Southall, N.; Cho, M. H.; Xia, M.; Inglese, J.; Austin, C. P. Characterization of diversity in toxicity mechanism using *in vitro* cytotoxicity assays in quantitative high throughput screening. *Chem. Res. Toxicol.* **2008**, *21*, 659–667.
54. Krishnaja, A. P.; Sharma, N. K. Heterogeneity in chemical mutagen-induced chromosome damage after G2 phase exposure to bleomycin, Ara-C and gentian violet in cultured lymphocytes of β-thalassemia traits. *Mutat. Res.* **1995**, *331*, 143–148.
55. McCarroll, N. E.; Piper, C. E.; Keech, B. H. An *E. coli* microsuspension assay for the detection of DNA damage induced by direct-acting agents and promutagens. *Environ. Mutagen.* **1981**, *3*, 429–444.
56. Walsh, L.; Hastwell, P. W.; Keenan, P. O.; Knight, A. W.; Billinton, N.; Walmsley, R. M. Genetic modification and variations in solvent increase the sensitivity of the yeast RAD54-GFP genotoxicity assay. *Mutagenesis* **2005**, *20*, 317–327.
57. Choudhary, E.; Capalash, N.; Sharma, P. Genotoxicity of degradation products of textile dyes evaluated with rec-assay after photo Fenton and ligninase treatment. *J. Environ. Pathol., Toxicol. Oncol.* **2004**, *23*, 279–285.
58. Sakagami, Y.; Yamazaki, H.; Ogasawara, N.; Yokoyama, H.; Ose, Y.; Sato, T. The evaluation of genotoxic activities of disinfectants and their metabolites by umu test. *Mutat. Res. Lett.* **1988**, *209*, 155–160.
59. Serafimova, R.; Todorov, M.; Pavlov, T.; Kotov, S.; Jacob, E.; Aptula, A.; Mekenyan, O. Identification of the structural requirements for mutagenicity, by incorporating molecular flexibility and metabolic activation of chemicals. II. General Ames mutagenicity model. *Chem. Res. Toxicol.* **2007**, *20*, 662–676.
60. Malachova, K.; Pavlickova, Z.; Novotny, C.; Svobodova, K.; Lednicka, D.; Musilkova, E. Reduction in the mutagenicity of synthetic dyes by successive treatment with activated sludge and the ligninolytic fungus, *Irpex lacteus*. *Environ. Mol. Mutagen.* **2006**, *47*, 533–540.
61. Klopman, G.; Frierson, M. R.; Rosenkranz, H. S. The structural basis of the mutagenicity of chemicals in *Salmonella typhimurium*: the Gene-Tox data base. *Mutat. Res.* **1990**, *228*, 1–50.
62. Ferguson, L. R.; Baguley, B. C. Verapamil as a co-mutagen in the *Salmonella*/mammalian microsome mutagenicity test. *Mutat. Res. Lett.* **1988**, *209*, 57–62.
63. Tom, L. W. C. Ototoxicity of common topical antimycotic preparations. *Laryngoscope* **2000**, *110*, 509–516.
64. Burnett, C.; Goldenthal, E. I.; Harris, S. B.; Wazeter, F. X.; Strausburg, J.; Kapp, R.; Voelker, R. Teratology and perc-utaneous toxicity studies on hair dyes. *J. Toxicol. Environ. Health* **1976**, *1*, 1027–1040.
65. Kandela, R. K.; Bartlett, J. A.; Indig, G. L. Effect of molecular structure on the selective phototoxicity of triarylmethane dyes towards tumor cells. *Photochem. Photobiol. Sci.* **2002**, *1*, 309–314.

CYANOTOLYL TETRAZOLIUM CHLORIDE (CTC)

CAS Registry Number 90217-02-0

Chemical Structure

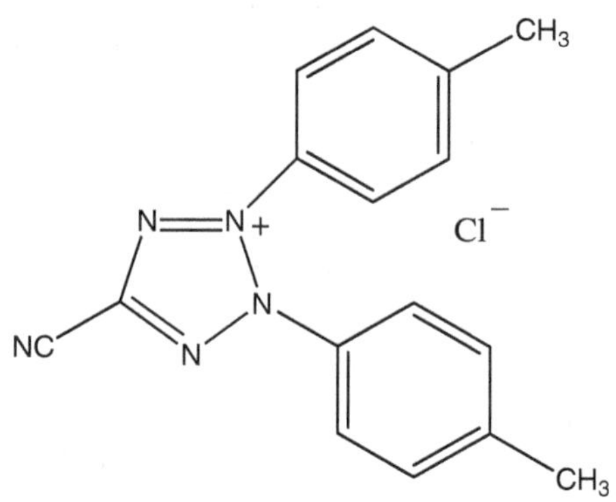

CA Index Name 2*H*-Tetrazolium, 5-cyano-2,3-bis(4-methylphenyl)-, chloride (1:1)

Other Names 2*H*-Tetrazolium, 5-cyano-2,3-bis(4-methylphenyl)-, chloride; 5-Cyano-2,3-di-(*p*-tolyl)tetrazolium chloride; cyanotolyl tetrazolium, CTC; CTC (biological stain)

Merck Index Number Not listed

Chemical/Dye Class Tetrazolium salt

Molecular Formula $C_{16}H_{14}ClN_5$

Molecular Weight 311.77

Physical Form Off-white powder

Solubility Soluble in water

Melting Point 225°C

Absorption (**λ_{max}**) 450 nm (of formazan)

Synthesis Synthetic method[1]

Staining Applications Bacteria;[2–12] cells;[13] microorganisms;[14,15] plasma membrane[16]

Biological Applications NADPH-cytochrome P 450 reductase activity assays;[17] detecting gamma-hydroxybutyric acid (GHB),[18] metabolic activity,[19] microorganisms;[20] localizing glucose-6-phosphate dehydrogenase activity;[21] monitoring respiratory activity;[22–31] treating cancer[32]

Industrial Applications Not reported

Safety/Toxicity Bacterial toxicity[25,33]

REFERENCES

1. Stellmach, J.; Severin, E. Fluorescent redox indicators and their use in cytodiagnostics. Ger. Offen. DE 3418852, 1985; *Chem. Abstr.* **1986**, *105*, 3077.
2. Leung, H. D.; Chen, G.; Sharma, K. Effect of detached/re-suspended solids from sewer sediment on the sewage phase bacterial activity. *Water Sci. Technol.* **2005**, *52*, 147–152.
3. Maraha, N.; Backman, A.; Jansson, J. K. Monitoring physiological status of GFP-tagged *Pseudomonas fluorescens* SBW25 under different nutrient conditions and in soil by flow cytometry. *FEMS Microbiol. Ecol.* **2004**, *51*, 123–132.
4. Mangoni, M. L.; Papo, N.; Barra, D.; Simmaco, M.; Bozzi, A.; Di Giulio, A.; Rinaldi, A. C. Effects of the antimicrobial peptide temporin L on cell morphology, membrane perme-ability and viability of *Escherichia coli*. *Biochem. J.* **2004**, *380*, 859–865.
5. Caruso, G.; Mancuso, M.; Crisafi, E. Combined fluorescent antibody assay and viability staining for the assessment of the physiological states of *Escherichia coli* in seawaters. *J. Appl. Microbiol.* **2003**, *95*, 225–233.
6. Bartosch, S.; Mansch, R.; Knotzsch, K.; Bock, E. CTC staining and counting of actively respiring bacteria in natural stone using confocal laser scanning microscopy. *J. Microbiol. Methods* **2003**, *52*, 75–84.
7. Creach, V.; Baudoux, A. C.; Bertru, G.; Le Rouzic, B. Direct estimate of active bacteria: CTC use and limitations. *J. Microbiol. Methods* **2003**, *52*, 19–28.
8. Fuller, M. E.; Streger, S. H.; Rothmel, R. K.; Mailloux, B. J.; Hall, J. A.; Onstott, T. C.; Fredrickson, J. K.; Balkwill, D. L.; DeFlaun, M. F. Development of a vital fluorescent staining method for monitoring bacterial transport in subsurface environments. *Appl. Environ. Microbiol.* **2000**, *66*, 4486–4496.
9. Gavin, J.; Button, N. F.; Watson-Craik, I. A.; Logan, N. A. Observation of soft contact lens disinfection with fluorescent metabolic stains. *Appl. Environ. Microbiol.* **2000**, *66*, 874–875.
10. Bhupathiraju, V. K.; Hernandez, M.; Landfear, D.; Alvarez-Cohen, L. Application of a tetrazolium dye as an indicator of viability in anaerobic bacteria. *J. Microbiol. Methods* **1999**, *37*, 231–243.

11. Pyle, B. H.; Broadaway, S. C.; McFeters, G. A. Sensitive detection of *Escherichia coli* O157:H7 in food and water by immunomagnetic separation and solid-phase laser cytometry. *Appl. Environ. Microbiol.* **1999**, *65*, 1966–1972.

12. Lopez-Amoros, R.; Comas, J.; Garcia, M. T.; Vives-Rego, J. Use of the 5-cyano-2,3-ditolyl tetrazolium chloride reduction test to assess respiring marine bacteria and grazing effects by flow cytometry during linear alkylbenzene sulfonate degradation. *FEMS Microbiol. Ecol.* **1998**, *27*, 33–42.

13. Berridge, M. V.; Herst, P. M.; Tan, A. S. Tetrazolium dyes as tools in cell biology: new insights into their cellular reduction. *Biotechnol. Annu. Rev.* **2005**, *11*, 127–152.

14. Noda, N.; Mizutani, T. Microorganism-measuring method using multiple staining. Jpn. Kokai Tokkyo Koho. JP 2006340684, 2006; *Chem. Abstr.* **2006**, *146*, 77563.

15. Gruden, C. L.; Khijniak, A.; Adriaens, P. Activity assessment of microorganisms eluted from sediments using 5-cyano-2,3-ditolyl tetrazolium chloride: a quantitative comparison of flow cytometry to epifluorescent microscopy. *J. Microbiol. Methods* **2003**, *55*, 865–874.

16. Bernas, T.; Dobrucki, J. W. The role of plasma membrane in bioreduction of two tetrazolium salts, MTT, and CTC. *Arch. Biochem. Biophys.* **2000**, *380*, 108–116.

17. Kim, D. H.; Yim, S. K.; Kim, K. H.; Ahn, T.; Yun, C. H. Continuous spectrofluorometric and spectrophotometric assays for NADPH-cytochrome P 450 reductase activity using 5-cyano-2,3-ditolyl tetrazolium chloride. *Biotechnol. Lett.* **2009**, *31*, 271–275.

18. Parsons, S. M.; Harris, D. O.; Bravo, D. T. Methods, compositions and apparatuses for detection of gamma-hydroxybutyric acid (GHB). U.S. Pat. Appl. Publ. US 2003175846, 2003; *Chem. Abstr.* **2003**, *139*, 256601.

19. Bovill, R. A.; Shallcross, J. A.; Mackey, B. M. Comparison of the fluorescent redox dye 5-cyano-2,3-ditolyltetrazolium chloride with *p*-iodonitrotetrazolium violet to detect metabolic activity in heat-stressed *Listeria monocytogenes* cells. *J. Appl. Bacteriol.* **1994**, *77*, 353–358.

20. Pyle, B. H.; Mcfeters, G. A.; Jutila, J. W.; Schiemann, D. A.; Bargatze, R. F.; Jutila, M. A. Fluorochrome dye and antibody for detection, identification and enumeration of specific viable microorganisms. PCT Int. Appl. WO 9531481, 1995; *Chem. Abstr.* **1996**, *124*, 115461.

21. Frederiks, W. M.; van Marle, J.; van Oven, C.; Comin-Anduix, B.; Cascante, M. Improved localization of glucose-6-phosphate dehydrogenase activity in cells with 5-cyano-2,3-ditolyl-tetrazolium chloride as fluorescent redox dye reveals its cell cycle-dependent regulation. *J. Histochem. Cytochem.* **2006**, *54*, 47–52.

22. Kodama, T.; Ara, K.; Sekiguchi, J. Analyzing method and kit of cell respiration activity and screening method and kit of strain with high protein productivity. Jpn. Kokai Tokkyo Koho. JP 2009028011, 2009; *Chem. Abstr.* **2009**, *150*, 186339.

23. Yamaguchi, N.; Sasada, M.; Yamanaka, M.; Nasu, M. Rapid detection of respiring *Escherichia coli* O157: H7 in apple juice, milk, and ground beef by flow cytometry. *Cytometry, Part A* **2003**, *54*, 27–35.

24. Ogawa, M.; Tani, K.; Yamaguchi, N.; Nasu, M. Development of multicolour digital image analysis system to enumerate actively respiring bacteria in natural river water. *J. Appl. Microbiol.* **2003**, *95*, 120–128.

25. Hatzinger, P. B.; Palmer, P.; Smith, R. L.; Penarrieta, C. T.; Yoshinari, T. Applicability of tetrazolium salts for the measurement of respiratory activity and viability of groundwater bacteria. *J. Microbiol. Methods* **2003**, *52*, 47–58.

26. Caro, A.; Got, P.; Baleux, B. Physiological changes of *Salmonella typhimurium* cells under osmotic and starvation conditions by image analysis. *FEMS Microbiol. Lett.* **1999**, *179*, 265–273.

27. Yamaguchi, N.; Nasu, M. Flow cytometric analysis of bacterial respiratory and enzymic activity in the natural aquatic environment. *J. Appl. Microbiol.* **1997**, *83*, 43–52.

28. Pyle, B. H.; Broadaway, S. C.; McFeters, G. A. A rapid, direct method for enumerating respiring enterohemorrhagic *Escherichia coli* O157:H7 in water. *Appl. Environ. Microbiol.* **1995**, *61*, 2614–2619.

29. Kaprelyants, A. S.; Kell, D. B. The use of 5-cyano-2,3-ditolyl tetrazolium chloride and flow cytometry for the visualization of respiratory activity in individual cells of *Micrococcus luteus*. *J. Microbiol. Methods* **1993**, *17*, 115–122.

30. Schaule, G.; Flemming, H. C.; Ridgway, H. F. Use of 5-cyano-2,3-ditolyl tetrazolium chloride for quantifying planktonic and sessile respiring bacteria in drinking water. *Appl. Environ. Microbiol.* **1993**, *59*, 3850–3857.

31. Rodriguez, G. G.; Phipps, D.; Ishiguro, K.; Ridgway, H. F. Use of a fluorescent redox probe for direct visualization of actively respiring bacteria. *Appl. Environ. Microbiol.* **1992**, *58*, 1801–1808.

32. Kong, Q. Method and composition for treating cancer using cellular organelle crystallizing agents. U.S. Pat. Appl. Publ. US 2006286170, 2006; *Chem. Abstr.* **2006**, *146*, 55482.

33. Ullrich, S.; Karrasch, B.; Hoppe, H. G.; Jeskulke, K.; Mehrens, M. Toxic effects on bacterial metabolism of the redox dye 5-cyano-2,3-ditolyl tetrazolium chloride. *Appl. Environ. Microbiol.* **1996**, *62*, 4587–4593.

DAF FM

CAS Registry Number 254109-20-1

Chemical Structure

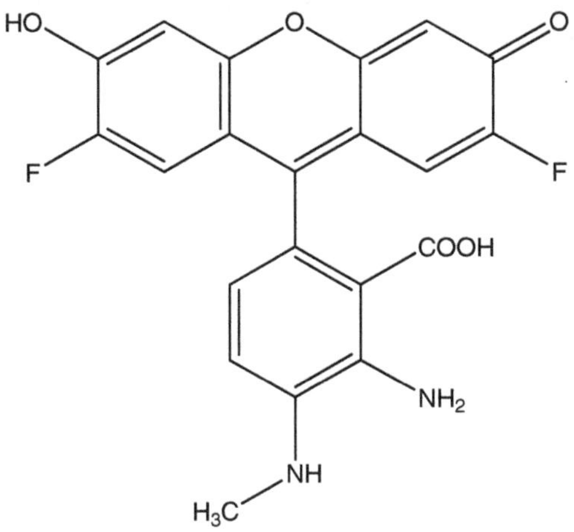

CA Index Name Spiro[isobenzofuran-1(3*H*),9′-[9*H*] xanthen]-3-one, 4-amino-2′,7′-difluoro-3′,6′-dihydroxy-5-(methylamino)-

Other Names 4-Amino-5-methylamino-2′,7′-difluorofluorescein; DAF-FM

Merck Index Number Not listed

Chemical/Dye Class Xanthene

Molecular Formula $C_{21}H_{14}F_2N_2O_5$

Molecular Weight 412.34

Physical Form Solid

Solubility Soluble in dimethyl sulfoxide, methanol

Melting Point 265 °C

Boiling Point (Calcd.) 693.7 ± 55.0 °C, pressure: 760 Torr

pK_a (Calcd.) 6.88 ± 0.20, most acidic, temperature: 25 °C; 3.99 ± 0.20, most basic, temperature: 25 °C

Absorption (λ_{max}) 487 nm, 495 nm

Emission (λ_{max}) 515 nm

Synthesis Synthetic method[1]

Staining Applications Nitric oxide ions[1–10]

Biological Applications Nitric oxide indicator;[1–10] detecting microorganisms;[11] measuring Gα i-coupled or Gα o-coupled receptors activity[12]

Industrial Applications Not reported

Safety/Toxicity No data available

REFERENCES

1. Kojima, H.; Urano, Y.; Kikuchi, K.; Higuchi, T.; Hirata, Y.; Nagano, T. Fluorescent indicators for imaging nitric oxide production. *Angew. Chem. Int. Ed.* **1999**, *38*, 3209–3212.
2. Vitecek, J.; Reinohl, V.; Jones, R. L. Measuring NO production by plant tissues and suspension cultured cells. *Mol. Plant* **2008**, *1*, 270–284.
3. Yang, Q.; Zhang, X.; Bao, X.; Lu, H.; Zhang, W.; Wu, W.; Miao, H.; Jiao, B. Single cell determination of nitric oxide release using capillary electrophoresis with laser-induced fluorescence detection. *J. Chromatogr., A* **2008**, *1201*, 120–127.
4. Kolonics, A. Tory, K. Literati Nagy, P.; Koranyi, L.; Egri, J. A process and a reagent kit for the examination of the metabolic state related to the carbohydrate and lipid metabolism of a human organism. PCT Int. Appl. WO 2007088446, 2007; *Chem. Abstr.* **2007**, *147*, 230168.
5. Zguris, J.; Pishko, M. V. Nitric oxide sensitive fluorescent poly(ethylene glycol) hydrogel microstructures. *Sens. Actuators, B* **2006**, *B115*, 503–509.
6. Kim, W. S.; Ye, X.; Rubakhin, S. S.; Sweedler, J. V. Measuring nitric oxide in single neurons by capillary electrophoresis with laser-induced fluorescence: use of ascorbate oxidase in diaminofluorescein measurements. *Anal. Chem.* **2006**, *78*, 1859–1865.
7. Balcerczyk, A.; Soszynski, M.; Bartosz, G. On the specificity of 4-amino-5-methylamino-2′,7′-difluorofluorescein as a probe for nitric oxide. *Free Radical Biol. Med.* **2005**, *39*, 327–335.
8. Broillet, M. C.; Randin, O.; Chatton, J. Y. Photoactivation and calcium sensitivity of the fluorescent NO indicator 4,5-diaminofluorescein (DAF-2): implications for cellular NO imaging. *FEBS Lett.* **2001**, *491*, 227–232.

Handbook of Biological Dyes and Stains By R. W. Sabnis

9. Itoh, Y.; Ma, F. H.; Hoshi, H.; Oka, M.; Noda, K.; Ukai, Y.; Kojima, H.; Nagano, T.; Toda, N. Determination and bioimaging method for nitric oxide in biological specimens by diaminofluorescein fluorometry. *Anal. Biochem.* **2000**, *287*, 203–209.
10. Kojima, H.; Nagano, T. Fluorescent indicators for nitric oxide. *Adv. Mater.* **2000**, *12*, 763–765.
11. Yoshimi, K.; Ogawa, N. Method and apparatus for detecting microorganism by fluorometry. Jpn. Kokai Tokkyo Koho JP 2003144193, 2003; *Chem. Abstr.* **2003**, *138*, 381689.
12. Wade, E. J.; Janocha, E.; Germann, T. Methods for measuring the activity of Gα i-coupled or Gα o-coupled receptors using cellular calcium ion influx, and use in agonist and antagonist identification. PCT Int. Appl. WO 2004051264, 2004; *Chem. Abstr.* **2004**, *141*, 33742.

DANSYL CADAVERINE

CAS Registry Number 10121-91-2

Chemical Structure

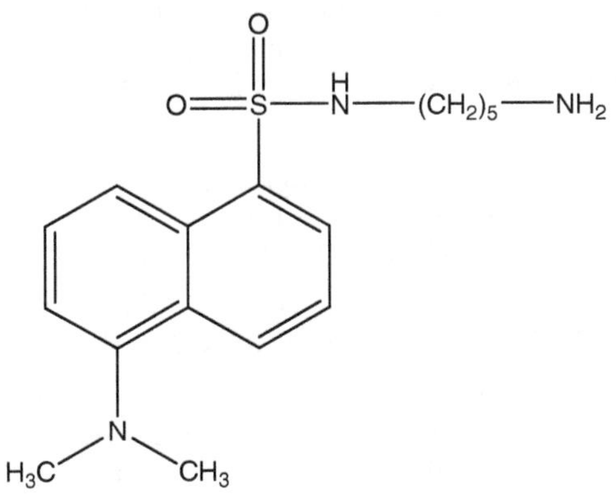

CA Index Name 1-Naphthalenesulfonamide, *N*-(5-aminopentyl)-5-(dimethylamino)-

Other Names Dansylcadaverine; Monodansylcadaverine; *N*-(5-Aminopentyl)-5-dimethylamino-1-naphthalenesulfonamide; 5-Dimethylaminonaphthalene-1-(*N*-(5-aminopentyl))sulfonamide

Merck Index Number Not listed

Chemical/Dye Class Naphthalene

Molecular Formula $C_{17}H_{25}N_3O_2S$

Molecular Weight 335.46

Physical Form Yellow to yellowish green powder

Solubility Soluble in ethanol, *N,N*-dimethylformamide

Melting Point 137 °C

Boiling Point (Calcd.) 505.5 ± 60.0 °C, pressure: 760 Torr

pK_a (Calcd.) 11.96 ± 0.50, most acidic, temperature: 25 °C; 10.49 ± 0.10, most basic, temperature: 25 °C

Absorption (λ_{max}) 335 nm

Emission (λ_{max}) 518 nm

Synthesis Synthetic methods[1–4]

Staining Applications Autophagy/autophagic vacuoles;[5–10] transglutaminase activity[11–15]

Biological Applications Measuring cardiac autophagic flux;[16] treating cataract,[17] fibrosis,[18] Parkinson's disease,[19] peritoneal ovarian tumor dissemination,[20] respiratory diseases,[21] lung diseases,[21] Huntington's disease,[22] spinobulbar atrophy,[22] spinocerebellar ataxia,[22] dentatorubralpallidoluysian atrophy,[22] multiple sclerosis,[22] rheumatoid arthritis,[22] insulin-dependent diabetes mellitus[22]

Industrial Applications Not reported

Safety/Toxicity Cytotoxicity;[23,24] neurotoxicity[25]

REFERENCES

1. Takashi, R. A novel actin label: a fluorescent probe at glutamine-41 and its consequences. *Biochemistry* **1988**, *27*, 938–943.
2. Narayanan, R.; Balaram, P. Synthesis and fluorescence properties of a probe for membrane anionic sites. *Biochem. Biophys. Res. Commun.* **1976**, *70*, 1122–1128.
3. Nilsson, J.; Lars, G.; Stenberg, P.; Ljunggren, C. Fibrin-stabilizing factor inhibitors. III. Sulfonamides related to monodansylcadaverine. *Acta Pharm. Suecica* **1971**, *8*, 497–504.
4. Lorand, L.; Rule, N. G.; Ong, H. H.; Furlanetto, R.; Jacobsen, A.; Downey, J.; Oner, N.; Bruner-Lorand, J. Amine specificity in transpeptidation: inhibition of fibrin cross-linking. *Biochemistry* **1968**, *7*, 1214–1223.
5. Contento, A. L.; Xiong, Y.; Bassham, D. C. Visualization of autophagy in Arabidopsis using the fluorescent dye monodansylcadaverine and a GFP-AtATG8e fusion protein. *Plant J.* **2005**, *42*, 598–608.
6. Mizushima, N. Methods for monitoring autophagy. *Int. J. Biochem. Cell Biol.* **2004**, *36*, 2491–2502.
7. Niemann, A.; Baltes, J.; Elsasser, H. P. Fluorescence properties and staining behavior of monodansylpentane, a structural homologue of the lysosomotropic agent monodansylcadaverine. *J. Histochem. Cytochem.* **2001**, *49*, 177–185.
8. Munafo, D. B.; Colombo, M. I. A novel assay to study autophagy: regulation of autophagosome vacuole size by amino acid deprivation. *J. Cell Sci.* **2001**, *114*, 3619–3629.
9. Niemann, A.; Takatsuki, A.; Elsasser, H. P. The lysosomotropic agent monodansylcadaverine also acts as a solvent polarity probe. *J. Histochem. Cytochem.* **2000**, *48*, 251–258.
10. Biederbick, A.; Kern, H. F.; Elsaesser, H. P. Monodansylcadaverine (MDC) is a specific *in vivo* marker for autophagic vacuoles. *Eur. J. Cell Biol.* **1995**, *66*, 3–14.

11. De Young, L.; Ballaron, S.; Epstein, W. Transglutaminase activity in human and rabbit ear comedogenesis: a histochemical study. *J. Invest. Dermatol.* **1984**, *82*, 275–279.

12. Lee, N. H.; Takeuchi, A.; Konno, K. Inhibition of transglutaminase and microbial transglutaminase activity by garlic. *Food Sci. Biotechnol.* **2007**, *16*, 223–227.

13. Brobey, R. K. B.; Soong, L. Leishmania species: evidence for transglutaminase activity and its role in parasite proliferation. *Exp. Parasitol.* **2006**, *114*, 94–102.

14. Sessa, A.; Tunici, P.; Rabellotti, E.; Bardocz, S.; Grant, G.; Pusztai, A.; Perin, A. Response of intestinal transglutaminase activity to dietary phytohaemagglutinin. *Biochim. Biophys. Acta* **1996**, *1314*, 66–70.

15. Zirvi, K. A.; Keogh, J. P.; Slomiany, A.; Slomiany, B. L. Transglutaminase activity in human colorectal carcinomas of differing metastatic potential. *Cancer Lett.* **1991**, *60*, 85–92.

16. Iwai-Kanai, E.; Yuan, H.; Huang, C.; Sayen, M. R.; Perry-Garza, C. N.; Kim, L.; Gottlieb, R. A. A method to measure cardiac autophagic flux *in vivo*. *Autophagy* **2008**, *4*, 322–329.

17. Kim, I. G.; Jeon, J. H.; Shin, D. M.; Cho, S. Y.; Jeong, E. M.; Jang, G. Y.; Park, S. C. Pharmaceutical compositions comprising transglutaminase 2 inhibitor or *N*-acetylcysteine, for preventing or treating cataract. PCT Int. Appl. WO 2009038288, 2009; *Chem. Abstr.* **2009**, *150*, 359766.

18. Kim, I. G.; Shin, D. M.; Cho, S. Y.; Jeong, E. M.; Jang, G. Y.; Park, S. C.; Lee, D. S.; Oh, K. H.; Kim, Y. W. Pharmaceutical compositions for preventing or treating fibrosis. PCT Int. Appl. WO 2009038289, 2009; *Chem. Abstr.* **2009**, *150*, 345559.

19. Mouradian, M. M.; Junn, E. Methods and compositions using transglutaminase 2 inhibitors for the treatment of Parkinson's disease and other α-synucleinopathies. PCT Int. Appl. WO 2004069175, 2004; *Chem. Abstr.* **2004**, *141*, 200207.

20. Matei, D. Tissue transglutaminase modulators for treatment of peritoneal ovarian tumor dissemination. PCT Int. Appl. WO 2008098129, 2008; *Chem. Abstr.* **2008**, *149*, 259456.

21. Nyce, J. W.; Tang, L.; Sandrasagra, A.; Aguilar, D.; Miller, S.; Shahabuddin, S.; Lu, H.; Cong, H. Compositions, formulations & kit for treatment of respiratory & lung diseases. PCT Int. Appl. WO 2004011613, 2004; *Chem. Abstr.* **2004**, *140*, 175174.

22. Steinman, L.; Karpuj, M. V. Methods and compositions for treating diseases mediated by transglutaminase activity. PCT Int. Appl. WO 9965516, 1999; *Chem. Abstr.* **1999**, *132*, 45002.

23. Monier, S.; Samadi, M.; Prunet, C.; Denance, M.; Laubriet, A.; Athias, A.; Berthier, A.; Steinmetz, E.; Jurgens, G.; Negre-Salvayre, A.; Bessede, G.; Lemaire-Ewing, S.; Neel, D.; Gambert, P.; Lizard, G. Impairment of the cytotoxic and oxidative activities of 7 beta-hydroxycholesterol and 7-ketocholesterol by esterification with oleate. *Biochem. Biophys. Res. Commun.* **2003**, *303*, 814–824.

24. Naim, R.; Iida, T.; Takahashi, A.; Honda, T. Monodansylcadaverine inhibits cytotoxicity of *Vibrio parahaemo-lyticus*thermostable direct hemolysin on cultured rat embryonic fibroblast cells. *FEMS Microbiol. Lett.* **2001**, *196*, 99–105.

25. Gilad, G. M.; Gilad, V. H. Cytotoxic effects of monodansylcadaverine and methylamine in primary cultures of rat cerebellar neurons. *Int. J. Dev. Neurosci.* **1986**, *4*, 401–405.

DAPI

CAS Registry Number 28718-90-3

Chemical Structure

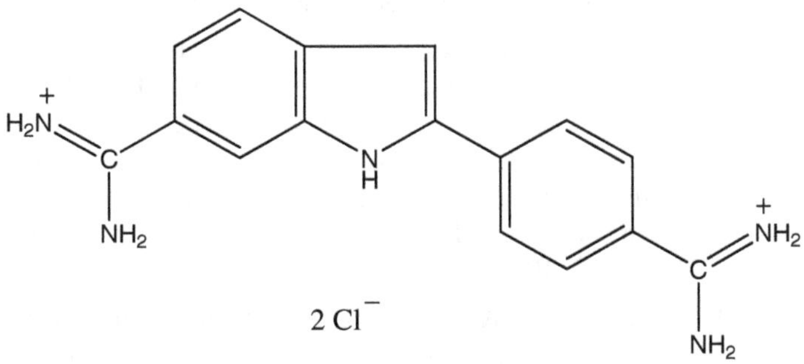

CA Index Name 1*H*-Indole-6-carboximidamide, 2-[4-(aminoiminomethyl)phenyl]-, dihydrochloride

Other Names Indole-6-carboxamidine, 2-(*p*-amidinophenyl)-, dihydrochloride; 4′,6-Diamidino-2-phenylindole dihydrochloride; 6-Amidino-2-(4-amidinophenyl) indole dihydrochloride; DAPI

Merck Index Number Not listed

Chemical/Dye Class Indole

Molecular Formula $C_{16}H_{17}Cl_2N_5$

Molecular Weight 350.25

Physical Form Yellow powder

Solubility Soluble in water, *N,N*-dimethylformamide

Melting Point >200 °C

Absorption (λ_{max}) 358 nm, 342 nm

Emission (λ_{max}) 461 nm, 450 nm

Synthesis Synthetic methods[1–6]

Staining Applications Nucleic acids;[3,7–10] bacteria;[11,12] cells;[13] chromosomes;[14] microorganisms;[15] neuron-specific nuclear protein NeuN;[16] polynucleotide;[17] proteins[18]

Biological Applications Nuclear apoptosis assay;[19] targeting drug to cerebral neuron;[20] treating amyloidosis,[21] neurodegenerative diseases,[21] Alzheimer's disease,[22] Down's syndrome,[22] type II diabetes;[22] monitoring bioaerosols[23]

Industrial Applications Not reported

Safety/Toxicity No data available

REFERENCES

1. Teller, J.; Gruttner, C.; Rudershausen, S.; Westphal, F. Preparation of homogeneous colored or fluorescent silicic acid particles by controlled hydrolysis and polycondensation. Eur. Pat. Appl. EP 1036763, 2000; *Chem. Abstr.* **2000**, *133*, 225158.
2. Zhang, X.; Chen, G.; Li, G.; Chen, L.; Dai, Z. Antimalarial agents. Part 14. Synthesis of 4′,6-diamidino-2-phenylindole and its antimalarial effect. *Yiyao Gongye* **1983**, 2–4; *Chem. Abstr.* **1984**, *100*, 174592.
3. Kapuscinski, J.; Skoczylas, B. Fluorescent complexes of DNA with DAPI (4′,6-diamidine-2-phenyl indole dihydrochloride) or DCI (4′,6-dicarboxyamide-2-phenyl indole). *Nucleic Acids Res.* **1978**, *5*, 3775–3799.
4. Dann, O.; Bergen, G.; Demant, E.; Vol, G. Trypanosomicidal diamidines of 2-phenylbenzofuran, 2-phenylindene, and 2-phenylindole. *Justus Liebigs Ann. Chem.* **1971**, *749*, 68–89.
5. Dann, O. Tripanosomicidal diamidine compounds. Brit. GB 1236812, 1971; *Chem. Abstr.* **1971**, *75*, 110182.
6. Dann, O. Trypanocidal diamidine compounds. Fr. Demande FR 1586113, 1970; *Chem. Abstr.* **1970**, *73*, 109674.
7. Noirot, M.; Barre, P.; Louarn, J.; Duperray, C.; Hamon, S. Consequences of stoichiometric error on nuclear DNA content evaluation in *Coffea liberica* var. dewevrei using DAPI and propidium iodide. *Ann. Bot.* **2002**, *89*, 385–389.
8. Kapuscinski, J.; Skoczylas, B. Simple and rapid fluorimetric method for DNA microassay. *Anal. Biochem.* **1977**, *83*, 252–257.
9. Vollenweider, I.; Groscurth, P. Comparison of four DNA staining fluorescence dyes for measuring cell proliferation of lymphokine-activated killer (LAK) cells. *J. Immunol. Methods* **1992**, *149*, 133–135.
10. Mabuchi, T.; Nishikawa, S. Selective staining with two fluorochromes of DNA fragments on gels depending on their AT content. *Nucleic Acids Res.* **1990**, *18*, 7461–7462.
11. Kawaharasaki, M.; Tanaka, H.; Kanagawa, T.; Nakamura, K. *In situ* identification of polyphosphate-accumulating bacteria in activated

sludge by dual staining with rRNA-targeted oligonucleotide probes and 4′,6-diamidino-2-phenylindol (DAPI) at a polyphosphate-probing concentration. *Water Res.* **1998**, *33*, 257–265.

12. Kawaharasaki, M.; Kanagawa, T.; Tanaka, H.; Nakamura, K. Development and application of 16S rRNA-targeted oligonucleotide probe for detection of phosphate-accumulating bacterium, *Microlunatus phosphovorus* in an enhanced biological phosphorus removal process. *Water Sci. Technol.* **1998**, *37*, 481–484.
13. Takahama, M. Reagents and method for determination of protein/DNA ratios in cells. Jpn. Kokai Tokkyo Koho JP 62135769, 1987; *Chem. Abstr.* **1987**, *107*, 214487.
14. Gosden, J. R.; Spowart, G.; Lawrie, S. S. Satellite DNA and cytological staining patterns in heterochromatic inversions of human chromosome 9. *Hum. Genet.* **1981**, *58*, 276–278.
15. Li, C. S.; Chia, W. C.; Chen, P. S. Fluorochrome and flow cytometry to monitor microorganisms in treated hospital wastewater. *J. Environ. Sci. Health, Part A* **2007**, *42*, 195–203.
16. Gill, S. K.; Ishak, M.; Rylett, R. J. Exposure of nuclear antigens in formalin-fixed, paraffin-embedded necropsy human spinal cord tissue: detection of NeuN. *J. Neurosci. Methods* **2005**, *148*, 26–35.
17. Cavatorta, P.; Masotti, L.; Szabo, A. G. A time-resolved fluorescence study of 4′,6′-diamidine-2-phenylindole dihydrochloride binding to polynucleotides. *Biophys. Chem.* **1985**, *22*, 11–16.
18. Stoehr, M.; Vogt-Schaden, M.; Knobloch, M.; Vogel, R.; Futterman, G. Evaluation of eight fluorochrome combinations for simultaneous DNA–protein flow analyses. *Stain Technol.* **1978**, *53*, 205–215.
19. Susin, S. A.; Zamzami, N.; Larochette, N.; Dallaporta, B.; Marzo, I.; Brenner, C.; Hirsch, T.; Petit, P. X.; Geuskens, M.; Kroemer, G. A cytofluorometric assay of nuclear apoptosis induced in a cell-free system: application to ceramide-induced apoptosis. *Exp. Cell Res.* **1997**, *236*, 397–403.
20. Kuchiiwa, S.; Kuchiiwa, T. Agent for targeting drug to cerebral neuron. PCT Int. Appl. WO 2007086587, 2007; *Chem. Abstr.* **2007**, *147*, 220081.
21. Chalifour, R. J.; Kong, X.; Wu, X.; Lu, W.; Tidwell, R. R.; Boykin, D. Amidine derivatives for treating amyloidosis and neurodegenerative diseases. PCT Int. Appl. WO 2003103598, 2003; *Chem. Abstr.* **2003**, *140*, 35966.
22. Chalifour, R. J.; Kong, X.; Wu, X.; Lu, W. Amidine derivatives for treating amyloid-related diseases. PCT Int. Appl. WO 2003017994, 2003; *Chem. Abstr.* **2003**, *138*, 215350.
23. Chi, M.; Li, C. Fluorochrome in monitoring atmospheric bioaerosols and correlations with meteorological factors and air pollutants. *Aerosol Sci. Technol.* **2007**, *41*, 672–678.

DARROW RED

CAS Registry Number 15391-59-0

Chemical Structure

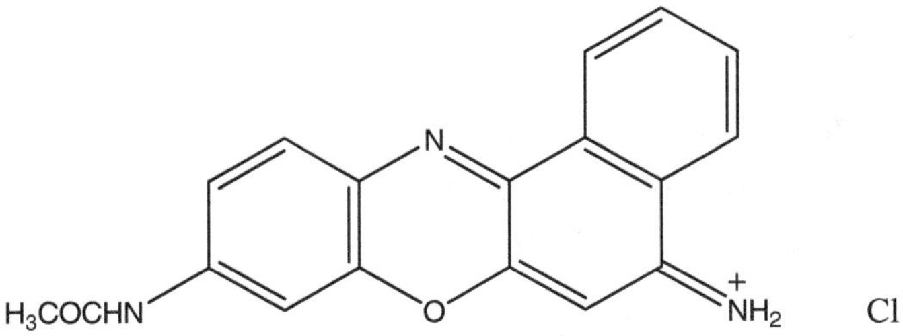

CA Index Name Acetamide, *N*-(5-imino-5*H*-benzo[*a*]phenoxazin-9-yl)-chloride

Other Names 9-Acetylamino-5-aminobenzo[α]phenoxazonium chloride; 5*H*-Benzo[*a*]phenoxazine, acetamide derivative; Darrow red; NSC 364163

Merck Index Number Not listed

Chemical/Dye Class Phenoxazine

Molecular Formula $C_{18}H_{14}ClN_3O_2$

Molecular Weight 339.79

Physical Form Brown powder

Solubility Soluble in water, ethanol

Boiling Point (Calcd.) 536.3 ± 50.0 °C, pressure: 760 Torr

pK_a (Calcd.) 14.07 ± 0.20, most acidic, temperature: 25 °C; 3.33 ± 0.20, most basic, temperature: 25 °C

Absorption (λ_{max}) 502 nm

Synthesis Synthetic methods[1,2]

Staining Applications Brain;[3,4] spinal cord;[3,4] cells;[5,6] leukocytes[7]

Biological Applications Treatment of diabetes-associated pain[8]

Industrial Applications Pesticides[9]

Safety/Toxicity No data available

REFERENCES

1. Powers, M. M.; Clark, G. Darrow Red, a new basic dye. *Stain Technol.* **1960**, *35*, 19–21.
2. Green, F. J. Polychromasia of Cresyl Violet dyes; a chemical explanation of variations in pre-World War II products. *Stain Technol.* **1966**, *41*, 223–228.
3. Augulis, V.; Sigg, E. B. Supravital staining and fixation of brain and spinal cord by intravascular perfusion. *Stain Technol.* **1971**, *46*, 183–190.
4. Augulis, V.; Sepinwall, J. Brazilin-Toluidine Blue O and Hematoxylin-Darrow Red methods for brain and spinal cord. *Stain Technol.* **1969**, *44*, 131–137.
5. Nakamoto, H.; Fujiwara, C. Reagent and method for analyzing cells in urine. Eur. Pat. Appl. EP 513762, 1992; *Chem. Abstr.* **1993**, *118*, 35449.
6. Sievers, J. Basic two-dye stains for epoxy-embedded 0.3–1 μ sections. *Stain Technol.* **1971**, *46*, 195–199.
7. Maekawa, Y. Classification of leukocytes by flow cytometry using ethidium bromide and other dyes. Jpn. Kokai Tokkyo Koho JP 04184168, 1992; *Chem. Abstr.* **1993**, *118*, 3410.
8. Buschmann, H. H. Use of compounds binding to the sigma receptor for the treatment of diabetes-associated pain. PCT Int. Appl. WO 2007025613, 2007; *Chem. Abstr.* **2007**, *146*, 309358.
9. Nadian, A. K. Photostabilization of pesticides in microsphere formulations. PCT Int. Appl. WO 2006077394, 2006; *Chem. Abstr.* **2006**, *145*, 139211.

DASPEI

CAS Registry Number 3785-01-1

Chemical Structure

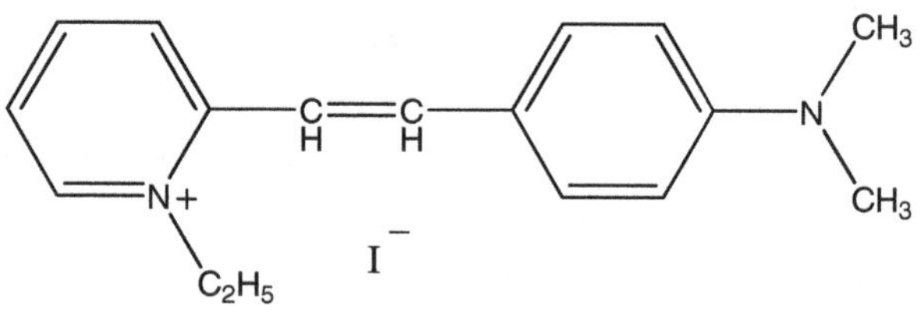

CA Index Name Pyridinium, 2-[2-[4-(dimethylamino) phenyl]ethenyl]-1-ethyl-, iodide (1 : 1)

Other Names 2-[*p*-(Dimethylamino)styryl]-1-ethylpyridinium iodide; Pyridinium, 2-[2-[4-(dimethylamino) phenyl]ethenyl]-1-ethyl-, iodide; Pyridinium, 2-[*p*-(dimethylamino)styryl]-1-ethyl-, iodide; 2-[4-(Dimethylamino)styryl]-1-ethylpyridinium iodide; D 426; DASPEI; NK 557; Pinaflavol

Merck Index Number Not listed

Chemical/Dye Class Styryl

Molecular Formula $C_{17}H_{21}IN_2$

Molecular Weight 380.27

Physical Form Red powder

Solubility Soluble in methanol, *N,N*-dimethylformamide, dimethyl sulfoxide

Melting Point 267 °C (decompose)

Absorption (λ_{max}) 461 nm

Emission (λ_{max}) 589 nm

Synthesis Synthetic method[1–7]

Staining Applications Mitochondria;[8–10,19] amyloid plaques;[11] bacteria;[12] epidermal cells;[13] nerve terminals;[14,15] phospholipid vesicles;[16] hairs[17,18]

Biological Applications Detecting prostate cancer;[19] drug screening assays;[20] treating amyloidosis disorder[11]

Industrial Applications Photoresists;[21] photography[22]

Safety/Toxicity No data available

REFERENCES

1. Antonious, M. S.; Mahmoud, M. R.; Guirguis, D. B. Solvent polarity indicators: mass spectral properties of some styryl pyridinium and quinolinium salts. *Ann. Chim.* **1993**, *83*, 457–460.
2. Zhmurova, I. N.; Yurchenko, R. I.; Kirsanov, A. V. Auxochromic action of a phosphazo group. III. *Zh. Obshch. Khim.* **1970**, *40*, 982–986; *Chem. Abstr.* **1970**, *73*, 99985.
3. Phillips, A. P. Synthetic curare substitutes from stilbazoline bisquaternary ammonium salts. *J. Am. Chem. Soc.* **1952**, *74*, 3683–3685.
4. Takahashi, T.; Satake, K.; Nomura, N.; Yoshikawa, K.; Sawata, M. Syntheses of heterocyclic compounds of nitrogen. LXII. *Yakugaku Zasshi* **1952**, *72*, 42–45; *Chem. Abstr.* **1952**, *46*, 67054.
5. Brooker, L. G. S.; Keyes, G. H.; Sprague, R. H.; VanDyke, R. H.; VanLare, E.; VanZandt, G.; White, F. L.; Cressman, H. W. J.; Dent, S. G., Jr. Color and constitution. X. Absorption of the merocyanines. *J. Am. Chem. Soc.* **1951**, *73*, 5332–5350.
6. Crippa, G. B.; Maffei, S. The reactivity of picoline alkiodides. *Gazz. Chim. Ital.* **1947**, *77*, 416–421.
7. Doja, M. Q.; Prasad, D. J. The cyanine dyes of the pyridine series. II. *J. Indian Chem. Soc.* **1942**, *19*, 125–129.
8. Hattori, F.; Fukuda, K. Method for selecting myocardial cells using intracellular mitochondria labeled with fluorescent indicator. PCT Int. Appl. WO 2006022377, 2006; *Chem. Abstr.* **2006**, *144*, 270175.
9. Sakamoto, T.; Yokota, S.; Ando, M. Rapid morphological oscillation of mitochondrion-rich cell in estuarine mudskipper following salinity changes. *J. Exp. Zool.* **2000**, *286*, 666–669.
10. Jezek, P.; Borecký, J. Mitochondrial uncoupling protein may participate in futile cycling of pyruvate and other monocarboxylates. *Am. J. Physiol.* **1998**, *275*, C496–C504.
11. Gervais, F.; Kong, X.; Chalifour, R.; Migneault, D. Amyloid targeting imaging agents and uses thereof. U.S. Pat. Appl. Publ. US 2005048000, 2005; *Chem. Abstr.* **2005**, *142*, 285149.
12. Sedgwick, E. G.; Bragg, P. D. The role of efflux systems and the cell envelope in fluorescence changes of the lipophilic cation 2-(4-dimethylaminostyryl)-1-ethylpyridinium in *Escherichia coli*. *Biochim. Biophys. Acta* **1996**, *1278*, 205–212.
13. Leise, E. M. Selective retention of the fluorescent dye DASPEI in a larval gastropod mollusc after paraformaldehyde fixation. *Microsc. Res. Tech.* **1996**, *33*, 496–500.

14. Herrera, A. A.; Banner, L. R. The use and effects of vital fluorescent dyes: observation of motor nerve terminals and satellite cells in living frog muscles. *J. Neurocytol.* **1990**, *19*, 67–83.

15. Magrassi, L.; Purves, D.; Lichtman, J. W. Fluorescent probes that stain living nerve terminals. *J. Neurosci.* **1987**, *7*, 1207–1214.

16. Sedgwick, E. G.; Bragg, P. D. Mechanism of uptake of the fluorescent dye 2-(4-dimethylaminostyryl)-1-ethylpyridinium cation (DMP+) by phospholipid vesicles. *Biochim. Biophys. Acta* **1993**, *1146*, 113–120.

17. Pastore, F.; Gourlaouen, L.; Lagrange, A. Hair dye brightener composition for human keratinic fibers. Fr. Demande FR 2830189, 2003; *Chem. Abstr.* **2003**, *138*, 292373.

18. Moeller, H.; Oberkobusch, D.; Hoeffkes, H. Hair dye compositions containing aromatic aldehydes or ketones. Ger. Offen. DE 19936911, 2001; *Chem. Abstr.* **2001**, *134*, 152366.

19. Dickman, D. Methods of detecting prostate cancer. PCT Int. Appl. WO 2006054296, 2006; *Chem. Abstr.* **2006**, *145*, 3346.

20. Dykens, J. A.; Velicelebi, G.; Ghosh, S. S. Compositions and methods for assaying subcellular conditions and processes using energy transfer for drug screening. PCT Int. Appl. WO 2000079274, 2000; *Chem. Abstr.* **2000**, *134*, 66121.

21. Toba, Y.; Yamaguchi, T.; Yasuike, M. Photopolymerizable composition. Jpn. Kokai Tokkyo Koho JP 06348011, 1994; *Chem. Abstr.* **1995**, *122*, 303020.

22. Thiry, H. Preparation and properties of ultrafine grain silver bromide emulsions. *J. Photogr. Sci.* **1987**, *35*, 150–154.

DASPMI

CAS Registry Number 959-81-9

Chemical Structure

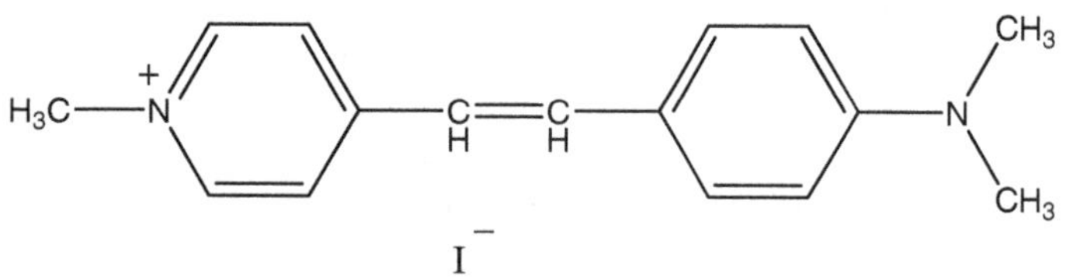

CA Index Name Pyridinium, 4-[2-[4-(dimethylamino) phenyl]ethenyl]-1-methyl-, iodide (1 : 1)

Other Names 4-[*p*-(Dimethylamino)styryl]-1-methylpyridinium iodide; Pyridinium, 4-[2-[4-(dimethylamino) phenyl]ethenyl]-1-methyl-, iodide; Pyridinium, 4-[*p*-(dimethylamino)styryl]-1-methyl-, iodide; 4-[2-[4-(Dimethylamino)phenyl]ethenyl]-1-methylpyridinium iodide; 4-[4-(Dimethylamino)-α-styryl]-1-methylpyridinium iodide; 4-[4-(Dimethylamino)styryl]-1-methylpyridinium iodide; 4-[4-(Dimethylamino)styryl]-*N*-methylpyridinium iodide; 4′-Dimethylamino-1-methylstilbazolium iodide; 4′-Dimethylamino-*N*-methyl-4-stilbazolium iodide; D 288; DASPMI; DASPI; 4-Di-1-ASP; ω-(*N*′-Methylpyridyl-4′)-4-dimethylaminostyrene iodide

Merck Index Number Not listed

Chemical/Dye Class Styryl

Molecular Formula $C_{16}H_{19}IN_2$

Molecular Weight 366.24

Physical Form Solid

Solubility Soluble in methanol, *N,N*-dimethylformamide

Melting Point 261–262 °C

Absorption (λ_{max}) 475 nm

Emission (λ_{max}) 605 nm

Synthesis Synthetic method[1–21]

Staining Applications Mitochondria;[22–24] cells;[1] amyloid plaques[25]

Biological Applications Detecting nucleic acids,[26] prostate cancer;[27] treating amyloidosis disorders,[25] bacterial infection[28]

Industrial Applications Organic thin films;[29] nonlinear optical materials;[13] photoconductors;[30] glass[31]

Safety/Toxicity No data available

REFERENCES

1. Chang, Y. T.; Li, Q.; Rosania, G. RNA-selective styryl probes for live cell imaging of nuclear structure and function. U.S. Pat. Appl. Publ. US 2008064037, 2008; *Chem. Abstr.* **2008**, *148*, 350590.
2. Vasilev, A.; Deligeorgiev, T.; Gadjev, N.; Kaloyanova, S.; Vaquero, J. J.; Alvarez-Builla, J.; Baeza, A. G. Novel environmentally benign procedures for the synthesis of styryl dyes. *Dyes Pigm.* **2008**, *77*, 550–555.
3. Balanda, A. O.; Volkova, K. D.; Kovalska, V. B.; Losytskyy, M. Yu.; Lukashov, S. S.; Yarmoluk, S. M. Novel styryl cyanines and their dimers as fluorescent dyes for nucleic acids detection: synthesis and spectral-luminescent studies. *Ukr. Bioorg. Acta* **2006**, *4*, 17–29.
4. Chang, Y.; Rosania, G. Combinatorial fluorescent library based on the styryl scaffold. U.S. Pat. Appl. Publ. US 2005054006, 2005; *Chem. Abstr.* **2005**, *142*, 294281.
5. Wang, L.; Zhang, X.; Shi, Y.; Zhang, Z. Microwave-assisted solvent-free synthesis of some hemicyanine dyes. *Dyes Pigm.* **2004**, *62*, 21–25.
6. Umezawa, H.; Tsuji, K.; Okada, S.; Oikawa, H.; Matsuda, H.; Nakanishi, H. Molecular design on substituted DAST derivatives for second-order nonlinear optics. *Opt. Mater.* **2003**, *21*, 75–78.
7. Berneth, H.; Bruder, F.; Haese, W.; Hagen, R.; Hassenrueck, K.; Kostromine, S.; Landenberger, P.; Oser, R.; Sommermann, T.; Stawitz, J.; Bieringer, T. Optical data carrier comprising a hemicyanine dye in the information layer as light-absorbing compound. PCT Int. Appl. WO 2002086879, 2002; *Chem. Abstr.* **2002**, *137*, 343938.
8. Zhang, J.; Wu, Y.; Wang, Z.; Li, F.; Jin, L. Electrochemical properties of 4-[2-4(dimethylamino)phenyl] ethenylpyridine *N*-oxide. *Wuli Huaxue Xuebao* **2000**, *16*, 362–365; *Chem. Abstr.* **2000**, *133*, 44889.
9. Shah, S. S.; Ahmad, R.; Shah, S. W. H.; Asif, K. M.; Naeem, K. Synthesis of cationic hemicyanine dyes and their interactions with ionic surfactants. *Colloids Surf., A* **1998**, *137*, 301–305.
10. Stewart, K. R. Preparation of high melting point stilbazolium salts as second harmonic generators. U.S. Patent 5292888, 1994; *Chem. Abstr.* **1994**, *120*, 270132.
11. Jing, H.; Wang, X.; Hou, Z.; Chen, S.; Li, D. Synthesis, characterization and structure of

complexes of 4-[2-(4-dimethylamino phenyl)-vinyl]-1-methylpyridinium penta- and hexanitrato lanthanides. *Polyhedron* **1994**, *13*, 1035–1038.

12. Matsui, M.; Kawamura, S.; Shibata, K.; Muramatsu, H. Synthesis and characterization of mono-, bis-, and tris-substituted pyridinium and pyrylium dyes. *Bull. Chem. Soc. Jpn.* **1992**, *65*, 71–74.
13. Okada, S.; Matsuda, H.; Nakanishi, H.; Kato, M.; Muramatsu, R. Organic nonlinear optical materials. Jpn. Kokai Tokkyo Koho JP 63048265, 1988; *Chem. Abstr.* **1988**, *109*, 219268.
14. Barni, E.; Savarino, P.; Larovere, R.; Viscardi, G.; Pelizzetti, E. Long-chain heterocyclic dyes. Part I. Hydrophobic structures. *J. Heterocyclic Chem.* **1986**, *23*, 209–221.
15. Kramer, D. N.; Bisauta, L. P.; Bato, R. Kinetics of the condensation of *N*-methyl-4-picolinium iodide with *p*-dimethylaminobenzaldehyde in aqueous ethanol. *J. Org. Chem.* **1974**, *39*, 3132–3136.
16. Sheinkman, A. K.; Prilepskaya, A. N.; Kolomoitsev, L. R.; Kost, A. N. 4-*p*-Dialkylaminophenylpyri-dinium salts: a new group of bactericides, fungi-cides, and herbicides. *Vestnik Moskov. Univ., Ser. 2, Khim.* **1964**, *19*, 74–82; *Chem. Abstr.* **1965**, *62*, 58806
17. Sheinkman, A. K.; Rudenko, N. Z.; Kazarinova, N. F.; Lysenko, V. B. Structure of quaternary salts of 4-(*p*-dimethylaminophenyl) and 4-(*p*-dimethylaminostyryl)pyridines. *Zh. Obshch. Khim.* **1963**, *33*, 1964–1969; *Chem. Abstr.* **1963**, *59*, 69042.
18. Freifelder, M. Hydrogenation of pyridinecarboxylic acids with platinum catalyst. *J. Org. Chem.* **1962**, *27*, 4046.
19. Finkelstein, J.; Lee, J. Piperidine derivatives. U.S. Patent 2686784, 1954; *Chem. Abstr.* **1955**, *49*, 56932.
20. Phillips, A. P. Condensation of aromatic aldehydes with 4-picoline methiodide. *J. Org. Chem.* **1949**, *14*, 302–305.
21. Doja, M. Q.; Prasad, K. B. Cyanine dyes of the pyridine series. *V. J. Indian Chem. Soc.* **1947**, *24*, 301–306.
22. Swayne, T. C.; Gay, A. C.; Pon, L. A. Visualization of mitochondria in budding yeast. *Methods Cell Biol.* **2007**, *80*, 591–626.
23. Hattori, F.; Fukuda, K. Method for selecting myocardial cells using intracellular mitochondria labeled with fluorescent indicator. PCT Int. Appl. WO 2006022377, 2006; *Chem. Abstr.* **2006**, *144*, 270175.
24. Villa, A. M.; Doglia, S. M. Mitochondria in tumor cells studied by laser scanning confocal microscopy. *J. Biomed. Opt.* **2004**, *9*, 385–394.
25. Gervais, F.; Kong, X.; Chalifour, R.; Migneault, D. Amyloid targeting imaging agents and uses thereof. U.S. Pat. Appl. Publ. US 2005048000, 2005; *Chem. Abstr.* **2005**, *142*, 285149.
26. Yashchuk, V. M.; Kudrya, V. Y.; Losytskyy, M. Y.; Tokar, V. P.; Yarmoluk, S. M.; Dmytruk, I. M.; Prokopets, V. M.; Kovalska, V. B.; Balanda, A. O.; Kryvorotenko, D. V.; Ogul'chansky, T. Y. The optical biomedical sensors for the DNA detection and imaging based on two-photon excited luminescent styryl dyes: phototoxic influence on the DNA. *Proc. SPIE-Int. Soc. Opt. Eng.* **2007**, *6796*, 67960M/1–67960M/14.
27. Dickman, D. Methods of detecting prostate cancer. PCT Int. Appl. WO 2006054296, 2006; *Chem. Abstr.* **2006**, *145*, 3346.
28. Mendz, G. L.; Hazell, S. L. Inhibitors of fumarate reductase for treating *Helicobacter pylori* infection. PCT Int. Appl. WO 9427606, 1994; *Chem. Abstr.* **1995**, *122*, 96487.
29. Forrest, S. R.; Burrows, P.; Ban, V. S. Low pressure vapor phase deposition of organic thin films. PCT Int. Appl. WO 9925894, 1999; *Chem. Abstr.* **1999**, *130*, 359585.
30. Narasimharaghavan, P. K.; Yadav, H. O.; Varadarajan, T. S.; Patnaik, L. N.; Das, S. Organic photoconductors: dark and photoconduction studies in two *p*-dimethylamino styryl dyes derived from pyridine-2 and pyridine-4. *J. Mater. Sci.* **1991**, *26*, 4774–4786.
31. Imashita, K.; Yokokura, S. Glass containing photofunctional organic compound and its manufacture. Jpn. Kokai Tokkyo Koho JP 07064135, 1995; *Chem. Abstr.* **1995**, *122*, 302623.

Di-2-ANEPEQ

CAS Registry Number 160605-94-7

Chemical Structure

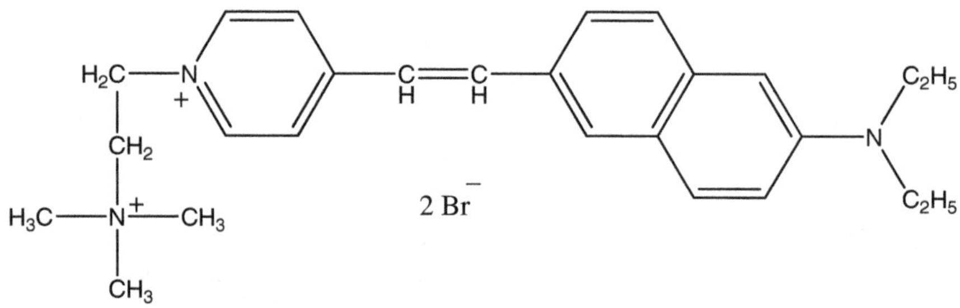

CA Index Name Pyridinium, 4-[2-[6-(diethylamino)-2-naphthalenyl]ethenyl]-1-[2-(trimethylammonio)ethyl]-, bromide (1 : 2)

Other Names Pyridinium, 4-[2-[6-(diethylamino)-2-naphthalenyl]ethenyl]-1-[2-(trimethylammonio)ethyl]-, dibromide; Di-2-ANEPEQ; JPW 1114

Merck Index Number Not listed

Chemical/Dye Class Styryl

Molecular Formula $C_{26}H_{35}Br_2N_3$

Molecular Weight 549.39

Physical Form Solid

Solubility Soluble in water, ethanol, dimethyl sulfoxide

Melting Point >200 °C

Absorption (λ_{max}) 517 nm

Emission (λ_{max}) 721 nm

Synthesis Synthetic method[1]

Staining Applications Dorsal cochlear nucleus;[2] neurons[3]

Biological Applications Measuring membrane potential;[4] examining activity of ion channels[5]

Industrial Applications Not reported

Safety/Toxicity No data available

REFERENCES

1. Loew, L. M. Voltage-sensitive dyes and imaging neuronal activity. *NeuroProtocols* **1994**, *5*, 72–79.
2. Kaltenbach, J. A.; Zhang, J. S. *In vivo* optical imaging of tone-evoked activity in the dorsal cochlear nucleus with a voltage sensitive dye. *J. Neurosci. Res.* **2004**, *78*, 908–917.
3. Tsau, Y.; Wenner, P.; O'Donovan, M. J.; Cohen, L. B.; Loew, L. M.; Wuskell, J. P. Dye screening and signal-to-noise ratio for retrogradely transported voltage-sensitive dyes. *J. Neurosci. Methods* **1996**, *70*, 121–129.
4. Bullen, A.; Saggau, P. Indicators and optical configuration for simultaneous high-resolution recording of membrane potential and intracellular calcium using laser scanning microscopy. *Pfluegers Arch.* **1998**, *436*, 788–796.
5. Ebneth, A.; Netzer, R.; Hahn, U. Method for examining the activity of ion channels. PCT Int. Appl. WO 2005069008, 2005; *Chem. Abstr.* **2005**, *143*, 149507.

Di-4-ANEPPS

CAS Registry Number 90134-00-2

Chemical Structure

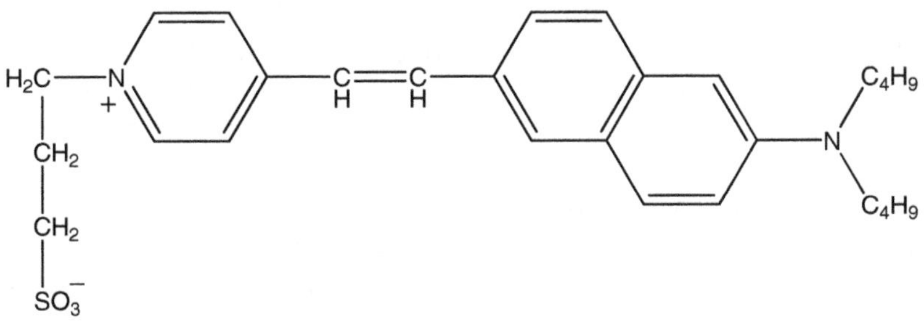

CA Index Name Pyridinium, 4-[2-[6-(dibutylamino)-2-naphthalenyl]ethenyl]-1-(3-sulfopropyl)-, inner salt

Other Names 6-[2-(*N*,*N*-Dibutylamino)naphthyl]ethenyl-4′-pyridinium propanesulfonate; 4-(2-(6-(Dibutylamino)-2-naphthalenyl)ethenyl)-1-(3-sulfopropyl)pyridinium hydroxide inner salt; Di-4-ANEPPS; JPW 211; 1-(3-Sulfonatopropyl)4-[β-[2-(di-*n*-butylamino)-6-naphthyl] vinyl]pyridinium betaine

Merck Index Number Not listed

Chemical/Dye Class Styryl

Molecular Formula $C_{28}H_{36}N_2O_3S$

Molecular Weight 480.66

Physical Form Orange to red powder

Solubility Soluble in ethanol, methanol, *N*,*N*-dimethylformamide, dimethyl sulfoxide

Melting Point 122–124 °C

Absorption (λ_{max}) 495 nm

Emission (λ_{max}) 705 nm

Synthesis Synthetic method[1]

Staining Applications Bacteria;[2] cardiac tissues;[3–6] myocardium;[7] neurons;[8–11] proteins[12]

Biological Applications Measuring membrane potential;[13–16] preventing arrhythmias;[17] probes for Na,K-ATPase reaction mechanism;[18] assays for identifying taste-specific genes;[19] quantum dots[20]

Industrial Applications Monitoring polymerization[21]

Safety/Toxicity Cardiac toxicity;[7] photodynamic damage[22]

REFERENCES

1. Hassner, A.; Birnbaum, D.; Loew, L. M. Charge-shift probes of membrane potential: synthesis. *J. Org. Chem.* **1984**, *49*, 2546–2551.
2. Teuber, M.; Rogner, M.; Berry, S. Fluorescent probes for non-invasive bioenergetic studies of whole cyanobacterial cells. *Biochim. Biophys. Acta Bioenerg.* **2001**, *1506*, 31–46.
3. Knisley, S. B.; Pollard, A. E. Use of translucent indium tin oxide to measure stimulatory effects of a passive conductor during field stimulation of rabbit hearts. *Am. J. Physiol.* **2005**, *289*, H1137–H1146.
4. Dobrzynski, H.; Nikolski, V. P.; Sambelashvili, A. T.; Greener, I. D.; Yamamoto, M.; Boyett, M. R.; Efimov, I. R. Site of origin and molecular substrate of atrioventricular junctional rhythm in the rabbit heart. *Circ. Res.* **2003**, *93*, 1102–1110.
5. Nygren, A.; Kondo, C.; Clark, R. B.; Giles, W. R. Voltage-sensitive dye mapping in Langendorff-perfused rat hearts. *Am. J. Physiol.* **2003**, *284*, H892–H902.
6. Hooks, D. A.; LeGrice, I. J.; Harvey, J. D.; Smaill, B. H. Intramural multisite recording of transmembrane potential in the heart. *Biophys. J.* **2001**, *81*, 2671–2680.
7. Novakova, M.; Bardonova, J.; Provaznik, I.; Taborska, E.; Bochorakova, H.; Paulova, H.; Horky, D. Effects of voltage sensitive dye di-4-ANEPPS on guinea pig and rabbit myocardium. *Gen. Physiol. Biophys.* **2008**, *27*, 45–54.
8. Ito, Y.; Oyamada, Y.; Okada, Y.; Hakuno, H.; Aoyama, R.; Yamaguchi, K. Optical mapping of pontine chemosensitive regions of neonatal rat. *Neurosci. Lett.* **2004**, *366*, 103–106.

9. Tsau, Y.; Wenner, P.; O'Donovan, M. J.; Cohen, L. B.; Loew, L. M.; Wuskell, J. P. Dye screening and signal-to-noise ratio for retrogradely transported voltage-sensitive dyes. *J. Neurosci. Methods* **1996**, *70*, 121–129.
10. Kent, P. F.; Mozell, M. M.; Murphy, S. J.; Hornung, D. E. The interaction of imposed and inherent olfactory mucosal activity patterns and their composite representation in a mammalian species using voltage-sensitive dyes. *J. Neurosci.* **1996**, *16*, 345–353.
11. Loew, L. M. Voltage-sensitive dyes and imaging neuronal activity. *NeuroProtocols* **1994**, *5*, 72–79.
12. Haugland, R. P.; Singer, V. L.; Jones, L. J.; Steinberg, T. H. Merocyanine dye protein stains. PCT Int. Appl. WO 9636882, 1996; *Chem. Abstr.* **1997**, *126*, 44635.
13. Millard, A. C; Jin, L.; Wei, M. D.; Wuskell, J. P.; Lewis, A.; Loew, L. M. Sensitivity of second harmonic generation from styryl dyes to transmembrane potential. *Biophys. J.* **2004**, *86*, 1169–1176.
14. Xu, C.; Loew, L. M. The effect of asymmetric surface potentials on the intramembrane electric field measured with voltage-sensitive dyes. *Biophys. J.* **2003**, *84*, 2768–2780.
15. Laurita, K. R.; Singal, A. Mapping action potentials and calcium transients simultaneously from the intact heart. *Am. J. Physiol.* **2001**, *280*, H2053–H2060.
16. Murakami, T. Cytotoxicity test method by measuring membrane electric potential. Jpn. Kokai Tokkyo Koho JP 2000300290, 2000; *Chem. Abstr.* **2000**, *133*, 330852.
17. Marban, E.; Abraham, M. R. Preventing arrhythmias associated with cell transplantation. PCT Int. Appl. WO 2005092033, 2005; *Chem. Abstr.* **2005**, *143*, 319156.
18. Fedosova, N. U.; Cornelius, F.; Klodos, I. Fluorescent styryl dyes as probes for Na,K-ATPase reaction mechanism: Significance of the charge of the hydrophilic moiety of RH dyes. *Biochemistry* **1995**, *34*, 16806–16814.
19. Moyer, B.; Zlotnik, A.; Hevezi, P.; Soto, H.; Kalabat, D.; Lu, M.; Gao, N.; White, E. Rationale, methods, and assays for identifying human and non-human primate taste-specific genes and use thereof in taste modulator and therapeutic screening assays. PCT Int. Appl. WO 2008153924, 2008; *Chem. Abstr.* **2008**, *150*, 49063.
20. Nadeau, J. L.; Cohen, N. Use of quantum dots for biological labels and sensors. PCT Int. Appl. WO 2006037226, 2006; *Chem. Abstr.* **2006**, *144*, 385738.
21. Wang, F. W.; Lowry, R. E. Wavelength-shift fluorescent probes for monitoring of polymerization. *J. Appl. Polym. Sci.* **2006**, *101*, 747–750.
22. Schaffer, P.; Ahammer, H.; Muller, W.; Koidl, B.; Windisch, H. Di-4-ANEPPS causes photodynamic damage to isolated cardiomyocytes. *Eur. J. Physiol.* **1994**, *426*, 548–551.

Di-8-ANEPPS

CAS Registry Number 157134-53-7

Chemical Structure

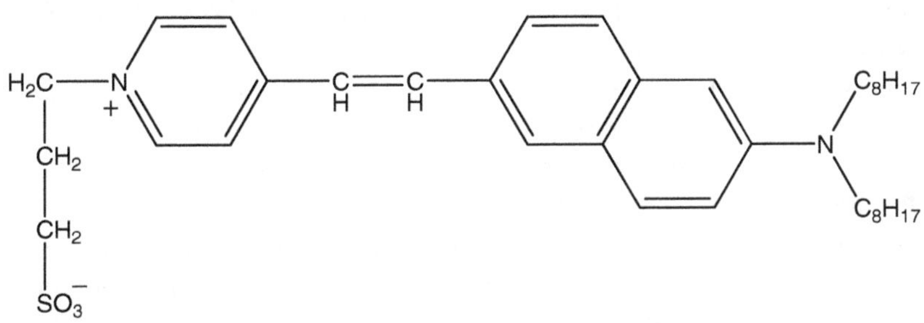

CA Index Name Pyridinium, 4-[2-[6-(dioctylamino)-2-naphthalenyl]ethenyl]-1-(3-sulfopropyl)-, inner salt

Other Names Di-8-ANEPPS; 4-(2-[6-(Dioctylamino)-2-naphthalenyl]ethenyl)-1-(3-sulfopropyl)pyridinium inner salt; JPW 1153

Merck Index Number Not listed

Chemical/Dye Class Styryl

Molecular Formula $C_{36}H_{52}N_2O_3S$

Molecular Weight 592.88

Physical Form Orange to red powder

Solubility Soluble in ethanol, methanol, dimethyl sulfoxide

Melting Point >200 °C

Absorption (λ_{max}) 498 nm

Emission (λ_{max}) 713 nm

Synthesis Synthetic methods[1,2]

Staining Applications Cardiac myocytes;[3] lipid membrane surfaces;[4] lipid particles;[5] neurons;[6,7] prestin;[8] proteins;[9] sea urchin eggs[19]

Biological Applications Measuring membrane potential;[10–18] imaging exocytosis at fertilization[19]

Industrial Applications Not reported

Safety/Toxicity No data available

REFERENCES

1. Gross, E.; Bedlack, R. S., Jr.; Loew, L. M. Dual-wavelength ratiometric fluorescence measurement of the membrane dipole potential. *Biophys. J.* **1994**, *67*, 208–216.
2. Loew, L. M. Voltage-sensitive dyes and imaging neuronal activity. *NeuroProtocols* **1994**, *5*, 72–79.
3. Hardy, M. E. L.; Lawrence, C. L.; Standen, N. B.; Rodrigo, G. C. Can optical recordings of membrane potential be used to screen for drug-induced action potential prolongation in single cardiac myocytes? *J. Pharmacol. Toxicol. Methods* **2006**, *54*, 173–182.
4. Le Goff, G.; Vitha, M. F.; Clarke, R. J. Orientational polarizability of lipid membrane surfaces. *Biochim. Biophys. Acta, Biomembr.* **2007**, *1768*, 562–570.
5. Doranz, B. J. Lipid particles containing ion channels and membrane potential-sensitive dyes and their use in screening for effectors of ion channels. PCT Int. Appl. WO 2007089582, 2007; *Chem. Abstr.* **2007**, *147*, 250560.
6. Fisher, J. A. N.; Salzberg, B. M.; Yodh, A. G. Near infrared two-photon excitation cross-sections of voltage-sensitive dyes. *J. Neurosci. Methods* **2005**, *148*, 94–102.
7. Tsau, Y.; Wenner, P.; O'Donovan, M. J.; Cohen, L. B.; Loew, L. M.; Wuskell, J. P. Dye screening and signal-to-noise ratio for retrogradely transported voltage-sensitive dyes. *J. Neurosci. Methods* **1996**, *70*, 121–129.
8. Santos-Sacchi, J.; Zhao, H. B. Excitation of fluorescent dyes inactivates the outer hair cell integral membrane motor protein prestin and betrays its lateral mobility. *Pfluegers Arch.* **2003**, *446*, 617–622.
9. Haugland, R. P.; Singer, V. L.; Jones, L. J.; Steinberg, T. H. Merocyanine dye protein stains. PCT Int. Appl. WO 9636882, 1996; *Chem. Abstr.* **1997**, *126*, 44635.
10. Rusu, C. F.; Lanig, H.; Othersen, O. G.; Kryschi, C.; Clark, T. Monitoring biological membrane-potential changes: a CI QM/MM Study. *J. Phys. Chem. B* **2008**, *112*, 2445–2455.

11. Vitha, M. F.; Clarke, R. J. Comparison of excitation and emission ratiometric fluorescence methods for quantifying the membrane dipole potential. *Biochim. Biophys. Acta, Biomembr.* **2007**, *1768*, 107–114.
12. Kataoka, T.; Matsubara, M.; Hisada, M. Membrane-potential change detecting method, drug screening method, and well plate manufacture. Jpn. Kokai Tokkyo Koho JP 2006126073, 2006; *Chem. Abstr.* **2006**, *144*, 447614.
13. Chen, Y.; Rivers, R. J. Measurement of membrane potential and intracellular Ca^{2+} of arteriolar endothelium and smooth muscle *in vivo*. *Microvasc. Res.* **2001**, *62*, 55–62.
14. Murakami, T. Cytotoxicity test method by measuring membrane electric potential. Jpn. Kokai Tokkyo Koho JP 2000300290, 2000; *Chem. Abstr.* **2000**, *133*, 330852.
15. Koyano, T. Fluorescent method for monitoring change of membrane potential and intracellular ions. Jpn. Kokai Tokkyo Koho JP 09021800, 1997; *Chem. Abstr.* **1997**, *126*, 183503.
16. Koyano, T. Simultaneous measurement of cellular ion concentration and membrane potential with fluorescent pigments. Jpn. Kokai Tokkyo Koho JP 09005243, 1997; *Chem. Abstr.* **1997**, *126*, 168805.
17. Clarke, R. J.; Kane, D. J. Optical detection of membrane dipole potential: avoidance of fluidity and dye-induced effects. *Biochim. Biophys. Acta, Biomembr.* **1997**, *1323*, 223–239.
18. Hayashi, Y.; Zviman, M. M.; Brand, J. G.; Tetter, J. H.; Restrepo, D. Measurement of membrane potential and [Ca^{2+}] in cell ensembles: application to the study of glutamate taste in mice. *Biophys. J.* **1996**, *71*, 1057–1070.
19. Millard, A. C.; Terasaki, M.; Loew, L. M. Second harmonic imaging of exocytosis at fertilization. *Biophys. J.* **2005**, *88*, L46–L48.

DiBAC$_4$(3)

CAS Registry Number 70363-83-6

Chemical Structure

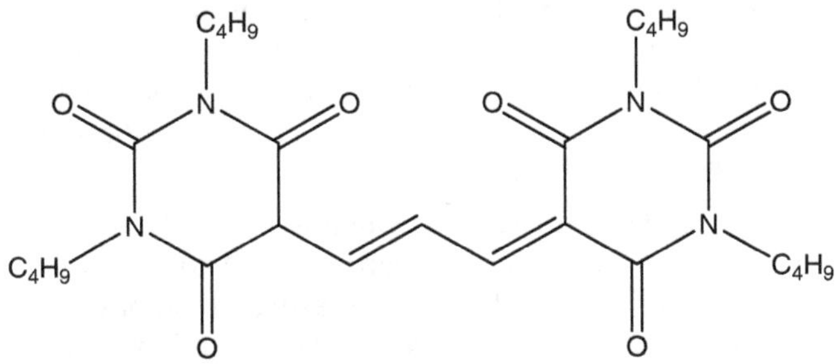

CA Index Name 2,4,6(1*H*,3*H*,5*H*)-Pyrimidinetrione, 1,3-dibutyl-5-[3-(1,3-dibutylhexahydro-2,4,6-trioxo-5-pyrimidinyl)-2-propen-1-ylidene]-

Other Names 2,4,6(1*H*,3*H*,5*H*)-Pyrimidinetrione, 1,3-dibutyl-5-[3-(1,3-dibutylhexahydro-2,4,6-trioxo-5-pyrimidinyl)-2-propenylidene]-; Bis(1,3-dibutylbarbituric acid) trimethine oxonol; DiBAC$_4$(3)

Merck Index Number Not listed

Chemical/Dye Class Oxonol

Molecular Formula $C_{27}H_{40}N_4O_6$

Molecular Weight 516.63

Physical Form Orange or brown-red powder

Solubility Soluble in ethanol, methanol, *N,N*-dimethylformamide, dimethyl sulfoxide

Melting Point >200 °C

Boiling Point (Calcd.) 614.3 ± 65.0 °C, pressure: 760 Torr

pK$_a$ (Calcd.) 4.20 ± 0.20, most acidic, temperature: 25 °C; −2.17 ± 0.20, most basic, temperature: 25 °C

Absorption (λ_{max}) 493 nm

Emission (λ_{max}) 516 nm

Synthesis Synthetic methods[1–8]

Staining Applications Bacteria;[9–13] α- and β-cells in islets of Langerhans;[14] *Bacillus* spores;[15] cells;[16] fungi;[17,18] yeast[19,20]

Biological Applications Measuring membrane potential;[21–28] HCN channel modulators;[29] BK channel activator;[30] treating Alzheimer's disease;[31] antimicrobial susceptibility test;[2] ion channel screening assays;[32] hERG channel assay[33]

Industrial Applications Photographic materials[1]

Safety/Toxicity Neurotoxicity[34]

REFERENCES

1. Suzuki, K.; Wariishi, K. Silver halide photographic material containing (poly)methineoxonol solid dispersed dye in colloidal layer. Jpn. Kokai Tokkyo Koho JP 06332112, 1994; *Chem. Abstr.* **1995**, *122*, 201116.
2. Gauthier, C.; St-Pierre, Y.; Villemur, R. Rapid antimicrobial susceptibility testing of urinary tract isolates and samples by flow cytometry. *J. Med. Microbiol.* **2002**, *51*, 192–200.
3. Connell, C.; Rutter, A.; Hill, B.; Suller, M.; Lloyd, D. Encystation of *Acanthamoeba castellanii*: dye uptake for assessment by flow cytometry and confocal laser scanning microscopy. *J. Appl. Microbiol.* **2001**, *90*, 706–712.
4. Peter, M., Jr.; Varga, Z.; Panyi, G.; Bene, L.; Damjanovich, S.; Pieri, C.; Possani, L. D.; Gaspar, R., Jr. Pandinus imperator scorpion venom blocks voltage-gated K^+ channels in human lymphocytes. *Biochem. Biophys. Res. Commun.* **1998**, *242*, 621–625.
5. Breittmayer, J. P.; Pelassy, C.; Aussel, C. Effect of membrane potential on phosphatidylserine synthesis and calcium movements in control and CD3-activated Jurkat T cells. *J. Lipid Mediators Cell Signalling* **1996**, *13*, 151–161.
6. Civitelli, R.; Reid, I. R.; Halstead, L. R.; Avioli, L. V.; Hruska, K. A. Membrane potential and cation content of osteoblast-like cells (UMR 106) assessed by fluorescent dyes. *J. Cell. Physiol.* **1987**, *131*, 434–441.
7. Brauner, T.; Huelser, D. F.; Strasser, R. J. Comparative measurements of membrane potentials with microelectrodes and voltage-sensitive dyes. *Biochim. Biophys. Acta, Biomembr.* **1984**, *771*, 208–216.
8. Russell, J. T.; Beeler, T.; Martonosi, A. Optical probe responses on sarcoplasmic reticulum: merocyanine and oxonol dyes. *J. Biol. Chem.* **1979**, *254*, 2047–2052.
9. Alakomi, H. L.; Maettoe, J.; Virkajaervi, I.; Saarela, M. Application of a microplate scale fluorochrome staining assay for the assessment of viability of probiotic preparations. *J. Microbiol. Methods* **2005**, *62*, 25–35.

10. Haidinger, W.; Szostak, M. P.; Jechlinger, W.; Lubitz, W. Online monitoring of *Escherichia coli* ghost production. *Appl. Environ. Microbiol.* **2003**, *69*, 468–474.
11. Jepras, R. I.; Paul, F. E.; Pearson, S. C.; Wilkinson, M. J. Rapid assessment of antibiotic effects on *Escherichia coli* by bis-(1,3-dibutylbarbituric acid) trimethine oxonol and flow cytometry. *Antimicrob. Agents Chemother.* **1997**, *41*, 2001–2005.
12. Deere, D.; Porter, J.; Edwards, C.; Pickup, R. Evaluation of the suitability of bis-(1,3-dibutyl-barbituric acid)trimethine oxonol, (diBA-C$_4$(3)-), for the flow cytometric assessment of bacterial viability. *FEMS Microbiol. Lett.* **1995**, *130*, 165–170.
13. Mason, D. J.; Allman, R.; Stark, J. M.; Lloyd, D. Rapid estimation of bacterial antibiotic susceptibility with flow cytometry. *J. Microsc.* **1994**, *176*, 8–16.
14. Hjortoe, G. M.; Hagel, G. M.; Terry, B. R.; Thastrup, O.; Arkhammar, P. O. G. Functional identification and monitoring of individual α and β cells in cultured mouse islets of Langerhans. *Acta Diabetol.* **2004**, *41*, 185–193.
15. Laflamme, C.; Ho, J.; Veillette, M.; Latremoille, M. C.; Verreault, D.; Meriaux, A.; Duchaine, C. Flow cytometry analysis of germinating *Bacillus* spores, using membrane potential dye. *Arch. Microbiol.* **2005**, *183*, 107–112.
16. Watabe, E.; Nakai, T.; Matsumoto, S.; Ikeda, F.; Hatano, K. Killing activity of micafungin against *Aspergillus fumigatus* hyphae assessed by specific fluorescent staining for cell viability. *Antimicrob. Agents Chemother.* **2003**, *47*, 1995–1998.
17. Bowman, J. C.; Hicks, P. S.; Kurtz, M. B.; Rosen, H.; Schmatz, D. M.; Liberator, P. A.; Douglas, C. M. The antifungal echinocandin caspofungin acetate kills growing cells of *Aspergillus fumigatus in vitro*. *Antimicrob. Agents Chemother.* **2002**, *46*, 3001–3012.
18. Liao, R. S.; Rennie, R. P.; Talbot, J. A. Assessment of the effect of amphotericin B on the vitality of *Candida albicans*. *Antimicrob. Agents Chemother.* **1999**, *43*, 1034–1041.
19. Boyd, A. R.; Gunasekera, T. S.; Attfield, P. V.; Simic, K.; Vincent, S. F.; Veal, D. A. A flow-cytometric method for determination of yeast viability and cell number in a brewery. *FEMS Yeast Res.* **2003**, *3*, 11–16.
20. Deere, D.; Shen, J.; Vesey, G.; Bell, P.; Bissinger, P.; Veal, D. Flow cytometry and cell sorting for yeast viability assessment and cell selection. *Yeast* **1998**, *14*, 147–160.
21. Konrad, K. R.; Hedrich, R. The use of voltage-sensitive dyes to monitor signal-induced changes in membrane potential-ABA triggered membrane depolarization in guard cells. *Plant J.* **2008**, *55*, 161–173.
22. Dumas, D.; Stoltz, J. F. New tool to monitor membrane potential by FRET voltage sensitive dye (FRET-VSD) using spectral and fluorescence lifetime imaging microscopy. *Clin. Hemorheol. Microcirc.* **2005**, *33*, 293–302.
23. Lloyd, D.; Harris, J. C.; Biagini, G. A.; Hughes, M. R.; Maroulis, S.; Bernard, C.; Wadley, R. B.; Edwards, M. R. The plasma membrane of microaerophilic protists: oxidative and nitrosative stress. *Microbiology* **2004**, *150*, 1183–1190.
24. de Poorter, L. M. I.; Keltjens, J. T. Convenient fluorescence-based methods to measure membrane potential and intracellular pH in the Archaeon, *Methanobacterium thermoautotrophicum*. *J. Microbiol. Methods* **2001**, *47*, 233–241.
25. Okun, I.; Okun, A.; Kaler, G. Identifying compounds that alter membrane biological potentials with fluorescent ionic dyes. U.S. Patent 6287758, 2001; *Chem. Abstr.* **2001**, *135*, 223771.
26. Maric, D.; Maric, I.; Barker, J. L. Dual video microscopic imaging of membrane potential and cytosolic calcium of immunoidentified embryonic rat cortical cells. *Methods* **2000**, *21*, 335–347.
27. Tanner, M. K.; Wellhausen, S. R. Flow cytometric detection of fluorescent redistributional dyes for measurement of cell transmembrane potential. *Methods Mol. Biol.* **1998**, *91*, 85–95.
28. Seamer, L. C.; Mandler, R. N. Method to improve the sensitivity of flow cytometric membrane potential measurements in mouse spinal cord cells. *Cytometry* **1992**, *13*, 545–552.
29. Vasylyev, D.; Lu, Q.; Mark, B.; Edward, K.; Shan, Q. High-throughput screening of modulators of hyperpolarization-activated cyclic nucleotide-gated (HCN) channels. PCT Int. Appl. WO 2008121735, 2008; *Chem. Abstr.* **2008**, *149*, 455063.
30. Morimoto, T.; Sakamoto, K.; Sade, H.; Ohya, S.; Muraki, K.; Imaizumi, Y. Voltage-sensitive oxonol dyes are novel large-conductance Ca^{2+}-activated K^+ channel activators selective for β 1 and β 4 but not for β 2 subunits. *Mol. Pharmacol.* **2007**, *71*, 1075–1088.
31. Ingram, V. M.; Blanchard, B. J.; Stockwell, B. R. Treatments for neurotoxicity in Alzheimer's disease. U.S. Pat. Appl. Publ. US 2003105152, 2003; *Chem. Abstr.* **2003**, *139*, 30831.

32. Wolff, C.; Fuks, B.; Chatelain, P. Comparative study of membrane potential-sensitive fluorescent probes and their use in ion channel screening assays. *J. Biomol. Screening* **2003**, *8*, 533–543.

33. Tang, W.; Kang, J.; Wu, X.; Rampe, D.; Wang, L.; Shen, H.; Li, Z.; Dunnington, D.; Garyantes, T. Development and evaluation of high throughput functional assay methods for hERG potassium channel. *J. Biomol. Screening* **2001**, *6*, 325–331.

34. Okada, J.; Shimokawa, N.; Koibuchi, N. Polychlorinated biphenyl (PCB) alters acid-sensitivity of cultured neurons derived from the medulla oblongata. *Int. J. Biochem. Cell Biol.* **2005**, *37*, 1368–1374.

DiBAC$_4$(5)

CAS Registry Number 63560-89-4

Chemical Structure

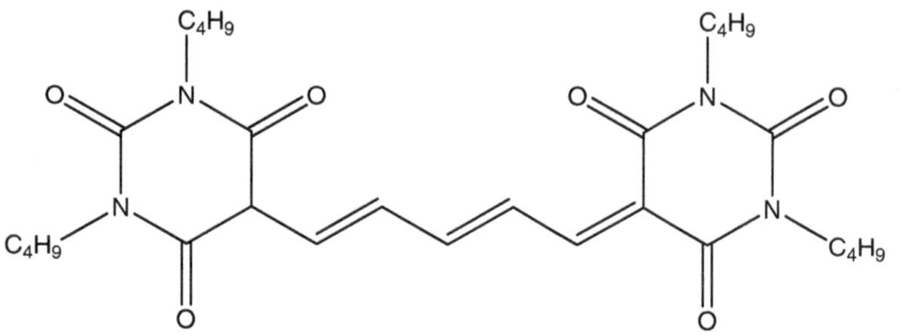

CA Index Name 2,4,6(1*H*,3*H*,5*H*)-Pyrimidinetrione, 1,3-dibutyl-5-[5-(1,3-dibutylhexahydro-2,4,6-trioxo-5-pyrimidinyl)-2,4-pentadien-1-ylidene]-

Other Names Bis(1,3-dibutylbarbituric acid)pentamethine oxonol; 2,4,6(1*H*,3*H*,5*H*)-Pyrimidinetrione, 1,3-dibutyl-5-[5-(1,3-dibutylhexahydro-2,4,6-trioxo-5-pyrimidinyl)-2,4-pentadienylidene]-; DiBAC$_4$(5)

Merck Index Number Not listed

Chemical/Dye Class Oxonol

Molecular Formula $C_{29}H_{42}N_4O_6$

Molecular Weight 542.67

Physical Form Grey powder

Solubility Soluble in ethanol, methanol, dimethyl sulfoxide

Melting Point >200 °C

Boiling Point (Calcd.) 642.6 ± 65.0 °C, pressure: 760 Torr

pK$_a$ (Calcd.) 4.79 ± 0.20, most acidic, temperature: 25 °C; − 2.16 ± 0.20, most basic, temperature: 25 °C

Absorption (λ_{max}) 590 nm

Emission (λ_{max}) 616 nm

Synthesis Synthetic methods[1–4]

Staining Applications Liposomes[5]

Biological Applications Measuring membrane potential;[6–9] HCN channel modulators;[10] potassium channel openers[11]

Industrial Applications Not reported

Safety/Toxicity No data available

REFERENCES

1. Molotkovskii, Yu. G.; Yakovleva, G. A. Oxonol dye diBa-C$_4$-(5) as a fluorescent probe for membrane potential in chloroplasts and its interaction with valinomycin. *Photosynthetica* **1985**, *19*, 493–499.
2. Russell, J. T.; Beeler, T.; Martonosi, A. Optical probe responses on sarcoplasmic reticulum: merocyanine and oxonol dyes. *J. Biol. Chem.* **1979**, *254*, 2047–2052.
3. Walsh Kinnally, K.; Tedeschi, H.; Maloff, B. L. Use of dyes to estimate the electrical potential of the mitochondrial membrane. *Biochemistry* **1978**, *17*, 3419–3428.
4. Kinnally, K. W.; Tedeschi, H. Metabolic effects of some electrofluorimetric dyes. *Biochim. Biophys. Acta, Bioenerg.* **1978**, *503*, 380–388.
5. Rajendra, J.; Damianoglou, A.; Hicks, M.; Booth, P.; Rodger, P. M.; Rodger, A. Quantitation of protein orientation in flow-oriented unilamellar liposomes by linear dichroism. *Chem. Phys.* **2006**, *326*, 210–220.
6. Steinberg, B. E.; Touret, N.; Vargas-Caballero, M.; Grinstein, S. *In situ* measurement of the electrical potential across the phagosomal membrane using FRET and its contribution to the proton-motive force. *Proc. Natl. Acad. Sci. U.S.A.* **2007**, *104*, 9523–9528.
7. Okun, I.; Okun, A.; Kaler, G. Identifying compounds that alter membrane biological potentials with fluorescent ionic dyes. U.S. Patent 6287758, 2001; *Chem. Abstr.* **2001**, *135*, 223771.
8. Farinas, J. A.; Wada, H. G. Microfluidic devices and use of Nernstein voltage sensitive dyes in measuring transmembrane voltage. PCT Int. Appl. WO 2001027253, 2001; *Chem. Abstr.* **2001**, *134*, 307569.

9. Ross, W. N.; Salzberg, B. M.; Cohen, L. B.; Grinvald, A.; Davila, H. V.; Waggoner, A. S.; Wang, C. H. Changes in absorption, fluorescence, dichroism, and birefringence in stained giant axons: optical measurement of membrane potential. *J. Membr. Biol.* **1977**, *33*, 141–183.

10. Vasylyev, D.; Lu, Q.; Mark, B.; Edward, K.; Shan, Q. High-throughput screening of modulators of hyperpolarization-activated cyclic nucleotide-gated (HCN) channels. PCT Int. Appl. WO 2008121735, 2008; *Chem. Abstr.* **2008**, *149*, 455063.

11. Imaizumi, Y.; Oya, S.; Morimoto, T. Compounds with oligo methine chain containing oxooxoanion stable stereostructure as potassium channel openers. Jpn. Kokai Tokkyo Koho JP 2008266161, 2008; *Chem. Abstr.* **2008**, *149*, 525486.

DiD

CAS Registry Number 127274-91-3

Chemical Structure

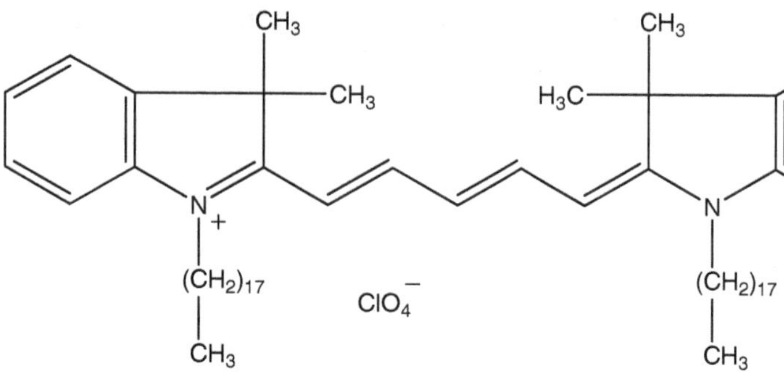

CA Index Name 3*H*-Indolium, 2-[5-(1,3-dihydro-3,3-dimethyl-1-octadecyl-2*H*-indol-2-ylidene)-1,3-pentadien-1-yl]-3,3-dimethyl-1-octadecyl-, perchlorate (1 : 1)

Other Names 3*H*-Indolium, 2-[5-(1,3-dihydro-3,3-dimethyl-1-octadecyl-2*H*-indol-2-ylidene)-1,3-pentadienyl]-3,3-dimethyl-1-octadecyl-, perchlorate; 1,1′-Dioctadecyl-3,3,3′,3′-tetramethylindodicarbocyanine perchlorate; D 307; DiD; DiD oil; DiIC18(5); Lipophilic dye DiD

Merck Index Number Not listed

Chemical/Dye Class Cyanine

Molecular Formula $C_{61}H_{99}ClN_2O_4$

Molecular Weight 959.90

Physical Form Dark brown crystals

Solubility Soluble in ethanol, methanol, dimethyl sulfoxide

Absorption (λ_{max}) 644 nm

Emission (λ_{max}) 665 nm

Synthesis Synthetic method[1]

Staining Applications Cells;[2] circulating apoptotic cells;[3] liposomes;[4,5] membranes;[6–9] neurons[10–12]

Biological Applications Detecting cell lysis,[13] microorganisms,[14] nucleic acids,[15] lipid diffusion in giant unilamellar vesicles,[16] lipid diffusion in phospholipid bilayers,[16] lipids and proteins diffusion in membranes;[17,18] monitoring cell migration[19]

Industrial Applications Optical recording materials;[20] thin films[21]

Safety/Toxicity No data available

REFERENCES

1. Adier, C.; Thomas, M.; Turpin, F.; Brigand, C.; Cenatiempo, Y.; Belgsir, E. M. Method for obtaining water-soluble fluorescent complexes from lipophilic dyes and cyclodextrin derivatives. Fr. Demande FR 2878853, 2006; *Chem. Abstr.* **2006**, *145*, 23859.
2. Lichtman, J. W.; Wong, W. T.; Wong, R.; Gan, W. B.; Grutzendler, J. Method for labeling individual cells. PCT Int. Appl. WO 2001069244, 2001; *Chem. Abstr.* **2001**, *135*, 238975.
3. Lin, C. P.; Wei, X.; Sipkins, D. *In-vivo* monitoring of circulating apoptotic cells by flow cytometry. U.S. Pat. Appl. Publ. US 2006134005, 2006; *Chem. Abstr.* **2006**, *145*, 42686.
4. Ise, H.; Kobayashi, S.; Takahashi, M.; Aso, S.; Ikeda, Y.; Goto, M. *N*-acetylglucosamine sugar chain-recognizing proteins and drug delivery systems having *N*-acetylglucosamine compounds. Jpn. Kokai Tokkyo Koho JP 2009046413, 2009; *Chem. Abstr.* **2009**, *150*, 290771.
5. Ngo, A. T.; Karam, P.; Fuller, E.; Burger, M.; Cosa, G. Liposome encapsulation of conjugated polyelectrolytes: toward a liposome beacon. *J. Am. Chem. Soc.* **2008**, *130*, 457–459.
6. Chiantia, S.; Ries, J.; Kahya, N.; Schwille, P. Combined AFM and two-focus SFCS study of raft-exhibiting model membranes. *ChemPhysChem* **2006**, *7*, 2409–2418.
7. Bacia, K.; Scherfeld, D.; Kahya, N.; Schwille, P. Fluorescence correlation spectroscopy relates rafts in model and native membranes. *Biophys. J.* **2004**, *87*, 1034–1043.
8. Krishna, M. M. G.; Periasamy, N. Fluorescence of organic dyes in lipid membranes: site of solubilization

and effects of viscosity and refractive index on lifetimes. *J. Fluoresc.* **1998**, *8*, 81–91.

9. Bloom, J. A.; Webb, W. W. Photodamage to intact erythrocyte membranes at high laser intensities: methods of assay and suppression. *J. Histochem. Cytochem.* **1984**, *32*, 608–616.
10. Sumner, J. P.; Conroy, R.; Shapiro, E. M.; Moreland, J.; Koretsky, A. P. Delivery of fluorescent probes using iron oxide particles as carriers enables *in-vivo* labeling of migrating neural precursors for magnetic resonance imaging and optical imaging. *J. Biomed. Opt.* **2007**, *12*, 051504/1–051504/6.
11. Mangat, G.; Brzozowski, L. Intra-operative head and neck nerve mapping. PCT Int. Appl. WO 2007016790, 2007; *Chem. Abstr.* **2007**, *146*, 223533.
12. Hutter, H. Five-colour *in vivo* imaging of neurons in *Caenorhabditis elegans*. *J. Microsc.* **2004**, *215*, 213–218.
13. Nixon, D.; McDermott, A. B.; Furlan, S.; Bigos, M.; Sheehy, M.; Klenerman, P. Methods of detecting specific cell lysis. U.S. Pat. Appl. Publ. US 2002146746, 2002; *Chem. Abstr.* **2002**, *137*, 291273.
14. Pindzola, B. A.; Nguyen, A. T.; Reppy, M. A. Antibody-functionalized polydiacetylene coatings on nanoporous membranes for microorganism detection. *Chem. Commun.* **2006**, 906–908.
15. Miller, B. L.; Krauss, T. D.; Du, H.; Crnkovich, N.; Strohsahl, C. M. Use of sensor arrays containing hairpin probes for detecting nucleic acids of pathogens. PCT Int. Appl. WO 2004061127, 2004; *Chem. Abstr.* **2004**, *141*, 118285.
16. Przybylo, M.; Sykora, J.; Humpolickova, J.; Benda, A.; Zan, A.; Hof, M. Lipid diffusion in giant unilamellar vesicles is more than 2 times faster than in supported phospholipid bilayers under identical conditions. *Langmuir* **2006**, *22*, 9096–9099.
17. Bloom, J. A.; Webb, W. W. Lipid diffusibility in the intact erythrocyte membrane. *Biophys. J.* **1983**, *42*, 295–305.
18. Vaz, W. L. C.; Goodsaid-Zalduondo, F.; Jacobson, K. Lateral diffusion of lipids and proteins in bilayer membranes. *FEBS Lett.* **1984**, *174*, 199–207.
19. Kaiser, J. P.; Bruinink, A. Investigating cell-material interactions by monitoring and analysing cell migration. *J. Mater. Sci.: Mater. Med.* **2004**, *15*, 429–435.
20. Noguchi, S.; Nagano, S.; Onuma, Y.; Matsushita, Y.; Seki, T. Optical recording material using amphipathic block copolymer phase-separated monolayer, and its manufacturing method. Jpn. Kokai Tokkyo Koho JP 2006312253, 2006; *Chem. Abstr.* **2006**, *145*, 497726.
21. Ishihara, M. Functional thin film containing photochromic compound and organic dye, and its manufacture. Jpn. Kokai Tokkyo Koho JP 2009042438, 2009; *Chem. Abstr.* **2009**, *150*, 272115.

DIHYDROETHIDIUM

CAS Registry Number 104821-25-2

Chemical Structure

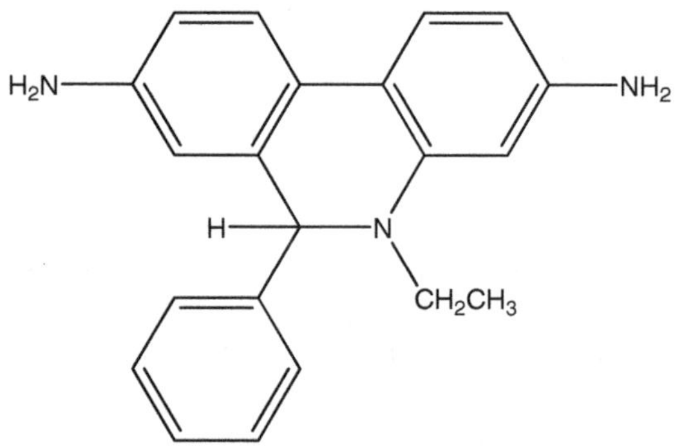

CA Index Name 3,8-Phenanthridinediamine, 5-ethyl-5,6-dihydro-6-phenyl-

Other Names 3,8-Phenanthridinediamine, 5-ethyl-5,6-dihydro-6-phenyl-, (±)-; Dihydroethidium; Hydroethidine; HE; HYD; PD-MY 003

Merck Index Number Not listed

Chemical/Dye Class Phenanthridine

Molecular Formula $C_{21}H_{21}N_3$

Molecular Weight 315.41

Physical Form Pink to dark brown powder

Solubility Soluble in *N,N*-dimethylformamide, dimethyl sulfoxide, acetonitrile, methanol, chloroform

Boiling Point (Calcd.) 580.4 ± 50.0 °C, pressure: 760 Torr

pK_a (Calcd.) 5.30 ± 0.40, most basic, temperature: 25 °C

Absorption (λ_{max}) 355 nm

Emission (λ_{max}) 420 nm

Synthesis Synthetic methods[1–4]

Staining Applications Superoxide ions;[5–15] nucleic acids;[16–18] cells;[19,20,29] myocardium;[21] bacteria;[22] prokaryotes[23]

Biological Applications Apoptosis assay;[24,25] generating and detecting reactive oxygen species (ROS);[5–15,26,27] detecting nucleic acids,[16–18] cells;[19,20,29] measuring respiratory burst;[28] superoxide indicator;[5–15,26,27] viability assay[29,30]

Industrial Applications Assaying reactive oxidants in smoke[31]

Safety/Toxicity Metabolic toxicity;[32] neutotoxicity;[33–36] parasite toxicity;[37] vascular toxicity[38]

REFERENCES

1. Loccufier, J.; Schacht, E. Regioselective acylation of 3,8-diamino-5-ethyl-6-phenylphenanthridinium bromide. Preparation of potential new trypanocides. *Tetrahedron* **1989**, *45*, 3385–3396.
2. Loccufier, J.; Schacht, E. Synthesis of anionic and hydrophobic derivatives of ethidium bromide as potential new trypanocides. *Bull. Soc. Chim. Belg.* **1989**, *98*, 145–152.
3. Kinsey, B. M.; Kassis, A. I.; Fayad, F.; Layne, W. W.; Adelstein, S. J. Synthesis and biological studies of iodinated (127/125I) derivatives of rhodamine 123. *J. Med. Chem.* **1987**, *30*, 1757–1761.
4. Mori, Y.; Yoshida, M. 3,8-Diamino-5-alkyl-6-phenyl-5,6-dihydrophenanthridines. Jpn. Kokai Tokkyo Koho JP 49045077, 1974; *Chem. Abstr.* **1974**, *81*, 120342.
5. Lam, C.; Liu, Y.; Tseng, F.; Sung, Y.; Huang, C.; Jiang, M.; Tsai, Y. High-dose morphine impairs vascular endothelial function by increased production of superoxide anions. *Anesthesiology* **2007**, *106*, 532–537.
6. Fernandes, D. C.; Wosniak, J., Jr.; Pescatore, L. A.; Bertoline, M. A.; Liberman, M.; Laurindo, F. R. M.; Santos, C. X. C. Analysis of DHE-derived oxidation products by HPLC in the assessment of superoxide production and NADPH oxidase activity in vascular systems. *Am. J. Physiol.* **2007**, *292*, C413–C422.
7. Burnaugh, L.; Sabeur, K.; Ball, B. A. Generation of superoxide anion by equine spermatozoa as detected by dihydroethidium. *Theriogenology* **2007**, *67*, 580–589.
8. Zielonka, J.; Vasquez-Vivar, J.; Kalyanaraman, B. The confounding effects of light, sonication, and Mn(III)TBAP on quantitation of superoxide using hydroethidine. *Free Radical Biol. Med.* **2006**, *41*, 1050–1057.
9. Georgiou, C. D.; Papapostolou, I.; Patsoukis, N.; Tsegenidis, T.; Sideris, T. An ultrasensitive fluorescent assay for the *in vivo* quantification of superoxide radical in organisms. *Anal. Biochem.* **2005**, *347*, 144–151.

10. Zanetti, M.; d'Uscio, L. V.; Peterson, T. E.; Katusic, Z. S.; O'Brien, T. Analysis of superoxide anion production in tissue. *Methods Mol. Med.* **2005**, *108*, 65–72.

11. Patsoukis, N.; Papapostolou, I.; Georgiou, C. D. Interference of non-specific peroxidases in the fluorescence detection of superoxide radical by hydroethidine oxidation: a new assay for H_2O_2. *Anal. Bioanal. Chem.* **2005**, *381*, 1065–1072.

12. Bindokas, V. P.; Kuznetsov, A.; Sreenan, S.; Polonsky, K. S.; Roe, M. W.; Philipson, L. H. Visualizing superoxide production in normal and diabetic rat Islets of Langerhans. *J. Biol. Chem.* **2003**, *278*, 9796–9801.

13. Szucs, S.; Vamosi, G.; Poka, R.; Sarvary, A.; Bardos, H.; Balazs, M.; Kappelmayer, J.; Toth, L.; Szollosi, J.; Adany, R. Single-cell measurement of superoxide anion and hydrogen peroxide production by human neutrophils with digital imaging fluorescence microscopy. *Cytometry* **1998**, *33*, 19–31.

14. Budd, S. L.; Castilho, R. F.; Nicholls, D. G. Mitochondrial membrane potential and hydroethidine-monitored superoxide generation in cultured cerebellar granule cells. *FEBS Lett.* **1997**, *415*, 21–24.

15. Bindokas, V. P.; Jordan, J.; Lee, C. C.; Miller, R. J. Superoxide production in rat hippocampal neurons: selective imaging with hydroethidine. *J. Neurosci.* **1996**, *16*, 1324–1336.

16. Park, H. O.; Kim, H. B.; Chi, S. M. Detection method of DNA amplification using probes labeled with intercalating dyes. PCT Int. Appl. WO 2006004267, 2006; *Chem. Abstr.* **2006**, *144*, 122707.

17. Ashley, N.; Harris, D.; Poulton, J. Detection of mitochondrial DNA depletion in living human cells using PicoGreen staining. *Exp. Cell Res.* **2005**, *303*, 432–446.

18. Kim, G. W.; Noshita, N.; Sugawara, T.; Chan, P. H. Early decrease in DNA repair proteins, Ku70 and Ku86, and subsequent DNA fragmentation after transient focal cerebral ischemia in mice. *Stroke* **2001**, *32*, 1401–1407.

19. Anderson, A. L.; Knutson, C. R.; Mueth, D.; Plewa, J.; Tanner, E. Methods for staining cells for identification and sorting. U.S. Pat. Appl. Publ. US 2006172315, 2006; *Chem. Abstr.* **2006**, *145*, 183714.

20. Posen, Y.; Kalchenko, V.; Seger, R.; Brandis, A.; Scherz, A.; Salomon, Y. Manipulation of redox signaling in mammalian cells enabled by controlled photogeneration of reactive oxygen species. *J. Cell Sci.* **2005**, *118*, 1957–1969.

21. Lu, L.; Liu, Y.; Sun, C.; Lin, L.; Su, M.; Wu, C. Optical mapping of myocardial reactive oxygen species production throughout the reperfusion of global ischemia. *J. Biomed. Opt.* **2006**, *11*, 021012/1–021012/7.

22. Yano, R.; Nogami, S.; Meiga, H.; Toba, Y.; Hiraoka, M.; Noda, N.; Oto, T.; Tada, H. Method and apparatus for detecting bacteria by fluorescent staining using bacteriophage. Jpn. Kokai Tokkyo Koho JP 2000342300, 2000; *Chem. Abstr.* **2000**, *134*, 27268.

23. Swannell, R. P. J.; Caplin, R.; Nedwell, D. B.; Williamson, F. A. An investigation of hydroethidine as a fluorescent vital stain for prokaryotes. *FEMS Microbiol. Ecol.* **1992**, *101*, 173–182.

24. Molina-Jimenez, M. F.; Sanchez-Reus, M. I.; Andres, D.; Cascales, M.; Benedi, J. Neuroprotective effect of fraxetin and myricetin against rotenone-induced apoptosis in neuroblastoma cells. *Brain Res.* **2004**, *1009*, 9–16.

25. Valente, E.; Assis, M. C.; Alvim, I. M.; Pereira, G. M.; Plotkowski, M. C. Microbial *Pseudomonas aeruginosa* induces apoptosis in human endothelial cells. *Pathogenesis* **2000**, *29*, 345–356.

26. Rinaldi, M.; Moroni, P.; Paape, M. J.; Bannerman, D. D. Evaluation of assays for the measurement of bovine neutrophil reactive oxygen species. *Vet. Immunol. Immunopathol.* **2007**, *115*, 107–125.

27. Luo, J.; Li, N.; Paul Robinson, J.; Shi, R. Detection of reactive oxygen species by flow cytometry after spinal cord injury. *J. Neurosci. Methods* **2002**, *120*, 105–112.

28. Filatov, M. V.; Varfolomeeva, E. Y.; Ivanov, E. I. Flow cytofluorometric detection of inflammatory processes by measuring respiratory burst reaction of peripheral blood neutrophils. *Biochem. Mol. Med.* **1995**, *55*, 116–121.

29. Millard, P. J.; Roth, B. L.; Yue, S. T.; Haugland, R. P. Fluorescent viability assay using cyclic-substituted unsymmetrical cyanine dyes. U.S. Patent 5534416, 1996; *Chem. Abstr.* **1996**, *125*, 162751.

30. Breeuwer, P.; Drocourt, J. Method for assessing microorganism viability. PCT Int. Appl. WO 9500660, 1995; *Chem. Abstr.* **1995**, *122*, 155727.

31. Huang, D.; Ou, B. Method for assaying reactive oxidants in smoke. U.S. Pat. Appl. Publ. US 2004126891, 2004; *Chem. Abstr.* **2004**, *141*, 66653.

32. Downing, T. W.; Garner, D. L.; Ericsson, S. A.; Redelman, D. Metabolic toxicity of fluorescent stains on thawed cryopreserved bovine sperm cells. *J. Histochem. Cytochem.* **1991**, *39*, 485–489.

33. Berry, E. V.; Toms, N. J. Pyruvate and oxaloacetate limit zinc-induced oxidative HT-22 neuronal cell injury. *Neurotoxicology* **2006**, *27*, 1043–1051.

34. Qiao, S.; Li, W.; Tsubouchi, R.; Haneda, M.; Murakami, K.; Yoshino, M. Involvement of peroxynitrite in capsaicin-induced apoptosis of C6 glioma cells. *Neurosci. Res.* **2005**, *51*, 175–183.

35. Sheline, C. T.; Wang, H.; Cai, A.; Dawson, V. L.; Choi, D. W. Involvement of poly ADP ribosyl polymerase-1 in acute but not chronic zinc toxicity. *Eur. J. Neurosci.* **2003**, *18*, 1402–1409.

36. Howard, S.; Bottino, C.; Brooke, S.; Cheng, E.; Giffard, R. G.; Sapolsky, R. Neuroprotective effects of bcl-2 overexpression in hippocampal cultures: interactions with pathways of oxidative damage. *J. Neurochem.* **2002**, *83*, 914–923.

37. Vivas, L.; Easton, A.; Kendrick, H.; Cameron, A.; Lavandera, J.; Barros, D.; de las Heras, F. G.; Brady, R. L.; Croft, S. L. *Plasmodium falciparum*: stage specific effects of a selective inhibitor of lactate dehydrogenase. *Exp. Parasitol.* **2005**, *111*, 105–114.

38. Matsui, R.; Xu, S.; Maitland, K. A.; Hayes, A.; Leopold, J. A.; Handy, D. E.; Loscalzo, J.; Cohen, R. A. Glucose-6 phosphate dehydrogenase deficiency decreases the vascular response to angiotensin II. *Circulation* **2005**, *112*, 257–263.

DIHYDRORHODAMINE 6G

CAS Registry Number 217176-83-5

Chemical Structure

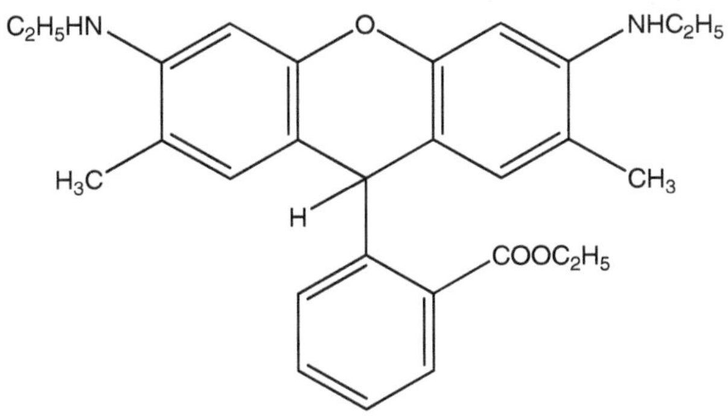

CA Index Name Benzoic acid, 2-[3,6-bis(ethylamino)-2,7-dimethyl-9*H*-xanthen-9-yl]-, ethyl ester

Other Names D 633; Dihydrorhodamine 6G; d-R 6G

Merck Index Number Not listed

Chemical/Dye Class Xanthene

Molecular Formula $C_{28}H_{32}N_2O_3$

Molecular Weight 444.57

Physical Form Solid

Solubility Soluble in *N,N*-dimethylformamide, dimethyl sulfoxide, methanol

Boiling Point (Calcd.) 572.4 ± 50.0 °C, pressure: 760 Torr

pK_a (Calcd.) 4.97 ± 0.40, most basic, temperature: 25 °C

Absorption (λ_{max}) 296 nm

Synthesis Synthetic method[1]

Staining Applications Mitochondria[2]

Biological Applications Detecting genetic anomalies,[3] methylation in DNA,[4] intracellular hydrogen peroxide in tumor cells,[5] intracellular active oxygen and reduced glutathione,[6] gene expression,[7] reactive oxygen species (ROS),[8] small RNAs;[9] nucleic acid ligation assays[10]

Industrial Applications Solar cells[11]

Safety/Toxicity No data available

REFERENCES

1. Haugland, R. P. *The Handbook: A Guide to Fluorescent Probes and Labeling Technologies*; Molecular Probes Inc.: Eugene, **2005**; pp 569–570.
2. Hattori, F.; Fukuda, K. Method for selecting myocardial cells using intracellular mitochondria labeled with fluorescent indicator. PCT Int. Appl. WO 2006022377, 2006; *Chem. Abstr.* **2006**, *144*, 270175.
3. Livak, K. J.; Stevens, J.; Lazaruk, K. D.; Ziegle, J. S.; Wong, L. Y. Detection of genetic anomalies, particularly duplications, using end-point fluorogenic real-time PCR assays and diagnostic uses. U.S. Pat. Appl. Publ. US 2005255485, 2005; *Chem. Abstr.* **2005**, *143*, 454005.
4. Zon, G. Cytosine and 5-methylcytosine selective base analogs for the detection of methylation in DNA. U.S. Patent 7399614, 2008; *Chem. Abstr.* **2008**, *149*, 97037.
5. Qin, Y.; Lu, M.; Gong, X. Dihydrorhodamine 123 is superior to 2,7-dichlorodihydrofluorescein diacetate and dihydro-rhodamine 6G in detecting intracellular hydrogen peroxide in tumor cells. *Cell Biol. Int.* **2008**, *32*, 224–228.
6. Qin, J.; Ye, N.; Jiang, L.; Xie, M.; Lin, B. Method for synchronously detecting intracellular active oxygen and reduced glutathione with high sensitivity. Faming Zhuanli Shenqing Gongkai Shuomingshu CN 1940550, 2007; *Chem. Abstr.* **2007**, *146*, 449286.
7. Lao, K. Q.; Reed, M. Detection of gene expression by PCR using primers with common and target-specific sequences. U.S. Pat. Appl. Publ. US 2006141518, 2006; *Chem. Abstr.* **2006**, *145*, 97399.
8. Ou, B.; Huang, D. Fluorescent approach to quantitation of reactive oxygen species in mainstream cigarette smoke. *Anal. Chem.* **2006**, *78*, 3097–3103.

9. Brandis, J.; Bolchakova, E. V.; Karger, A. E. Detection of small RNAs based on RNA-templated ligation and PCR amplification. U.S. Pat. Appl. Publ. US 2006003337, 2006; *Chem. Abstr.* **2006**, *144*, 82005.
10. Wenz, H. M.; Day, J. P. Methods and controls for determining reaction performance in nucleic acid ligation assays. U.S. Pat. Appl. Publ. US 2006014189, 2006; *Chem. Abstr.* **2006**, *144*, 122737.
11. He, Z.; Xu, Z. Fabrication of energy-saving light-transmitting construction boards integrated with solar cells. Faming Zhuanli Shenqing Gongkai Shuomingshu CN 101294435, 2008; *Chem. Abstr.* **2008**, *149*, 579543.

DIHYDRORHODAMINE 123

CAS Registry Number 109244-58-8

Chemical Structure

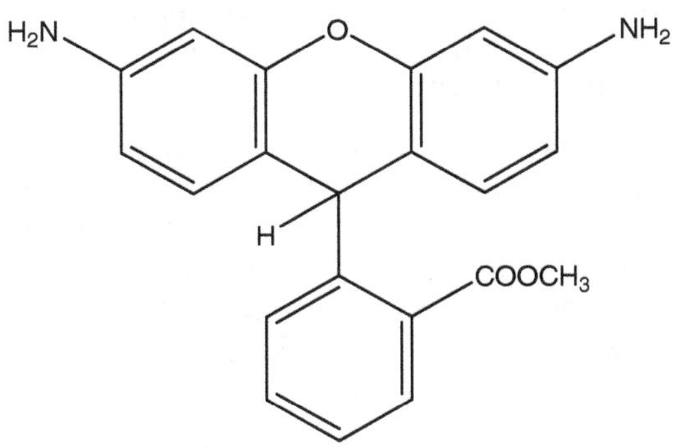

CA Index Name Benzoic acid, 2-(3,6-diamino-9*H*-xanthen-9-yl)-, methyl ester

Other Names D 23806; D 632; Dihydrorhodamine 123

Merck Index Number Not listed

Chemical/Dye Class Xanthene

Molecular Formula $C_{21}H_{18}N_2O_3$

Molecular Weight 346.38

Physical Form Light red or pinkish-white powder

Solubility Soluble in *N,N*-dimethylformamide, dimethyl sulfoxide, methanol

Melting Point 163–165 °C

Boiling Point (Calcd.) 526.9 ± 50.0 °C, pressure: 760 Torr

pK_a (Calcd.) 4.70 ± 0.40, most basic, temperature: 25 °C

Absorption (λ_{max}) 289 nm

Synthesis Synthetic methods[1–3]

Staining Applications Mitochondria;[4,5] blood samples;[6] leukocytes;[6] cells;[7] microbes;[8] granulocytes;[9] epithelial cells;[10] myeloid cells;[11] neutrophils[12]

Biological Applications Generating, detecting & measuring reactive oxygen species;[13–18] generating, detecting & measuring reactive nitrogen species;[18,19] detecting phosphates,[20] peroxynitrile;[21] measuring respiratory burst;[22] chronic granulomatous disease DHR assay;[23] screening antioxidant activity assay;[24] peroxidase substrate[25]

Industrial Applications Not reported

Safety/Toxicity Cytotoxicity;[26] hepatotoxicity;[27] nephrotoxicity;[28] neurotoxicity;[29,30] skin toxicity[31]

REFERENCES

1. Kinsey, B. M.; Kassis, A. I.; Adelstein, S. J. Preparation of dihydrorhodamines and their radiohalogenated derivatives as tumor imaging and therapeutic agents. PCT Int. Appl. WO 8706138, 1987; *Chem. Abstr.* **1988**, *108*, 204497.
2. Kinsey, B. M.; Kassis, A. I.; Fayad, F.; Layne, W. W.; Adelstein, S. J. Synthesis and biological studies of iodinated (127/125I) derivatives of rhodamine 123. *J. Med. Chem.* **1987**, *30*, 1757–1761.
3. Millard, P. J.; Roth, B. L.; Yue, S. T.; Haugland, R. P. Fluorescent viability assay using cyclic-substituted unsymmetrical cyanine dyes. U.S. Patent 5534416, 1996; *Chem. Abstr.* **1996**, *125*, 162751.
4. Hattori, F.; Fukuda, K. Method for selecting myocardial cells using intracellular mitochondria labeled with fluorescent indicator. PCT Int. Appl. WO 2006022377, 2006; *Chem. Abstr.* **2006**, *144*, 270175.
5. Sobreira, C.; Davidson, M.; King, M. P.; Miranda, A. F. Dihydrorhodamine 123 identifies impaired mitochondrial respiratory chain function in cultured cells harboring mitochondrial DNA mutations. *J. Histochem. Cytochem.* **1996**, *44*, 571–579.
6. Dambaeva, S. V.; Mazurov, D. V.; Pinegin, B. V. Assessment of active oxygen form production in human peripheral blood using laser flow cytometry. *Immunologiya* **2001**, 58–61.
7. Zeigler, F. C. Determination of cell viability and phenotype by FACS. U.S. Patent 7018804, 2006; *Chem. Abstr.* **2006**, *144*, 307914.
8. Little, R. G., II; Abrahamson, S.; Wong, P. Identification of novel antimicrobial agents using membrane potential indicator dyes. PCT Int. Appl. WO 2000018951, 2000; *Chem. Abstr.* **2000**, *132*, 260670.
9. Prince, H. E.; Lape-Nixon, M. Influence of specimen age and anticoagulant on flow cytometric evaluation of granulocyte oxidative burst generation. *J. Immunol. Methods* **1995**, *188*, 129–138.

10. Wersto, R. P.; Rosenthal, E. R.; Crystal, R. G.; Spring, K. R. Uptake of fluorescent dyes associated with the functional expression of the cystic fibrosis transmembrane conductance regulator in epithelial cells. *Proc. Natl. Acad. Sci. U.S.A.* **1996**, *93*, 1167–1172.

11. Trayner, I. D.; Rayner, A. P.; Freeman, G. E.; Farzaneh, F. Quantitative multiwell myeloid differentiation assay using dichlorodihydrofluorescein diacetate (H2DCF-DA) or dihydrorhodamine 123 (H2R123). *J. Immunol. Methods* **1995**, *186*, 275–284.

12. Szucs, S.; Vamosi, G.; Poka, R.; Sarvary, A.; Bardos, H.; Balazs, M.; Kappelmayer, J.; Toth, L.; Szollosi, J.; Adany, R. Single-cell measurement of superoxide anion and hydrogen peroxide production by human neutrophils with digital imaging fluorescence microscopy. *Cytometry* **1998**, *33*, 19–31.

13. Rinaldi, M.; Moroni, P.; Paape, M. J.; Bannerman, D. D. Evaluation of assays for the measurement of bovine neutrophil reactive oxygen species. *Vet. Immunol. Immunopathol.* **2007**, *115*, 107–125.

14. Ling, Y. Y.; Yin, X. F.; Fang, Z. L. Simultaneous determination of glutathione and reactive oxygen species in individual cells by microchip electrophoresis. *Electrophoresis* **2005**, *26*, 4759–4766.

15. Sun, Y.; Yin, X. F.; Ling, Y. Y.; Fang, Z. L. Determination of reactive oxygen species in single human erythrocytes using microfluidic chip electrophoresis. *Anal. Bioanal. Chem.* **2005**, *382*, 1472–1476.

16. Hanson, K. M.; Clegg, R. M. Two-photon fluorescence imaging and reactive oxygen species detection within the epidermis. *Methods Mol. Biol.* **2004**, *289*, 413–421.

17. Grzelak, A.; Rychlik, B.; Bartosz, G. Light-dependent generation of reactive oxygen species in cell culture media. *Free Radical Biol. Med.* **2001**, *30*, 1418–1425.

18. Crow, J. P. Dichlorodihydrofluorescein and dihydrorhodamine 123 are sensitive indicators of peroxynitrite *in vitro*: implications for intracellular measurement of reactive nitrogen and oxygen species. *Nitric Oxide* **1997**, *1*, 145–157.

19. Ischiropoulos, H.; Gow, A.; Thom, S. R.; Kooy, N. W.; Royall, J. A.; Crow, J. P. Detection of reactive nitrogen species using 2,7-dichlorodihydrofluorescein and dihydrorhodamine 123. *Methods Enzymol.* **1999**, *301*, 367–373.

20. Zhou, M.; Haugland, R. P. Detection of phosphate using coupled enzymatic reactions. U.S. Patent 6265179, 2001; *Chem. Abstr.* **2001**, *135*, 119239.

21. Malcolm, S.; Foust, R.; Hertkom, C.; Ischiropoulos, H. Detection of peroxynitrite in biological fluids. *Methods Mol. Med.* **2000**, *36*, 171–177.

22. Richardson, M. P.; Ayliffe, M. J.; Helbert, M.; Davies, E. G. A simple flow cytometry assay using dihydrorhodamine for the measurement of the neutrophil respiratory burst in whole blood: comparison with the quantitative nitrobluetetrazolium test. *J. Immunol. Methods* **1998**, *219*, 187–193.

23. Jirapongsananuruk, O.; Malech, H. L.; Kuhns, D. B.; Niemela, J. E.; Brown, M. R.; Anderson-Cohen, M.; Fleisher, T. A. Diagnostic paradigm for evaluation of male patients with chronic granulomatous disease, based on the dihydrorhodamine 123 assay. *J. Allergy Clin. Immunol.* **2003**, *111*, 374–379.

24. Dunlap, W.; Llewellyn, L.; Doyle, J.; Yamamoto, Y. A microtiter plate assay for screening antioxidant activity in extracts of marine organisms. *Mar. Biotechnol.* **2003**, *5*, 294–301.

25. Whitaker, J. E.; Moore, P. L.; Haugland, R. P.; Haugland, R. P. Dihydrotetramethylrosamine: a long wave-length, fluorogenic peroxidase substrate evaluated *in vitro* and in a model phagocyte. *Biochem. Biophys. Res. Commun.* **1991**, *175*, 387–393.

26. Choi, J. J.; Kong, M. Y.; Lee, S. J.; Kim, H. C.; Ko, K. H.; Kim, W. K. Ciclopirox prevents peroxynitrite toxicity in astrocytes by maintaining their mitochondrial function: a novel mechanism for cytoprotection by ciclopirox. *Neuropharmacology* **2002**, *43*, 408–417.

27. Weng, D.; Lu, Y.; Wei, Y.; Liu, Y.; Shen, P. The role of ROS in microcystin-LR-induced hepatocyte apoptosis and liver injury in mice. *Toxicology* **2007**, *232*, 15–23.

28. Khand, F. D.; Gordge, M. P.; Robertson, W. G.; Noronha-Dutra, A. A.; Hothersall, J. S. Mitochondrial superoxide production during oxalate-mediated oxidative stress in renal epithelial cells. *Free Radical Biol. Med.* **2002**, *32*, 1339–1350.

29. Vanderveldt, G. M.; Regan, R. F. The neurotoxic effect of sickle cell hemoglobin. *Free Radical Res.* **2004**, *38*, 431–437.

30. Rogers, B.; Yakopson, V.; Teng, Z.; Guo, Y.; Regan, R. F. Heme oxygenase-2 knockout neurons are less vulnerable to hemoglobin toxicity. *Free Radical Biol. Med.* **2003**, *35*, 872–881.

31. Hanson, K. M.; Clegg, R. M. Observation and quantification of ultraviolet-induced reactive oxygen species in *ex vivo* human skin. *Photochem. Photobiol.* **2002**, *76*, 57–63.

DiI

CAS Registry Number 41085-99-8

Chemical Structure

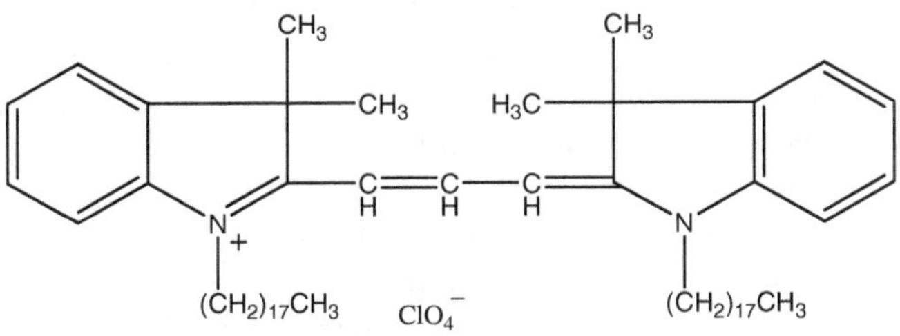

CA Index Name 3*H*-Indolium, 2-[3-(1,3-dihydro-3,3-dimethyl-1-octadecyl-2*H*-indol-2-ylidene)-1-propen-1-yl]-3,3-dimethyl-1-octadecyl-, perchlorate (1 : 1)

Other Names 3*H*-Indolium, 2-[3-(1,3-dihydro-3,3-di-methyl-1-octadecyl-2*H*-indol-2-ylidene)-1-propenyl]-3,3-dimethyl-1-octadecyl, perchlorate; 1,1′-Dioctadecyl-3,3,3′,3′-tetramethylindocarbocyanine perchlorate; D 282; D 282 (dye); DiI; DiI (dye); DiIC18(3)

Merck Index Number Not listed

Chemical/Dye Class Cyanine

Molecular Formula $C_{59}H_{97}ClN_2O_4$

Molecular Weight 933.87

Physical Form Burgundy to purples-violet crystals or powder

Solubility Soluble in ethanol, methanol, *N,N*-dimethylformamide, dimethyl sulfoxide

Melting Point 68 °C (decompose)

Absorption (λ_{max}) 549 nm

Emission (λ_{max}) 565 nm

Synthesis Synthetic methods[1–5]

Staining Applications Axons;[6] bovine brain tissues;[7] mitral/tufted cells;[8] retinal ganglion cells;[9] bacteria;[10] cells;[11] lipid bilayers;[12] lipid monolayers;[13] lipid membranes;[14] lipoproteins;[15,16] liposomes;[17–19] membranes;[20,21] neurons;[22,23] neural tracers;[24,25] vessel;[26] peptides;[3,4] proteins;[3,4] antibodies[3,4]

Biological Applications Detecting cell fusion,[27] cell adhesion,[28] cell migration,[28] cell proliferation;[28] membrane fusion assay;[29] exocytosis and/or endocytosis assay[29]

Industrial Applications Silicon-chip[30]

Safety/Toxicity Neurotoxicity[31]

REFERENCES

1. Zheng, G.; Chen, J.; Li, H.; Glickson, J. D. Rerouting lipoprotein nanoparticles to selected alternate receptors for the targeted delivery of cancer diagnostic and therapeutic agents. *Proc. Natl. Acad. Sci. U.S.A.* **2005**, *102*, 17757–17762.
2. Campbell, A. I.; Bartlett, P. Fluorescent hard-sphere polymer colloids for confocal microscopy. *J. Colloid Interface Sci.* **2002**, *256*, 325–330.
3. Hahn, K. M.; Toutchkine, A.; Muthyala, R.; Kraynov, V.; Bark, S. J.; Burton, D. R.; Chamberlain, C. Labeled peptides, proteins and antibodies and processes and intermediates useful for their preparation. U.S. Pat. Appl. Publ. US 2002055133, 2002; *Chem. Abstr.* **2002**, *136*, 366139.
4. Hahn, K. M.; Toutchkine, A.; Muthyala, R.; Kraynov, V.; Bark, S. J.; Burton, D. R.; Chamberlain, C. Labeled peptides, proteins and antibodies and processes and intermediates useful for their preparation. PCT Int. Appl. WO 2002008245, 2002; *Chem. Abstr.* **2002**, *136*, 163716.
5. Minciacchi, D.; Granato, A.; Antonini, A.; Sbriccoli, A.; Macchi, G. A procedure for the simultaneous visualization of two anterograde and different retrograde fluorescent tracers: application to the study of the afferent–efferent organization of thalamic anterior intralaminar nuclei. *J. Neurosci. Methods* **1991**, *38*, 183–191.
6. Chen, B. K.; Miller, S. M.; Mantilla, C. B.; Gross, L.; Yaszemski, M. J.; Windebank, A. J. Optimizing conditions and avoiding pitfalls for prolonged axonal tracing with carbocyanine dyes in fixed rat spinal cords. *J. Neurosci. Methods* **2006**, *154*, 256–263.
7. Takenouchi, T.; Iwamaru, Y.; Sato, M.; Yokoyama, T.; Shinagawa, M.; Kitani, H. Establishment and characterization of SV40 large T antigen-immortalized cell lines derived from fetal bovine brain tissues after prolonged cryopreservation. *Cell Biol. Int.* **2007**, *31*, 57–64.

8. Salazar, I.; Brennan, P. A. Retrograde labeling of mitral/tufted cells in the mouse accessory olfactory bulb following local injections of the lipophilic tracer DiI into the vomeronasal amygdala. *Brain Res.* **2001**, *896*, 198–203.
9. Butowt, R.; Jeffrey, P. L.; von Bartheld, C. S. Purification of chick retinal ganglion cells for molecular analysis: combining retrograde labeling and immunopanning yields 100% purity. *J. Neurosci. Methods* **2000**, *95*, 29–38.
10. Fuller, M. E.; Streger, S. H.; Rothmel, R. K.; Mailloux, B. J.; Hall, J. A.; Onstott, T. C.; Fredrickson, J. K.; Balkwill, D. L.; DeFlaun, M. F. Development of a vital fluorescent staining method for monitoring bacterial transport in subsurface environments. *Appl. Environ. Microbiol.* **2000**, *66*, 4486–4496.
11. Lichtman, J. W.; Wong, W. T.; Wong, R.; Gan, W. B.; Grutzendler, J. Method for labeling individual cells. PCT Int. Appl. WO 2001069244, 2001; *Chem. Abstr.* **2001**, *135*, 238975.
12. Orth, R. N.; Kameoka, J.; Zipfel, W. R.; Ilic, B.; Webb, W. W.; Clark, T. G.; Craighead, H. G. Creating biological membranes on the micron scale: forming patterned lipid bilayers using a polymer lift-off technique. *Biophys. J.* **2003**, *85*, 3066–3073.
13. Hollars, C. W.; Dunn, R. C. Submicron fluorescence, topology, and compliance measurements of phase-separated lipid monolayers using tapping-mode near-field scanning optical microscopy. *J. Phys. Chem. B* **1997**, *101*, 6313–6317.
14. Talley, C. E.; Dunn, R. C. Single molecules as probes of lipid membrane microenvironments. *J. Phys. Chem. B* **1999**, *103*, 10214–10220.
15. Okaji, Y.; Tsuno, N. H.; Kitayama, J.; Saito, S.; Takahashi, T.; Kawai, K.; Yazawa, K.; Asakage, M.; Tsuchiya, T.; Sakurai, D.; Tsuchiya, N.; Tokunaga, K.; Takahashi, K.; Nagawa, H. A novel method for isolation of endothelial cells and macrophages from murine tumors based on Ac-LDL uptake and CD16 expression. *J. Immunol. Methods* **2004**, *295*, 183–193.
16. Dantuma, N. P.; Pijnenburg, M. A. P.; Diederen, J. H. B.; Van Der Horst, D. J. Electron microscopic visualization of receptor-mediated endocytosis of DiI-labeled lipoproteins by diaminobenzidine photoconversion. *J. Histochem. Cytochem.* **1998**, *46*, 1085–1089.
17. Verma, D. D.; Verma, S.; Blume, G.; Fahr, A. Particle size of liposomes influences dermal delivery of substances into skin. *Int. J. Pharm.* **2003**, *258*, 141–151.
18. Mueller, A.; Bondurant, B.; O'Brien, D. F. Visible-light-stimulated destabilization of PEG-liposomes. *Macromolecules* **2000**, *33*, 4799–4804.
19. Choice, E.; Ayyobi, A. F.; Pritchard, P. H.; Madden, T. D. Separation of liposomes from plasma components using fast protein liquid chromatography. *Anal. Biochem.* **1999**, *270*, 1–8.
20. Ries, J.; Schwille, P. Studying slow membrane dynamics with continuous wave scanning fluorescence correlation spectroscopy. *Biophys. J.* **2006**, *91*, 1915–1924.
21. Schneckenburger, H.; Wagner, M.; Kretzschmar, M.; Strauss, W. S. L.; Sailer, R. Laser-assisted fluorescence microscopy for measuring cell membrane dynamics. *Photochem. Photobiol. Sci.* **2004**, *3*, 817–822.
22. Gan, W. B.; Grutzendler, J.; Wong, W. T.; Wong, R. O. L.; Lichtman, J. W. Multicolor "DiOlistic" labeling of the nervous system using lipophilic dye combinations. *Neuron* **2000**, *27*, 219–225.
23. Vogalis, F.; Hillsley, K.; Smith, T. Recording ionic events from cultured, DiI-labelled myenteric neurons in the guinea-pig proximal colon. *J. Neurosci. Methods* **2000**, *96*, 25–34.
24. Choi, D.; Li, D.; Raisman, G. Fluorescent retrograde neuronal tracers that label the rat facial nucleus: a comparison of Fast Blue, Fluoro-ruby, Fluoro-emerald, Fluoro-Gold and DiI. *J. Neurosci. Methods* **2002**, *117*, 167–172.
25. Marin, G.; Henny, P.; Letelier, J. C.; Sentis, E.; Karten, H.; Mrosko, B.; Mpodozis, J. A simple method to microinject solid neural tracers into deep structures of the brain. *J. Neurosci. Methods* **2001**, *106*, 121–129.
26. Ravnic, D. J.; Jiang, X.; Wolloscheck, T.; Pratt, J. P.; Huss, H.; Mentzer, S. J.; Konerding, M. A. Vessel painting of the microcirculation using fluorescent lipophilic tracers. *Microvasc. Res.* **2005**, *70*, 90–96.
27. Huerta, L.; Lopez-Balderas, N.; Larralde, C.; Lamoyi, E. Discriminating *in vitro* cell fusion from cell aggregation by flow cytometry combined with fluorescence resonance energy transfer. *J. Virol. Methods* **2006**, *138*, 17–23.
28. Goldbrunner, R. H.; Bouterfa, H.; Vince, G. H.; Bernstein, J. J.; Roosen, K.; Tonn, J. C. Transfection and dye premarking of human and rat

glioma cell lines affects adhesion, migration and proliferation. *Anticancer Res.* **1997**, *17*, 4467–4471.

29. Duncan, R. R.; Cousin, M. A. Membrane fusion assay using FRET in combination with fluorescence microscopy and time resolved fluorometry. PCT Int. Appl. WO 2006003397, 2006; *Chem. Abstr.* **2006**, *144*, 103480.

30. Fromherz, P.; Kiessling, V.; Kottig, K.; Zeck, G. Membrane transistor with giant lipid vesicle touching a silicon chip. *Appl. Phys. A* **1999**, *69*, 571–576.

31. St. John, P. A. Toxicity of "DiI" for embryonic rat motoneurons and sensory neurons *in vitro*. *Life Sci.* **1991**, *49*, 2013–2021.

DiIC$_1$(5)

CAS Registry Number 36536-22-8

Chemical Structure

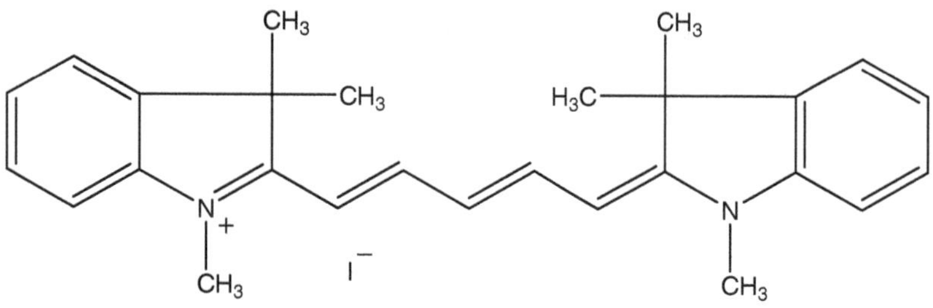

Physical Form Blue green crystals

Solubility Soluble in methanol, dimethyl sulfoxide; partially soluble in water

Melting Point 266–267 °C (decompose)

CA Index Name 3*H*-Indolium, 2-[5-(1,3-dihydro-1,3,3-trimethyl-2*H*-indol-2-ylidene)-1,3-pentadien-1-yl]-1,3,3-trimethyl-, iodide (1 : 1)

Other Names 1,3,3-Trimethyl-2-[5-(1,3,3-trimethyl-2-indolinylidene)-1,3-pentadienyl]-3*H*-indolium iodide; 3*H*-Indolium, 2-[5-(1,3-dihydro-1,3,3-trimethyl-2*H*-indol-2-ylidene)-1,3-pentadienyl]-1,3,3-trimethyl-, iodide; 1,1′,3,3,3′,3′-Hexamethylindodicarbocyanine iodide; 1,1′,3,3,3′,3′-Hexamethylindoledicarbocyanine iodide; 1,3,3,1′,3′,3′-Hexamethylindoledicarbocyanine iodide; HIDC; HIDCI; HIDC iodide; Hexacyanine 2; NK 529; PK 643; SNC 6

Merck Index Number Not listed

Chemical/Dye Class Cyanine

Molecular Formula $C_{27}H_{31}IN_2$

Molecular Weight 510.45

Absorption (λ_{max}) 638 nm

Emission (λ_{max}) 658 nm

Synthesis Synthetic methods[1–8]

Staining Applications Fungi;[9] sperms;[9] microorganism;[10] proteoliposomes;[11] analytical samples;[12] reticulocytes;[13] leukocytes[13]

Biological Applications Measuring membrane potential;[11,14] detecting prostate cancer;[15] analyzing urine;[9] apoptosis assays;[16] *in vivo* hematotoxicity assays[17]

Industrial Applications Dye laser;[18] photoresists;[19] liquid crystals;[20] semiconductor integrated circuits;[21] thin films;[22] recording materials;[23,24] inks;[24,25] toners;[24] photographic materials[4,5,26]

Safety/Toxicity No data available

REFERENCES

1. Sato, S.; Tsunoda, M.; Suzuki, M.; Kutsuna, M.; Takido-uchi, K.; Shindo, M.; Mizuguchi, H.; Obara, H.; Ohya, H. Synthesis and spectral properties of polymethine-cyanine dye–nitroxide radical hybrid compounds for use as fluorescence probes to monitor reducing species and radicals. *Spectrochim. Acta* **2009**, *71A*, 2030–2039.
2. Miyagawa, T.; Yamamoto, M.; Muraki, R.; Onouchi, H.; Yashima, E. Supramolecular helical assembly of an achiral cyanine dye in an induced helical amphiphilic poly(phenylacetylene) interior in water. *J. Am. Chem. Soc.* **2007**, *129*, 3676–3682.
3. Jung, J. H.; Lee, K. H.; Yang, C. S. Synthesis and physical properties of cyanine dyes as optical information recording medium. *J. Korean Chem. Soc.* **1998**, *42*, 570–574.
4. Kimura, S.; Homma, S.; Kobayashi, S. Dicarbocyanine dyes for photographic sensitizers. JP 31006437, 1956; *Chem. Abstr.* **1958**, *52*, 29004.
5. Kendall, J. D.; Edwards, Harry D. Dicarbocyanine dyes for use as sensitizers in photographic emulsions. U.S. Patent 2412815, 1946; *Chem. Abstr.* **1947**, *41*, 25284.
6. Ogata, T. Cyanine dyes. VII. Synthesis of pentamethinecyanine. *Proc. Imperial Acad.* **1934**, *10*, 572–574; *Chem. Abstr.* **1935**, *29*, 14023.
7. Ogata, T. Synthesis of polymethine dyes. *Nippon Kagaku Kaishi* **1934**, *55*, 394–436; *Chem. Abstr.* **1934**, *28*, 48038.

8. Ogata, T. Cyanine dyestuffs. IV. Pentamethine- and heptamethine-cyanines. *Proc. Imperial Acad.* **1932**, *8*, 421–423; *Chem. Abstr.* **1933**, *27*, 10023.
9. Kawashima, Y. Reagent and method for analyzing urine in which reagent contains a fungus membrane damaging agent and fluorescent dyes. Eur. Pat. Appl. EP 1837402, 2007; *Chem. Abstr.* **2007**, *147*, 360036.
10. Yeh, M. H. Sensor for monitoring metabolic activity of an anaerobic or aerobic microorganism. Eur. Pat. Appl. EP 1327874, 2003; *Chem. Abstr.* **2003**, *139*, 97604.
11. Ivashchuk-Kienbaum, Y. A. Monitoring of the membrane potential in proteoliposomes with incorporated cytochrome-*c* oxidase using the fluorescent dye indocyanine. *J. Membr. Biol.* **1996**, *151*, 247–259.
12. Iketaki, Y.; Fujii, M.; Omatsu, T.; Suzuki, T.; Minakata, M.; Yamamoto, K.; Nakaya, K. Method for dyeing or labeling analytical samples with quantum dots or dendrimers containing fluorescent groups. Jpn. Kokai Tokkyo Koho JP 2001305030, 2001; *Chem. Abstr.* **2001**, *135*, 366150.
13. Akai, Y.; Hatanaka, K.; Itose, Y.; Sakata, T. A reagent for measuring reticulocytes and a method of measuring them. Eur. Pat. Appl. EP 806664, 1997; *Chem. Abstr.* **1997**, *128*, 11616.
14. Murakami, T. Cytotoxicity test method by measuring membrane electric potential. Jpn. Kokai Tokkyo Koho JP 2000300290, 2000; *Chem. Abstr.* **2000**, *133*, 330852.
15. Dickman, D. Methods of detecting prostate cancer. PCT Int. Appl. WO 2006054296, 2006; *Chem. Abstr.* **2006**, *145*, 3346.
16. Li, X.; Darzynkiewicz, Z. The Schrodinger's cat quandary in cell biology: integration of live cell functional assays with measurements of fixed cells in analysis of apoptosis. *Exp. Cell Res.* **1999**, *249*, 404–412.
17. Dertinger, S. D.; Bemis, J. C.; Bryce, S. M. Method for measuring *in vivo* hematotoxicity with an emphasis on radiation exposure assessment. U.S. Pat. Appl. Publ. US 2008311586, 2008; *Chem. Abstr.* **2008**, *150*, 50134.
18. Kessler, W. J.; Davis, S. J.; Ferguson, D. R.; Pugh, E. R. Solid-state dye laser host. U.S. Patent 5610932, 1997; *Chem. Abstr.* **1997**, *126*, 256922.
19. Farid, S. Y.; Haley, N. F.; Moody, R. E.; Specht, D. P. Negative working photoresists responsive to longer wavelengths and novel coated articles. U.S. Patent 4743530, 1988; *Chem. Abstr.* **1988**, *109*, 180445.
20. Kurihara, S.; Nomiyama, S.; Nonaka, T. Photochemical control of the macrostructure of cholesteric liquid crystals by means of photoisomerization of chiral azobenzene molecules. *Chem. Mater.* **2001**, *13*, 1992–1997.
21. Hashiba, T.; Itabashi, T.; Akaboshi, H.; Fukada, S. Copper electroplating baths and method for manufacturing semiconductor integrated circuits with no voids and seams by damascene method using them. Jpn. Kokai Tokkyo Koho JP 2002155390, 2002; *Chem. Abstr.* **2002**, *136*, 410023.
22. Xu, L.; Gao, Y.; Shen, Y.; You, R. Preparation of photosensitive thin films containing photosensitive coating layer for holographic recording. Faming Zhuanli Shenqing Gongkai Shuomingshu CN 101034257, 2007; *Chem. Abstr.* **2007**, *147*, 436872.
23. Ito, H.; Kume, M.; Ohe, Y. Hologram recording material with light and heat resistance. Jpn. Kokai Tokkyo Koho JP 11161140, 1999; *Chem. Abstr.* **1999**, *131*, 80821.
24. Murofushi, K.; Hosada, Y. Light-decolorizable recording material, ink, or toner. U.S. Patent 5846682, 1998; *Chem. Abstr.* **1998**, *130*, 59181.
25. Kunimatsu, M.; Nakatsu, K.; Sugawa, T.; Shirota, T. Fluorescent dye-containing ink-jet inks. Jpn. Kokai Tokkyo Koho JP 2001026732, 2001; *Chem. Abstr.* **2001**, *134*, 133041.
26. Farid, S. Y.; Haley, N. F.; Moody, R. E.; Specht, D. P. Dye-sensitized photographic imaging system. U.S. Patent 4743531, 1988; *Chem. Abstr.* **1988**, *109*, 139163.

$DiOC_2(3)$

CAS Registry Number 905-96-4

Chemical Structure

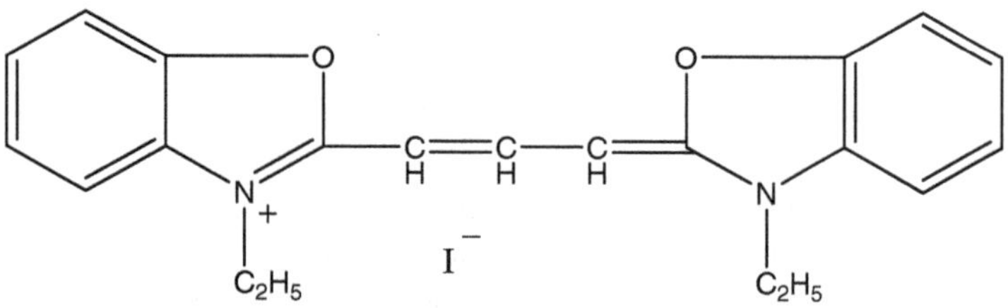

CA Index Name Benzoxazolium, 3-ethyl-2-[3-(3-ethyl-2(3*H*)-benzoxazolylidene)-1-propen-1-yl]-, iodide (1 : 1)

Other Names 3-Ethyl-2-[3-(3-ethyl-2-benzoxazolinylidene)propenyl]benzoxazolium iodide; Benzoxazolium, 3-ethyl-2-[3-(3-ethyl-2(3*H*)-benzoxazolylidene)-1-propenyl]-, iodide; Benzoxazolium, 3-ethyl-2-[3-(3-ethyl-2-benzoxazolinylidene)propenyl]-, iodide; 3,3′-Diethyl-2,2′-oxacarbocyanine iodide; 3,3′-Diethyloxacarbocyanine iodide; 3-Ethyl-2-[3-ethyl-2(3*H*)-benzoxazolylidene)-1-propenyl]benzoxazolium iodide); DOC; DOC (dye); DOC iodide; DOCI; $DiOC_2$; DiOC2(3); G 1745; NK 85

Merck Index Number Not listed

Chemical/Dye Class Cyanine

Molecular Formula $C_{21}H_{21}IN_2O_2$

Molecular Weight 460.31

Physical Form Dark red crystals or powder

Solubility Soluble in methanol, dimethyl sulfoxide

Melting Point 278 °C (decompose); 210 °C (decompose) (impure)

Absorption (λ_{max}) 482 nm

Emission (λ_{max}) 497 nm

Synthesis Synthetic methods[1–9]

Staining Applications Hepatocytes;[10] P-glycoprotein;[11] malaria infected cells;[12] nucleic acids;[13] sperms;[14] hairs[15]

Biological Applications Measuring membrane potential;[16] detecting prostate cancer,[17] nucleic acid hybridization,[18] polynucleotides;[19] treating acute myelogenous leukemia,[20] metastatic breast cancer,[20] cancers expressing P-glycoprotein[20]

Industrial Applications Dye laser;[21] displays;[22] recording materials;[23] light emitting diode lens;[24] thin films;[25] photographic materials;[9,26,27] nonlinear optical materials;[28] solar panels[29]

Safety/Toxicity Carcinogenicity[30]

REFERENCES

1. Glazer, A. N.; Mathies, R. A.; Hung, S. C.; Jue, J. Cyanine dyes with high-absorbance cross section as donor chromophores in energy transfer labels. PCT Int. Appl. WO 9814612, 1998; *Chem. Abstr.* **1998**, *128*, 280583.
2. Akins, D. L.; Kumar, V. T. High-performance liquid chromatography of cyanine dyes: multiphase separation, purification, and substitution of the counter ion. *J. Chromatogr., A* **1995**, *689*, 269–273.
3. Sims, P. J.; Waggoner, A. S.; Wang, C. H.; Hoffman, J. F. Mechanism by which cyanine dyes measure membrane potential in red blood cells and phosphatidylcholine vesicles. *Biochemistry* **1974**, *13*, 3315–3330.
4. Ciernik, J. V. Formylation of nitrogen-containing heterocycles and their quaternary salts. *Collect. Czech. Chem. Commun.* **1972**, *37*, 2273–2281.
5. Pokrovskaya, K. I.; Levkoev, I. I.; Natanson, S. V. Complex compounds of polymethine dyes with silver ions. I. The formation of silver ions of carbo- and polycarbocyanines. *Zh. Fiz. Khim.* **1956**, *30*, 161–171; *Chem. Abstr.* **1956**, *50*, 51973.
6. Hishiki, Y. J. Trinuclear cyanine dyes. V. Decomposition of oxacyanine. *J. Sci. Res. Inst.* **1954**, *48*, 130–142.
7. Dent, S. G., Jr.; Brooker, L. G. S. Polymethine dyes. U.S. Patent 2537880, 1951; *Chem. Abstr.* **1951**, *45*, 21354.
8. Hofer, L. J. E.; Grabenstetter, R. J.; Wiig, E. O. The fluorescence of cyanine and related dyes in the monomeric state. *J. Am. Chem. Soc.* **1950**, *72*, 203–209.

9. Hamer, F. M.; Rathbone, R. J. Photographic layers. Kodak Ltd. GB 541330, 1941; *Chem. Abstr.* **1942**, *36*, 30854.

10. Li, M.; Yuan, H.; Li, N.; Song, G.; Zheng, Y.; Baratta, M.; Hua, F.; Thurston, A.; Wang, J.; Lai, Y. Identification of interspecies difference in efflux transporters of hepatocytes from dog, rat, monkey and human. *Eur. J. Pharm. Sci.* **2008**, *35*, 114–126.

11. Broxterman, H. J.; Lankelma, J.; Pinedo, H. M.; Eekman, C. A.; Wahrer, D. C. R.; Ossenkoppele, G. J.; Schuurhuis, G. J. Theoretical and practical considerations for the measurement of P-glycoprotein function in acute myeloid leukemia. *Leukemia* **1997**, *11*, 1110–1118.

12. Sakata, T.; Matsumoto, H. A staining reagent and method for detecting malaria infected cells using auramine analog and condensed benzene derivative. Eur. Pat. Appl. EP 613003, 1994; *Chem. Abstr.* **1994**, *121*, 225851.

13. Kerwin, S. M.; Sun, D.; Kern, J. T.; Rangan, A.; Thomas, P. W. G-Quadruplex DNA binding by a series of carbocyanine dyes. *Bioorg. Med. Chem. Lett.* **2001**, *11*, 2411–2414.

14. Nascimento, J. M.; Shi, L. Z.; Chandsawangbhuwana, C.; Tam, J.; Durrant, B.; Botvinick, E. L.; Berns, M. W. Use of laser tweezers to analyze sperm motility and mitochondrial membrane potential. *J. Biomed. Opt.* **2008**, *13*, 014002/1–014002/7.

15. Ohashi, Y.; Miyabe, H.; Matsunaga, K. Hair dye composition. Eur. Pat. Appl. EP 1166753, 2002; *Chem. Abstr.* **2002**, *136*, 74280.

16. Murakami, T. Cytotoxicity test method by measuring membrane electric potential. Jpn. Kokai Tokkyo Koho JP 2000300290, 2000; *Chem. Abstr.* **2000**, *133*, 330852.

17. Dickman, D. Methods of detecting prostate cancer. PCT Int. Appl. WO 2006054296, 2006; *Chem. Abstr.* **2006**, *145*, 3346.

18. Bupp, C. R., II; Choi, K. Y.; Holmes-Davis, R. A.; Izmailov, A.; Koshinsky, H.; Nulf, C. J.; Urdea, M.; Wang, M.; Warner, B. D.; Zwick, M. Carbocyanine dye dimers linked by a conjugated alkenyl chain for use in detection of nucleic acid hybridization. PCT Int. Appl. WO 2007100711, 2007; *Chem. Abstr.* **2007**, *147*, 358263.

19. Bupp, C. R.; Choi, K. Y.; Koshinsky, H.; Nulf, C.; Urdea, M.; Wang, M.; Zwick, M. Measurement of rate of light-induced change of dye optical property in presence of probe-target hybrids for detecting polynucleotides. U.S. Pat. Appl. Publ. US 2008096193, 2008; *Chem. Abstr.* **2008**, *148*, 419118.

20. Sikic, B.; Hoth, D.; Socks, D.; Glenn, S.; Marcelletti, J.; Walsh, M. J.; Multani, P. S. Zosuquidar–chemotherapeutic agent combination for the treatment of cancer patients exhibiting activation of the P-glycoprotein efflux pump mechanism, and methods for the determination of P-glycoprotein expression and/or function. PCT Int. Appl. WO 2007008499, 2007; *Chem. Abstr.* **2007**, *146*, 135535.

21. Kessler, W. J.; Davis, S. J.; Ferguson, D. R.; Pugh, E. R. Solid-state dye laser host. U.S. Patent 5610932, 1997; *Chem. Abstr.* **1997**, *126*, 256922.

22. Hajto, J.; Hindle, C.; Graham, A. Displays. PCT Int. Appl. WO 2000007039, 2000; *Chem. Abstr.* **2000**, *132*, 144506.

23. Takano, H.; Yamashita, N. Manufacture of optical recording material. Jpn. Kokai Tokkyo Koho JP 2007199125, 2007; *Chem. Abstr.* **2007**, *147*, 246859.

24. Butterworth, M. M.; Helbing, R. P. Fluorescent dye added to epoxy of light emitting diode lens. U.S. Patent 5847507, 1998; *Chem. Abstr.* **1998**, *130*, 58886.

25. O'Regan, B.; Schwartz, D. T. Efficient photo-hole injection from adsorbed cyanine dyes into electrodeposited copper(I) thiocyanate thin films. *Chem. Mater.* **1995**, *7*, 1349–1354.

26. Hermans, T.; Vrielynck, M. Static pressure effects on photographic materials. Part III. The influence of dyes. *J. Photogr. Sci.* **1984**, *32*, 153–157.

27. Kitova, S.; Malinovski, I. Dye sensitized photoprocess in silver bromide evaporated layers. *Photogr. Sci. Eng.* **1980**, *24*, 50–54.

28. Morita, K.; Suehiro, T.; Yokoh, Y.; Ashitaka, H. The development of organic third-order nonlinear optical materials. *J. Photopolym. Sci. Technol.* **1993**, *6*, 229–238.

29. Krokoszinski, H. J.; Mayer, O.; Stromberger, J.; Korman, C. S. Fabrication of solar panels consisting of photovoltaic cells. U.S. Pat. Appl. Publ. US 2007137696, 2007; *Chem. Abstr.* **2007**, *147*, 98621.

30. Vasconcelos, F. C.; Cavalcanti, G. B.; Silva, K. L.; de Meis, E.; Kwee, J. K.; Rumjanek, V. M.; Maia, R. C. Contrasting features of MDR phenotype in leukemias by using two fluorochromes: implications for clinical practice. *Leuk. Res.* **2007**, *31*, 445–454.

DiOC$_5$(3)

CAS Registry Number 53213-81-3
Chemical Structure

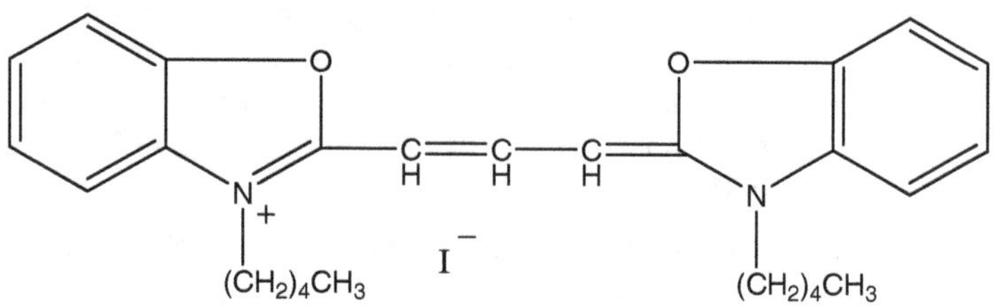

CA Index Name Benzoxazolium, 3-pentyl-2-[3-(3-pentyl-2(3*H*)-benzoxazolylidene)-1-propenyl]-, iodide
Other Names 3,3′-Dipentyloxacarbocyanine iodide, DiOC$_5$(3)
Merck Index Number Not listed
Chemical/Dye Class Cyanine
Molecular Formula $C_{27}H_{33}IN_2O_2$
Molecular Weight 544.47
Physical Form Red powder
Solubility Soluble in methanol, dimethyl sulfoxide
Melting Point 214–215 °C
Absorption (λ_{max}) 484 nm
Emission (λ_{max}) 500 nm
Synthesis Synthetic method[1]
Staining Applications Cells;[2] leukocytes;[3,4] squamous epithelial cells;[4] microorganisms;[4,5] proteins;[6] sperms;[7] hairs[8]
Biological Applications Detecting prostate cancer;[9] monitoring membrane potential;[10–15] cytotoxicity assay;[15,16] drug screening assay;[17] susceptibility assay[18]
Industrial Applications Photoresists[19]
Safety/Toxicity No data available

REFERENCES

1. Sims, P. J.; Waggoner, A. S.; Wang, C.; Hoffman, J. F. Mechanism by which cyanine dyes measure membrane potential in red blood cells and phosphatidylcholine vesicles. *Biochemistry* **1974**, *13*, 3315–3330.
2. Crissman, H. A.; Hofland, M. H.; Stevenson, A. P.; Wilder, M. E.; Tobey, R. A. Supravital cell staining with Hoechst 33342 and DiOC$_5$(3). *Methods Cell Biol.* **1990**, *33*, 89–95.
3. Sakata, T.; Kuroda, T. Fluorochrome-containing reagent for classifying leukocytes by flow cytometry. U.S. Patent 5175109, 1992; *Chem. Abstr.* **1993**, *118*, 142984.
4. Mansour, J. D.; Schulte, T. H.; Sage, B. H. Detection and quantitation of microorganisms, leukocytes and squamous epithelial cells in urine by fluorescent dye staining. U.S. Patent 4622298, 1986; *Chem. Abstr.* **1987**, *106*, 29715.
5. Mansour, J. D. Fluorescent gram stain. U.S. Patent 4665024, 1987; *Chem. Abstr.* **1987**, *107*, 55329.
6. Aebersold, R. H.; Teplow, D. B.; Hood, L. E.; Kent, S. B. H. Electroblotting onto activated glass. High efficiency preparation of proteins from analytical sodium dodecyl sulfate-polyacrylamide gels for direct sequence analysis. *J. Biol. Chem.* **1986**, *261*, 4229–4238.
7. Peterson, R. N.; Bundman, D.; Freund, M. Use of a fluorescent dye to measure drug-induced changes in the membrane potential of boar spermatozoa. *Life Sci.* **1978**, *22*, 659–666.
8. Ohashi, Y.; Miyabe, H.; Matsunaga, K. Hair dye composition. Eur. Pat. Appl. EP 1166753, 2002; *Chem. Abstr.* **2002**, *136*, 74280.
9. Dickman, D. Methods of detecting prostate cancer. PCT Int. Appl. WO 2006054296, 2006; *Chem. Abstr.* **2006**, *145*, 3346.
10. Remani, P.; Ostapenko, V. V.; Akagi, K.; Bhattathiri, V. N.; Nair, M. K.; Tanaka, Y. Relation of transmembrane potential to cell survival following hyperthermia in HeLa cells. *Cancer Lett.* **1999**, *144*, 117–123.
11. Bramhall, J. S.; Morgan, J. I.; Perris, A. D.; Britten, A. Z. The use of a fluorescent probe to monitor alterations in trans-membrane potential in single cell suspensions. *Biochem. Biophys. Res. Commun.* **1976**, *72*, 654–662.

12. Miller, J. B.; Koshland, D. E., Jr. Effects of cyanine dye membrane probes on cellular properties. *Nature* **1978**, *272*, 83–84.
13. Sapiro, H. M. Cell membrane potential analysis. *Methods Cell Biol.* **1994**, *41*, 121–133.
14. Ordonez, J. V.; Wehman, N. M. Amphotericin B susceptibility of *Candida* species assessed by rapid flow cytometric membrane potential assay. *Cytometry* **1995**, *22*, 154–157.
15. Murakami, T. Cytotoxicity test method by measuring membrane electric potential. Jpn. Kokai Tokkyo Koho JP 2000300290, 2000; *Chem. Abstr.* **2000**, *133*, 330852.
16. Wadkins, R. M.; Houghton, P. J. Kinetics of transport of dialkyloxacarbocyanines in multidrug-resistant cell lines overexpressing p-glycoprotein: interrelationship of dye alkyl chain length, cellular flux, and drug resistance. *Biochemistry* **1995**, *34*, 3858–3872.
17. Dykens, J. A.; Velicelebi, G.; Ghosh, S. S. Compositions and methods for assaying subcellular conditions and processes using energy transfer for drug screening. PCT Int. Appl. WO 2000079274, 2000; *Chem. Abstr.* **2000**, *134*, 66121.
18. Ordonez, J. V.; Wehman, N. M. Rapid flow cytometric antibiotic susceptibility assay for *Staphylococcus aureus*. *Cytometry* **1993**, *14*, 811–818.
19. Noppakundilograt, S.; Miyagawa, N.; Takahara, S.; Yamaoka, T. Visible light-sensitive positive-working photopolymer based on poly(*p*-hydroxystyrene) and vinyl ether crosslinker. *J. Photopolym. Sci. Technol.* **2000**, *13*, 719–721.

DiOC$_6$(3)

CAS Registry Number 53213-82-4

Chemical Structure

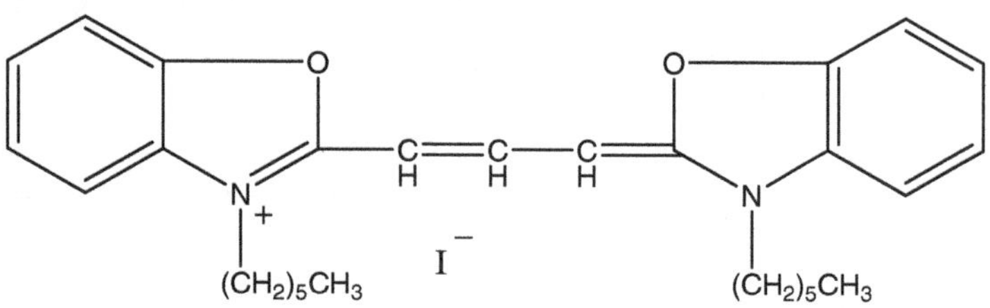

CA Index Name Benzoxazolium, 3-hexyl-2-[3-(3-hexyl-2(3*H*)-benzoxazolylidene)-1-propen-1-yl]-, iodide (1 : 1)

Other Names Benzoxazolium, 3-hexyl-2-[3-(3-hexyl-2(3*H*)-benzoxazolylidene)-1-propenyl]-, iodide; 3,3′-Dihexyloxacarbocyanine iodide; D 273; DiOC$_6$(3); NK 2280

Merck Index Number Not listed

Chemical/Dye Class Cyanine

Molecular Formula $C_{29}H_{37}IN_2O_2$

Molecular Weight 572.52

Physical Form Red powder

Solubility Soluble in ethanol, methanol, dimethyl sulfoxide

Melting Point 222–224 °C

Absorption (λ_{max}) 484 nm

Emission (λ_{max}) 501 nm

Synthesis Synthetic methods[1–3,16]

Staining Applications Mitochondria;[4–17] endoplasmic reticulum;[16–37] bacteria;[38,39] fungi;[40] nucleic acids;[41] leukocytes;[42] lymphocytes;[43] microtubules;[59] myelin;[44] presynaptic nerve terminals;[45] sperms;[46] spores[51]

Biological Applications Detecting mitochondrial membrane potential,[3,10–16,47,48] nucleic acid hybridization,[49] prostate cancer,[50] germination *Bacillus anthracis* spores;[51] apoptosis assays;[52,53] diagnostic assays;[54] in photodynamic therapy;[55] treating malaria,[56] leishmaniasis, trypanosomiasis,[56] trichomoniasis,[56] neosporosis,[56] coccidiosis[56]

Industrial Applications Photoresists;[57] photography[58]

Safety/Toxicity Photodamage of microtubules and inactivation of organelle motility[59]

REFERENCES

1. Adier, C.; Thomas, M.; Turpin, F.; Brigand, C.; Cenatiempo, Y.; Belgsir, E. M. Method for obtaining water-soluble fluorescent complexes from lipophilic dyes and cyclodextrin derivatives. Fr. Demande FR 2878853, 2006; *Chem. Abstr.* **2006**, *145*, 23859.
2. Akins, D. L.; Kumar, V. T. High-performance liquid chromatography of cyanine dyes: multiphase separation, purification, and substitution of the counter ion. *J. Chromatogr., A* **1995**, *689*, 269–273.
3. Sims, P. J.; Waggoner, A. S.; Wang, C.; Hoffman, J. F. Mechanism by which cyanine dyes measure membrane potential in red blood cells and phosphatidylcholine vesicles. *Biochemistry* **1974**, *13*, 3315–3330.
4. Swayne, T. C.; Gay, A. C.; Pon, L. A. Visualization of mitochondria in budding yeast. *Methods Cell Biol.* **2007**, *80*, 591–626.
5. Hattori, F.; Fukuda, K. Method for selecting myocardial cells using intracellular mitochondria labeled with fluorescent indicator. PCT Int. Appl. WO 2006022377, 2006; *Chem. Abstr.* **2006**, *144*, 270175.
6. Salvioli, S.; Ardizzoni, A.; Franceschi, C.; Cossarizza, A. JC-1, but not DiOC$_6$(3) or rhodamine 123, is a reliable fluorescent probe to assess $\Delta\Psi$ changes in intact cellsimplications for studies on mitochondrial functionality during apoptosis. *FEBS Lett.* **1997**, *411*, 77–82.
7. Yang, H. C.; Taguchi, H.; Nishimura, K.; Miyaji, M. Effect of miconazole on diO-C$_6$-(3) accumulation in mitochondria of *Candida albicans. Mycoscience* **1996**, *37*, 243–248.
8. Miyakawa, I.; Higo, K.; Osaki, F.; Sando, N. Double staining of mitochondria and mitochondrial nucleoids in the living yeast during the life cycle. *J. Gen. Appl. Microbiol.* **1994**, *40*, 1–14.
9. Korchak, H. M.; Rich, A. M.; Wilkenfeld, C.; Rutherford, L. E.; Weissmann, G. A carbocyanine dye, DiOC$_6$(3), acts as a mitochondrial probe in

human neutrophils. *Biochem. Biophys. Res. Commun.* **1982**, *108*, 1495–1501.

10. Le, S. B.; Holmuhamedov, E. L.; Narayanan, V. L.; Sausville, E. A.; Kaufmann, S. H. Adaphostin and other anticancer drugs quench the fluorescence of mitochondrial potential probes. *Cell Death Differ.* **2006**, *13*, 151–159.
11. Kataoka, M.; Fukura, Y.; Shinohara, Y.; Baba, Y. Analysis of mitochondrial membrane potential in the cells by microchip flow cytometry. *Electrophoresis* **2005**, *26*, 3025–3031.
12. Petit, P. X.; O'Connor, J. E.; Grunwald, D.; Brown, S. C. Analysis of the membrane potential of rat- and mouse-liver mitochondria by flow cytometry and possible applications. *Eur. J. Biochem.* **1990**, *194*, 389–397.
13. Kalbacova, M.; Vrbacky, M.; Drahota, Z.; Melkova, Z. Comparison of the effect of mitochondrial inhibitors on mitochondrial membrane potential in 2 different cell lines using flow cytometry and spectrofluorometry. *Cytometry* **2003**, *52A*, 110–116.
14. Zamzami, N.; Metivier, D.; Kroemer, G. Quantitation of mitochondrial transmembrane potential in cells and in isolated mitochondria. *Methods Enzymol.* **2000**, *322*, 208–213.
15. Rottenberg, H.; Wu, S. Quantitative assay by flow cytometry of the mitochondrial membrane potential in intact cells. *Biochim. Biophys. Acta* **1998**, *1404*, 393–404.
16. Sabnis, R. W.; Deligeorgiev, T. G.; Jachak, M. N.; Dalvi, T. S. $DiOC_6(3)$: a useful dye for staining the endoplasmic reticulum, *Biotech. Histochem.* **1997**, *72*, 253–258.
17. Soltys, B. J.; Gupta, R. S. Interrelationships of endoplasmic reticulum, mitochondria, intermediate filaments, and microtubules: a quadruple fluorescence labeling study. *Biochem. Cell Biol.* **1992**, *70*, 1174–1186.
18. Ramoino, P.; Diaspro, A.; Fato, M.; Beltrame, F.; Robello, M. Changes in the endoplasmic reticulum structure of *Paramecium primaurelia* in relation to different cellular physiological states. *J. Photochem. Photobiol., B* **2000**, *54*, 35–42.
19. Terasaki, M.; Reese, T. S. Interaction among endoplasmic reticulum, microtubules and retrograde movements of the cell surface. *Cell Motil. Cytoskeleton* **1994**, *29*, 291–300.
20. Terasaki, M. Probes for the endoplasmic reticulum. In *Fluorescent and Luminescent Probes for Biological Activity*; Mason, W. T., Ed.; Academic Press: London, **1993**; pp 120–123.
21. Terasaki, M.; Reese, T. S. Characterization of endoplasmic reticulum by co-localization of BiP and dicarbocyanine dyes. *J. Cell Sci.* **1992**, *101*, 315–322.
22. Toyoshima, I.; Yu, H.; Steuer, E. R.; Sheetz, M. P. Kinectin, a major kinesin-binding protein on ER. *J. Cell Biol.* **1992**, *118*, 1121–1131.
23. Chen, L. B.; Lee, C. Probing endoplasmic reticulum in living cells by epifluorescence and digitized video microscopy. In *Optical Microscopy for Biology*; Herman, B.; Jacobson, E., Eds.; Wiley-Liss: New York, **1990**; pp 409–418.
24. Terasaki, M. Recent progress on structural interactions of the endoplasmic reticulum. *Cell Motil. Cytoskeleton* **1990**, *15*, 71–75.
25. Quader, H. Formation and disintegration of cisternae of the endoplasmic reticulum visualized in live cells by conventional fluorescence and confocal laser scanning microscopy: evidence for the involvement of calcium and the cytoskeleton. *Protoplasma* **1990**, *155*, 166–175.
26. Quader, H.; Fast, H. Influence of cytosolic pH changes on the organization of the endoplasmic reticulum in epidermal cells of onion bulb scales: acidification by loading with weak organic acids. *Protoplasma* **1990**, *157*, 216–224.
27. Dailey, M. E.; Bridgman, P. C. Dynamics of the endoplasmic reticulum and other membranous organelles in growth cones of cultured neurons. *J. Neurosci.* **1989**, *9*, 1897–1909.
28. Terasaki, M. Fluorescent labeling of endoplasmic reticulum. *Methods Cell Biol.* **1989**, *29*, 125–135.
29. Lee, C.; Ferguson, M.; Chen, L. B. Construction of the endoplasmic reticulum. *J. Cell Biol.* **1989**, *109*, 2045–2055.
30. Quader, H.; Hofmann, A.; Schnepf, E. Reorganization of the endoplasmic reticulum in epidermis cells of onion bulb scales after cold stress: involvement of cytoskeletal elements. *Planta* **1989**, *177*, 273–280.
31. Sanger, J. M.; Dome, J. S.; Mittal, B.; Somlyo, A. V.; Sanger, J. W. Dynamics of the endoplasmic reticulum in living non-muscle and muscle cells. *Cell Motil. Cytoskeleton* **1989**, *13*, 301–319.
32. Lee, C.; Chen, L. B. Dynamic behavior of endoplasmic reticulum in living cells. *Cell* **1988**, *54*, 37–46.

33. Allen, N. S.; Brown, D. T. Dynamics of the endoplasmic reticulum in living onion epidermal cells in relation to microtubules, microfilaments and intracellular particle movement. *Cell Motil. Cytoskeleton* **1988**, *10*, 153–163.
34. Quader, H.; Hofmann, A.; Schnepf, E. Shape and movement of the endoplasmic reticulum in onion bulb epidermis cells: possible involvement of actin. *Eur. J. Cell Biol.* **1987**, *44*, 17–26.
35. Terasaki, M.; Chen, L. B.; Fujiwara, K. Microtubules and the endoplasmic reticulum are highly interdependent structures. *J. Cell Biol.* **1986**, *103*, 1557–1568.
36. Quader, H.; Schnepf, E. Endoplasmic reticulum and cytoplasmic streaming: fluorescence microscopical observations in adaxial epidermis cells of onion bulb scales. *Protoplasma* **1986**, *131*, 250–252.
37. Terasaki, M.; Song, J.; Wong, J. R.; Weiss, M. J.; Chen, L. B. Localization of endoplasmic reticulum in living and glutaraldehyde-fixed cells with fluorescent dyes. *Cell* **1984**, *38*, 101–108.
38. Mason, D. J.; Lopez-Amoros, R.; Allman, R.; Stark, J. M.; Lloyd, D. The ability of membrane potential dyes and calcofluor white to distinguish between viable and non-viable bacteria. *J. Appl. Bacteriol.* **1995**, *78*, 309–315.
39. Diaper, J. P.; Tither, K.; Edwards, C. Rapid assessment of bacterial viability by flow cytometry. *Appl. Microbiol. Biotechnol.* **1992**, *38*, 268–272.
40. Duckett, J. G.; Read, D. J. The use of the fluorescent dye, 3,3′-dihexyloxacarbocyanine iodide, for selective staining of ascomycete fungi associated with liverwort rhizoids and ericoid mycorrhizal roots. *New Phytol.* **1991**, *118*, 259–272.
41. Honda, M.; Nakai, N.; Fukuda, M.; Kawabe, Y. Optical amplification and laser action in cyanine dyes doped in DNA complex. *Proc. SPIE-Int. Soc. Opt. Eng.* **2007**, *6646*, 664609/1–664609/8.
42. Sakata, T.; Kuroda, T. Fluorochrome-containing reagent for classifying leukocytes by flow cytometry. U.S. Patent 5175109, 1992; *Chem. Abstr.* **1993**, *118*, 142984.
43. Shapiro, H. M.; Natale, P. J.; Kamentsky, L. A. Estimation of membrane potentials of individual lymphocytes by flow cytometry. *Proc. Natl. Acad. Sci. U.S.A.* **1979**, *76*, 5728–5730.
44. Micu, I.; Ridsdale, A.; Zhang, L.; Woulfe, J.; McClintock, J.; Brantner, C. A.; Andrews, S. B.; Stys, P. K. Real-time measurement of free Ca2+ changes in CNS myelin by two-photon microscopy. *Nat. Med.* **2007**, *13*, 874–879.
45. Yoshikami, D.; Okun, L. M. Staining of living presynaptic nerve terminals with selective fluorescent dyes. *Nature* **1984**, *310*, 53–56.
46. Matsumoto, T.; Okada, H.; Hamaguchi, Y. Method and reagent for counting sperm by flow cytometry. Jpn. Kokai Tokkyo Koho JP 2001242168, 2001; *Chem. Abstr.* **2001**, *135*, 207869.
47. Mathur, A.; Hong, Y.; Kemp, B. K.; Barrientos, A. A.; Erusalimsky, J. D. Evaluation of fluorescent dyes for the detection of mitochondrial membrane potential changes in cultured cardiomyocytes. *Cardiovasc. Res.* **2000**, *46*, 126–138.
48. Johnson, L. V.; Wash, M. L.; Bockus, B. J.; Chen, L. B. Monitoring of relative mitochondrial membrane potential in living cells by fluorescence microscopy. *J. Cell Biol.* **1981**, *88*, 526–535.
49. Bupp, C. R., II; Choi, K. Y.; Holmes-Davis, R. A.; Izmailov, A.; Koshinsky, H.; Nulf, C. J.; Urdea, M.; Wang, M.; Warner, B. D.; Zwick, M. Carbocyanine dye dimers linked by a conjugated alkenyl chain for use in detection of nucleic acid hybridization. PCT Int. Appl. WO 2007100711, 2007; *Chem. Abstr.* **2007**, *147*, 358263.
50. Dickman, D. Methods of detecting prostate cancer. PCT Int. Appl. WO 2006054296, 2006; *Chem. Abstr.* **2006**, *145*, 3346.
51. Laflamme, C.; Ho, J.; Veillette, M.; Latremoille, M. C.; Verreault, D.; Meriaux, A.; Duchaine, C. Flow cytometry analysis of germinating *Bacillus* spores, using membrane potential dye. *Arch. Microbiol.* **2005**, *183*, 107–112.
52. Buenz, E. J.; Limburg, P. J.; Howe, C. L. A high-throughput 3-parameter flow cytometry-based cell death assay. *Cytometry* **2007**, *71A*, 170–173.
53. Li, X.; Darzynkiewicz, Z. The Schrodinger's cat quandary in cell biology: integration of live cell functional assays with measurements of fixed cells in analysis of apoptosis. *Exp. Cell Res.* **1999**, *249*, 404–412.
54. Thiruvengadam, A. P.; Chandrasekaran, K. Diagnosis of bipolar disorder and ADHD based on regulation of Na^+,K^+-ATPase and membrane potential of lymphoblasts or whole blood cells. PCT Int. Appl. WO 2006113095, 2006; *Chem. Abstr.* **2006**, *145*, 451297.
55. Lipshutz, G. S.; Castro, D. J.; Saxton, R. E.; Haugland, R. P.; Soudant, J. Evaluation of four new carbocyanine dyes for photodynamic

therapy with lasers. *Laryngoscope* **1994**, *104*, 996–1002.

56. Zeligs, M. A. Antiparasitic methods and compositions using diindolylmethane-related indoles. PCT Int. Appl. WO 2008057253, 2008; *Chem. Abstr.* **2008**, *148*, 554046.
57. Farid, S. Y.; Haley, N. F.; Moody, R. E.; Specht, D. P. Negative working photoresists responsive to shorter wavelength visible light and novel coated articles. U.S. Patent 4743529, 1988; *Chem. Abstr.* **1988**, *109*, 160608.
58. Farid, S. Y.; Haley, N. F.; Moody, R. E.; Specht, D. P. Dye-sensitized photographic imaging system. U.S. Patent 4743531, 1988; *Chem. Abstr.* **1988**, *109*, 139163.
59. Lee, C.; Wu, S. S.; Chen, L. B. Photosensitization by 3,3′-dihexyloxacarbocyanine iodide: specific disruption of microtubules and inactivation of organelle motility. *Cancer Res.* **1995**, *55*, 2063–2069.

DiOC$_7$(3)

CAS Registry Number 53213-83-5

Chemical Structure

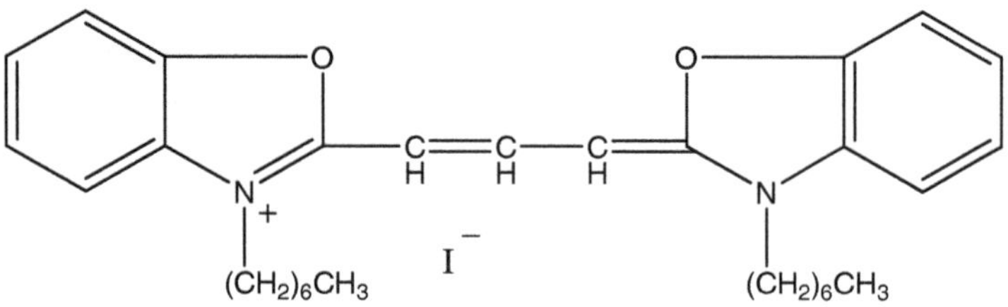

Solubility Soluble in ethanol, methanol, dimethyl sulfoxide

Melting Point 194–197 °C

Absorption (λ_{max}) 482 nm

Emission (λ_{max}) 504 nm

CA Index Name Benzoxazolium, 3-heptyl-2-[3-(3-heptyl-2(3*H*)-benzoxazolylidene)-1-propenyl]-, iodide

Other Names D 378; DiOC$_7$(3)

Merck Index Number Not listed

Chemical/Dye Class Cyanine

Molecular Formula $C_{31}H_{41}IN_2O_2$

Molecular Weight 600.57

Physical Form Red crystals

Synthesis Synthetic method[1]

Staining Applications Mitochondria;[2] cells;[3,4] fungi;[5] tumors[6]

Biological Applications Detecting cells,[3] fungal stress,[5] membrane potential;[1,2] cytotoxicity assays[7,8]

Industrial Applications Not reported

Safety/Toxicity No data available

REFERENCES

1. Sims, P. J.; Waggoner, A. S.; Wang, C.; Hoffman, J. F. Mechanism by which cyanine dyes measure membrane potential in red blood cells and phosphatidylcholine vesicles. *Biochemistry* **1974**, *13*, 3315–3330.
2. Hattori, F.; Fukuda, K. Method for selecting myocardial cells using intracellular mitochondria labeled with fluorescent indicator. PCT Int. Appl. WO 2006022377, 2006; *Chem. Abstr.* **2006**, *144*, 270175.
3. Garini, Y.; Mcnamara, G.; Soenksen, D. G.; Cabib, D.; Buckwald, R. A. *In situ* method of analyzing cells by staining with multiple stains and using a spectral data collection device. PCT Int. Appl. WO 2000031534, 2000; *Chem. Abstr.* **2000**, *133*, 14300.
4. Olive, P. L.; Durand, R. E. Characterization of a carbocyanine derivative as a fluorescent penetration probe. *Cytometry* **1987**, *8*, 571–575.
5. Thrane, C.; Olsson, S.; Harder Nielsen, T.; Sorensen, J. Vital fluorescent stains for detection of stress in *Pythium ultimum* and *Rhizoctonia solani* challenged with viscosinamide from *Pseudomonas fluorescens* DR54. *FEMS Microbiol. Ecol.* **1999**, *30*, 11–23.
6. Trotter, M. J.; Chaplin, D. J.; Olive, P. L. Use of a carbocyanine dye as a marker of functional vasculature in murine tumors. *Br. J. Cancer* **1989**, *59*, 706–709.
7. Murakami, T. Cytotoxicity test method by measuring membrane electric potential. Jpn. Kokai Tokkyo Koho JP 2000300290, 2000; *Chem. Abstr.* **2000**, *133*, 330852.
8. Wadkins, R. M.; Houghton, P. J. Kinetics of transport of dialkyloxacarbocyanines in multidrug-resistant cell lines overexpressing p-glycoprotein: interrelationship of dye alkyl chain length, cellular flux, and drug resistance. *Biochemistry* **1995**, *34*, 3858–3872.

DiSBAC$_2$(3)

CAS Registry Number 47623-98-3

Chemical Structure

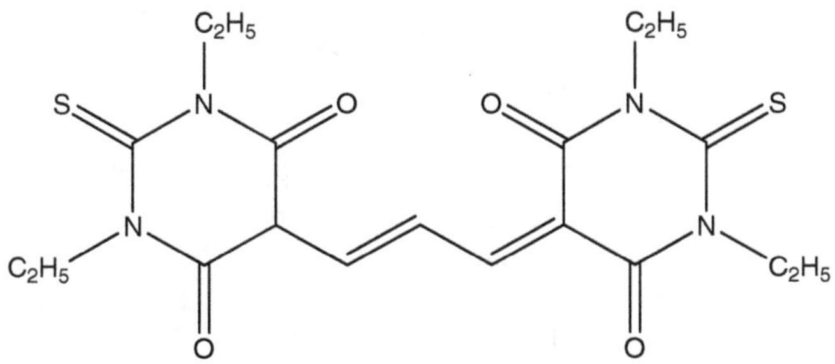

CA Index Name 4,6(1*H*,5*H*)-Pyrimidinedione, 5-[3-(1,3-diethylhexahydro-4,6-dioxo-2-thioxo-5-pyrimidinyl)-2-propen-1-ylidene]-1,3-diethyldihydro-2-thioxo-

Other Names 4,6(1*H*,5*H*)-Pyrimidinedione, 5-[3-(1,3-diethylhexahydro-4,6-dioxo-2-thioxo-5-pyrimidinyl)-2-propenylidene]-1,3-diethyldihydro-2-thioxo-; Bis(1,3-diethylthiobarbituric acid)trimethine oxonol; DiSBAC$_2$(3)

Merck Index Number Not listed

Chemical/Dye Class Oxonol

Molecular Formula $C_{19}H_{24}N_4O_4S_2$

Molecular Weight 436.55

Physical Form Purple powder

Solubility Soluble in ethanol, methanol, dimethyl sulfoxide

Boiling Point (Calcd.) $545.0 \pm 60.0\,°C$, pressure 760 Torr

pK_a (Calcd.) 3.72 ± 0.20, most acidic, temperature: 25 °C; -2.49 ± 0.20, most basic, temperature: 25 °C

Absorption (λ_{max}) 535 nm

Emission (λ_{max}) 560 nm

Synthesis Synthetic methods[1,2]

Staining Applications Cells;[3] clostridial toxin substrate;[4] endothelial cells;[5] lipid particles;[6] myocardium;[7] neurons;[8] islets of Langerhans;[9] epithelial cells[14]

Biological Applications Measuring membrane potential;[3,5,7,10,11] ion channel screening assays;[12] taste modulators screening assays;[13] detecting cystic fibrosis transmembrane conductance regulator (CFTR);[14] HCN channel modulators;[15] BK channel openers[16]

Industrial Applications Photographic materials[17]

Safety/Toxicity No data available

REFERENCES

1. Tanaka, A.; Yamashita, K.; Horii, S.; Haino, K. Polymethines. Jpn. Kokai Tokkyo Koho JP 52104537, 1977; *Chem. Abstr.* **1978**, *88*, 38963.
2. Ohkuma, S.; Katsura, M.; Higo, A.; Shirotani, K.; Hara, A.; Tarumi, C.; Ohgi, T. Peroxynitrite affects Ca^{2+} influx through voltage-dependent calcium channels. *J. Neurochem.* **2001**, *76*, 341–350.
3. Dall'Asta, V.; Gatti, R.; Orlandini, G.; Rossi, P. A.; Rotoli, B. M.; Sala, R.; Bussolati, O.; Gazzola, G. C. Membrane potential changes visualized in complete growth media through confocal laser scanning microscopy of bis-oxonol-loaded cells. *Exp. Cell Res.* **1997**, *231*, 260–268.
4. Fernandez-Salas, E.; Steward, L. E.; Aoki, K. R. Lipophilic dye-containing cells producing fluorophore-labeled clostridial toxin substrate for FRET detection of toxin activity. PCT Int. Appl. WO 2006107921, 2006; *Chem. Abstr.* **2006**, *145*, 390749.
5. He, P.; Curry, F. E. Measurement of membrane potential of endothelial cells in single perfused microvessels. *Microvasc. Res.* **1995**, *50*, 183–198.
6. Doranz, B. J. Lipid particles containing ion channels and membrane potential-sensitive dyes and their use in screening for effectors of ion channels. PCT Int. Appl. WO 2007089582, 2007; *Chem. Abstr.* **2007**, *147*, 250560.
7. Brasch, F.; Neckel, M.; Volkmann, R.; Schmidt, G.; Hellige, G.; Vetterlein, F. Mapping of capillary flow, cellular redox state, and resting membrane potential in hypoperfused rat myocardium. *Am. J. Physiol.* **1999**, *277*, H2050–H2064.
8. Manger, R.; Woodle, D.; Berger, A.; Hungerford, J. Flow cytometric detection of saxitoxins using fluorescent voltage-sensitive dyes. *Anal. Biochem.* **2007**, *366*, 149–155.
9. Kuznetsov, A.; Bindokas, V. P.; Marks, J. D.; Philipson, L. H. FRET-based voltage probes for confocal imaging: membrane potential oscillations throughout pancreatic islets. *Am. J. Physiol.* **2005**, *289*, C224–C229.
10. Kataoka, T.; Matsubara, M.; Hisada, M. Membrane-potential change detecting method, drug screening

method, and well plate manufacture. Jpn. Kokai Tokkyo Koho JP 2006126073, 2006; *Chem. Abstr.* **2006**, *144*, 447614.

11. Okun, I.; Okun, A.; Kaler, G. Identifying compounds that alter membrane biological potentials with fluorescent ionic dyes. U.S. Patent 6287758, 2001; *Chem. Abstr.* **2001**, *135*, 223771.
12. Wolff, C.; Fuks, B.; Chatelain, P. Comparative study of membrane potential-sensitive fluorescent probes and their use in ion channel screening assays. *J. Biomol. Screen.* **2003**, *8*, 533–543.
13. Servant, G.; Chang, H.; Redcrow, C.; Ray, S.; Clark, I. Improved electrophysiological screening assays for taste modulators using oocytes that express human ENaC and the use of phenamil to improve the effect of ENaC enhancers in assays using membrane potential reporting dyes. PCT Int. Appl. WO 2005014848, 2005; *Chem. Abstr.* **2005**, *142*, 236026.
14. Coclet-Ninin, J.; Rochat, T.; Poitry, S.; Chanson, M. Discrimination between cystic fibrosis and CFTR-corrected epithelial cells by a membrane potential-sensitive probe. *Exp. Lung Res.* **2002**, *28*, 181–199.
15. Vasylyev, D.; Lu, Q.; Mark, B.; Edward, K.; Shan, Q. High-throughput screening of modulators of hyperpolarization-activated cyclic nucleotide-gated (HCN) channels. PCT Int. Appl. WO 2008121735, 2008; *Chem. Abstr.* **2008**, *149*, 455063.
16. Morimoto, T.; Sakamoto, K.; Sade, H.; Ohya, S.; Muraki, K.; Imaizumi, Y. Voltage-sensitive oxonol dyes are novel large-conductance Ca^{2+}-activated K^+ channel activators selective for β1 and β4 but not for β2 subunits. *Mol. Pharmacol.* **2007**, V 71, 1075–1088.
17. Nishide, R. Silver halide photographic photosensitive materials. Jpn. Kokai Tokkyo Koho JP 61267756, 1986; *Chem. Abstr.* **1987**, *107*, 31140.

DiSC$_3$(5)

CAS Registry Number 53213-94-8

Chemical Structure

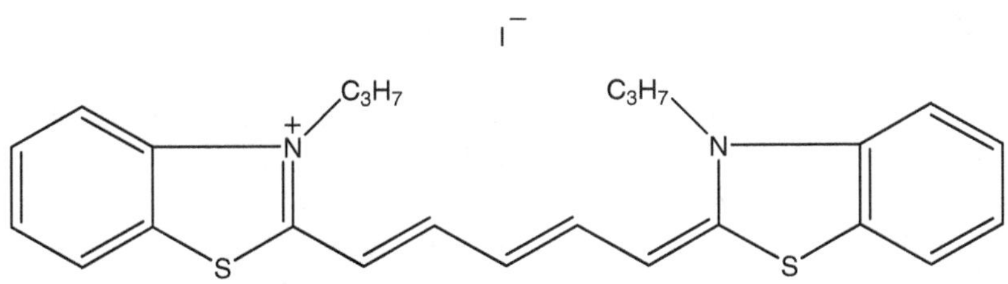

Melting Point 248–249 °C

Absorption (λ_{max}) 651 nm

Emission (λ_{max}) 675 nm

Synthesis Synthetic method[1]

CA Index Name Benzothiazolium, 3-propyl-2-[5-(3-propyl-2(3*H*)-benzothiazolylidene)-1,3-pentadien-1-yl]-, iodide (1:1)

Other Names Benzothiazolium, 3-propyl-2-[5-(3-propyl-2(3*H*)-benzothiazolylidene)-1,3-pentadienyl]-, iodide; 3,3′-Dipropyl-2,2′-thiadicarbocyanine iodide; 3,3′-Dipropylthiadicarbocyanine iodide; Di-S-C3-5; NK 2251

Merck Index Number Not listed

Chemical/Dye Class Cyanine

Molecular Formula $C_{25}H_{27}IN_2S_2$

Molecular Weight 546.53

Physical Form Dark green powder

Solubility Soluble in methanol, *N,N*-dimethylformamide, dimethyl sulfoxide

Staining Applications Bacteria;[2] cells;[3] cytochrome *c* oxidase-reconstituted proteoliposomes;[4] fungus;[5] sperms;[5] epithelial cells;[6] erythrocytes;[7] lipid bilayer;[8] lipid membranes;[9] lipid particles;[10] liposome;[11] lymphocytes;[12] lysosomes;[13] microorganisms;[14] mitochondrial neuroblastoma cells;[15,16] neutrophils;[17] red blood cells;[18] tumor cells;[19] yeast[20,21]

Biological Applications Measuring membrane potential;[1–3,15–25] measuring membrane fusion;[26,27] detecting prostate cancer,[28] nucleic acid hybridization;[29] immunoassays;[30] nucleic acid assays;[30] as anticancer agents[31]

Industrial Applications Not reported

Safety/Toxicity No data available

REFERENCES

1. Sims, P. J.; Waggoner, A. S.; Wang, C. H.; Hoffman, J. F. Mechanism by which cyanine dyes measure membrane potential in red blood cells and phosphatidylcholine vesicles. *Biochemistry* **1974**, *13*, 3315–3330.
2. Goulbourne, E. A., Jr.; Greenberg, E. P. Chemotaxis of *Spirochaeta aurantia*: involvement of membrane potential in chemosensory signal transduction. *J. Bacteriol.* **1981**, *148*, 837–844.
3. Plasek, J.; Dale, R. E.; Sigler, K.; Laskay, G. Transmembrane potentials in cells: a diS-C3(3) assay for relative potentials as an indicator of real changes. *Biochim. Biophys. Acta, Biomembr.* **1994**, *1196*, 181–190.
4. Singh, A. P.; Chanady, G. A.; Nicholls, P. Interactions involving the cyanine dye, diS-C3-(5), cytochrome *c* and liposomes and their implications for estimations of $\Delta\psi$ in cytochrome *c* oxidase-reconstituted proteoliposomes. *J. Membr. Biol.* **1985**, *84*, 183–190.
5. Kawashima, Y. Reagent and method for analyzing urine in which reagent contains a fungus membrane damaging agent and fluorescent dyes. Eur. Pat. Appl. EP 1837402, 2007; *Chem. Abstr.* **2007**, *147*, 360036.
6. Kaunitz, J. D. Preparation and characterization of viable epithelial cells from rabbit distal colon. *Am. J. Physiol.* **1988**, *254*, G502–G512.
7. Hladky, S. B.; Rink, T. J. pH changes in human erythrocytes reported by 3,3′-dipropylthiadicarbocyanine. *J. Physiol.* **1976**, *263*, 213P–214P.
8. Matylevich, N. P. Fluctuation accumulations of cholesterol molecules in lipid bilayer determine substance distribution between the membrane and water phases. *Biofizika* **1986**, *31*, 714–716; *Chem. Abstr.* **1986**, *105*, 130214.

9. Krishna, M. M. G.; Periasamy, N. Fluorescence of organic dyes in lipid membranes: site of solubilization and effects of viscosity and refractive index on lifetimes. *J. Fluoresc.* **1998**, *8*, 81–91.
10. Doranz, B. J. Lipid particles containing ion channels and membrane potential-sensitive dyes and their use in screening for effectors of ion channels. PCT Int. Appl. WO 2007089582, 2007; *Chem. Abstr.* **2007**, *147*, 250560.
11. Koyano, T.; Saito, M.; Myamoto, H.; Umibe, K.; Kato, M. Biochip with liposome-encapsulated with light-responsive chemical and liposome-encapsulated with fluorescent dye. Jpn. Kokai Tokkyo Koho JP 05175574, 1993; *Chem. Abstr.* **1993**, *119*, 221117.
12. Bakeeva, L. E.; Derevyanchenko, I. G.; Konoshenko, G. I.; Mokhova, E. N. Interaction of diS-C3-(5) and ethylrhodamine with lymphocyte mitochondria. *Biokhimiya* **1983**, *48*, 1463–1470; *Chem. Abstr.* **1983**, *99*, 192529.
13. Harikumar, P.; Reeves, J. P. The lysosomal proton pump is electrogenic. *J. Biol. Chem.* **1983**, *258*, 10403–10410.
14. Lykov, V. P.; Khovrychev, M. P.; Polin, A. N. Determination of the sensitivity of microorganisms to antibiotics. U.S.S.R. SU 1337411, 1987; *Chem. Abstr.* **1988**, *108*, 164429.
15. Yamamoto, T.; Tachikawa, A.; Terauchi, S.; Yamashita, K.; Kataoka, M.; Terada, H.; Shinohara, Y. Multiple effects of DiS-C3(5) on mitochondrial structure and function. *Eur. J. Biochem.* **2004**, *271*, 3573–3579.
16. Milligan, G.; Strange, P. G. Biochemical estimation of membrane potential in neuroblastoma cells. *Biochem. Soc. Trans.* **1981**, *9*, 414–415.
17. Tatham, P. E. R.; Delves, P. J.; Shen, L.; Roitt, I. M. Chemotactic factor-induced membrane potential changes in rabbit neutrophils monitored by the fluorescent dye 3,3′-dipropylthiadicarbocyanine iodide. *Biochim. Biophys. Acta, Biomembr.* **1980**, *602*, 285–298.
18. Tsien, R. Y.; Hladky, S. B. A quantitative resolution of the spectra of a membrane potential indicator, diS-C3-(5), bound to cell components and to red blood cells. *J. Membr. Biol.* **1978**, *38*, 73–97.
19. Smith, T. C.; Robinson, S. C. The effect of the fluorescent probe, 3,3′-dipropylthiadicarbocyanine iodide, on the membrane potential of Ehrlich ascites tumor cells. *Biochem. Biophys. Res. Commun.* **1980**, *95*, 722–727.
20. Farrelly, E.; Amaral, M. C.; Marshall, L.; Huang, S. G. A high-throughput assay for mitochondrial membrane potential in permeabilized yeast cells. *Anal. Biochem.* **2001**, *293*, 269–276.
21. Kovac, L.; Poliachova, V. Membrane potential monitoring cyanine dyes uncouple respiration and induce respiration-deficient mutants in intact yeast cells. *Biochem. Int.* **1981**, *2*, 503–507.
22. Murakami, T. Cytotoxicity test method by measuring membrane electric potential. Jpn. Kokai Tokkyo Koho JP 2000300290, 2000; *Chem. Abstr.* **2000**, *133*, 330852.
23. Plasek, J.; Hrouda, V. Assessment of membrane potential changes using the carbocyanine dye diS-C3-(5): Synchronous excitation spectroscopy studies. *Eur. Biophys. J.* **1991**, *19*, 183–188.
24. Cabrini, G.; Verkman, A. S. Potential-sensitive response mechanism of 3,3′-dipropylthiadicarbocyanine iodide (DiS-C3-(5)) in biological membranes. *J. Membr. Biol.* **1986**, *92*, 171–182.
25. Schummer, U.; Schiefer, H. G. Ion diffusion potentials across mycoplasma membranes determined by a novel method using a carbocyanine dye. *Arch. Biochem. Biophys.* **1986**, *244*, 553–562.
26. Nakae, T.; Ryo, E. Measuring membrane fusion. Eur. Pat. Appl. EP 148551, 1985; *Chem. Abstr.* **1985**, *103*, 156880.
27. Yoshihara, E.; Nakae, T. Quantitative measurement of membrane fusions induced by calcium and polyethylene glycol using the porin function. *FEBS Lett.* **1984**, *166*, 49–52.
28. Dickman, D. Methods of detecting prostate cancer. PCT Int. Appl. WO 2006054296, 2006; *Chem. Abstr.* **2006**, *145*, 3346.
29. Bupp, C. R., II; Choi, K. Y.; Holmes-Davis, R. A.; Izmailov, A.; Koshinsky, H.; Nulf, C. J.; Urdea, M.; Wang, M.; Warner, B. D.; Zwick, M. Carbocyanine dye dimers linked by a conjugated alkenyl chain for use in detection of nucleic acid hybridization. PCT Int. Appl. WO 2007100711, 2007; *Chem. Abstr.* **2007**, *147*, 358263.
30. Buechler, K. F.; Noar, J. B.; Tadesse, L. Hybrid phthalocyanine derivatives and their uses in immunoassays and nucleic acid assays. U.S. Patent 5824799, 1998; *Chem. Abstr.* **1998**, *129*, 317583.
31. Uchiumi, K.; Yasui, S.; Hara, H. Cyanine heterocycles as anticancer agents. Jpn. Kokai Tokkyo Koho JP 54157839, 1979; *Chem. Abstr.* **1980**, *93*, 19532.

EOSIN B

CAS Registry Number 548-24-3

Chemical Structure

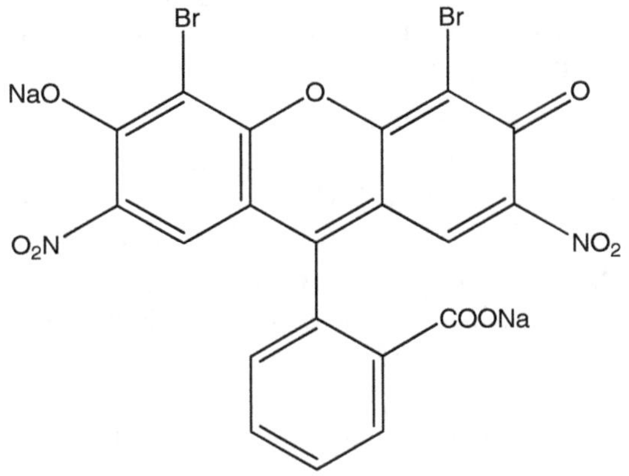

CA Index Name Spiro[isobenzofuran-1(3*H*),9′-[9*H*]xanthen]-3-one, 4′,5′-dibromo-3′,6′-dihydroxy-2′,7′-dinitro-, sodium salt (1 : 2)

Other Names Fluorescein, 4′,5′-dibromo-2′,7′-dinitro-, disodium salt; Spiro[isobenzofuran-1(3*H*),9′-[9*H*]xanthen]-3-one, 4′,5′-dibromo-3′,6′-dihydroxy-2′,7′-dinitro-, disodium salt; C.I. 45400; C.I. Acid Red 91; Dibromodinitrofluorescein sodium; Eosin B; Eosin BA; Eosine B; Eosine BA; Eosine BN; Eosine BNX; Eosine I Bluish; Saffrosine; Simacid Nitro Eosin 24669

Merck Index Number 3602

Chemical/Dye Class Xanthene

Molecular Formula $C_{20}H_6Br_2N_2Na_2O_9$

Molecular Weight 624.06

Physical Form Red-brown to green crystals or powder

Solubility Freely soluble in water; soluble in ethanol

Melting Point 295 °C

Absorption (λ_{max}) 514 nm, 395 nm

Emission (λ_{max}) 544 nm

Synthesis Synthetic method[1–4]

Staining Applications Brain;[5] cells;[6] microorganisms;[7] nucleic acids;[8] peptides;[9] proteins;[9–13] enzyme substrates;[14,15] hairs[16]

Biological Applications Antimalarial agent;[17] protein assay;[18] detecting enzyme activity;[14] treating cancer,[19] malaria,[19] diabetes,[20] a variety of conditions affecting skin,[21] mouth,[21] digestive tract,[21] urinary tract,[21] reproductive tract,[21] respiratory tract,[21] circulatory system,[21] head,[21] neck,[21] endocrine system,[21] lymphoreticular system;[21] dental materials[22,23]

Industrial Applications Color filters;[24,25] liquid crystal displays;[25] inks;[26–28] NLO materials;[29] photographic materials;[30] laundry detergent;[31] textiles[32]

Safety/Toxicity No data available

Certification/Approval Certified by Biological Stain Commission (BSC)

REFERENCES

1. Shimidzu, T.; Iyoda, T.; Koide, Y. An advanced visible-light-induced water reduction with dye-sensitized semiconductor powder catalyst. *J. Am. Chem. Soc.* **1985**, *107*, 35–41.
2. Gandin, E.; Lion, Y.; Van de Vorst, A. Quantum yield of singlet oxygen production by xanthene derivatives. *Photochem. Photobiol.* **1983**, *37*, 271–278.
3. Maiti, S.; Saha, M. K.; Palit, S. R. Dye-sensitized photopolymerization of methyl methacrylate in nonaqueous media. *Makromol. Chem.* **1969**, *127*, 224–239.
4. Holmes, W. C.; Melin, C. G.; Peterson, A. R.; Eosin, B. *Stain Technol.* **1932**, *7*, 121–127.
5. Smith, T. M.; Squires, R. F. Differential inhibition of brain specific [^{3}H]flunitrazepam binding by several types of dyes. *Neurochem. Res.* **1983**, *8*, 1177–1183.
6. Drocourt, J. L.; Levesque, P. Method for counting viable cells by dual staining. PCT Int. Appl. WO 9855861, 1998; *Chem. Abstr.* **1998**, *130*, 35353.
7. Chong, N. M. Development of a tool for measuring the degradation capacity of microorganisms for a xenobiotic. *Enzyme Microb. Technol.* **2005**, *37*, 467–471.
8. Gagna, C. E.; Kuo, H. R.; Agostino, N.; Rizzo, D.; Isquith, I. R.; Mathew, J.; Mohammed, J.; Hoo, S.; Lambert, W. C. Novel use of bovine zeta-crystallin as a conformational DNA probe to characterize a phase

Handbook of Biological Dyes and Stains By R. W. Sabnis

transition zone and terminally differentiating fiber cells in the adult canine ocular lens. *Arch. Histol. Cytol.* **2001**, *64*, 379–391.

9. Ben-Sasson, S.; Eilat, D. Pharmaceutical compositions containing polymers of aromatic compounds which affect tissue distribution of bioactive peptides and proteins. Eur. Pat. Appl. EP 354714, 1990; *Chem. Abstr.* **1991**, *114*, 115104.
10. Kosaka, H.; Sakamoto, H. Development of a colorimetric quantitation method of protein in solution by excluding the influence of creatinine. PCT Int. Appl. WO 2005071420, 2005; *Chem. Abstr.* **2005**, *143*, 188838.
11. Waheed, A. A.; Rao, K. S.; Gupta, P. D. Mechanism of dye binding in the protein assay using eosin dyes. *Anal. Biochem.* **2000**, *287*, 73–79.
12. Waheed, A. A.; Gupta, P. D. Application of an eosin B dye method for estimating a wide range of proteins. *J. Biochem. Biophys. Methods* **1996**, *33*, 187–196.
13. Waheed, A. A.; Gupta, P. D. Estimation of protein using eosin B dye. *Anal. Biochem.* **1996**, *233*, 249–252.
14. Lihme, A. O. F.; Wikborg, M. Pre-stained 3,3′,5,5′-tetrame-thylbenzidine substrates for the detection of enzyme activity. PCT Int. Appl. WO 9904261, 1999; *Chem. Abstr.* **1999**, *130*, 150342.
15. Sumiya, M. Dyeing of substrates. Jpn. Kokai Tokkyo Koho JP 62021884, 1987; *Chem. Abstr.* **1987**, *107*, 117243.
16. Moeller, H.; Meinigke, B. Hair dye containing nitroxanthene derivatives. Eur. Pat. Appl. EP 1199065, 2002; *Chem. Abstr.* **2002**, *136*, 345464.
17. Massimine, K. M.; McIntosh, M. T.; Doan, L. T.; Atreya, C. E.; Gromer, S.; Sirawaraporn, W.; Elliott, D. A.; Joiner, K. A.; Schirmer, R. H.; Anderson, K. S. Eosin B as a novel antimalarial agent for drug-resistant *Plasmodium falciparum*. *Antimicrob. Agents Chemother.* **2006**, *50*, 3132–3141.
18. Das Gupta, P.; Waheed, A. L. A.; A process for the preparation of a formulation useful for the detection and estimation of proteins present in biological fluids. Indian IN 185275, 2000; *Chem. Abstr.* **2004**, *142*, 3086.
19. Becker-Brandenburg, K.; Zimmermann, H.; Fritz-Wolf, K. Ellagic acid-related compound and polyaromatic phenol inhibitors of glutathione-*S*-transferase, and their therapeutic use. PCT Int. Appl. WO 2006097472, 2006; *Chem. Abstr.* **2006**, *145*, 328416.
20. Geier, A.; Kanety, H.; Karasik, A. Biologically active synthetic dye compounds mimicking the physiological activity of insulin or insulin-like growth factor, and their therapeutic use. U.S. Patent 6020374, 2000; *Chem. Abstr.* **2000**, *132*, 117549.
21. Dees, H. C.; Scott, T. C.; Wachter, E. A.; Fisher, W. G.; Smolik, J. Intracorporeal medicaments for high energy phototherapeutic treatment of disease. U.S. Pat. Appl. Publ. US 2007208076, 2007; *Chem. Abstr.* **2007**, *147*, 317211.
22. Piergallini, R.; Loupis, N. Teeth whitening composition comprising activating agents, such as Eosin B and Erythrosin B. U.S. Pat. Appl. Publ. US 2007128132, 2007; *Chem. Abstr.* **2007**, *147*, 38007.
23. Trom, M. C.; Oxman, J. D. Color-changing dental compositions. U. S. Patent 6444725, 2002; *Chem. Abstr.* **2002**, *137*, 206597.
24. Machiguchi, K.; Ueda, Y. Colored light-sensitive resin composition containing specific dyes for manufacturing color filters. Jpn. Kokai Tokkyo Koho JP 2002278056, 2002; *Chem. Abstr.* **2002**, *137*, 270519.
25. Shirota, K.; Myazaki, T.; Nakazawa, K.; Yamada, S. Ink-jet printing ink for manufacture of liquid crystal display color filter. Jpn. Kokai Tokkyo Koho JP 08327812, 1996; *Chem. Abstr.* **1997**, *126*, 164272.
26. Hasemann, L. Dye mixtures and their use in jet printing inks. PCT Int. Appl. WO 2003052007, 2003; *Chem. Abstr.* **2003**, *139*, 54286.
27. Sano, H.; Yamada, M. Ink-jet recording method using magenta inks to give light-resistant images. Jpn. Kokai Tokkyo Koho JP 2000265099, 2000; *Chem. Abstr.* **2000**, *133*, 268387.
28. Kubota, N.; Kanetani, M. Ink sets for light-resistant ink-jet recording. Jpn. Kokai Tokkyo Koho JP 11315230, 1999; *Chem. Abstr.* **1999**, *131*, 338433.
29. Yitzchaik, S.; Kakkar, A. K.; Roscoe, S. B.; Marks, T. J.; Lundquist, P. M.; Lin, W.; Wong, G. K. Descriptive chemistry of self-assembled multilayer second-order NLO materials: chemical, microstructural and performance interrelationships. *Mater. Res. Soc. Symp. Proc.* **1994**, *328*, 27–35.
30. Inaba, Y.; Hasegawa, T.; Hamada, N. Electrophotographic liquid developer. Jpn. Kokai Tokkyo Koho JP 63266463, 1988; *Chem. Abstr.* **1989**, *111*, 87362.
31. Batchelor, S. N.; Bird, J. M.; Chapple, A. P. Laundry treatment compositions containing nonionic surfactants and dyes. PCT Int. Appl. WO 2006053598, 2006; *Chem. Abstr.* **2006**, *144*, 490675.
32. Deeds, S. M. Fabric repair coloring device and using the device to color unwanted bleached out spots on fabric. U.S. Patent 6739779, 2004; *Chem. Abstr.* **2004**, *140*, 424935.

EOSIN Y

CAS Registry Number 17372-87-1

Chemical Structure

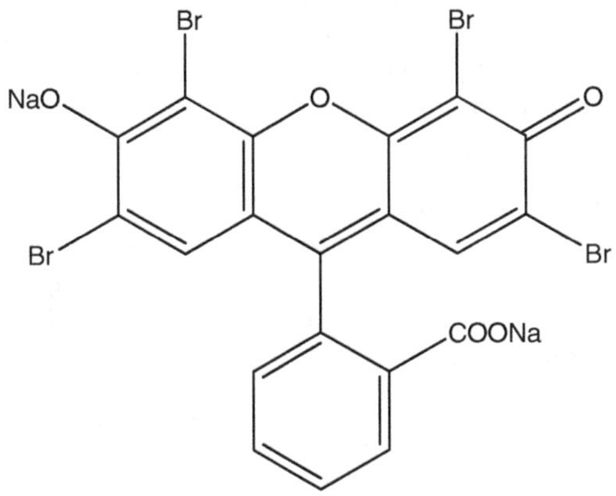

CA Index Name Spiro[isobenzofuran-1(3*H*),9′-[9*H*]xanthen]-3-one, 2′,4′,5′,7′-tetrabromo-3′,6′-dihydroxy-, sodium salt (1 : 2)

Other Names Eosin Yellowish; Fluorescein, 2′,4′,5′,7′-tetrabromo-, disodium salt; Spiro[isobenzofuran-1(3*H*),9′-[9*H*]xanthen]-3-one, 2′,4′,5′,7′-tetrabromo-3′,6′-dihydroxy-, disodium salt; 11445 Red; 11731 Red; 1903 Yellow Pink; ABCO Eosin; Abbey Eosin; Acid Red 87; Acid Red 87-39014; Aizen Eosine; Aizen Eosine GH; Basacid Red 316; Bromo 4D; Bromo 4DC; Bromo 4DL; Bromo B; Bromo FL; Bromo JPS; Bromo TS; Bromo X 100; Bromo XX; Bromo acid; Bromoeosine; Bromofluoresceic acid; Bromofluorescein; Bronze Bromo; C.I. 45380; C.I. Acid Red 87; Certiqual Eosine; Chugai Ink Red AM 5; Cogilor Orange 212.10; Conacid Red ML; D and C Red No. 22; D&C Red 22; D&C Red No. 22; D&C Red No. 22-90133; Daiwa Red 103WB; Dawn red; Dinacid Eosine Red; Disodium eosin; Dycosweak Acid Red A; Eosin; Eosin A; Eosin OJ; Eosin Y; Eosin Y 17255; Eosin YS; Eosin yellow; Eosine; Eosine 3G; Eosine 3Y; Eosine A; Eosine AG; Eosine BPC; Eosine BS; Eosine BS-SF; Eosine DA; Eosine DWC 73; Eosine Extra Conc. A. Export; Eosine Extra Yellowish; Eosine FA; Eosine G; Eosine GF; Eosine GH; Eosine J; Eosine K; Eosine K Salt Free; Eosine Lake Red Y; Eosine OJ; Eosine S; Eosine S 13; Eosine SSO; Eosine Salt Free; Eosine Y; Eosine Y 90133; Eosine YB; Eosine YS; Eosine Yellow; Eosine Yellowish; Eosine sodium; Eosine sodium salt; Eosine w/s; Eosine water-soluble; Fenazo Eosine XG; Food Red 103; Hidacid Boiling Bromo; Hidacid Bromo acid regular; Hidacid Dibromo Fluorescein; Hidacid Eosine Soda Salt; Hidacid White Bromo; Irgalite Bronze Red CL; Japan Red 103; Japan Red 230-1; Japan Red 230-2; Japan Red No. 103; Japan Red No. 230-1; Morning red; Neelicol Eosine; Orient Water Red 2; Orient Watger Red 2; Phlox Red Toner X 1354; Phloxine Red 20-7600; Phloxine Toner B; Pure Eosine YY; Red 103; Red 17255; Red No. 103; Red No. 230-1; Silk Discharge Orange 3R; Simacid Eosine Y; Sodium eosin; Sodium eosinate; Sodium eosine; Spiro[isobenzofuran-1(3*H*),9′-[9*H*]xanthen]-3-one, 2′,4′,5′,7′-tetrabromo-3′,6′-dihydroxy-, disodium salt; Symuler Eosin Toner; Tetrabromofluorescein; Tetrabromofluorescein D; Tetrabromofluorescein S; Tetrabromofluorescein soluble; Toyo Eosine G; Triacid Eosine; Vicoacid Eosine; Water Red 2

Merck Index Number 3603

Chemical/Dye Class Xanthene

Molecular Formula $C_{20}H_6Br_4Na_2O_5$

Molecular Weight 691.85

Physical Form Red-brown crystals or powder

Solubility Freely soluble in water; slightly soluble in ethanol, methanol; insoluble in ether

Melting Point 295.5 °C

pH Range 0.0–3.0

Color Change at pH Non-fluorescence (0.0) to green fluorescence (3.0)

pK_a 2.9, 4.5

Absorption (λ_{max}) 517 nm

Emission (λ_{max}) 538 nm

Synthesis Synthetic methods[1–11]

Staining Applications Blood;[12] blood smears;[13] bone marrow;[12] cells;[14] nucleus;[15] cytoplasm;[15] membrane;[15] candies;[16] drinks;[16] keloid;[17] orthodontic adhesives;[18] proteins;[19] tissues;[20] thrombocytes;[21] eye lens;[22] eye shadow;[23] lips;[24] skin;[24,25] hairs;[1,26] keratin fibers[27]

Biological Applications Treating age-related macular degeneration,[28] burns,[1,29] cancer,[1,30] diabetes,[1,31] obesity,[1,31] dental bone defects,[1,32] gastroesophageal reflux disease,[33] prostate cancer,[34] viral diseases;[35] stents;[1,36] wound-healing materials[1,37]

Industrial Applications Solar cell;[38] semiconductor devices;[39] color filters;[1,40] Light-emitting devices;[41] photovoltaic devices;[1,42] electrochromic devices;[43] thin films;[44] sol–gel materials;[45] inks;[1,46,47] colored bubbles[48]

Safety/Toxicity Acute toxicity;[1,49] carcinogenicity;[50] cardiotoxicity;[51] ecotoxicity;[52] environmental toxicity;[53,54] microbial toxicity;[55] mutagenicity;[1,56] neurotoxicity;[57] nucleic acid damage;[58] phototoxicity;[1,59] pulmonary toxicity;[60] reproductive toxicity;[61] skin toxicity[62]

Certification/Approval Certified by Biological Stain Commission (BSC)

REFERENCES

1. Sabnis, R.W.; *Handbook of Acid-Base Indicators*; CRC Press: Boca Ratan, **2008**; pp 139–141.
2. Anantharaman, P. N.; Vasudevan, D.; Chellammal, S. An improved process for the preparation of eosin by electrolytically brominating fluorescein. Indian IN 185585, 2001; *Chem. Abstr.* **2005**, *142*, 95715.
3. Vasudevan, D.; Basha, C. A. A kinetic study on the electrochemical bromination of fluorescein to eosin in a batch reactor. *J. Indian Chem. Soc.* **1998**, *75*, 165–166.
4. Vasudevan, D.; Anantharaman, P. N. Electrochemical synthesis of eosin from fluorescein. *J. Appl. Electrochem.* **1993**, *23*, 808–812.
5. Anantharaman, P. N.; Noel, M. Electrochemical manufacture of erythrosine or eosine from fluorescein. Indian IN 168346, 1991; *Chem. Abstr.* **1992**, *116*, 237372.
6. Amt-Guerri, F.; Lopez-Gonzalez, M. M. C.; Martinez-Utrilla, R.; Sastra, R. Synthesis and spectroscopic properties of new Rose Bengal and Eosin Y derivatives. *Dyes Pigments* **1990**, *12*, 249–272.
7. Jagannathan, E.; Anantharaman, P. N. Electrochemical preparation of erythrosine and eosine. *Bull. Electrochem.* **1987**, *3*, 29–31.
8. Fompeydie, D.; Onur, F.; Levillain, P. Synthesis of pure eosine. *Bull. Soc. Chim. Fr.* **1982**, 5–6; *Chem. Abstr.* **1982**, *96*, 201245.
9. Paucescu, S. D.; Ionescu-Ioan, C. Microscopy xanthine dyes. II. Fluorescein bromination to tetrabromo compounds. *Rev. Chim.* **1980**, *31*, 339–341; *Chem. Abstr.* **1980**, *93*, 96775.
10. Dolinsky, M.; Jones, J. H. Studies on coal-tar colors. IX. D&C Yellow No. 7, D&C Orange No. 5, D&C Red No. 21, tetrachlorofluorescein, D&C Red No. 27, and FD&C Red No. 3. *J. Assoc. off. Agric. Chem.* **1951**, *34*, 114–126.
11. Welcher, F. J. *Organic Analytical Reagents*; D. Van Nostrand Company: New York, **1948**; Vol. *4*, p 486.
12. Dagiral, R. L. G. Cell staining method for blood and bone marrow. PCT Int. Appl. WO 2008043963, 2008; *Chem. Abstr.* **2008**, *148*, 466507.
13. Trukhachev, V. I.; Rodin, V. V.; Mikhailenko, V. V.; Dergunov, A. A. Staining of blood smears by using the Erlich's hematoxylin and sodium eosin. Russ. RU 2304776, 2007; *Chem. Abstr.* **2007**, *147*, 253438.
14. Korchagina, E.; Bovin, N.; Henry, S. Fluorescent cell markers containing fluorophore and diacyl lipid. PCT Int. Appl. WO 2008030115, 2008; *Chem. Abstr.* **2008**, *148*, 326200.
15. Gholap, A. S.; Gholap, G. A.; Joshi, A.; Basu, A.; Rao, C. V. K.; Jadhav, P. Method and system for digital image based tissue independent simultaneous nucleus, cytoplasm and membrane quantitation. PCT Int. Appl. WO 2007024264, 2007; *Chem. Abstr.* **2007**, *146*, 269702.
16. Yoshioka, N.; Ichihashi, K. Determination of 40 synthetic food colors in drinks and candies by high-performance liquid chromatography using a short column with photodiode array detection. *Talanta* **2008**, *74*, 1408–1413.
17. Butler, P. D.; Ly, D. P.; Longaker, M. T.; Yang, G. P. Use of organotypic coculture to study keloid biology. *Am. J. Surg.* **2008**, *195*, 144–148.
18. Brennan, J. V.; Soo, P. P.; Kalgutkar, R. S.; Suhrwardy, R. A. Methods of identifying orthodontic adhesives on teeth by selective staining with pigments and/or food dyes. U.S. Pat. Appl. Publ. US 2007148609, 2007; *Chem. Abstr.* **2007**, *147*, 102288.
19. Becker, K. F.; Schott, C.; Becker, I.; Hoefler, H. Guided protein extraction from formalin-fixed tissues for quantitative multiplex analysis avoids detrimental effects of histological stains. *Proteomics* **2008**, *2*, 737–743.
20. Brusilovskiy, A. I. Tissue staining protocol using eosin and hematoxylin solutions. U.S. Pat. Appl. Publ. US 2005202524, 2005; *Chem. Abstr.* **2005**, *143*, 282169.
21. Lindberg, S. Enumeration of thrombocytes. PCT Int. Appl. WO 2008010760, 2008; *Chem. Abstr.* **2008**, *148*, 163110.
22. Korsakova, N. V.; Sergeeva, V. E. Cytochemical analysis of cellular structures in eye lens of laboratory animals including vital–supravital staining by methylene blue. Russ. RU 2319132, 2008; *Chem. Abstr.* **2008**, *148*, 302857.
23. Bouchard, F. Eye shadow comprising least 50% colorants. Eur. Pat. Appl. EP 1757262, 2007; *Chem. Abstr.* **2007**, *146*, 280374.
24. Dumousseaux, C. Cosmetic composition for application to the skin, lips, and body surface. Fr. Demande FR 2868694, 2005; *Chem. Abstr.* **2005**, *143*, 392520.
25. Teboul, K. Method for coloring dark skins. PCT Int. Appl. WO 2008087308, 2008; *Chem. Abstr.* **2008**, *149*, 160219.

26. Hercouet, L. Hair dye composition comprising a derivative of diamino-*N,N*-dihydropyrazolone, a coupler, and a heterocyclic direct dye. Fr. Demande FR 2886132, 2006; *Chem. Abstr.* **2006**, *146*, 49738.
27. Kravtchenko, S. Agent for simultaneously bleaching and coloring of keratin fibers comprising an anionic or nonionic dye and an inert organic liquid. Eur. Pat. Appl. EP 1738801, 2007; *Chem. Abstr.* **2007**, *146*, 128004.
28. Hofmann, R. F. Use of targeted oxidative therapeutic formulation in treatment of age-related macular degeneration. U.S. Pat. Appl. Publ. US 2005250756, 2005; *Chem. Abstr.* **2005**, *143*, 416314.
29. Hofmann, R. F. Use of targeted oxidative therapeutic formulation in treatment of burns. PCT Int. Appl. WO 2006002302, 2006; *Chem. Abstr.* **2006**, *144*, 114433.
30. Hofmann, R. F. Use of targeted oxidative therapeutic formulation in treatment of cancer. U.S. Pat. Appl. Publ. US 2005250757, 2005; *Chem. Abstr.* **2005**, *143*, 16223.
31. Hofmann, R. F. Use of targeted oxidative therapeutic formulation in treatment of diabetes and obesity. PCT Int. Appl. WO 2005107728, 2005; *Chem. Abstr.* **2005**, *143*, 466228.
32. Walline, K. S.; Atkinson, B. L. Light-curable bone growth material for treating dental bone defects. PCT Int. Appl. WO 2007079053, 2007; *Chem. Abstr.* **2007**, *147*, 173744.
33. Chandrasoma, P. Methodologies for the diagnosis and treatment of gastroesophageal reflux disease. U.S. Pat. Appl. Publ. US 2007178510, 2007; *Chem. Abstr.* **2007**, *147*, 184153.
34. Datta, M. W.; Datta, S.; Altaba, A. R. I. Methods for diagnosis and treatment of prostate and other cancers by inhibition of proteoglycans perlecan and patched and smoothened receptors. PCT Int. Appl. WO 2006007450, 2006; *Chem. Abstr.* **2006**, *144*, 143034.
35. Hofmann, R. F. Targeted oxidative therapeutic formulation in treatment of viral diseases. U.S. Pat. Appl. Publ. US 2005192267, 2005; *Chem. Abstr.* **2005**, *143*, 272504.
36. Nakayama, Y.; Nemoto, Y. Stents having soft polymer coating layers with fine pores, and production thereof. Jpn. Kokai Tokkyo Koho JP 2006333940, 2006; *Chem. Abstr.* **2006**, *146*, 50430.
37. Nakayama, Y.; Nemoto, Y. Visible light-curable materials and their use for wound-healing promotion materials and hemostatics. Jpn. Kokai Tokkyo Koho JP 2008073443, 2008; *Chem. Abstr.* **2008**, *148*, 410835.
38. Okamoto, T.; Ito, S. Dye-sensitized solar cell and its manufacture by electrodepositing zinc oxide with template compound. Jpn. Kokai Tokkyo Koho JP 2008177099, 2008; *Chem. Abstr.* **2008**, *149*, 180291.
39. Latchinov, A. N.; Salazkin, S. N. Anisotropic electrically conductive adhesives for semiconductor devices. PCT Int. Appl. WO 2007120078, 2007; *Chem. Abstr.* **2007**, *147*, 487258.
40. Deno, T.; Eichenberger, T.; Voigt, J. Use of pyrimido [5,4-g]pteridines as shading component in color filter colorant compositions. PCT Int. Appl. WO 2006037728, 2006; *Chem. Abstr.* **2006**, *144*, 400791.
41. Uehara, M. Light-emitting device, method for manufacturing light-emitting device, electronic device, and electronic instrument. PCT Int. Appl. WO 2006132374, 2006; *Chem. Abstr.* **2006**, *146*, 35756.
42. Ando, H.; Takeuchi, H.; Minoura, H.; Yoshida, T. Manufacture of porous metal oxide/dye composite film for photovoltaic devices. Jpn. Kokai Tokkyo Koho JP 2006283048, 2006; *Chem. Abstr.* **2006**, *145*, 445190.
43. Kanitz, A.; Roth, W. Electrochromic active formulation and electrochromic device manufactured therewith. Eur. Pat. Appl. EP 1892566, 2008; *Chem. Abstr.* **2008**, *148*, 273017.
44. Verma, A.; Goyal, A.; Sharma, R. K. Microstructural, photocatalysis and electrochemical investigations on $CeTi_2O_6$ thin films. *Thin Solid Films* **2008**, *516*, 4925–4933.
45. Murciano, A.; Blaya, S.; Carretero, L.; Madrigal, R. F.; Fimia, A. Holographic reflection gratings on photopolymerizable solgel materials. *Opt. Lett.* **2006**, *31*, 2317–2319.
46. Asahino, K. Water-erasable oil-based inks applied on thermally reversible recording sheets. Jpn. Kokai Tokkyo Koho JP 2007169319, 2007; *Chem. Abstr.* **2007**, *147*, 119863.
47. Nakamura, S. Water-thinned writing ink compositions with good lightfastness, transparent cartridges for them, and pens using them. Jpn. Kokai Tokkyo Koho JP 2006321861, 2006; *Chem. Abstr.* **2006**, *146*, 9427.
48. Sabnis, R. W.; Kehoe, T. D. Composition and method for producing colored bubbles. U.S. Pat. Appl. Publ. US 2006004110, 2006; *Chem. Abstr.* **2006**, *144*, 110164.
49. Tonogai, Y.; Ito, Y.; Iwaida, M.; Tati, M.; Ose, Y.; Hori, M. Studies on the toxicity of coal-tar dyes III. Reason of acute toxicity to fish caused by coal-tar dyes and their industrial effluents. *J. Toxicol. Sci.* **1980**, *5*, 23–33.

50. Oliveira, P. A.; Colaco, A. A.; Palmeira, C. A.; De la Cruz, P. L. F.; Lopes, C. A. A phyllodes tumor of the urinary bladder in a rat. *Exp. Oncol.* **2007**, *29*, 159–161.

51. Miyazato, H.; Biro, S.; Setoguchi, M.; Maeda, M.; Tashiro, T.; Nakao, S.; Tanaka, H. Abnormal immunostaining for dystrophin in isoproterenol-induced acute myocardial injury in rats: evidence for change in dystrophin in the absence of genetic defect. *J. Mol. Cell. Cardiol.* **1997**, *29*, 1217–1223.

52. Benoit-Guyod, J. L.; Rochat, J.; Alary, J.; Andre, C.; Taillandier, G. Correlations between physico-chemical properties and ecotoxicity of fluorescent xanthenic water tracers. *Toxicol. Eur. Res.* **1979**, *2*, 241–246.

53. Rastogi, S. C.; Pritzl, G. Red lipstick: a source of barium to humans and the environment. *Bull. Environ. Contam. Toxicol.* **1998**, *60*, 507–510.

54. Tonogai, Y.; Ogawa, S.; Ito, Y.; Iwaida, M. Actual survey on TLm (median tolerance limit) values of environmental pollutants, especially on amines, nitriles, aromatic nitrogen compounds and artificial dyes. *J. Toxicol. Sci.* **1982**, *7*, 193–203.

55. Oros, G.; Cserhati, T.; Forgacs, E. Separation of the strength and selectivity of the microbiological effect of synthetic dyes by spectral mapping technique. *Chemosphere* **2003**, *52*, 185–193.

56. Serafimova, R.; Todorov, M.; Pavlov, T.; Kotov, S.; Jacob, E.; Aptula, A.; Mekenyan, O. Identification of the structural requirements for mutagenicity, by incorporating molecular flexibility and metabolic activation of chemicals. II. General Ames mutagenicity model. *Chem. Res. Toxicol.* **2007**, *20*, 662–676.

57. Zettler, C.; Head, R. J.; Rush, R. A. Chronic nerve growth factor treatment of normotensive rats. *Brain Res.* **1991**, *538*, 251–262.

58. Rhee, Y.; Valentine, M. R.; Termini, J. Oxidative base damage in RNA detected by reverse transcriptase. *Nucleic Acids Res.* **1995**, *23*, 3275–3282.

59. Wang, H.; Lu, L.; Zhu, S.; Li, Y.; Cai, W. The phototoxicity of xanthene derivatives against *Escherichia coli*, *Staphylococcus aureus*, and *Saccharomyces cerevisiae*. *Curr. Microbiol.* **2006**, *52*, 1–5.

60. Ferguson, J. S.; Alarie, Y. Long term pulmonary impairment following a single exposure to methyl isocyanate. *Toxicol. Appl. Pharmacol.* **1991**, *107*, 253–268.

61. Uzumcu, M.; Suzuki, H.; Skinner, M. K. Effect of the anti-androgenic endocrine disruptor vinclozolin on embryonic testis cord formation and postnatal testis development and function. *Reprod. Toxicol.* **2004**, *18*, 765–774.

62. Smith, H. R.; Orchard, G. E.; Calonje, E.; Basketter, D. A.; McFadden, J. P. Irritant threshold and histological response of epidermis to irritant application. *Contact Dermatitis* **2004**, *51*, 227–230.

ERYTHROSIN

CAS Registry Number 16423-68-0

Chemical Structure

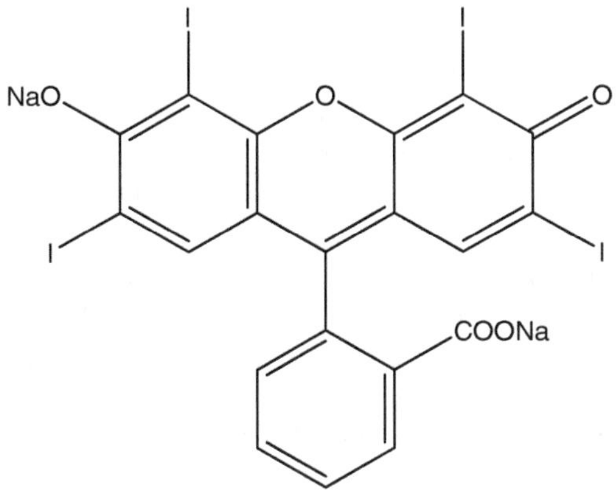

CA Index Name Spiro[isobenzofuran-1(3*H*),9′-[9*H*]xanthen]-3-one, 3′,6′-dihydroxy-2′,4′,5′,7′-tetraiodo-, sodium salt (1 : 2)

Other Names Erythrosine B; Fluorescein, 2′,4′,5′,7′-tetraiodo-, disodium salt; Spiro[isobenzofuran-1(3*H*),9′-[9*H*]xanthen]-3-one, 3′,6′-dihydroxy-2′,4′,5′,7′-tetraiodo-, disodium salt; 1427 Red; 1671 Red; 2′,4′,5′,7′-Tetraiodofluorescein disodium salt; 2,4,5,7-Tetraiodofluorescein disodium salt; Acid Red 51; Aizen Erythrosine; Aizen Food Red 3; Basovit Red 425E; C.I. 45430; C.I. Acid Red 51; C.I. Food Red 14; Calcocid Erythrosine N; Canacert Erythrosine BS; Ceplac; Cilefa Pink B; Cogilor Red 312.10; D and C Red No. 3; D&C Red No. 3; Dolkwal Erythrosine; E 127; Edicol Supra Erythrosin AS; Edicol Supra Erythrosine A; Erythrosin; Erythrosin B; Erythrosin B sodium salt; Erythrosin BS; Erythrosine; Erythrosine 307046; Erythrosine 36003; Erythrosine 37003; Erythrosine 3B; Erythrosine B-FO; Erythrosine BS; Erythrosine Bluish; Erythrosine Extra; Erythrosine Extra Conc. A Export; Erythrosine Extra Pure A; Erythrosine I; Erythrosine K-FO; Erythrosine TB; Erythrosine TB Extra; Erythrosine Extra Bluish; Eurocert Erythrosine 311807; FD & C Red No. 3-307020; FD and C Red 3; FD and C Red No. 3; FD&C Red No. 3; FD&C Red No. 3-37003; FDC Red 3; FDC Red 3 dye; Food Color Red 3; Food Dye Red 3; Food Red 14; Food Red 3; Food Red No. 3; Hexacert Red No. 3; Hexacol Erythrosine BS; Japan Food Red No. 3; Japan Red 3; Japan Red No. 3; Maple Erythrosine; Necol Erythrosine; Neelicol Erythrosine; New Ink Bluish Geigy; Red 1799; Red 3; Red No. 3; S 887; Simacid Pink 24107; Spiro[isobenzofuran-1(3*H*),9′-[9*H*]xanthen]-3-one, 3′,6′-dihydroxy-2′,4′,5′,7′-tetraiodo-, disodium salt; Synerid; Tetraiodofluorescein sodium salt; Usacert FD & C Red No. 3-310116; Usacert Red No. 3; Water Pink 176575

Merck Index Number 3693

Chemical/Dye Class Xanthene

Molecular Formula $C_{20}H_6I_4Na_2O_5$

Molecular Weight 879.86

Physical Form Red to brown powder

Solubility Soluble in water, ethanol

Melting Point >250 °C

pH Range 2.5–4.0

Color Change at pH Non-fluorescence (2.5) to light green or reddish fluorescence (4.0)

pK_a 4.1

Absorption (λ_{max}) 525 nm

Emission (λ_{max}) 555 nm

Synthesis Synthetic methods[1–8]

Staining Applications Blood;[9] bone marrow;[9] bacterial plaque;[10] cancer cells;[11] dental plaque;[12,13] human serum albumin (HSA);[19] lymph node;[14] microorganisms;[15] neurons;[16] nucleic acids;[17,18] prions;[20] spores;[11] animal feeds;[21] alcohol;[22] baked food;[23] beverages;[24] candies;[25] caramels;[26] confectionery;[27,28] cotton candy;[29] drinks;[25,27] fish;[30] olives;[31] orange juices;[32] papaya fruit;[33] soft drinks;[28] sport drink;[22] sweetener;[34] capsules;[35,36] tablets;[37,38] sunscreen;[39] eyelids;[41] lips;[40,41] skin;[41–43] tattoos;[44] teeth;[45,46] hairs;[1,47–49] keratin fibers[49,50]

Biological Applications Detecting gene expression,[51] phosphoproteins,[52] protease,[53] stress biomarkers;[54] treating age-related macular degeneration,[55] arteriosclerosis,[56] bone metabolic diseases,[57] burns,[1,58] cancer,[1,59] diabetes,[1,60] human immunodeficiency virus infection,[61] obesity,[1,60] viral diseases;[1,62] medical devices;[63] photodynamic therapy[1,64,65]

Industrial Applications Solar cells;[66] photoelectric device;[67] light emitting diodes;[1,68] color filters;[1,69] liquid-crystal displays;[69] thin films;[70] inks;[1,71–74] lithographic plate;[75] photographic materials;[76] recording materials;[77] sol-gel materials;[78] photonics;[78] adhesive;[79] paints;[1,80] thermoplastics;[81] colored bubbles;[1,82] textiles,[83] entertainment products;[84] toys[85]

Safety/Toxicity Acute toxicity;[1,86] carcinogenicity;[1,87,88,95,96] cytotoxicity;[1,89] developmental toxicity;[90] DNA-damage;[98] embryotoxicity;[1,91] genotoxicity;[1,92–94] lifetime toxicity;[95,96] mutagenicity;[1,97–100] neurotoxicity;[1,101,102] phototoxicity;[103] psychotoxicity;[90] reproductive toxicity[1,104–106]

Certification/Approval Certified by Biological Stain Commission (BSC); Approved by Food & Drugs Administration (FDA)

REFERENCES

1. Sabnis, R. W. *Handbook of Acid–Base Indicators*; CRC Press: Boca Raton, **2008**; pp 142–144.
2. Vasudevan, D.; Anantharaman, P. N. Electrochemical synthesis of erythrosin from fluorescein. *J. Appl. Electrochem.* **1994**, *24*, 1188–1190.
3. Anantharaman, P. N.; Noel, M. Electrochemical manufacture of erythrosine or eosine from fluorescein. Indian IN 168346, 1991; *Chem. Abstr.* **1992**, *116*, 237372.
4. Jagannathan, E.; Anantharaman, P. N. Electrochemical synthesis of erythrosine. *Bull. Electrochem.* **1988**, *4*, 481–483.
5. Jagannathan, E.; Anantharaman, P. N. Electrochemical preparation of erythrosine and eosine. *Bull. Electrochem.* **1987**, *3*, 29–31.
6. Lopez Sanchez, M. Process for manufacture of erythrosine. Span. ES 548295, 1986; *Chem. Abstr.* **1987**, *106*, 178113.
7. Welcher, F. J. *Organic Analytical Reagents*; D. Van Nostrand Company: New York, **1948**; Vol. *4*, p 360.
8. Pukirev, A. Preparation and purification of iodoeosin and erythrosin. *Khim. Farm. Prom.* **1933**, *2*, 58–63; *Chem. Abstr.* **1933**, *27*, 55004.
9. Dagiral, R. L. G. Cell staining method for blood and bone marrow. PCT Int. Appl. WO 2008043963, 2008; *Chem. Abstr.* **2008**, *148*, 466507.
10. Eduardo da Cruz, L. Dentifrice containing an antibacterial agent that can show the presence of bacterial plaque in the oral cavity. Braz. Pedido PI BR 2001002757, 2005; *Chem. Abstr.* **2006**, *145*, 255597.
11. Levy, M.; Ellington, A. D.; Pai, S. Proximity ligation assays with peptide conjugate 'burrs' and aptamers for the sensitive detection of spores and cancer cells. PCT Int. Appl. WO 2007027796, 2007; *Chem. Abstr.* **2007**, *146*, 312239.
12. Lecharny, M. Oral hygiene kits containing dyes as plaque-disclosing agents. PCT Int. Appl. WO 2003020230, 2003; *Chem. Abstr.* **2003**, *138*, 226421.
13. Morishima, S.; Umeda, S.; Miyoshi, N.; Yuta, S. Encapsulated dye solutions for detection of dental plaque. Jpn. Kokai Tokkyo Koho. JP 2002138056, 2002; *Chem. Abstr.* **2002**, *136*, 374840.
14. Morton, D. L. Methods for lymph node identification. U.S. Patent 6815170, 2004; *Chem. Abstr.* **2004**, *141*, 391557.
15. Kozawa, K.; Ozawa, A. Antimicrobial sheet supporting dye or fullerene and method for controlling microorganisms and decay. Jpn. Kokai Tokkyo Koho JP 2006316037, 2006; *Chem. Abstr.* **2006**, *145*, 501071.
16. Mahley, R. W.; Weisgraber, K. H.; Pitas, R. E. Compounds effecting neuron remodeling and assays for same. U.S. Pat. Appl. Publ. US 2002009439, 2002; *Chem. Abstr.* **2002**, *136*, 112683.
17. Cruceanu, M.; Stephen, A. G.; Beuning, P. J.; Gorelick, R. J.; Fisher, R. J.; Williams, M. C. Single DNA molecule stretching measures the activity of chemicals that target the HIV-1 nucleocapsid protein. *Anal. Biochem.* **2006**, *358*, 159–170.
18. Collis, M. P. Methods for detecting nucleic acids detection method using multiple pairs of donor flurophores and quencher molecules in the same probe. Eur. Pat. Appl. EP 1586662, 2005; *Chem. Abstr.* **2005**, *143*, 400770.
19. Isoe, J.; Kaneko, E. A new spectrophotometric method for determination of urinary protein using erythrosin B. *Chem. Lett.* **2006**, *35*, 922–923.
20. Orser, C. S.; Pan, T.; Sethi, J. Detecting and removing pathogenic misfolded proteins, such as prions, using conformational capture peptide probes and double-labeled detection-amplification peptides. PCT Int. Appl. WO 2006088823, 2006; *Chem. Abstr.* **2006**, *145*, 265232.
21. Heindl, F.; Drexel, C. P.; Aul, C. Silicic acid, silica gels or silicates coated with wax, oil or fats for use in foods and animal feeds. Ger. Offen. DE 10253193, 2004; *Chem. Abstr.* **2004**, *141*, 6176.
22. Donhowe, E. T. Product and process of making an alcohol containing sport drink. U.S. Pat. Appl. Publ. US 2003134007, 2003; *Chem. Abstr.* **2003**, *139*, 100270.
23. Stewart, D. R. Method of making and using an edible cellulose-based polymeric film for decorating foods, especially baked goods. U.S. Patent 6616958, 2003; *Chem. Abstr.* **2003**, *139*, 229693.
24. Roy, G.; Berardi, R.; Chan, W.; Lee, T. Prevention of synthetic color fading in beverages using botanically derived color stabilizers such as phenylpropenoic carbonyl compounds. U.S. Pat. Appl. Publ. US 2004091589, 2004; *Chem. Abstr.* **2004**, *140*, 405940.

25. Yoshioka, N.; Ichihashi, K. Determination of 40 synthetic food colors in drinks and candies by high-performance liquid chromatography using a short column with photodiode array detection. *Talanta* **2008**, *74*, 1408–1413.

26. Bernard, J.; Kowalczyk, J. Gelatin-free isomaltulose-containing soft caramels. U.S. Pat. Appl. Publ. US 2005089618, 2005; *Chem. Abstr.* **2005**, *142*, 410200.

27. Gueler, Z. Determination of synthetic colorants in confectionery and instant drink powders consumed in Turkey using UV/VIS spectrophotometry. *J. Food Qual.* **2005**, *28*, 98–108.

28. Chou, S. S.; Lin, Y. H.; Cheng, C. C.; Hwang, D. F. Determination of synthetic colors in soft drinks and confectioneries by micellar electrokinetic capillary chromatography. *J. Food Sci.* **2002**, *67*, 1314–1318.

29. Litvin, A. Sugar-substitute cotton candy and procedure for its production. Ger. Offen. DE 102008037185, 2009; *Chem. Abstr.* **2009**, *150*, 258878.

30. Moen, V.; Nordtug, T.; Zahlsen, K. Mixture and procedure for marking fish. Norw. NO 317169, 2004; *Chem. Abstr.* **2007**, *146*, 141497.

31. Cirillo, F. Process for dyeing olives by use of erythrosine. U.S. Pat. Appl. Publ. US 2006240158, 2006; *Chem. Abstr.* **2006**, *145*, 437697.

32. Avila, A. Y. V.; Ramirez, C. H. H.; Quiros, M. B. Physicochemical evaluation of quality of orange juices. *Ingen. Ciencia Quim.* **2004**, *21*, 42–48; *Chem. Abstr.* **2005**, *144*, 330179.

33. Sangwanangkul, P.; Paull, R. E. The role of hexose transporter in sugar accumulation of papaya fruit during maturation and ripening. *Acta Hortic.* **2007**, *740*, 313–316.

34. Catani, S. J.; Liao, S.; Loades, M. Self-mixing tabletop sweetener. U.S. Pat. Appl. Publ. US 2007059418, 2007; *Chem. Abstr.* **2007**, *146*, 315560.

35. Fujii, T.; Noami, M.; Kida, A. Colored capsules in which discoloration is prevented. Jpn. Kokai Tokkyo Koho JP 2007091670, 2007; *Chem. Abstr.* **2007**, *146*, 387140.

36. Hochrainer, D.; Wachtel, H. Method and device to seal inhalation capsules. PCT Int. Appl. WO 2003011574, 2003; *Chem. Abstr.* **2003**, *138*, 158844.

37. Vichkanova, S. A.; Kolkhir, V. K.; Sokol'skaya, T. A.; Okhotnikova, V. F.; Krutikova, N. M.; Fateeva, T. V.; Adgina, V. V.; Kachalina, T. V.; Bykov, V. A.; Val, E. V. Enterosoluble tablets of the drug sanguiritrine. Russ. RU 2202341, 2003; *Chem. Abstr.* **2003**, *139*, 219263.

38. Liu, S.; Feng, P. Resonance Rayleigh scattering for the determination of berberine in tablet form with some acidic xanthene fluorescent dyes original paper. *Mikrochim. Acta* **2002**, *140*, 189–193.

39. Ross, J. S.; Morefield, E. M. Sunscreen with disappearing color indicator. U.S. Patent 6290936, 2001; *Chem. Abstr.* **2001**, *135*, 215789.

40. Bobka, E.; Gibtner, T.; Mara, M. Water-free formulations for application to the lips. U.S. Pat. Appl. Publ. US 2005226832, 2005; *Chem. Abstr.* **2005**, *143*, 372863.

41. Emig, S.; Engelhardt, S.; Lober, S. Decorative cosmetic pastes for skin, lips and eyelids and their preparation. PCT Int. Appl. WO 2006027151, 2006; *Chem. Abstr.* **2006**, *144*, 317935.

42. Magallon, A. G.; Boone, K. W.; Bardwell, J. R. Aqueous antiseptic solution and compatible dye for staining skin. U.S. Pat. Appl. Publ. US 2007253909, 2007; *Chem. Abstr.* **2007**, *147*, 491676.

43. Lee, H.; Soistier, N.; Simon, J. C. Cosmetic makeup foundations for darkened skins. Fr. Demande FR 2848822, 2004; *Chem. Abstr.* **2004**, *141*, 76373.

44. Kundakovic, L.; Klitzman, B.; Koger, K. E.; Mathiowitz, E.; Drill, C. A. Cellular or organelle-entrapped nanoparticles containing dyes for tattoos. PCT Int. Appl. WO 2007024429, 2007; *Chem. Abstr.* **2007**, *146*, 280376.

45. Brennan, J. V.; Soo, P. P.; Kalgutkar, R. S.; Suhrwardy, R. A. Methods of identifying orthodontic adhesives on teeth by selective staining with pigments and/or food dyes. U.S. Pat. Appl. Publ. US 2007148609, 2007; *Chem. Abstr.* **2007**, *147*, 102288.

46. Hoic, D. A.; Curtis, J. P. Tooth coating compositions with resin, colorant, and pearlescent particles. U.S. Pat. Appl. Publ. US 2005175552, 2005; *Chem. Abstr.* **2005**, *143*, 216750.

47. Hercouet, L. Hair dye composition comprising a derivative of diamino-*N*,*N*-dihydropyrazolone, a coupler, and a heterocyclic direct dye. Fr. Demande FR 2886132, 2006; *Chem. Abstr.* **2006**, *146*, 49738.

48. Kanda, N. Acidic hair dye compositions containing specified acidic dyes, acids, and aromatic alcohols. Jpn. Kokai Tokkyo Koho JP 2004018505, 2004; *Chem. Abstr.* **2004**, *140*, 116929.

49. Sallwey, A.; Schmitt, M.; Dieckow, A. Dyestuff for hair and fibers. Eur. Pat. Appl. EP 976385, 2000; *Chem. Abstr.* **2000**, *132*, 127457.

50. Javet, M.; Mueller, C.; Weber, I. Bleaching keratin fibers with compositions containing an agent having a basic pH and sulfamate bleach booster. Eur. Pat. Appl. EP 1759684, 2007; *Chem. Abstr.* **2007**, *146*, 301740.

51. Fu, X. D.; Rosenfeld, M. G. Methods for detection of genome-wide gene expression in digital format by coupling DNA annealing, selection and ligation with sequencing. PCT Int. Appl. WO 2008069906, 2008; *Chem. Abstr.* **2008**, *149*, 46241.

52. Stevens, S. M.; Chung, A. Y. Fluorescent affinity tag to enhance phosphoprotein detection and characterization. U.S. Pat. Appl. Publ. US 2008050736, 2008; *Chem. Abstr.* **2008**, *148*, 302909.

53. Liu, G. L.; Ellman, J. A.; Lee, L. P.; Chen, F. F. Detection of protease using a single peptide-nanocrescent hybrid SERS probe, and diagnostic applications. PCT Int. Appl. WO 2008018933, 2008; *Chem. Abstr.* **2008**, *148*, 256536.

54. Cullum, M. E.; Duplessis, C. A.; Crepeau, L. J. Method for the detection of stress biomarkers including cortisol by fluorescence polarization. U.S. Pat. Appl. Publ. US 2006105397, 2006; *Chem. Abstr.* **2006**, *144*, 463810.

55. Hofmann, R. F. Use of targeted oxidative therapeutic formulation in treatment of age-related macular degeneration. U.S. Pat. Appl. Publ. US 2005250756, 2005; *Chem. Abstr.* **2005**, *143*, 416314.

56. Carpenter, R. H. Targeted oxidative therapeutic formulation for arteriosclerosis treatment. PCT Int. Appl. WO 2002078623, 2002; *Chem. Abstr.* **2002**, *137*, 284357.

57. Diaz, L. E. Liquid composition comprising alendronic acid for prevention and/or treatment of different bone metabolic diseases, uses thereof, and preparation process therefore. Eur. Pat. Appl. EP 1894557, 2008; *Chem. Abstr.* **2008**, *148*, 292170.

58. Hofmann, R. F. Use of targeted oxidative therapeutic formulation in treatment of burns. PCT Int. Appl. WO 2006002302, 2006; *Chem. Abstr.* **2006**, *144*, 114433.

59. Hofmann, R. F. Use of targeted oxidative therapeutic formulation in treatment of cancer. U.S. Pat. Appl. Publ. US 2005250757, 2005; *Chem. Abstr.* **2005**, *143*, 416223.

60. Hofmann, R. F. Use of targeted oxidative therapeutic formulation in treatment of diabetes and obesity. PCT Int. Appl. WO 2005107728, 2005; *Chem. Abstr.* **2005**, *143*, 466228.

61. Goldenberg, D. M.; Chang, C. H.; Rossi, E. A.; McBride, W. J. Methods and compositions for treatment of human immunodeficiency virus infection with conjugated antibodies or antibody fragments. U.S. Pat. Appl. Publ. US 2007264265, 2007; *Chem. Abstr.* **2007**, *147*, 534630.

62. Hofmann, R. F. Targeted oxidative therapeutic formulation in treatment of viral diseases. U.S. Pat. Appl. Publ. US 2005192267, 2005; *Chem. Abstr.* **2005**, *143*, 272504.

63. Raad, I. Coating medical device surfaces with antimicrobial agents. PCT Int. Appl. WO 2007062306, 2007; *Chem. Abstr.* **2007**, *147*, 39321.

64. Metcalf, D.; Robinson, C.; Devine, D.; Wood, S. Enhancement of erythrosine-mediated photodynamic therapy of *Streptococcus mutans* biofilms by light fractionation. *J. Antimicrob. Chemother.* **2006**, *58*, 190–192.

65. Wood, S.; Metcalf, D.; Devine, D.; Robinson, C. Erythrosine is a potential photosensitizer for the photodynamic therapy of oral plaque biofilms. *J. Antimicrob. Chemother.* **2006**, *57*, 680–684.

66. Nanba, N.; Kadota, A.; Tanabe, J. Dye-sensitized solar cells having low internal resistance. Jpn. Kokai Tokkyo Koho JP 2001076777, 2001; *Chem. Abstr.* **2001**, *134*, 254612.

67. Jung, W. C.; Lee, E. S.; Park, S. C.; Park, Y. J.; Sohn, B. H.; Nam, J. G.; Jo, H. S. Dye for photoelectric device and photoelectric device comprising the dye. U.S. Pat. Appl. Publ. US 2008110496, 2008; *Chem. Abstr.* **2008**, *148*, 565353.

68. Leatherdale, C. A.; Thompson, D. S.; Boardman, L. D.; Kalgutkar, R. S. Encapsulated light emitting diodes and methods of making. U.S. Pat. Appl. Publ. US 2006105483, 2006; *Chem. Abstr.* **2006**, *144*, 477453.

69. Shirota, K.; Yamashita, Y. Color filters, their manufacture by ink-jet printing, and liquid-crystal displays using them. Jpn. Kokai Tokkyo Koho JP 2001159710, 2001; *Chem. Abstr.* **2001**, *135*, 38954.

70. Okabe, K.; Yoshida, T.; Sugiura, T.; Minoura, H. Electrodeposition of photoactive ZnO/xanthene dye hybrid thin films. *Trans. Mater. Res. Soc. Jpn.* **2001**, *26*, 523–526.

71. Nakamura, S. Water-thinned writing ink compositions with good lightfastness, transparent cartridges for them, and pens using them. Jpn. Kokai Tokkyo Koho JP 2006321861, 2006; *Chem. Abstr.* **2006**, *146*, 9427.

72. Mukai, H. Polyoxyalkylene surfactant-containing water-thinned jet-printing inks with good breakage prevention of heads, their manufacture, and containers, heads, and printers containing them. Jpn. Kokai Tokkyo Koho JP 2006016532, 2006; *Chem. Abstr.* **2006**, *144*, 130597.

73. Saito, S. Aqueous black ink compositions. Jpn. Kokai Tokkyo Koho JP 2004123798, 2004; *Chem. Abstr.* **2004**, *140*, 323006.

74. Hasemann, L. Dye mixtures and their use in jet printing inks. PCT Int. Appl. WO 2003052007, 2003; *Chem. Abstr.* **2003**, *139*, 54286.

75. Toko, S.; Mochizuki, A.; Urushibata, K. Lithographic plate master for trimming by red light emitting diode. Jpn. Kokai Tokkyo Koho JP 2008033137, 2008; *Chem. Abstr.* **2008**, *148*, 225579.

76. Nagami, T. Photographic materials containing microparticle-supported filter dyes and their processing. Jpn. Kokai Tokkyo Koho JP 2001100354, 2001; *Chem. Abstr.* **2001**, *134*, 287807.

77. Lawrence, J. R.; O'Neill, F. T.; Sheridan, J. T. Photopolymer holographic recording material. *Optik* **2001**, *112*, 449–463.

78. Zhu, X. L.; Lam, S. K.; Chan, M. S. A.; Tsang, Y. C.; Lo, D. Recent development of organic dye-doped sol–gel materials for photonics. *Proc. SPIE-Int. Soc. Opt. Eng.* **2000**, *3943*, 210–217.

79. Cole, B. M.; Dower, W. V.; Oxman, J. D. Optical fiber connector using colored photocurable adhesive. PCT Int. Appl. WO 2000022060, 2000; *Chem. Abstr.* **2000**, *132*, 294806.

80. Sabnis, R. W.; Kehoe, T. D.; Balchunis, R. J. Paint compositions comprising color changing phthalein acid–base indicator. PCT Int. Appl. WO 2006105191, 2006; *Chem. Abstr.* **2006**, *145*, 403489.

81. Van Den Oever, M. J. A.; Boeriu, C. G.; Blaauw, R.; Van Haveren, J. Colorants based on renewable resources and food-grade colorants for application in thermoplastics. *J. Appl. Polym. Sci.* **2004**, *92*, 2961–2969.

82. Sabnis, R. W.; Kehoe, T. D. Composition and method for producing colored bubbles. U.S. Pat. Appl. Publ. US 2006004110, 2006; *Chem. Abstr.* **2006**, *144*, 110164.

83. Deeds, S. M. Fabric repair coloring device and using the device to color unwanted bleached out spots on fabric. U.S. Patent 6739779, 2004; *Chem. Abstr.* **2004**, *140*, 424935.

84. Sundaram, M.; Ihatsu, P.; Bourhis, Y. Dry ice products for the entertainment industry. PCT Int. Appl. WO 2007088442, 2007; *Chem. Abstr.* **2007**, *147*, 191690.

85. Ono, Y. Water-erasable markers for toys and figure toy sets therewith. Jpn. Kokai Tokkyo Koho JP 2002129085, 2002; *Chem. Abstr.* **2002**, *136*, 371209.

86. Tonogai, Y.; Ito, Y.; Iwaida, M.; Tati, M.; Ose, Y.; Hori, M. Studies on the toxicity of coal-tar dyes. III. Reason of acute toxicity to fish caused by coal-tar dyes and their industrial effluents. *J. Toxicol. Sci.* **1980**, *5*, 23–33.

87. Helguera, A. M.; Cabrera Perez, M. A.; Gonzalez, M. P.; Ruiz, R. M.; Gonzalez Diaz, H. A topological substructural approach applied to the computational prediction of rodent carcinogenicity. *Bioorg. Med. Chem.* **2005**, *13*, 2477–2488.

88. Benigni, R.; Cotta-Ramusino, M.; Andreoli, C.; Giuliani, A. Electrophilicity as measured by K_e: molecular determinants, relationship with other physical–chemical and quantum mechanical parameters, and ability to predict rodent carcinogenicity. *Carcinogenesis* **1992**, *13*, 547–553.

89. Yamashoji, S.; Isshiki, K. Rapid detection of cytotoxicity of food additives and contaminants by a novel cytotoxicity test, menadione-catalyzed H_2O_2 production assay. *Cytotechnology* **2001**, *37*, 171–178.

90. Vorhees, C. V.; Butcher, R. E.; Brunner, R. L.; Wootten, V.; Sobotka, T. J. A developmental toxicity and psychotoxicity evaluation of FD and C red dye No. 3 (erythrosine) in rats. *Arch. Toxicol.* **1983**, *53*, 253–264.

91. Pereira, L.; Ali, B.; Mohite, K.; Arora, P.; Rao, C. V. A study of embryotoxicity and teratogenesis of spurious lipsticks chemicals in the chick embryo. *J. Environ. Biol.* **1999**, *20*, 177–181.

92. Hagiwara, M.; Watanabe, E.; Barrett, J. C.; Tsutsui, T. Assessment of genotoxicity of 14 chemical agents used in dental practice: ability to induce chromosome aberrations in Syrian hamster embryo cells. *Mutat. Res.* **2006**, *603*, 111–120.

93. Kirkland, D.; Aardema, M.; Henderson, L.; Mueller, L. Evaluation of the ability of a battery of three *in vitro* genotoxicity tests to discriminate rodent carcinogens and non-carcinogens. I. Sensitivity, specificity and

relative predictivity. *Mutat. Res.* **2005**, *584*, 1–256.

94. Haveland-Smith, R. B.; Combes, R. D. Screening of food dyes for genotoxic activity. *Food Cosmet. Toxicol.* **1980**, *18*, 215–221.
95. Borzelleca, J. F.; Hallagan, J. B. Lifetime toxicity/carcinogenicity study of FD & C Red No. 3 (erythrosine) in mice. *Food Chem. Toxicol.* **1987**, *25*, 735–737.
96. Borzelleca, J. F.; Capen, C. C.; Hallagan, J. B. Lifetime toxicity/carcinogenicity study of FD & C Red No. 3 (erythrosine) in rats. *Food Chem. Toxicol.* **1987**, *25*, 723–733.
97. Pereira, L.; Ali, B.; Mohite, K.; Arora, P.; Rao, C. V. A study of mutagenic effects of spurious lipsticks chemicals in *Drosophila melanogaster*. *J. Environ. Biol.* **2000**, *21*, 309–315.
98. Ozaki, A.; Kitano, M.; Itoh, N.; Kuroda, K.; Furusawa, N.; Masuda, T.; Yamaguchi, H. Mutagenicity and DNA-damaging activity of decomposed products of food colors under UV irradiation. *Food Chem. Toxicol.* **1998**, *36*, 811–817.
99. Klopman, G.; Frierson, M. R.; Rosenkranz, H. S. The structural basis of the mutagenicity of chemicals in *Salmonella typhimurium*: the Gene-Tox data base. *Mutat. Res.* **1990**, *228*, 1–50.
100. Lin, G. H. Y.; Brusick, D. J.; Joseph, C. W. Mutagenicity studies on FD&C Red No. 3. *Mutagenesis* **1986**, *1*, 253–259.
101. Mailman, R, B.; Lewis, M. H. Neurotoxicants and central catecholamine systems. *Neurotoxicology* **1987**, *8*, 123–139.
102. Anderson, R. J. Relative refractory period as a measure of peripheral nerve neurotoxicity. *Toxicol. Appl. Pharmacol.* **1983**, *71*, 391–397.
103. Wang, H.; Lu, L.; Zhu, S.; Li, Y.; Cai, W. The phototoxicity of xanthene derivatives against *Escherichia coli*, *Staphylococcus aureus*, and *Saccharomyces cerevisiae*. *Curr. Microbiol.* **2006**, *52*, 1–5.
104. Vivekanandhi, J.; Devi, C. P. A.; Jayaraman, K.; Raghavan, L. Effect of erythrosine on testicular function of mice. *Toxicol. Int.* **2006**, *13*, 119–125.
105. Tanaka, T. Reproductive and neurobehavioral toxicity study of erythrosine administered to mice in the diet. *Food Chem. Toxicol.* **2001**, *39*, 447–454.
106. Abdel Aziz, A. H.; Shouman, S. A.; Attia, A. S.; Saad, S. F. A study on the reproductive toxicity of erythrosine in male mice. *Pharmacol. Res.* **1997**, *35*, 457–462.

ETHIDIUM BROMIDE

CAS Registry Number 1239-45-8

Chemical Structure

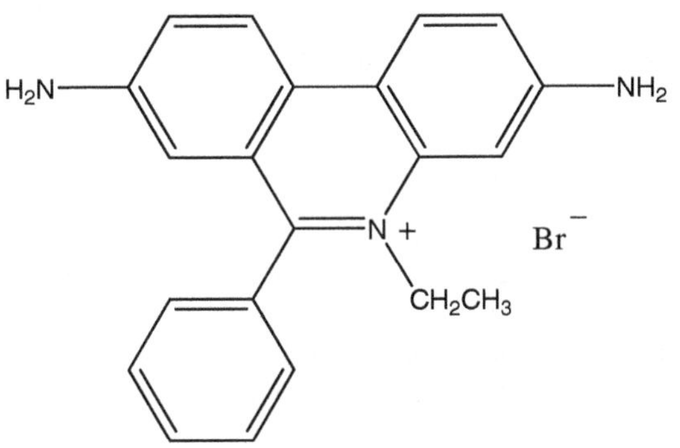

CA Index Name Phenanthridinium, 3,8-diamino-5-ethyl-6-phenyl-, bromide (1:1)

Other Names 3,8-Diamino-5-ethyl-6-phenylphenanthridinium bromide; Phenanthridinium, 3,8-diamino-5-ethyl-6-phenyl-, bromide; 2,7-Diamino-10-ethyl-9-phenylphenanthridinium bromide; 2,7-Diamino-9-phenyl-10-ethylphenanthridinium bromide; 2,7-Diamino-9-phenylphenanthridine ethobromide; Dromilac; Ethidium bromide; Homidium bromide

Merck Index Number 4731

Chemical/Dye Class Phenanthridine

Molecular Formula $C_{21}H_{20}BrN_3$

Molecular Weight 394.31

Physical Form Dark red crystals

Solubility Soluble in water, chloroform, dimethyl sulfoxide, ethanol

Melting Point 238–240°C

Absorption (λ_{max}) 518 nm, 210 nm, 285 nm, 316 nm, 343 nm, 480 nm, 525 nm

Emission (λ_{max}) 605 nm, 620 nm

Synthesis Synthetic methods[1–11]

Staining Applications Nucleic acids;[12–16] cells;[17] HeLa cells;[18] bacteria[19]

Biological Applications Nucleic acid hybridization;[20] detecting nucleic acids,[12–16] cells,[17] cancer cells,[21] human cytomegalovirus,[22] hydrogenase A (hydA) of *Clostridia*,[23] influenza A virus,[24] oligonucleotides,[25] viable *Plesiomonas shigelloides*;[26] apoptosis assay;[27] nucleic acid amplification;[28] nucleic acid quantification;[29] treating diseases associated with androgens,[30] cancer,[31] plasma cell disorders,[32] neurodegenerative diseases[33]

Industrial Applications Not reported

Safety/Toxicity Carcinogenicity;[34] cytotoxicity;[35] DNA damage;[36] embryotoxicity;[37] genotoxicity;[38,39] mutagenicity;[40–42] neurotoxicity;[43] phototoxicity[44]

REFERENCES

1. Lu, T.; Guo, C.; Ni, P. Design, synthesis and evaluation of phenanthridine derivatives interacting with G-quadruplex as telomerase. *Zhongguo Yaoke Daxue Xuebao* **2004**, *35*, 99–105; *Chem. Abstr.* **2005**, *144*, 31987.
2. Hahn, K. M.; Toutchkine, A.; Muthyala, R.; Kraynov, V.; Bark, S. J.; Burton, D. R.; Chamberlain, C. Labeled peptides, proteins and antibodies and processes and intermediates useful for their preparation. U.S. Pat. Appl. Publ. US 2002055133, 2002; *Chem. Abstr.* **2002**, *136*, 366139.
3. Hahn, K. M.; Toutchkine, A.; Muthyala, R.; Kraynov, V.; Bark, S. J.; Burton, D. R.; Chamberlain, C. Labeled peptides, proteins and antibodies and processes and intermediates useful for their preparation. PCT Int. Appl. WO 2002008245, 2002; *Chem. Abstr.* **2002**, *136*, 163716.
4. Yen, F. W.; Liu, J. M.; Chen, C. H. The design and synthesis of bisazo series compound used in organophotoconductor. *MRL Bull. Res. Dev.* **1992**, *6*, 21–27.
5. Lion, C.; Boukou-Poba, J. P.; Charvy, C. Synthesis and chemistry of phenanthridines. III. Preparation of some model ω-(3,8-diamino-6-phenanthridinyl) alkanoic acids. *Bull. Soc. Chim. Belg.* **1990**, *99*, 171–181.
6. Malinge, J. M.; Schwartz, A.; Leng, M. Characterization of the ternary complexes formed in the reaction of cis-diamminedichloroplatinum (II), ethidium bromide and nucleic acids. *Nucleic Acids Res.* **1987**, *15*, 1779–1797.
7. Vacek, A. T.; Bourque, D. P.; Hewlett, N. G. An ethidium-acrylamide affinity medium for recovery of nucleic acids from free solution and from polyacrylamide and agarose gels. *Anal. Biochem.* **1982**, *124*, 414–420.

8. El-Gewely, M. R.; Helling, R. B. Preparative separation of DNA–ethidium bromide complexes by zonal density gradient centrifugation. *Anal. Biochem.* **1980**, *102*, 423–428.

9. Jacquemin-Sablon, H.; Jacquemin-Sablon, A.; Paoletti, C. Yeast mitochondrial deoxyribonuclease stimulated by ethidium bromide. 1. Purification and properties. *Biochemistry* **1979**, *18*, 119–127.

10. Minor, P. D.; Dimmock, N. J. Selective inhibition of influenza virus protein synthesis by inhibitors of DNA function. *Virology* **1977**, *78*, 393–406.

11. Patel, D. J.; Canuel, L. L. Ethidium bromide (dC-dG-dC-dG)$_2$ complex in solution: intercalation and sequence specificity of drug binding at the tetranucleotide duplex level. *Proc. Natl. Acad. Sci. U.S.A* **1976**, *73*, 3343–3347.

12. Exner, M.; Rogers, A. Methods for identifying nucleic acids and determining melting temperature using multiple fluorophores and FRET. U.S. Pat. Appl. Publ. US 2007172836, 2007; *Chem. Abstr.* **2007**, *147*, 182489.

13. Ragazzon, P. A.; Garbett, N. C.; Chaires, J. B. Competition dialysis: a method for the study of structural selective nucleic acid binding. *Methods* **2007**, *42*, 173–182.

14. Rajendran, A.; Magesh, C. J.; Perumal, P. T. DNA–DNA cross-linking mediated by bifunctional [SalenAlIII]$^+$ complex. *Biochim. Biophys. Acta* **2008**, *1780*, 282–288.

15. Dolezel, J.; Greilhuber, J.; Suda, J. Estimation of nuclear DNA content in plants using flow cytometry. *Nat. Protoc.* **2007**, *2*, 2233–2244.

16. Hilal, H.; Taylor, J. A. Determination of the stoichiometry of DNA–dye interaction and application to the study of a bis-cyanine dye–DNA complex. *Dyes Pigments* **2007**, *75*, 483–490.

17. Guda, K.; Natale, L.; Markowitz, S. D. An improved method for staining cell colonies in clonogenic assays. *Cytotechnology* **2007**, *54*, 85–88.

18. Sobczak, T. J.; Bourgain, G. F.; Gendron, M. C.; Gasnereau, I.; Ganier, O.; De, G. A. Fluorescent detection, sorting, and enumeration of HeLa cells in cytokinesis. Fr. Demande FR 2902440, 2007; *Chem. Abstr.* **2007**, *148*, 47032.

19. Hannig, C.; Hannig, M.; Rehmer, O.; Braun, G.; Hellwig, E.; Al-Ahmad, A. Fluorescence microscopic visualization and quantification of initial bacterial colonization on enamel *in situ*. *Arch. Oral Biol.* **2007**, *52*, 1048–1056.

20. Qi, H.; Li, X.; Chen, P.; Zhang, C. Electrochemical detection of DNA hybridization based on polypyrrole/ss-DNA/multi-wall carbon nanotubes paste electrode. *Talanta* **2007**, *72*, 1030–1035.

21. Fong, Y.; Adusumilli, P. Replication-competent conditional oncolytic herpes and vaccinia viruses for detection of cancer cells. PCT Int. Appl. WO 2007075879, 2007; *Chem. Abstr.* **2007**, *147*, 93393.

22. Chan, Y.; Lin, J.; Liu, C.; Ho, M. Method and kits for detecting human cytomegalovirus using real-time PCR. U.S. Pat. Appl. Publ.US 2008032280, 2008; *Chem. Abstr.* **2008**, *148*, 231410.

23. Wang, M.; Olson, B. H.; Chang, J. Improving PCR and qPCR detection of hydrogenase A (hydA) associated with *Clostridia* in pure cultures and environmental sludges using bovine serum albumin. *Appl. Microbiol. Biotechnol.* **2007**, *77*, 645–656.

24. Espy, M. J.; Smith, T. F. Detection of influenza A virus RNA by real-time PCR. U.S. Pat. Appl. Publ. US 2007238093, 2007; *Chem. Abstr.* **2007**, *147*, 420744.

25. Tan, W.; Shangguan, D.; Li, Y. Cell-based SELEX process for identifying and isolating oligonucleotides (probes) having high affinity for extracellular or cell surface markers on target cells, such as CCRF-CEM cells. PCT Int. Appl.WO 2007142713, 2007; *Chem. Abstr.* **2007**, *148*, 47020.

26. Gu, W.; Levin, R. E. Quantification of viable *Plesiomonas shigelloides* in a mixture of viable and dead cells using ethidium bromide monoazide and conventional PCR. *Food Biotechnol.* **2007**, *21*, 145–159.

27. Ferlini, C.; Scambia, G. Assay for apoptosis using the mitochondrial probes, Rhodamine123 and 10-N-nonyl acridine orange. *Nat. Protoc.* **2007**, *2*, 3111–3114.

28. Kermekchiev, M.; Kirilova, L. Taq polymerase mutants for DNA amplification in the presence of dye, blood, and soil PCR inhibitors. PCT Int. Appl. WO 2008034110, 2008; *Chem. Abstr.* **2008**, *148*, 348542.

29. Bonasera, V.; Alberti, S.; Sacchetti, A. Protocol for high-sensitivity/long linear-range spectrofluorimetric DNA quantification using ethidium bromide. *BioTechniques* **2007**, *43* 173–174, 176.

30. Singh, M.; Gatson, J. W. Membrane androgen receptor as a therapeutic target for the prevention/promotion of cell death. U.S. Pat. Appl. Publ. US 2007141581, 2007; *Chem. Abstr.* **2007**, *147*, 88350.

31. Reiter, R. E. N-cadherin and Ly6-E: targets for cancer diagnosis and therapy. PCT Int. Appl. WO 2007109347, 2007; *Chem. Abstr.* **2007**, *147*, 398645.

32. Boruchov, A. M.; Comenzo, R. L. Methods of diagnosing, treating, or preventing plasma cell disorders. PCT Int. Appl. WO 2008033571, 2008; *Chem. Abstr.* **2008**, *148*, 377546.

33. Stockwell, B. R.; Hoffstrom, B.; Varma, H. Agents for treating neurodegenerative diseases. U.S. Pat. Appl. Publ. US 2007149543, 2007; *Chem. Abstr.* **2007**, *147*, 110249.

34. Na, M. R.; Koo, S. K.; Kim, D. Y.; Park, S. D.; Rhee, S. K.; Kang, K. W.; Joe, C. O. *In vitro* inhibition of gap junctional intercellular communication by chemical carcinogens. *Toxicology* **1995**, *98*, 199–206.

35. Huang, R.; Southall, N.; Cho, M.; Xia, M.; Inglese, J.; Austin, C. P. Characterization of diversity in toxicity mechanism using *in vitro* cytotoxicity assays in quantitative high throughput screening. *Chem. Res. Toxicol.* **2008**, *21*, 659–667.

36. Barclay, B. J.; DeHaan, C. L.; Hennig, U. G. G.; Iavorovska, O.; Von Borstel, R. W.; Von Borstel, R. C. A rapid assay for mitochondrial DNA damage and respiratory chain inhibition in the yeast *Saccharomyces cerevisiae. Environ. Mol. Mutagen.* **2001**, *38*, 153–158.

37. Kohler, M.; Kundig, A.; Reist, H. W.; Michel, C. Modification of *in vitro* mouse embryogenesis by X-rays and fluorochromes. *Radiat. Environ. Biophys.* **1994**, *33*, 341–351.

38. Cahill, P. A.; Knight, A. W.; Billinton, N.; Barker, M. G.; Walsh, L.; Keenan, P. O.; Williams, C. V.; Tweats, D. J.; Walmsley, R. M. The Green Screen genotoxicity assay: a screening validation programme. *Muta-genesis* **2004**, *19*, 105–119.

39. Fernandez, M.; Gauthier, L.; Jaylet, A. Use of newt larvae for *in vivo* genotoxicity testing of water: results on 19 compounds evaluated by the micronucleus test. *Mutagenesis* **1989**, *4*, 17–26.

40. Singer, V. L.; Lawlor, T. E.; Yue, S. Comparison of SYBR Green I nucleic acid gel stain mutagenicity and ethidium bromide mutagenicity in the *Salmonella*/mammalian microsome reverse mutation assay (Ames test). *Mutat. Res.* **1999**, *439*, 37–47.

41. Buchholz, S. E.; Omer, C. A.; Viitanen, P. V. Induction of cytochrome P-450 synthesis in *Streptomycetes* and determining the mutagenicity of chemicals. Eur. Pat. Appl. EP 347226, 1989; *Chem. Abstr.* **1990**, *113*, 72908.

42. Ferguson, L. R.; Baguley, B. C. Verapamil as a co-mutagen in the *Salmonella*/mammalian microsome mutagenicity test. *Mutat. Res. Lett.* **1988**, *209*, 57–62.

43. Guazzo, E. P. A technique for producing demyelination of the rat optic nerves. *J. Clin. Neurosci.* **2005**, *12*, 54–58.

44. Dobrucki, J. W.; Feret, D.; Noatynska, A. Scattering of exciting light by live cells in fluorescence confocal imaging: phototoxic effects and relevance for FRAP studies. *Biophys. J.* **2007**, *93*, 1778–1786.

ETHIDIUM HOMODIMER-1 (EthD-1)

CAS Registry Number 61926-22-5

Chemical Structure

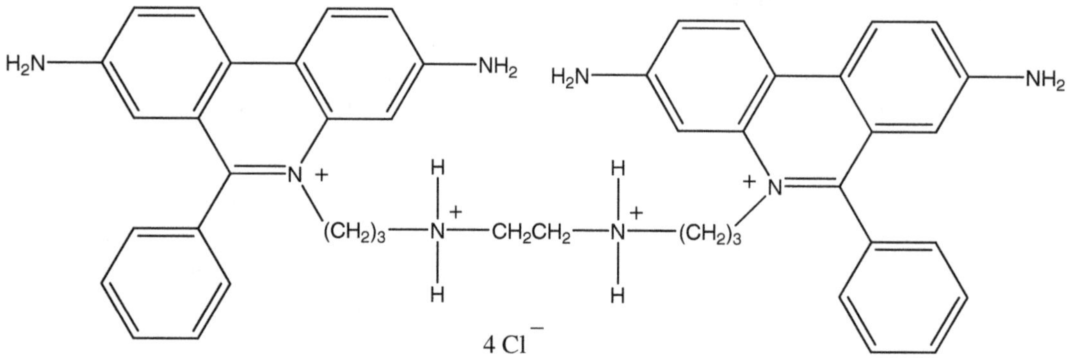

CA Index Name Phenanthridinium, 5,5′-[1,2-ethanediylbis(imino-3,1-propanediyl)]bis[3,8-diamino-6-phenyl-, chloride, hydrochloride (1:2:2)

Other Names Phenanthridinium, 5,5′-[1,2-ethanediylbis(imino-3,1-propanediyl)]bis[3,8-diamino-6-phenyl-, dichloride, dihydrochloride; 5,5′-(4,7-Diazadecamethylene)bis(3,8-diamino-6-phenylphenanthridinium) dichloride dihydrochloride; EB2; ETDI; EthD 1; EtDi; Ethidium dimer; Ethidium homodimer; Ethidium homodimer 1

Merck Index Number Not listed

Chemical/Dye Class Phenanthridine

Molecular Formula $C_{46}H_{50}Cl_4N_8$

Molecular Weight 856.77

Physical Form Red powder

Solubility Soluble in water, *N,N*-dimethylformamide, dimethyl sulfoxide

Melting Point 275°C

Absorption (λ_{max}) 528 nm

Emission (λ_{max}) 617 nm

Synthesis Synthetic methods[1–5]

Staining Applications Nucleic acids;[6–9,18,19] cells;[10,20–22] leukocytes;[11,29] nuclei;[29] megakaryocytes;[12] microorganisms;[13] nucleated red blood cells;[14] sperms;[15,16] Schwann cells (SCs) in whole nerves[17]

Biological Applications Nucleic acid hybridization;[18] detecting nucleic acids,[6–9,18,19] cells,[10,20–22] retinoblastoma susceptibility gene protein (pRB),[23] enzymes,[24] PCR products,[25] viral infection;[26] nucleic acid amplification;[27] nucleic acid fragment sizing;[28] counting embryoblasts;[29] treating cancer[30]

Industrial Applications For authentification[31]

Safety/Toxicity Cytotoxicity;[32] gastric toxicity;[33] neurotoxicity;[34] retinal toxicity[35,36]

REFERENCES

1. Millard, P. J.; Roth, B. L.; Yue, S. T.; Haugland, R. P. Fluorescent viability assay using cyclic-substituted unsymmetrical cyanine dyes. U.S. Patent 5534416, 1996; *Chem. Abstr.* **1996**, *125*, 162751.
2. Haugland, R. P.; MacCoubrey, I. C.; Moore, P. L. Dual-fluorescence cell viability assay using ethidium homodimer and calcein AM. U.S. Patent 5314805, 1994; *Chem. Abstr.* **1994**, *121*, 53487.
3. Gaugain, B.; Barbet, J.; Oberlin, R.; Roques, B. P.; Le Pecq, J. B. DNA bifunctional intercalators. 1. Synthesis and conformational properties of an ethidium homodimer and of an acridine ethidium heterodimer. *Biochemistry* **1978**, *17*, 5071–5078.
4. Gaugain, B.; Barbet, J.; Capelle, N.; Roques, B. P.; Le Pecq, J. B.; Le Bret, M. DNA bifunctional intercalators. 2. Fluorescence properties and DNA binding interaction of an ethidium homodimer and an acridine ethidium heterodimer. Appendix: numerical solution of McGhee and von Hippel equations for competing ligands. *Biochemistry* **1978**, *17*, 5078–5088.
5. Roques, B. P.; Barbet, J.; Oberlin, R.; Le Pecq, J. B. DNA intercalating drugs: synthesis of phenanthridinium monomers and of one dimer with aminoalkyl chains. *Sci. Naturelles* **1976**, *283*, 1365–1367.

6. Exner, M.; Rogers, A. Methods for identifying nucleic acids and determining melting temperature using multiple fluorophores and FRET. U.S. Pat. Appl. Publ. US 2007172836, 2007; *Chem. Abstr.* **2007**, *147*, 182489.
7. Cui, H. H.; Valdez, J. G.; Steinkamp, J. A.; Crissman, H. A. Fluorescence lifetime-based discrimination and quantification of cellular DNA and RNA with phase-sensitive flow cytometry. *Cytometry* **2003**, *52A*, 46–55.
8. Scheinert, P. Electrophoresis system for nucleic acids and gel staining methods. *BioTec (Germany)* **1996**, *8*, 47–49.
9. Rye, H. S.; Glazer, A. N. Interaction of dimeric intercalating dyes with single-stranded DNA. *Nucleic Acids Res.* **1995**, *23*, 1215–1222.
10. Anderson, A. L.; Knutson, C. R.; Mueth, D.; Plewa, J.; Tanner, E. Methods for staining cells for identification and sorting. U.S. Pat. Appl. Publ. US 2006172315, 2006; *Chem. Abstr.* **2006**, *145*, 183714.
11. Sakata, T.; Mizukami, T.; Hatanaka, K. Method for classifying and counting immature leukocytes using cell hemolysis, staining and flow cytometry. Eur. Pat. Appl. EP 844481, 1998; *Chem. Abstr.* **1998**, *129*, 38386.
12. Minakami, T.; Mori, Y.; Tsuji, T.; Ikeuchi, Y. Megakaryocyte classification/counting method by double fluorescent staining and flow cytometry. Jpn. Kokai Tokkyo Koho JP 2006275985, 2006; *Chem. Abstr.* **2006**, *145*, 413679.
13. Noda, N.; Mizutani, T. Microorganism-measuring method using multiple staining. Jpn. Kokai Tokkyo Koho JP 2006340684, 2006; *Chem. Abstr.* **2006**, *146*, 77563.
14. Kim, Y. R.; Yee, M. W.; Mehta, S. N.; Sagala, J. C.; Kantor, J. Method for rapid and simultaneous analysis of nucleated red blood cells. PCT Int. Appl. WO 9618878, 1996; *Chem. Abstr.* **1996**, *125*, 109652.
15. Kato, M.; Makino, S.; Kimura, H.; Ota, T.; Furuhashi, T.; Nagamura, Y.; Hirano, K. *In vitro* evaluation of acrosomal status and motility in rat epididymal spermatozoa treated with α-chlorohydrin for predicting their fertilizing capacity. *J. Reprod. Dev.* **2002**, *48*, 461–468.
16. Matsumoto, T.; Okada, H.; Hamaguchi, Y. Method and reagent for counting sperm by flow cytometry. Jpn. Kokai Tokkyo Koho JP 2001242168, 2001; *Chem. Abstr.* **2001**, *135*, 207869.
17. Decherchi, P.; Cochard, P.; Gauthier, P. Dual staining assessment of Schwann cell viability within whole peripheral nerves using calcein-AM and ethidium homodimer. *J. Neurosci. Methods* **1997**, *71*, 205–213.
18. Erikson, G. H.; Daksis, J. I. Fluorescent intensity assay for triplex and quadruplex nucleic acid hybridization in solution utilizing fluorescent intercalating dyes for detection of target nucleic acids. U.S. Patent 6911536, 2005; *Chem. Abstr.* **2005**, *143*, 72739.
19. Erikson, G. H.; Daksis, J. I.; Kandic, I.; Picard, P. Conditions for formation of three- and four-stranded nucleic acid complexes and their detection and analytical use. PCT Int. Appl. WO 2002103051, 2002; *Chem. Abstr.* **2002**, *138*, 50812.
20. Fukutome, K. Method for evaluating microorganism cell activity by flow cytometry analysis. Jpn. Kokai Tokkyo Koho JP 2006238771, 2006; *Chem. Abstr.* **2006**, *145*, 331200.
21. Horikiri, S. Microorganism cell detection method using fluorescent indicator. Jpn. Kokai Tokkyo Koho JP 2006262775, 2006; *Chem. Abstr.* **2006**, *145*, 392006.
22. Oberhardt, B. J. Cell analysis methods and apparatus. U.S. Patent 6251615, 2001; *Chem. Abstr.* **2001**, *135*, 58132.
23. Darzynkiewicz, Z.; Traganos, F.; Juan, G.; Gruenwald, S. Flow cytometric detection of conformations of pRB in single cells. PCT Int. Appl. WO 9944067, 1999; *Chem. Abstr.* **1999**, *131*, 181965.
24. Chavan, S. J.; Prochaska, H. J. Method and kit for fluorometric analysis of enzymes catalyzing synthesis of nucleic acids. PCT Int. Appl. WO 9624694, 1996; *Chem. Abstr.* **1996**, *125*, 241775.
25. Chia, D.; Terasaki, P.; Chan, H.; Tonai, R.; Siauw, P. A. Direct detection of PCR products for HLA class II typing. *Tissue Antigens* **1993**, *42*, 146–149.
26. Phimister, A. J.; Day, K. C.; Gunderson, A. D.; Wong, V. J.; Lawson, G. W.; Fanucchi, M. V.; Van, W. L. S.; Kendall, L. V.; Plopper, C. G. Detection of viral infection in the respiratory tract of virus antibody free mice: advantages of high-resolution imaging for respiratory toxicology. *Toxicol. Appl. Pharmacol.* **2003**, *190*, 286–293.
27. Park, H. O.; Kim, H. B.; Chi, S. M. Detection method of DNA amplification using probes labeled with intercalating dyes. PCT Int. Appl. WO 2006004267, 2006; *Chem. Abstr.* **2006**, *144*, 122707.
28. Yan, X.; Grace, W. K.; Yoshida, T. M.; Habbersett, R. C.; Velappan, N.; Jett, J. H.; Keller, R. A.; Marrone, B. L. Characteristics of different nucleic acid staining

dyes for DNA fragment sizing by flow cytometry. *Anal. Chem.* **1999**, *71*, 5470–5480.

29. Heuven, B.; Wong, F.; Tsuji, T.; Sakata, T.; Hamaguchi, I. Method for classifying and counting erythroblasts by flow cytometry. Jpn. Kokai Tokkyo Koho JP 11326323, 1999; *Chem. Abstr.* **1999**, *132*, 1804.
30. Reiter, Y.; Klechevsky, E.; Denkberg, G. Antibodies specific to MHC–peptide complexes for diagnosis and treatment of cancer and infection and for drug screening. PCT Int. Appl. WO 2005120166, 2005; *Chem. Abstr.* **2005**, *144*, 86569.
31. Isler, U.; Hoehener, K.; Meier, W.; Poux, S. Procedure for the treatment of yarn with encapsulated marking substance for authentification and device for proof of authenticity. Ger. Offen. DE 102005047786, 2007; *Chem. Abstr.* **2007**, *146*, 423457.
32. Dias N.; Lima, N. A comparative study using a fluorescence-based and a direct-count assay to determine cytotoxicity in *Tetrahymena pyriformis*. *Res. Microbiol.* **2002**, *153*, 313–322.
33. Kokoska, E. R.; Smith, G. S.; Deshpande, Y.; Wolff, A. B.; Rieckenberg, C.; Miller, T. A. Calcium accentuates injury induced by ethanol in human gastric cells. *J. Gastrointest. Surg.* **1999**, *3*, 308–318.
34. Nattie, E. E.; Erlichman, J. S.; Li, A. Brain stem lesion size determined by DEAD red or conjugation of neurotoxin to fluorescent beads. *J. Appl. Physiol.* **1998**, *85*, 2370–2375.
35. Jackson, T. L.; Vote, B.; Knight, B. C.; El-Amir, A.; Stanford, M. R.; Marshall, J. Safety testing of infracyanine green using retinal pigment epithelium and glial cell cultures. *Invest. Ophthalmol. Vis. Sci.* **2004**, *45*, 3697–3703.
36. Ho, J.; Chen, H.; Chen, S.; Tsai, R. Reduction of indocyanine green-associated photosensitizing toxicity in retinal pigment epithelium by sodium elimination. *Arch. Ophthalmol.* **2004**, *122*, 871–878.

ETHIDIUM HOMODIMER-2 (EthD-2)

CAS Registry Number 180389-01-9

Chemical Structure

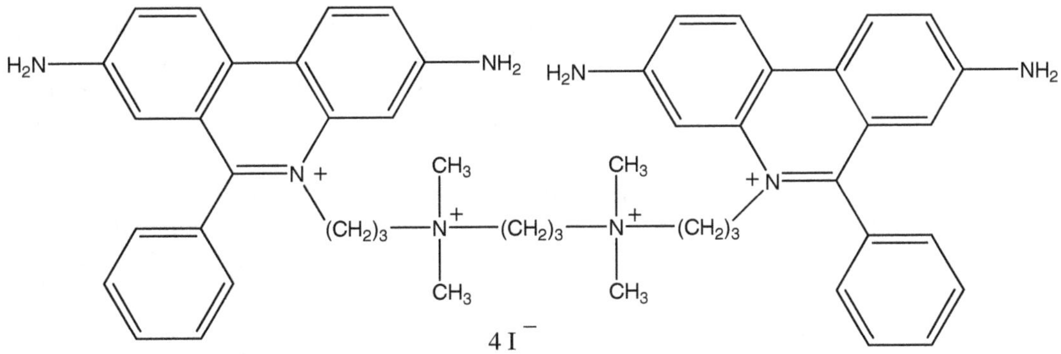

CA Index Name Phenanthridinium, 5,5′-[1,3-propanediylbis[(dimethyliminio)-3,1-propanediyl]]bis[3,8-diamino-6-phenyl]-, iodide (1:4)

Other Names Phenanthridinium, 5,5′-[1,3-propanediylbis[(dimethyliminio)-3,1-propanediyl]]bis[3,8-diamino-6-phenyl]-, tetraiodide; EthD 2; Ethidium homodimer 2

Merck Index Number Not listed

Chemical/Dye Class Phenanthridine

Molecular Formula $C_{51}H_{60}I_4N_8$

Molecular Weight 1292.71

Physical Form Red powder

Solubility Soluble in dimethyl sulfoxide, water

Melting Point >200°C

Absorption (λ_{max}) 535 nm

Emission (λ_{max}) 624 nm

Synthesis Synthetic method[1]

Staining Applications Nucleic acids;[2–4] cells;[1,5] leukocytes;[6,14] megakaryocyte;[7] bone marrow cells;[8] sperms[9]

Biological Applications Nucleic acid hybridization;[10,15] detecting nucleic acids,[2–4,10,11] proteins,[11] cells,[1,5] spores,[12] cancer cells,[12] stress biomarkers;[13] counting embryoblasts;[14] hybridization assay;[15] viability assay[1]

Industrial Applications Not reported

Safety/Toxicity No data available

REFERENCES

1. Millard, P. J.; Roth, B. L.; Yue, S. T.; Haugland, R. P. Fluorescent viability assay using cyclic-substituted unsymmetrical cyanine dyes. U.S. Patent 5534416, 1996; *Chem. Abstr.* **1996**, *125*, 162751.
2. Exner, M.; Rogers, A. Methods for identifying nucleic acids and determining melting temperature using multiple fluorophores and FRET. U.S. Pat. Appl. Publ. US 2007172836, 2007; *Chem. Abstr.* **2007**, *147*, 182489.
3. Park, H. O.; Kim, H. B.; Chi, S. M. Detection method of DNA amplification using probes labeled with intercalating dyes. PCT Int. Appl. WO 2006004267, 2006; *Chem. Abstr.* **2006**, *144*, 122707.
4. Erikson, G. H.; Daksis, J. I.; Kandic, I.; Picard, P. Conditions for formation of three- and four-stranded nucleic acid complexes and their detection and analytical use. PCT Int. Appl. WO 2002103051, 2002; *Chem. Abstr.* **2002**, *138*, 50812.
5. Anderson, A. L.; Knutson, C. R.; Mueth, D.; Plewa, J.; Tanner, E. Methods for staining cells for identification and sorting. U.S. Pat. Appl. Publ. US 2006172315, 2006; *Chem. Abstr.* **2006**, *145*, 183714.
6. Sakata, T.; Mizukami, T.; Hatanaka, K. Method for classifying and counting immature leukocytes using cell hemolysis, staining and flow cytometry. Eur. Pat. Appl. EP 844481, 1998; *Chem. Abstr.* **1998**, *129*, 38386.
7. Minakami, T.; Mori, Y.; Tsuji, T.; Ikeuchi, Y. Megakaryocyte classification/counting method by double fluorescent staining and flow cytometry. Jpn. Kokai Tokkyo Koho JP 2006275985, 2006; *Chem. Abstr.* **2006**, *145*, 413679.

8. Tsuji, T.; Itose, H.; Konishi, A. Method for automatically analyzing nucleated bone marrow cell. Jpn. Kokai Tokkyo Koho JP 2003329668, 2003; *Chem. Abstr.* **2003**, *139*, 393149.
9. Matsumoto, T.; Okada, H.; Hamaguchi, Y. Method and reagent for counting sperm by flow cytometry. Jpn. Kokai Tokkyo Koho JP 2001242168, 2001; *Chem. Abstr.* **2001**, *135*, 207869.
10. Erikson, G. H.; Daksis, J. I. Fluorescent intensity assay for triplex and quadruplex nucleic acid hybridization in solution utilizing fluorescent intercalating dyes for detection of target nucleic acids. U.S. Patent 6911536, 2005; *Chem. Abstr.* **2005**, *143*, 72739.
11. Alba, F. J.; Bermudez, A.; Daban, J. R. Green-light transilluminator for the detection without photodamage of proteins and DNA labeled with different fluorescent dyes. *Electrophoresis* **2001**, *22*, 399–403.
12. Levy, M.; Ellington, A. D.; Pai, S. Proximity ligation assays with peptide conjugate 'burrs' and aptamers for the sensitive detection of spores and cancer cells. PCT Int. Appl. WO 2007027796, 2007; *Chem. Abstr.* **2007**, *146*, 312239.
13. Cullum, M. E.; Duplessis, C. A.; Crepeau, L. J. Method for the detection of stress biomarkers including cortisol by fluorescence polarization. U.S. Pat. Appl. Publ. US 2006105397, 2006; *Chem. Abstr.* **2006**, *144*, 463810.
14. Heuven, B.; Wong, F. S.; Tsuji, T.; Sakata, T.; Hamaguchi, I. Method for classifying and counting erythroblasts by flow cytometry. Jpn. Kokai Tokkyo Koho JP 11326323, 1999; *Chem. Abstr.* **1999**, *132*, 1804.
15. Daksis, J. I.; Picard, P.; Erikson, G. H. Cation mediated nucleic acid triplex hybridization assay. PCT Int. Appl. WO 2002004655, 2002; *Chem. Abstr.* **2002**, *136*, 97269.

ETHIDIUM MONOAZIDE (EMA)

CAS Registry Number 58880-05-0

Chemical Structure

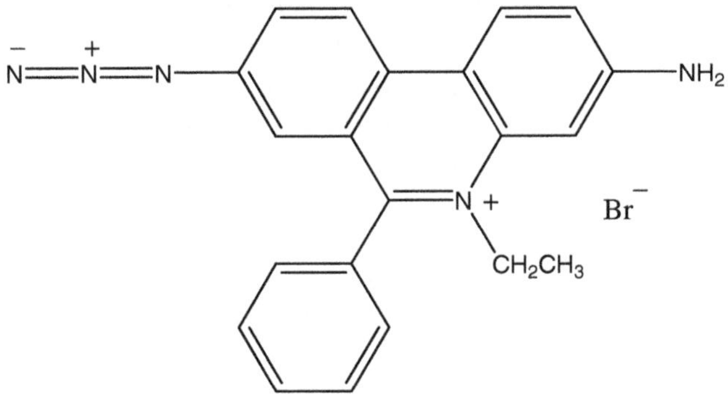

CA Index Name Phenanthridinium, 3-amino-8-azido-5-ethyl-6-phenyl-, bromide (1:1)

Other Names Phenanthridinium, 3-amino-8-azido-5-ethyl-6-phenyl-, bromide; Ethidium monoazide; Ethidium monoazide bromide

Merck Index Number Not listed

Chemical/Dye Class Phenanthridine

Molecular Formula $C_{21}H_{18}BrN_5$

Molecular Weight 420.31

Physical Form Orange powder

Solubility Soluble in ethanol, *N,N*-dimethylformamide

Melting Point >200°C

Absorption (λ_{max}) 462 nm

Emission (λ_{max}) 625 nm

Synthesis Synthetic methods[1,2]

Staining Applications Nucleic acids;[3–6] polynucleotide;[6] bacteria;[7,8] leukocyte;[9] micronuclei;[10,11] chromatin;[10,11] nucleus[12]

Biological Applications Biochips;[13] DNA chips;[14] nucleic acid hybridization;[15,16] detecting nucleic acids,[3–6,17,18] cells,[19] bacteria,[7,8,20] spore contents[21]

Industrial Applications Not reported

Safety/Toxicity Mutagenicity;[22] nucleic acid damage[23]

REFERENCES

1. Graves, D. E.; Yielding, L. W.; Watkins, C. L.; Yielding, K. L. Synthesis, separation and characterization of the mono- and diazide analogs of ethidium bromide. *Biochim. Biophys. Acta* **1977**, *479*, 98–104.
2. Firth, W. J.; Watkins, C. L.; Graves, D. E.; Yielding, L. W. Synthesis and characterization of ethidium analogs: emphasis on amino and azido substituents. *J. Heterocycl. Chem.* **1983**, *20*, 759–765.
3. Exner, M.; Rogers, A. Methods for identifying nucleic acids and determining melting temperature using multiple fluorophores and FRET. U.S. Pat. Appl. Publ. US 2007172836, 2007; *Chem. Abstr.* **2007**, *147*, 182489.
4. Rueckert, A.; Morgan, H. W. Removal of contaminating DNA from polymerase chain reaction using ethidium monoazide. *J. Microbiol. Methods* **2007**, *68*, 596–600.
5. Hein, I.; Flekna, G.; Wagner, M. Possible errors in the interpretation of ethidium bromide and PicoGreen DNA staining results from ethidium monoazide-treated DNA. Comments. *Appl. Environ. Microbiol.* **2006**, *72*, 6860–6861.
6. Cantrell, C. E.; Yielding, K. L.; Pruitt, K. M. Efficiency of photolytic binding of ethidium monoazide to nucleic acids and synthetic polynucleotides. *Mol. Pharmacol.* **1979**, *15*, 322–330.
7. Regan, J. M.; Oldenburg, P. S.; Park, H. D.; Harrington, G. W.; Noguera, D. R. Simultaneous determination of bacterial viability and identity in biofilms using ethidium monoazide and fluorescent *in situ* hybridization. *Water Sci. Technol.* **2003**, *47*, 123–128.
8. Nogva, H. K.; Dromtorp, S. M.; Nissen, H.; Rudi, K. Ethidium monoazide for DNA-based differentiation of viable and dead bacteria by 5′-nuclease PCR. *BioTechniques* **2003**, *34* 804–806, 808, 810, 812–813.
9. Sakata, T.; Mizukami, T.; Hatanaka, K. Method for classifying and counting immature leukocytes using cell hemolysis, staining and flow cytometry. Eur. Pat. Appl. EP 844481, 1998; *Chem. Abstr.* **1998**, *129*, 38386.
10. Avlasevich, S. L.; Bryce, S. M.; Cairns, S. E.; Dertinger, S. D. *In vitro* micronucleus scoring by flow cytometry: differential staining of micronuclei versus apoptotic and necrotic chromatin enhances assay reliability. *Environ. Mol. Mutagen.* **2006**, *47*, 56–66.
11. Dertinger, S. D.; Cairns, S. E.; Avlasevich, S. L.; Torous, D. K. Method for enumerating and

differentiating between mammalian cell micronuclei by differentially staining micronuclei and the chromatin of dead, dying and viable cells. PCT Int. Appl. WO 2006007479, 2006; *Chem. Abstr.* **2006**, *144*, 124521.

12. Yoshida, S.; Soejima, T. Microorganism detection method by nuclear staining and flow cytometry, and microorganism detection kit. PCT Int. Appl. WO 2007010641, 2007; *Chem. Abstr.* **2007**, *146*, 138259.
13. Chen, S. H.; Yang, C. S.; Chang, S. C.; Lin, C. C.; Sun, S. Y. Organic electroluminescence (OEL)-based biochips. U.S. Pat. Appl. Publ. US 2003035755, 2003; *Chem. Abstr.* **2003**, *138*, 166193.
14. Yamana, K.; Kumamoto, S.; Hasegawa, T.; Nakano, H.; Matsuo, Y.; Sugie, T. DNA chip, and assay method. Jpn. Kokai Tokkyo Koho JP 2003083968, 2003; *Chem. Abstr.* **2003**, *138*, 234410.
15. Erikson, G. H.; Daksis, J. I. Improving the signal/noise ratio of nucleic acid hybridization assays by preincubation of primer and target with nucleic acid binding agents. U.S. Pat. Appl. Publ. US 2004180345, 2004; *Chem. Abstr.* **2004**, *141*, 255469.
16. Dattagupta, N. Hybridization assays using target enhanced signal amplification for detection of *Mycobacterium tuberculosis*. U.S. Pat. Appl. Publ. US 2003219755, 2003; *Chem. Abstr.* **2003**, *140*, 1537.
17. Erikson, G. H.; Daksis, J. I.; Kandic, I.; Picard, P. Conditions for formation of three- and four-stranded nucleic acid complexes and their detection and analytical use. PCT Int. Appl.WO 2002103051, 2002; *Chem. Abstr.* **2002**, *138*, 50812.
18. Park, H. O.; Kim, H. B.; Chi, S. M. Detection method of DNA amplification using probes labeled with intercalating dyes. PCT Int. Appl. WO 2006004267, 2006; *Chem. Abstr.* **2006**, *144*, 122707.
19. Rudi, K. A method of cell detection in a biological sample which distinguishes between living and dead cells. PCT Int. Appl. WO 2001077379, 2001; *Chem. Abstr.* **2001**, *135*, 283953.
20. Nocker, A.; Cheung, C. Y.; Camper, A. K. Comparison of propidium monoazide with ethidium monoazide for differentiation of live vs. dead bacteria by selective removal of DNA from dead cells. *J. Microbiol. Methods* **2006**, *67*, 310–320.
21. Rueckert, A.; Ronimus, R. S.; Morgan, H. W. Rapid differentiation and enumeration of the total, viable vegetative cell and spore content of thermophilic bacilli in milk powders with reference to *Anoxybacillus flavithermus*. *J. Appl. Microbiol.* **2005**, *99*, 1246–1255.
22. Klopman, G.; Frierson, M. R.; Rosenkranz, H. S. The structural basis of the mutagenicity of chemicals in *Salmonella typhimurium*: the Gene-Tox data base. *Mutat. Res.* **1990**, *228*, 1–50.
23. Kulkarni, M. S.; Yielding, K. L. Alkali lability and rapid initiation of excision repair following photoaffinity damage by ethidium azide. *Chem.-Biol. Interact.* **1985**, *56*, 89–99.

ETHYL EOSIN

CAS Registry Number 6359-05-3

Chemical Structure

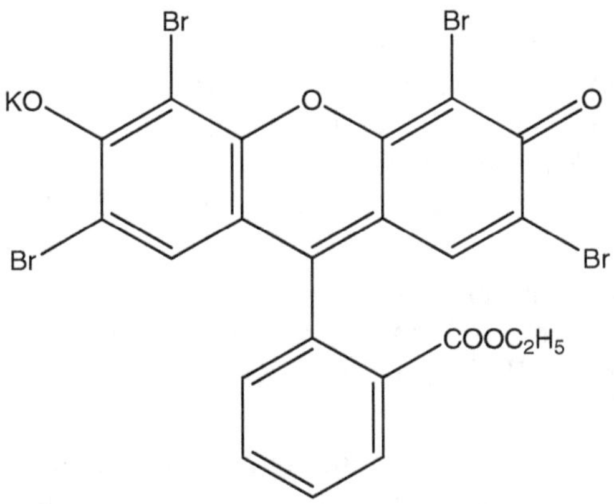

CA Index Name Benzoic acid, 2-(2,4,5,7-tetrabromo-6-hydroxy-3-oxo-3*H*-xanthen-9-yl)-, ethyl ester, potassium salt

Other Names Ethyl eosin; Fluorescein, 2′,4′,5′,7′-tetrabromo-, ethyl ester, potassium derivative; Fluorescein, 2′,4′,5′,7′-tetrabromo-, ethyl ester, potassium salt; C.I. 45386; C.I. Solvent Red 45; Ethyl eosine Y; NSC 8670; Spirit Eosine; Spirit Primrose DH

Merck Index Number Not listed

Chemical/Dye Class Xanthene

Molecular Formula $C_{22}H_{11}Br_4KO_5$

Molecular Weight 714.03

Physical Form Red-brown crystals or powder

Solubility Soluble in hot water; slightly soluble in ethanol

Melting Point >200°C

Absorption (λ_{max}) 532 nm

Synthesis Synthetic method[1–3]

Staining Applications Cells;[4] gelatin;[5] lysosomes;[6] mammalian tissues;[7] mitochondria;[8] neurons;[9] nucleic acids;[10,11] proteins;[12] urine[13]

Biological Applications Drug delivery and tissue engineering;[14] photodynamic therapy;[15] dental materials;[16] treating cancer,[17] diabetes;[18] wound dressing materials[19]

Industrial Applications Liquid crystal display device;[20] lithographic printing plates;[21–23] inks;[24] recording materials;[25] photographic materials;[26] polymer waveguides[27]

Safety/Toxicity No data available

Certification/Approval Certified by Biological Stain Commission (BSC)

REFERENCES

1. Hayatsu, H. Pigment molecules linked to polymer support: blue rayon, blue chitin, and green chitosan-synthesis and applications. *Yakugaku Zasshi* **2000**, *120*, 534–547; *Chem. Abstr.* **2000**, *133*, 160755.
2. Mchedlov-Petrosyan, N. O.; Egorova, S. I.; Arias-Cordova, E. Structure and extractability of xanthene and triphenylmethane dye anions in the form of ionic associates. *Dokl. Akad. Nauk* **1992**, *323*, 696–701; *Chem. Abstr.* **1993**, *118*, 126477.
3. Holmes, W. C.; Scanlon, J. T. The evaluation of ethyl eosin. *Stain Technol.* **1927**, *2*, 101–103.
4. Drocourt, J. L.; Levesque, P. Method for counting viable cells by dual staining. PCT Int. Appl.WO 9855861, 1998; *Chem. Abstr.* **1998**, *130*, 35353.
5. Birkedal-Hansen, H. Eosin staining of gelatine. *Histochemie* **1973**, *36*, 73–87.
6. Rashid, F.; Horobin, R. W.; Williams, M. A. Predicting the behavior and selectivity of fluorescent probes for lysosomes and related structures by means of structure–activity models. *Histochem. J.* **1991**, *23*, 450–459.
7. Hine, I. F. Block staining of mammalian tissues with hematoxylin and eosin. *Stain Technol.* **1981**, *56*, 119–123.
8. Rashid, F.; Horobin, R. W. Accumulation of fluorescent non-cationic probes in mitochondria of cultured cells: observations, a proposed mechanism, and some implications. *J. Microsc.* **1991**, *163*, 233–241.
9. Holemans, S.; Feron, O.; Octave, J. N.; Maloteaux, J. M. Interaction of fluorescein derivatives with glibenclamide binding sites in rat brain. *Neurosci. Lett.* **1995**, *183*, 183–186.
10. Stephen, A. G.; Worthy, K. M.; Towler, E.; Mikovits, J. A.; Sei, S.; Roberts, P.; Yang, Q.; Akee, R. K.; Klausmeyer, P.; McCloud, T. G.; Henderson, L.; Rein, A.; Covell, D. G.; Currens, M.; Shoemaker, R. H.; Fisher, R. J. Identification of HIV-1 nucleocapsid protein:nucleic acid antagonists with

cellular anti-HIV activity. *Biochem. Biophys. Res. Commun.* **2002**, *296*, 1228–1237.

11. Ginsberg, S. D.; Che, S. Amplification of RNA through a cDNA intermediate with bacteriophage promoters incorporated at the 3′-end of the cDNA. PCT Int. Appl. WO 2002065093, 2002; *Chem. Abstr.* **2002**, *137*, 180748.
12. Muckle, T. J.; Bradford, J. Dye induced alterations in lipoproteins and other plasma proteins. *Acta Histochem.* **1980**, *67*, 32–39.
13. Inoue, J.; Nishizaki, M. Analytical method for formative components in urine. Jpn. Kokai Tokkyo Koho JP 11083849, 1999; *Chem. Abstr.* **1999**, *130*, 293613.
14. Langer, R. S.; Elisseeff, J. H.; Anseth, K.; Sims, D. Semi-interpenetrating or interpenetrating polymer networks for drug delivery and tissue engineering. U.S. Patent 6224893, 2001; *Chem. Abstr.* **2001**, *134*, 331625.
15. Dees, H. C.; Scott, T. C.; Smolik, J.; Wachter, E. A.; Fisher, W. G. Topical halogenated xanthenes for photodynamic treatment. U.S. Pat. Appl. Publ.US 2008118578, 2008; *Chem. Abstr.* **2008**, *148*, 568986.
16. Trom, M. C.; Oxman, J. D. Color-changing dental compositions. U.S. Patent 6444725, 2002; *Chem. Abstr.* **2002**, *137*, 206597.
17. Zheng, J.; Chu, F. Polymer gel for cancer treatment. U.S. Pat. Appl. Publ. US 2002192289, 2002; *Chem. Abstr.* **2002**, *138*, 44708.
18. Scharp, D.; Latta, P.; Yu, X.; Yue, C.; Hubbell, J. Implantation of encapsulated biological materials for treating diseases such as diabetes. PCT Int. Appl. WO 2004032881, 2004; *Chem. Abstr.* **2004**, *140*, 344912.
19. Soon-Shiong, P.; Desai, N. P.; Sandford, P. A.; Heintz, R. A.; Sojomihardjo, S. Crosslinkable polysaccharides, polycations and lipids useful for encapsulation of drugs and cells and manufacture of wound dressings. PCT Int. Appl. WO 9309176, 1993; *Chem. Abstr.* **1993**, *119*, 125199.
20. Hughes, C. J.; Coates, D.; Alkins, A. O. Cholesteric liquid crystal display device. PCT Int. Appl. WO 2006054050, 2006; *Chem. Abstr.* **2006**, *145*, 17884.
21. Takagi, K. Lithographic plates with high sensitivity, printability, and clean ability, method for laser platemaking thereof, aluminum supports thereof, and manufacture thereof. Jpn. Kokai Tokkyo Koho JP 2005070267, 2005; *Chem. Abstr.* **2005**, *142*, 325965.
22. Suzuki, K.; Kuroki, T. Lithographic printing plates, their treatment, and image formation. Jpn. Kokai Tokkyo Koho JP 2004347676, 2004; *Chem. Abstr.* **2004**, *142*, 45933.
23. Ota, T. Presensitized lithographic printing plate having oxygen-blocking layer and image-forming method. Jpn. Kokai Tokkyo Koho JP 2004317937, 2004; *Chem. Abstr.* **2004**, *141*, 417950.
24. Yoshisawa, T.; Ninomiya, H.; Abe, T. Fluorescent color inks for ink jet-printing. Jpn. Kokai Tokkyo Koho JP 2003026964, 2003; *Chem. Abstr.* **2003**, *138*, 124042.
25. Zertani, R.; Mohr, D.; Frass, W. Photopolymerizable recording material with dye-containing cover layer. Ger. Offen. DE 3827245, 1990; *Chem. Abstr.* **1990**, *113*, 49877.
26. Konstantinov, I.; Mednikarov, B.; Sakhatchiev, M.; Burov, A. Photographic system and photographic material for use in it. Eur. Pat. Appl. EP 34490, 1981; *Chem. Abstr.* **1982**, *96*, 133198.
27. Samoc, A.; Samoc, M.; Luther-Davies, B. Upconversion of He–Ne laser light in xanthene dye-doped polymer waveguides. *Pol. J. Chem.* **2002**, *76*, 345–358.

FAST GREEN FCF

CAS Registry Number 2353-45-9

Chemical Structure

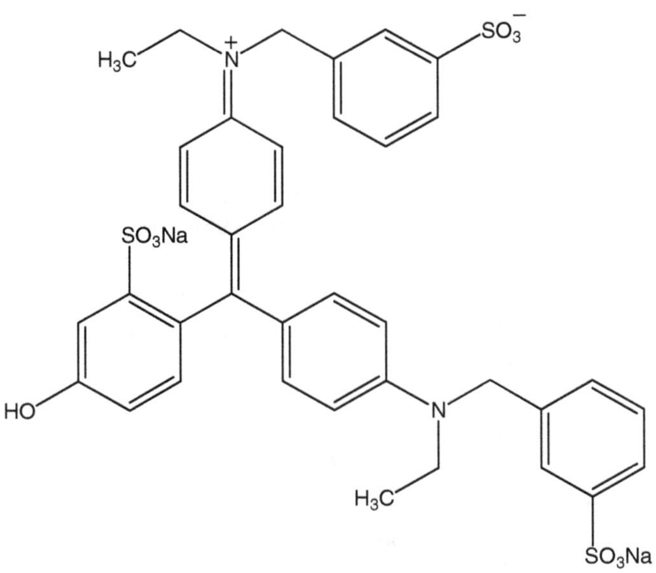

CA Index Name Benzenemethanaminium, *N*-ethyl-*N*-[4-[[4-[ethyl[(3-sulfophenyl)methyl]amino]phenyl](4-hydroxy-2-sulfophenyl)methylene]-2,5-cyclohexadien-1-ylidene]-3-sulfo-, inner salt, sodium salt (1:2)

Other Names Benzenemethanaminium, *N*-ethyl-*N*-[4-[[4-[ethyl[(3-sulfophenyl)methyl]amino]phenyl](4-hydroxy-2-sulfophenyl)methylene]-2,5-cyclohexadien-1-ylidene]-3-sulfo-, hydroxide, inner salt, disodium salt; Benzenemethanaminium, *N*-ethyl-*N*-[4-[[4-[ethyl[(3-sulfophenyl)methyl]amino]phenyl](4-hydroxy-2-sulfophenyl)methylene]-2,5-cyclohexadien-1-ylidene]-3-sulfo-, inner salt, disodium salt; C.I. Food Green 3; C.I. Food Green 3, disodium salt; Fast Green FCF; C.I. 42053; FD and C Green No. 3; FD&C Green No. 3; FDC Green 3; Food Green 3; Food Green No. 3; Green No. 3; Japan Food Green No. 3; Japan Green 3; Japan Green No. 3; NSC 379443

Merck Index Number 3941

Chemical/Dye Class Triphenylmethane

Molecular Formula $C_{37}H_{34}N_2Na_2O_{10}S_3$

Molecular Weight 808.85

Physical Form Dark green or red-brown powder

Solubility Soluble in water, ethanol; insoluble in xylene

Melting Point 290°C (decompose)

Absorption (λ_{max}) 624 nm

Synthesis Synthetic method[1–7]

Staining Applications Brain;[8] cytoplasm;[9] cytoskeleton;[9] nuclei;[9] nucleoli;[9] aluminum;[10] cells;[9,11] microorganisms;[12] stain for endoscopy;[13] beverages;[14,15] candies;[16] drinks;[16] fish;[17] sweeteners;[18,19] capsules;[20] tablets;[21] skin;[22–24] lips;[22,23] tattoos;[25] tooth;[26] hairs[27–29]

Biological Applications Detecting proteins;[30] medical devices[31]

Industrial Applications Inks;[32,33] highlighters;[34] paints;[35] detergents;[36] colored bubbles;[37] entertainment products[38]

Safety/Toxicity Acute toxicity;[39] carcinogenicity;[40,41] chronic toxicity;[42] cytotoxicity;[43] effects on chromosomes;[44] genotoxicity;[45,46] mutagenicity:[47,48] neurotoxicity[49]

Certification/Approval Certified by Biological Stain Commission (BSC); Approved by Food & Drugs Administration (FDA)

REFERENCES

1. Gonzalez-Hernandez, J.; Mendoza-Galvan, A. Study of SiO_2 sol–gel derived glasses containing inorganic and organic nanostructures. *Recent Res. Dev. Vac. Sci. Technol.* **2003**, *4*, 213–276.
2. Terpetschnig, E. A.; Yang, D.; Owicki, J. C. Luminescent metal–ligand complexes. PCT Int. Appl. WO 2000047693, 2000; *Chem. Abstr.* **2000**, *133*, 174260.
3. Clark, H. W.; Chapman, T. E.; Yates, R. L. Insoluble pigments. Eur. Pat. Appl. EP 206798, 1986; *Chem. Abstr.* **1987**, *107*, 178132.

Handbook of Biological Dyes and Stains By R. W. Sabnis

4. Walles, W. E.; Keskey, W. H.; Young, R. G. Pigments for toiletries. U.S. Patent. 4533484, 1985; *Chem. Abstr.* **1986**, *104*, 10368.
5. Kase, T. Electrolytic oxidation of triphenylmethane derivatives. Jpn. Kokai Tokkyo Koho JP 54130481, 1979; *Chem. Abstr.* **1980**, *92*, 163717.
6. Imai, F.; Ogawa, Y.; Kagawa, K. Colors for cosmetics. Jpn. Kokai Tokkyo Koho JP 53071132, 1978; *Chem. Abstr.* **1978**, *89*, 185924.
7. Kraus, H. Histochemical detection of histones at pH 8. *Acta Histochem., Suppl.* **1965**, *4*, 349–361.
8. Mericle, R. A.; Batich, C. D.; Watkins, C. S.; Burry, M. V.; Richter, E. O.; Swandeshmukul, S. Method of endovascular brain mapping. PCT Int. Appl. WO 2003090604, 2003; *Chem. Abstr.* **2003**, *139*, 347551.
9. Reischle, H. Preparation of a stain and a staining method for biological specimens. Ger. Offen. DE 4422250, 1996; *Chem. Abstr.* **1996**, *124*, 170007.
10. Walton, J. R.; Diamond, T. H.; Kumar, S.; Murrell, G. A. C. A sensitive stain for aluminum in undecalcified cancellous bone. *J. Inorg. Biochem.* **2007**, *101*, 1285–1290.
11. Gelvan, D. J.; Goltsman, L.; Chausovsky, A. Staining methods and compositions for identifying a cell phenotype. PCT Int. Appl. WO 2007102146, 2007; *Chem. Abstr.* **2007**, *147*, 339010.
12. Noda, N.; Mizutani, T. Microorganism-measuring method using multiple staining. Jpn. Kokai Tokkyo Koho JP 2006340684, 2006; *Chem. Abstr.* **2006**, *146*, 77563.
13. Yamamoto, A.; Iimori, Y.; Sase, M.; Ishiguro, M.; Li, B.; Sasaki, H. Histochemical fluorescent staining agent composition for endoscope. Jpn. Kokai Tokkyo Koho JP 2007326789, 2007; *Chem. Abstr.* **2007**, *148*, 49094.
14. Smith, L. C. Carbonated beverage national school lunch meal. U.S. Pat. Appl. Publ. US 2007281059, 2007; *Chem. Abstr.* **2007**, *148*, 10213.
15. Roy, G.; Berardi, R.; Chan, W.; Lee, T. Prevention of synthetic color fading in beverages using botanically derived color stabilizers such as phenylpropenoic carbonyl compounds. U.S. Pat. Appl. Publ. US 2004091589, 2004; *Chem. Abstr.* **2004**, *140*, 405940.
16. Yoshioka, N.; Ichihashi, K. Determination of 40 synthetic food colors in drinks and candies by high-performance liquid chromatography using a short column with photodiode array detection. *Talanta* **2008**, *74*, 1408–1413.
17. Moen, V.; Nordtug, T.; Zahlsen, K. Mixture and procedure for marking fish. Norw. NO 317169, 2004; *Chem. Abstr.* **2007**, *146*, 141497.
18. Catani, S. J.; Liao, S.; Loades, M. Self-mixing tabletop sweetener. U.S. Pat. Appl. Publ. US 2007059418, 2007; *Chem. Abstr.* **2007**, *146*, 315560.
19. Mori, K. Sweetener composition. U.S. Pat. Appl. Publ. US 2006257550, 2006; *Chem. Abstr.* **2006**, *145*, 488169.
20. Mori, T.; Ito, T.; Sano, Y.; Ito, M.; Enomoto, I. Capsules in which content comprises visibly distinguishable multiple component phases and their manufacture. Jpn. Kokai Tokkyo Koho JP 2000239159, 2000; *Chem. Abstr.* **2000**, *133*, 227780.
21. Meloy, G. K. Color display means. U.S. Patent. 4587069, 1986; *Chem. Abstr.* **1986**, *105*, 26193.
22. Blin, X. Range of compositions for application to the skin, lips, and/or superficial body growths. Fr. Demande FR 2868697, 2005; *Chem. Abstr.* **2005**, *143*, 392523.
23. Dumousseaux, C. Cosmetic composition for application to the skin, lips, and/or body surface. Fr. Demande FR 2868696, 2005; *Chem. Abstr.* **2005**, *143*, 392522.
24. Teboul, K. Method for coloring dark skins. PCT Int. Appl. WO 2008087308, 2008; *Chem. Abstr.* **2008**, *149*, 160219.
25. Kundakovic, L.; Klitzman, B.; Koger, K. E.; Mathiowitz, E.; Drill, C. A. Cellular or organelle-entrapped nanoparticles containing dyes for tattoos. PCT Int. Appl. WO 2007024429, 2007; *Chem. Abstr.* **2007**, *146*, 280376.
26. Hoic, D. A.; Curtis, J. P. Tooth coating compositions with resin, colorant, and pearlescent particles. U.S. Pat. Appl. Publ. US 2005175552, 2005; *Chem. Abstr.* **2005**, *143*, 216750.
27. Iwata, S. Scratch-resistant film-forming hair dyes with reduced dye migration. Jpn. Kokai Tokkyo Koho JP 2005247794, 2005; *Chem. Abstr.* **2005**, *143*, 271942.
28. Sugimoto, K. Scratch-resistant film-forming hair dyes with reduced dye migration. Jpn. Kokai Tokkyo Koho JP 2005232021, 2005; *Chem. Abstr.* **2005**, *143*, 253457.
29. Pollack, G. Hair dye compositions containing polymers. U.S. Pat. Appl. Publ. US 2004025264, 2004; *Chem. Abstr.* **2004**, *140*, 169236.
30. Albarella, J. P.; Cahill, S. E.; Johnson, G. M.; Pugia, M. J. Improved method for the detection of protein. Eur. Pat. Appl. EP 793099, 1997; *Chem. Abstr.* **1997**, *127*, 231593.
31. Raad, I. Coating medical device surfaces with antimicrobial agents. PCT Int. Appl. WO 2007062306, 2007; *Chem. Abstr.* **2007**, *147*, 39321.

32. Wen, L.; Lanner, D. A.; Lin, P. Y. T. Ink jetting inks for food application. U.S. Pat. Appl. Publ. US 2006038866, 2006; *Chem. Abstr.* **2006**, *144*, 234782.
33. Yui, I.; Sakurai, N. Color-fading aqueous ink compositions. Jpn. Kokai Tokkyo Koho JP 2003073593, 2003; *Chem. Abstr.* **2003**, *138*, 223088.
34. Davies-Smith, L.; Sum, V. W. Highlightable and highlighted mixtures, marking instruments, eradicator solution and kits, using the mixtures in highlighting. U.S. Pat. Appl. Publ. US 2005120919, 2005; *Chem. Abstr.* **2005**, *143*, 9331.
35. Franer, V. R. Color dissipating paint. U.S. Patent. 5548010, 1996; *Chem. Abstr.* **1996**, *125*, 250497.
36. Macdonald, J. G.; Huang, Y.; Yang, K.; Kim, J.; Wei, N. Single phase color change agents for liquid detergents for cleaning surfaces. U.S. Pat. Appl. Publ. US 2005049157, 2005; *Chem. Abstr.* **2005**, *142*, 263578.
37. Sabnis, R. W.; Kehoe, T. D. Composition and method for producing colored bubbles. U.S. Pat. Appl. Publ. US 2006004110, 2006; *Chem. Abstr.* **2006**, *144*, 110164.
38. Sundaram, M.; Ihatsu, P.; Bourhis, Y. Dry ice products for the entertainment industry. PCT Int. Appl. WO 2007088442, 2007; *Chem. Abstr.* **2007**, *147*, 191690.
39. Tonogai, Y.; Iwaida, M.; Tati, M.; Ose, Y.; Sato, T. Biochemical decomposition of coal-tar dyes. II. Acute toxicity of coal-tar dyes and their decomposed products. *J. Toxicol. Sci.* **1978**, *3*, 205–214.
40. Helguera, A. M.; Cabrera Perez, M. A.; Gonzalez, M. P.; Ruiz, R. M.; Gonzalez Diaz, H. A topological substructural approach applied to the computational prediction of rodent carcinogenicity. *Bioorg. Med. Chem.* **2005**, *13*, 2477–2488.
41. Haseman, J. K.; Winbush, J. S.; O'Donnell, M. W. Jr. Use of dual control groups to estimate false positive rates in laboratory animal carcinogenicity studies. *Fundam. Appl. Toxicol.* **1986**, *7*, 573–584.
42. Hansen, W. H.; Long, E. L.; Davis, K. J.; Nelson, A. A.; Fitzhugh, O. G. Chronic toxicity of three food colorings: Guinea Green B, Light Green SF Yellowish, and Fast Green FCF in rats, dogs and mice. *Food Cosmet. Toxicol.* **1966**, *4*, 389–410.
43. Sako, F.; Kobayashi, N.; Watabe, H.; Taniguchi, N. Cytotoxicity of food dyes on cultured fetal rat hepatocytes. *Toxicol. Appl. Pharmacol.* **1980**, *54*, 285–292.
44. Roychoudhury, A.; Giri, A. K. Effects of certain food dyes on chromosomes of *Allium cepa*. *Mutat. Res.* **1989**, *223*, 313–319.
45. Knight, A. W.; Billinton, N.; Cahill, P. A.; Scott, A.; Harvey, J. S.; Roberts, K. J.; Tweats, D. J.; Keenan, P. O.; Walmsley, R. M. An analysis of results from 305 compounds tested with the yeast RAD54-GFP genotoxicity assay (GreenScreen GC)—including relative predictivity of regulatory tests and rodent carcinogenesis and performance with autofluorescent and colored compounds. *Mutagenesis* **2007**, *22*, 409–416.
46. Ohno, K.; Tanaka-Azuma, Y.; Yoneda, Y.; Yamada, T. Genotoxicity test system based on p53R2 gene expression in human cells: examination with 80 chemicals. *Mutat. Res.* **2005**, *588*, 47–57.
47. Klopman, G.; Frierson, M. R.; Rosenkranz, H. S. The structural basis of the mutagenicity of chemicals in *Salmonella typhimurium*: the Gene-Tox data base. *Mutat. Res.* **1990**, *228*, 1–50.
48. Ishidate, M. Jr.; Sofuni, T.; Yoshikawa, K.; Hayashi, M.; Nohmi, T.; Sawada, M.; Matsuoka, A. Primary mutagenicity screening of food additives currently used in Japan. *Food Chem. Toxicol.* **1984**, *22*, 623–636.
49. van Hooft, J. A. Fast Green FCF (Food Green 3) inhibits synaptic activity in rat hippocampal interneurons. *Neurosci. Lett.* **2002**, *318*, 163–165.

FLUO 3

CAS Registry Number 123632-39-3

Chemical Structure

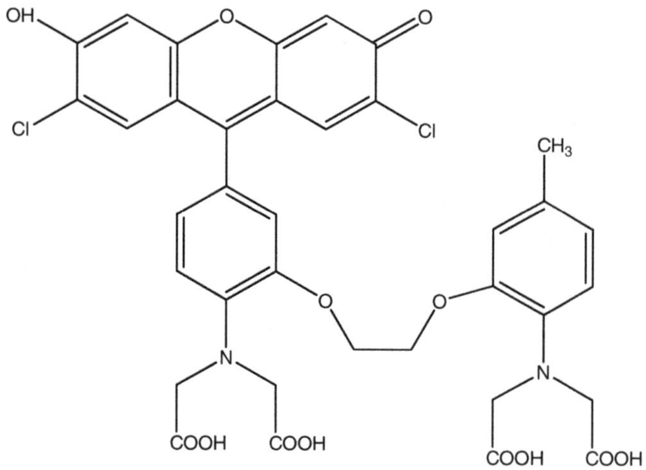

CA Index Name Glycine, *N*-[2-[2-[2-[bis(carboxymethyl)amino]-5-(2,7-dichloro-6-hydroxy-3-oxo-3*H*-xanthen-9-yl)phenoxy]ethoxy]-4-methylphenyl]-*N*-(carboxymethyl)-

Other Names Fluo 3

Merck Index Number Not listed

Chemical/Dye Class Xanthene

Molecular Formula $C_{36}H_{30}Cl_2N_2O_{13}$

Molecular Weight 769.53

Physical Form Dark reddish-brown powder

Solubility Soluble in dimethyl sulfoxide

Melting Point >250°C

Boiling Point (Calcd.) 1056.0 ± 65.0°C, pressure: 760 Torr

pK_a (Calcd.) 1.66 ± 0.10, most acidic temperature: 25°C; 6.18 ± 0.50, most basic temperature: 25°C

Absorption (λ_{max}) 506 nm

Emission (λ_{max}) 526 nm

Synthesis Synthetic methods[1–3]

Staining Applications Calcium ions;[2,3,8–22] zinc ions;[23] bone;[4] cells;[5,6] neurons;[7] peptides;[1] proteins;[1] antibodies[1]

Biological Applications Calcium indicator;[2,3,8–22] zinc indicator;[23] identifying taste modulators;[24] measuring membrane potential;[25] treating defective skeletal muscle function during heart failure,[26] pain[27]

Industrial Applications Not reported

Safety/Toxicity Cardiac toxicity;[28] cytotoxicity;[29,30] excitotoxicity;[31] genotoxicity;[32] neurotoxicity[33,34]

REFERENCES

1. Hahn, K. M.; Toutchkine, A.; Muthyala, R.; Kraynov, V.; Bark, S. J.; Burton, D. R.; Chamberlain, C. Labeled peptides, proteins and antibodies and processes and intermediates useful for their preparation. U.S. Pat. Appl. Publ. US 2002055133, 2002; *Chem. Abstr.* **2002**, *136*, 366139.
2. Tsien, R. Y.; Minta, A. Preparation and properties of calcium-specific, long-wavelength indicator dyes. Eur. Pat. Appl. EP 314480, 1989; *Chem. Abstr.* **1990**, *112*, 135620.
3. Minta, A.; Kao, J. P. Y.; Tsien, R. Y. Fluorescent indicators for cytosolic calcium based on rhodamine and fluorescein chromophores. *J. Biol. Chem.* **1989**, *264*, 8171–8178.
4. Parkesh, R.; Mohsin, S.; Lee, T. C.; Gunnlaugsson, T. Histological, spectroscopic, and surface analysis of microdamage in bone: toward real-time analysis using fluorescent sensors. *Chem. Mater.* **2007**, *19*, 1656–1663.
5. Lacombe, F.; Belloc, F.; Veriac, S.; Lefevre, D. Method of discrimination of at least two cellular populations by immunofluorescence flow cytometry and application to oncopathology. Fr. Demande FR 2895087, 2007; *Chem. Abstr.* **2007**, *147*, 93934.
6. Chinn, K. S.; Jim, K. T. C.; Yen, B.; Wuh, H. Apparatus and method for detecting activity of living cells. PCT Int. Appl. WO 2007041308, 2007; *Chem. Abstr.* **2007**, *146*, 397842.
7. Goldman, S. A.; Roy, N. S.; Nakano, T. Neuron-specific reporter gene expression for isolation of motor neurons from a population of embryonic stem cells undergoing induced differentiation. U.S. Pat. Appl. Publ. US 2005003544, 2005; *Chem. Abstr.* **2005**, *142*, 108461.
8. Parkesh, R.; Lewis, A. M.; Aley, P. K.; Arredouani, A.; Rossi, S.; Tavares, R.; Vasudevan, S. R.; Rosen, D.; Galione, A.; Dowden, J.; Churchill, G. C. Cell-permeant NAADP: a novel chemical tool enabling the study of Ca^{2+} signalling in intact cells. *Cell Calcium* **2008**, *43*, 531–538.
9. Swietach, P.; Spitzer, K. W.; Vaughan-Jones, R. D. Ca^{2+}-mobility in the sarcoplasmic reticulum of

ventricular myocytes is low. *Biophys. J.* **2008**, *95*, 1412–1427.

10. Sobie, E. A.; Kao, J. P. Y.; Lederer, W. J. Novel approach to real-time flash photolysis and confocal [Ca^{2+}] imaging. *Pfluegers. Arch.* **2007**, *454*, 663–673.

11. Ishibashi, K.; Okazaki, S.; Hiramatsu, M. Simultaneous measurement of superoxide generation and intracellular Ca^{2+} concentration reveals the effect of extracellular Ca^{2+} on rapid and transient contents of superoxide generation in differentiated THP-1 cells. *Biochem. Biophys. Res. Commun.* **2006**, *344*, 571–580.

12. Bailey, S.; Macardle, P. J. A flow cytometric comparison of Indo-1 to fluo-3 and Fura Red excited with low power lasers for detecting Ca^{2+} flux. *J. Immunol. Methods* **2006**, *311*, 220–225.

13. Oikawa, H.; Nakamichi, N.; Kambe, Y.; Ogura, M.; Yoneda, Y. An increase in intracellular free calcium ions by nicotinic acetylcholine receptors in a single cultured rat cortical astrocyte. *J. Neurosci. Res.* **2005**, *79*, 535–544.

14. Zou, H.; Lifshitz, L. M.; Tuft, R. A.; Fogarty, K. E.; Singer, J. J. Using total fluorescence increase (signal mass) to determine the Ca^{2+} current underlying localized Ca^{2+} events. *J. Gen. Physiol.* **2004**, *124*, 259–272.

15. Scaduto, R. C., Jr.; Grotyohann, L. W. Hydrolysis of Ca^{2+}-sensitive fluorescent probes by perfused rat heart. *Am. J. Physiol.* **2003**, *285*, H2118–H2124.

16. Powell, J. A.; Molgo, J.; Adams, D. S.; Colasante, C.; Williams, A.; Bohlen, M.; Jaimovich, E. IP3 receptors and associated Ca^{2+} signals localize to satellite cells and to components of the neuromuscular junction in skeletal muscle. *J. Neurosci.* **2003**, *23*, 8185–8192.

17. Rintoul, G. L.; Baimbridge, K. G. Effects of calcium buffers and calbindin-D28k upon histamine-induced calcium oscillations and calcium waves in HeLa cells. *Cell Calcium* **2003**, *34*, 131–144.

18. De Maria, R.; Palmieri, G.; Santoni, A. Induction of Ca^{2+} flux by adhesion molecules in lymphocytes. *Methods Mol. Biol.* **1999**, *96*, 199–204.

19. Klauke, N.; Plattner, H. Imaging of Ca^{2+} transients induced in *Paramecium* cells by a polyamine secretagogue. *J. Cell Sci.* **1997**, *110*, 975–983.

20. Schnetkamp, P. P. M. Functional expression of Na–Ca exchanger clones measured with the fluorescent Ca^{2+}-indicating dye fluo-3. *Biochem. Cell Biol.* **1996**, *74*, 535–539.

21. Atsumi, T.; Sugita, K.; Kohno, M.; Takahashi, T.; Ueha, T. Simultaneous measurement of Ca^{2+} and pH by laser cytometry using fluo-3 and SNARF-1. *Cytometry* **1996**, *24*, 99–105.

22. Wilcox, R. A.; Strupish, J.; Nahorski, S. R. Measurement of Ca^{2+} fluxes in permeabilized cells using $45Ca^{2+}$ and fluo-3. *Methods Mol. Biol.* **1995**, *41*, 215–227.

23. Martin, J. L.; Stork, C. J.; Li, Y. V. Determining zinc with commonly used calcium and zinc fluorescent indicators, a question on calcium signals. *Cell Calcium* **2006**, *40*, 393–402.

24. Li, X.; Xu, H.; Li, Q.; Tang, H.; Pronin, A. Identification of bitter ligands that specifically activate human T2R receptors and related assays for identifying human bitter taste modulators. U.S. Pat. Appl. Publ. US 2008187936, 2008; *Chem. Abstr.* **2008**, *149*, 241823.

25. Klaubert, D.; Diwu, Z.; Yi, G.; Kirk, M. Improved method for measuring membrane potential. PCT Int. Appl. WO 2003014701, 2003; *Chem. Abstr.* **2003**, *138*, 166188.

26. Marks, A. R. Type 1 ryanodine receptor-based methods for treating defective skeletal muscle function during heart failure and for drug screening. PCT Int. Appl. WO 2004080283, 2004; *Chem. Abstr.* **2004**, *141*, 271568.

27. Benjamin, E. R.; Chen, Z.; Sha, D.; Tafesse, L.; Victory, S. F.; Whitehead, J. W. F.; Zhou, X. Piperidinyl compounds and their preparation, pharmaceutical compositions, and use as N-type calcium channel modulators or blockers for treatment or prevention of pain. PCT Int. Appl. WO 2006040181, 2006; *Chem. Abstr.* **2006**, *144*, 412381.

28. Tanaka, H.; Oyamada, M.; Tsujii, E.; Nakajo, T.; Takamatsu, T. Excitation-dependent intracellular Ca^{2+} waves at the border zone of the cryo-injured rat heart revealed by real-time confocal microscopy. *J. Mol. Cell. Cardiol.* **2002**, *34*, 1501–1512.

29. Li, Y.; King, M. A.; Grimes, J.; Smith, N.; de Fiebre, C. M.; Meyer, E. M. Alpha7 nicotinic receptor mediated protection against ethanol-induced cytotoxicity in PC12 cells. *Brain Res.* **1999**, *816*, 225–228.

30. Jiang, T.; Grant, R. L.; Acosta, D. A digitized fluorescence imaging study of intracellular free calcium, mitochondrial integrity and cytotoxicity in rat renal cells exposed to ionomycin, a calcium ionophore. *Toxicology* **1993**, *85*, 41–65.

31. Ahern, K. V.; Lustig, H. S.; Chan, J.; Greenberg, D. A. Calcium indicators and excitotoxicity in cultured cortical neurons. *Neurosci. Lett.* **1993**, *162*, 169–172.

32. Dopp, E.; Muller, J.; Hahnel, C.; Schiffmann, D. Induction of genotoxic effects and modulation of the intracellular calcium level in Syrian hamster embryo (SHE) fibroblasts caused by ochratoxin A. *Food Chem. Toxicol.* **1999**, *37*, 713–721.
33. Shang, L.; Liu, J.; Zhu, Q.; Zhao, L.; Feng, Y.; Wang, X.; Cao, W.; Xin, H. Gypenosides protect primary cultures of rat cortical cells against oxidative neurotoxicity. *Brain Res.* **2006**, *1102*, 163–174.
34. Diaz-Trelles, R.; Novelli, A.; Fernandez-Sanchez, M. T. RNA synthesis-dependent potentiation of alpha-amino-3-hydroxy-5-methyl-4-isoxazole propionate receptor-mediated toxicity by antihistamine terfenadine in cultured rat cerebellar neurons. *Neurosci. Lett.* **2003**, *345*, 136–140.

FLUO 3 AM

CAS Registry Number 121714-22-5

Chemical Structure

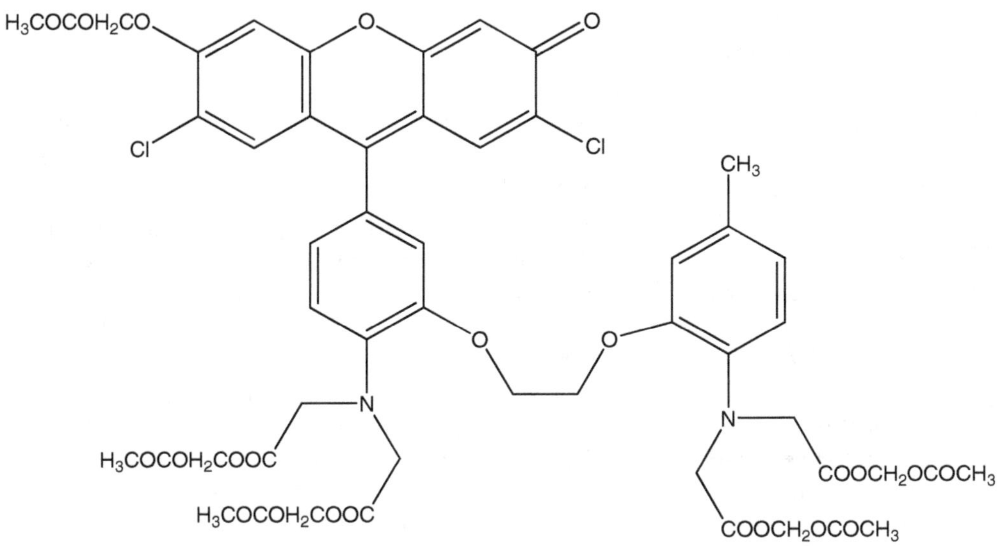

Melting Point >250°C

Boiling Point (Calcd.) 1090.9 ± 65.0°C, pressure: 760 Torr

pK_a (Calcd.) 2.03 ± 0.50, most basic, temperature: 25°C

CA Index Name Glycine, *N*-[4-[6-[(acetyloxy)methoxy]-2,7-dichloro-3-oxo-3*H*-xanthen-9-yl]-2-[2-[2-[bis[2-[(acetyloxy)methoxy]-2-oxoethyl]amino]-5-methylphenoxy]ethoxy]phenyl]-*N*-[2-[(acetyloxy)methoxy]-2-oxoethyl]-, (acetyloxy)methyl ester

Other Names Fluo 3 pentaacetoxymethyl ester; Fluo 3AM

Merck Index Number Not listed

Chemical/Dye Class Xanthene

Molecular Formula $C_{51}H_{50}Cl_2N_2O_{23}$

Molecular Weight 1129.85

Physical Form Dark red crystals

Solubility Soluble in dimethyl sulfoxide, methanol

Absorption (λ_{max}) 464 nm

Emission (λ_{max}) Fluorescence is very weak

Synthesis Synthetic methods[1,2]

Staining Applications Calcium ions;[1,2,7–16] cells;[3,4] sperms;[5] neurons[6]

Biological Applications Calcium indicator;[1,2,7–16] detecting leukocyte tumor cells;[17] identifying taste modulators;[18] treating pain[19]

Industrial Applications Light emitting diodes[20]

Safety/Toxicity Cellular toxicity;[21] phototoxicity[20]

REFERENCES

1. Tsien, R. Y.; Minta, A. Preparation and properties of calcium-specific, long-wavelength indicator dyes. Eur. Pat. Appl. EP 314480, 1989; *Chem. Abstr.* **1990**, *112*, 135620.
2. Minta, A.; Kao, J. P. Y.; Tsien, R. Y. Fluorescent indicators for cytosolic calcium based on rhodamine and fluorescein chromophores. *J. Biol. Chem.* **1989**, *264*, 8171–8178.
3. Ghosh, R. N.; Debiasio, R.; Chen, Y. T.; Bellutta, P.; Giuliano, K.; Pasley, J. W. A system for cell-based screening. PCT Int. Appl. WO 2001035072, 2001; *Chem. Abstr.* **2001**, *134*, 363630.
4. Taylor, D. L. Miniaturized cell array methods and apparatus for cell-based screening. U.S. Patent 6103479, 2000; *Chem. Abstr.* **2000**, *133*, 161565.

5. Niu, Y.; Greube, A.; Ji, W.; Jewgenow, K. The application of *in vitro* sperm competition test to evaluate the impact of ZP-derived peptides on fertilization capacity of cat sperm. *Theriogenology* **2006**, *66*, 989–995.
6. Chan, G. K. Y.; Richards, G. R.; Peters, M.; Simpson, P. B. High content kinetic assays of neuronal signaling implemented on BDTM Pathway HT. *Assay Drug Dev. Technol.* **2005**, *3*, 623–636.
7. Wilson, J. M.; Dombeck, D. A.; Diaz-Rios, M.; Harris-Warrick, R. M.; Brownstone, R. M. Two-photon calcium imaging of network activity in XFP-expressing neurons in the mouse. *J. Neurophysiol.* **2007**, *97*, 3118–3125.
8. Richards, G. R.; Jack, A. D.; Platts, A.; Simpson, P. B. Measurement and analysis of calcium signaling in heterogeneous cell culture. *Methods Enzymol.* **2006**, *414*, 335–347.
9. Plank, D. M.; Sussman, M. A. Intracellular Ca^{2+} measurements in live cells by rapid line scan confocal microscopy: simplified calibration methodology. *Methods Cell Sci.* **2003**, *25*, 123–133.
10. Rintoul, G. L.; Baimbridge, K. G. Effects of calcium buffers and calbindin-D28k upon histamine-induced calcium oscillations and calcium waves in HeLa cells. *Cell Calcium* **2003**, *34*, 131–144.
11. Gensburger, C.; Freyermuth, S.; Klein, C.; Malviya, A. N. *In vivo* nuclear Ca^{2+}-ATPase phosphorylation triggers intermediate size molecular transport to the nucleus. *Biochem. Biophys. Res. Commun.* **2003**, *303*, 1225–1228.
12. Katoh, H.; Sasano, C.; Niwa, R.; Honjo, H.; Hayashi, H. Imaging of mitochondrial Ca^{2+} concentration in saponin permeabilized rat ventricular myocytes. *Environ. Med.* **2001**, *45*, 48–50.
13. Zhang, W. H.; Rengel, Z.; Kuo, J. Determination of intracellular Ca^{2+} in cells of intact wheat roots: loading of acetoxymethyl ester of Fluo-3 under low temperature. *Plant J.* **1998**, *15*, 147–151.
14. Brezden, C. B.; Hedley, D. W.; Rauth, A. M. Constitutive expression of P-glycoprotein as a determinant of loading with fluorescent calcium probes. *Cytometry* **1994**, *17*, 343–348.
15. Burnier, M.; Centeno, G.; Burki, E.; Brunner, H. R. Confocal microscopy to analyze cytosolic and nuclear calcium in cultured vascular cells. *Am. J. Physiol.* **1994**, *266*, C1118–C1127.
16. Kao, J. P. Y.; Harootunian, A. T.; Tsien, R. Y. Photochemically generated cytosolic calcium pulses and their detection by fluo-3. *J. Biol. Chem.* **1989**, *264*, 8179–8184.
17. Minakami, T.; Tsuji, T.; Oguni, S.; Hamaguchi, Y.; Tsuruta, K.; Kamihiro, T. Method for detecting leukocyte tumor cells. Jpn. Kokai Tokkyo Koho JP 2002207036, 2002; *Chem. Abstr.* **2002**, *137*, 106038.
18. Servant, G.; Kamdar, P.; Rivadeneyra, A. Novel cell-based assays for identifying enhancers or inhibitors of T1R taste receptors (T1R2/T1R3 sweet) and umami (T1R1/T1R3 umami) taste receptors. U.S. Pat. Appl. Publ. US 2008305500, 2008; *Chem. Abstr.* **2008**, *150*, 31118.
19. Davis, J. B.; Smith, G. D. Protein and cDNA sequences of human vanilloid receptor 1 (VR1) ion channel subunits and vanilloid receptor like 3 (VRL3) ion channel subunits and their uses for pain treatments. PCT Int. Appl. WO 2003074562, 2003; *Chem. Abstr.* **2003**, *139*, 256316.
20. Nishigaki, T.; Wood, C. D.; Shiba, K.; Baba, S. A.; Darszon, A. Stroboscopic illumination using light-emitting diodes reduces phototoxicity in fluorescence cell imaging. *BioTechniques* **2006**, *41*, 191–197.
21. Burghardt, R. C.; Barhoumi, R.; Lewis, E. H.; Bailey, H.; Pyle, K. A.; Clement, B. A.; Phillips, T. D. Patulin-induced cellular toxicity: a vital fluorescence study. *Toxicol. Appl. Pharmacol.* **1992**, *112*, 235–244.

FLUO 4 AM

CAS Registry Number 273221-67-3

Chemical Structure

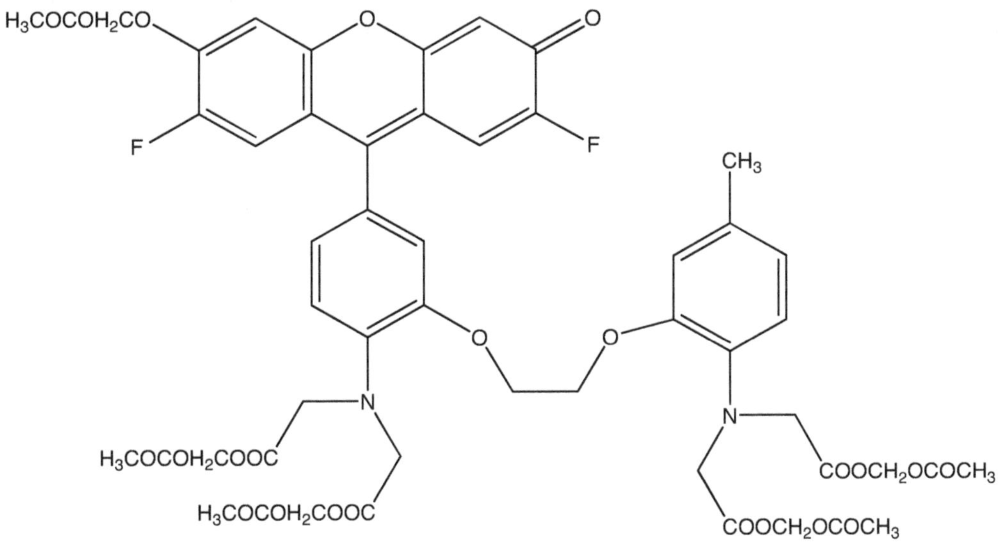

CA Index Name Glycine, *N*-[4-[6-[(acetyloxy)methoxy]-2,7-difluoro-3-oxo-3*H*-xanthen-9-yl]-2-[2-[2-[bis[2-[(acetyloxy)methoxy]-2-oxoethyl]amino]-5-methylphenoxy]ethoxy]phenyl]-*N*-[2-[(acetyloxy)methoxy]-2-oxoethyl]-, (acetyloxy)methyl ester

Other Names Fluo 4AM; Fluo 4AM ester; Fluo 4 acetoxymethyl ester

Merck Index Number Not listed

Chemical/Dye Class Xanthene

Molecular Formula $C_{51}H_{50}F_2N_2O_{23}$

Molecular Weight 1096.94

Physical Form Solid

Solubility Soluble in dimethyl sulfoxide, methanol

Melting Point >200°C

Boiling Point (Calcd.) 1063.2 ± 65.0°C, pressure: 760 Torr

pK_a (Calcd.) 2.02 ± 0.50, most basic, temperature: 25°C

Absorption (λ_{max}) 456 nm

Emission (λ_{max}) Fluorescence is very weak

Synthesis Synthetic methods[1,2]

Staining Applications Calcium ions;[1,2,5–8] cells;[3] neurons;[4]

Biological Applications Calcium indicator[1,2,5–8]

Industrial Applications Not reported

Safety/Toxicity Cytotoxicity;[9] hepatotoxicity[9]

REFERENCES

1. Gee, K. R.; Brown, K. A.; Chen, W. N. U.; Bishop-Stewart, J.; Gray, D.; Johnson, I. Chemical and physiological characterization of fluo-4 Ca^{2+}-indicator dyes. *Cell Calcium* **2000**, *27*, 97–106.
2. Kreitzer, A. C.; Gee, K. R.; Archer, E. A.; Regehr, W. G. Monitoring presynaptic calcium dynamics in projection fibers by *in vivo* loading of a novel calcium indicator. *Neuron* **2000**, *27*, 25–32.
3. Chen, T. W.; Lin, B. J.; Brunner, E.; Schild, D. *In situ* background estimation in quantitative fluorescence imaging. *Biophys. J.* **2006**, *90*, 2534–2547.
4. Garaschuk, O.; Milos, R. I.; Konnerth, A. Targeted bulk-loading of fluorescent indicators for two-photon brain imaging *in vivo*. *Nat. Protoc.* **2006**, *1*, 380–386.
5. Picht, E.; Zima, A. V.; Blatter, L. A.; Bers, D. M. SparkMaster: automated calcium spark analysis with ImageJ. *Am. J. Physiol.* **2007**, *293*, C1073–C1081.
6. Hashimoto, T.; Ohata, H.; Nobe, K.; Sakai, Y.; Honda, K. A novel approach for the determination of contractile and calcium responses of the basilar artery employing real-time confocal laser microscopy. *J. Pharmacol. Toxicol. Methods* **2007**, *56*, 79–86.

7. Zhang, X.; Yin, H.; Cooper, J. M.; Haswell, S. J. A microfluidic-based system for analysis of single cells based on Ca^{2+} flux. *Electrophoresis* **2006**, *27*, 5093–5100.
8. Liu, E. C. K.; Abell, L. M. Development and validation of a platelet calcium flux assay using a fluorescent imaging plate reader. *Anal. Biochem.* **2006**, *357*, 216–224.
9. O'Brien, P. J.; Irwin, W.; Diaz, D.; Howard-Cofield, E.; Krejsa, C. M.; Slaughter, M. R.; Gao, B.; Kaludercic, N.; Angeline, A.; Bernardi, P.; Brain, P.; Hougham, C. High concordance of drug-induced human hepatoto-xicity with *in vitro* cytotoxicity measured in a novel cell-based model using high content screening. *Arch. Toxicol.* **2006**, *80*, 580–604.

FLUOZIN 1

CAS Registry Number 411209-53-5

Chemical Structure

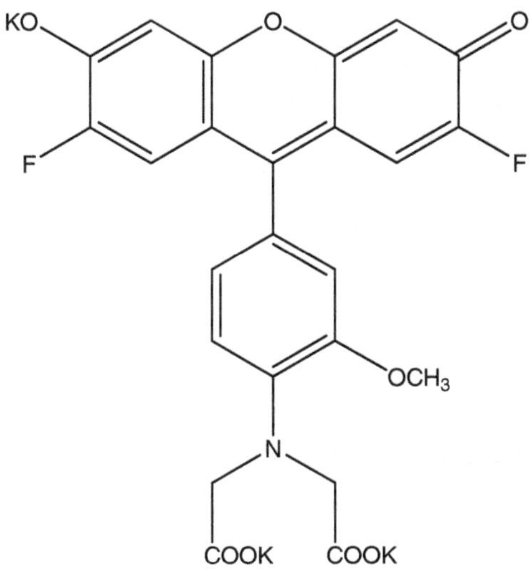

CA Index Name Glycine, *N*-(carboxymethyl)-*N*-[4-(2,7-difluoro-6-hydroxy-3-oxo-3*H*-xanthen-9-yl)-2-methoxyphenyl]-, tripotassium salt

Other Names FluoZin 1; FluoZin 1 tripotassium salt

Merck Index Number Not listed

Chemical/Dye Class Xanthene

Molecular Formula $C_{24}H_{14}F_2K_3NO_8$

Molecular Weight 599.67

Physical Form Solid

Solubility Soluble in water

Melting Point >200°C

Absorption (λ_{max}) 495 nm

Emission (λ_{max}) 517 nm

Synthesis Synthetic method[1]

Staining Applications Zinc ions;[1–4] copper ions[5]

Biological Applications Zinc indicator;[1–4] copper indicator;[5] detecting modulators of ion channels;[6] identifying genes[7]

Industrial Applications Not reported

Safety/Toxicity No data available

REFERENCES

1. Gee, K. R.; Zhou, Z. L.; Ton-That, D.; Sensi, S. L.; Weiss, J. H. Measuring zinc in living cells. A new generation of sensitive and selective fluorescent probes. *Cell Calcium* **2002**, *31*, 245–251.
2. Wei, Y.; Fu, D. Selective metal binding to a membrane-embedded aspartate in the *Escherichia coli* metal transporter YiiP (FieF). *J. Biol. Chem.* **2005**, *280*, 33716–33724.
3. Chao, Y.; Fu, D. Kinetic study of the antiport mechanism of an *Escherichia coli* zinc transporter, ZitB. *J. Biol. Chem.* **2004**, *279*, 12043–12050.
4. Dineley, K. E.; Malaiyandi, L. M.; Reynolds, I. J. A reevaluation of neuronal zinc measurements: artifacts associated with high intracellular dye concentration. *Mol. Pharmacol.* **2002**, *62*, 618–627.
5. Royzen, M.; Dai, Z.; Canary, J. W. Ratiometric displacement approach to Cu(II) sensing by fluorescence. *J. Am. Chem. Soc.* **2005**, *127*, 1612–1613.
6. Weaver, C. D. Methods for detecting modulators of ion channels using thallium (I) sensitive assays. PCT Int. Appl. WO 2002031508, 2002; *Chem. Abstr.* **2002**, *136*, 321701.
7. Kim, E. J.; Barker, L.; Burnet, M.; Guse, J. H.; Luyten, K.; Tsotsou, G. Use of a yeast expression host to identify genes for transport proteins using fluorescent dye reporters. PCT Int. Appl. WO 2003038092, 2003; *Chem. Abstr.* **2003**, *138*, 363834.

FLUOZIN 2

CAS Registry Number 411209-54-6

Chemical Structure

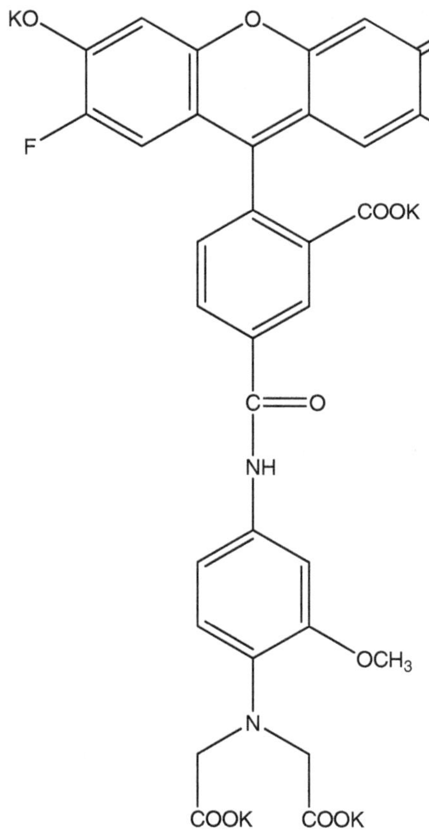

CA Index Name Glycine, *N*-(carboxymethyl)-*N*-[4-[[(2′,7′-difluoro-3′,6′-dihydroxy-3-oxospiro[isobenzofuran-1(3*H*),9′-[9*H*]xanthen]-5-yl)carbonyl]amino]-2-methoxyphenyl]-, tetrapotassium salt

Other Names FluoZin 2; FluoZin 2 tetrapotassium salt

Merck Index Number Not listed

Chemical/Dye Class Xanthene

Molecular Formula $C_{32}H_{18}F_2K_4N_2O_{11}$

Molecular Weight 800.89

Physical Form Solid

Solubility Soluble in water

Melting Point >200°C

Absorption (λ_{max}) 494 nm

Emission (λ_{max}) 522 nm

Synthesis Synthetic method[1]

Staining Applications Zinc ions[1,2]

Biological Applications Zinc indicator;[1,2] detecting modulators of ion channels[3]

Industrial Applications Not reported

Safety/Toxicity No data available

REFERENCES

1. Gee, K. R.; Zhou, Z. L.; Ton-That, D.; Sensi, S. L.; Weiss, J. H. Measuring zinc in living cells. A new generation of sensitive and selective fluorescent probes. *Cell Calcium* **2002**, *31*, 245–251.
2. Dineley, K. E.; Malaiyandi, L. M.; Reynolds, I. J. A reevaluation of neuronal zinc measurements: artifacts associated with high intracellular dye concentration. *Mol. Pharmacol.* **2002**, *62*, 618–627.
3. Weaver, C. D. Methods for detecting modulators of ion channels using thallium (I) sensitive assays. PCT Int. Appl. WO 2002031508, 2002; *Chem. Abstr.* **2002**, *136*, 321701.

FLUOZIN 3

CAS Registry Number 404335-95-1

Chemical Structure

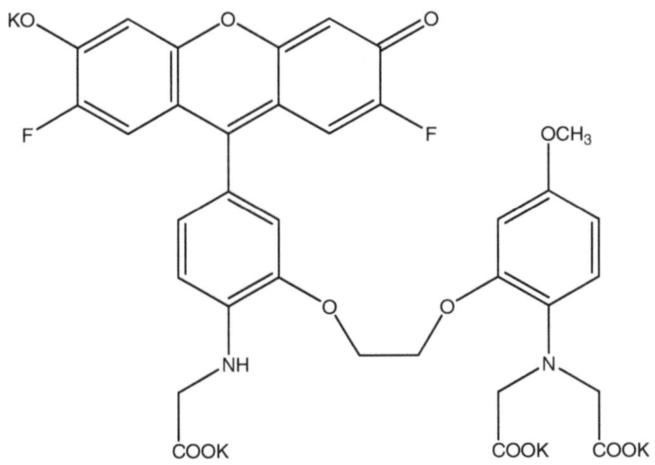

CA Index Name Glycine, *N*-(carboxymethyl)-*N*-[2-[2-[2-[(carboxymethyl)amino]-5-(2,7-difluoro-6-hydroxy-3-oxo-3*H*-xanthen-9-yl)phenoxy]ethoxy]-4-methoxyphenyl]-, potassium salt (1:4)

Other Names Glycine, *N*-(carboxymethyl)-*N*-[2-[2-[2-[(carboxymethyl)amino]-5-(2,7-difluoro-6-hydroxy-3-oxo-3*H*-xanthen-9-yl)phenoxy]ethoxy]-4-methoxyphenyl]-, tetrapotassium salt; FluoZin 3

Merck Index Number Not listed

Chemical/Dye Class Xanthene

Molecular Formula $C_{34}H_{24}F_2K_4N_2O_{12}$

Molecular Weight 846.96

Physical Form Solid

Solubility Soluble in water

Melting Point >200°C

Absorption (λ_{max}) 494 nm

Emission (λ_{max}) 516 nm

Synthesis Synthetic method[1]

Staining Applications Zinc;[1–15] chromium;[3] manganese;[3] iron;[3] cobalt;[3] copper;[3] nickel;[3] cadmium ions[3]

Biological Applications Zinc indicator[1–15]

Industrial Applications Not reported

Safety/Toxicity No data available

REFERENCES

1. Gee, K. R.; Zhou, Z. L.; Qian, W. J.; Kennedy, R. Detection and imaging of zinc secretion from pancreatic β-cells using a new fluorescent zinc indicator. *J. Am. Chem. Soc.* **2002**, *124*, 776–778.
2. Gee, K. R.; Zhou, Z. L.; Ton-That, D.; Sensi, S. L.; Weiss, J. H. Measuring zinc in living cells: a new generation of sensitive and selective fluorescent probes. *Cell Calcium* **2002**, *31*, 245–251.
3. Zhao, J.; Bertoglio, B. A.; Devinney, M. J., Jr.; Dineley, K. E.; Kay, A. R. The interaction of biological and noxious transition metals with the zinc probes FluoZin-3 and Newport Green. *Anal. Biochem.* **2009**, *384*, 34–41.
4. Zhao, J.; Bertoglio, B. A.; Gee, K. R.; Kay, A. R. The zinc indicator FluoZin-3 is not perturbed significantly by physiological levels of calcium or magnesium. *Cell Calcium* **2008**, *44*, 422–426.
5. Krezel, A.; Maret, W. Dual nanomolar and picomolar Zn(II) binding properties of metallothionein. *J. Am. Chem. Soc.* **2007**, *129*, 10911–10921.
6. Datki, Z. L.; Hunya, A.; Penke, B. A novel and simple fluorescence method for the measurement of presynaptic vesicular zinc release in acute hippocampal slices with a fluorescence plate reader. *Brain Res. Bull.* **2007**, *74*, 183–187.
7. Dineley, K. E. On the use of fluorescent probes to distinguish Ca^{2+} from Zn^{2+} in models of excitotoxicity. *Cell Calcium* **2007**, *42*, 341–342.
8. Henares, T. G.; Takaishi, M.; Yoshida, N.; Terabe, S.; Mizutani, F.; Sekizawa, R.; Hisamoto, H. Integration of multianalyte sensing functions on a capillary-assembled microchip: simultaneous determination of ion concentrations and enzymatic activities by a "Drop-and-Sip" technique. *Anal. Chem.* **2007**, *79*, 908–915.
9. Stork, C. J.; Li, Y. V. Measuring cell viability with membrane impermeable zinc fluorescent indicator. *J. Neurosci. Methods* **2006**, *155*, 180–186.
10. Haase, H.; Hebel, S.; Engelhardt, G.; Rink, L. Flow cytometric measurement of labile zinc in peripheral blood mononuclear cells. *Anal. Biochem.* **2006**, *352*, 222–230.
11. St. Croix, C. M.; Stitt, M. S.; Watkins, S. C.; Pitt, B. R. Fluorescence resonance energy transfer based assays for the real-time detection of nitric oxide signaling. *Methods Enzymol.* **2005**, *396*, 317–326.

12. Malaiyandi, L. M.; Vergun, O.; Dineley, K. E.; Reynolds, I. J. Direct visualization of mitochondrial zinc accumulation reveals uniporter-dependent and -independent transport mechanisms. *J. Neurochem.* **2005**, *93*, 1242–1250.
13. Devinney, M. J.; Reynolds, I. J.; Dineley, K. E. Simultaneous detection of intracellular free calcium and zinc using fura-2FF and FluoZin-3. *Cell Calcium* **2005**, *37*, 225–232.
14. Qian, W. J.; Peters, J. L.; Dahlgren, G. M.; Gee, K. R.; Kennedy, R. T. Simultaneous monitoring of Zn^{2+} secretion and intracellular Ca^{2+} from islets and islet cells by fluorescence microscopy. *BioTechniques* **2004**, *37*, 922–924, 926, 928–930, 932–933.
15. Qian, W. J.; Gee, K. R.; Kennedy, R. T. Imaging of Zn^{2+} release from pancreatic β-cells at the level of single exocytotic events. *Anal. Chem.* **2003**, *75*, 3468–3475.

FM 4-64

CAS Registry Number 162112-35-8

Chemical Structure

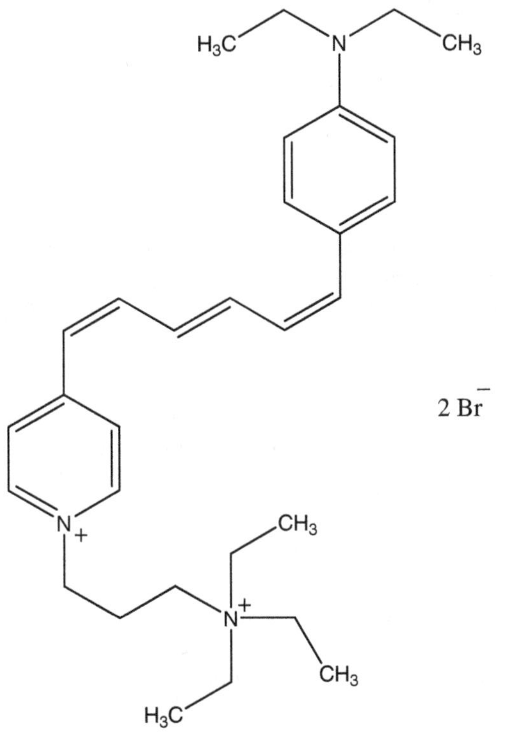

CA Index Name Pyridinium, 4-[6-[4-(diethylamino) phenyl]-1,3,5-hexatrien-1-yl]-1-[3-(triethylammonio) propyl]-, bromide (1:2)

Other Names Pyridinium, 4-[6-[4-(diethylamino)phenyl]-1,3,5-hexatrienyl]-1-[3-(triethylammonio)propyl]-, dibromide; FM 4-64; SynaptoRed C2; *N*-(3-Triethylammoniopropyl)-4-(6-(4-(diethylamino)phenyl)hexatrienyl) pyridinium dibromide

Merck Index Number Not listed

Chemical/Dye Class Styryl

Molecular Formula $C_{30}H_{45}Br_2N_3$

Molecular Weight 607.51

Physical Form Dark purple powder

Solubility Soluble in water, dimethyl sulfoxide, methanol

Melting Point >200°C

Absorption (λ_{max}) 505 nm

Emission (λ_{max}) 725 nm

Synthesis Synthetic method[1]

Staining Applications Vacuolar membrane;[2,3] plasma membrane;[8,20] bacterial membrane;[4,5] lipid membrane;[6,7] plasma membrane-bound flavoproteins;[8,20]nuclear envelope (NE);[9] synaptic vesicles;[10] secretory vesicles;[11] lactotroph vesicles;[12] synaptic terminals;[13] neurons;[14,15] endocytosis;[2,23–26] exocytosis;[27]smooth-muscle-associated airway receptors (SMARs) in lungs;[16] embryos[17]

Biological Applications Detecting FRET in cells;[18] monitoring fast neuronal activity and signaling;[19] quantifying plasma membrane expression;[20] cytotoxicity assay;[21] membrane fusion assay;[22] probe for endocytosis;[23–26] probe for exocytosis[27]

Industrial Applications Not reported

Safety/Toxicity No data available

REFERENCES

1. Haugland, R. P. *The Handbook: A Guide to Fluorescent Probes and Labeling Technologies*; Molecular Probes Inc.: Eugene, **2005**; pp 585–588.
2. Vida, T. A.; Emr, S. D. A new vital stain for visualizing vacuolar membrane dynamics and endocytosis in yeast. *J. Cell Biol.* **1995**, *128*, 779–792.
3. Zheng, B.; Wu, J. N.; Schober, W.; Lewis, D. E.; Vida, T. Isolation of yeast mutants defective for localization of vacuolar vital dyes. *Proc. Natl. Acad. Sci. U.S.A.* **1998**, *95*, 11721–11726.
4. Fishov, I.; Woldringh, C. L. Visualization of membrane domains in *Escherichia coli*. *Mol. Microbiol.* **1999**, *32*, 1166–1172.
5. Pogliano, J.; Osborne, N.; Sharp, M. D.; Mello, A. A. D.; Perez, A.; Sun, Y. L.; Pogliano, K. A vital stain for studying membrane dynamics in bacteria: a novel mechanism controlling septation during *Bacillus subtilis* sporulation. *Mol. Microbiol.* **1999**, *31*, 1149–1159.
6. Krishna, M. M. G.; Periasamy, N. Fluorescence of organic dyes in lipid membranes: site of solubilization and effects of viscosity and refractive index on lifetimes. *J. Fluoresc.* **1998**, *8*, 81–91.
7. Haertel, S.; Tykhonova, S.; Haas, M.; Diehl, H. A. The susceptibility of non-UV fluorescent membrane dyes to dynamical properties of lipid membranes. *J. Fluoresc.* **2002**, *12*, 465–479.
8. Samhan-Arias, A. K.; Garcia-Bereguiain, M. A.; Martin-Romero, F. J.; Gutierrez-Merino, C. Regionalization of plasma membrane-bound flavoproteins of cerebellar granule neurons in

culture by fluorescence energy transfer imaging. *J. Fluoresc.* **2006**, *16*, 393–401.

9. Zal, T.; Zal, M. A.; Lotz, C.; Goergen, C. J.; Gascoigne, N. R. J. Spectral shift of fluorescent dye FM4-64 reveals distinct microenvironment of nuclear envelope in living cells. *Traffic* **2006**, *7*, 1607–1613.
10. Gaffield, M. A.; Betz, W. J. Imaging synaptic vesicle exocytosis and endocytosis with FM dyes. *Nat. Protoc.* **2006**, *1*, 2916–2921.
11. Wang, X.; Teng, Y.; Wang, Q.; Li, X.; Sheng, X.; Zheng, M.; Samaj, J.; Baluska, F.; Lin, J. Imaging of dynamic secretory vesicles in living pollen tubes of *Picea meyeri* using evanescent wave microscopy. *Plant Physiol.* **2006**, *141*, 1591–1603.
12. Stenovec, M.; Poberaj, I.; Kreft, M.; Zorec, R. Concentration-dependent staining of lactotroph vesicles by FM 4-64. *Biophys. J.* **2005**, *88*, 2607–2613.
13. Cali, C.; Marchaland, J.; Regazzi, R.; Bezzi, P. SDF 1-alpha (CXCL12) triggers glutamate exocytosis from astrocytes on a millisecond time scale: imaging analysis at the single-vesicle level with TIRF microscopy. *J. Neuroimmunol.* **2008**, *198*, 82–91.
14. Jomphe, C.; Bourque, M. J.; Fortin, G. D.; St-Gelais, F.; Okano, H.; Kobayashi, K.; Trudeau, L. E. Use of TH-EGFP transgenic mice as a source of identified dopaminergic neurons for physiological studies in postnatal cell culture. *J. Neurosci. Methods* **2005**, *146*, 1–12.
15. Jiang, J.; Eisenthal, K. B.; Yuste, R. Second harmonic generation in neurons: electro-optic mechanism of membrane potential sensitivity. *Biophys. J.* **2007**, *93*, L26–L28.
16. De Proost, I.; Pintelon, I.; Brouns, I.; Timmermans, J. P.; Adriaensen, D. Selective visualization of sensory receptors in the smooth muscle layer of *ex-vivo* airway whole-mounts by styryl pyridinium dyes. *Cell Tissue Res.* **2007**, *329*, 421–431.
17. Mohler, W. A.; White, J. G. Stereo-4-D reconstruction and animation from living fluorescence specimens. *BioTechniques* **1998**, *24*, 1006–1008, 1010, 1011.
18. Duncan, R. R.; Bergmann, A.; Cousin, M. A.; Apps, D. K.; Shipston, M. J. Multi-dimensional time-correlated single photon counting (TCSPC) fluorescence lifetime imaging microscopy (FLIM) to detect FRET in cells. *J. Microsc.* **2004**, *215*, 1–12.
19. Pages, S.; Veilleux, I.; De Koninck, P.; Cote, D. Multimodal optical microscopy for monitoring fast neuronal activity and signaling. *Proc. SPIE-Int. Soc. Opt. Eng.* **2008**, *6860*, 68601X/1–68601X/8.
20. Janecki, A. J.; Janecki, M.; Akhter, S.; Donowitz, M. Quantitation of plasma membrane expression of a fusion protein of Na/H exchanger NHE3 and green fluorescence protein (GFP) in living PS120 fibroblasts. *J. Histochem. Cytochem.* **2000**, *48*, 1479–1491.
21. Murakami, T. Cytotoxicity test method by measuring membrane electric potential. Jpn. Kokai Tokkyo Koho JP 2000300290, 2000; *Chem. Abstr.* **2000**, *133*, 330852.
22. Duncan, R. R.; Cousin, M. A. Membrane fusion assay using FRET in combination with fluorescence microscopy and time resolved fluorometry. PCT Int. Appl. WO 2006003397, 2006; *Chem. Abstr.* **2006**, *144*, 103480.
23. Fischer-Parton, S.; Parton, R. M.; Hickey, P. C.; Dijksterhuis, J.; Atkinson, H. A.; Read, N. D. Confocal microscopy of FM4-64 as a tool for analyzing endocytosis and vesicle trafficking in living fungal hyphae. *J. Microsc.* **2000**, *198*, 246–259.
24. Wendland, B.; McCaffery, J. M.; Xiao, Q.; Emr, S. D. A novel fluorescence-activated cell sorter-based screen for yeast endocytosis mutants identifies a yeast homolog of mammalian eps15. *J. Cell Biol.* **1996**, *135*, 1485–1500.
25. Collings, D. A.; Gebbie, L. K.; Howles, P. A.; Hurley, U. A.; Birch, R. J.; Cork, A. H.; Hocart, C. H.; Arioli, T.; Williamson, R. E. *Arabidopsis* dynamin-like protein DRP1A: a null mutant with widespread defects in endocytosis, cellulose synthesis, cytokinesis, and cell expansion. *J. Exp. Bot.* **2008**, *59*, 361–376.
26. Morishita, M.; Engebrecht, J. End3p-mediated endocytosis is required for spore wall formation in *Saccharomyces cerevisiae*. *Genetics* **2005**, *170*, 1561–1574.
27. Vardjan, N.; Stenovec, M.; Jorgacevski, J.; Kreft, M.; Zorec, R. Subnanometer fusion pores in spontaneous exocytosis of peptidergic vesicles. *J. Neurosci.* **2007**, *27*, 4737–4746.

FURA 2

CAS Registry Number 113694-64-7

Chemical Structure

Physical Form Yellow powder

Solubility Soluble in water

Melting Point >250°C

Absorption (λ_{max}) 363 nm; 335 nm

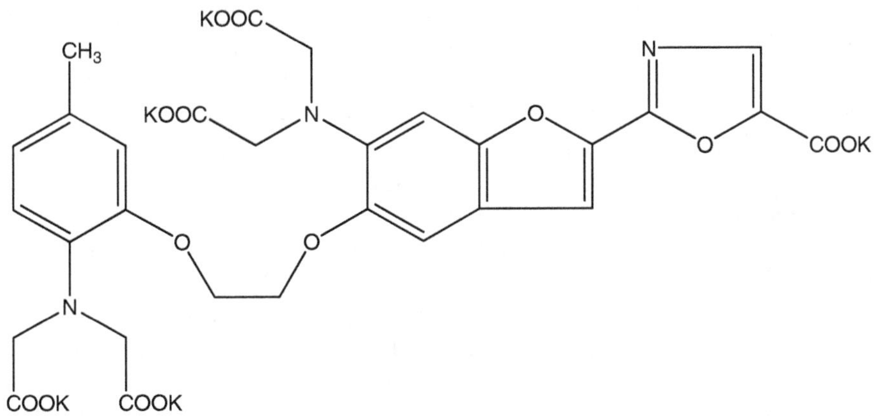

CA Index Name 5-Oxazolecarboxylic acid, 2-[6-[bis(carboxymethyl)amino]-5-[2-[2-[bis(carboxymethyl)amino]-5-methylphenoxy]ethoxy]-2-benzofuranyl]-, potassium salt (1:5)

Other Names 5-Oxazolecarboxylic acid, 2-[6-[bis(carboxymethyl)amino]-5-[2-[2-[bis(carboxymethyl)amino]-5-methylphenoxy]ethoxy]-2-benzofuranyl]-, pentapotassium salt; Fura 2; Fura 2 pentapotassium salt

Merck Index Number Not listed; 4293 for Free Acid

Chemical/Dye Class Benzofuran

Molecular Formula $C_{29}H_{22}K_5N_3O_{14}$

Molecular Weight 831.99

Emission (λ_{max}) 512 nm; 505 nm

Synthesis Synthetic methods[1,2]

Staining Applications Calcium ions;[4,5,8–15] zinc ions;[6,16–18] cells;[3] neurons;[4–6] peptides;[7] proteins;[7] antibodies[7]

Biological Applications Calcium indicator;[4,5,8–15] zinc indicator;[6,16–18] identifying taste modulators;[19,20] measuring membrane potential;[21] treating epilepsy;[22] neurological disorders[23]

Industrial Applications Not reported

Safety/Toxicity No data available

REFERENCES

1. Gee, K. R.; Archer, E. A.; Lapham, L. A.; Leonard, M. E.; Zhou, Z. L.; Bingham, J.; Diwu, Z. New ratiometric fluorescent calcium indicators with moderately attenuated binding affinities. *Bioorg. Med. Chem. Lett.* **2000**, *10*, 1515–1518.
2. Grynkiewicz, G.; Poenie, M.; Tsien, R. Y. A new generation of calcium indicators with greatly improved fluorescence properties. *J. Biol. Chem.* **1985**, *260*, 3440–3450.
3. Lacombe, F.; Belloc, F.; Veriac, S.; Lefevre, D. Method of discrimination of at least two cellular populations by immunofluorescence flow cytometry and application to oncopathology. Fr. Demande FR 2895087, 2007; *Chem. Abstr.* **2007**, *147*, 93934.
4. Ohshiro, H.; Ogawa, S.; Shinjo, K. Visualizing sensory transmission between dorsal root ganglion and dorsal horn neurons in co-culture with calcium imaging. *J. Neurosci. Methods* **2007**, *165*, 49–54.
5. Raza, M.; Deshpande, L. S.; Blair, R. E.; Carter, D. S.; Sombati, S.; DeLorenzo, R. J. Aging is associated with elevated intracellular calcium levels and altered calcium homeostatic mechanisms in hippocampal neurons. *Neurosci. Lett.* **2007**, *418*, 77–81.
6. Cheng, C.; Reynolds, I. J. Calcium-sensitive fluorescent dyes can report increases in intracellular free zinc concentration in cultured forebrain neurons. *J. Neurochem.* **1998**, *71*, 2401–2410.

7. Hahn, K. M.; Toutchkine, A.; Muthyala, R.; Kraynov, V.; Bark, S. J.; Burton, D. R.; Chamberlain, C. Labeled peptides, proteins and antibodies and processes and intermediates useful for their preparation. U.S. Pat. Appl. Publ. US 2002055133, 2002; *Chem. Abstr.* **2002**, *136*, 366139.
8. Freisinger, C. M.; Houston, D. W.; Slusarski, D. C. Image analysis of calcium release dynamics. *Methods Mol. Biol.* **2008**, *468*, 145–156.
9. McAinsh, M. R.; Ng, C. K. Y. Measurement of cytosolic-free Ca^{2+} in plant tissue. *Methods Mol. Biol.* **2005**, *312*, 289–302.
10. Hirst, R. A.; Harrison, C.; Hirota, K.; Lambert, D. G. Measurement of $[Ca^{2+}]i$ in whole cell suspensions using fura-2. *Methods Mol. Biol.* **2005**, *312*, 37–45.
11. Robinson, J. A.; Jenkins, N. S.; Holman, N. A.; Roberts-Thomson, S. J.; Monteith, G. R. Ratiometric and nonratiometric Ca^{2+} indicators for the assessment of intracellular free Ca^{2+} in a breast cancer cell line using a fluorescence microplate reader. *J. Biochem. Biophys. Methods* **2004**, *58*, 227–237.
12. Marrelli, S. P. Selective measurement of endothelial or smooth muscle $[Ca^{2+}]_i$ in pressurized/perfused cerebral arteries with fura-2. *J. Neurosci. Methods* **2000**, *97*, 145–155.
13. Lansley, A. B.; Sanderson, M. J. Regulation of airway ciliary activity by Ca^{2+}: simultaneous measurement of beat frequency and intracellular Ca^{2+}. *Biophys. J.* **1999**, *77*, 629–638.
14. Partridge, L. D.; Zeilhofer, H. U.; Swandulla, D. Combined whole-cell and single-channel current measurement with quantitative Ca^{2+} injection or Fura-2 measurement of Ca^{2+}. *Methods Enzymol.* **1998**, *293*, 371–383.
15. Gasalla-Herraiz, J.; Rhee, S.; Isales, C. M. Calcium-sensitive probes for the measurement of intracellular calcium: effects of buffer system and magnesium concentration. *Biochem. Biophys. Res. Commun.* **1995**, *214*, 373–388.
16. Haase, H.; Hebel, S.; Engelhardt, G.; Rink, L. Zinc ions cause the thimerosal-induced signal of fluorescent calcium probes in lymphocytes. *Cell Calcium* **2009**, *45*, 185–191.
17. Martin, J. L.; Stork, C. J.; Li, Y. V. Determining zinc with commonly used calcium and zinc fluorescent indicators, a question on calcium signals. *Cell Calcium* **2006**, *40*, 393–402.
18. Thompson, R. B.; Peterson, D.; Mahoney, W.; Cramer, M.; Maliwal, B. P.; Suh, S. W.; Frederickson, C.; Fierke, C.; Herman, P. Fluorescent zinc indicators for neurobiology. *J. Neurosci. Methods* **2002**, *118*, 63–75.
19. Li, X.; Xu, H.; Zhang, L.; Liu, H.; Wu, K. Identification of bitter ligands that activate human bitter receptors T2R1 & T2R67 and related assays, for modulating bitter taste of hydrolyzed soy protein. U.S. Pat. Appl. Publ. US 2009074894, 2009; *Chem. Abstr.* **2009**, *150*, 348870.
20. Servant, G.; Kamdar, P.; Rivadeneyra, A. Novel cell-based assays for identifying enhancers or inhibitors of T1R taste receptors (T1R2/T1R3 sweet) and umami (T1R1/T1R3 umami) taste receptors. U.S. Pat. Appl. Publ. US 2008305500, 2008; *Chem. Abstr.* **2008**, *150*, 31118.
21. Klaubert, D.; Diwu, Z.; Yi, G.; Kirk, M. Improved method for measuring membrane potential. PCT Int. Appl. WO 2003014701, 2003; *Chem. Abstr.* **2003**, *138*, 166188.
22. Nedergaard, M.; Tian, G. F. Method for treatment and prevention of epilepsy. PCT Int. Appl. WO 2006062683, 2006; *Chem. Abstr.* **2006**, *145*, 55958.
23. Haydon, P. G.; Halassa, M. M.; Fellin, T.; Ding, S.; Zhu, Y. Methods for treating neurological and psychiatric conditions, and test compound screening methods. PCT Int. Appl. WO 2007002285, 2007; *Chem. Abstr.* **2007**, *146*, 93591.

FURA 2 AM

CAS Registry Number 108964-32-5

Chemical Structure

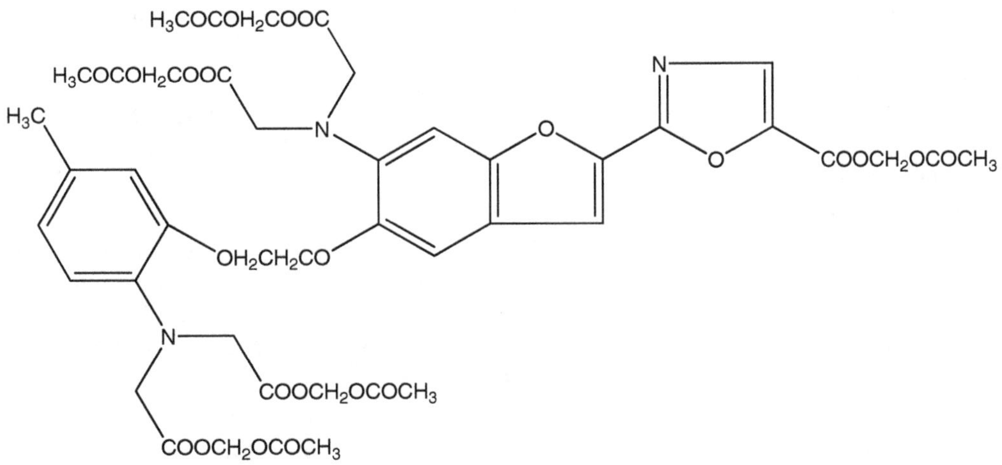

CA Index Name 5-Oxazolecarboxylic acid, 2-[6-[bis[2-[(acetyloxy)methoxy]-2-oxoethyl]amino]-5-[2-[2-[bis[2-[(acetyloxy)methoxy]-2-oxoethyl]amino]-5-methylphenoxy]ethoxy]-2-benzofuranyl]-, (acetyloxy)methyl ester

Other Names Fura 2 acetoxymethyl ester; Fura 2AM; Fura 2 AM ester

Merck Index Number Not listed

Chemical/Dye Class Benzofuran

Molecular Formula $C_{44}H_{47}N_3O_{24}$

Molecular Weight 1001.85

Physical Form Yellow powder

Solubility Insoluble in water; soluble in dimethyl sulfoxide, ethyl acetate

Melting Point >250°C

Boiling Point (Calcd.) 975.9 ± 75.0°C, pressure: 760 Torr

pK_a (Calcd.) 1.91 ± 0.50, most basic, temperature: 25°C

Absorption (λ_{max}) 370 nm

Emission (λ_{max}) 476 nm

Synthesis Synthetic method[1]

Staining Applications Calcium ions;[7–15] cells;[2,3] leukocyte tumor cells;[4] neurons;[5,6] sperms[7]

Biological Applications Calcium indicator;[7–15] monitoring membrane potential;[16] treating ischemia[17]

Industrial Applications Not reported

Safety/Toxicity No data available

REFERENCES

1. Grynkiewicz, G.; Poenie, M.; Tsien, R. Y. A new generation of calcium indicators with greatly improved fluorescence properties. *J. Biol. Chem.* **1985**, *260*, 3440–3450.
2. O'Riordan, T. C.; Fitzgerald, K.; Ponomarev, G. V.; Mackrill, J.; Hynes, J.; Taylor, C.; Papkovsky, D. B. Sensing intracellular oxygen using near-infrared phosphorescent probes and live-cell fluorescence imaging. *Am. J. Physiol.* **2007**, *292*, R1613–R1620.
3. Okun, I.; Okun, A. Apparatus and method for real-time measurement of cellular response. U.S. Patent 5919646, 1999; *Chem. Abstr.* **1999**, *131*, 56118.
4. Minakami, T.; Tsuji, T.; Oguni, S.; Hamaguchi, Y.; Tsuruta, K.; Kamihiro, T. Method for detecting leukocyte tumor cells. Jpn. Kokai Tokkyo Koho JP 2002207036, 2002; *Chem. Abstr.* **2002**, *137*, 106038.
5. Agasse, F.; Bernardino, L. I.; Silva, B. A. C.; Malva, J. J. O. Method for the functional identification of new neurons, neural progenitors, astrocytes and immature cells from stem cell cultures, and uses thereof. PCT Int. Appl. WO 2008100168, 2008; *Chem. Abstr.* **2008**, *149*, 284871.

6. Garaschuk, O.; Milos, R. I.; Konnerth, A. Targeted bulk-loading of fluorescent indicators for two-photon brain imaging *in vivo*. *Nat. Protoc.* **2006**, *1*, 380–386.
7. Micera, E.; Albrizio, M.; Surdo, N. C.; Zarrilli, A. A semi-immobilizing system associated with microspectrofluorimetric and videoimaging analysis for intracytoplasmic calcium measurement in individual viable spermatozoa. *J. Biotechnol.* **2008**, *133*, 90–95.
8. Rehberg, M.; Lepier, A.; Solchenberger, B.; Osten, P.; Blum, R. A new non-disruptive strategy to target calcium indicator dyes to the endoplasmic reticulum. *Cell Calcium* **2008**, *44*, 386–399.
9. Hirst, R. A.; Harrison, C.; Hirota, K.; Lambert, D. G. Measurement of $[Ca^{2+}]i$ in whole cell suspensions using fura-2. *Methods Mol. Biol.* **2005**, *312*, 37–45.
10. Quintana, A.; Hoth, M. Apparent cytosolic calcium gradients in T-lymphocytes due to fura-2 accumulation in mitochondria. *Cell Calcium* **2004**, *36*, 99–109.
11. Terada, S.; Muraoka, I.; Tabata, I. Changes in $[Ca^{2+}]_i$ induced by several glucose transport-enhancing stimuli in rat epitrochlearis muscle. *J. Appl. Physiol.* **2003**, *94*, 1813–1820.
12. Ricken, S.; Leipziger, J.; Greger, R.; Nitschke, R. Simultaneous measurements of cytosolic and mitochondrial Ca^{2+} transients in HT29 cells. *J. Biol. Chem.* **1998**, *273*, 34961–34969.
13. Xu, Y. J.; Shao, Q.; Dhalla, N. S. Fura-2 fluorescent technique for the assessment of Ca^{2+} homeostasis in cardiomyocytes. *Mol. Cell. Biochem.* **1997**, *172*, 149–157.
14. Zhao, M.; Hollingworth, S.; Baylor, S. M. AM-loading of fluorescent Ca^{2+} indicators into intact single fibers of frog muscle. *Biophys. J.* **1997**, *72*, 2736–2747.
15. Brooke, S. M.; Trafton, J. A.; Sapolsky, R. M. Autofluorescence as a confound in the determination of calcium levels in hippocampal slices using fura-2AM dye. *Brain Res.* **1996**, *706*, 283–288.
16. Koyano, T. Fluorescent method for monitoring change of membrane potential and intracellular ions. Jpn. Kokai Tokkyo Koho JP 09021800, 1997; *Chem. Abstr.* **1997**, *126*, 183503.
17. Simon, R. P.; Xiong, Z. G. Acid-sensing ion channel 1a (ASIC1a) inhibitors for the treatment of ischemia, and anti-ischemic drug screening methods. PCT Int. Appl. WO 2006034035, 2006; *Chem. Abstr.* **2006**, *144*, 324831.

FURA 2FF

CAS Registry Number 192140-58-2

Chemical Structure

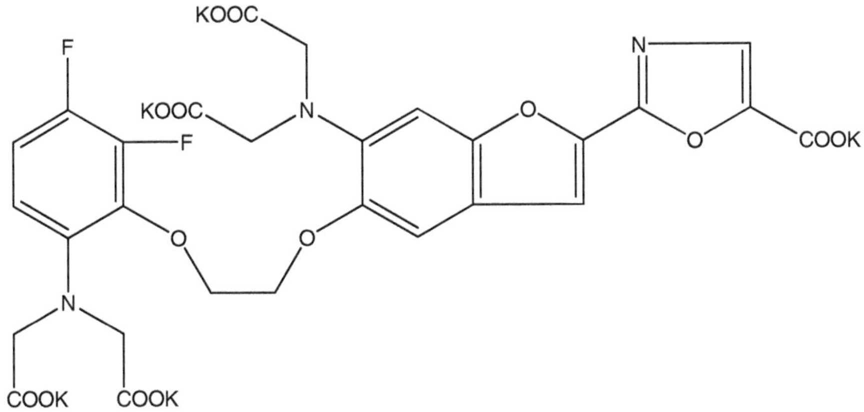

CA Index Name 5-Oxazolecarboxylic acid, 2-[6-[bis(carboxymethyl)amino]-5-[2-[6-[bis(carboxymethyl)amino]-2,3-difluorophenoxy]ethoxy]-2-benzofuranyl]-, potassium salt (1:5)

Other Names 5-Oxazolecarboxylic acid, 2-[6-[bis(carboxymethyl)amino]-5-[2-[6-[bis(carboxymethyl)amino]-2,3-difluorophenoxy]ethoxy]-2-benzofuranyl]-, pentapotassium salt; Fura 2FF; Fura FF pentapotassium salt

Merck Index Number Not listed

Chemical/Dye Class Benzofuran

Molecular Formula $C_{28}H_{18}F_2K_5N_3O_{14}$

Molecular Weight 853.94

Physical Form Solid

Solubility Soluble in water

Melting Point >250°C

Absorption (λ_{max}) 364 nm, 335 nm

Emission (λ_{max}) 510 nm, 506 nm

Synthesis Synthetic methods[1,2]

Staining Applications Calcium ions;[2,4–9] proteins[1,3]

Biological Applications Calcium indicator;[2,4–9] treating pain[9]

Industrial Applications Not reported

Safety/Toxicity No data available

REFERENCES

1. Gee, K. R.; Hart, C. R.; Haugland, R.; Patton, W. F.; Whitney, S. Site-specific labeling of affinity tags in fusion proteins. U.S. Pat. Appl. Publ. US 2006141554, 2006; *Chem. Abstr.* **2006**, *145*, 120046.
2. London, R. E.; Rhee, C. K.; Murphy, E.; Gabel, S.; Levy, L. A. NMR-sensitive fluorinated and fluorescent intracellular calcium ion indicators with high dissociation constants. *Am. J. Physiol.* **1994**, *266*, C1313–C1322.
3. Diwu, Z.; Gee, K.; Hart, C.; Haugland, R.; Leung, W. Y.; Patton, W.; Rukavishnikov, A. Site-specific fluorescent labeling of affinity tags in fusion proteins. PCT Int. Appl. WO 2004025259, 2004; *Chem. Abstr.* **2004**, *140*, 300064.
4. Yoshida, Y.; Tsuchiya, R.; Matsumoto, N.; Morita, M.; Miyakawa, H.; Kudo, Y. Ca^{2+}-dependent induction of intracellular Ca^{2+} oscillation in hippocampal astrocytes during metabotropic glutamate receptor activation. *J. Pharmacol. Sci.* **2005**, *97*, 212–218.
5. Devinney, M. J.; Reynolds, I. J.; Dineley, K. E. Simultaneous detection of intracellular free calcium and zinc using fura-2FF and FluoZin-3. *Cell Calcium* **2005**, *37*, 225–232.
6. Montero, M.; Alonso, M. T.; Albillos, A.; Cuchillo-Ibanez, I.; Olivares, R.; Garcia, A.; Garcia-Sancho, J.; Alvarez, J. Control of secretion by mitochondria depends on the size of the local [Ca^{2+}] after chromaffin cell stimulation. *Eur. J. Neurosci.* **2001**, *13*, 2247–2254.
7. Hyrc, K. L.; Bownik, J. M.; Goldberg, M. P. Ionic selectivity of low-affinity ratiometric calcium indicators: mag-Fura-2, Fura-2FF and BTC. *Cell Calcium* **2000**, *27*, 75–86.

8. Zhao, M.; Hollingworth, S.; Baylor, S. M. AM-loading of fluorescent Ca^{2+} indicators into intact single fibers of frog muscle. *Biophys. J.* **1997**, *72*, 2736–2747.
9. Benjamin, E. R.; Chen, Z.; Sha, D.; Tafesse, L.; Victory, S. F.; Whitehead, J. W. F.; Zhou, X. Piperidinyl compounds and their preparation, pharmaceutical compositions, and use as *N*-type calcium channel modulators or blockers for treatment or prevention of pain. PCT Int. Appl. WO 2006040181, 2006; *Chem. Abstr.* **2006**, *144*, 412381.

GIEMSA STAIN

CAS Registry Number 51811-82-6

Chemical Structure

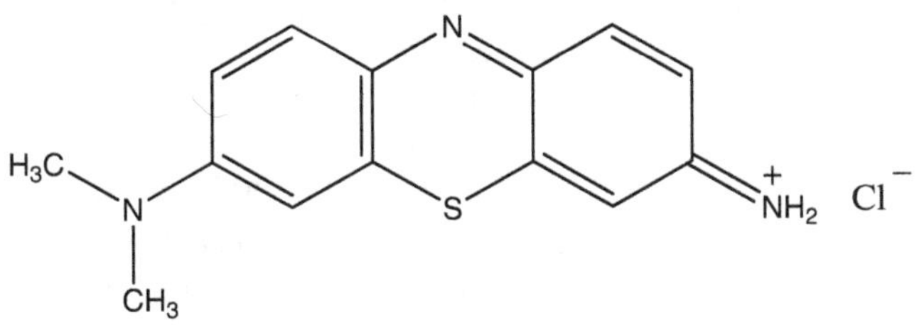

CA Index Name Giemsa's stain

Other Names Azure Mixture sicc.; 3-(Dimethylamino)-7-aminophenothiazin-5-ium chloride; Giemsa; Giemsa stain

Merck Index Number Not listed

Chemical/Dye Class Phenothiazine

Molecular Formula $C_{14}H_{14}ClN_3S$

Molecular Weight 291.80

Physical Form Dark green to black crystals or powder

Solubility Soluble in water; ethanol, methanol

Melting Point 300 °C

Absorption (λ_{max}) 521 nm, 648 nm

Synthesis Synthetic methods[1–6]

Staining Applications Blood smears;[1–8] bone marrow cells;[9,10] adrenaline- and noradrenaline-containing cells;[11] cancer cells;[12] cells;[13–15] chromosomes;[16–22] DNA;[23–25] embryos;[26] germ cells;[27] lipids;[28] nucleus;[29] nucleous;[30] oral lesions;[31] oral mucosa;[32] parasites;[1–6,8,33–35] sperms;[36] spinal fluid;[37] tissues[38]

Biological Applications Diagnosis of fetal aneuploidies,[39] cancer cells,[12] herpes simplex virus-1 keratitis,[40] malaria,[1–6,8,33–35] spermatogenesis disorders;[27] detection of INK4 gene and cell proliferation markers,[41] lymphocyte subgroups,[42] eperythrozoon,[43] mouse bone marrow micronucleus;[44] treating gastritis,[45] lupus,[9] microorganisms in the oral cavity,[46,47] pathogens[48]

Industrial Applications Not reported

Safety/Toxicity Genotoxicity;[49] immunotoxicity[50,51]

Certification/Approval Certified by Biological Stain Commission (BSC)

REFERENCES

1. Giemsa, G. Farbemethoden fur malariaparasiten. *Zbl. Bakteriol.* **1902**, *31*, 429–430.
2. Giemsa, G. Farbemethoden fur malariaparasiten. *Zbl. Bakteriol.* **1902**, *32*, 307–313.
3. Giemsa, G. Eine vereinfachung und vervollkommung meiner methylenazur-methylenblau-eosinfarbung zur erzielung der Romanowsky–Nochtschen chromatinfarbung. *Zbl. Bakteriol.* **1904**, *37*, 308–311.
4. Green, F. J. *The Sigma–Aldrich Handbook of Stains, Dyes and Indicators*; Aldrich Chemical Company Inc.: Milwaukee, **1991**; pp 388–389.
5. Wittekind, D. H. On the nature of Romanowsky–Giemsa staining and its significance for cytochemistry and histochemistry: an overall view. *Histochem. J.* **1983**, *15*, 1029–1047.
6. Barcia, J. J. The Giemsa stain: its history and applications. *Int. J. Surg. Pathol.* **2007**, *15*, 292–296.
7. Zhou, H.; Huang, G.; Yan, X. Method for manufacturing blood smear leukocyte stain. Faming Zhuanli Shenqing Gongkai Shuomingshu CN 1924543, 2007; *Chem. Abstr.* **2007**, *146*, 354216.
8. Aubouy, A.; Carme, B. Plasmodium DNA contamination between blood smears during Giemsa staining and microscopic examination. *J. Infect. Dis.* **2004**, *190*, 1335–1337.
9. Kallick, C. A. Specific bacterial inclusions in bone marrow cells indicate systematic lupus erythematosus, and treatment for lupus. U.S. Pat. Appl. Publ. US 2007031820, 2007; *Chem. Abstr.* **2007**, *146*, 201609.
10. Naresh, K. N.; Lampert, I.; Hasserjian, R.; Lykidis, D.; Elderfield, K.; Horncastle, D.; Smith, N.; Murray-Brown, W.; Stamp, G. W. Optimal processing of bone marrow trephine biopsy: the Hammersmith protocol. *J. Clin. Pathol.* **2006**, *59*, 903–911.
11. Ortiz-Butron, R.; Pinon-Lopez, M. J.; Juarez-Santiago, L.; Racotta, R. An improved histochemical technique for differentiating

Handbook of Biological Dyes and Stains By R. W. Sabnis

adrenaline- and noradrenaline-containing cells. *Arch. Med. Res.* **1998**, *29*, 75–77.

12. Daniely, M.; Kaplan, T.; Kaplan, E.; Freiberger, A. Methods of detecting cancer cells in biological samples using double staining and dual imaging. U.S. Pat. Appl. Publ. US 2004197839, 2004; *Chem. Abstr.* **2004**, *141*, 310232.
13. Daniely, M.; Reichart, M.; Kaplan, E.; Zilberstein, Y. Kits and methods for preparing cell samples optimized for dual staining. PCT Int. Appl. WO 2003040064, 2003; *Chem. Abstr.* **2003**, *138*, 365157.
14. Fernandes, T. S.; Lloyd, D.; Amaral, A. A comparison of different cytological stains for biological dosimetry. *Int. J. Radiat. Biol.* **2008**, *84*, 703–711.
15. Garini, Y.; Mcnamara, G.; Soenksen, D. G.; Cabib, D.; Buckwald, R. A. *In situ* method of analyzing cells by staining with multiple stains and using a spectral data collection device. PCT Int. Appl. WO 2000031534, 2000; *Chem. Abstr.* **2000**, *133*, 14300.
16. Qin, H.; Zhu, X. F.; Shi, A. M. Effects of adefovir dipivoxil on lung cell chromosome aberration of *Cricetulus griseus*. *Jiangsu Yufang Yixue* **2006**, *17*, 66–67; *Chem. Abstr.* **2006**, *147*, 23294.
17. Swarca, A. C.; Fenocchio, A. S.; Cestari, M. M.; Dias, A. L. First chromosome data on *Steindachneridion scripta* (Pisces, Siluriformes, Pimelodidae) from Brazilian rivers: Giemsa, CBG, G-, and RE banding. *Genet. Mol. Res.* **2005**, *4*, 734–741.
18. Gotoh, E.; Tanno, Y. Simple biodosimetry method for cases of high-dose radiation exposure using the ratio of the longest/shortest length of Giemsa-stained drug-induced prematurely condensed chromosomes (PCC). *Int. J. Radiat. Biol.* **2005**, *81*, 379–385.
19. Volobouev, V. T.; Aniskin, V. M. Comparative chromosome banding analysis of three South American species of rice rats of the genus *Oryzomys (rodentia, sigmodontinae)*. *Chromosome Res.* **2000**, *8*, 295–304.
20. Stein, C. K. Modified Giemsa-11 staining protocol for chromosomes of human and hybrid cells. *Somatic Cell Mol. Genet.* **1998**, *24*, 191–195.
21. Mosesso, P.; Darroudi, F.; Van Den Berg, M.; Vermeulen, S.; Palitti, F.; Natarajan, A. T. Induction of chromosomal aberrations (unstable and stable) by inhibitors of topoisomerase II, m-AMSA and VP16, using conventional Giemsa staining and chromosome painting techniques. *Mutagenesis* **1998**, *13*, 39–43.
22. Ellard, S.; Toper, S.; Stemp, G.; Parry, E. M.; Wilcox, P.; Parry, J. M. A comparison of conventional metaphase analysis of Giemsa-stained chromosomes with multi-color fluorescence *in situ* hybridization. *Mutagenesis* **1996**, *11*, 537–546.
23. Motazedian, H.; Karamian, M.; Noyes, H. A.; Ardehali, S. DNA extraction and amplification of *Leishmania* from archived, Giemsa-stained slides, for the diagnosis of cutaneous leishmaniasis by PCR. *Ann. Trop. Med. Parasitol.* **2002**, *96*, 31–34.
24. Serth, J.; Kuczyk, M. A.; Paeslack, U.; Lichtinghagen, R.; Jonas, U. Quantitation of DNA extracted after micro-preparation of cells from frozen and formalin-fixed tissue sections. *Am. J. Pathol.* **2000**, *156*, 1189–1196.
25. Dutt, M. K. Basic dyes for the staining of DNA in mammalian tissues and absorption spectra of stained nuclei in visible light. *Microsc. Acta* **1982**, *86*, 59–68.
26. Tian, Y.; Shen, L.; Gao, Y.; Yamauchi, T.; Shen, X. M.; Ma, N. Comparison of 4′,6′-diamidino-2-phenylindole and Giemsa stainings in preimplantation mouse embryos micronucleus assay including a triple dose study. *Ind. Health* **2007**, *45*, 343–347.
27. Kurilo, L. F. Method for diagnosis of spermatogenesis disorders involving microscopic evaluation of germ cell cytological specimens treated with potassium chloride at temp. 37.0–37.1°, fixed by methanol–glacial acetate–chloroform mixture, and colored with Giemsa stain. Russ. RU 2328736, 2008; *Chem. Abstr.* **2008**, *149*, 102070.
28. Farrell, M.; Bieniarz, C.; Reinhardt, K.; Ward, G.; Kosmeder, J. W.; Ghusson, A.; Walk, E. E.; Manriquez, G. H.; Grogan, T. M. Biological sample processing composition and method for staining the sample and contacting the sample with a lipid compound composition. PCT Int. Appl. WO 2007084429, 2007; *Chem. Abstr.* **2007**, *147*, 184152.
29. Sumner, A. T. Some observations on the mechanisms of blocking of nuclear staining by cisplatin. *Histochem. J.* **1982**, *14*, 283–299.
30. Das, J. K.; Khuda-Bukhsh, A. R. Preponderance of GC-rich sites in silver-stained nucleolus organizing regions of *Rita rita* (Hamilton) and *Mystus gulio* (Hamilton) (Bagridae, Pisces), as revealed by chromomycin A3-staining technique and scanning electron microscopic studies. *Genet. Mol. Res.* **2007**, *6*, 184–191.

31. Hirshberg, A.; Yarom, N.; Amariglio, N.; Yahalom, R.; Adam, I.; Stanchescu, R.; Ben-Dov, I.; Taicher, S.; Rechavi, G.; Trakhtenbrot, L. Detection of non-diploid cells in premalig-nant and malignant oral lesions using combined morphological and FISH analysis—a new method for early detection of suspicious oral lesions. *Cancer Lett.* **2007**, *253*, 282–290.

32. Nersesyan, A.; Kundi, M.; Atefie, K.; Schulte-Hermann, R.; Knasmueller, S. Effect of staining procedures on the results of micronucleus assays with exfoliated oral mucosa cells. *Cancer Epidemiol. Biomarkers Prev.* **2006**, *15*, 1835–1840.

33. Sio, S. W. S.; Sun, W.; Kumar, S.; Bin, W. Z.; Tan, S. S.; Ong, S. H.; Kikuchi, H.; Oshima, Y.; Tan, K. S. W. MalariaCount: an image analysis-based program for the accurate determination of parasitemia. *J. Microbiol. Methods* **2007**, *68*, 11–18.

34. Xu, L.; Chaudhuri, A. *Plasmodium yoelii*: a differential fluorescent technique using acridine orange to identify infected erythrocytes and reticulocytes in Duffy knockout mouse. *Exp. Parasitol.* **2005**, *110*, 80–87.

35. Tham, J. M.; Lee, S. H.; Tan, T. M. C.; Ting, R. C. Y.; Kara, U. A. K. Detection and species determination of malaria parasites by PCR: comparison with microscopy and with ParaSight-F and ICT Malaria Pf tests in a clinical environment. *J. Clin. Microbiol.* **1999**, *37*, 1269–1273.

36. Santiani, A.; Huanca, W.; Sapana, R.; Huanca, T.; Sepulveda, N.; Sanchez, R. Effects on the quality of frozen-thawed alpaca (*Lama pacos*) semen using two different cryoprotectants and extenders. *Asian J. Androl.* **2005**, *7*, 303–309.

37. Masuoka, S.; Yoneda, K.; Oka, M. Apparatus and method for analyzing spinal fluid by staining and imaging analysis. Jpn. Kokai Tokkyo Koho JP 2004132787, 2004; *Chem. Abstr.* **2004**, *140*, 371466.

38. Gough, A. H.; Giuliano, K. A.; Taylor, D. L. Method for automated tissue analysis using immuno-histochemical staining in combination with fluorometry. PCT Int. Appl. WO 2007136724, 2007; *Chem. Abstr.* **2007**, *148*, 4477.

39. Parano, E.; Falcidia, E.; Grillo, A. A non invasive method for screening and diagnosis fetal aneuploidies. PCT Int. Appl. WO 2002101387, 2002; *Chem. Abstr.* **2002**, *138*, 21819.

40. Subhan, S.; Jose, R. J.; Duggirala, A.; Hari, R.; Krishna, P. V.; Reddy, S. B.; Sharma, S. Diagnosis of herpes simplex virus-1 keratitis: comparison of Giemsa stain, immunofluorescence assay and polymerase chain reaction. *Curr. Eye Res.* **2004**, *29*, 209–213.

41. Trunk-Gehmacher, M.; Reichert, A.; Batrla, R.; Ridder, R. Detection of INK4 gene and cell proliferation markers for improved diagnosis of dysplasias using cytological and histological methods. Eur. Pat. Appl. EP 1416278, 2004; *Chem. Abstr.* **2004**, *140*, 371467.

42. Yao, B. Method for detecting lymphocyte subgroups by staphylococcal protein a (spa)-coated erythrocyte rosette test using monoclonal antibody. Faming Zhuanli Shenqing Gongkai Shuomingshu CN 101109752, 2008; *Chem. Abstr.* **2008**, *148*, 138992.

43. Zheng, L.; Zhang, Y.; Chang, W. Comparison of staining method with Wright, Giemsa and acridine stains with PCR assay for the detection of *Eperythrozoon. Zhongguo Renshou Gonghuanbing Xuebao* **2006**, *22*, 243–245; *Chem. Abstr.* **2007**, *148*, 489889.

44. Kolanko, C. J.; Pyle, M. D.; Loats, H.; Parton, J.; Blakely, W. F.; Nath, J. Fast-*in situ* hybridization and immunoenzymic color pigment detection of mouse bone marrow micronucleus. *Biotech. Histochem.* **1999**, *74*, 111–115.

45. Shi, T.; Gao, F. Application of proanthocyanidins in preparing medical preparations or health food for treating *Helicobacter pylori*-associated gastritis. Faming Zhuanli Shenqing Gongkai Shuomingshu CN 101239058, 2008; *Chem. Abstr.* **2008**, *149*, 293664.

46. Kennedy, J. Treatment of microorganisms in the oral cavity by photodynamic therapy. Can. Pat. Appl. CA 2486475, 2006; *Chem. Abstr.* **2006**, *144*, 440202.

47. Kennedy, J. Topical photosensitizers for treating microorganisms in the oral cavity by photodynamic therapy. U.S. Pat. Appl. Publ. US 2006093561, 2006; *Chem. Abstr.* **2006**, *144*, 419749.

48. Lurie, R. Phototherapy-based method and composition for treating pathogens. PCT Int. Appl. WO 9904628, 1999; *Chem. Abstr.* **1999**, *130*, 150435.

49. Rozgaj, R.; Kasuba, V.; Fucic, A. Genotoxicity of cadmium chloride in human lymphocytes evaluated by the comet assay and cytogenetic tests. *J. Trace Elem. Med. Biol.* **2002**, *16*, 187–192.

50. Mao, Z.; Ma, Q.; Yuan, X.; Han, J. Effects of Giemsa on the immunological function of peritoneal

macrophage in mice. *Xinxiang Yixueyuan Xuebao* **2006**, *23*, 231–232; *Chem. Abstr.* **2007**, *147*, 421030.

51. Suzuki, M.; Katsuyama, K.; Adachi, K.; Ogawa, Y.; Yorozu, K.; Fujii, E.; Misawa, Y.; Sugimoto, T. Combination of fixation using PLP fixative and embedding in paraffin by the AMeX method is useful for histochemical studies in assessment of immunotoxicity. *J. Toxicol. Sci.* **2002**, *27*, 165–172.

HEMATOXYLIN

CAS Registry Number 517-28-2

Chemical Structure

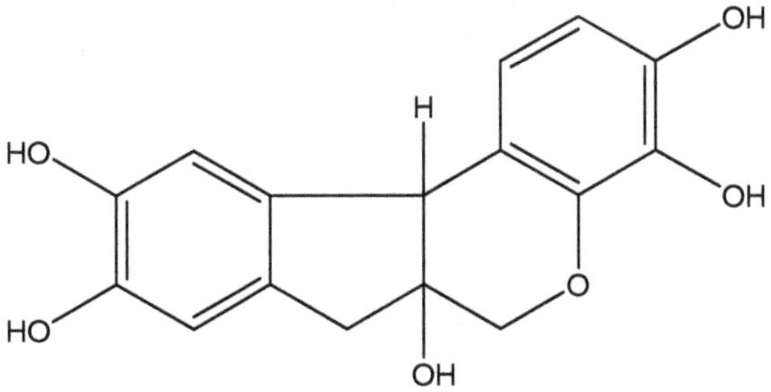

CA Index Name Benz[*b*]indeno[1,2-*d*]pyran-3,4,6a,9,10(6*H*)-pentol, 7,11b-dihydro, (6aS,11bR)-

Other Names *cis*-(+)-7,11b-Dihydrobenz[*b*]indeno[1,2-*d*]pyran-3,4,6a,9,10(6*H*)-pentol; Benz[*b*]indeno[1,2-*d*]pyran-3,4,6a,9,10(6*H*)-pentol, 7,11b-dihydro; Benz[*b*]indeno[1,2-*d*]pyran-3,4,6a,9,10(6*H*)-pentol, C.I. 75290; C.I. Natural Black 1; 7,11b-Dihydro, (6aS-*cis*)-; Hematoxiline; (+)-Hematoxylin; Haematoxylin; Hematoxylin; Hematoxyline; Hydroxybrasilin; Hydroxybrazilin; NSC 270085

Merck Index Number 4637

Chemical/Dye Class Flavone

Molecular Formula $C_{16}H_{14}O_6$

Molecular Weight 302.28

Physical Form White to yellowish crystals turns red on exposure to light

Solubility Soluble in water, ethanol, ethylene glycol, methyl cellosolve

Melting Point 140°C; 200°C (decompose)

Boiling Point (Calcd.) 579.9 ± 50.0°C, pressure: 760 Torr

pH Range 0.0–1.0; 5.0–6.0

Color Change at pH Red (0.0) to yellow (1.0); pale yellow (5.0) to violet (6.0)

Absorption (λ_{max}) 292 nm

Synthesis Synthetic methods[1–17]

Staining Applications Aluminum;[34] antigen;[1,18] blood smears;[19] cells;[20,21] collagen;[22] epithelial cells;[23] eye lens;[24] fish;[25] genes;[32] gluten structure;[26] horny cells;[27] lipid;[26,28] liver tissues;[29] malignant melanoma;[30] myocardial biopsies;[31] neurons;[32] nucleic acids;[32] nucleus;[33] oil droplets;[26] proteins;[26,35] starch granules;[26] tissues;[36] hairs;[1,37,38] keratin fibers[39]

Biological Applications Detecting breast cancer,[1,40] collagen,[1,41] genes,[1,42] microorganism;[43] treating age-related macular degeneration,[1,44] burns,[1,45] cancer,[1,46] diabetes,[1,47] obesity,[1,47] gastroesophageal reflux disease,[48] peripheral neural and vascular ailments,[1,49] prostate cancers,[1,50] skin disorders,[1,51] viral diseases[1,52]

Industrial Applications Plasma display panel;[1,53] textiles[1,54]

Safety/Toxicity Carcinogenicity;[1,55,56] cytotoxicity;[57] genotoxicity;[1,58] mutagenicity;[1,59] neurotoxicity;[1,60,61] pulmonary toxicity[62]

Certification/Approval Certified by Biological Stain Commission (BSC)

REFERENCES

1. Sabnis, R. W. *Handbook of Acid–Base Indicators*; CRC Press: Boca Raton, **2008**; pp 180–182.
2. Bosma, R. B.; Elias, E. A. Environmentally friendly mercury-free hematoxylin. *J. Histotechnol.* **1993**, *16*, 371–374.
3. Morsingh, F.; Robinson, R. Syntheses of brazilin and hematoxylin. *Tetrahedron* **1970**, *26*, 281–289.
4. Hahn, F. L. Vegetable coloring materials of Guatamala. *Ciencia* **1945**, *6*, 385–391; *Chem. Abstr.* **1946**, *40*, 35369.
5. Drew, R. L. Hematoxylin. U.S. Patent 2369343, 1945; *Chem. Abstr.* **1945**, *39*, 22578.
6. Drew, R. L. Extraction of hematoxylin. U.S. Patent 2360114, 1944; *Chem. Abstr.* **1945**, *39*, 20124.
7. Pfeiffer, P.; Epler, H. Brazilin and hematoxylin. XVII. Synthesis of μ-aminochromindan. *Justus Liebigs Ann. Chem.* **1940**, *545*, 263–286; *Chem. Abstr.* **1941**, *35*, 2676.

Handbook of Biological Dyes and Stains By R. W. Sabnis

8. Appel, V. H.; Baker, W.; Hagenbach, H.; Robinson, R. Synthesis of brazilin and hematoxylin. *J. Chem. Soc.* **1937**, 738–744.
9. Perkin, W. H., Jr.; Pollard, A.; Robinson, R. Synthesis of brazilin and of hematoxylin. IV. Synthesis of *O*-diethylenehematoxylone. *J. Chem. Soc.* **1937**, 49–53.
10. Conn, J. The history of staining. Logwood dyes. II. *Stain Technol.* **1929**, *4*, 37–48.
11. Perkin, W. H., Jr.; Ray, J. N.; Robinson, R. Synthesis of brazilin and hematoxylin and their derivatives. III. *J. Chem. Soc.* **1928**, 1504–1513.
12. Perkin, W. H., Jr.; Ray, J. N.; Robinson, R. Synthesis of brazilin and hematoxylin and their derivatives. II. Synthesis of deoxytrimethylbrazilone and of isobrazilein ferrichloride trimethyl ether. *J. Chem. Soc.* **1927**, 2094–2100.
13. Perkin, W. H., Jr.; Ray, J. N.; Robinson, R. Synthesis of brazilin and hematoxylin and their derivatives. I. Veratrylidene-7-methoxychromanone and an account of a new synthesis of some benzopyrylium salts. *J. Chem. Soc.* **1926**, 941–953.
14. Drew, A. H. The laboratory preparation of a purified hematoxylin. *Br. J. Exp. Pathol.* **1922**, *3*, 307.
15. Houseman, P. A.; Swift, C. K. The preparation of hematoxylin. *J. Ind. Eng. Chem.* **1920**, *12*, 173–174.
16. Perkin, W. H., Jr.; Robinson, R. Experiments on the synthesis of brasilin and hematoxylin and their derivatives: preliminary note. *Proc. Chem. Soc.* **1913**, *28*, 7–8.
17. Perkin, W. H., Jr.; Robinson, R. Brazilin and haematoxylin. Part VII. Synthesis of derivatives of hydrindene closely allied to brazilin and haematoxylin. *J. Chem. Soc., Trans.* **1907**, *91*, 1073–1103.
18. Hasui, K. Method for detecting antigen by immunohistochemical staining. Jpn. Kokai Tokkyo Koho JP 2005017133, 2005; *Chem. Abstr.* **2005**, *142*, 130326.
19. Trukhachev, V. I.; Rodin, V. V.; Mikhailenko, V. V.; Dergunov, A. A. Staining of blood smears by using the Erlich's hematoxylin and sodium eosin. Russ. RU 2304776, 2007; *Chem. Abstr.* **2007**, *147*, 253438.
20. Vega, C. J. Laser microdissection sample preparation for RNA analyses. *Methods Mol. Biol.* **2008**, *414*, 241–252.
21. Gelvan, D. J.; Goltsman, L.; Chausovsky, A. Staining methods and compositions for identifying a cell phenotype. PCT Int. Appl. WO 2007102146, 2007; *Chem. Abstr.* **2007**, *147*, 339010.
22. Borges, L. F.; Gutierrez, P. S.; Marana, H. R. C.; Taboga, S. R. Picrosirius-polarization staining method as an efficient histopathological tool for collagenolysis detection in vesical prolapse lesions. *Micron* **2007**, *38*, 580–583.
23. Jensen, C.; French, C. E. V.; McGlashan, S. R.; Elliott, D. A.; Vintiner, S. K. Method of differentiating between epithelial cells by fixation in methanol and staining using the Dane's technique. PCT Int. Appl. WO 2007018438, 2007; *Chem. Abstr.* **2007**, *146*, 247078.
24. Korsakova, N. V.; Sergeeva, V. E. Cytochemical analysis of cellular structures in eye lens of laboratory animals including vital–supravital staining by methylene blue. Russ. RU 2319132, 2008; *Chem. Abstr.* **2008**, *148*, 302857.
25. Moen, V.; Nordtug, T.; Zahlsen, K. Mixture and procedure for marking fish. Norw. NO 317169, 2004; *Chem. Abstr.* **2007**, *146*, 141497.
26. Ashida, Y. Examination of food containing wheat flour and fats/oils by double staining and staining solution kits thereof. Jpn. Kokai Tokkyo Koho JP 2008122401, 2008; *Chem. Abstr.* **2008**, *148*, 584596.
27. Yamazaki, K.; Fujiwara, N.; Seto, T.; Yang, P. Horny cell-staining liquid, and horny cell-staining method. PCT Int. Appl. WO 2007029437, 2007; *Chem. Abstr.* **2007**, *146*, 312248.
28. Farrell, M.; Bieniarz, C.; Reinhardt, K.; Ward, G.; Kosmeder, J. W.; Ghusson, A.; Walk, E. E.; Manriquez, G. H.; Grogan, T. M. Biological sample processing composition and method for staining the sample and contacting the sample with a lipid compound composition. PCT Int. Appl. WO 2007084429, 2007; *Chem. Abstr.* **2007**, *147*, 184152.
29. Galy, O.; Petit, M. A.; Benjelloun, S.; Chevallier, P.; Chevallier, M.; Srivatanakul, P.; Karalak, A.; Carreira, C.; Lyandrat, N.; Essaid, A.; Trepo, C.; Hainaut, P.; Chemin, I. Efficient hepatitis C antigen immunohistological staining in sections of normal, cirrhotic and tumoral liver using a new monoclonal antibody directed against serum-derived HCV E2 glycoproteins. *Cancer Lett.* **2007**, *248*, 81–88.
30. Kimyai-Asadi, A.; Ayala, G. B.; Goldberg, L. H.; Vujevich, J.; Jih, M. H. The 20-minute rapid MART-1 immunostain for malignant melanoma frozen sections. *Dermatol. Surg.* **2008**, *34*, 498–500.
31. Shipulin, V. M.; Kazakov, V. A.; Sukhodolo, I. V.; Krivoshchekov, E. V.; Lezhnev, A. A.; Vaizov, V. K.; Stasev, A. N.; Kozhevnikov, M. L. Prognosis of post-surgical cardiac remodeling in patients with ischemic cardiomyopathy based on histochemical examination. Russ. RU 2310372, 2007; *Chem. Abstr.* **2007**, *147*, 539653.
32. Ginsberg, S. D.; Che, S. Combined histochemical staining, RNA amplification, regional, and single cell

cDNA analysis with the hippocampus. *Lab. Invest.* **2004**, *84*, 952–962.

33. Foster, P. Method for improving the shelf-life of hematoxylin staining solutions. U.S. Pat. Appl. Publ. US 2008139827, 2008; *Chem. Abstr.* **2008**, *149*, 48203.
34. Walton, J. R. A bright field/fluorescent stain for aluminum: its specificity, validation, and staining characteristics. *Biotech. Histochem.* **2004**, *79*, 169–176.
35. Becker, K. F.; Schott, C.; Becker, I.; Hoefler, H. Guided protein extraction from formalin-fixed tissues for quantitative multiplex analysis avoids detrimental effects of histological stains. *Proteomics* **2008**, *2*, 737–743.
36. van der Loos, C. M. Multiple immunoenzyme staining: methods and visualizations for the observation with spectral imaging. *J. Histochem. Cytochem.* **2008**, *56*, 313–328.
37. Hercouet, L. Hair dye composition comprising a derivative of diamino-*N,N*-dihydropyrazolone, a coupler, and a heterocyclic direct dye. Fr. Demande FR 2886132, 2006; *Chem. Abstr.* **2006**, *146*, 49738.
38. Machida, G. Natural products for hair dyes. Jpn. Kokai Tokkyo Koho JP 2005060293, 2005; *Chem. Abstr.* **2005**, *142*, 266272.
39. Kainz, S.; Kruegermann, I.; Hoeffkes, H. Oxidation colorant for coloring keratin-containing fibers. Ger. Offen. DE 102005062830, 2007; *Chem. Abstr.* **2007**, *146*, 106792.
40. Abdul-Rasool, S.; Kidson, S. H.; Panieri, E.; Dent, D.; Pillay, K.; Hanekom, G. S. An evaluation of molecular markers for improved detection of breast cancer metastases in sentinel nodes. *J. Clin. Pathol.* **2006**, *59*, 289–297.
41. D'Andrea, M. R. Method for immunohistochemical detection of collagen in a tissue sample. U.S. Pat. Appl. Publ. US 2005227297, 2005; *Chem. Abstr.* **2005**, *143*, 362848.
42. Hoon, D. S. B. Methods for detecting genes associated with melanoma or carcinoma in body fluids and monitoring efficacy of therapy using quantitative RT-PCR. PCT Int. Appl. WO 2006031843, 2006; *Chem. Abstr.* **2006**, *144*, 310063.
43. Martin, S. M.; MacDonald, J. G.; Lye, J.; Sayre, C.; Thompson, K. Array for rapid detection of a microorganism based on dyes. PCT Int. Appl. WO 2008026104, 2008; *Chem. Abstr.* **2008**, *148*, 302867.
44. Hofmann, R. F. Use of targeted oxidative therapeutic formulation in treatment of age-related macular degeneration. U.S. Pat. Appl. Publ. US 2005250756, 2005; *Chem. Abstr.* **2005**, *143*, 416314.
45. Hofmann, R. F. Use of targeted oxidative therapeutic formulation in treatment of burns. PCT Int. Appl. WO 2006002302, 2006; *Chem. Abstr.* **2006**, *144*, 114433.
46. Hofmann, R. F. Use of targeted oxidative therapeutic formulation in treatment of cancer. U.S. Pat. Appl. Publ. US 2005250757, 2005; *Chem. Abstr.* **2005**, *143*, 416223.
47. Hofmann, R. F. Use of targeted oxidative therapeutic formulation in treatment of diabetes and obesity. PCT Int. Appl. WO 2005107728, 2005; *Chem. Abstr.* **2005**, *143*, 466228.
48. Chandrasoma, P. Methodologies for the diagnosis and treatment of gastroesophageal reflux disease. U.S. Pat. Appl. Publ. US 2007178510, 2007; *Chem. Abstr.* **2007**, *147*, 184153.
49. Rosenbloom, R. A. Methods using antioxidant flavonoid compounds for the treatment of peripheral neural and vascular ailments. U.S. Pat. Appl. Publ. US 2004087516, 2004; *Chem. Abstr.* **2004**, *140*, 386058.
50. Datta, M. W.; Datta, S.; Altaba, A. R. I. Methods for diagnosis and treatment of prostate and other cancers by inhibition of proteoglycans perlecan and patched and smoothened receptors. PCT Int. Appl. WO 2006007450, 2006; *Chem. Abstr.* **2006**, *144*, 143034.
51. Rosenbloom, R. A. Flavonoid compositions for the treatment of skin disorders. U.S. Pat. Appl. Publ. US 2003105031, 2003; *Chem. Abstr.* **2003**, *139*, 12323.
52. Hofmann, R. F. Targeted oxidative therapeutic formulation in treatment of viral diseases. U.S. Pat. Appl. Publ. US 2005192267, 2005; *Chem. Abstr.* **2005**, *143*, 272504.
53. Yamaguchi, N.; Suzuki, M. Plasma display panel, glass substrates with dielectric layer containing color material, and its manufacture. Jpn. Kokai Tokkyo Koho JP 2005327542, 2005; *Chem. Abstr.* **2005**, *143*, 485917.
54. Park, J. H.; Gatewood, B. M.; Ramaswamy, G. N. Naturally occurring quinones and flavonoid dyes for wool: insect feeding deterrents. *J. Appl. Polym. Sci.* **2005**, *98*, 322–328.
55. Oliveira, P. A.; Colaco, A. A.; Palmeira, C. A.; De la Cruz, P. L. F.; Lope, C. A. A phyllodes tumor of the urinary bladder in a rat. *Exp. Oncol.* **2007**, *29*, 159–161.
56. Ashby, J.; Paton, D. The Influence of chemical structure on the extent and sites of carcinogenesis for 522 rodent carcinogens and 55 different human carcinogen exposures. *Mutat. Res.* **1993**, *286*, 3–74.
57. Huang, R.; Southall, N.; Cho, M. H.; Xia, M.; Inglese, J.; Austin, C. P. Characterization of diversity in toxicity

mechanism using *in vitro* cytotoxicity assays in quantitative high throughput screening. *Chem. Res. Toxicol.* **2008**, *21*, 659–667.

58. Kirkland, D.; Aardema, M.; Henderson, L.; Mueller, L. Evaluation of the ability of a battery of three *in vitro* genotoxicity tests to discriminate rodent carcinogens and non-carcinogens. I. Sensitivity, specificity and relative predictivity. *Mutat. Res.* **2005**, *584*, 1–256.

59. Serafimova, R.; Todorov, M.; Pavlov, T.; Kotov, S.; Jacob, E.; Aptula, A.; Mekenyan, O. Identification of the structural require-ments for mutagenicity, by incorporating molecular flexibility and metabolic activation of chemicals. II. General Ames mutagenicity model. *Chem. Res. Toxicol.* **2007**, *20*, 662–676.

60. Bernocchi, G.; Gerzeli, G.; Scherini, E.; Vignola, C. Neuroprotective effects of alpha-dihydro-ergocryptine against damages in the substantia nigra caused by severe treatment with 1-methyl-4-phenyl-1,2,3,6-tetrahydropyridine. *Acta Neuropathol.* **1993**, *85*, 404–413.

61. Zettler, C.; Head, R. J.; Rush, R. A. Chronic nerve growth factor treatment of normotensive rats. *Brain Res.* **1991**, *538*, 251–262.

62. Ferguson, J. S.; Alarie, Y. Long term pulmonary impairment following a single exposure to methyl isocyanate. *Toxicol. Appl. Pharmacol.* **1991**, *107*, 253–268.

HEXIDIUM IODIDE

CAS Registry Number 211566-66-4

Chemical Structure

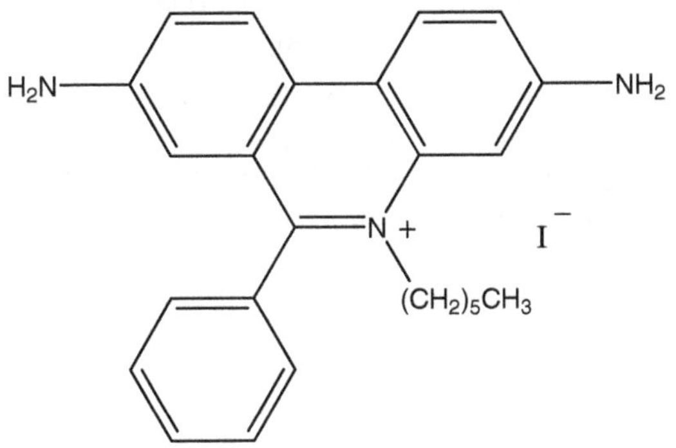

CA Index Name Phenanthridinium, 3,8-diamino-5-hexyl-6-phenyl-, iodide (1:1)

Other Names Phenanthridinium, 3,8-diamino-5-hexyl-6-phenyl-, iodide; Hexidium iodide

Merck Index Number Not listed

Chemical/Dye Class Phenanthridine

Molecular Formula $C_{25}H_{28}IN_3$

Molecular Weight 497.42

Physical Form Solid

Solubility Soluble in dimethyl sulfoxide, water

Melting Point >250 °C

Absorption (λ_{max}) 518 nm. 482 nm

Emission (λ_{max}) 600 nm, 625 nm

Synthesis Synthetic method[1]

Staining Applications Nucleic acids;[2–4] bacteria;[5,6] fungi;[7] microorganisms;[8] nuclei[9]

Biological Applications Detecting nucleic acids;[3] microbes;[10] molds;[11] nucleic acid binding proteins;[12] polynucleotides;[13] pRB in single cells;[14] carrying out polymerase chain reaction (PCR);[15] measuring membrane potential[16]

Industrial Applications Not reported

Safety/Toxicity No data available

REFERENCES

1. Schneider, K.; Naujok, A.; Zimmermann, H. W. Influence of trans-membrane potential and of hydrophobic interactions on dye accumulation in mitochondria of living cells: photoaffinity labeling of mitochondrial proteins, action of potential dissipating drugs, and competitive staining. *Histochemistry* **1994**, *101*, 455–461.
2. Weber, J.; Brendler, A.; Bley, T. Method and kit for the analysis of the DNA content in cells, especially plant cells using barium hydroxide treatment prior staining. Ger. Offen. DE 102006046079, 2008; *Chem. Abstr.* **2008**, *148*, 373652.
3. Park, H. O.; Kim, H. B.; Chi, S. M. Detection method of DNA amplification using probes labeled with intercalating dyes. PCT Int. Appl. WO 2006004267, 2006; *Chem. Abstr.* **2006**, *144*, 122707.
4. McNally, A. J.; Wu, R. S.; Li, Z. Immunoassay based on DNA replication using labeled primer. U.S. Pat. Appl. Publ. US 2002072053, 2002; *Chem. Abstr.* **2002**, *137*, 30238.
5. Holm, C.; Jespersen, L. A flow-cytometric gram-staining technique for milk-associated bacteria. *Appl. Environ. Microbiol.* **2003**, *69*, 2857–2863.
6. Forster, S.; Snape, J. R.; Lappin-Scott, H. M.; Porter, J. Simultaneous fluorescent gram staining and activity assessment of activated sludge bacteria. *Appl. Environ. Microbiol.* **2002**, *68*, 4772–4779.
7. Bradner, J. R.; Nevalainen, K. M. H. Metabolic activity in filamentous fungi can be analyzed by flow cytometry. *J. Microbiol. Methods* **2003**, *54*, 193–201.
8. Horikiri, S. Microorganism cell detection method using fluorescent indicator. Jpn. Kokai Tokkyo Koho JP 2006262775, 2006; *Chem. Abstr.* **2006**, *145*, 392006.
9. Gunasekera, N.; Musier-Forsyth, K.; Arriaga, E. Electrophoretic behavior of individual nuclear species as determined by capillary electrophoresis with laser-induced fluorescence detection. *Electrophoresis* **2002**, *23*, 2110–2116.
10. Oppedahl, A. M.; Lasky, S. J.; Baker, D. D.; Buttry, D. A.; Steger, A. M. Simultaneous rapid detection of more than one microbe class using labeling agents including fluorescent labels. U.S. Pat. Appl. Publ. US 2008261229, 2008; *Chem. Abstr.* **2008**, *149*, 465856.
11. Oppedahl, A. M.; Lasky, S. J.; Baker, D. D.; Buttry, D.; Steger, A. M. Rapid detection of mold by accelerated growth and detection. U.S. Pat. Appl. Publ. US 2007231852, 2007; *Chem. Abstr.* **2007**, *147*, 401104.

12. Loewy, Z.; Chaung, W.; Pottathil, R. Methods for high throughput screening and characterization of nucleic acid binding proteins. PCT Int. Appl. WO 2004011606, 2004; *Chem. Abstr.* **2004**, *140*, 158522.
13. Hyldig-Nielsen, J. J.; Fiandaca, M. J.; Coull, J. M. Fluorescent signal and quencher oligonucleotide probe design and methods for multiplex analysis of polynucleotides. PCT Int. Appl. WO 2004074447, 2004; *Chem. Abstr.* **2004**, *141*, 237692.
14. Darzynkiewicz, Z.; Traganos, F.; Juan, G.; Gruenwald, S. Flow cytometric detection of conformations of pRB in single cells. PCT Int. Appl. WO 9944067, 1999; *Chem. Abstr.* **1999**, *131*, 181965.
15. Gauer, C.; Mann, W.; Alunni-Fabbroni, M. Methods for carrying out enzymic reactions, such as PCR, in single cells. PCT Int. Appl. WO 2008064730, 2008; *Chem. Abstr.* **2008**, *148*, 578429.
16. Deutsch, M.; Namer, Y. A method and system for measuring membrane potential based on fluorescence polarization. PCT Int. Appl. WO 2007122602, 2007; *Chem. Abstr.* **2007**, *147*, 498288.

HOECHST 33258

CAS Registry Number 23491-45-4

Chemical Structure

Molecular Formula $C_{25}H_{27}Cl_3N_6O$

Molecular Weight 533.88

Physical Form Dark yellow to tan powder with green cast

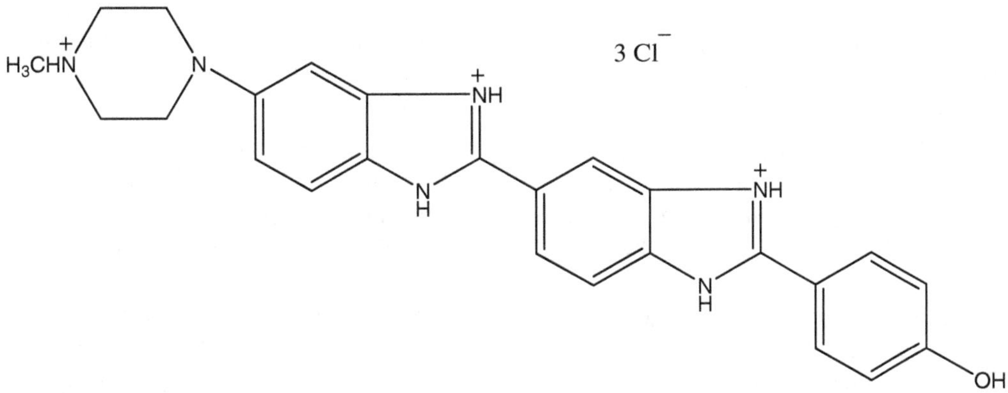

CA Index Name Phenol, 4-[5-(4-methyl-1-piperazinyl) [2,5′-bi-1*H*-benzimidazol]-2′-yl]-, hydrochloride (1:3)

Other Names Phenol, 4-[5-(4-methyl-1-piperazinyl) [2,5′-bi-1*H*-benzimidazol]-2′-yl]-, trihydrochloride; Phenol, *p*-[5-[5-(4-methyl-1-piperazinyl)-2-benzimidazolyl]-2-benzimidazolyl]-, trihydrochloride; 2′-(4-Hydroxyphenyl)-5-(4-methyl-1-piperazinyl)-2,5′-bi-1*H*-benzimidazole; 2-[2-(4-Hydroxyphenyl)-6-benzimidazolyl]-6-(1-methyl-4-piperazinyl)benzimidazole trihydrochloride; 2-[2-(4-Hydroxyphenyl)-6-benzimidazolyl]-6-(1-methyl-4-piperazyl)-benzimidazole trihydrochloride; 2-[2-(4-Hydroxyphenyl)-6-benzimidazolyl]-6-(1-methyl-4-piperazyl)benzimidazole trichloride; 2-[2-(4-Hydroxyphenyl)-6-benzimidazolyl]-6-(1-methyl-4-piperazyl)benzimidazole-3HCl; 4-[5-[5-(4-Methyl-1-piperazinyl)-2-benzimidazolyl]-2-benzimidazolyl]phenol trihydrochloride; Bisbenzimide; Bisbenzimide (quenchant); Bisbenzimide trihydrochloride; H 33258; HOE 33258; Ho 33258; Hoechst 33258

Merck Index Number Not listed

Chemical/Dye Class Benzimidazole

Solubility Soluble in water, *N,N*-dimethyl formamide

Melting Point >300 °C

Absorption (λ_{max}) 352 nm, 343 nm

Emission (λ_{max}) 461 nm

Synthesis Synthetic methods[1–16]

Staining Applications Nucleic acids;[17–23] cells;[24] embryos;[25] microorganisms;[26] nuclei;[27] sperms[28,29]

Biological Applications Nucleic acid hybridization;[30] detecting nucleic acids,[17–23,31,32] cancer cells,[33] spores,[33] hepatitis C virus,[34] human papilloma virus (HPV),[35] single nucleotide polymorphism (SNP),[36] stress biomarkers;[37] nucleic acid amplification;[38] nucleic acid quantification;[39] nucleic acid sequencing;[40] treating of cancer,[41,42] allergy,[42] autoimmune disease,[42] neoplasia,[43] sexually transmitted diseases[44]

Industrial Applications Not reported

Safety/Toxicity Carcinogenicity;[45] cytotoxicity;[46] DNA damage;[47,48] genotoxicity;[49] neurotoxicity;[50] oral toxicity;[51,52] radiotoxicity;[53] retinal toxicity[54]

REFERENCES

1. Bielawski, K.; Bielawska, A.; Anchim, T.; Wolczynski, S. Synthesis, DNA binding, topoisomerase inhibition and cytotoxic properties of 2-chloroethylnitrosourea derivatives of Hoechst 33258. *Biol. Pharm. Bull.* **2005**, *28*, 1004–1009.
2. Karlsson, H. J.; Lincoln, P.; Westman, G. Synthesis and DNA binding studies of a new asymmetric cyanine dye binding in the minor groove of [poly (dA-dT)]$_2$. *Bioorg. Med. Chem.* **2003**, *11*, 1035–1040.
3. Tawar, U.; Jain, A. K.; Dwarakanath, B. S.; Chandra, R.; Singh, Y.; Chaudhury, N. K.; Khaitan, D.; Tandon, V. Influence of phenyl ring disubstitution on bisbenzimidazole and terbenzimidazole cytotoxicity: synthesis and biological evaluation as radioprotectors. *J. Med. Chem.* **2003**, *46*, 3785–3792.
4. Bielawski, K.; Bielawska, A.; Wolczynski, S. Synthesis, DNA-binding activity and cytotoxicity of carbamate derivatives of Hoechst 33258 in

breast cancer MCF-7 cells. *Biol. Pharm. Bull.* **2002**, *25*, 916–919.

5. Satyanarayana, Y.; Lown, J. Design and synthesis of a new class of pyrrolobenzimidazole based agents to target human tumor helicases. *Heterocycl. Commun.* **2000**, *6*, 199–208.
6. Singh, A. K.; Lown, J. W. Synthesis of Hoechst 33258 analogues designed to target human tumor helicases. *Heterocycl. Commun.* **1999**, *5*, 11–18.
7. Soderlind, K. J.; Gorodetsky, B.; Singh, A. K.; Bachur, N. R.; Miller, G. G.; Lown, J. W. Bis-benzimidazole anticancer agents: targeting human tumour helicases. *Anti-Cancer Drug Des.* **1999**, *14*, 19–36.
8. Harapanhalli, R. S.; McLaughlin, L. W.; Howell, R. W.; Rao, D. V.; Adelstein, S. J.; Kassis, A. I. [^{125}I/^{127}I] IodoHoechst 33342: synthesis, DNA binding, and biodistribution. *J. Med. Chem.* **1996**, *39*, 4804–4809.
9. Czarny, A.; Wilson, W. D.; Boykin, D. W. Synthesis of mono-cationic and dicationic analogs of Hoechst 33258. *J. Heterocycl. Chem.* **1996**, *33*, 1393–1397.
10. Wiederholt, K.; Rajur, S. B.; Giuliano, J., Jr.; O'Donnell, M. J.; McLaughlin, L. W. DNA-tethered Hoechst groove-binding agents: duplex stabilization and fluorescence characteristics. *J. Am. Chem. Soc.* **1996**, *118*, 7055–7062.
11. Gupta, R.; Wang, H.; Huang, L.; Lown, J. Design, synthesis, DNA sequence preferential alkylation and biological evaluation of N-mustard derivatives of Hoechst 33258 analogs. *Anti-Cancer Drug Des.* **1995**, *10*, 25–41.
12. Harapanhalli, R. S.; Howell, R. W.; Rao, D. V. Bis-benzimidazole dyes, Hoechst 33258 and Hoechst 33342: radioiodination, facile purification and subcellular distribution. *Nucl. Med. Biol.* **1994**, *21*, 641–647.
13. Kelly, D. P.; Bateman, S. A.; Martin, R. F.; Reum, M. E.; Rose, M.; Whittaker, A. R. D. DNA binding compounds. V. Synthesis and characterization of boron-containing bibenzimidazoles related to the DNA minor groove binder, Hoechst 33258. *Aust. J. Chem.* **1994**, *47*, 247–262.
14. Lee, M.; Walker, C. D.; Eckert, J. M.; Bowers, S. K.; Montague, D.; McAdams, S.; Hartley, J. A. DNA sequence selective alkylation and cytotoxicity of monoheterocyclic analogs of Hoechst 33258. *Med. Chem. Res.* **1993**, *3*, 79–86.
15. Martin, R. F.; Pardee, M. Preparation of carrier free [^{125}I]IodoHoechst 33258. *Int. J. Appl. Radiat. Isotopes* **1985**, *36*, 745–747.
16. Farbwerke Hoechst AG. Piperazino bisbenzimidazoles. Fr. Demande FR 1519964, 1968; *Chem. Abstr.* **1969**, *71*, 81418.
17. Ramsing, N. B.; Mouritzen, P.; Echwald, S. M.; Tolstrup, N. Probes, libraries and kits for analysis of mixtures of nucleic acids, such as the human transcriptome, and methods for their construction. PCT Int. Appl. WO 2006066592, 2006; *Chem. Abstr.* **2006**, *145*, 118232.
18. Martin, R. M.; Leonhardt, H.; Cardoso, M. C. DNA labeling in living cells. *Cytometry* **2005**, *67A*, 45–52.
19. Stockert, J. C.; Pinna-Senn, E.; Bella, J. L.; Lisanti, J. A. DNA-binding fluorochromes: correlation between C-banding of mouse metaphase chromosomes and hydrogen bonding to adenine–thymine base pairs. *Acta Histochem.* **2005**, *106*, 413–420.
20. Chaudhury, N. K.; Bhardwaj, R. Structural stabilization by Hoechst 33258 in γ-irradiated DNA: evidenced by spectroscopic studies. *Curr. Sci.* **2004**, *87*, 1256–1262.
21. Gromyko, A. V.; Streltsov, S. A.; Zhuze, A. L. A DNA-specific dimeric bisbenzimidazole. *Russ. J. Bioorg. Chem.* **2004**, *30*, 400–402.
22. Baraldi, P. G.; Bovero, A.; Fruttarolo, F.; Preti, D.; Tabrizi, M. A.; Pavani, M. G.; Romagnoli, R. DNA minor groove binders as potential antitumor and antimicrobial agents. *Med. Res. Rev.* **2004**, *24*, 475–528.
23. Dominick, P. K.; Keppler, B. R.; Legassie, J. D.; Moon, I. K.; Jarstfer, M. B. Nucleic acid-binding ligands identify new mechanisms to inhibit telomerase. *Bioorg. Med. Chem. Lett.* **2004**, *14*, 3467–3471.
24. Anderson, A. L.; Knutson, C. R.; Mueth, D.; Plewa, J.; Tanner, E. Methods for staining cells for identification and sorting. U.S. Pat. Appl. Publ. US 2006172315, 2006; *Chem. Abstr.* **2006**, *145*, 183714.
25. Masci, J.; Monteiro, A. Visualization of early embryos of the butterfly *Bicyclus anynana. Zygote* **2005**, *13*, 139–144.
26. Noda, N.; Mizutani, T. Microorganism-measuring method using multiple staining. Jpn. Kokai Tokkyo Koho JP 2006340684, 2006; *Chem. Abstr.* **2006**, *146*, 77563.
27. Horobin, R. W.; Stockert, J. C.; Rashid-Doubell, F. Fluorescent cationic probes for nuclei of living cells: why are they selective? A quantitative structure–activity relations analysis. *Histochem. Cell Biol.* **2006**, *126*, 165–175.
28. Marco-Jimenez, F.; Garzon, D. L.; Penaranda, D. S.; Perez, L.; Viudes-de-Castro, M. P.; Vicente, J. S.;

Jover, M.; Asturiano, J. F. Cryopreservation of European eel (*Anguilla anguilla*) spermatozoa: effect of dilution ratio, fetal bovine serum supplementation, and cryoprotectants. *Cryobiology* **2006**, *53*, 51–57.

29. Vadnais, M. L.; Kirkwood, R. N.; Tempelman, R. J.; Sprecher, D. J.; Chou, K. Effect of cooling and seminal plasma on the capacitation status of fresh boar sperm as determined using chlortetracycline assay. *Anim. Reprod. Sci.* **2005**, *87*, 121–132.

30. Kandimalla, E. R.; Agrawal, S. Hybridization-based fluorescence assay. U.S. Pat. Appl. Publ. US 2005089915, 2005; *Chem. Abstr.* **2005**, *142*, 424827.

31. Guan, Y.; Zhou, W.; Yao, X.; Zhao, M.; Li, Y. Determination of nucleic acids based on the fluorescence quenching of Hoechst 33258 at pH 4.5. *Anal. Chim. Acta* **2006**, *570*, 21–28.

32. Horstkotte, B.; Rehbein, H. Determination of DNA content of whole fish. *Fish. Sci.* **2006**, *72*, 429–436.

33. Levy, M.; Ellington, A. D.; Pai, S. Proximity ligation assays with peptide conjugate 'burrs' and aptamers for the sensitive detection of spores and cancer cells. PCT Int. Appl. WO 2007027796, 2007; *Chem. Abstr.* **2007**, *146*, 312239.

34. Hashimoto, K.; Ito, K. Development of DNA detection electrode system based on amplification and hybridization and application to detection of hepatitis C virus genomic sequence and ras gene variants. Jpn. Kokai Tokkyo Koho JP 2004154121, 2004; *Chem. Abstr.* **2004**, *141*, 18697.

35. Poetter, K.; Gould, T. Human papilloma virus (HPV) detection using nucleic acid probes, microbeads, and fluorescence-activated cell sorter (FACS). PCT Int. Appl. WO 2006060872, 2006; *Chem. Abstr.* **2006**, *145*, 41223.

36. Choi, Y. S.; Park, D. H. Electrochemical single nucleotide polymorphism (SNP) detection using a microelectrode array biochip by Hoechst 33258. *J. Korean Phys. Soc.* **2005**, *46*, 1445–1451.

37. Cullum, M. E.; Duplessis, C. A.; Crepeau, L. J. Method for the detection of stress biomarkers including cortisol by fluorescence polarization. U. S. Pat. Appl. Publ. US 2006105397, 2006; *Chem. Abstr.* **2006**, *144*, 463810.

38. Park, H. O.; Kim, H. B.; Chi, S. M. Detection method of DNA amplification using probes labeled with intercalating dyes. PCT Int. Appl. WO 2006004267, 2006; *Chem. Abstr.* **2006**, *144*, 122707.

39. Georgiou, C. D.; Papapostolou, I. Assay for the quantification of intact/fragmented genomic DNA. *Anal. Biochem.* **2006**, *358*, 247–256.

40. Hoser, M. J. Methods for nucleic acid sequencing. PCT Int. Appl. WO 2004074503, 2004; *Chem. Abstr.* **2004**, *141*, 237698.

41. Zeldis, J. B.; Zeitlin, A.; Barer, S. Compositions for the treatment of cancer comprising a topoisomerase inhibitor and thalidomide. Eur. Pat. Appl. EP 1676577, 2006; *Chem. Abstr.* **2006**, *145*, 96425.

42. Vollmer, J.; Krieg, A. M.; Uhlmann, E. Compositions comprising immunostimulatory nucleic acid-lipophilic conjugates and antigen or therapeutic agent for treating allergy, cancer, infection and autoimmune disease. PCT Int. Appl. WO 2005030259, 2005; *Chem. Abstr.* **2005**, *142*, 372464.

43. Masferrer, J. L. Combination of a cyclooxygenase 2 (COX-2) inhibitor and a DNA topoisomerase 1 inhibitor for treatment of neoplasia. U.S. Pat. Appl. Publ. US 2005187172, 2005; *Chem. Abstr.* **2005**, *143*, 241966.

44. Marcucci, F. Use of nitrogen heterocyclic compounds as microbicides for the treatment of sexually transmitted diseases. PCT Int. Appl. WO 2006000863, 2006; *Chem. Abstr.* **2006**, *144*, 94404.

45. Huang, M.; Liu, G. The study of innate drug resistance of human hepatocellular carcinoma Bel7402 cell line. *Cancer Lett.* **1999**, *135*, 97–105.

46. Mizutani, Y.; Yoshida, O.; Miki, T.; Bonavida, B. Synergistic cytotoxicity and apoptosis by Apo-2 ligand and adriamycin against bladder cancer cells. *Clin. Cancer Res.* **1999**, *5*, 2605–2612.

47. Suzuki, H.; Mori, M.; Suzuki, M.; Sakurai, K.; Miura, S.; Ishii, H. Extensive DNA damage induced by monochloramine in gastric cells. *Cancer Lett.* **1997**, *115*, 243–248.

48. Limoli, C. L.; Ward, J. F. DNA damage in bromodeoxyuridine substituted SV40 DNA and minichromosomes following UVA irradiation in the presence of Hoechst dye 33258. *Int. J. Radiat. Biol.* **1994**, *66*, 717–728.

49. He, L.; Jurs, P. C.; Custer, L. L.; Durham, S. K.; Pearl, G. M. Predicting the genotoxicity of polycyclic aromatic compounds from molecular structure with different classifiers. *Chem. Res. Toxicol.* **2003**, *16*, 1567–1580.

50. Fernandez, M.; Pirondi, S.; Antonelli, T.; Ferraro, L.; Giardino, L.; Calza, L. Role of c-Fos protein on glutamate toxicity in primary neural hippocampal cells. *J. Neurosci. Res.* **2005**, *82*, 115–125.

51. Chang, Y. C.; Hu, C. C.; Lii, C. K.; Tai, K. W.; Yang, S. H.; Chou, M. Y. Cytotoxicity and arecoline mechanisms in human gingival fibroblasts *in vitro*. *Clin. Oral Invest.* **2001**, *5*, 51–56.

52. Chang, Y. C.; Tai, K. W.; Huang, F. M.; Huang, M. F. Cytotoxic and nongenotoxic effects of phenolic compounds in human pulp cell cultures. *J. Endod.* **2000**, *26*, 440–443.

53. Singh, S. P.; Jayanth, V. R.; Chandna, S.; Dwarakanath, B. S.; Singh, S.; Adhikari, J. S.; Jain, V. Radioprotective effects of DNA ligands Hoechst-33342 and 33258 in whole body irradiated mice. *Indian J. Exp. Biol.* **1998**, *36*, 375–384.

54. Oh, H.; Kwon, O. W.; Chung, I.; Lee, S. C.; Koh, H. J.; Lee, S.; Lee, J. H. Retinal toxicity of commercial tissue plasminogen activator is mediated by the induction of nitric oxide in the mouse retinal primary cells. *Curr. Eye Res.* **2005**, *30*, 291–297.

HOECHST 33342

CAS Registry Number 23491-52-3

Chemical Structure

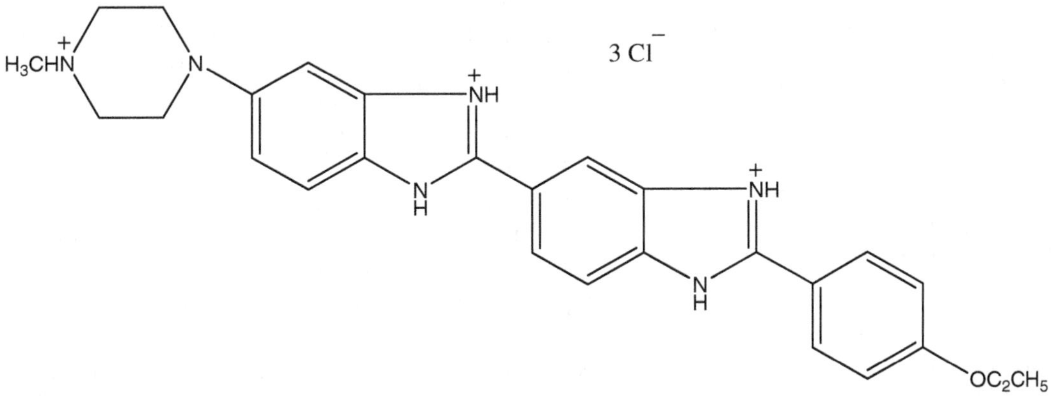

Absorption (λ_{max}) 350 nm

Emission (λ_{max}) 461 nm

Synthesis Synthetic methods[1–7]

CA Index Name 2,5′-Bi-1*H*-benzimidazole, 2′-(4-ethoxyphenyl)-5-(4-methyl-1-piperazinyl)-,hydrochloride (1:3)

Other Names 2,5′-Bibenzimidazole, 2′-(*p*-ethoxyphenyl)-5-(4-methyl-1-piperazinyl)-; 2′-(4-Ethoxyphenyl)-5-(4-methyl-1-piperazinyl)-2,5′-bi-1*H*-benzimidazole; 2-[2-(4-Ethoxyphenyl)-6-benzimidazolyl]-6-(1-methyl-4-piperazinyl)benzimidazole; Bisbenzimide; HOE 33342; Ho 342; Hoechst 33342; NSC 334072

Merck Index Number Not listed

Chemical/Dye Class Benzimidazole

Molecular Formula $C_{27}H_{31}Cl_3N_6O$

Molecular Weight 561.93

Physical Form Yellow to green powder

Solubility Soluble in water, *N,N*-dimethyl formamide

Melting Point >300 °C

Boiling Point (Calcd.) 725.9 ± 70.0 °C, pressure: 760 Torr

pK_a (Calcd.) 11.08 ± 0.69, most acidic, temperature: 25 °C; 7.66 ± 0.42, most basic, temperature: 25 °C

Staining Applications Nucleic acids;[7–10] cells;[11,12] neurons;[13] endothelial cells;[13] inflammatory cells;[13] cancer stem cells;[11,14] animal mammary gland stem cells;[15] hematopoietic stem cells;[16–18] hepatocytes stem cells;[19] human tumor cell;[20] SV40 human corneal epithelial cell;[21] microorganisms;[22] nuclei;[23,24] parasites;[25] chromosomes;[26] sperms[27]

Biological Applications Detecting abnormal gametes,[27] apoptosis,[28] human papilloma virus (HPV),[29] single nucleotide polymorphism (SNP),[30] mitochondrial membrane potential change,[31] polynucleotides,[32] proteins,[33] peptides;[33] nucleic acid amplification;[34] nucleic acid quantification;[35] nucleic acid sequencing;[36] treating of cancer,[37,38] Alzheimers disease,[38] multiple sclerosis,[38] epilepsy[39]

Industrial Applications Not reported

Safety/Toxicity Carcinogenicity;[40,41] cytotoxicity;[42–44] DNA damage;[41,43,44] genotoxicity;[45,46] metabolic toxicity;[47] mutagenicity;[44] neurotoxicity;[48–50] radiotoxicity;[51] reproductive toxicity[52,53]

REFERENCES

1. Tawar, U.; Jain, A. K.; Dwarakanath, B. S.; Chandra, R.; Singh, Y.; Chaudhury, N. K.; Khaitan, D.; Tandon, V. Influence of phenyl ring disubstitution on bisbenzimidazole and terbenzimidazole cytotoxicity: synthesis and biological evaluation as radioprotectors. *J. Med. Chem.* **2003**, *46*, 3785–3792.
2. Ji, Y. H.; Bur, D.; Hasler, W.; Runtz, S. V.; Dorn, A.; Bailly, C.; Waring, M. J.; Hochstrasser, R.; Leupin, W. F. Tris–benzimidazole derivatives: design, synthesis and DNA sequence recognition. *Bioorg. Med. Chem.* **2001**, *9*, 2905–2919.
3. Harapanhalli, R. S.; McLaughlin, L. W.; Howell, R. W.; Rao, D. V.; Adelstein, S. J.; Kassis, A. I. [^{125}I/^{127}I]IodoHoechst 33342: synthesis, DNA binding, and biodistribution. *J. Med. Chem.* **1996**, *39*, 4804–4809.

4. Sun, Q.; Gatto, B.; Yu, C.; Liu, A.; Liu, L. F.; LaVoie, E. J. Synthesis and evaluation of terbenzimidazoles as topoisomerase I inhibitors. *J. Med. Chem.* **1995**, *38*, 3638–3644.
5. Harapanhalli, R. S.; Howell, R. W.; Rao, D. V. Bis-benzimidazole dyes, Hoechst 33258 and Hoechst 33342: radioiodination, facile purification and subcellular distribution. *Nucl. Med. Biol.* **1994**, *21*, 641–647.
6. Loewe, H.; Urbanietz, J. Basic substituted 2,6-bisbenzimidazole derivatives, a novel series of substances with chemotherapeutic activity. *Arzneim.-Forsch.* **1974**, *24*, 1927–1933; *Chem. Abstr.* **1975**, *82*, 112032.
7. Farbwerke Hoechst AG. Piperazino bisbenzimidazoles. Fr. Demande FR 1519964, 1968; *Chem. Abstr.* **1969**, *71*, 81418.
8. Ramsing, N. B.; Mouritzen, P.; Echwald, S. M.; Tolstrup, N. Probes, libraries and kits for analysis of mixtures of nucleic acids, such as the human transcriptome, and methods for their construction. PCT Int. Appl. WO 2006066592, 2006; *Chem. Abstr.* **2006**, *145*, 118232.
9. Wu, J.; Apontes, P.; Song, L.; Liang, P.; Yang, L.; Li, F. Molecular mechanism of upregulation of survivin transcription by the AT-rich DNA-binding ligand, Hoechst33342: evidence for survivin involvement in drug resistance. *Nucleic Acids Res.* **2007**, *35*, 2390–2402.
10. Evans, D. A.; Neidle, S. Virtual screening of DNA minor groove binders. *J. Med. Chem.* **2006**, *49*, 4232–4238.
11. Adamski, D.; Mayol, J. F.; Platet, N.; Berger, F.; Herodin, F.; Wion, D. Effects of Hoechst 33342 on C2C12 and PC12 cell differentiation. *FEBS Lett.* **2007**, *581*, 3076–3080.
12. Anderson, A. L.; Knutson, C. R.; Mueth, D.; Plewa, J.; Tanner, E. Methods for staining cells for identification and sorting. U.S. Pat. Appl. Publ. US 2006172315, 2006; *Chem. Abstr.* **2006**, *145*, 183714.
13. Casella, G. T. B.; Bunge, M. B.; Wood, P. M. Improved immunocytochemical identification of neural, endothelial, and inflammatory cell types in paraffin-embedded injured adult rat spinal cord. *J. Neurosci. Methods* **2004**, *139*, 1–11.
14. Zheng, X.; Shen, G.; Yang, X.; Liu, W. Most C6 cells are cancer stem cells: evidence from clonal and population analyses. *Cancer Res.* **2007**, *67*, 3691–3697.
15. Grezina, N. M.; Zinov'eva, N. A. *In vitro* microscopic determination of animal mammary gland stem cells. Russ. RU 2295565, 2007; *Chem. Abstr.* **2007**, *146*, 333660.
16. Lin, K. K.; Goodell, M. A. Purification of hematopoietic stem cells using the side population. *Methods Enzymol.* **2006**, *420*, 255–264.
17. Uchida, N.; Dykstra, B.; Lyons, K.; Leung, F.; Kristiansen, M.; Eaves, C. ABC transporter activities of murine hematopoietic stem cells vary according to their developmental and activation status. *Blood* **2004**, *103*, 4487–4495.
18. Bertoncello, I.; Williams, B. Hematopoietic stem cell characterization by Hoechst 33342 and rhodamine 123 staining. *Methods Mol. Biol.* **2004**, *263*, 181–200.
19. Uryvaeva, I. V.; Tsitrin, E. B.; Gorodetsky, S. I.; Tsvetkova, I. A.; Delone, G. V.; Gulyaev, D. V.; Khrushchov, N. G. The phenotypic characters of the stem cells in hepatocytes during liver regeneration: the expression of the Bcrp1/Abcg2 membrane transporter and the Hoechst 33342 export. *Dokl. Biol. Sci.* **2004**, *398*, 413–416.
20. Adhikari, J. S.; Khaitan, D.; Arya, M. B.; Dwarakanath, B. S. Heterogeneity in the radiosensitizing effects of the DNA ligand Hoechst-33342 in human tumor cell lines. *J. Cancer Res. Ther.* **2005**, *1*, 151–161.
21. Liliensiek, S. J.; Schell, K.; Howard, E.; Nealey, P.; Murphy, C. J. Cell sorting but not serum starvation is effective for SV40 human corneal epithelial cell cycle synchronization. *Exp. Eye Res.* **2006**, *83*, 61–68.
22. Noda, N.; Mizutani, T. Microorganism-measuring method using multiple staining. Jpn. Kokai Tokkyo Koho JP 2006340684, 2006; *Chem. Abstr.* **2006**, *146*, 77563.
23. Yasui, L. S.; Chen, K.; Wang, K.; Jones, T. P.; Caldwell, J.; Guse, D.; Kassis, A. I. Using Hoechst 33342 to target radioactivity to the cell nucleus. *Radiat. Res.* **2007**, *167*, 167–175.
24. Horobin, R. W.; Stockert, J. C.; Rashid-Doubell, F. Fluorescent cationic probes for nuclei of living cells: why are they selective? A quantitative structure–activity relations analysis. *Histochem. Cell Biol.* **2006**, *126*, 165–175.
25. Jean-Moreno, V.; Rojas, R.; Goyeneche, D.; Coombs, G. H.; Walker, J. *Leishmania donovani*: differential activities of classical topoisomerase inhibitors and antileishmanials against parasite and host cells at the level of DNA topoisomerase I and in cytotoxicity assays. *Exp. Parasitol.* **2006**, *112*, 21–30.
26. Lu, Y. Q.; Zhang, M.; Meng, B.; Lu, S. S.; Wei, Y. M.; Lu, K. H. Identification of X- and Y-chromosome

bearing buffalo (*Bubalus bubalis*) sperm. *Reprod. Sci.* **2006**, *95*, 158–164.

27. Harton, G.; Karabinus, D.; Stern, H. Enrichment of euploid sperms using Hoechst 33342 for detection of abnormal gametes. PCT Int. Appl. WO 2006116627, 2006; *Chem. Abstr.* **2006**, *145*, 478095.

28. Wlodkowic, D.; Skommer, J.; Pelkonen, J. Towards an understanding of apoptosis detection by SYTO dyes. *Cytometry* **2007**, *71A*, 61–72.

29. Poetter, K.; Gould, T. Human papilloma virus (HPV) detection using nucleic acid probes, microbeads, and fluorescence-activated cell sorter (FACS). PCT Int. Appl. WO 2006060872, 2006; *Chem. Abstr.* **2006**, *145*, 41223.

30. Sakata, T.; Miyahara, Y. Potentiometric detection of single nucleotide polymorphism by using a genetic field-effect transistor. *ChemBioChem* **2005**, *6*, 703–710.

31. Chen, J. C.; Zhang, X.; Singleton, T. P.; Kiechle, F. L. Mitochondrial membrane potential change induced by Hoechst 33342 in myelogenous leukemia cell line HL-60. *Ann. Clin. Lab. Sci.* **2004**, *34*, 458–466.

32. Hyldig-Nielsen, J. J.; Fiandaca, M. J.; Coull, J. M. Fluorescent signal and quencher oligonucleotide probe design and methods for multiplex analysis of polynucleotides. PCT Int. Appl. WO 2004074447, 2004; *Chem. Abstr.* **2004**, *141*, 237692.

33. Sawyer, J. R.; Li, C.; Choong, V.; Maracas, G.; Zhang, P. Protein and peptide sensors using electrical detection methods. U.S. Pat. Appl. Publ. US 2005023155, 2005; *Chem. Abstr.* **2005**, *142*, 193892.

34. Park, H. O.; Kim, H. B.; Chi, S. M. Detection method of DNA amplification using probes labeled with intercalating dyes. PCT Int. Appl. WO 2006004267, 2006; *Chem. Abstr.* **2006**, *144*, 122707.

35. Kobayashi, M.; Kaji, S.; Omi, M.; Tamiya, E. Electrochemical DNA quantification based on aggregation induced by phosphate group-binding substance. Jpn. Kokai Tokkyo Koho JP 2006145342, 2006; *Chem. Abstr.* **2006**, *145*, 2140.

36. Hoser, M. J. Methods for nucleic acid sequencing. PCT Int. Appl. WO 2004074503, 2004; *Chem. Abstr.* **2004**, *141*, 237698.

37. Zeldis, J. B.; Zeitlin, A.; Barer, S. Compositions for the treatment of cancer comprising a topoisomerase inhibitor and thalidomide. Eur. Pat. Appl. EP 1676577, 2006; *Chem. Abstr.* **2006**, *145*, 96425.

38. Churcher, I.; Simpson, P. B.; Nanthakumar, C. B. Method for identifying modulators of γ-secretase or notch for treatment of Alzheimers disease, multiple sclerosis, and cancer. PCT Int. Appl. WO 2007029030, 2007; *Chem. Abstr.* **2007**, *146*, 288481.

39. Nedergaard, M.; Tian, G. F. Method for treatment and prevention of epilepsy. PCT Int. Appl. WO 2006062683, 2006; *Chem. Abstr.* **2006**, *145*, 55958.

40. Balachandran, R.; ter Haar, E.; Yalowich, J. C.; Welsh, M. J.; Grant, S. G.; Day, B. W. Induction of human breast cancer cell apoptosis from G2/M preceded by stimulation into the cell cycle by *Z*-1,1-dichloro-2,3-diphenylcyclopropane. *Biochem. Pharmacol.* **1999**, *57*, 97–110.

41. Olive, P. L. Detection of hypoxia by measurement of DNA damage in individual cells from spheroids and murine tumours exposed to bioreductive drugs. I. Tirapazamine. *Br. J. Cancer* **1995**, *71*, 529–536.

42. Wiezorek, C. Cell cycle dependence of Hoechst 33342 dye cytotoxicity on sorted living cells. *Histochemistry* **1984**, *81*, 493–495.

43. Erba, E.; Ubezio, P.; Broggini, M.; Ponti, M.; D'Incalci, M. DNA damage, cytotoxic effect and cell-cycle perturbation of Hoechst 33342 on L1210 cells *in vitro*. *Cytometry* **1988**, *9*, 1–6.

44. Durand, R. E.; Olive, P. L. Cytotoxicity, mutagenicity and DNA damage by Hoechst 33342. *J. Histochem. Cytochem.* **1982**, *30*, 111–116.

45. Argentin, G.; Cicchetti, R. Genotoxic and antiapoptotic effect of nicotine on human gingival fibroblasts. *Toxicol. Sci.* **2004**, *79*, 75–81.

46. He, L.; Jurs, P. C.; Custer, L. L.; Durham, S. K.; Pearl, G. M. Predicting the genotoxicity of polycyclic aromatic compounds from molecular structure with different classifiers. *Chem. Res. Toxicol.* **2003**, *16*, 1567–1580.

47. Downing, T. W.; Garner, D. L.; Ericsson, S. A.; Redelman, D. Metabolic toxicity of fluorescent stains on thawed cryopreserved bovine sperm cells. *J. Histochem. Cytochem.* **1991**, *39*, 485–489.

48. Pin, S.; Chen, H.; Lein, P. J.; Wang, M. M. Nucleic acid binding agents exert local toxic effects on neurites via a non-nuclear mechanism. *J. Neurochem.* **2006**, *96*, 1253–1266.

49. Shimazawa, M.; Yamashima, T.; Agarwal, N.; Hara, H. Neuroprotective effects of minocycline against *in vitro* and *in vivo* retinal ganglion cell damage. *Brain Res.* **2005**, *1053*, 185–194.

50. Celsi, F.; Ferri, A.; Casciati, A.; D'Ambrosi, N.; Rotilio, G.; Costa, A.; Volonte, C.; Carri, M. T. Overexpression of superoxide dismutase 1 protects against beta-amyloid peptide toxicity: effect of estrogen and copper chelators. *Neurochem. Int.* **2004**, *44*, 25–33.

51. Singh, S. P.; Jayanth, V. R.; Chandna, S.; Dwarakanath, B. S.; Singh, S.; Adhikari, J. S.; Jain, V. Radioprotective effects of DNA ligands Hoechst-33342 and 33258 in whole body irradiated mice. *Indian J. Exp. Biol.* **1998**, *36*, 375–384.

52. Velilla, E.; Lopez-Bejar, M.; Rodriguez-Gonzalez, E.; Vidal, F.; Paramio, M. Effect of Hoechst 33342 staining on developmental competence of prepubertal goat oocytes. *Zygote* **2002**, *10*, 201–208.

53. Watkins, A. M.; Chan, P. J.; Kalugdan, T. H.; Patton, W. C.; Jacobson, J. D.; King, A. Analysis of the flow cytometer stain Hoechst 33342 on human spermatozoa. *Mol. Hum. Reprod.* **1996**, *2*, 709–712.

HOECHST 34580

CAS Registry Number 23555-00-2

Chemical Structure

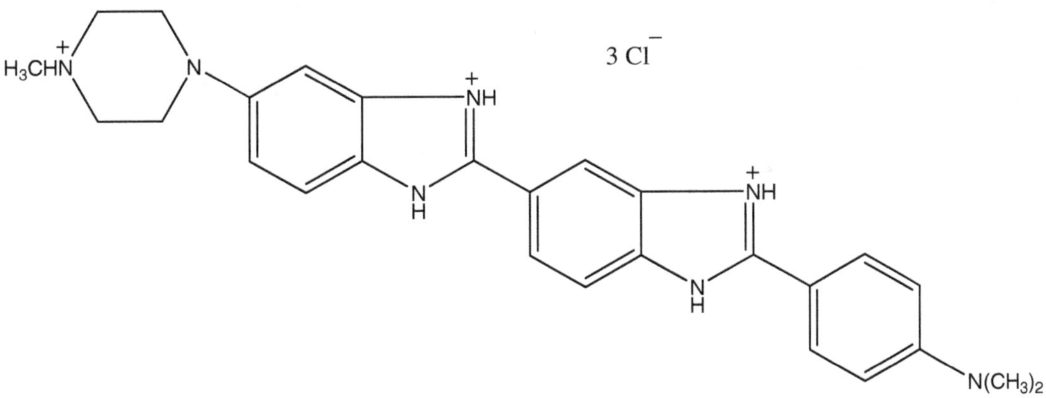

CA Index Name Benzenamine, *N,N*-dimethyl-4-[5-(4-methyl-1-piperazinyl)[2,5′-bi-1*H*-benzimidazol]-2′-yl]-, hydrochloride (1:3)

Other Names 2,5′-Bibenzimidazole, 2′-[*p*-(dimethylamino)phenyl]-5-(4-methyl-1-piperazinyl)-; HOE 34580; Hoechst 34580; Proamine

Merck Index Number Not listed

Chemical/Dye Class Benzimidazole

Molecular Formula $C_{27}H_{32}Cl_3N_7$

Molecular Weight 560.96

Physical Form Yellow to green powder

Solubility Soluble in dimethyl sulfoxide, water

Melting Point >300°C

Boiling Point (Calcd.) 733.2 ± 70.0°C, pressure: 760 Torr

pK_a (Calcd.) 12.46 ± 0.69, most acidic, temperature: 25 °C; 7.66 ± 0.42, most basic, temperature: 25 °C

Absorption (λ_{max}) 392 nm

Emission (λ_{max}) 440 nm

Synthesis Synthetic methods[1–5]

Staining Applications Nucleic acids;[3,6–8] cells;[9] granules[8]

Biological Applications Detecting malaria infected red blood cells,[10] polynucleotides;[11] nucleic acid amplification;[12] nucleic acid sequencing[13]

Industrial Applications Not reported

Safety/Toxicity No data available

REFERENCES

1. Martin, R. F.; Kelly, D. P.; White, J. M. Radioprotectants. U.S. Patent 6548505, 2003; *Chem. Abstr.* **2003**, *138*, 299892.
2. Martin, R. F.; Kelly, D. P.; White, J. M. Preparation of benzazoles as radioprotectors. PCT Int. Appl. WO 9704776, 1997; *Chem. Abstr.* **1997**, *126*, 225300.
3. Kelly, D. P.; Bateman, S. A.; Hook, R. J.; Martin, R. F.; Reum, M. E.; Rose, M.; Whittaker, A. R. D. DNA binding compounds. VI. Synthesis and characterization of 2,5′-disubstituted bibenzimidazoles related to the DNA minor groove binder Hoechst 33258. *Aust. J. Chem.* **1994**, *47*, 1751–1769.
4. Loewe, H.; Urbanietz, J. Basic substituted 2,6-bisbenzimidazole derivatives, a novel series of substances with chemotherapeutic activity. *Arzneim.-Forsch.* **1974**, *24*, 1927–1933; *Chem. Abstr.* **1975**, *82*, 112032.
5. Farbwerke Hoechst AG. Piperazino bisbenzimidazoles. Fr. Demande FR 1519964, 1968; *Chem. Abstr.* **1969**, *71*, 81418.
6. Evans, D. A.; Neidle, S. Virtual screening of DNA minor groove binders. *J. Med. Chem.* **2006**, *49*, 4232–4238.
7. Martin, R. F.; Broadhurst, S.; D'Abrew, S.; Budd, R.; Sephton, R.; Reum, M.; Kelly, D. P. Radioprotection by DNA ligands. *Br. J. Cancer, Suppl.* **1996**, *74*, S99–S101.
8. Maples, J. A.; Lopez, L. L.; Torke, N. Dye compositions which provide enhanced differential fluorescence and light scatter characteristics. U.S.

Pat. Appl. Publ. US 2004185447, 2004; *Chem. Abstr.* **2004**, *141*, 291842.

9. Anderson, A. L.; Knutson, C. R.; Mueth, D.; Plewa, J.; Tanner, E. Methods for staining cells for identification and sorting. U.S. Pat. Appl. Publ. US 2006172315, 2006; *Chem. Abstr.* **2006**, *145*, 183714.
10. Yoshida, A.; Uchihashi, K.; Itose, Y.; Konishi, A.; Iizuka, H. Reagent for partially lysing a cell membrane of a red blood cell, a reagent for detecting malaria infected red blood cells, and a sample analyzing method for detecting malaria infected red blood cells. U.S. Pat. Appl. Publ. US 2006223137, 2006; *Chem. Abstr.* **2006**, *145*, 372400.
11. Hyldig-Nielsen, J. J.; Fiandaca, M. J.; Coull, J. M. Fluorescent signal and quencher oligonucleotide probe design and methods for multiplex analysis of polynucleotides. PCT Int. Appl. WO 2004074447, 2004; *Chem. Abstr.* **2004**, *141*, 237692.
12. Park, H. O.; Kim, H. B.; Chi, S. M. Detection method of DNA amplification using probes labeled with intercalating dyes. PCT Int. Appl. WO 2006004267, 2006; *Chem. Abstr.* **2006**, *144*, 122707.
13. Hoser, M. J. Methods for nucleic acid sequencing. PCT Int. Appl. WO 2004074503, 2004; *Chem. Abstr.* **2004**, *141*, 237698.

INDIGO CARMINE

CAS Registry Number 860-22-0

Chemical Structure

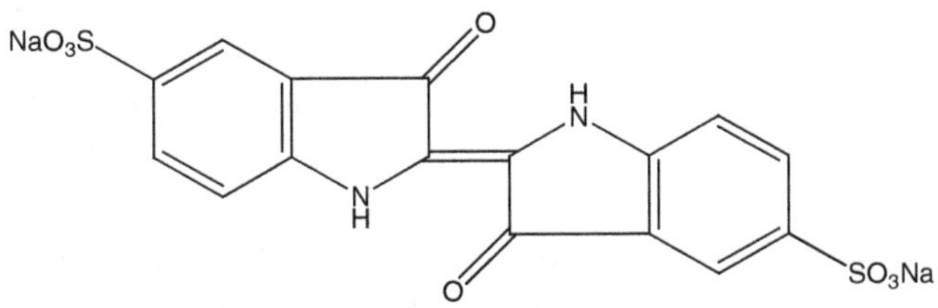

CA Index Name 1*H*-Indole-5-sulfonic acid, 2-(1,3-dihydro-3-oxo-5-sulfo-2*H*-indol-2-ylidene)-2,3-dihydro-3-oxo-, sodium salt

Other Names 1*H*-Indole-5-sulfonic acid, 2-(1,3-dihydro-3-oxo-5-sulfo-2*H*-indol-2-ylidene)-2,3-dihydro-3-oxo-, disodium salt; C.I. Acid Blue 74; [Δ2,2′-Biindoline]-5,5′-disulfonic acid, 3,3′-dioxo-, disodium salt; 12070 Blue; 1311 Blue; 5,5′-Indigodisulfonic acid disodium salt; A.F. Blue No. 2; Acid Blue 74; Acid Blue W; Acid Leather Blue IC; Airedale Blue IN; Amacid Brilliant Blue; Aniline Carmine Powder; Ariavit Indigo Carmine; Atul Indigo Carmine; Basovit Blue 665E; Blue 2; Bucacid Indigotine B; C.I. 73015; C.I. 75781; C.I. Food Blue 1; C.I. Natural Blue 2; Canacert Indigo Carmine; Carmine Blue; Cilefa Blue R; Cogilor Blue 511.11; Disodium 5,5′-indigodisulfonate; Disodium 5,5′-indigotin disulfonate; Dolkwal Indigo Carmine; E 132; Edicol Supra Blue X; Eurocert Indigo Carmine 311811; FD & C Blue 2; FD & C Blue No. 2-307045; FD and C Blue 2; FD and C Blue No. 2; FD&C Blue No. 2; FD&C Blue No. 2-37006; Food Blue 1; Food Blue 2; Food Blue No. 1; Food Blue No. 2; Grape Blue A; HD Indigo Carmine; HD Indigo Carmine Supra; Hexacert Blue No. 2; Hexacol Indigo Carmine Supra; Indigo Carmine 307019; Indigo Carmine 36009; Indigo Carmine 37006; Indigo Carmine A; Indigo Carmine AC; Indigo Carmine BP; Indigo Carmine Conc. FQ; Indigo Carmine Powder; Indigo Carmine X; Indigo Extract; Indigo carmine; Indigo carmine NB; Indigotin; Indigotin (solubilized); Indigotine; Indigotine B; Indigotine Blue LZ; Indigotine Carmine; Indigotine Extra Pure A; Indigotine I; Indigotine IA; Indigotine Lake; Indigotine N; Indigotine disodium salt; Indocarmine F; Intense Blue; Japan Blue 2; Japan Food Blue No. 2; L Blue 5010; Maple Indigo Carmine; Mitsui Indigo Carmine; Necol Indigo Carmine; Neelicol Indigo Carmine; San-ei Indigo Carmine; Sepisperse Dry 1003; Sodium 5,5′-indigodisulfonate; Sodium 5,5′-indigotindisulfonate; Soluble indigo; Soluble indigo blue; Sumitomo Wool Blue SBC; Usacert Blue No. 2; Usacert FD & C Blue No. 2-310118; Usacert FD and C Blue No. 2; WAS 35; Water Blue 177557; Water Blue 177558

Merck Index Number 4944

Chemical/Dye Class Indole

Molecular Formula $C_{16}H_8N_2Na_2O_8S_2$

Molecular Weight 466.35

Physical Form Dark blue-purple powder, sensitive to light

Solubility Soluble in water; slightly soluble in ethanol

Melting Point >250 °C

pH Range 11.5–14.0

Color Change at pH Blue (11.5) to yellow (14.0)

Absorption (λ_{max}) 608 nm

Synthesis Synthetic methods[1–23]

Staining Applications Cells;[24,25] beverages;[26–28] chewing gum;[29] candies;[30] drinks;[30] frozen products;[31] sweetener;[32] tablets;[33,34] dosage form;[35] sunscreen;[36] skin;[37] hairs[1,38–40]

Biological Applications Detecting microorganisms;[41] treating amyloidosis,[42] testicular cancer;[43] medical devices;[44] drug delivery system[45]

Industrial Applications Color filters;[46] display devices;[1,47] inks;[1,48,49] toners;[50] lithographic printing plates;[51,52] paints;[53] photographic material;[1,54] colored bubbles;[1,55] toys[56]

Safety/Toxicity Acute toxicity;[57] carcinogenicity;[1,58,59,61] chromosome aberrations;[60] chronic toxicity;[61] cytotoxicity;[1,62] genotoxicity;[1,63] hypertension;[1,64] hypotension;[1,65] mutagenicity;[1,66,68] nucleic acid damage[66,67]

Certification/Approval Certified by Biological Stain Commission (BSC); Approved by Food & Drugs Administration (FDA)

Handbook of Biological Dyes and Stains By R. W. Sabnis

REFERENCES

1. Sabnis, R. W. *Handbook of Acid–Base Indicators*; CRC Press: Boca Raton, **2008**; pp 193–195.
2. Welcher, F. J. *Organic Analytical Reagents*; D. Van Nostrand Company: New York, **1948**; Vol. 4, pp 505–511.
3. Shadi, I. T.; Chowdhry, B. Z.; Snowden, M. J.; Withnall, R. Analysis of the conversion of indigo into indigo carmine dye using SERRS. *Chem. Commun.* **2004**, 1436–1437.
4. Kabasawa, Y.; Shinomiya, K.; Nakazawa, H.; Tanimura, T.; Fujita, M. Purification of Food Blue No. 1 by rotation locular countercurrent chromatography. *Anal. Sci.* **1991**, *7*, 273–276.
5. Salas-Peregrin, J. M.; Suarez-Varela, J. Synthesis, characterization and thermal behavior of some metal indigodisulfonates. *J. Therm. Anal.* **1984**, *29*, 515–521.
6. Ogawa, Y. Purification of cosmetic colors from coal tar. Jpn. Kokai Tokkyo Koho JP 53041439, 1978; *Chem. Abstr.* **1978**, *89*, 135674.
7. Fridman, S. A.; Kenin, S. L.; Antonova, E. G.; Balunina, L. P. Removal of impurities from the synthetic dye indigo carmine. U.S.S.R. SU 274284, 1970; *Chem. Abstr.* **1971**, *74*, 43521.
8. Jones, J. H.; Harrow, L. S.; Heine, K. S., Jr. Coal-tar colors. XX. FD & C Blue No. 2. *J. Assoc. Off. Agric. Chem.* **1955**, *38*, 949–977.
9. Jewett, J. E. Treating molten materials with liquids as in indigo manufacture. U.S. Patent 1934716, 1933; *Chem. Abstr.* **1934**, *28*, 7236.
10. Shibata, R. N.; Okuyama, M.; Okamura, K. Sandmeyer's indigo synthesis. *Kogyo Kagaku Zasshi* **1933**, *36*, 569–571; *Chem. Abstr.* **1934**, *28*, 3609.
11. Spalding, W. L. Treatment of indoxyl for indigo production. U.S. Patent 1827828, 1931; *Chem. Abstr.* **1932**, *26*, 7438.
12. Overmyer, C. J. Synthesis of substitution derivatives of indigo. II. Ethyl nitrotrimethylgallyl acetate and related compounds. *J. Am. Chem. Soc.* **1927**, *49*, 499–509.
13. Gupta, S. N. Indigo. U.S. Patent 1588960, 1926; *Chem. Abstr.* **1926**, *20*, 21299.
14. Overmyer, C. J. Synthesis of substitution derivatives of indigo. I. *o*-Nitrobenzoylacetic acid and related compounds. *J. Am. Chem. Soc.* **1926**, *48*, 454–460.
15. Martinet, J. The chemical synthesis of artificial indigo. *Chim. Ind.* (*Paris*) **1925**, *13*, 531–536; *Chem. Abstr.* **1925**, *19*, 16447.
16. Nanson, W. B. Indigo. *Cotton* **1922**, *86*, 695–697, 786.
17. Bailey, G. C.; Potter, R. S. Synthesis of indigo from fumaric acid and aniline. *J. Am. Chem. Soc.* **1922**, *44*, 215–216.
18. Bergve, E. Alkaline melts for chemical reactions. U. S. Patent 1354451, 1920; *Chem. Abstr.* **1921**, *15*, 176.
19. Shibata, T. J. Synthesis of indigo and 6,6′-dimethylindigo. *Chem. Ind. Tokyo* **1920**, *23*, 373–379; *Chem. Abstr.* **1920**, *14*, 15693.
20. Matthews, J. M. Indigo carmine. II. *Color Trade J.* **1920**, *7*, 8–11.
21. Holland, P. Continuous scheme for preparation of indigotin froth phthalimide. *Chem. News J. Ind. Sci.* **1920**, *120*, 208–209.
22. Machida, S.; Irisawa, J. Preparation of indigo from wild indigo plants. JP 34841, 1919; *Chem. Abstr.* **1920**, *14*, 12172.
23. Matthews, J. M. Indigo carmine. *Color Trade J.* **1920**, *6*, 96–99.
24. Yamamoto, A.; Iimori, Y.; Koyama, M. Compositions and method for staining of cell surfaces or tissue surfaces. Jpn. Kokai Tokkyo Koho JP 2007223966, 2007; *Chem. Abstr.* **2007**, *147*, 317775.
25. Yamamoto, A.; Iimori, Y.; Koyama, M. Cell or tissue inside staining agent composition, and staining method using it. Jpn. Kokai Tokkyo Koho JP 2007222100, 2007; *Chem. Abstr.* **2007**, *147*, 317774.
26. Tennant, D. R. Screening potential intakes of colour additives used in non-alcoholic beverages. *Food Chem. Toxicol.* **2008**, *46*, 1985–1993.
27. Komissarchik, S. M.; Nyanikova, G. G. Method of determination of synthetic dyes in alcoholic beverages. Russ. RU 2324179, 2008; *Chem. Abstr.* **2008**, *148*, 536939.
28. Huang, H. Y.; Shih, Y. C.; Chen, Y. C. Determining eight colorants in milk beverages by capillary electrophoresis. *J. Chromatogr. A* **2002**, *959*, 317–325.
29. Salini, A. Carrageenan-based chewing gum. Eur. Pat. Appl. EP 1946751, 2008; *Chem. Abstr.* **2008**, *149*, 174989.
30. Yoshioka, N.; Ichihashi, K. Determination of 40 synthetic food colors in drinks and candies by high-performance liquid chromatography using a short column with photodiode array detection. *Talanta* **2008**, *74*, 1408–1413.

31. Stanley, S. A.; Pachivas, J. A.; Annis, J. R. Frozen product and method of oral delivery of active ingredients. U.S. Patent 6258384, 2001; *Chem. Abstr.* **2001**, *135*, 97458.

32. Catani, S. J.; Liao, S.; Loades, M. Self-mixing tabletop sweetener. U.S. Pat. Appl. Publ. US 2007059418, 2007; *Chem. Abstr.* **2007**, *146*, 315560.

33. Lobo, M.; Patel, J.; Kamins, G.; Francis, R.; Breza, B.; Jerzewski, R. Interaction of omapatrilat with FD&C Blue No. 2 lake during dissolution of modified release tablets. *Int. J. Pharm.* **2007**, *339*, 168–174.

34. Moroni, A.; Drefko, W. J. Pharmaceutical tablet coating composition. U.S. Pat. Appl. Publ. US 2004001884, 2004; *Chem. Abstr.* **2004**, *140*, 65216.

35. Nadkarni, S. S.; Vaya, N.; Karan, R. S.; Gupta, V. K. Novel dosage form. Indian Pat. Appl. IN 2005MU01013, 2007; *Chem. Abstr.* **2007**, *148*, 387155.

36. Ross, J. S.; Morefield, E. M. Sunscreen with disappearing color indicator. U.S. Patent 5747011, 1998; *Chem. Abstr.* **1998**, *129*, 8406.

37. Magallon, A. G.; Boone, K. W.; Bardwell, J. R. Aqueous antiseptic solution and compatible dye for staining skin. U.S. Pat. Appl. Publ. US 2007253909, 2007; *Chem. Abstr.* **2007**, *147*, 491676.

38. Greaves, E.; Greaves, J. T. Hair coloring compositions containing plant dyes and metal salts or mineral salts. U.S. Pat. Appl. Publ. US 2007251024, 2007; *Chem. Abstr.* **2007**, *147*, 508055.

39. Sallwey, A.; Schmitt, M.; Dieckow, A. Dyestuff for hair and fibers. Eur. Pat. Appl. EP 976385, 2000; *Chem. Abstr.* **2000**, *132*, 127457.

40. Kunz, M.; Le Cruer, D.; Balzer, W. R. Direct hair dye. Ger. Offen. DE 19618595, 1997; *Chem. Abstr.* **1997**, *128*, 39385.

41. Martin, S. M.; MacDonald, J. G.; Lye, J.; Sayre, C.; Thompson, K. Array for rapid detection of a microorganism based on dyes. PCT Int. Appl. WO 2008026104, 2008; *Chem. Abstr.* **2008**, *148*, 302867.

42. Kisilevsky, R.; Szarek, W.; Weaver, D. Method for treating amyloidosis. U.S. Patent 5840294, 1998; *Chem. Abstr.* **1998**, *130*, 20587.

43. Sidorenko, Y. S.; Raevskii, V. A. Method for treating testicular cancer. Russ. RU 2195876, 2003; *Chem. Abstr.* **2003**, *138*, 297641.

44. Raad, I. Coating medical device surfaces with antimicrobial agents. PCT Int. Appl. WO 2007062306, 2007; *Chem. Abstr.* **2007**, *147*, 39321.

45. Nadkarni, S. S.; Vaya, N.; Karan, R. S.; Gupta, V. K. Novel drug delivery system. Indian Pat. Appl. IN 2005MU01012, 2007; *Chem. Abstr.* **2007**, *148*, 503081.

46. Sakamoto, M.; Shimohata, Y.; Uehara, K.; Aoki, K. Self-dispersible color fine particles, water-thinned dispersions containing them, and their inks with good water and light resistance for ink-jet printing or color filters. Jpn. Kokai Tokkyo Koho JP 2008031194, 2008; *Chem. Abstr.* **2008**, *148*, 216986.

47. Vincent, J. B.; Flick, D. W. Electrochromic display device. U.S. Pat. Appl. Publ. US 2003179432, 2003; *Chem. Abstr.* **2003**, *139*, 283537.

48. Iida, Y.; Higo, S.; Furukawa, K. Polyglycerin fatty ester-containing screen inks and pressure-sensitive transfer sheets printed therewith. Jpn. Kokai Tokkyo Koho JP 2008143992, 2008; *Chem. Abstr.* **2008**, *149*, 106640.

49. Nishio, A.; Mukoda, S.; Okamoto, E. Aqueous inks with long-lasting fragrance. Jpn. Kokai Tokkyo Koho JP 2001040262, 2001; *Chem. Abstr.* **2001**, *134*, 164644.

50. Ono, H.; Maekawa, K. Electrophotographic toner. Jpn. Kokai Tokkyo Koho JP 08123094, 1996; *Chem. Abstr.* **1996**, *125*, 100086.

51. Kawamura, T. Lithographic printing plate material and its manufacturing method. Jpn. Kokai Tokkyo Koho JP 2008018593, 2008; *Chem. Abstr.* **2008**, *148*, 179367.

52. Takagi, M. Lithographic printing plate with good blue semiconductor laser sensitivity. Jpn. Kokai Tokkyo Koho JP 2003107716, 2003; *Chem. Abstr.* **2003**, *138*, 311605.

53. Franer, V. R. Color dissipating paint. U.S. Patent 5548010, 1996; *Chem. Abstr.* **1996**, *125*, 250497.

54. Tsukada, K. Processing of silver halide photographic material. Jpn. Kokai Tokkyo Koho JP 06258786, 1994; *Chem. Abstr.* **1995**, *122*, 174189.

55. Sabnis, R. W.; Kehoe, T. D. Composition and method for producing colored bubbles. U.S. Pat. Appl. Publ. US 2006004110, 2006; *Chem. Abstr.* **2006**, *144*, 110164.

56. Ono, Y. Water-erasable markers for toys and figure toy sets therewith. Jpn. Kokai Tokkyo Koho JP 2002129085, 2002; *Chem. Abstr.* **2002**, *136*, 371209.

57. Tonogai, Y.; Ito, Y.; Iwaida, M.; Tati, M.; Ose, Y.; Hori, M. Studies on the toxicity of coal-tar dyes. III. Reason of acute toxicity to fish caused by coal-tar

dyes and their industrial effluents. *J. Toxicol. Sci.* **1980**, *5*, 23–33.

58. Helguera, A. M.; Cabrera Perez, M. A.; Gonzalez, M. P.; Ruiz, R. M.; Gonzalez Diaz, H. A topological substructural approach applied to the computational prediction of rodent carcinogenicity. *Bioorg. Med. Chem.* **2005**, *13*, 2477–2488.
59. Haseman, J. K.; Winbush, J. S.; O'Donnell, M. W., Jr. Use of dual control groups to estimate false positive rates in laboratory animal carcinogenicity studies. *Fundam. Appl. Toxicol.* **1986**, *7*, 573–584.
60. Serra, J. R.; Thompson, E. D.; Jurs, P. C. Development of binary classification of structural chromosome aberrations for a diverse set of organic compounds from molecular structure. *Chem. Res. Toxicol.* **2003**, *16*, 153–163.
61. Borzelleca, J. F.; Hogan, G. K.; Koestner, A. Chronic toxicity/carcinogenicity study of FD and C Blue No. 2 in rats. *Food Chem. Toxicol.* **1985**, *23*, 551–558.
62. Sako, F.; Kobayashi, N.; Watabe, H.; Taniguchi, N. Cytotoxicity of food dyes on cultured fetal rat hepatocytes. *Toxicol. Appl. Pharmacol.* **1980**, *54*, 285–292.
63. Haveland-Smith, R. B.; Combes, R. D. Screening of food dyes for genotoxic activity. *Food Cosmet. Toxicol.* **1980**, *18*, 215–221.
64. Jeffords, D. L.; Lange, P. H.; DeWolf, W. C. Severe hypertensive reaction to indigo carmine. *Urology* **1977**, *9*, 180–181.
65. Shir, Y.; Raja, S. N. Indigo carmine-induced severe hypotension in patients undergoing radical prostatectomy. *Anesthesiology* **1993**, *79*, 378–381.
66. Ozaki, A.; Kitano, M.; Itoh, N.; Kuroda, K.; Furusawa, N.; Masuda, T.; Yamaguchi, H. Mutagenicity and DNA-damaging activity of decomposed products of food colors under UV irradiation. *Food Chem. Toxicol.* **1998**, *36*, 811–817.
67. Rhee, Y.; Valentine, M. R.; Termini, J. Oxidative base damage in RNA detected by reverse transcriptase. *Nucleic Acids Res.* **1995**, *23*, 3275–3282.
68. Ishidate, M., Jr.; Sofuni, T.; Yoshikawa, K.; Hayashi, M.; Nohmi, T.; Sawada, M.; Matsuoka, A. Primary mutagenicity screening of food additives currently used in Japan. *Food Chem. Toxicol.* **1984**, *22*, 623–636.

INDO 1

CAS Registry Number 132319-56-3

Chemical Structure

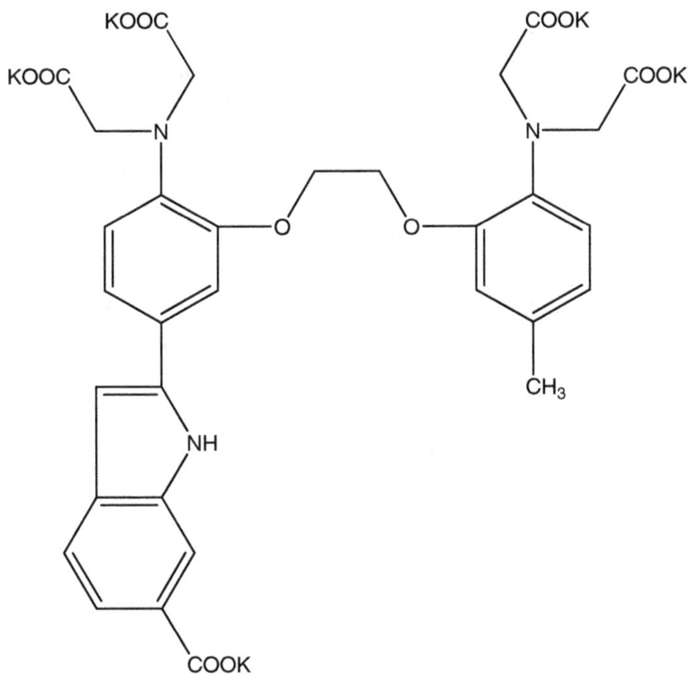

CA Index Name 1*H*-Indole-6-carboxylic acid, 2-[4-[bis(carboxymethyl)amino]-3-[2-[2-[bis(carboxymethyl)amino]-5-methylphenoxy]ethoxy]phenyl]-, potassium salt (1 : 5)

Other Names 1*H*-Indole-6-carboxylic acid, 2-[4-[bis(carboxymethyl)amino]-3-[2-[2-[bis(carboxymethyl)amino]-5-methylphenoxy]ethoxy]phenyl]-, pentapotassium salt

Merck Index Number Not listed (Free acid: 4960)

Chemical/Dye Class Indole

Molecular Formula $C_{32}H_{26}K_5N_3O_{12}$

Molecular Weight 840.05

Physical Form Light grey crystals

Solubility Soluble in water

Melting Point >250 °C

Absorption (λ_{max}) 346 nm, 330 nm

Emission (λ_{max}) 475 nm, 401 nm

Synthesis Synthetic methods[1–3]

Staining Applications Calcium ions;[2–11] cadmium ions;[12] lead ions;[13] zinc ions;[14] peptides;[1] proteins;[1] antibodies[1]

Biological Applications Calcium indicator;[2–11] cadmium indicator;[12] lead indicator;[13] zinc indicator;[14] identifying taste modulators;[15] measuring membrane potential;[16] nucleic acid sequencing;[17] preventing arrhythmias[18]

Industrial Applications Not reported

Safety/Toxicity Neurotoxicity[19,20]

REFERENCES

1. Hahn, K. M.; Toutchkine, A.; Muthyala, R.; Kraynov, V.; Bark, S. J.; Burton, D. R.; Chamberlain, C. Labeled peptides, proteins and antibodies and processes and intermediates useful for their preparation. U.S. Pat. Appl. Publ. US 2002055133, 2002; *Chem. Abstr.* **2002**, *136*, 366139.
2. Vo-Dinh, T.; Viallet, P.; Ramirez, L.; Pal, A. Gel-based Indo-1 probe for monitoring calcium(II) ions. *Anal. Chem.* **1994**, *66*, 813–817.
3. Grynkiewicz, G.; Poenie, M.; Tsien, R. Y. A new generation of calcium indicators with greatly improved fluorescence properties. *J. Biol. Chem.* **1985**, *260*, 3440–3450.
4. Bailey, S.; Macardle, P. J. A flow cytometric comparison of Indo-1 to fluo-3 and Fura Red excited with low power lasers for detecting Ca^{2+} flux. *J. Immunol. Methods* **2006**, *311*, 220–225.
5. Nelemans, A. Measurement of $[Ca^{2+}]_i$ in cell suspensions using Indo-1. *Methods Mol. Biol.* **2005**, *312*, 47–53.
6. Eerbeek, O.; Mik, E. G.; Zuurbier, C. J.; van't Loo, M.; Donkersloot, C.; Ince, C. Ratiometric intracellular calcium imaging in the isolated beating rat heart using indo-1 fluorescence. *J. Appl. Physiol.* **2004**, *97*, 2042–2050.
7. Black, D. J.; Tran, Q. K.; Persechini, A. Monitoring the total available calmodulin concentration in intact cells over the physiological range in free Ca^{2+}. *Cell Calcium* **2004**, *35*, 415–425.
8. Brustein, E.; Marandi, N.; Kovalchuk, Y.; Drapeau, P.; Konnerth, A. *In vivo* monitoring of neuronal network activity in zebrafish by two-photon Ca^{2+} imaging. *Pfluegers Arch.* **2003**, *446*, 766–773.

9. Scheenen, W. J. J. M.; Makings, L. R.; Gross, L. R.; Pozzan, T.; Tsien, R. Y. Photodegradation of indo-1 and its effect on apparent Ca^{2+} concentrations. *Chem. Biol.* **1996**, *3*, 765–774.
10. Levi, A. J.; Li, J.; Litwin, S. E.; Spitzer, K. W. Effect of internal sodium and cellular calcium load on voltage-dependence of the Indo-1 transient in guinea pig ventricular myocytes. *Cardiovasc. Res.* **1996**, *32*, 534–550.
11. Levi, A. J.; Li, J.; Spitzer, K. W.; Bridge, J. H. B. Effect on the indo-1 transient of applying Ca^{2+} channel blocker for a single beat in voltage-clamped guinea pig cardiac myocytes. *J. Physiol.* **1996**, *494*, 653–673.
12. Vo-Dinh, T.; Viallet, P.; Ramirez, L.; Pal, A.; Vigo, J. Detection of cadmium ion using the fluorescence probe Indo-1. *Anal. Chim. Acta* **1994**, *295*, 67–72.
13. Esposito, A.; Robello, M.; Pellistri, F.; Marchetti, C. Two-photon analysis of lead accumulation in rat cerebellar granule neurons. *Neurochem. Res.* **2005**, *30*, 949–954.
14. Jefferson, J. R.; Hunt, J. B.; Ginsburg, A. Characterization of indo-1 and quin-2 as spectroscopic probes for $zinc^{(2+)}$-protein interactions. *Anal. Biochem.* **1990**, *187*, 328–336.
15. Servant, G.; Kamdar, P.; Rivadeneyra, A. Novel cell-based assays for identifying enhancers or inhibitors of T1R taste receptors (T1R2/T1R3 sweet) and umami (T1R1/T1R3 umami) taste receptors. U.S. Pat. Appl. Publ. US 2008305500, 2008; *Chem. Abstr.* **2008**, *150*, 31118.
16. Klaubert, D.; Diwu, Z.; Yi, G.; Kirk, M. Improved method for measuring membrane potential. PCT Int. Appl. WO 2003014701, 2003; *Chem. Abstr.* **2003**, *138*, 166188.
17. Hoser, M. J. Methods for nucleic acid sequencing. PCT Int. Appl. WO 2004074503, 2004; *Chem. Abstr.* **2004**, *141*, 237698.
18. Marban, E.; Abraham, M. R. Preventing arrhythmias associated with cell transplantation. PCT Int. Appl. WO 2005092033, 2005; *Chem. Abstr.* **2005**, *143*, 319156.
19. Marin, P.; Israel, M.; Glowinski, J.; Premont, J. Routes of zinc entry in mouse cortical neurons: role in zinc-induced neurotoxicity. *Eur. J. Neurosci.* **2000**, *12*, 8–18.
20. Legare, M. E.; Barhoumi, R.; Hebert, E.; Bratton, G. R.; Burghardt, R. C.; Tiffany-Castiglioni, E. Analysis of Pb^{2+} entry into cultured astroglia. *Toxicol. Sci.* **1998**, *46*, 90–100.

INDO 1 AM

CAS Registry Number 112926-02-0

Chemical Structure

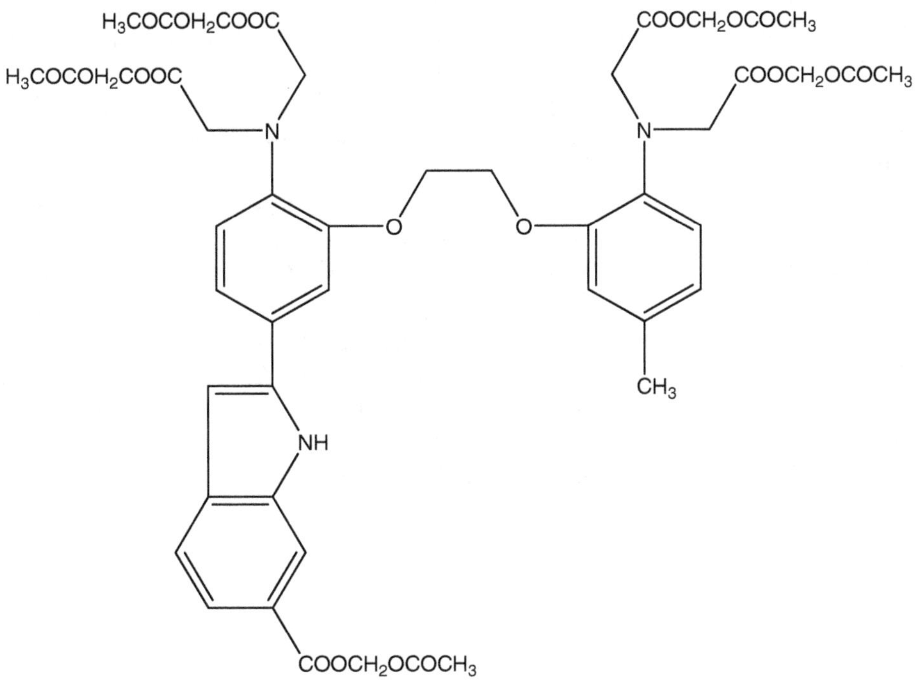

CA Index Name 1*H*-Indole-6-carboxylic acid, 2-[4-[bis[2-[(acetyloxy)methoxy]-2-oxoethyl]amino]-3-[2-[2-[bis[2-[(acetyloxy)methoxy]-2-oxoethyl]amino]-5-methylphenoxy]ethoxy]phenyl]-, (acetyloxy)methyl ester

Other Names Indo 1 acetoxymethyl ester; Indo 1 AM; Indo 1 AM ester

Merck Index Number Not listed

Chemical/Dye Class Indole

Molecular Formula $C_{47}H_{51}N_3O_{22}$

Molecular Weight 1009.91

Physical Form Light yellow powder

Solubility Soluble in dimethyl sulfoxide, methanol

Melting Point >250 °C

Boiling Point (Calcd.) 1009.2 ± 65.0 °C, pressure: 760 Torr

pK_a (Calcd.) 14.85 ± 0.30, most acidic temperature: 25 °C; 1.92 ± 0.50, most basic temperature: 25 °C

Absorption (λ_{max}) 356 nm

Emission (λ_{max}) 478 nm

Synthesis Synthetic methods[1,2]

Staining Applications Calcium ions;[2,4–11] leukocyte tumor cells[3]

Biological Applications Calcium indicator;[2,4–11] preventing arrhythmias[12]

Industrial Applications Not reported

Safety/Toxicity Cytotoxicity[13]

REFERENCES

1. Yang, L.; Zhang, X.; You, F.; Tu, H.; Sheng, M.; Zhu, P. Synthesis of fluorescence indicator indo-1 AM. *Hecheng Huaxue* **1993**, *1*, 341–347; *Chem. Abstr.* **1994**, *121*, 82940.
2. Grynkiewicz, G.; Poenie, M.; Tsien, R. Y. A new generation of calcium indicators with greatly improved fluorescence properties. *J. Biol. Chem.* **1985**, *260*, 3440–3450.
3. Minakami, T.; Tsuji, T.; Oguni, S.; Hamaguchi, Y.; Tsuruta, K.; Kamihiro, T. Method for detecting leukocyte tumor cells. Jpn. Kokai Tokkyo Koho JP 2002207036, 2002; *Chem. Abstr.* **2002**, *137*, 106038.
4. Nelemans, A. Measurement of $[Ca^{2+}]_i$ in cell suspensions using Indo-1. *Methods Mol. Biol.* **2005**, *312*, 47–53.

5. Brewis, I. A.; Morton, I. E.; Mohammad, S. N.; Browes, C. E.; Moore, H. D. M. Measurement of intracellular calcium concentration and plasma membrane potential in human spermatozoa using flow cytometry. *J. Androl.* **2000**, *21*, 238–249.
6. Toth, A.; Ivanics, T.; Ruttner, Z.; Slaaf, D. W.; Reneman, R. S.; Ligeti, L. Quantitative assessment of $[Ca^{2+}]_i$ levels in rat skeletal muscle *in vivo*. *Am. J. Physiol.* **1998**, *275*, H1652–H1662.
7. Zhou, Z.; Matlib, M. A.; Bers, D. M. Cytosolic and mitochondrial Ca^{2+} signals in patch clamped mammalian ventricular myocytes. *J. Physiol.* **1998**, *507*, 379–403.
8. Schreur, J. H. M.; Figueredo, V. M.; Miyamae, M.; Shames, D. M.; Baker, A. J.; Camacho, S. A. Cytosolic and mitochondrial $[Ca^{2+}]$ in whole hearts using indo-1 acetoxymethyl ester: effects of high extracellular Ca^{2+}. *Biophys. J.* **1996**, *70*, 2571–2580.
9. Lorell, B. H.; Apstein, C. S.; Cunningham, M. J.; Schoen, F. J.; Weinberg, E. O.; Peeters, G. A.; Barry, W. H. Contribution of endothelial cells to calcium-dependent fluorescence transients in rabbit hearts loaded with indo 1. *Circ. Res.* **1990**, *67*, 415–425.
10. Elliott, D. C.; Petkoff, H. S. Measurement of cytoplasmic free calcium in plant protoplasts. *Plant Sci.* **1990**, *67*, 125–131.
11. Lopez, M.; Olive, D.; Mannoni, P. Analysis of cytosolic ionized calcium variation in polymorphonuclear leukocytes using flow cytometry and Indo-1 AM. *Cytometry* **1989**, *10*, 165–173.
12. Marban, E.; Abraham, M. R. Preventing arrhythmias associated with cell transplantation. PCT Int. Appl. WO 2005092033, 2005; *Chem. Abstr.* **2005**, *143*, 319156.
13. Aubry, J. P.; Blaecke, A.; Lecoanet-Henchoz, S.; Jeannin, P.; Herbault, N.; Caron, G.; Moine, V.; Bonnefoy, J. Y. Annexin V used for measuring apoptosis in the early events of cellular cytotoxicity. *Cytometry* **1999**, *37*, 197–204.

IODONITRO TETRAZOLIUM (INT)

CAS Registry Number 146-68-9

Chemical Structure

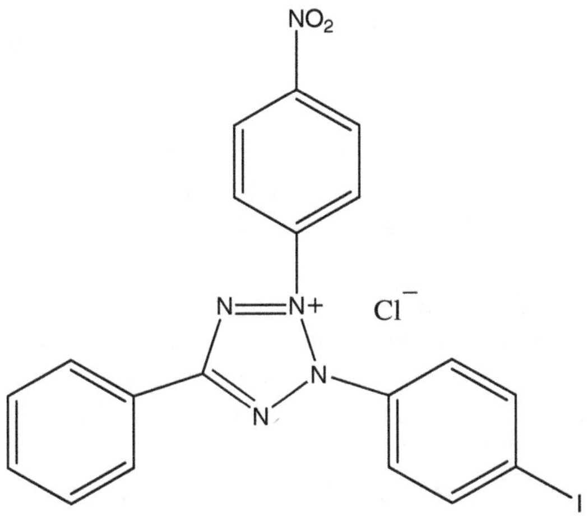

CA Index Name 2*H*-Tetrazolium, 2-(4-iodophenyl)-3-(4-nitrophenyl)-5-phenyl-, chloride (1 : 1)

Other Names 2*H*-Tetrazolium, 2-(4-iodophenyl)-3-(4-nitrophenyl)-5-phenyl-, chloride; 2*H*-Tetrazolium, 2-(*p*-iodophenyl)-3-(*p*-nitrophenyl)-5-phenyl-, chloride; [2-(*p*-Iodophenyl)-3-(*p*-nitrophenyl)-5-phenyl-2*H*-tetrazolium chloride]; 2-(4-Iodophenyl)-3-(4-nitrophenyl)-5-phenyl-2*H*-tetrazolium chloride; 2-(4-Iodophenyl)-3-(4-nitrophenyl)-5-phenyltetrazolium chloride; 2-(*p*-Iodophenyl)-3-(*p*-nitrophenyl)-5-phenyltetrazolium chloride; 3-(*p*-Nitrophenyl)-2-(*p*-iodophenyl)-5-phenyltetrazolium chloride; INT; Iodonitro tetrazolium; Iodonitrotetrazolium purple; Iodonitrotetrazolium violet; NSC 27620; *p*-Iodonitrotetrazolium violet

Merck Index Number Not listed

Chemical/Dye Class Tetrazolium salt

Molecular Formula $C_{19}H_{13}ClIN_5O_2$

Molecular Weight 505.70

Physical Form Light yellow powder

Solubility Soluble in water, ethanol, dimethyl sulfoxide

Melting Point 229–231 °C (decompose)

Absorption (λ_{max}) 248 nm

Synthesis Synthetic methods[1–7]

Staining Applications Cells;[8] oxidized nicotinamide adenine dinucleotide kinase;[9] sperms[10]

Biological Applications Bacterial vaginosis diagnosis assay;[11] dehydrogenase enzyme assay;[12–14] diagnostic assay;[15] food and beverage analytes assays;[16] microbial growth assay;[17] detecting bacteria,[18] yeast,[18] fungi,[18] gamma-hydroxybutyric acid (GHB),[19] microbial growth;[20] measuring niacin,[21] bacterial respiratory activity,[22] superoxide dismutase;[23] treating cancer[24]

Industrial Applications Steel products[25]

Safety/Toxicity Bacterial toxicity;[22] heavy metal inhibition;[26] microbial toxicity[27]

REFERENCES

1. Cui, X.; Vlahakis, J. Z.; Crandall, I. E.; Szarek, W. A. Anti-*Plasmodium* activity of tetrazolium salts. *Bioorg. Med. Chem.* **2008**, *16*, 1927–1947.
2. James, P. R. Tetrazolium salts as steam sterilization process indicators. PCT Int. Appl. WO 9837157, 1998; *Chem. Abstr.* **1998**, *129*, 218237.
3. Malev, M. N. Preparation of 2-(*p*-iodophenyl)-3-(*p*-nitrophenyl)-5-phenyltetrazole chloride. U.S.S.R. SU 196874, 1967; *Chem. Abstr.* **1968**, *68*, 95826.
4. Ostrovskaya, V. M.; Pryanishnikov, A. A. Iodonitrotetrazolium chloride: 3-(4-iodophenyl)-2-(4-nitrophenyl)-5-phenyltetrazo-lium chloride. *Metody Poluch. Khim. Reakt. Prep.* **1964**, 18–20; *Chem. Abstr.* **1966**, *65*, 99352.
5. Pryanishnikov, A. A.; Ostrovskaya, V. M.; Raikhlin, N. T. Synthesis, properties, and histochemical application of salts of tetrazole with electron-accepting substituents. *Tr. Vses. Nauchn.-Issled. Inst. Khim. Reakt.* **1963**, 139–148; *Chem. Abstr.* **1964**, *61*, 4214.
6. Moeckel, P.; Staerk, G. Substituted tetrazolium salts. German (East) DD 22410, 1961; *Chem. Abstr.* **1963**, *58*, 46792.
7. Fox, S. W.; Atkinson, E. H. Synthesis of some substituted tetrazolium chlorides. *J. Am. Chem. Soc.* **1950**, *72*, 3629–3631.
8. Berridge, M. V.; Herst, P. M.; Tan, A. S. Tetrazolium dyes as tools in cell biology: new insights into their cellular reduction. *Biotechnol. Annu. Rev.* **2005**, *11*, 127–152.
9. Mailloux, R. J.; Singh, R.; Appanna, V. D. In-gel activity staining of oxidized nicotinamide adenine dinucleotide kinase by blue native polyacrylamide gel electrophoresis. *Anal. Biochem.* **2006**, *359*, 210–215.

10. Hirata, R.; Kawagoe, M.; Aono, H. Reagent and method for enzymatically measuring spermatozoon number. Jpn. Kokai Tokkyo Koho JP 2004129620, 2004; *Chem. Abstr.* **2004**, *140*, 353210.

11. Dwir, O. S.; Keren, T.; Fish, F. Dry format sialidase colorimetric assay for bacterial vaginosis diagnosis. PCT Int. Appl. WO 2008018074, 2008; *Chem. Abstr.* **2008**, *148*, 232647.

12. Mosher, J. J.; Levison, B. S.; Johnston, C. G. A simplified dehydrogenase enzyme assay in contaminated sediment using 2-(*p*-iodophenyl)-3 (*p*-nitrophenyl)-5-phenyl tetrazolium chloride. *J. Microbiol. Methods* **2003**, *53*, 411–415.

13. Mathew, M.; Obbard, J. P. Optimisation of the dehydrogenase assay for measurement of indigenous microbial activity in beach sediments contaminated with petroleum. *Biotechnol. Lett.* **2001**, *23*, 227–230.

14. Gong, P. Dehydrogenase activity in soil: a comparison between the TTC and INT assay under their optimum conditions. *Soil Biol. Biochem.* **1997**, *29*, 211–214.

15. Wong, S. T.; Hiltibran, R. G.; Huang, T. M.; Calfin, B. B.; Pope, M. R.; Schapira, T. G.; Shain, E. B.; Young, D. F. Diagnostic assay and multilayer assay apparatus for a sample of biological fluid. U.S. Patent 6312888, 2001; *Chem. Abstr.* **2001**, *135*, 341131.

16. McCleary, B. V.; Charnock, S. A kit for colorimetric assays of food and beverage analytes, especially those in wine. PCT Int. Appl. WO 2006064488, 2006; *Chem. Abstr.* **2006**, *145*, 61798.

17. Storey, D.; Field, T.; Ramage, G. Microbial growth assay and assay device. PCT Int. Appl. WO 2005098023, 2005; *Chem. Abstr.* **2005**, *143*, 402159.

18. Horn, J. Improved growth and colorimetric detection of bacteria, yeast, fungi, or cocci. Ger. Offen. DE 19602345, 1997; *Chem. Abstr.* **1997**, *127*, 158788.

19. Parsons, S. M.; Harris, D. O.; Bravo, D. T. Methods, compositions and apparatuses for detection of gamma-hydroxybutyric acid (GHB). U.S. Pat. Appl. Publ. US 2003175846, 2003; *Chem. Abstr.* **2003**, *139*, 256601.

20. Ushiyama, M. Tetrazolium salts and polyhydric alcohols for microbial growth and detection. Jpn. Kokai Tokkyo Koho JP 2005287452, 2005; *Chem. Abstr.* **2005**, *143*, 363250.

21. Jacobson, E. L.; Jacobson, M. K. Biochemical method to measure niacin status in a biological sample. U.S. Patent 6287796, 2001; *Chem. Abstr.* **2001**, *135*, 223772.

22. Hatzinger, P. B.; Palmer, P.; Smith, R. L.; Penarrieta, C. T.; Yoshinari, T. Applicability of tetrazolium salts for the measurement of respiratory activity and viability of groundwater bacteria. *J. Microbiol. Methods* **2003**, *52*, 47–58.

23. Haghighi, A. Z.; Wei, R. Measurement of superoxide dismutase in erythrocytes and whole blood using iodonitrotetrazolium violet. *Anal. Lett.* **1998**, *31*, 981–990.

24. Kong, Q. Method and composition for treating cancer using cellular organelle crystallizing agents. U.S. Pat. Appl. Publ. US 2006286170, 2006; *Chem. Abstr.* **2006**, *146*, 55482.

25. Cheng, L. Corrosion inhibitors with tetrazolium salts for steel protection in aqueous systems. PCT Int. Appl. WO 2000066809, 2000; *Chem. Abstr.* **2000**, *133*, 353308.

26. Yin, J.; Tan, X. J.; Ren, N. Q.; Cui, Y. B.; Tang, L. Evaluation of heavy metal inhibition of activated sludge by TTC and INT-electron transport system activity tests. *Water Sci. Technol.* **2005**, *52*, 231–239.

27. Mosher, J. J.; Findlay, R. H.; Johnston, C. G. Physical and chemical factors affecting microbial biomass and activity in contaminated subsurface riverine sediments. *Can. J. Microbiol.* **2006**, *52*, 397–403.

JANUS GREEN B

CAS Registry Number 2869-83-2

Chemical Structure

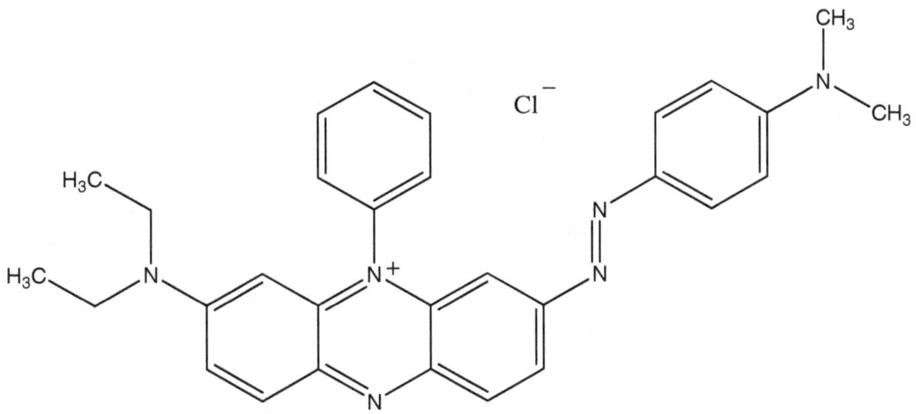

CA Index Name Phenazinium, 3-(diethylamino)-7-[2-[4-(dimethylamino)phenyl]diazenyl]-5-phenyl-, chloride (1 : 1)

Other Names Diazin Green S; Diazine Green S; 3-Diethylamino-7-(4-dimethylaminophenylazo)-5-phenylphenazinium chloride; 3-Diethylamino-7-(*p*-dimethylaminophenylazo)-5-phenylphenazinium chloride; Janus Green B; Phenazinium, 3-(diethylamino)-7-[[4-(dimethylamino)phenyl]azo]-5-phenyl-, chloride; Phenazinium, 3-(diethylamino)-7-[[*p*-(dimethylamino)phenyl]azo]-5-phenyl-, chloride; C.I. 11050; Janus Green V; Union Green B

Merck Index Number 5255

Chemical/Dye Class Phenazine

Molecular Formula $C_{30}H_{31}ClN_6$

Molecular Weight 511.06

Physical Form Dark green to dark brown to dark black powder

Solubility Soluble in water; slightly soluble in ethanol

Melting Point >200 °C

Absorption (λ_{max}) 660 nm, 395 nm

Synthesis Synthetic methods[1,2]

Staining Applications Biomolecules;[3] brain;[4,24] spinal cord;[4] chromosomes;[5] DNA;[6–8] embryos;[9,10] fungi;[11] lymph vessels;[12] mitochondria;[13–21] neurons;[22–24] nucleic acids;[25] sperms;[26] tissue culture monolayers;[27] yeast cell;[28] hairs[29,30]

Biological Applications Antimalarial agents;[31] diagnosis of diseases related to amyloid accumulation;[32] diagnostic assays;[33] detecting fungi;[11] nucleic acids;[34] sugars[35]

Industrial Applications Copper electroplating;[36–39] electronic devices;[40] semiconductor chips;[41] adhesives;[42,43] paints[44]

Safety/Toxicity Bacterial toxicity;[45] cytotoxicity;[46] germ morphogenesis[47]

Certification/Approval Certified by Biological Stain Commission (BSC)

REFERENCES

1. Ismailov, N. I.; Rustamov, N. K. Derivatives of chlorogallates with azo derivatives of safranin T and application in analysis. *Azerb. Khim. Zh.* **2003**, 81–84; *Chem. Abstr.* **2004**, *142*, 85366.
2. Toma, H.; Kashimura, N.; Hisamura, M.; Sumino, F.; Tanaka, S. Alcohol-soluble resin interlayer containing monoazo compound for electrophotographic photoreceptor. Jpn. Kokai Tokkyo

Koho JP 62269966, 1987; *Chem. Abstr.* **1988**, *109*, 83326.

3. Chirikjian, J. G. Articles of manufacture and methods for staining and labeling biomolecules. PCT Int. Appl. WO 9942620, 1999; *Chem. Abstr.* **1999**, *131*, 167357.
4. Augulis, V.; Sigg, E. B. Supravital staining and fixation of brain and spinal cord by intravascular perfusion. *Stain Technol.* **1971**, *46*, 183–190.
5. Curtis, D.; Horobin, R. W. Chromosome banding: specification of structural features of dyes giving rise to G-banding. *Histochem. J.* **1982**, *14*, 911–928.
6. Huang, C. Z.; Li, Y. F.; Huang, X. H.; Li, M. Interactions of Janus Green B with double stranded DNA and the determination of DNA based on the measurement of enhanced resonance light scattering. *Analyst* **2000**, *125*, 1267–1272.
7. Chen, L.; Liu, L.; Shen, H. Fluorescence quenching investigation for Janus Green B and used as probe in determination of nucleic acids. *Chin. J. Chem.* **2005**, *23*, 291–296.
8. Dutt, M. K. Basic dyes for the staining of DNA in mammalian tissues and absorption spectra of stained nuclei in visible light. *Microsc. Acta* **1982**, *86*, 59–68.
9. Fernandez-Teran, M. A.; Hurle, J. M. Syndactyly induced by Janus Green B in the embryonic chick leg bud: a reexamination. *J. Embryol. Exp. Morphol.* **1984**, *84*, 159–175.
10. Stein, E.; Iditoiu, C.; Deleanu, M. Janus Green B and experimental syndactyly in chick embryos. *Experientia* **1969**, *25*, 916–917.
11. Capps, C. L. Detection of *Candida*. U.S. Patent 5372935, 1994; *Chem. Abstr.* **1995**, *122*, 101134.
12. Lee, B. C.; Yoo, J. S.; Baik, K. Y.; Kim, K. W.; Soh, K. S. Novel threadlike structures (Bonghan ducts) inside lymphatic vessels of rabbits visualized with a Janus Green B staining method. *Anatom. Record Part B* **2005**, *286*, 1–7.
13. Yang, J.; Ma, L.; Zhang, Y.; Fang, F.; Li, L. Flow cytometric identification of two different rhodamine-123-stained mitochondrial populations in maize leaves. *Protoplasma* **2007**, *231*, 249–252.
14. Ghazi-Khansari, M.; Mohammadi-Bardbori, A.; Hosseini, M. J. Using Janus Green B to study paraquat toxicity in rat liver mitochondria: role of ACE inhibitors (thiol and nonthiol ACEi). *Ann. N.Y. Acad. Sci.* **2006**, *1090*, 98–107.
15. Rashid, F.; Horobin, R. W. Interaction of molecular probes with living cells and tissues. Part 2. A structure–activity analysis of mitochondrial staining by cationic probes, and a discussion of the synergistic nature of image-based and biochemical approaches. *Histochemistry* **1990**, *94*, 303–308.
16. Koide, T.; Baba, K.; Watanabe, K.; Ikeda, M.; Shimosato, Y. Mitochondrial injury produced by Janus Green B. Enzyme-morphological and ultrastructural observation. *Acta Histochem. Cytochem.* **1971**, *4*, 137–152.
17. Tanaka, Y. Deposition of Janus Green B and pinocyanol in mitochondria of supravitally stained KB cells. *Exp. Cell Res.* **1968**, *52*, 338–348.
18. Showacre, J. L.; duBuy, H. G. The enzymic nature of mitochondrial characterization by Janus Green B and the detection of Krebs-cycle dehydrogenases with Janus Green B. *J. Natl. Cancer Inst.* **1955**, *16*, 173–194.
19. Cooperstein, S. J.; Lazarow, A.; Patterson, J. W. The mechanism of Janus Green B staining of mitochondria. II. Reactions and properties of Janus Green B and its derivatives. *Exp. Cell Res., Suppl.* **1953**, *5*, 69–82.
20. Lazarow, A.; Cooperstein, S. J. The mechanism of Janus Green B staining of mitochondria. I. A review. *Exp. Cell Res., Suppl.* **1953**, *5*, 56–69.
21. Showacre, J. L. A critical study of Janus Green B coloration as a tool for characterizing mitochondria. *J. Natl. Cancer Inst.* **1953**, *13*, 829–845.
22. Yack, J. E. Janus Green B as a rapid, vital stain for peripheral nerves and chordotonal organs in insects. *J. Neurosci. Methods* **1993**, *49*, 17–22.
23. Uzdensky, A. B.; Zhavoronkova, A. A.; Mironov, A. F.; Kuz'min, S. G. A study of photodynamic effect of new photosensitizers on a single nerve cell. *Biol. Bull.* **2000**, *27*, 186–193.
24. Smith, T. M.; Squires, R. F. Differential inhibition of brain specific [^{3}H]flunitrazepam binding by several types of dyes. *Neurochem. Res.* **1983**, *8*, 1177–1183.
25. Goldstein, D. J. Mechanism of differential staining of nucleic acids. *Nature* **1961**, *191*, 407–408.
26. Michael, A. Y.; Drejer, J. O.; Bagger, P. V.; Detlefsen, G. U.; Stakemann, G. Complete staining of human spermatozoa and immature germ cells combined with phase contrast microscopy. *Arch. Androl.* **1987**, *19*, 217–221.
27. Everett, M. M.; Miller, W. A. Differential dichrome staining of tissue culture monolayers: alternate dyes and possible mechanism. *Stain Technol.* **1978**, *53*, 315–320.

28. Kwon-Chung, K. J.; Tewari, R. P. Determination of viability of *Histoplasma capsulatum* yeast cells grown *in vitro*: comparison between dye and colony count methods. *J. Med. Vet. Mycol.* **1987**, *25*, 107–114.

29. Manneck, H.; Schwartz, S.; Schwart, S.; Kleen, A. Hair bleaching and coloring agents containing imidazoles and amino alcohols. PCT Int. Appl. WO 2007124927, 2007; *Chem. Abstr.* **2007**, *147*, 527705.

30. Hoeffkes, H.; Mucha, T. Preparation for simultaneous dyeing and lightening of hair containing a direct dye and a carbonate or carbonate precursor. Ger. Offen. DE 102005013488, 2006; *Chem. Abstr.* **2006**, *145*, 362835.

31. Vennerstrom, J. L.; Makler, M. T.; Angerhofer, C. K.; Williams, J. A. Antimalarial dyes revisited: xanthenes, azines, oxazines, and thiazines. *Antimicrob. Agents Chemother.* **1995**, *39*, 2671–2677.

32. Kudo, K.; Suemoto, T.; Suzuki, M.; Tojo, H.; Shimazu, H. Azure A analogs as imaging agents and probes for diagnosis of diseases related to amyloid accumulation. Jpn. Kokai Tokkyo Koho JP 2000344685, 2000; *Chem. Abstr.* **2000**, *134*, 27007.

33. Wong, S. T.; Hiltibran, R. G.; Huang, T. M.; Calfin, B. B.; Pope, M. R.; Schapira, T. G.; Shain, E. B.; Young, D. F. Diagnostic assay and multilayer assay apparatus for a sample of biological fluid. U.S. Patent 6312888, 2001; *Chem. Abstr.* **2001**, *135*, 341131.

34. Gupta, A.; Will, S. G. Light emission modifiers and their uses in nucleic acid detection, amplification and analysis. Eur. Pat. Appl. EP 1739190, 2007; *Chem. Abstr.* **2007**, *146*, 115979.

35. Friedman, A. J. Method of detecting reducing sugars by using phenazines as colorimetric indicators. U.S. Patent 6586195, 2003; *Chem. Abstr.* **2003**, *139*, 68250.

36. Dow, W. P.; Yen, M. Y.; Liao, S. Z.; Chiu, Y. D.; Huang, H. C. Filling mechanism in microvia metallization by copper electroplating. *Electrochim. Acta* **2008**, *53*, 8228–8237.

37. Mori, M. Manufacture of copper alloy ribbons and foils, and electronic parts manufactured by using them. Jpn. Kokai Tokkyo Koho JP 2008081836, 2008; *Chem. Abstr.* **2008**, *148*, 438421.

38. Burke, L. D.; Sharna, R. AC impedance investigation of copper in acid solution. II. Effect of bath additives on copper electrodeposition. *J. Electrochem. Soc.* **2008**, *155*, D285–D297.

39. Lee, Y. Y.; Park, Y. J.; Lee, J. B.; Cho, B. W. Effects of leveler on the trench filling during damascene copper plating. *J. Korean Electrochem. Soc.* **2002**, *5*, 153–158.

40. Mori, M. Copper wires or foils with excellent strength, conductivity, and bending processability, their manufacture, and electronic devices therewith. Jpn. Kokai Tokkyo Koho JP 2007294923, 2007; *Chem. Abstr.* **2007**, *147*, 513127.

41. Kim, B. C.; Lee, I. H.; Park, Y. H. Acid copper electroplating solution, especially useful in manufacture of semiconductor chip. Repub. Korean Kongkae Taeho Kongbo KR 2004088322, 2004; *Chem. Abstr.* **2006**, *145*, 365094.

42. Bauer, H.; Piestert, G.; Gilch, H. G. Adhesive compositions. U.S. Patent 4395301, 1983; *Chem. Abstr.* **1983**, *99*, 123839.

43. Piestert, G.; Gilch, H. G.; Bauer, H. Adhesives and activators for them. Ger. Offen. DE 3039287, 1981; *Chem. Abstr.* **1981**, *95*, 26141.

44. Fasano, D. M. Method for increasing the hiding power of a wet paint film. Eur. Pat. Appl. EP 549145, 1993; *Chem. Abstr.* **1993**, *119*, 252219.

45. Oros, G.; Cserhati, T.; Forgacs, E. Separation of the strength and selectivity of the microbiological effect of synthetic dyes by spectral mapping technique. *Chemosphere* **2003**, *52*, 185–193.

46. Mohammadi-Bardbori, A.; Nejati, M.; Esmaeili, J.; Ghafari, H.; Ghazi-Khansari, M. Comparative measurement of *in vitro* paraquat and aflatoxin B1 cytotoxicity using three different cytotoxicity assays in pheochromocytoma cells (PC12). *Toxicol. Mech. Methods* **2008**, *18*, 685–689.

47. Mayerson, P. L.; Fallon, J. F. The effects of Janus Green B on the temporal and spatial pattern of feather germ morphogenesis. *J. Exp. Zool.* **1984**, *232*, 231–242.

JC 1

CAS Registry Number 3520-43-2

Chemical Structure

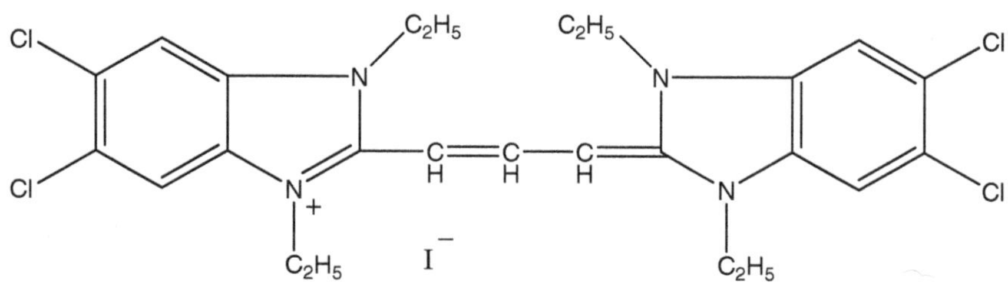

CA Index Name 1*H*-Benzimidazolium, 5,6-dichloro-2-[3-(5,6-dichloro-1,3-diethyl-1,3-dihydro-2*H*-benzimidazol-2-ylidene)-1-propen-1-yl]-1,3-diethyl-, iodide (1 : 1)

Other Names 1*H*-Benzimidazolium, 5,6-dichloro-2-[3-(5,6-dichloro-1,3-diethyl-1,3-dihydro-2*H*-benzimidazol-2-ylidene)-1-propenyl]-1,3-diethyl-, iodide; 5,6-Dichloro-2-[3-(5,6-dichloro-1,3-diethyl-2-benzimidazolinylidene)propenyl]-1,3-diethylbenzimidazolium iodide; Benzimidazolium, 5,6-dichloro-2-[3-(5,6-dichloro-1,3-diethyl-2-benzimidazolinylidene)propenyl]-1,3-diethyl-, iodide; Benzimidazolocarbocyanine iodide, 5,5′,6,6′-tetrachloro-1,1′,3,3′-tetraethyl-; Imidacarbocyanine iodide, 1,1′,3,3′-tetraethyl-5,5′,6,6′-tetrachloro-; 1,1′,3,3′-Tetraethyl-5,5′,6,6′-tetrachlorobenzimidazolocarbocyanine iodide; 1,1′,3,3′-Tetraethyl-5,5′,6,6′-tetrachloroimidacarbocyanine iodide; 5,5′,6,6′-Tetrachloro-1,1′,3,3′-tetraethylbenzimidazolocarbocyanine iodide; Bis(5,6-dichloro-1,3-diethyl-2-benzimidazole)trimethinecyanine iodide; CBIC2; JC 1; NK 1420

Merck Index Number Not listed

Chemical/Dye Class Cyanine

Molecular Formula $C_{25}H_{27}Cl_4IN_4$

Molecular Weight 652.23

Physical Form Deep red crystals

Solubility Soluble in methanol, *N,N*-dimethyl formamide, dimethyl sulfoxide

Melting Point 275–278 °C

Absorption (λ_{max}) 514 nm

Emission (λ_{max}) 529 nm

Synthesis Synthetic methods[1–10]

Staining Applications Mitochondria;[11–20] blood cells;[21] fungi;[22] microbes;[23] proteins;[24] pulmonary neuroepithelial body;[25] sperms;[26,27] hairs[28]

Biological Applications Detecting mitochondrial membrane potential,[20,29–35] ABCB1, ABCC1, and ABCG2 transporters inhibitors,[36] nucleic acid hybridization,[37] prostate cancer;[38] treating cellular death,[39] Alzheimer's disease;[40] apoptosis assay;[41] cytotoxicity assay;[42] hematotoxicity assay;[21] drug screening assay;[43] P-glycoprotein (P-gp) activity acute myeloid leukemia (AML) assay;[44,45] multidrug resistance assay[46]

Industrial Applications Nonlinear optical material;[47] photographic material;[1,4,7–10,48] Semiconductors[49]

Safety/Toxicity Hepatotoxicity[50]

REFERENCES

1. Peng, Z. H.; Geise, H. J.; Zhou, X. F.; Peng, B. X.; Carleer, R.; Dommisse, R. The structure of benzimidazole cyanine dyes, their spectroscopy, and their performance in photographic emulsions. *Liebigs Ann./Rec.* **1997**, 27–33.
2. Gandino, M.; Baldassarri, A. Benzimidazacarbocyanines. Fr. FR 1525450, 1968; *Chem. Abstr.* **1969**, *71*, 40236.
3. Yagupol'skii, L. M.; Troitskaya, V. I.; Levkoev, I. I.; Lifshits, E. B.; Yufa, P. A.; Barvyn, N. S. Cyanine dyes containing fluorine. XIV. Some ditri- and tetrasubstituted benzimidazolocyanines. *Zh. Obshch. Khim.* **1967**, *37*, 191–198; *Chem. Abstr.* **1967**, *66*, 96202.
4. Kodak, Soc. Anon. Photographic emulsions. BE 659415, 1965; *Chem. Abstr.* **1966**, *64*, 36434.
5. Gevaert Photo-Producten N.V. Benzimidazole cyanines. BE 510948, 1952; *Chem. Abstr.* **1958**, *52*, 60247.

6. Heseltine, D. W.; Brooker, L. G. S. Trinuclear polymethine dyes containing a pyrrolyl, indolyl, or pyrrocolyl nucleus. GB 756226, 1956; *Chem. Abstr.* **1957**, *51*, 20894.

7. Van Lare, E. J. Symmetrical carbocyanine dyes for optically sensitizing gelatin silver halide emulsions. GB 754546, 1956; *Chem. Abstr.* **1957**, *51*, 20893.

8. Allen, C. F. H.; Kennard, K. C. Stabilized photographic silver halide emulsions. U.S. Patent 2776211, 1957; *Chem. Abstr.* **1957**, *51*, 20892.

9. Van Lare, E. J. Symmetrical carbocyanin dyes for optically sensitizing gelatin silver halide emulsions. U.S. Patent 2739149, 1956; *Chem. Abstr.* **1957**, *51*, 4571.

10. Carroll, B. H.; Jones, J. E. Supersensitization of photographic emulsions. U.S. Patent 2688545, 1954; *Chem. Abstr.* **1955**, *49*, 601.

11. Ona, T.; Kosaihira, A. Intracellular mitochondria polarization-monitoring method. PCT Int. Appl. WO 2007069692, 2007; *Chem. Abstr.* **2007**, *147*, 48184.

12. Dressler, C.; Beuthan, J.; Mueller, G.; Zabarylo, U.; Minet, O. Fluorescence imaging of heat-stress induced mitochondrial long-term depolarization in breast cancer cells. *J. Fluoresc.* **2006**, *16*, 689–695.

13. Szilagyi, G.; Simon, L.; Koska, P.; Telek, G.; Nagy, Z. Visualization of mitochondrial membrane potential and reactive oxygen species via double staining. *Neurosci. Lett.* **2006**, *399*, 206–209.

14. Lecoeur, H.; Langonne, A.; Baux, L.; Rebouillat, D.; Rustin, P.; Prevost, M. C.; Brenner, C.; Edelman, L.; Jacotot, E. Real-time flow cytometry analysis of permeability transition in isolated mitochondria. *Exp. Cell Res.* **2004**, *294*, 106–117.

15. Safiulina, D.; Kaasik, A.; Seppet, E.; Peet, N.; Zharkovsky, A.; Seppet, E. Method for *in situ* detection of the mitochondrial function in neurons. *J. Neurosci. Methods* **2004**, *137*, 87–95.

16. Bernas, T.; Dobrucki, J. Mitochondrial and nonmitochondrial reduction of MTT: interaction of MTT with TMRE, JC-1, and NAO mitochondrial fluorescent probes. *Cytometry* **2002**, 47, 236–242.

17. Nuydens, R.; Novalbos, J.; Dispersyn, G.; Weber, C.; Borgers, M.; Geerts, H. A rapid method for the evaluation of compounds with mitochondria-protective properties. *J. Neurosci. Methods* **1999**, *92*, 153–159.

18. Poot, M.; Zhang, Y. Z.; Kraemer, J. A.; Wells, K. S.; Jones, L. J.; Hanzel, D. K.; Lugade, A. G.; Singer, V. L.; Haugland, R. P. Analysis of mitochondrial morphology and function with novel fixable fluorescent stains. *J. Histochem. Cytochem.* **1996**, *44*, 1363–1372.

19. Cossarizza, A.; Ceccarelli, D.; Masini, A. Functional heterogeneity of an isolated mitochondrial population revealed by cytofluorometric analysis at the single organelle level. *Exp. Cell Res.* **1996**, *222*, 84–94.

20. Zamzami, N.; Metivier, D.; Kroemer, G. Quantitation of mitochondrial transmembrane potential in cells and in isolated mitochondria. *Methods Enzymol.* **2000**, *322*, 208–213.

21. Dertinger, S. D.; Bemis, J. C.; Bryce, S. M. Method for measuring *in vivo* hematotoxicity with an emphasis on radiation exposure assessment. U.S. Pat. Appl. Publ. US 2008311586, 2008; *Chem. Abstr.* **2008**, *150*, 50134.

22. Pina-Vaz, C.; Sansonetty, F.; Rodrigues, A. G.; Costa-Oliveira, S.; Tavares, C.; Martinez-De-Oliveira, J. Cytometric approach for a rapid evaluation of susceptibility of *Candida* strains to antifungals. *Clin. Microbiol. Infect.* **2001**, *7*, 609–618.

23. Little, R. G., II; Abrahamson, S.; Wong, P. Identification of novel antimicrobial agents using membrane potential indicator dyes. PCT Int. Appl. WO 2000018951, 2000; *Chem. Abstr.* **2000**, *132*, 260670.

24. Kim, E. J.; Barker, L.; Burnet, M.; Guse, J. H.; Luyten, K.; Tsotsou, G. Use of a yeast expression host to identify genes for transport proteins using fluorescent dye reporters. PCT Int. Appl. WO 2003038092, 2003; *Chem. Abstr.* **2003**, *138*, 363834.

25. De Proost, I.; Pintelon, I.; Brouns, I.; Kroese, A. B. A.; Riccardi, D.; Kemp, P. J.; Timmermans, J. P.; Adriaensen, D. Functional live cell imaging of the pulmonary neuroepithelial body microenvironment. *Am. J. Respir. Cell Mol. Biol.* **2008**, *39*, 180–189.

26. Martinez-Pastor, F.; Johannisson, A.; Gil, J.; Kaabi, M.; Anel, L.; Paz, P.; Rodriguez-Martinez, H. Use of chromatin stability assay, mitochondrial stain JC-1, and fluorometric assessment of plasma membrane to evaluate frozen-thawed ram semen. *Anim. Reprod. Sci.* **2004**, *84*, 121–133.

27. De Pauw, I. M. C.; Van Soom, A.; Laevens, H.; Verberckmoes, S.; De Kruif, A. Sperm binding to epithelial oviduct explants in bulls with different nonreturn rates investigated with a new *in vitro* model. *Biol. Reprod.* **2002**, *67*, 1073–1079.

28. Ohashi, Y.; Miyabe, H.; Matsunaga, K. Hair dye composition. Eur. Pat. Appl. EP 1166753, 2002; *Chem. Abstr.* **2002**, *136*, 74280.

29. Troiano, L.; Ferraresi, R.; Lugli, E.; Nemes, E.; Roat, E.; Nasi, M.; Pinti, M.; Cossarizza, A. Multiparametric analysis of cells with different mitochondrial membrane potential during apoptosis by polychromatic flow cytometry. *Nat. Protoc.* **2007**, *2*, 2719–2727.

30. Morrison, M. L.; Williamson, K.; Arthur, K.; Price, G. J.; Hamilton, P. W.; Maxwell, P. Phenotypic changes in mitochondrial membrane potential (Δψ m) during valinomycin-induced depolarisation and apoptosis. *Cell. Oncol.* **2005**, *27*, 231–236.

31. Feeney, C. J.; Pennefather, P. S.; Gyulkhandanyan, A. V. A cuvette-based fluorometric analysis of mitochondrial membrane potential measured in cultured astrocyte monolayers. *J. Neurosci. Methods* **2003**, *125*, 13–25.

32. Mathur, A.; Hong, Y.; Kemp, B. K.; Barrientos, A. A.; Erusalimsky, J. D. Evaluation of fluorescent dyes for the detection of mitochondrial membrane potential changes in cultured cardiomyocytes. *Cardiovasc. Res.* **2000**, *46*, 126–138.

33. Troiano, L.; Granata, A. R. M.; Cossarizza, A.; Kalashnikova, G.; Bianchi, R.; Pini, G.; Tropea, F.; Carani, C.; Franceschi, C. Mitochondrial membrane potential and DNA stainability in human sperm cells: a flow cytometry analysis with implications for male infertility. *Exp. Cell Res.* **1998**, *241*, 384–393.

34. Reers, M.; Smiley, S. T.; Mottola-Hartshorn, C.; Chen, A.; Lin, M.; Chen, L. B. Mitochondrial membrane potential monitored by JC-1 dye. *Methods Enzymol.* **1995**, *260*, 406–417.

35. Cossarizza, A.; Cooper, E. L.; Quaglino, D.; Salvioli, S.; Kalachnikova, G.; Franceschi, C. Mitochondrial mass and membrane potential in coelomocytes from the earthworm *Eisenia foetida*: studies with fluorescent probes in single intact cells. *Biochem. Biophys. Res. Commun.* **1995**, *214*, 503–510.

36. Ivnitski-Steele, I.; Larson, R. S.; Lovato, D. M.; Khawaja, H. M.; Winter, S. S.; Oprea, T. I.; Sklar, L. A.; Edwards, B. S. High-throughput flow cytometry to detect selective inhibitors of ABCB1, ABCC1, and ABCG2 transporters. *Assay Drug Dev. Technol.* **2008**, *6*, 263–276.

37. Bupp, C. R., II; Choi, K. Y.; Holmes-Davis, R. A.; Izmailov, A.; Koshinsky, H.; Nulf, C. J.; Urdea, M.; Wang, M.; Warner, B. D.; Zwick, M. Carbocyanine dye dimers linked by a conjugated alkenyl chain for use in detection of nucleic acid hybridization. PCT Int. Appl. WO 2007100711, 2007; *Chem. Abstr.* **2007**, *147*, 358263.

38. Dickman, D. Methods of detecting prostate cancer. PCT Int. Appl. WO 2006054296, 2006; *Chem. Abstr.* **2006**, *145*, 3346.

39. Chauvier, D.; Borgne, A.; Jacotot, E.; Langonne, A.; Lecoeur, H.; Rebouillat, D. Caspase-2 inhibitors and pharmaceutical compositions for preventing and treating cellular death and their biological applications. PCT Int. Appl. WO 2004103389, 2004; *Chem. Abstr.* **2004,** *142*, 33018.

40. Herrnstadt, C.; Parker, W. D. Alleles of cytochrome *c* oxidase genes associated with Alzheimer's disease and their use in diagnosis and treatment of the disease. U.S. Patent 6171859, 2001; *Chem. Abstr.* **2001**, *134*, 84625.

41. Lecoeur, H.; Chauvier, D.; Langonne, A.; Rebouillat, D.; Brugg, B.; Mariani, J.; Edelman, L.; Jacotot, E. Dynamic analysis of apoptosis in primary cortical neurons by fixed- and real-time cytofluorometry. *Apoptosis* **2004**, *9*, 157–169.

42. Murakami, T. Cytotoxicity test method by measuring membrane electric potential. Jpn. Kokai Tokkyo Koho JP 2000300290, 2000; *Chem. Abstr.* **2000**, *133*, 330852.

43. Dykens, J. A.; Velicelebi, G.; Ghosh, S. S. Compositions and methods for assaying subcellular conditions and processes using energy transfer for drug screening. PCT Int. Appl. WO 2000079274, 2000; *Chem. Abstr.* **2000**, *134*, 66121.

44. Swerts, K.; De Moerloose, B.; Dhooge, C.; Noens, L.; Laureys, G.; Benoit, Y.; Philippe, J. Comparison of two functional flow cytometric assays to assess P-gp activity in acute leukemia. *Leuk. Lymphoma* **2004**, *45*, 2221–2228.

45. Legrand, O.; Perrot, J. Y.; Simonin, G.; Baudard, M.; Marie, J. P. JC-1: a very sensitive fluorescent probe to test Pgp activity in adult acute myeloid leukemia. *Blood* **2001**, *97*, 502–508.

46. Loken, M. R. Immunocytometric tumor cell multidrug resistance assay. U.S. Pat. Appl. Publ. US 2006263834, 2006; *Chem. Abstr.* **2006**, *145*, 500017.

47. Kato, T. Composition used as nonlinear optical material. Jpn. Kokai Tokkyo Koho JP 2005128152, 2005; *Chem. Abstr.* **2005**, *142*, 454036.

48. Ikeda, T. Silver halide emulsion containing cyanine dye for improved storage stability and high sensitivity

and heat-developable photographic material. Jpn. Kokai Tokkyo Koho JP 2004094015, 2004; *Chem. Abstr.* **2004**, *140*, 278336.

49. Alivisatos, A. P.; Larabell, C. A.; Parak, W. J.; Le Gros, M.; Boudreau, R. Semiconductor nanocrystal-based cellular imaging. PCT Int. Appl. WO 2003038401, 2003; *Chem. Abstr.* **2003**, *138*, 334028.

50. Oshima, R.; Nakano, H.; Katayama, M.; Sakurai, J.; Wu, W.; Koizumi, S.; Asano, T.; Watanabe, T.; Asakura, T.; Ohta, T.; Otsubo, T. Modification of the hepatic mitochondrial proteome in response to ischemic preconditioning following ischemia-reperfusion injury of the rat liver. *Eur. Surg. Res.* **2008**, *40*, 247–255.

JC 9

CAS Registry Number 522592-13-8
Chemical Structure

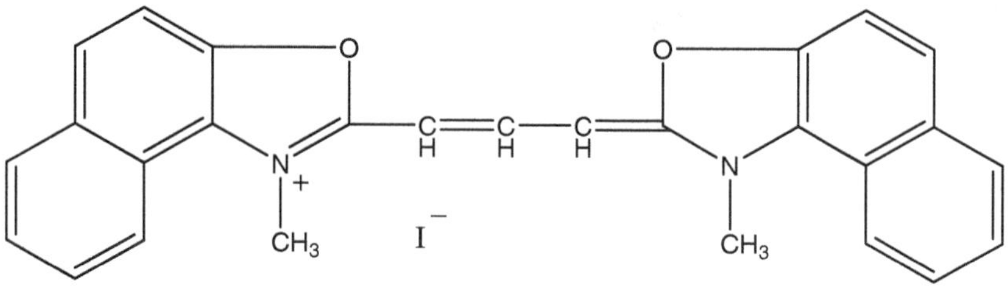

CA Index Name Naphth[1,2-*d*]oxazolium, 1-methyl-2-[3-(1-methylnaphth[1,2-*d*]oxazol-2(1*H*)-ylidene)-1-propen-1-yl], iodide (1 : 1)

Other Names D 22421; DiNOC$_1$(3); 3,3′-Dimethyl-α-naphthoxacarbocyanine iodide; JC 9; Naphth[1,2-*d*]oxazolium, 1-methyl-2-[3-(1-methylnaphth[1,2-*d*]oxazol-2(1*H*)-ylidene)-1-propenyl]-, iodide

Merck Index Number Not listed

Chemical/Dye Class Cyanine

Molecular Formula $C_{27}H_{21}IN_2O_2$

Molecular Weight 532.38

Physical Form Solid

Solubility Soluble in *N,N*-dimethyl formamide, dimethyl sulfoxide

Melting Point >200 °C

Absorption (λ_{max}) 522 nm

Emission (λ_{max}) 535 nm

Synthesis Synthetic method[1]

Staining Applications Mitochondria;[2] blood cells;[3] proteins[4]

Biological Applications Identifying genes for transport proteins;[4] treating male infertility,[5] Kennedy disease,[5] prostate cancer,[5] breast cancer,[5] liver cancer,[5] bladder cancer,[5] benign prostate hyperplasia,[5] acne,[5] baldness,[5] hirsutism,[5] exposed wounds,[5] unwanted pregnancy;[5] apoptosis assay;[6] hematotoxicity assay[3]

Industrial Applications Not reported

Safety/Toxicity No data available

REFERENCES

1. Haugland, R. P. *The Handbook: A Guide to Fluorescent Probes and Labeling Technologies*; Molecular Probes Inc.: Eugene, **2005**; pp 568–569.
2. Hattori, F.; Fukuda, K. Method for selecting myocardial cells using intracellular mitochondria labeled with fluorescent indicator. PCT Int. Appl. WO 2006022377, 2006; *Chem. Abstr.* **2006**, *144*, 270175.
3. Dertinger, S. D.; Bemis, J. C.; Bryce, S. M. Method for measuring *in vivo* hematotoxicity with an emphasis on radiation exposure assessment. U.S. Pat. Appl. Publ. US 2008311586, 2008; *Chem. Abstr.* **2008**, *150*, 50134.
4. Kim, E. J.; Barker, L.; Burnet, M.; Guse, J. H.; Luyten, K.; Tsotsou, G. Use of a yeast expression host to identify genes for transport proteins using fluorescent dye reporters. PCT Int. Appl. WO 2003038092, 2003; *Chem. Abstr.* **2003**, *138*, 363834.
5. Shih, C. C. Y.; Su, C. Y. Methods and compositions for enhancing degradation of nuclear receptor transcription factors and therapeutic uses thereof. PCT Int. Appl. WO 2006029040, 2006; *Chem. Abstr.* **2006**, *144*, 267320.
6. Smyth, P. G.; Berman, S. A. Markers of apoptosis: methods for elucidating the mechanism of apoptotic cell death from the nervous system. *Biotechniques* **2002**, *32*, 648–650, 652, 654.

JOJO 1

CAS Registry Number 305801-87-0

Chemical Structure

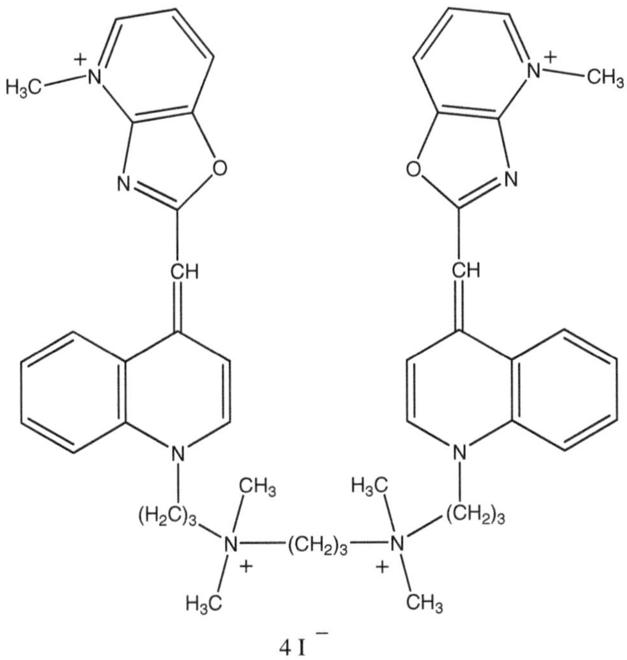

CA Index Name Oxazolo[4,5-*b*]pyridinium, 2,2′-[1,3-propanediylbis[(dimethyliminio)-3,1-propanediyl-1(4*H*)-quinolinyl-4-ylidenemethylidyne]]bis[4-methyl]-, tetraiodide

Other Names JOJO 1; JOJO 1 iodide

Merck Index Number Not listed

Chemical/Dye Class Cyanine

Molecular Formula $C_{47}H_{56}I_4N_8O_2$

Molecular Weight 1272.63

Physical Form Yellow-brown powder

Solubility Soluble in dimethyl sulfoxide

Melting Point >250 °C

Absorption (λ_{max}) 529 nm

Emission (λ_{max}) 545 nm

Synthesis Synthetic method[1]

Staining Applications Nucleic acids;[2–5] cells;[6] hairs[7]

Biological Applications Nucleic acid hybridization;[4,8,9] detecting nucleic acids,[2–5] cells;[6] nucleic acid sequencing[3]

Industrial Applications Not reported

Safety/Toxicity No data available

REFERENCES

1. Haugland, R. P.; Yue, S. T. Aza-benzazolium-containing cyanine dyes and their use in fluorescent biological stains. PCT Int. Appl. WO 2000066664, 2000; *Chem. Abstr.* **2000**, *133*, 351506.
2. Exner, M.; Rogers, A. Methods for identifying nucleic acids and determining melting temperature using multiple fluorophores and FRET. U.S. Pat. Appl. Publ. US 2007172836, 2007; *Chem. Abstr.* **2007**, *147*, 182489.
3. Williams, J. G. K.; Anderson, J. P. Field-switch single molecule DNA sequencing in a two-electrode chamber. PCT Int. Appl. WO 2005111240, 2005; *Chem. Abstr.* **2005**, *143*, 474548.
4. Atkinson, I. J.; Erikson, G. H.; Daksis, J. I.; Picard, P. Kits and methods for purification of nucleic acids using heteropolymeric capture probes and duplex, triplex or quadruplex hybridization in solution utilizing fluorescent intercalating dyes. U.S. Pat. Appl. Publ. US 2003049673, 2003; *Chem. Abstr.* **2003**, *138*, 232955.
5. Erikson, G. H.; Daksis, J. I.; Kandic, I.; Picard, P. Conditions for formation of three- and four-stranded nucleic acid complexes and their detection and analytical use. PCT Int. Appl. WO 2002103051, 2002; *Chem. Abstr.* **2002**, *138*, 50812.
6. Anderson, A. L.; Knutson, C. R.; Mueth, D.; Plewa, J.; Tanner, E. Methods for staining cells for identification and sorting. U.S. Pat. Appl. Publ. US 2006172315, 2006; *Chem. Abstr.* **2006**, *145*, 183714.
7. Lagrange, A. Hair dye compositions containing a polycationic direct dye. Fr. Demande FR 2848840, 2004; *Chem. Abstr.* **2004**, *141*, 76344.

8. Erikson, G. H.; Daksis, J. I. Improving the signal/noise ratio of nucleic acid hybridization assays by preincubation of primer and target with nucleic acid binding agents. U.S. Pat. Appl. Publ. US 2004180345, 2004; *Chem. Abstr.* **2004**, *141*, 255469.

9. Erikson, G. H. Method for modifying transcription and/or translation in an organism by heteropolymeric probes and duplex, triplex or quadruplex hybridization for therapeutic, prophylactic and/or analytic uses. U.S. Pat. Appl. Publ. US 2003181412, 2003; *Chem. Abstr.* **2003**, *139*, 272000.

JO-PRO 1

CAS Registry Number 305801-86-9

Chemical Structure

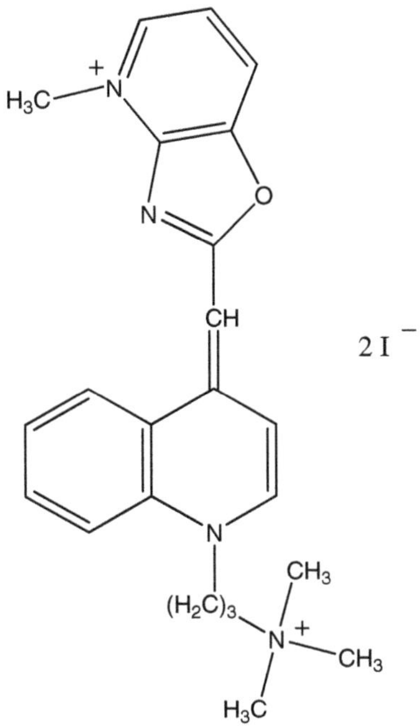

CA Index Name Quinolinium, 4-[(4-methyloxazolo [4,5-*b*]pyridin-2(4*H*)-ylidene)methyl]-1-[3-(trimethylammonio)propyl]-, diiodide

Other Names JO-PRO 1, JO-PRO 1 iodide

Merck Index Number Not listed

Chemical/Dye Class Cyanine

Molecular Formula $C_{23}H_{28}I_2N_4O$

Molecular Weight 630.31

Physical Form Yellow-brown powder

Solubility Soluble in dimethyl sulfoxide

Melting Point >250 °C

Absorption (λ_{max}) 530 nm

Emission (λ_{max}) 546 nm

Synthesis Synthetic method[1]

Staining Applications Nucleic acids;[2–5] cells[6]

Biological Applications Nucleic acid hybridization;[7,8] detecting nucleic acids,[2–5] cells;[6] nucleic acid sequencing[4]

Industrial Applications Not reported

Safety/Toxicity No data available

REFERENCES

1. Haugland, R. P.; Yue, S. T. Aza-benzazolium-containing cyanine dyes and their use in fluorescent biological stains. PCT Int. Appl. WO 2000066664, 2000; *Chem. Abstr*. **2000**, *133*, 351506.
2. Exner, M.; Rogers, A. Methods for identifying nucleic acids and determining melting temperature using multiple fluorophores and FRET. U.S. Pat. Appl. Publ. US 2007172836, 2007; *Chem. Abstr*. **2007**, *147*, 182489.
3. Wittwer, C. T.; Dujols, V. E.; Reed, G.; Zhou, L. Genotyping by thermal denaturation analysis using loss of fluorescence of double-stranded DNA-specific dyes as a reporter to monitor denaturation. PCT Int. Appl. WO 2004038038, 2004; *Chem. Abstr*. **2004**, *140*, 386988.
4. Hoser, M. J. Methods for nucleic acid sequencing. PCT Int. Appl. WO 2004074503, 2004; *Chem. Abstr*. **2004**, *141*, 237698.
5. Erikson, G. H.; Daksis, J. I.; Kandic, I.; Picard, P. Conditions for formation of three- and four-stranded nucleic acid complexes and their detection and analytical use. PCT Int. Appl. WO 2002103051, 2002; *Chem. Abstr*. **2002**, *138*, 50812.
6. Anderson, A. L.; Knutson, C. R.; Mueth, D.; Plewa, J.; Tanner, E. Methods for staining cells for identification and sorting. U.S. Pat. Appl. Publ. US 2006172315, 2006; *Chem. Abstr*. **2006**, *145*, 183714.
7. Erikson, G. H.; Daksis, J. I. Improving the signal/noise ratio of nucleic acid hybridization assays by preincubation of primer and target with nucleic acid binding agents. U.S. Pat. Appl. Publ. US 2004180345, 2004; *Chem. Abstr*. **2004**, *141*, 255469.
8. Erikson, G. H. Method for modifying transcription and/or translation in an organism by heteropolymeric probes and duplex, triplex or quadruplex hybridization for therapeutic, prophylactic and/or analytic uses. U.S. Pat. Appl. Publ. US 2003181412, 2003; *Chem. Abstr*. **2003**, *139*, 272000.

LIGHT GREEN SF YELLOWISH

CAS Registry Number 5141-20-8

Chemical Structure

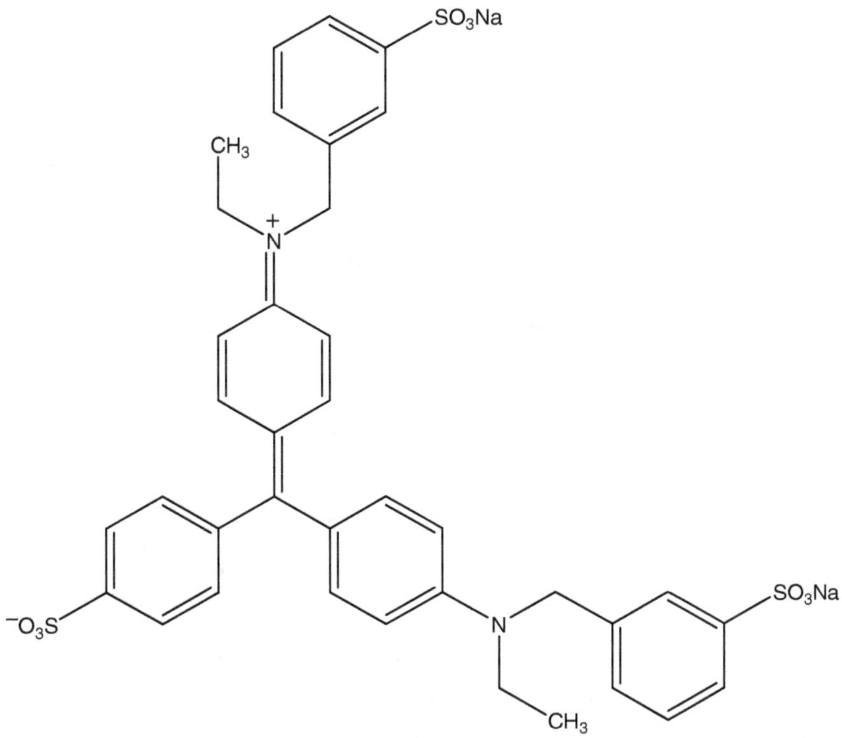

CA Index Name Benzenemethanaminium, *N*-ethyl-*N*-[4-[[4-[ethyl[(3-sulfophenyl)methyl]amino]phenyl](4-sulfophenyl)methylene]-2,5-cyclohexadien-1-ylidene]-3-sulfo-, inner salt, sodium salt (1 : 2)

Other Names Benzenemethanaminium, *N*-ethyl-*N*-[4-[[4-[ethyl[(3-sulfophenyl)methyl]amino]phenyl](4-sulfophenyl)methylene]-2,5-cyclohexadien-1-ylidene]-3-sulfo-, hydroxide, inner salt, disodium salt; Benzenemethanaminium, *N*-ethyl-*N*-[4-[[4-[ethyl[(3-sulfophenyl) methyl]amino]phenyl](4-sulfophenyl)methylene]-2,5-cyclohexadien-1-ylidene]-3-sulfo-, inner salt, disodium salt; C.I. Acid Green 5; C.I. Acid Green 5, disodium salt; Light Green SF; A F Green No. 2; Acid Brilliant Green SF; Acid Green 5; Acid Green A; Acidal Light Green SF; Acilan Green SFG; Acilan Light Green SFG; Amacid Green G; C.I. 42095; C.I. Food Green 2; D and C Green No. 4; FD and C Green No. 2; Fenazo Green 7G; Food Green 2; Green No. 203; Japan Green 205; Japan Green No. 205; Leather Green SF; Light Green Lake; Light Green SF Yellowish; Light Green SFA; Light Green SFD; Light Green Yellowish; Light SF Yellowish; Lissamine Green SF; Lissamine Lake Green SF; MY/68; Merantine Green SF; NSC 9619; Pencil Green SF; Sulfo Green J; Sumitomo Light Green SF Yellowish

Merck Index Number 5485

Chemical/Dye Class Triphenylmethane

Molecular Formula $C_{37}H_{34}N_2Na_2O_9S_3$

Molecular Weight 792.85

Physical Form Reddish-brown powder or crystals

Solubility Soluble in water; slightly soluble in ethanol; insoluble in xylene

Melting Point 288 °C (decompose)

Absorption (λ_{max}) 630 nm, 422 nm

Synthesis Synthetic methods[1–5]

Staining Applications Cell;[6,7] cytoplasm;[8] endoscope;[9] microorganisms;[10] eye membranes;[11] retina;[12–14] proteins;[15] hairs[16]

Biological Applications Cosmetics;[17] oral hygiene products;[18] sunscreen;[19] detecting proteins;[20] treating apolipoprotein E-related diseases[21]

Industrial Applications Color filters;[22] recording materials;[23] inks;[24,25] highlighters;[26] adhesives;[27] photographic materials;[28] detergents;[29] textiles;[30,31] leather[32]

Safety/Toxicity Acute toxicity;[33,34] carcinogenicity;[35–38] chronic toxicity;[39] genotoxicity;[40] mutagenicity;[41,42] retinal toxicity[12–14]

Certification/Approval Certified by Biological Stain Commission (BSC)

REFERENCES

1. Tijssen, P. M. H. P.; Stanssens, D. A. W.; Muscat, D. Enhancement of dye solubility using branched poly (amide esters). Eur. Pat. Appl. EP 1321493, 2003; *Chem. Abstr.* **2003**, *139*, 54283.
2. Wang, Y.; Ni, Y.; Li, Z.; Li, Z. Synthesis of LG-SFY as biological stains. *Zhejiang Gongye Daxue Xuebao* **1995**, *23*, 212–218; *Chem. Abstr.* **1995**, *124*, 120098.
3. Liao, W.; Cai, X.; Wang, M.; Lu, X. Synthesis of Brilliant Green SF Light Yellow. *Huaxue Shiji* **1995**, *17*, 184–185; *Chem. Abstr.* **1995**, *123*, 172614.
4. Sakar, L.; Chaloupka, J.; Hlinovska, Z. Sulfonated leuco base of a green triphenylmethane dye. Czech. CS 213158, 1984; *Chem. Abstr.* **1984**, *100*, 211665.
5. Bodforss, S. Sulfonated triphenylmethane dyes. *Kungliga Fysiografiska Saellskapets Lund, Foerhandlingar* **1954**, *24*, 1–6; *Chem. Abstr.* **1955**, *49*, 25844.
6. Gelvan, D. J.; Goltsman, L.; Chausovsky, A. Staining methods and compositions for identifying a cell phenotype. PCT Int. Appl. WO 2007102146, 2007; *Chem. Abstr.* **2007**, *147*, 339010.
7. Garini, Y.; Mcnamara, G.; Soenksen, D. G.; Cabib, D.; Buckwald, R. A. *In situ* method of analyzing cells by staining with multiple stains and using a spectral data collection device. PCT Int. Appl. WO 2000031534, 2000; *Chem. Abstr.* **2000**, *133*, 14300.
8. Zahniser, D. J.; Isenstein, L. M.; Soule, N. W.; Mui, K. K.; Lapen, D. C. Cytological stain composition. PCT Int. Appl. WO 2001033192, 2001; *Chem. Abstr.* **2001**, *134*, 337923.
9. Yamamoto, A.; Iimori, Y.; Sase, M.; Ishiguro, M.; Li, B.; Sasaki, H. Histochemical fluorescent staining agent composition for endoscope. Jpn. Kokai Tokkyo Koho JP 2007326789, 2007; *Chem. Abstr.* **2007**, *148*, 49094.
10. Noda, N.; Mizutani, T. Microorganism-measuring method using multiple staining. Jpn. Kokai Tokkyo Koho JP 2006340684, 2006; *Chem. Abstr.* **2006**, *146*, 77563.
11. Haritoglou, C.; Freyer, W. Method, dye and medicament for staining the internal limiting membrane, epiretinal membrane, the vitreous and/or the capsule of an eye. U.S. Pat. Appl. Publ. US 2008206149, 2008; *Chem. Abstr.* **2008**, *149*, 315506.
12. Schuettauf, F.; Haritoglou, C.; May, C. A.; Rejdak, R.; Mankowska, A.; Freyer, W.; Eibl, K.; Zrenner, E.; Kampik, A.; Thaler, S. Administration of novel dyes for intraocular surgery: an *in vivo* toxicity animal study. *Invest. Ophthalmol. Vis. Sci.* **2006**, *47*, 3573–3578.
13. Haritoglou, C.; Tadayoni, R.; May, C. A.; Gass, C. A.; Freyer, W.; Priglinger, S. G.; Kampik, A. Short-term *in vivo* evaluation of novel vital dyes for intraocular surgery. *Retina* **2006**, *26*, 673–678.
14. Haritoglou, C.; Yu, A.; Freyer, W.; Priglinger, S. G.; Alge, C.; Eibl, K.; May, C. A.; Welge-Luessen, U.; Kampik, A. An evaluation of novel vital dyes for intraocular surgery. *Invest. Ophthalmol. Vis. Sci.* **2005**, *46*, 3315–3322.
15. De Moreno, M. R.; Smith, J. F.; Smith, R. V. Mechanism studies of Coomassie Blue and silver staining of proteins. *J. Pharm. Sci.* **1986**, *75*, 907–911.
16. Watanabe, K.; Ono, T.; Ota, T.; Minei, T.; Horikoshi, T. Wave-setting hair dye. Jpn. Kokai Tokkyo Koho JP 02076807, 1990; *Chem. Abstr.* **1991**, *114*, 49342.
17. Goto, H.; Taguchi, S.; Iida, N. Method for discoloration prevention of pigments in pharmaceutical and cosmetic compositions. Jpn. Kokai Tokkyo Koho JP 2000053522, 2000; *Chem. Abstr.* **2000**, *132*, 171121.
18. Maruoka, T. Disposable oral hygiene product comprising waterproof container and porous drug-holding material. Jpn. Kokai Tokkyo Koho JP 11197217, 1999; *Chem. Abstr.* **1999**, *131*, 149103.
19. Ross, J. S.; Morefield, E. M. Sunscreen with disappearing color indicator. U.S. Patent 6290936, 2001; *Chem. Abstr.* **2001**, *135*, 215789.
20. Albarella, J. P.; Cahill, S. E.; Johnson, G. M.; Pugia, M. J. Improved method for the detection of protein. Eur. Pat. Appl. EP 793099, 1997; *Chem. Abstr.* **1997**, *127*, 231593.

21. Crutcher, K. A.; Harmony, J. A. K. Methods for the treatment of apolipoprotein E-related diseases. PCT Int. Appl. WO 2000050042, 2000; *Chem. Abstr.* **2000**, *133*, 203003.

22. De Keyzer, G.; Lamatsch, B.; Muehlebach, A.; Rime, F.; Schmitt, G. Surface-modified nanoparticles comprising a cationic colorant for use in color filters. PCT Int. Appl. WO 2008107304, 2008; *Chem. Abstr.* **2008**, *149*, 357452.

23. Franke, W.; Brahm, R. Electrophotographic recording material. Ger. Offen. DE 3404365, 1985; *Chem. Abstr.* **1986**, *104*, 43146.

24. Godbout, D. A.; Vincent Kwan, W. S. Water-based, resin-free and solvent-free eradicable and re-write ball-pen inks or tinted fluid. U.S. Pat. Appl. Publ. US 2006032398, 2006; *Chem. Abstr.* **2006**, *144*, 214528.

25. Aoyama, M.; Tsuda, M.; Sago, H.; Umemura, M.; Yamazaki, H.; Higashiyama, S. Water-thinned fluorescent inks for ink-jet recording. Jpn. Kokai Tokkyo Koho JP 2005120367, 2005; *Chem. Abstr.* **2005**, *142*, 448372.

26. Davies-Smith, L.; Sum, V. W. Highlightable and highlighted mixtures, marking instruments, eradicator solution and kits, using the mixtures in highlighting. U.S. Pat. Appl. Publ. US 2005120919, 2005; *Chem. Abstr.* **2005**, *143*, 9331.

27. Rohowetz, S. E. Thermotropic adhesive tape. U.S. Patent 4188437, 1980; *Chem. Abstr.* **1980**, *92*, 164885.

28. Tschopp, P. Photographic silver halide material with at least one color-containing silver halide-free layer. Ger. Offen. DE 2855428, 1979; *Chem. Abstr.* **1980**, *92*, 31961.

29. Mito, K.; Gobayashi, T.; Shindo, K. Indicator system for use in a granular detergent composition. PCT Int. Appl. WO 9907817, 1999; *Chem. Abstr.* **1999**, *130*, 169869.

30. De La Torre, M. Method of treating textile carrying a stain. PCT Int. Appl. WO 2008122752, 2008; *Chem. Abstr.* **2008**, *149*, 473420.

31. Deeds, S. M. Fabric repair coloring device and using the device to color unwanted bleached out spots on fabric. U.S. Patent 6739779, 2004; *Chem. Abstr.* **2004**, *140*, 424935.

32. de la Maza, A.; Marsal, A.; Cot, J.; Manich, A.; Parra, J. L. Liposomes in leather dyeing: stability of dye–liposome systems and applications. *J. Am. Leather Chem. Assoc.* **1992**, *87*, 459–465.

33. Tonogai, Y.; Ito, Y.; Iwaida, M.; Tati, M.; Ose, Y.; Hori, M. Studies on the toxicity of coal-tar dyes. III. Reason of acute toxicity to fish caused by coal-tar dyes and their industrial effluents. *J. Toxicol. Sci.* **1980**, *5*, 23–33.

34. Tonogai, Y.; Iwaida, M.; Tati, M.; Ose, Y.; Sato, T. Biochemical decomposition of coal-tar dyes. II. Acute toxicity of coal-tar dyes and their decomposed products. *J. Toxicol. Sci.* **1978**, *3*, 205–214.

35. Matthews, E. J.; Contrera, J. F. A new highly specific method for predicting the carcinogenic potential of pharmaceuticals in rodents using enhanced MCASE QSAR-ES software. *Regul. Toxicol. Pharmacol.* **1998**, *28*, 242–264.

36. Fu, Z. D.; Chen, W. R.; Gu, L. J.; Gu, Z. W. The influence of the extent of target organs on sensitivities of methods for screening rodent carcinogens. *Mutat. Res.* **1995**, *331*, 99–117.

37. Ashby, J.; Paton, D. The influence of chemical structure on the extent and sites of carcinogenesis for 522 rodent carcinogens and 55 different human carcinogen exposures. *Mutat. Res.* **1993**, *286*, 3–74.

38. Gold, L. S.; Slone, T. H.; Stern, B. R.; Bernstein, L. Comparison of target organs of carcinogenicity for mutagenic and non-mutagenic chemicals. *Mutat. Res.* **1993**, *286*, 75–100.

39. Hansen, W. H.; Long, E. L.; Davis, K. J.; Nelson, A. A.; Fitzhugh, O. G. Chronic toxicity of three food colorings: Guinea Green B, Light Green SF Yellowish, and Fast Green FCF in rats, dogs and mice. *Food Cosmet. Toxicol.* **1966**, *4*, 389–410.

40. Kirkland, D.; Aardema, M.; Henderson, L.; Mueller, L. Evaluation of the ability of a battery of three *in vitro* genotoxicity tests to discriminate rodent carcinogens and non-carcinogens. I. Sensitivity, specificity and relative predictivity. *Mutat. Res.* **2005**, *584*, 1–256.

41. Klopman, G.; Frierson, M. R.; Rosenkranz, H. S. The structural basis of the mutagenicity of chemicals in *Salmonella typhimurium*: the Gene-Tox data base. *Mutat. Res.* **1990**, *228*, 1–50.

42. Cameron, T. P.; Hughes, T. J.; Kirby, P. E.; Fung, V. A.; Dunkel, V. C. Mutagenic activity of 27 dyes and related chemicals in the *Salmonella*/microsome and mouse lymphoma TK + /− assays. *Mutat. Res.* **1987**, *189*, 223–261.

LOLO 1

CAS Registry Number 305802-06-6

Chemical Structure

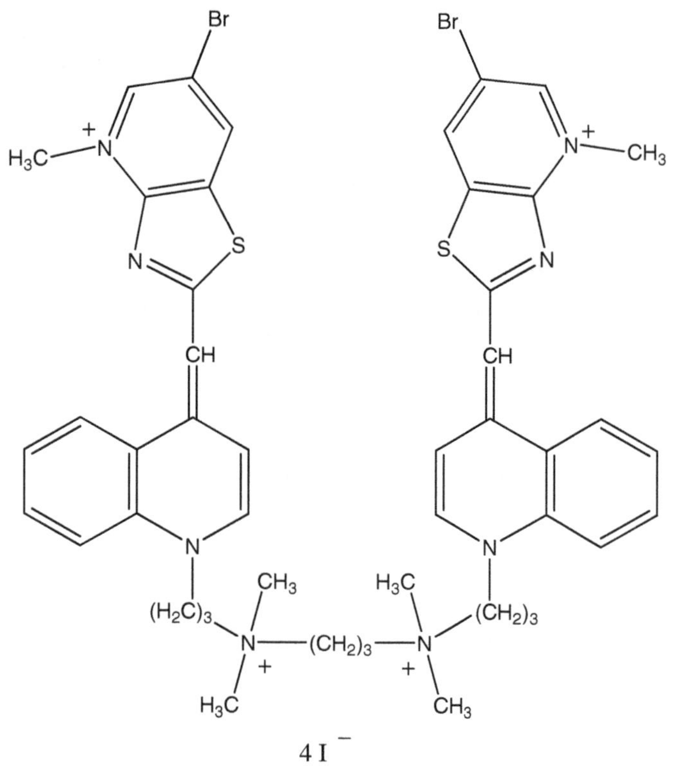

CA Index Name Thiazolo[4,5-*b*]pyridinium, 2,2′-[1,3-propanediylbis[(dimethyliminio)-3,1-propanediyl-1(4*H*)-quinolinyl-4-ylidenemethylidyne]]bis[6-bromo-4-methyl]-, tetraiodide

Other Names LOLO 1, LOLO iodide

Merck Index Number Not listed

Chemical/Dye Class Cyanine

Molecular Formula $C_{47}H_{54}Br_2I_4N_8S_2$

Molecular Weight 1462.54

Physical Form Yellow-brown powder

Solubility Soluble in dimethyl sulfoxide

Melting Point >250 °C

Absorption (λ_{max}) 565 nm

Emission (λ_{max}) 579 nm

Synthesis Synthetic method[1]

Staining Applications Nucleic acids;[2–5] cells;[6] hairs[7]

Biological Applications Nucleic acid hybridization;[3,8,9] detecting nucleic acids,[2–5] cells,[6] pathogens;[10] DNA sequencing[5]

Industrial Applications Not reported

Safety/Toxicity No data available

REFERENCES

1. Haugland, R. P.; Yue, S. T. Aza-benzazolium-containing cyanine dyes and their use in fluorescent biological stains. PCT Int. Appl. WO 2000066664, 2000; *Chem. Abstr.* **2000**, *133*, 351506.
2. Exner, M.; Rogers, A. Methods for identifying nucleic acids and determining melting temperature using multiple fluorophores and FRET. U.S. Pat. Appl. Publ. US 2007172836, 2007; *Chem. Abstr.* **2007**, *147*, 182489.
3. Atkinson, I. J.; Erikson, G. H.; Daksis, J. I.; Picard, P. Kits and methods for purification of nucleic acids using heteropolymeric capture probes and duplex, triplex or quadruplex hybridization in soln. utilizing fluorescent intercalating dyes. U.S. Pat. Appl. Publ. US 2003049673, 2003; *Chem. Abstr.* **2003**, *138*, 232955.
4. Erikson, G. H.; Daksis, J. I.; Kandic, I.; Picard, P. Conditions for formation of three- and four-stranded nucleic acid complexes and their detection and analytical use. PCT Int. Appl. WO 2002103051, 2002; *Chem. Abstr.* **2002**, *138*, 50812.
5. Williams, J. G. K.; Anderson, J. P. Field-switch single molecule DNA sequencing in a two-electrode chamber. PCT Int. Appl. WO 2005111240, 2005; *Chem. Abstr.* **2005**, *143*, 474548.
6. Anderson, A. L.; Knutson, C. R.; Mueth, D.; Plewa, J.; Tanner, E. Methods for staining cells for identification and sorting. U.S. Pat. Appl. Publ. US 2006172315, 2006; *Chem. Abstr.* **2006**, *145*, 183714.
7. Lagrange, A. Hair dye compositions containing a polycationic direct dye. Fr. Demande FR 2848840, 2004; *Chem. Abstr.* **2004**, *141*, 76344.
8. Erikson, G. H.; Daksis, J. I. Improving the signal/noise ratio of nucleic acid hybridization assays by preincubation of primer and target with nucleic acid binding agents. U.S. Pat. Appl. Publ. US

2004180345, 2004; *Chem. Abstr.* **2004**, *141*, 255469.

9. Erikson, G. H. Method for modifying transcription and/or translation in an organism by heteropolymeric probes and duplex, triplex or quadruplex hybridization for therapeutic, prophylactic and/or analytic uses. U.S. Pat. Appl. Publ. US 2003181412, 2003; *Chem. Abstr.* **2003**, *139*, 272000.
10. Vannier, E. Methods for detection of pathogens in red blood cells. PCT Int. Appl. WO 2006031544, 2006; *Chem. Abstr.* **2006**, *144*, 307966.

LUCIFER YELLOW CH

CAS Registry Number 67769-47-5

Chemical Structure

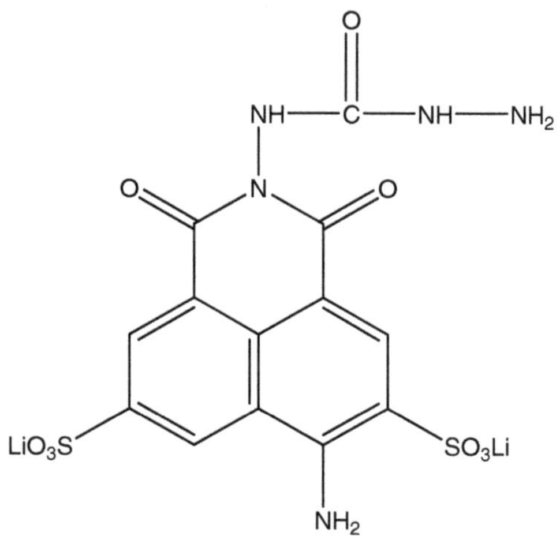

CA Index Name 1*H*-Benz[*de*]isoquinoline-5,8-disulfonic acid, 6-amino-2-[(hydrazinylcarbonyl)amino]-2,3-dihydro-1,3-dioxo-, lithium salt (1 : 2)

Other Names 1*H*-Benz[*de*]isoquinoline-5,8-disulfonic acid, 6-amino-2-[(hydrazinocarbonyl)amino]-2,3-dihydro-1,3-dioxo-, dilithium salt; Lucifer Yellow CH; Lucifer Yellow carbohydrazide

Merck Index Number 5594

Chemical/Dye Class Naphthalimide

Molecular Formula $C_{13}H_9Li_2N_5O_9S_2$

Molecular Weight 457.25

Physical Form Orange powder

Solubility Soluble in water; soluble in ethanol

Melting Point >200 °C

Absorption (λ_{max}) 280 nm, 428 nm

Emission (λ_{max}) 540 nm

Synthesis Synthetic methods[1,2]

Staining Applications Avidin;[4] bovine serum albumin;[4] bacteria;[5] cardiac fibers;[6] cells;[3,7,8] cell surface glycoconjugates;[9] cholesterol;[10] phospholipids;[10] exocytotic secretory processes;[11] gangliosides;[12] glycoprotein;[13] islet cells;[14] liposomes;[15] mitochondria;[16] neurons;[2,3] oxidized antibody;[17] plant cell;[18,19] protoplasts;[18,19] potato tuber storage tissues;[20] proteins;[21] retina;[22–24] saccharides;[25,26] skeletal muscle cells;[27] skin;[28] tissues;[29] plant vacuoles[30–34]

Biological Applications Antiviral agents[35,36]

Industrial Applications Optical nanosensors;[37] printing plates[38]

Safety/Toxicity Carcinogenicity[39]

REFERENCES

1. Stewart, W. W. Synthesis of 3,6-disulfonated 4-aminonaphthalimides. *J. Am. Chem. Soc.* **1981**, *103*, 7615–7620.
2. Stewart, W. W. Functional connections between cells as revealed by dye-coupling with a highly fluorescent naphthalimide tracer. *Cell* **1978**, *14*, 741–759.
3. Stewart, W. W. Lucifer dyes: highly fluorescent dyes for biological tracing. *Nature* **1981**, *292*, 17–21.
4. Heldt, J. M.; Fischer-Durand, N.; Salmain, M.; Vessieres, A.; Jaouen, G. The use of glycidol to introduce aldehyde functions into proteins—application to the fluorescent labelling of bovine serum albumin and avidin. *Eur. J. Org. Chem.* **2007**, 5429–5433.
5. Drevets, D. A.; Elliott, A. M. Fluorescence labeling of bacteria for studies of intracellular pathogenesis. *J. Immunol. Methods* **1995**, *187*, 69–79.
6. De Mello, W. C.; Gonzalez Castillo, M.; Van Loon, P. Intercellular diffusion of Lucifer Yellow CH in mammalian cardiac fibers. *J. Mol. Cell. Cardiol.* **1983**, *15*, 637–643.
7. O'Driscoll, D.; Wilson, G.; Steer, M. W. Lucifer Yellow and fluorescein isothiocyanate uptake by cells of *Morinda citrifolia* in suspension cultures is not confined to the endocytotic pathway. *J. Cell Sci.* **1991**, *100*, 237–241.
8. El-Fouly, M. H.; Trosko, J. E.; Chang, C. C. Scrape-loading and dye transfer. A rapid and simple technique to study gap junctional intercellular communication. *Exp. Cell Res.* **1987**, *168*, 422–430.
9. Spiegel, S.; Wilchek, M.; Fishman, P. H. Fluorescence labeling of cell surface glycoconjugates with Lucifer Yellow CH. *Biochem. Biophys. Res. Commun.* **1983**, *112*, 872–877.
10. Nothnagel, E. A. Synthesis and characterization of fluorescent Lucifer Yellow-lipid conjugates. *Biochim. Biophys. Acta, Biomembr.* **1989**, *980*, 209–219.

11. Kawasaki, Y.; Saitoh, T.; Okabe, T.; Kumakura, K.; Ohara-Imaizumi, M. Visualization of exocytotic secretory processes of mast cells by fluorescence techniques. *Biochim. Biophys. Acta, Biomembr.* **1991**, *1067*, 71–80.
12. Spiegel, S.; Kassis, S.; Wilchek, M.; Fishman, P. H. Direct visualization of redistribution and capping of fluorescent gangliosides on lymphocytes. *J. Cell Biol.* **1984**, *99*, 1575–1581.
13. Lee, J. A.; Fortes, P. A. G. Labeling of the glycoprotein subunit of sodium–potassium ATPase with fluorescent probes. *Biochemistry* **1985**, *24*, 322–330.
14. Meda, P. Tracer microinjections into islet cells. *Methods Diabet. Res.* **1984**, *1*, 193–204.
15. Scieszka, J. F.; Cho, M. J. Cellular uptake of a fluid-phase marker by human neutrophils from solutions and liposomes. *Pharm. Res.* **1988**, *5*, 352–358.
16. Bowman, C. L.; Tedeschi, H. Kinetics of Lucifer Yellow CH efflux in giant mitochondria. *Biochim. Biophys. Acta, Biomembr.* **1983**, *731*, 261–266.
17. Keener, C. R.; Wolfe, C. A. C.; Hage, D. S. Optimization of oxidized antibody labeling with Lucifer Yellow CH. *BioTechniques* **1994**, *16*, 894–895, 897.
18. Wright, K. M.; Oparka, K. J. Uptake of Lucifer Yellow CH into plant-cell protoplasts: a quantitative assessment of fluid-phase endocytosis. *Planta* **1989**, *179*, 257–264.
19. Hillmer, S.; Quader, H.; Robert-Nicoud, M.; Robinson, D. G. Lucifer Yellow uptake in cells and protoplasts of *Daucas carota* visualized by laser scanning microscopy. *J. Exp. Bot.* **1989**, *40*, 417–423.
20. Oparka, K. J.; Prior, D. A. M. Movement of Lucifer Yellow CH in potato tuber storage tissues: a comparison of symplastic and apoplastic transport. *Planta* **1988**, *176*, 533–540.
21. Chen, R. F.; Scott, C. H. Atlas of fluorescence spectra and lifetimes of dyes attached to protein. *Anal. Lett.* **1985**, *18*, 393–421.
22. Negishi, K.; Teranishi, T.; Kato, S. Opposite effects of ammonia and carbon dioxide on dye coupling between horizontal cells in the carp retina. *Brain Res.* **1985**, *342*, 330–339.
23. Layer, P. G.; Vollmer, G. Lucifer Yellow stains displaced amacrine cells of the chicken retina during embryonic development. *Neurosci. Lett.* **1982**, *31*, 99–104.
24. Detwiler, P. B.; Sarthy, P. V. Selective uptake of Lucifer Yellow by bipolar cells in the turtle retina. *Neurosci. Lett.* **1981**, *22*, 227–232.
25. Jackson, P. Analysis of carbohydrates and kits therefore. PCT Int. Appl. WO 9302356, 1993; *Chem. Abstr.* **1993**, *118*, 142985.
26. Caprioli, R. M.; Phoebe, C. H., Jr.; Jarrell, J. A. Derivatization and identification of saccharides using Lucifer Yellow CH and related compounds. PCT Int. Appl. WO 9118912, 1991; *Chem. Abstr.* **1992**, *116*, 231337.
27. Bondi, A. Y.; Chiarandini, D. J. Intracellular marking of skeletal muscle cells with horseradish peroxidase in combination with a stain for cholinesterase. *Stain Technol.* **1980**, *55*, 105–109.
28. Mansbridge, J. N.; Knapp, A. M. Penetration of Lucifer Yellow into human skin: a lateral diffusion channel in the stratum corneum. *J. Histochem. Cytochem.* **1993**, *41*, 909–914.
29. Rogers, R. A. Method for imaging tissue. PCT Int. Appl. WO 2000020846, 2000; *Chem. Abstr.* **2000**, *132*, 248259.
30. Klein, M.; Martinoia, E.; Weissenbock, G. Transport of Lucifer Yellow CH into plant vacuoles: evidence for direct energization of a sulfonated substance and implications for the design of new molecular probes. *FEBS Lett.* **1997**, *420*, 86–92.
31. Saito, M.; Ohi, A.; Matsuoka, H. Microinjection of fluorescent dye in a plant cell and its intercellular translocation using a multichannel microelectrode system. *Biochim. Biophys. Acta, Gen. Sub.* **1996**, *1289*, 1–4.
32. Wright, K. M.; Davies, T. G. E.; Steele, S. H.; Leigh, R. A.; Oparka, K. J. Development of a probenecid-sensitive Lucifer Yellow transport system in vacuolating oat aleurone protoplasts. *J. Cell Sci.* **1992**, *102*, 133–139.
33. Oparka, K. J.; Robinson, D.; Prior, D. A. M.; Derrick, P.; Wright, K. M. Uptake of Lucifer Yellow CH into intact barley roots: evidence for fluid-phase endocytosis. *Planta* **1988**, *176*, 541–547.
34. Madore, M. A.; Lucas, W. J. Characterization of the source leaf symplast by means of Lucifer Yellow CH. *Plant Biol.* **1986**, *1*, 129–133.
35. Rideout, D. C. Lucifer Yellow analogs as new antiviral agents. PCT Int. Appl. WO 9409773, 1994; *Chem. Abstr.* **1994**, *121*, 50082.
36. Rideout, D. C.; Elder, J. Antiviral Lucifer Yellow derivative conjugates and pharmaceuticals containing their self-assembling precursors. PCT

Int. Appl. WO 9004394, 1990; *Chem. Abstr.* **1991**, *114*, 75181.

37. Borisov, S. M.; Mayr, T.; Klimant, I. Poly(styrene-block-vinylpyrrolidone) beads as a versatile material for simple fabrication of optical nanosensors. *Anal. Chem.* **2008**, *80*, 573–582.
38. Naarmann, H.; Huemmer, W. Photopolymerizable system with conductive polymer support. Ger. Offen. DE 3844451, 1990; *Chem. Abstr.* **1991**, *114*, 72332.
39. Zeilmaker, M. J.; Yamasaki, H. Inhibition of junctional intercellular communication as a possible short-term test to detect tumor-promoting agents: results with nine chemicals tested by dye transfer assay in Chinese hamster V-79 cells. *Cancer Res.* **1986**, *46*, 6180–6186.

LUCIFER YELLOW VS

CAS Registry Number 71231-14-6

Chemical Structure

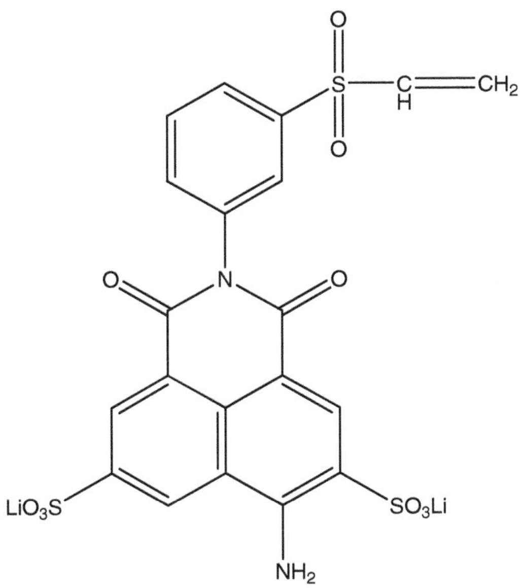

CA Index Name 1*H*-Benz[*de*]isoquinoline-5,8-disulfonic acid, 6-amino-2-[3-(ethenylsulfonyl)phenyl]-2,3-dihydro-1,3-dioxo-, lithium salt (1 : 2)

Other Names 1*H*-Benz[*de*]isoquinoline-5,8-disulfonic acid, 6-amino-2-[3-(ethenylsulfonyl)phenyl]-2,3-dihydro-1,3-dioxo-, dilithium salt; Lucifer Yellow VS

Merck Index Number 5594

Chemical/Dye Class Naphthalimide

Molecular Formula $C_{20}H_{12}Li_2N_2O_{10}S_3$

Molecular Weight 550.39

Physical Form Dark yellow powder

Solubility Soluble in water; insoluble in ethanol

Melting Point >200 °C

Absorption (λ_{max}) 280 nm, 428 nm

Emission (λ_{max}) 540 nm

Synthesis Synthetic methods[1,2]

Staining Applications Cells;[3] neurons;[3] albumin;[4,5] cholesterol;[6] phospholipids;[6] collagenase;[7] polynucleotides;[8] proteins;[9,10] progesterone derivative;[11] testosterone;[12] estriol[12]

Biological Applications Detecting nucleic acids;[8,13] lipid probes;[6] measuring collagenase,[7] target nucleic acid sequence;[13] fluorescent immunoassays[4,5,11,12]

Industrial Applications Not reported

Safety/Toxicity No data available

REFERENCES

1. Stewart, W. W. Synthesis of 3,6-disulfonated 4-aminonaphthalimides. *J. Am. Chem. Soc.* **1981**, *103*, 7615–7620.
2. Stewart, W. W. Functional connections between cells as revealed by dye-coupling with a highly fluorescent naphthalimide tracer. *Cell* **1978**, *14*, 741–759.
3. Stewart, W. W. Lucifer dyes. Highly fluorescent dyes for biological tracing. *Nature* **1981**, *292*, 17–21.
4. Bailey, M. P.; Rocks, B. F.; Riley, C. Homogeneous fluoroimmunoassay using Lucifer Yellow VS: determination of albumin plasma. *Ann. Clin. Biochem.* **1984**, *21*, 59–63.
5. Bailey, M. P.; Rocks, B. F.; Riley, C. Use of Lucifer yellow VS as a label in fluorescent immunoassays illustrated by the determination of albumin in serum. *Ann. Clin. Biochem.* **1983**, *20*, 213–216.
6. Nothnagel, E. A. Synthesis and characterization of fluorescent Lucifer Yellow-lipid conjugates. *Biochim. Biophys. Acta, Biomembr.* **1989**, *980*, 209–219.
7. Tang, L. X.; Rowell, F. J. Flow injection fluorescence measurement of collagenase using a mini-bioreactor with immobilized collagen labeled with Lucifer Yellow. *Anal. Proc.* **1995**, *32*, 255–256.
8. Heller, M. J. Polynucleotides conjugated with chromophores and fluorophores for determination of nucleic acid. PCT Int. Appl. WO 9309128, 1993; *Chem. Abstr.* **1993**, *119*, 242929.
9. Zvetkova, E.; Valet, G.; Katzarova, E.; Ianeva, E.; Neronov, A. Fluorescent and flow cytometric analysis of cellular biochemical content of basic (cationic) cytoplasmic proteins in granulocytes. *Acta Cytobiol. Morphol.* **1993**, *3*, 25–28.
10. Chen, R. F.; Scott, C. H. Atlas of fluorescence spectra and lifetimes of dyes attached to protein. *Anal. Lett.* **1985**, *18*, 393–421.

11. Kirk, D. N.; Miller, B. W. The synthesis of a conjugate of progesterone with Lucifer Yellow VS: a potential probe for fluoroimmunoassay of steroids. *J. Chem. Soc., Perkin Trans. 1* **1988**, 2979–2982.
12. Desfosses, B.; Urios, P.; Christeff, N.; Rajkowski, K. M.; Cittanova, N. The use of disulfonatonaphthalimide fluorescent dyes for the fluorescence polarization immunoassay of steroids. *Anal. Biochem.* **1986**, *159*, 179–186.
13. Di Cesare, J. L. Fluorescence detection assay for homogeneous PCR hybridization systems. PCT Int. Appl. WO 9729210, 1997; *Chem. Abstr.* **1997**, *127*, 215941.

LUCIGENIN

CAS Registry Number 2315-97-1

Chemical Structure

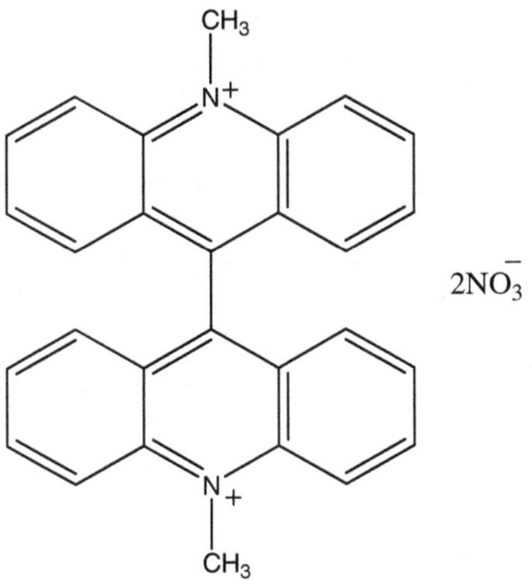

CA Index Name 9,9′-Biacridinium, 10,10′-dimethyl-, nitrate (1 : 2)

Other Names 10,10′-Dimethyl-9,9′-biacridinium dinitrate; *N,N′*-Dimethyl-9,9′-biacridinium dinitrate; 9,9′-Biacridinium, 10,10′-dimethyl-, dinitrate; 9,9′-Bis(*N*-methylacridinium nitrate); Bis-*N*-methylacridinium nitrate; L 6868; Lucigenin; Lucigenin nitrate; Lucigenine; *N,N′*-Dimethyl-9,9′-biacridinium dinitrate; NSC 151912

Merck Index Number Not listed

Chemical/Dye Class Acridine

Molecular Formula $C_{28}H_{22}N_4O_6$

Molecular Weight 510.50

Physical Form Yellow powder with orange to brown cast

Solubility Soluble in water, ethanol, dimethyl sulfoxide

Melting Point >330 °C

Absorption (λ_{max}) 455 nm

Emission (λ_{max}) 505 nm

Synthesis Synthetic methods[1–8]

Staining Applications Chloride ions;[14–18] mitochondria;[9–12] nuclei[13]

Biological Applications Chloride indicator;[14–18] diagnosis of hemostatic disorders;[19] detecting bacteria,[20,21] nucleic acids,[22] proteins,[22] pathogens;[23] identifying respiratory infections;[24] generating and detecting reactive oxygen species;[10–12,25–34] chemiluminescent indicator;[35–43] chemiluminescence determination of chromium,[44] cobalt,[45] arsenic,[46] iron,[47] vanadium,[48] molybdenum[49]

Industrial Applications Lamp;[50] optical nanosensor[51]

Safety/Toxicity Bacterial toxicity;[52] bone marrow toxicity;[53] carcinogenicity;[54,55] cytotoxicity;[56,57] hematotoxicity;[58] hepatotoxicity;[59] nephrotoxicity;[60] neurotoxicity;[61] immunotoxicity;[61] cardiovascular toxicity;[62,63] respiratory toxicity;[63,64] vascular toxicity[65]

REFERENCES

1. Suzuki, H.; Takahashi, T.; Aratani, G.; Katsuragi, H.; Hosogoe, M. Production method of chemiluminescence reagent. Jpn. Kokai Tokkyo Koho JP 2001115155, 2001; *Chem. Abstr.* **2001**, *134*, 318436.
2. Yamada, S.; Kubo, M.; Fuke, H.; Tsubaki, N.; Maeda, K. Photochemical ring closure of 10,10′-disubstituted 9,9′(10*H*,10′*H*)-biacridinylidenes followed by dehydrogenation. *Bull. Chem. Soc. Jpn.* **1993**, *66*, 1834–1836.
3. Shen, J.; Sun, T.; Chen, R.; Xu, X. Reagents for chemiluminescence—synthesis of luminol, isoluminol, lucigenin and ABEI. *Huaxue Shiji* **1988**, *10*, 178–179; *Chem. Abstr.* **1989**, *110*, 154249.
4. Amiet, R. G. The preparation of lucigenin: an experiment with charm. *J. Chem. Educ.* **1982**, *59*, 163–164.
5. Gleu, K.; Schaarschmidt, R. Further biacridenes and biacridylium salts. *Ber. Dtsch. Chem. Ges.* **1940**, *73B*, 909–915; *Chem. Abstr.* **1941**, *35*, 20345.
6. Gleu, K.; Schubert, A. Reaction of phosphorus oxychloride-acridones with Grignard reagents. *Ber. Dtsch. Chem. Ges.* **1940**, *73B*, 805–811; *Chem. Abstr.* **1941**, *35*, 20344.
7. Gleu, K.; Nitzsche, S. Methylated and methoxylated *N,N′*-dimethyldiacridines and *N,N′*-dimethyldiacridylium salts. *J. Prakt. Chem.* **1939**, *153*, 233–241.
8. Decker, H.; Dunant, G. Reduction of cyclaminones. II. Biacridyl. *Ber. Dtsch. Chem. Ges.* **1909**, *42*, 1176–1178; *Chem. Abstr.* **1909**, *3*, 11580.
9. Hattori, F.; Fukuda, K. Method for selecting myocardial cells using intracellular mitochondria labeled with fluorescent indicator. PCT Int. Appl.

WO 2006022377, 2006; *Chem. Abstr.* **2006**, *144*, 270175.

10. Li, Y.; Stansbury, K. H.; Zhu, H.; Trush, M. A. Biochemical characterization of lucigenin (bis-*N*-methylacridinium) as a chemiluminescent probe for detecting intramitochondrial superoxide anion radical production. *Biochem. Biophys. Res. Commun.* **1999**, *262*, 80–87.
11. Li, Y.; Zhu, H.; Trush, M. A. Detection of mitochondria-derived reactive oxygen species production by the chemilumigenic probes lucigenin and luminol. *Biochim. Biophys. Acta* **1999**, *1428*, 1–12.
12. Rembish, S. J.; Trush, M. A. Further evidence that lucigenin-derived chemiluminescence monitors mitochondrial superoxide generation in rat alveolar macrophages. *Free Radical Biol. Med.* **1994**, *17*, 117–126.
13. Horobin, R. W.; Stockert, J. C.; Rashid-Doubell, F. Fluorescent cationic probes for nuclei of living cells: why are they selective? A quantitative structure–activity relations analysis. *Histochem. Cell Biol.* **2006**, *126*, 165–175.
14. Ruedas-Rama, M. J.; Hall, E. A. H. A quantum dot-lucigenin probe for Cl^-. *Analyst* **2008**, *133*, 1556–1566.
15. Graefe, A.; Stanca, S. E.; Nietzsche, S.; Kubicova, L.; Beckert, R.; Biskup, C.; Mohr, G. J. Development and critical evaluation of fluorescent chloride nanosensors. *Anal. Chem.* **2008**, *80*, 6526–6531.
16. Jorg, G.; Bertau, M. Thiol-tolerant assay for quantitative colorimetric determination of chloride released from whole-cell biodehalogenations. *Anal. Biochem.* **2004**, *328*, 22–28.
17. Jiang, J.; Song, Y.; Bai, C.; Koller, B. H.; Matthay, M. A.; Verkman, A. S. Pleural surface fluorescence measurement of Na^+ and Cl^- transport across the air space–capillary barrier. *J. Appl. Physiol.* **2003**, *94*, 343–352.
18. Wissing, F.; Smith, J. A. C. Vacuolar chloride transport in *Mesembryanthemum crystallinum* L. measured using the fluorescent dye lucigenin. *J. Membr. Biol.* **2000**, *177*, 199–208.
19. Kraus, M.; Schelp, C.; Wiegand, A. Method for diagnosis of hemostatic disorders by fluorescence energy transfer spectroscopy. Eur. Pat. Appl. EP 924523, 1999; *Chem. Abstr.* **1999**, *131*, 41806.
20. Olstein, A. D.; Feirtag, J. M. Antibiotic–metal complexes in the detection of gram-negative bacteria and other biological analytes. PCT Int. Appl. WO 2001027628, 2001; *Chem. Abstr.* **2001**, *134*, 307594.
21. Manome, I.; Ikedo, M.; Tamura, S. Bacteria detection by chemiluminescence method. Jpn. Kokai Tokkyo Koho JP 10210998, 1998; *Chem. Abstr.* **1998**, *129*, 200168.
22. Levison, D. W. K.; Moller, U.; Levison, S. Chemiluminescent olefin probes and methods for detection of nucleic acids or proteins and their potential use in diagnosis of disease. U.S. Pat. Appl. Publ. US 2004014043, 2004; *Chem. Abstr.* **2004**, *140*, 107763.
23. Olstein, A. D.; Feirtag, J. Bacteriocin–metal complexes in the detection of pathogens and other biological analytes. U.S. Pat. Appl. Publ. US 2003175207, 2003; *Chem. Abstr.* **2003**, *139*, 260316.
24. Magrisso, M.; Marks, R. S. Chemiluminescent method for identifying respiratory infections of different origins. PCT Int. Appl. WO 2008026205, 2008; *Chem. Abstr.* **2008**, *148*, 302864.
25. Kopprasch, S.; Pietzsch, J.; Graessler, J. Validation of different chemilumigenic substrates for detecting extracellular generation of reactive oxygen species by phagocytes and endothelial cells. *Luminescence* **2003**, *18*, 268–273.
26. Myhre, O.; Andersen, J. M.; Aarnes, H.; Fonnum, F. Evaluation of the probes 2′,7′-dichlorofluorescein diacetate, luminol, and lucigenin as indicators of reactive species formation. *Biochem. Pharmacol.* **2003**, *65*, 1575–1582.
27. Lenaerts, I.; Braeckman, B. P.; Matthijssens, F.; Vanfleteren, J. R. A high-throughput microtiter plate assay for superoxide dismutase based on lucigenin chemiluminescence. *Anal. Biochem.* **2002**, *311*, 90–92.
28. Barbacanne, M. A.; Souchard, J. P.; Darblade, B.; Iliou, J. P.; Nepveu, F.; Pipy, B.; Bayard, F.; Arnal, J. F. Detection of superoxide anion released extracellularly by endothelial cells using cytochrome *c* reduction, ESR, fluorescence and lucigenin-enhanced chemiluminescence techniques. *Free Radical Biol. Med.* **2000**, *29*, 388–396.
29. Li, Y.; Zhu, H.; Kuppusamy, P.; Roubaud, V.; Zweier, J. L.; Trush, M. A. Validation of lucigenin (bis-*N*-methylacridinium) as a chemilumigenic probe for detecting superoxide anion radical production by enzymic and cellular systems. *J. Biol. Chem.* **1998**, *273*, 2015–2023.
30. Hasegawa, H.; Suzuki, K.; Nakaji, S.; Sugawara, K. Analysis and assessment of the capacity of neutrophils to produce reactive oxygen species in a 96-well microplate format using lucigenin- and

luminol-dependent chemiluminescence. *J. Immunol. Methods* **1997**, *210*, 1–10.

31. Brandes, R. P.; Barton, M.; Philippens, K. M. H.; Schweitzer, G.; Muegge, A. Endothelial-derived superoxide anions in pig coronary arteries: evidence from lucigenin chemiluminescence and histochemical techniques. *J. Physiol.* **1997**, *500*, 331–342.
32. Liochev, S. I.; Fridovich, I. Lucigenin luminescence as a measure of intracellular superoxide dismutase activity in *Escherichia coli*. *Proc. Natl. Acad. Sci. U.S.A.* **1997**, *94*, 2891–2896.
33. McKinney, K. A.; Lewis, S. E. M.; Thompson, W. Reactive oxygen species generation in human sperm: luminol and lucigenin chemiluminescence probes. *Arch. Androl.* **1996**, *36*, 119–125.
34. Supari, F.; Ungerer, T.; Harrison, D. G.; Williams, J. K. Fish oil treatment decreases superoxide anions in the myocardium and coronary arteries of atherosclerotic monkeys. *Circulation* **1995**, *91*, 1123–1128.
35. Vladimirov, Y. A.; Proskurnina, E. V.; Izmailov, D. Y. Chemiluminescence as a method for detection and study of free radicals in biological systems. *Bull. Exp. Biol. Med.* **2007**, *144*, 390–396.
36. Lee, J. M.; Karim, M. M.; Lee, S. H. Determination of catechin in aqueous solution by chemiluminescence method. *J. Fluoresc.* **2005**, *15*, 735–739.
37. Vaidya, N. A. Use of chemiluminescence in cosmetics & chromatography. U.S. Pat. Appl. Publ. US 2005118123, 2005; *Chem. Abstr.* **2005**, *142*, 487195.
38. Pavelkova, M.; Kubala, L. Luminol-, isoluminol- and lucigenin-enhanced chemiluminescence of rat blood phagocytes stimulated with different activators. *Luminescence* **2004**, *19*, 37–42.
39. Kitagawa, R. R.; Raddi, M. S. G.; Khalil, N. M.; Vilegas, W.; Marcos da Fonseca, L. Effect of the isocoumarin paepalantine on the luminol and lucigenin amplified chemiluminescence of rat neutrophils. *Biol. Pharm. Bull.* **2003**, *26*, 905–908.
40. Bunting, J. P.; Gray, D. A. Development of a flow injection chemiluminescent assay for the quantification of lipid hydroperoxides. *J. Am. Oil Chem. Soc.* **2003**, *80*, 951–955.
41. Perez-Ruiz, T.; Martinez-Lozano, C.; Tomas, V.; Fenoll, J. Chemiluminescence determination of glucose, fructose and their mixture by the stopped-flow mixing technique. *Mikrochim. Acta* **2003**, *141*, 73–78.
42. Kricka, L. J. Application of bioluminescence and chemiluminescence in biomedical sciences. *Methods Enzymol.* **2000**, *305*, 333–345.
43. Kournikakis, B.; Simpson, M. Optimization of a phagocyte microplate chemiluminescent assay. *J. Biolumin. Chemilumin.* **1995**, *10*, 63–67.
44. Du, J. X.; Li, Y. H.; Guan, R. Chemiluminescence determination of chromium(III) and total chromium in water samples using the periodate–lucigenin reaction. *Microchim. Acta* **2007**, *158*, 145–150.
45. Du, J.; Lu, J.; Zhang, X. Flow-injection chemiluminescence determination of cobalt using a cobalt(II) (1,10-phenanthroline)3 complex-catalyzed lucigenin–periodate reaction. *Microchim. Acta* **2006**, *153*, 21–25.
46. Li, M.; Lee, S. H. Determination of As(III) and As(V) ions by chemiluminescence method. *Microchem. J.* **2005**, *80*, 237–240.
47. Lee, S. H.; Nam, M. S. Determination of Fe(II) ion and Fe(III) ion by chemiluminescence method. *J. Korean Chem. Soc.* **2002**, *46*, 509–514.
48. Sukhan, V. V.; Zaporozhets, O. A.; Lipkovskaya, N. A.; Pogasii, L. B.; Chuiko, A. A. Solid-phase chemiluminescent reagent for determination of vanadium(IV) by flow methods. *Zh. Anal. Khim.* **1994**, *49*, 700–703; *Chem. Abstr.* **1994**, *121*, 194435.
49. Zhu, Z. J. Chemiluminescence determination of molybdenum by on-line reduction with a flow injection system. *Chin. Chem. Lett.* **2000**, *11*, 427–430.
50. Hou, R.; Zhao, X. High energy chemical lamp. Faming Zhuanli Shenqing Gongkai Shuomingshu CN 1157313, 1997; *Chem. Abstr.* **1999**, *132*, 16978.
51. Borisov, S. M.; Mayr, T.; Klimant, I. Poly(styrene-block-vinylpyrrolidone) beads as a versatile material for simple fabrication of optical nanosensors. *Anal. Chem.* **2008**, *80*, 573–582.
52. Barsukov, A. A.; Zhukhovitskii, V. G.; Shcherbakova, E. G.; Zemskov, V. M.; Diashev, A. N. New data on the effect of *Helicobacter pylori* on oxygen metabolism in human neutrophils. *Bull. Exp. Biol. Med.* **2005**, *139*, 70–72.
53. Twerdok, L. E.; Mosebrook, D. R.; Trush, M. A. Comparison of oxidant-generation and BP-diol activation by bone marrow cells from C57Bl/6 and DBA/2 mice: implications for risk of bone marrow toxicity induced by polycyclic hydrocarbons. *Toxicol. Appl. Pharmacol.* **1992**, *112*, 266–272.

54. Er, T. K.; Tsai, S. M.; Wu, S. H.; Chiang, W.; Lin, H. C.; Lin, S. F.; Wu, S. H.; Tsai, L. Y.; Liu, T. Z. Antioxidant status and superoxide anion radical generation in acute myeloid leukemia. *Clin. Biochem.* **2007**, *40*, 1015–1019.

55. Bouvier, G.; Hergenhahn, M.; Polack, A.; Bornkamm, G. W.; Bartsch, H. Validation of two test systems for detecting tumor promoters and EBV inducers: comparative responses of several agents in DR-CAT Raji cells and in human granulocytes. *Carcinogenesis* **1993**, *14*, 1573–1578.

56. Van, D. K.; Patel, S.; Vallyathan, V. Lucigenin chemiluminescence assay as an adjunctive tool for assessment of various stages of inflammation: a study of quiescent inflammatory cells. *J. Biosci.* **2003**, *28*, 115–119.

57. Blake, T.; Castranova, V.; Schwegler-Berry, D.; Baron, P.; Deye, G. J.; Li, C.; Jones, W. Effect of fiber length on glass microfiber cytotoxicity. *J. Toxicol. Environ. Health, Part A* **1998**, *54*, 243–259.

58. Miesel, R.; Dietrich, A.; Ulbrich, N.; Kroeger, H.; Mitchison, N. A. Assessment of collagen type II induced arthritis in mice by whole blood chemiluminescence. *Autoimmunity* **1994**, *19*, 153–159.

59. Caraceni, P.; Rosenblum, E. R.; Van, T. D. H.; Borle, A. B. Reoxygenation injury in isolated rat hepatocytes: relation to oxygen free radicals and lipid peroxidation. *Am. J. Physiol.* **1994**, *266*, G799–G806.

60. Zhang, J. G.; Lindup, W. E. Differential effects of cisplatin on the production of NADH-dependent superoxide and the activity of antioxidant enzymes in rat renal cortical slices *in vitro*. *Pharmacol. Toxicol.* **1996**, *79*, 191–198.

61. Aam, B. B.; Fonnum, F. (+/−)-2-Chloropropionic acid elevates reactive oxygen species formation in human neutrophil granulocytes. *Toxicology* **2006**, *228*, 124–134.

62. Borowiec, J. W.; Lahtinen, M.; Venge, P.; Henze, A.; Stiernstrom, H. Inflammatory response during simulated extracorporeal circulation with addition of nitric oxide. *J. Cardiovasc. Surg.* **2000**, *41*, 207–213.

63. Aam, B. B.; Fonnum, F. ROS scavenging effects of organic extract of diesel exhaust particles on human neutrophil granulocytes and rat alveolar macrophages. *Toxicology* **2007**, *230*, 207–218.

64. Brehm, M.; Schiller, E.; Zeller, W. J. Quantification of reactive oxygen species generated by alveolar macrophages using lucigenin-enhanced chemiluminescence—methodical aspects. *Toxicol. Lett.* **1996**, *87*, 131–138.

65. Skatchkov, M. P.; Sperling, D.; Hink, U.; Anggard, E.; Munzel, T. Quantification of superoxide radical formation in intact vascular tissue using a *Cypridina luciferin* analog as an alternative to lucigenin. *Biochem. Biophys. Res. Commun.* **1998**, *248*, 382–386.

MAG-FURA 2 (FURAPTRA)

CAS Registry Number 132319-57-4

Chemical Structure

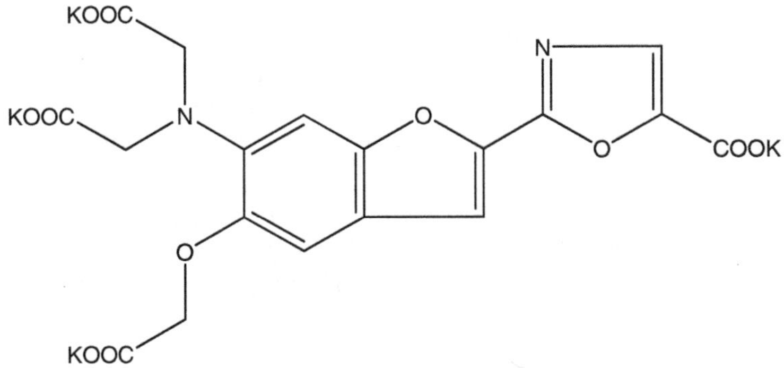

CA Index Name 5-Oxazolecarboxylic acid, 2-[6-[bis(carboxymethyl)amino]-5-(carboxymethoxy)-2-benzofuranyl]-, potassium salt (1:4)

Other Names 5-Oxazolecarboxylic acid, 2-[6-[bis(carboxymethyl)amino]-5-(carboxymethoxy)-2-benzofuranyl]-, tetrapotassium salt; Mag-Fura-2 tetrapotassium salt

Merck Index Number Not listed

Chemical/Dye Class Benzofuran

Molecular Formula $C_{18}H_{10}K_4N_2O_{11}$

Molecular Weight 586.68

Physical Form Light yellow powder

Solubility Soluble in water

Melting Point >200 °C

Absorption (λ_{max}) 369 nm, 330 nm

Emission (λ_{max}) 511 nm, 491 nm

Synthesis Synthetic method[1]

Staining Applications Magnesium ions;[1–27] calcium ions;[11,22,28–46] zinc ions[47–49]

Biological Applications Magnesium indicator;[1–27] calcium indicator;[11,22,28–46] zinc indicator[47–49]

Industrial Applications Not reported

Safety/Toxicity Neurotoxicity[47]

REFERENCES

1. Raju, B.; Murphy, E.; Levy, L. A.; Hall, R. D.; London, R. E. A fluorescent indicator for measuring cytosolic free magnesium. *Am. J. Physiol.* **1989**, *256*, C540–C548.
2. Malarkey, C. S.; Wang, G.; Ballicora, M. A.; Mota de Freitas, D. E. Evidence for two distinct Mg^{2+} binding sites in G(s alpha) and G(i alpha1) proteins. *Biochem. Biophys. Res. Commun.* **2008**, *372*, 866–869.
3. Gunther, T. Concentration, compartmentation and metabolic function of intracellular free Mg^{2+}. *Magnesium Res.* **2006**, *19*, 225–236.
4. Tashiro, M.; Tursun, P.; Konishi, M. Intracellular and extracellular concentrations of Na^+ modulate Mg^{2+} transport in rat ventricular myocytes. *Biophys. J.* **2005**, *89*, 3235–3247.
5. Tursun, P.; Tashiro, M.; Konishi, M. Modulation of Mg^{2+} efflux from rat ventricular myocytes studied with the fluorescent indicator Furaptra. *Biophys. J.* **2005**, *88*, 1911–1924.
6. Watanabe, M.; Konishi, M.; Ohkido, I.; Matsufuji, S. Enhanced sodium-dependent extrusion of magnesium in mutant cells established from a mouse renal tubular cell line. *Am. J. Physiol.* **2005**, *289*, F742–F748.
7. Delva, P.; Pastori, C.; Degan, M.; Montesi, G.; Lechi, A. Catecholamine-induced regulation *in vitro* and *ex vivo* of intralymphocyte ionized magnesium. *J. Membr. Biol.* **2004**, *199*, 163–171.
8. Froschauer, E. M.; Kolisek, M.; Dieterich, F.; Schweigel, M.; Schweyen, R. J. Fluorescence

Handbook of Biological Dyes and Stains By R. W. Sabnis

measurements of free [Mg^{2+}] by use of Mag-Fura 2 in *Salmonella enterica*. *FEMS Microbiol. Lett.* **2004**, *237*, 49–55.

9. Kolisek, M.; Zsurka, G.; Samaj, J.; Weghuber, J.; Schweyen, R. J.; Schweigel, M. Mrs2p is an essential component of the major electrophoretic Mg^{2+} influx system in mitochondria. *EMBO J.* **2003**, *22*, 1235–1244.
10. Watanabe, M.; Konishi, M. Intracellular calibration of the fluorescent Mg^{2+} indicator Furaptra in rat ventricular myocytes. *Pfluegers Arch.* **2001**, *442*, 35–40.
11. Tashiro, M.; Konishi, M.; Iwamoto, T.; Shigekawa, M.; Kurihara, S. Transport of magnesium by two isoforms of the Na^{+}–Ca^{2+} exchanger expressed in CCL39 fibroblasts. *Pfluegers Arch.* **2000**, *440*, 819–827.
12. Amari, L.; Layden, B.; Rong, Q.; Geraldes, C. F. G. C.; Mota de Freitas, D. Comparison of fluorescence, ^{31}P NMR, and ^{7}Li NMR spectroscopic methods for investigating Li^{+}/Mg^{2+} competition for biomolecules. *Anal. Biochem.* **1999**, *272*, 1–7.
13. Kennedy, H. J. Intracellular Mg^{2+} regulation in voltage-clamped *Helix aspersa* neurones measured with Mag-Fura-2 and $Mg^{(2+)}$-sensitive microelectrodes. *Exp. Physiol.* **1998**, *83*, 449–460.
14. Juttner, R.; Ebel, H. Characterization of Mg^{2+} transport in brush border membrane vesicles of rabbit ileum studied with Mag-Fura-2. *Biochim. Biophys. Acta* **1998**, *1370*, 51–63.
15. Kato, H.; Gotoh, H.; Kajikawa, M.; Suto, K. Depolarization triggers intracellular magnesium surge in cultured dorsal root ganglion neurons. *Brain Res.* **1998**, *779*, 329–333.
16. Sasaki, N.; Oshima, T.; Matsuura, H.; Yoshimura, M.; Yashiki, M.; Higashi, Y.; Ishioka, N.; Nakano, Y.; Kojima, R.; Kambe, M.; Kajiyama, G. Lack of effect of transmembrane gradient of magnesium and sodium on regulation of cytosolic free magnesium concentration in rat lymphocytes. *Biochim. Biophys. Acta* **1997**, *1329*, 169–173.
17. Wilson, W. R.; Phillips, K. S.; Leslie, S. W. Ethanol and glutamate effects on intracellular magnesium. *Alcohol* **1997**, *14*, 533–536.
18. Tashiro, M.; Konishi, M. Basal intracellular free Mg^{2+} concentration in smooth muscle cells of guinea pig tenia cecum: intracellular calibration of the fluorescent indicator Furaptra. *Biophys. J.* **1997**, *73*, 3358–3370.
19. Rodriguez-Zavala, J. S.; Saavedra-Molina, A.; Moreno-Sanchez, R. Effect of intramitochondrial Mg^{2+} on citrulline synthesis in rat liver mitochondria. *Biochem. Mol. Biol. Int.* **1997**, *41*, 179–187.
20. Niemela, J. E.; Snader, B. M.; Elin, R. J. Determination of ionized magnesium in platelets and correlation with selected variables. *Clin. Chem.* **1996**, *42*, 744–748.
21. Jung, D. W.; Chapman, C. J.; Baysal, K.; Pfeiffer, D. R.; Brierley, G. P. On the use of fluorescent probes to estimate free Mg^{2+} in the matrix of heart mitochondria. *Arch. Biochem. Biophys.* **1996**, *332*, 19–29.
22. Sugiyama, T.; Goldman, W. F. Measurement of SR free Ca^{2+} and Mg^{2+} in permeabilized smooth muscle cells with use of Furaptra. *Am. J. Physiol.* **1995**, *269*, C698–C705.
23. Garcia-Martin, E.; Martin-Romero, F. J.; Gutierrez-Merino, C. Intrasynaptosomal free Mg^{2+} concentration measured with the fluorescent indicator Mag-Fura-2: modulation by Na^{+} gradient and by extrasynaptosomal ATP. *J. Neurochem.* **1995**, *65*, 2757–2764.
24. Zhang, G. H.; Melvin, J. E. Regulation by extracellular Na^{+} of cytosolic Mg^{2+} concentration in $Mg^{(2+)}$-loaded rat sublingual acini. *FEBS Lett.* **1995**, *371*, 52–56.
25. Brocard, J. B.; Rajdev, S.; Reynolds, I. J. Glutamate-induced increases in intracellular free Mg^{2+} in cultured cortical neurons. *Neuron* **1993**, *11*, 751–757.
26. Schachter, M.; Gallagher, K. L.; Sever, P. S. Intracellular free magnesium in a vascular smooth muscle cell line. *Biochem. Soc. Trans.* **1990**, *18*, 1266.
27. Murphy, E.; Freudenrich, C. C.; Levy, L. A.; London, R. E.; Lieberman, M. Monitoring cytosolic free magnesium in cultured chicken heart cells by use of the fluorescent indicator Furaptra. *Proc. Natl. Acad. Sci. U.S.A.* **1989**, *86*, 2981–2984.
28. Paredes, R. M.; Etzler, J. C.; Watts, L. T.; Zheng, W.; Lechleiter, J. D. Chemical calcium indicators. *Methods* **2008**, *46*, 143–151.
29. Walsh, B. M.; Naik, H. B.; Dubach, J. M.; Beshire, M.; Wieland, A. M.; Soybel, D. I. Thiol-oxidant monochloramine mobilizes intracellular Ca^{2+} in parietal cells of rabbit gastric glands. *Am. J. Physiol.* **2007**, *293*, C1687–C1697.
30. Baylor, S. M.; Hollingworth, S. Simulation of Ca^{2+} movements within the sarcomere of fast-twitch mouse fibers stimulated by action potentials. *J. Gen. Physiol.* **2007**, *130*, 283–302.
31. Ukhanov, K.; Mills, S. J.; Potter, B. V.; Walz, B. InsP (3)-induced $Ca^{(2+)}$ release in permeabilized

invertebrate photoreceptors: a link between phototransduction and $Ca^{(2+)}$ stores. *Cell Calcium* **2001**, *29*, 335–345.

32. Gobel, A.; Krause, E.; Feick, P.; Schulz, I. IP(3) and cyclic ADP-ribose induced $Ca^{(2+)}$ release from intracellular stores of pancreatic acinar cells from rat in primary culture. *Cell Calcium* **2001**, *29*, 29–37.
33. Woon, L. A.; Holland, J. W.; Kable, E. P.; Roufogalis, B. D. Ca^{2+} sensitivity of phospholipid scrambling in human red cell ghosts. *Cell Calcium* **1999**, *25*, 313–320.
34. Braun, F. J.; Hegemann, P. Direct measurement of cytosolic calcium and pH in living *Chlamydomonas reinhardtii* cells. *Eur. J. Cell Biol.* **1999**, *78*, 199–208.
35. Stout, A. K.; Reynolds, I. J. High-affinity calcium indicators underestimate increases in intracellular calcium concentrations associated with excitotoxic glutamate stimulations. *Neuroscience* **1999**, *89*, 91–100.
36. Tengholm, A.; Hagman, C.; Gylfe, E.; Hellman, B. *In situ* characterization of nonmitochondrial Ca^{2+} stores in individual pancreatic beta-cells. *Diabetes* **1998**, *47*, 1224–1230.
37. Martinez-Zaguilan, R.; Parnami, J.; Martinez, G. M. Mag-Fura-2 (Furaptra) exhibits both low (μM) and high (nM) affinity for Ca^{2+}. *Cell. Physiol. Biochem.* **1998**, *8*, 158–174.
38. Naraghi, M. T-jump study of calcium binding kinetics of calcium chelators. *Cell Calcium* **1997**, *22*, 255–268.
39. Zhao, M.; Hollingworth, S.; Baylor, S. M. AM-loading of fluorescent Ca^{2+} indicators into intact single fibers of frog muscle. *Biophys. J.* **1997**, *72*, 2736–2747.
40. Sinha, S. R.; Wu, L. G.; Saggau, P. Presynaptic calcium dynamics and transmitter release evoked by single action potentials at mammalian central synapses. *Biophys. J.* **1997**, *72*, 637–651.
41. Regehr, W. G.; Atluri, P. P. Calcium transients in cerebellar granule cell presynaptic terminals. *Biophys. J.* **1995**, *68*, 2156–2170.
42. Tanimura, A.; Turner, R. J. Calcium release in HSY cells conforms to a steady-state mechanism involving regulation of the inositol 1,4,5-trisphosphate receptor Ca^{2+} channel by luminal [Ca^{2+}]. *J. Cell Biol.* **1996**, *132*, 607–616.
43. Ogden, D.; Khodakhah, K.; Carter, T.; Thomas, M.; Capiod, T. Analog computation of transient changes of intracellular free Ca^{2+} concentration with the low affinity Ca^{2+} indicator Furaptra during whole-cell patch-clamp recording. *Pfluegers Arch.* **1995**, *429*, 587–591.
44. Berlin, J. R.; Konishi, M. Ca^{2+} transients in cardiac myocytes measured with high and low affinity Ca^{2+} indicators. *Biophys. J.* **1993**, *65*, 1632–1647.
45. Konishi, M.; Berlin, J. R. Calcium transients in cardiac myocytes measured with a low affinity fluorescent indicator, Furaptra. *Biophys. J.* **1993**, *64*, 1331–1343.
46. Neher, E.; Zucker, R. S. Multiple calcium-dependent processes related to secretion in bovine chromaffin cells. *Neuron* **1993**, *10*, 21–30.
47. Dineley, K. E.; Scanlon, J. M.; Kress, G. J.; Stout, A. K.; Reynolds, I. J. Astrocytes are more resistant than neurons to the cytotoxic effects of increased [$Zn^{(2+)}$](i). *Neurobiol. Dis.* **2000**, *7*, 310–320.
48. Cheng, C.; Reynolds, I. J. Calcium-sensitive fluorescent dyes can report increases in intracellular free zinc concentration in cultured forebrain neurons. *J. Neurochem.* **1998**, *71*, 2401–2410.
49. Simons, T. J. Measurement of free Zn^{2+} ion concentration with the fluorescent probe Mag-Fura-2 (Furaptra). *J. Biochem. Biophys. Methods* **1993**, *27*, 25–37.

MAG-FURA 2 AM

CAS Registry Number 130100-20-8

Chemical Structure

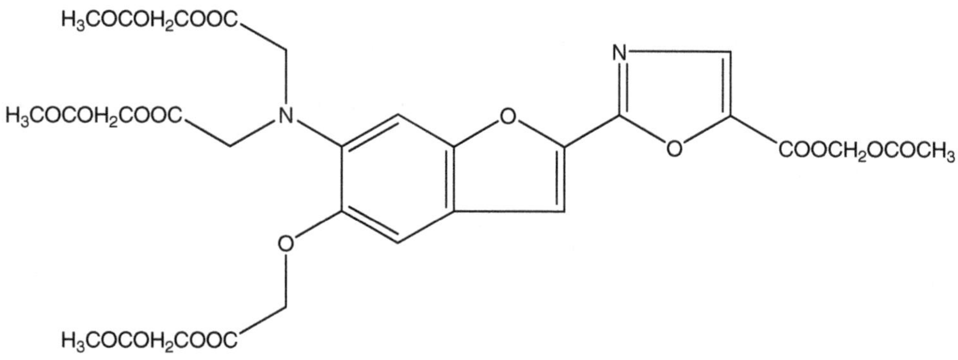

CA Index Name 5-Oxazolecarboxylic acid, 2-[5-[2-[(acetyloxy)methoxy]-2-oxoethoxy]-6-[bis[2-[(acetyloxy)methoxy]-2-oxoethyl]amino]-2-benzofuranyl]-, (acetyloxy)methyl ester

Other Names Mag-Fura-2; Mag-Fura-2 AM; Mag-Fura-2 AM ester; Mag-Fura-2 acetoxymethyl ester; Furaptra AM; Furaptra AM ester; Furaptra acetoxymethyl ester

Merck Index Number Not listed

Chemical/Dye Class Benzofuran

Molecular Formula $C_{30}H_{30}N_2O_{19}$

Molecular Weight 722.56

Physical Form Light yellow powder

Solubility Insoluble in water; soluble in dimethyl sulfoxide, ethyl acetate

Melting Point >200 °C

Boiling Point (Calcd.) 797.5 ± 70.0 °C, pressure: 760 Torr

pK_a (Calcd.) -2.88 ± 0.70, most basic, temperature: 25 °C

Absorption (λ_{max}) 366 nm

Emission (λ_{max}) 475 nm

Synthesis Synthetic method[1]

Staining Applications Magnesium ions;[1–15] calcium ions;[16–30] zinc ions[31–33]

Biological Applications Magnesium indicator;[1–15] calcium indicator;[16–30] zinc indicator[31–33]

Industrial Applications Not reported

Safety/Toxicity No data available

REFERENCES

1. Raju, B.; Murphy, E.; Levy, L. A.; Hall, R. D.; London, R. E. A fluorescent indicator for measuring cytosolic free magnesium. *Am. J. Physiol.* **1989**, *256*, C540–C548.
2. Froschauer, E. M.; Kolisek, M.; Dieterich, F.; Schweigel, M.; Schweyen, R. J. Fluorescence measurements of free [Mg^{2+}] by use of Mag-Fura 2 in *Salmonella enterica. FEMS Microbiol. Lett.* **2004**, *237*, 49–55.
3. Ishijima, S.; Uchibori, A.; Takagi, H.; Maki, R.; Ohnishi, M. Light-induced increase in free Mg^{2+} concentration in spinach chloroplasts: measurement of free Mg^{2+} by using a fluorescent probe and necessity of stromal alkalinization. *Arch. Biochem. Biophys.* **2003**, *412*, 126–132.
4. Touyz, R. M.; Mercure, C.; Reudelhuber, T. L. Angiotensin II type I receptor modulates intracellular free Mg^{2+} in renally derived cells via Na^+-dependent Ca^{2+}-independent mechanisms. *J. Biol. Chem.* **2001**, *276*, 13657–13663.
5. Murakami, T. Cytotoxicity test method using magnesium ion fluorescent dye. Jpn. Kokai Tokkyo Koho JP 2000189191, 2000; *Chem. Abstr.* **2000**, *133*, 100817.
6. Schweigel, M.; Lang, I.; Martens, H. $Mg^{(2+)}$ transport in sheep rumen epithelium: evidence for an electrodiffusive uptake mechanism. *Am. J. Physiol.* **1999**, *277*, G976–G982.
7. Howart, F. C.; Singh, J.; Waring, J. J.; Hustler, B. I.; Bailey, M. Effects of monovalent cations, pH and

temperature on the dissociation constant (K_D) for the fluorescent indicator Mag-Fura-2 at different excitation wavelengths. *Magnesium Res.* **1995**, *8*, 299–306.

8. Gow, I. F.; Latham, T.; Ellis, D.; Flatman, P. W. Measurement of intracellular ionized magnesium concentration in myocytes isolated from the septomarginal band of sheep hearts. *Magnesium Res.* **1995**, *8*, 223–232.
9. Garcia-Martin, E.; Martin-Romero, F. J.; Gutierrez-Merino, C. Intrasynaptosomal free Mg^{2+} concentration measured with the fluorescent indicator Mag-Fura-2: modulation by Na^+ gradient and by extrasynaptosomal ATP. *J. Neurochem.* **1995**, *65*, 2757–2764.
10. Willems, H. L.; Kruse, J.; Geven, W. B.; Vis, P.; Monnens, L. A. H. Defibrination essential in the assay of ionized magnesium in mononuclear cells. *Eur. J. Clin. Chem. Clin. Biochem.* **1995**, *33*, 103–104.
11. Grubbs, R. D.; Walter, A. Determination of cytosolic Mg^{2+} activity and buffering in BC3H-1 cells with Mag-Fura-2. *Mol. Cell. Biochem.* **1994**, *136*, 11–22.
12. Hurley, T. W.; Ryan, M. P.; Brinck, R. W. Changes of cytosolic calcium interfere with measurements of cytosolic magnesium using Mag-Fura-2. *Am. J. Physiol.* **1992**, *263*, C300–C307.
13. Rutter, G. A.; Osbaldeston, N. J.; McCormack, J. G.; Denton, R. M. Measurement of matrix free magnesium concentration in rat heart mitochondria by using entrapped fluorescent probes. *Biochem. J.* **1990**, *271*, 627–634.
14. Schachter, M.; Gallagher, K. L.; Sever, P. S. Measurement of intracellular magnesium in a vascular smooth muscle cell line using a fluorescent probe. *Biochim. Biophys. Acta, Gen. Sub.* **1990**, *1035*, 378–380.
15. Schachter, M.; Gallagher, K. L.; Sever, P. S. Intracellular free magnesium in a vascular smooth muscle cell line. *Biochem. Soc. Trans.* **1990**, *18*, 1266.
16. Narayanan, B.; Islam, M. N.; Bartelt, D.; Ochs, R. S. A direct mass-action mechanism explains capacitative calcium entry in Jurkat and skeletal L6 muscle cells. *J. Biol. Chem.* **2003**, *278*, 44188–44196.
17. Gurney, A. M.; Drummond, R. M.; Fay, F. S. Calcium signalling in sarcoplasmic reticulum, cytoplasm and mitochondria during activation of rabbit aorta myocytes. *Cell Calcium* **2000**, *27*, 339–351.
18. Hyrc, K. L.; Bownik, J. M.; Goldberg, M. P. Ionic selectivity of low-affinity ratiometric calcium indicators: Mag-Fura-2, Fura-2FF and BTC. *Cell Calcium* **2000**, *27*, 75–86.
19. Cheng, C.; Fass, D. M.; Reynolds, I. J. Emergence of excitotoxicity in cultured forebrain neurons coincides with larger glutamate-stimulated $[Ca^{2+}]_i$ increases and NMDA receptor mRNA levels. *Brain Res.* **1999**, *849*, 97–108.
20. Woon, L. A.; Holland, J. W.; Kable, E. P. W.; Roufogalis, B. D. Ca^{2+} sensitivity of phospholipid scrambling in human red cell ghosts. *Cell Calcium* **1999**, *25*, 313–320.
21. Khodorov, B.; Pinelis, V.; Storozhevykh, T.; Yuravichus, A.; Khaspekhov, L. Blockade of mitochondrial Ca^{2+} uptake by mitochondrial inhibitors amplifies the glutamate-induced calcium response in cultured cerebellar granule cells. *FEBS Lett.* **1999**, *458*, 162–166.
22. Churchill, G. C.; Louis, C. F. Imaging of intracellular calcium stores in single permeabilized lens cells. *Am. J. Physiol.* **1999**, *276*, C426–C434.
23. Stout, A. K.; Reynolds, I. J. High-affinity calcium indicators underestimate increases in intracellular calcium concentrations associated with excitotoxic glutamate stimulations. *Neuroscience* **1999**, *89*, 91–100.
24. Hofer, A. M.; Landolfi, B.; Debellis, L.; Pozzan, T.; Curci, S. Free $[Ca^{2+}]$ dynamics measured in agonist-sensitive stores of single living intact cells: a new look at the refilling process. *EMBO J.* **1998**, *17*, 1986–1995.
25. Saiki, Y.; Ikemoto, N. Fluorescence probe study of the lumenal Ca^{2+} of the sarcoplasmic reticulum vesicles during Ca^{2+} uptake and Ca^{2+} release. *Biochem. Biophys. Res. Commun.* **1997**, *241*, 181–186.
26. Tojyo, Y.; Tanimura, A.; Matsumoto, Y. Monitoring of Ca^{2+} release from intracellular stores in permeabilized rat parotid acinar cells using the fluorescent indicators Mag-Fura-2 and Calcium Green C18. *Biochem. Biophys. Res. Commun.* **1997**, *240*, 189–195.
27. Zhao, M.; Hollingworth, S.; Baylor, S. M. AM-loading of fluorescent Ca^{2+} indicators into intact single fibers of frog muscle. *Biophys. J.* **1997**, *72*, 2736–2747.
28. Hofer, A. M.; Schulz, I. Quantification of intraluminal free [Ca] in the agonist-sensitive internal calcium store using compartmentalized fluorescent indicators: some considerations. *Cell Calcium* **1996**, *20*, 235–242.
29. Boltz, R. C., Jr. A high capacity screen for immunoregulants using intracellular calcium concentration measurement. PCT Int. Appl. WO 9610090, 1996; *Chem. Abstr.* **1996**, *125*, 76326.

30. Martinez-Zaguilan, R.; Parnami, G.; Lynch, R. M. Selection of fluorescent ion indicators for simultaneous measurements of pH and Ca^{2+}. *Cell Calcium* **1996**, *19*, 337–349.

31. Dineley, K. E.; Malaiyandi, L. M.; Reynolds, I. J. A reevaluation of neuronal zinc measurements: artifacts associated with high intracellular dye concentration. *Mol. Pharmacol.* **2002**, *62*, 618–627.

32. Cheng, C.; Reynolds, I. J. Calcium-sensitive fluorescent dyes can report increases in intracellular free zinc concentration in cultured forebrain neurons. *J. Neurochem.* **1998**, *71*, 2401–2410.

33. Simons, T. J. B. Measurement of free zinc ion concentration with the fluorescent probe Mag-Fura-2 (Furaptra). *J. Biochem. Biophys. Methods* **1993**, *27*, 25–37.

MAG-INDO 1

CAS Registry Number 132299-21-9

Chemical Structure

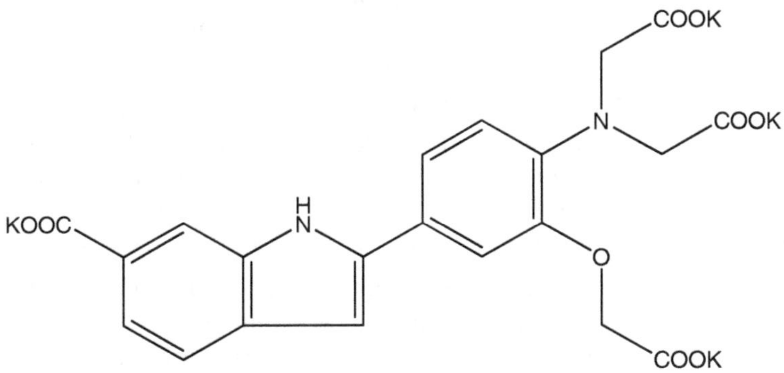

CA Index Name 1*H*-Indole-6-carboxylic acid, 2-[4-[bis(carboxymethyl)amino]-3-(carboxymethoxy)phenyl]-, potassium salt (1:4)

Other Names 1*H*-Indole-6-carboxylic acid, 2-[4-[bis (carboxymethyl)amino]-3-(carboxymethoxy)phenyl]-, tetrapotassium salt; Mag-indo-1; Mag-indo-1 tetrapotassium salt

Merck Index Number Not listed

Chemical/Dye Class Indole

Molecular Formula $C_{21}H_{14}K_4N_2O_9$

Molecular Weight 594.74

Physical Form Solid

Solubility Soluble in water

Melting Point >200 °C

Absorption (λ_{max}) 349 nm, 330 nm

Emission (λ_{max}) 480 nm, 417 nm

Synthesis Synthetic method[1]

Staining Applications Magnesium ions;[2–10] calcium ions[4,11–13]

Biological Applications Magnesium indicator;[2–10] calcium indicator;[4,11–13] identifying genes;[14] probing unfolding and refolding protein sequences,[15] protein subdomains[16,17]

Industrial Applications Not reported

Safety/Toxicity No data available

REFERENCES

1. Haugland, R. P. *The Handbook: A Guide to Fluorescent Probes and Labeling Technologies*; Molecular Probes Inc.: Eugene, **2005**; pp 910–912.
2. Park, E. J.; Brasuel, M.; Behrend, C.; Philbert, M. A.; Kopelman, R. Ratiometric optical PEBBLE nanosensors for real-time magnesium ion concentrations inside viable cells. *Anal. Chem.* **2003**, *75*, 3784–3791.
3. Suzuki, Y.; Komatsu, H.; Ikeda, T.; Saito, N.; Araki, S.; Citterio, D.; Hisamoto, H.; Kitamura, Y.; Kubota, T.; Nakagawa, J.; Oka, K.; Suzuki, K. Design and synthesis of Mg^{2+}-selective fluoroionophores based on a coumarin derivative and application for Mg^{2+} measurement in a living cell. *Anal. Chem.* **2002**, *74*, 1423–1428.
4. Pesco, J.; Salmon, J. M.; Vigo, J.; Viallet, P. Mag-indo1 affinity for Ca^{2+}, compartmentalization and binding to proteins: the challenge of measuring Mg^{2+} concentrations in living cells. *Anal. Biochem.* **2001**, *290*, 221–231.
5. Murakami, T. Cytotoxicity test method using magnesium ion fluorescent dye. Jpn. Kokai Tokkyo Koho JP 2000189191, 2000; *Chem. Abstr.* **2000**, *133*, 100817.
6. Csernoch, L.; Bernengo, J. C.; Szentesi, P.; Jacquemond, V. Measurements of intracellular Mg^{2+} concentration in mouse skeletal muscle fibers with the fluorescent indicator Mag-indo-1. *Biophys. J.* **1998**, *75*, 957–967.
7. Szmacinski, H.; Lakowicz, J. R. Fluorescence lifetime characterization of magnesium probes: improvement of Mg^{2+} dynamic range and sensitivity using phase-modulation fluorometry. *J. Fluoresc.* **1996**, *6*, 83–95.

8. Sebille, S.; Millot, J. M.; Maizieres, M.; Arnaud, M.; Delabroise, A. M.; Jacquot, J.; Manfait, M. Spatial and temporal Mg^{2+} signaling in single human tracheal gland cells. *Biochem. Biophys. Res. Commun.* **1996**, *227*, 743–749.

9. Morelle, B.; Salmon, J. M.; Vigo, J.; Viallet, P. Are intracellular ionic concentrations accessible using fluorescent probes? The example of Mag-indo-1. *Cell Biol. Toxicol.* **1994**, *10*, 339–344.

10. Morelle, B.; Salmon, J. M.; Vigo, J.; Viallet, P. Proton, Mg^{2+} and protein as competing ligands for the fluorescent probe, mag-indo-1: a first step to the quantification of intracellular Mg^{2+} concentration. *Photochem. Photobiol.* **1993**, *58*, 795–802.

11. Launikonis, B. S.; Zhou, J.; Royer, L.; Shannon, T. R.; Brum, G.; Rios, E. Confocal imaging of [Ca^{2+}] in cellular organelles by SEER, shifted excitation and emission ratioing of fluorescence. *J. Physiol.* **2005**, *567*, 523–543.

12. Ribou, A. C.; Vigo, J.; Viallet, P.; Salmon, J. M. Interaction of a protein, BSA, and a fluorescent probe, Mag-Indo-1, influence of EDTA and calcium on the equilibrium. *Biophys. Chem.* **1999**, *81*, 179–189.

13. Zhao, M.; Hollingworth, S.; Baylor, S. M. Properties of tri- and tetracarboxylate Ca^{2+} indicators in frog skeletal muscle fibers. *Biophys. J.* **1996**, *70*, 896–916.

14. Kim, E. J.; Barker, L.; Burnet, M.; Guse, J. H.; Luyten, K.; Tsotsou, G. Use of a yeast expression host to identify genes for transport proteins using fluorescent dye reporters. PCT Int. Appl. WO 2003038092, 2003; *Chem. Abstr.* **2003**, *138*, 363834.

15. Viallet, P. M.; Vo-Dinh, T.; Ribou, A. C.; Vigo, J.; Salmon, J. M. Native fluorescence and Mag-indo-1-protein interaction as tools for probing unfolding and refolding sequences of the bovine serum albumin subdomain in the presence of guanidine hydrochloride. *J. Protein Chem.* **2000**, *19*, 431–439.

16. Viallet, P. M.; Vo-Dinh, T.; Bunde, T.; Ribou, A. C.; Vigo, J.; Salmon, J. M. Fluorescent molecular reporter for the 3-D conformation of protein subdomains: the Mag-Indo system. *J. Fluoresc.* **1999**, *9*, 153–161.

17. Viallet, P. M.; Vo-Dinh, T.; Bunde, T.; Ribou, A. C.; Vigo, J.; Salmon, J. M. Mag-Indo-1 protein interaction as a tool for probing the 3D conformation of protein subdomains: influence of the chemical microenvironment of the histidine residue(s) on the parameters of the interaction. *Proc. SPIE-Int. Soc. Opt. Eng.* **1999**, *3595*, 231–242.

MAG-INDO 1 AM

CAS Registry Number 130926-94-2

Chemical Structure

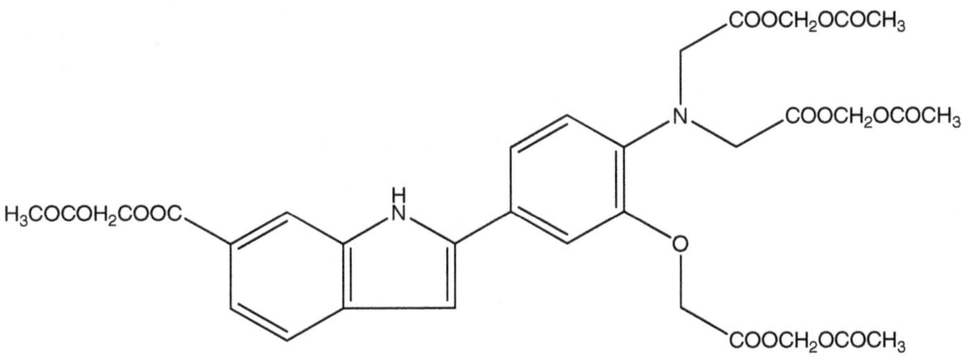

CA Index Name 1*H*-Indole-6-carboxylic acid, 2-[3-[2-[(acetyloxy)methoxy]-2-oxoethoxy]-4-[bis[2-[(acetyloxy)methoxy]-2-oxoethyl]amino]phenyl]-, (acetyloxy) methyl ester

Other Names Mag-indo-1/AM; Mag-indo-1 AM ester; Mag-indo-1 acetoxymethyl ester

Merck Index Number Not listed

Chemical/Dye Class Indole

Molecular Formula $C_{33}H_{34}N_2O_{17}$

Molecular Weight 730.63

Physical Form Pale yellow powder

Solubility Soluble in dimethyl sulfoxide, methanol

Melting Point >200 °C

Boiling Point 837.1 ± 65.0 °C, pressure: 760 Torr

pK_a 14.81 ± 0.30, most acidic, temperature: 25 °C; 1.37 ± 0.50, most basic, temperature: 25 °C

Absorption (λ_{max}) 354 nm

Emission (λ_{max}) 472 nm

Synthesis Synthetic method[1]

Staining Applications Magnesium ions;[2–4] calcium ions[5,6]

Biological Applications Magnesium indicator;[2–4] calcium indicator[5,6]

Industrial Applications Not reported

Safety/Toxicity No data available

REFERENCES

1. Haugland, R. P. *The Handbook: A Guide to Fluorescent Probes and Labeling Technologies*; Molecular Probes Inc.: Eugene, **2005**; pp 910–912.
2. Morelle, B.; Salmon, J. M.; Vigo, J.; Viallet, P. Measurement of intracellular magnesium concentration in 3T3 fibroblasts with the fluorescent indicator Mag-indo-1. *Anal. Biochem.* **1994**, *218*, 170–176.
3. Miyakawa, A.; Kamiya, K.; Hirano, M. Measurement of intracellular ion concentrations using fluorescent dye probes. Ger. Offen. DE 4239016, 1993; *Chem. Abstr.* **1993**, *119*, 112934.
4. Rutter, G. A.; Osbaldeston, N. J.; McCormack, J. G.; Denton, R. M. Measurement of matrix free magnesium concentration in rat heart mitochondria by using entrapped fluorescent probes. *Biochem. J.* **1990**, *271*, 627–634.
5. Zhao, M.; Hollingworth, S.; Baylor, S. M. AM-loading of fluorescent Ca^{2+} indicators into intact single fibers of frog muscle. *Biophys. J.* **1997**, *72*, 2736–2747.
6. Morris, S. J.; Beatty, D. M.; Chronwall, B. M. Simultaneous multiple wavelength fluorescence video microscopy shows Ca^{2+} regulation of pH in living cells. *Proc. SPIE-Int. Soc. Opt. Eng.* **1994**, *2137*, 93–104.

MAGNESIUM GREEN

CAS Registry Number 170516-41-3

Chemical Structure

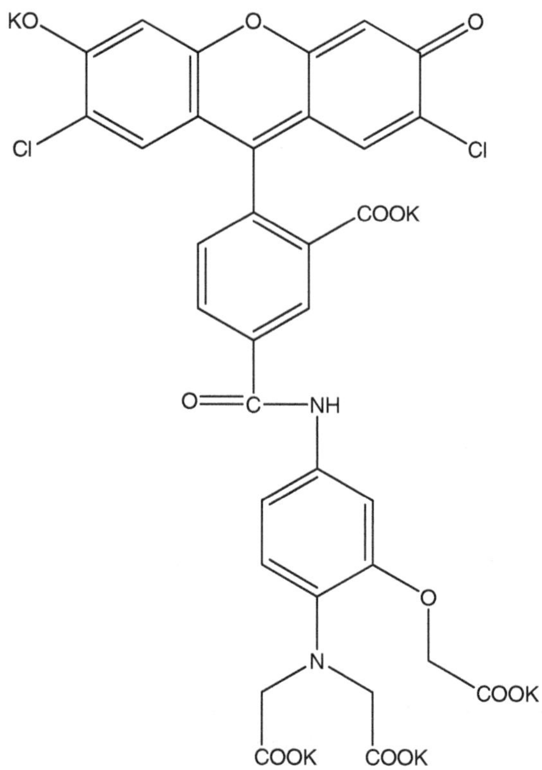

CA Index Name Glycine, *N*-[2-(carboxymethoxy)-4-[[(2′,7′-dichloro-3′,6′-dihydroxy-3-oxospiro[isobenzofuran-1(3*H*),9′-[9*H*]xanthen]-5-yl)carbonyl]amino]phenyl]-*N*-(carboxymethyl)-, potassium salt (1:5)

Other Names Glycine, *N*-[2-(carboxymethoxy)-4-[[(2′,7′-dichloro-3′,6′-dihydroxy-3-oxospiro[isobenzofuran-1(3*H*),9′-[9*H*]xanthen]-5-yl)carbonyl]amino]phenyl]-*N*-(carboxymethyl)-, pentapotassium salt; Magnesium Green; Magnesium green pentapotassium salt

Merck Index Number Not listed

Chemical/Dye Class Xanthene

Molecular Formula $C_{33}H_{17}Cl_2K_5N_2O_{13}$

Molecular Weight 915.90

Physical Form Solid

Solubility Soluble in water

Melting Point >250 °C

Absorption (λ_{max}) 506 nm

Emission (λ_{max}) 531 nm

Synthesis Synthetic methods[1,2]

Staining Applications Magnesium ions;[3–11] calcium ions;[12–23] zinc ions[1]

Biological Applications Magnesium indicator;[3–11] calcium indicator;[12–23] zinc indicator;[1] detecting nucleic acids[24]

Industrial Applications Not reported

Safety/Toxicity No data available

REFERENCES

1. Kuhn, M. A.; Hoyland, B.; Carter, S.; Zhang, C.; Haugland, R. P. Fluorescent ion indicators for detecting heavy metals. *Proc. SPIE-Int. Soc. Opt. Eng.* **1995**, *2388*, 238–244.
2. Hofmann, K. A.; Hoschele, K. Magnesium chloride as a mineralizer. II. Uranium-cerium-blue and the nature of constitutive coloring. Magnesium-red and magnesium-green. *Ber. Dtsch. Chem. Ges.* **1915**, *48*, 20–28; *Chem. Abstr.* **1915**, *9*, 6696.
3. Chinopoulos, C.; Vajda, S.; Csanady, L.; Adam-Vizi, V. Assay of adenine nucleotide translocase activity by fluorescent detection of magnesium concentration in a membrane compartment. PCT Int. Appl. WO 2009013555, 2009; *Chem. Abstr.* **2009**, *150*, 163067.
4. Erickson, J. R.; Moerland, T. S. A competition assay of magnesium affinity for EF-hand proteins based on the fluorescent indicator magnesium green. *Anal. Biochem.* **2005**, *345*, 343–345.
5. Park, E. J.; Brasuel, M.; Behrend, C.; Philbert, M. A.; Kopelman, R. Ratiometric optical PEBBLE nanosensors for real-time magnesium ion concentrations inside viable cells. *Anal. Chem.* **2003**, *75*, 3784–3791.
6. Suzuki, Y.; Komatsu, H.; Ikeda, T.; Saito, N.; Araki, S.; Citterio, D.; Hisamoto, H.; Kitamura, Y.; Kubota, T.; Nakagawa, J.; Oka, K.; Suzuki, K. Design and synthesis of Mg^{2+}-selective fluoroionophores based on a coumarin derivative and application for Mg^{2+} measurement in a living cell. *Anal. Chem.* **2002**, *74*, 1423–1428.
7. Baker, G. A.; Munson, C. A.; Bukowski, E. J.; Baker, S. N.; Bright, F. V. Assessment of one- and two-photon excited luminescence for directly measuring

O_2, pH, Na^+, Mg^{2+}, or Ca^{2+} in optically dense and biologically relevant samples. *Appl. Spectrosc.* **2002**, *56*, 455–463.

8. Mayr, T.; Liebsch, G.; Klimant, I.; Wolfbeis, O. S. Multi-ion imaging using fluorescent sensors in a microtiterplate array format. *Analyst* **2002**, *127*, 201–203.
9. Murakami, T. Cytotoxicity test method using magnesium ion fluorescent dye. Jpn. Kokai Tokkyo Koho JP 2000189191, 2000; *Chem. Abstr.* **2000**, *133*, 100817.
10. Denda, M.; Hosoi, J.; Asida, Y. Visual imaging of ion distribution in human epidermis. *Biochem. Biophys. Res. Commun.* **2000**, *272*, 134–137.
11. Szmacinski, H.; Lakowicz, J. R. Fluorescence lifetime characterization of magnesium probes: improvement of Mg^{2+} dynamic range and sensitivity using phase-modulation fluorometry. *J. Fluoresc.* **1996**, *6*, 83–95.
12. Paredes, R. M.; Etzler, J. C.; Watts, L. T.; Zheng, W.; Lechleiter, J. D. Chemical calcium indicators. *Methods* **2008**, *46*, 143–151.
13. Wilms, C. D.; Eilers, J. Photophysical properties of Ca^{2+}-indicator dyes suitable for two-photon fluorescence-lifetime recordings. *J. Microsc.* **2007**, *225*, 209–213.
14. Schiess, A. R. B.; Scullin, C. S.; Partridge, L. D. Neurosteroid-induced enhancement of short-term facilitation involves a component downstream from presynaptic calcium in hippocampal slices. *J. Physiol.* **2006**, *576*, 833–847.
15. Wade, E. J.; Janocha, E.; Germann, T. Methods for measuring the activity of Gα i-coupled or Gα o-coupled receptors using cellular calcium ion influx, and use in agonist and antagonist identification. PCT Int. Appl. WO 2004051264, 2004; *Chem. Abstr.* **2004**, *141*, 33742.
16. Brustein, E.; Marandi, N.; Kovalchuk, Y.; Drapeau, P.; Konnerth, A. *In vivo* monitoring of neuronal network activity in zebrafish by two-photon Ca^{2+} imaging *Pfluegers Arch.* **2003**, *446*, 766–773.
17. Kreitzer, A. C.; Gee, K. R.; Archer, E. A.; Regehr, W. G. Monitoring presynaptic calcium dynamics in projection fibers by *in vivo* loading of a novel calcium indicator. *Neuron* **2000**, *27*, 25–32.
18. Braun, F. J.; Hegemann, P. Direct measurement of cytosolic calcium and pH in living *Chlamydomonas reinhardtii* cells. *Eur. J. Cell Biol.* **1999**, *78*, 199–208.
19. Koester, H. J.; Baur, D.; Uhl, R.; Hell, S. W. Ca^{2+} fluorescence imaging with pico- and femtosecond two-photon excitation: signal and photodamage. *Biophys. J.* **1999**, *77*, 2226–2236.
20. Muschol, M.; Dasgupta, B. R.; Salzberg, B. M. Caffeine interaction with fluorescent calcium indicator dyes. *Biophys. J.* **1999**, *77*, 577–586.
21. Sabatini, B. L.; Regehr, W. G. Optical measurement of presynaptic calcium currents. *Biophys. J.* **1998**, *74*, 1549–1563.
22. Boltz, R. C., Jr. A high capacity screen for immunoregulants using intracellular calcium concentration measurement. PCT Int. Appl. WO 9610090, 1996; *Chem. Abstr.* **1996**, *125*, 76326.
23. Zhao, M.; Hollingworth, S.; Baylor, S. M. Properties of tri- and tetracarboxylate Ca^{2+} indicators in frog skeletal muscle fibers. *Biophys. J.* **1996**, *70*, 896–916.
24. Miller, B. L.; Krauss, T. D.; Du, H.; Crnkovich, N.; Strohsahl, C. M. Use of sensor arrays containing hairpin probes for detecting nucleic acids of pathogens. PCT Int. Appl. WO 2004061127, 2004; *Chem. Abstr.* **2004**, *141*, 118285.

MALACHITE GREEN

CAS Registry Number 569-64-2

Chemical Structure

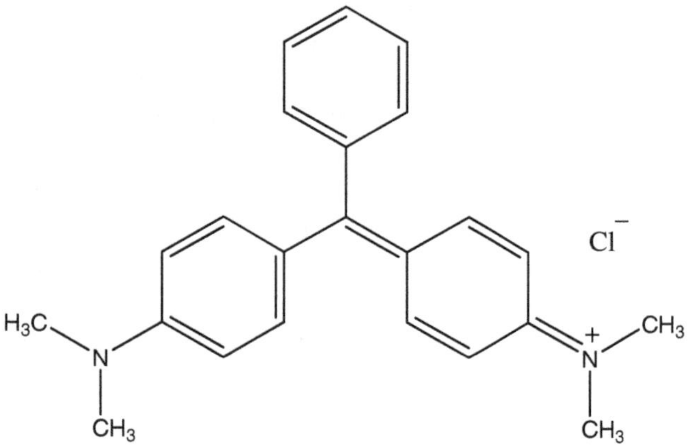

CA Index Name Methanaminium, *N*-[4-[[4-(dimethylamino)phenyl]phenylmethylene]-2,5-cyclohexadien-1-ylidene]-*N*-methyl-, chloride (1:1)

Other Names C.I. Basic Green 4; Methanaminium, *N*-[4-[[4-(dimethylamino)phenyl]phenylmethylene]-2,5-cyclohexadien-1-ylidene]-*N*-methyl-, chloride; Victoria Green WB; ADC Malachite Green Crystals; Acryl Brilliant Green B; Aizen Malachite Green; Aizen Malachite Green Crystals; Aizen Malachite Green Liquid; Aniline Green; Astra Malachite Green; Astra Malachite Green B; Astra Malachite Green BXX; Atlantic Malachite Green; Basacryl Green X-BF; Basic Green 4; Basonyl Green 830; Basonyl Green NB 832; Benzal Green; Benzaldehyde Green; Bronze Green Toner A 8002; Burma Green B; C.I. 42000; Calcozine Green V; China Green; Diabasic Malachite Green; Diamond Green B; Diamond Green B Extra; Diamond Green BX; Diamond Green P Extra; Green MX; Green Malaquite; Grenoble Green; Hidaco Malachite Green Base; Hidaco Malachite Green LC; Hidaco Malachite Green SC; LC 6220; Light Green N; Lincoln Green Toner B 15-2900; Malachite Green 4; Malachite Green A; Malachite Green AN; Malachite Green B; Malachite Green CP; Malachite Green Crystals; Malachite Green Crystals BPC; Malachite Green J 3E; Malachite Green Powder; Malachite Green WS; Malachite Green XLS; Malachite Lake Green A; Malachite Green; Malachite Green Chloride; Mitsui Malachite Green; New Victoria Green Extra I; New Victoria Green Extra II; New Victoria Green Extra O; Oji Malachite Green; Solid Green Crystals O; Solid Green O; Super Ick Cure; Tertrophene Green M; Tokyo Aniline Malachite Green; Verona Basic Green M; Victoria Green; Victoria Green (basic dye); Victoria Green B; Victoria Green S; Victoria Green WPB

Merck Index Number 5699

Chemical/Dye Class Triphenylmethane

Molecular Formula $C_{23}H_{25}ClN_2$

Molecular Weight 364.91

Physical Form Green crystals with metallic luster

Solubility Very soluble in water; soluble in ethanol, methanol, amyl alcohol

Melting Point 112–114 °C

pH Range 0.0–2.0; 11.6–14.0

Color Change at pH Yellow (0.0) to green (2.0); green (11.6) to colorless (14.0)

pK_a 6.90

Absorption (λ_{max}) 614 nm, 425 nm

Synthesis Synthetic methods[1–17]

Staining Applications Antigens;[18] antibodies;[18] bacteria;[19] bone cement;[20] collagen;[21] oxidized regenerated cellulose;[21] alginates;[21] chitosans;[21] galactomannans;[21] glycosaminoglycans;[21] erythroblast;[22] nucleic acids;[23,24] peptides;[25] proteins;[25] skin[26,27]

Biological Applications Antiseptic formulation;[28] detecting nucleic acids;[1,29] early diagnosis of tuberculosis;[30] identifying mammal genes;[1,31] treating cancers,[32] fungal diseases,[33] pulmonary tuberculosis;[1,34] medical device[35]

Industrial Applications Color filters;[1,36] light-emitting devices;[37] liquid crystal displays;[1,38] optical recording materials;[39] photoresists;[1,40,41] inks;[1,42] highlighters;[1,43] toners;[44] printed circuit board;[1,45] photographic materials;[46] adhesives;[47] textiles[48,49]

Safety/Toxicity Acute toxicity;[1,50–52] chronic toxicity;[52] carcinogenicity;[1,53,54] cytotoxicity;[1,55,56] genotoxicity;[1,57] mitochondrial toxicity;[58] mutagenicity;[1,59] nucleic acid damage[60]

REFERENCES

1. Sabnis, R. W. *Handbook of Acid–Base Indicators*; CRC Press: Boca Raton, **2008**; pp 212–214.
2. Welcher, F. J. *Organic Analytical Reagents*; D. Van Nostrand Company: New York, **1948**; Vol. *4*, pp 515–516.

3. Fierz-David, H. E.; Blangey, L.; Vittum, P. W. *Fundamental Processes of Dye Chemistry*; Interscience Publishers, Inc.: New York, **1949**; pp 299–301.
4. Agunwa, U. B.; Okonkwo, E. M. Production of malachite green by oxidation of its leuco base using potassium persulphate, potassium permanganate and manganese dioxide. *Global J. Pure Appl. Sci.* **2004**, *10*, 143–146.
5. Mukhortova, L. I.; Neboikov, P. L.; Devyatin, A. P.; Kazakov, V. V.; Andreev, O. K.; Gudkova, O. A. Composition for preparing liquid form of basic dye. Russ. RU 2222561, 2004; *Chem. Abstr.* **2004**, *141*, 158510.
6. Aizawa, T. Manufacture of triphenylmethane dye by electrochemical oxidation. Jpn. Kokai Tokkyo Koho JP 02194188, 1990; *Chem. Abstr.* **1991**, *114*, 52054.
7. Awad, W. I.; El-Sawi, E. A.; El-Sayed, W. A. Preparation of modified basic dyes and their direct application to cotton fabrics. *Tinctoria* **1985**, *82*, 357–360.
8. Kast, H.; Mayer, U. Basic dyes. Ger. Offen. DE 2736679, 1979; *Chem. Abstr.* **1979**, *90*, 170166.
9. Mayer, U.; Kast, H. Malachite green chloride. Ger. Offen. DE 2247346, 1974; *Chem. Abstr.* **1975**, *83*, 133388.
10. Kast, H. Preparation of basic dyes by catalytic oxidation. Ger. Offen. DE 2334918, 1975; *Chem. Abstr.* **1975**, *83*, 12202.
11. Kast, H.; Baumann, H.; Mayer, U.; Oberlinner, A. Basic dyes by oxidation. Ger. Offen. DE 2138931, 1973; *Chem. Abstr.* **1973**, *78*, 137964.
12. Feldman, M. L.; Gordon, J. E. Recovery and purification of basic triarylmethane dyes. U.S. Patent 3679713, 1972; *Chem. Abstr.* **1972**, *77*, 141488.
13. Tilak, B. D. Synthesis of dyes and intermediates involving hydride loss. *Ind. Chim. Belg.* **1967**, *32*, 50–53.
14. Minevitch, J. R. The manufacture of malachite green. *Color Trade J.* **1922**, *11*, 233–237.
15. Minevitch, J. R. The manufacture of malachite green. *Color Trade J.* **1921**, *9*, 181–184.
16. Minevitch, J. R. The manufacture of malachite green. *Color Trade J.* **1921**, *9*, 154–157.
17. Minevitch, J. R. The manufacture of malachite green. *Color Trade J.* **1921**, *9*, 114–118.
18. Kodaira, T.; Kato, M.; Kasai, D. Microplate-immobilized and labeled antigens or antibodies for detection of antibody, antigen or allergen in health, food or environmental sample. Jpn. Kokai Tokkyo Koho JP 2009031024, 2009; *Chem. Abstr.* **2009**, *150*, 235573.
19. Kocagoz, T. Z. Culture medium for rapid detection of mycobacterial growth by color change and for susceptibility testing. U.S. Pat. Appl. Publ. US 2005079570, 2005; *Chem. Abstr.* **2005**, *142*, 370332.
20. Kuehn, K. D.; Vogt, S. Colored polymethacrylate bone cement and method for preparation. Ger. DE 102005032110, 2006; *Chem. Abstr.* **2006**, *145*, 235931.
21. Cullen, B. M.; Addison, D.; Greenhalgh, D. Bioabsorbable wound dressing containing an antioxidant dye. Brit. UK Pat. Appl. GB 2402882, 2004; *Chem. Abstr.* **2004**, *142*, 62794.
22. Tsuji, T.; Sakata, T.; Ikeuchi, Y.; Oguni, S. Erythroblast diagnostic flow-cytometry method and reagents. Eur. Pat. Appl. EP 1004880, 2000; *Chem. Abstr.* **2000**, *133*, 14302.
23. Kolpashchikov, D. Binary probes for fluorescent analysis of nucleic acids. PCT Int. Appl. WO 2007115242, 2007; *Chem. Abstr.* **2007**, *147*, 441322.
24. Tyagi, S.; Kramer, F. R. Detection of probes in nucleic acid hybridization using non-FRET (fluorescence resonance energy transfer) pairs of chromophores. PCT Int. Appl. WO 9739008, 1997; *Chem. Abstr.* **1997**, *127*, 342638.
25. Ting, A. Y. Labeling target proteins with biotin analog of *Escherichia coli* biotin ligase mutant and their uses for labeling target proteins with biotin analog. U.S. Pat. Appl. Publ. US 2004209317, 2004; *Chem. Abstr.* **2004**, *141*, 362793.
26. Fulchand, S. C. A topical skin care composition. Indian Pat. Appl. IN 2004MU01051, 2007; *Chem. Abstr.* **2007**, *149*, 252004.
27. Kashibuchi, N.; Hirai, Y.; Fujiwara, T.; Miyazawa, M. A method for distinguishing sensitive skin. Jpn. Kokai Tokkyo Koho JP 11304798, 1999; *Chem. Abstr.* **1999**, *131*, 319876.
28. Magallon, A. G.; Boone, K. W.; Bardwell, J. R.; Hernandez, L. Y.; De la Rosa, M. P. Cationic antiseptic and dye formulation. U.S. Pat. Appl. Publ. US 2008108674, 2008; *Chem. Abstr.* **2008**, *148*, 503248.
29. Kolpashchikov, D. M. Binary malachite green aptamer for fluorescent detection of nucleic acids. *J. Am. Chem. Soc.* **2005**, *127*, 12442–12443.
30. Mordovskoi, G. G.; Zueva, M. N. Nutrient medium for *Mycobacteria tuberculosis* culture, and use for early diagnosis of tuberculosis. Russ. RU 2300571, 2007; *Chem. Abstr.* **2007**, *147*, 48199.

31. Sikela, J.; Pollack, J. R. Interspecies cDNA array-based comparative genomic hybridization for identifying mammal genes having lineage specific variations in copy number. PCT Int. Appl. WO 2005023983, 2005; *Chem. Abstr.* **2005**, *142*, 310878.
32. Mihail, A.; Mihail, N. Method for treatment of cancers and antineoplastic pharmaceutical formulations. U.S. Pat. Appl. Publ. US 2003114403, 2003; *Chem. Abstr.* **2003**, *139*, 26685.
33. Campbell, R. E.; Lilley, J. H.; Taukhid; Panyawachira, V.; Kanchanakhan, S. *In vitro* screening of novel treatments for *Aphanomyces invadans*. *Aquat. Res.* **2001**, *32*, 223–233.
34. Farnia, P.; Mohammadi, F.; Mirsaedi, M.; Zarife, A. Z.; Tabatabee, J.; Bahadori, K.; Bahadori, M.; Masjedi, M. R.; Velayati, A. A. Application of oxidation–reduction assay for monitoring treatment of patients with pulmonary tuberculosis. *J. Clin. Microbiol.* **2004**, *42*, 3324–3325.
35. Vyacheslav, D.; Yakeemovich, N.; DiTizio, V.; DiCosmo, F. Antimicrobial photo-stable coating composition and medical device coated thereby. PCT Int. Appl. WO 2009015476, 2009; *Chem. Abstr.* **2009**, *150*, 193534.
36. De Keyzer, G.; Lamatsch, B.; Muehlebach, A.; Rime, F.; Schmitt, G. Surface-modified nanoparticles comprising a cationic colorant for use in color filters. PCT Int. Appl. WO 2008107304, 2008; *Chem. Abstr.* **2008**, *149*, 357452.
37. Uehara, M. Light-emitting device, method for manufacturing light-emitting device, electronic device, and electronic instrument. PCT Int. Appl. WO 2006132374, 2006; *Chem. Abstr.* **2006**, *146*, 35756.
38. Fujiwara, T. Film having photosensitive color resin layer for manufacturing liquid crystal display parts. Jpn. Kokai Tokkyo Koho JP 2006078681, 2006; *Chem. Abstr.* **2006**, *144*, 321680.
39. Morinaka, A.; Funakoshi, N. Optical recording materials with information theft- and misuse-preventing layer. Jpn. Kokai Tokkyo Koho JP 61208647, 1986; *Chem. Abstr.* **1987**, *106*, 166313.
40. Sato, H.; Katsuma, K.; Tsujimoto, A. Photosensitive polymer compositions, photoresist films, and formation of resist patterns. Jpn. Kokai Tokkyo Koho JP 2005321587, 2005; *Chem. Abstr.* **2005**, *143*, 469619.
41. McMurdie, N. D. Positive photoresist with improved contrast ratio and photospeed. U.S. Patent 6100008, 2000; *Chem. Abstr.* **2000**, *133*, 170235.
42. Wakae, T.; Takagishi, I. Aqueous inks discolorable under pH change and writing tool sets therewith. Jpn. Kokai Tokkyo Koho JP 2004256706, 2004; *Chem. Abstr.* **2004**, *141*, 279186.
43. Davies-Smith, L.; Sum, V. W. Highlightable and highlighted mixtures, marking instruments, eradicator solution and kits, using the mixtures in highlighting. U.S. Pat. Appl. Publ. US 2005120919, 2005; *Chem. Abstr.* **2005**, *143*, 9331.
44. Hosfeld, L. K.; Lenox, R. S. Magnetic toner powders. Belg. BE 897918, 1984; *Chem. Abstr.* **1984**, *101*, 46285.
45. Sotomura, S. Resist ink for forming resist pattern to manufacture printed circuit board by ink jet printing. Jpn. Kokai Tokkyo Koho JP 2005057018, 2005; *Chem. Abstr.* **2005**, *142*, 269229.
46. Weyde, E.; Saleck, W.; Psaar, H.; Von Koenig, A.; Oehlschlaeger, H. Photographic color images and photographic recording materials. Ger. Offen. DE 3201557, 1983; *Chem. Abstr.* **1984**, *100*, 200854.
47. Zimmermann, D. Multicolor component adhesive system comprising polymerizable compositions, different coloring agents and a curing component. U.S. Pat. Appl. Publ. US 2008300321, 2008; *Chem. Abstr.* **2008**, *150*, 20872.
48. Mahltig, B.; Knittel, D.; Schollmeyer, E.; Boettcher, H. Incorporation of triarylmethane dyes into sol–gel matrices deposited on textiles. *J. Sol-Gel Sci. Technol.* **2004**, *31*, 293–297.
49. Mahltig, B.; Textor, T. Combination of silica sol and dyes on textiles. *J. Sol-Gel Sci. Technol.* **2006**, *39*, 111–118.
50. Kumar, P.; Roy, D. Acute toxicity of malachite green to the freshwater prawn, *Macrobrachium dayanum*. *J. Adv. Zool.* **2007**, *28*, 29–31.
51. Raizada, S.; Rana, K. S. Acute toxicity of malachite green to an air breathing teleost, *Clarias batrachus* (Linn.). *J. Environ. Biol.* **1998**, *19*, 237–241.
52. Srivastava, S. J.; Singh, N. D.; Sinha, R.; Srivastava, A. K. Acute and chronic toxicity of malachite green: Microscopic changes in the pituitary gonadotropic cells and gonads in a freshwater catfish, *Heteropneustes fossilis* (Bloch). *Proc. Natl. Acad. Sci., India, Sect. B: Biol. Sci.* **1998**, *68*, 253–256.
53. Culp, S. J.; Mellick, P. W.; Trotter, R. W.; Greenlees, K. J.; Kodell, R. L.; Beland, F. A. Carcinogenicity of malachite green chloride and leucomalachite green in B6C3F1 mice and F344 rats. *Food Chem. Toxicol.* **2006**, *44*, 1204–1212.

54. Vachalkova, A.; Novotny, L.; Blesova, M. Polarographic reduction of some triphenylmethane dyes and their potential carcinogenic activity. *Neoplasma* **1996**, *43*, 113–117.

55. Zahn, T.; Braunbeck, T. Cytotoxic effects of sublethal concentrations of malachite green in isolated hepatocytes from rainbow trout (*Oncorhynchus mykiss*). *Toxicol. In Vitro* **1995**, *9*, 729–741.

56. Panandiker, A.; Fernandes, C.; Rao, K. V. K. The cytotoxic properties of malachite green are associated with the increased demethylase, aryl hydrocarbon hydroxylase and lipid peroxidation in primary cultures of Syrian hamster embryo cells. *Cancer Lett.* **1992**, *67*, 93–101.

57. Mittelstaedt, R. A.; Mei, N.; Webb, P. J.; Shaddock, J. G.; Dobrovolsky, V. N.; McGarrity, L. J.; Morris, S. M.; Chen, T.; Beland, F. A.; Greenlees, K. J.; Heflich, R. H. Genotoxicity of malachite green and leucomalachite green in female Big Blue B6C3F1 mice. *Mutat. Res.* **2004**, *561*, 127–138.

58. Zhang, H.; Chen, Q. Y.; Xiang, M. L.; Ma, C. Y.; Huang, Q.; Yang, S. Y. *In silico* prediction of mitochondrial toxicity by using GA-CG-SVM approach. *Toxicol. In Vitro* **2009**, *23*, 134–140.

59. Fessard, V.; Godard, T.; Huet, S.; Mourot, A.; Poul, J. M. Mutagenicity of malachite green and leucomalachite green in *in vitro* tests. *J. Appl. Toxicol.* **1999**, *19*, 421–430.

60. Bose, B.; Motiwale, L.; Rao, K. V. K. DNA damage and G2/M arrest in Syrian hamster embryo cells during malachite green exposure are associated with elevated phosphorylation of ERK1 and JNK1. *Cancer Lett.* **2005**, *230*, 260–270.

MALACHITE GREEN OXALATE SALT

CAS Registry Number 2437-29-8

Chemical Structure

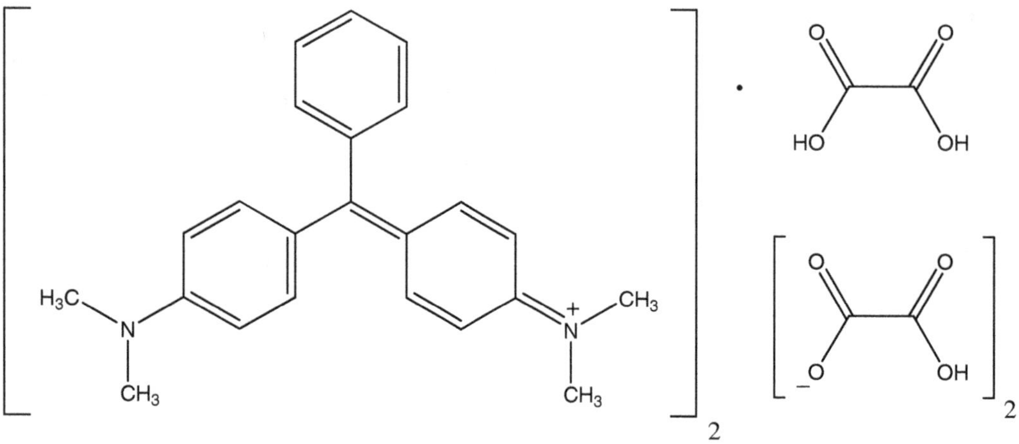

Melting Point 164 °C (decompose)

pH Range 0.0–2.0; 11.6–14.0

Color Change at pH Yellow (0.0) to green (2.0); green (11.6) to colorless (14.0)

pK_a 6.90

CA Index Name Methanaminium, *N*-[4-[[4-(dimethylamino)phenyl]phenylmethylene]-2,5-cyclohexadien-1-ylidene]-*N*-methyl-, ethanedioate, ethanedioate (2:2:1)

Other Names Ammonium, [4-[*p*-(dimethylamino)-α-phenylbenzylidene]-2,5-cyclohexadien-1-ylidene]dimethyl, oxalate (1:1), oxalate (1:1); Malachite green oxalate; Ethanedioic acid, ion(1-), *N*-[4-[[4-(dimethylamino)phenyl]phenylmethylene]-2,5-cyclohexadien-1-ylidene]-*N*-methylmethanaminium, ethanedioate (2:1); Malachite green oxalate salt

Merck Index Number 5699

Chemical/Dye Class Triphenylmethane

Molecular Formula $C_{52}H_{54}N_4O_{12}$

Molecular Weight 927.00

Physical Form Green crystals or powder

Solubility Soluble in water, ethanol; insoluble in xylene

Absorption (λ_{max}) 614 nm, 425 nm

Synthesis Synthetic methods[1–7]

Staining Applications Brain;[8] spinal cord;[8] phosphohydrolase[9]

Biological Applications Detecting bacterial growth;[10] treating fish diseases,[11] malignant neoplasm[12]

Industrial Applications Inks;[13] toners;[14] photoresists;[15,16] printing plates;[16] photosensitive resins;[17] recording materials;[18] printed circuit boards;[19] solar cells;[20] photoelectrographic materials;[21] laser dyes;[22] NLO dyes;[22] textiles[23]

Safety/Toxicity Cytotoxicity;[24,25] developmental abnormalities;[26] estrogenicity[27]

Certification/Approval Certified by Biological Stain Commission (BSC)

REFERENCES

1. Goff, T.; Wood, S. Production of malachite green oxalate and leucomalachite green reference materials certified for purity. *Anal. Bioanal. Chem.* **2008**, *391*, 2035–2045.
2. Nohr, R. S.; MacDonald, J. G. Novel colorant modifiers, and stabilization of colorants therewith. PCT Int. Appl. WO 9701605, 1997; *Chem. Abstr.* **1997**, *126*, 187485.
3. Berneth, H.; Raue, R. Oxidative manufacture of sulfo group-free triphenylmethane dyes. Ger. Offen. DE 3842014, 1990; *Chem. Abstr.* **1991**, *114*, 64262.
4. Kutzschbach, D.; Friedrich, A. Malachite green oxalate. Ger. Offen. DE 3036078, 1982; *Chem. Abstr.* **1982**, *97*, 40320.
5. Hermann, K. H. Cationic dyes. Ger. Offen. DE 2928466, 1981; *Chem. Abstr.* **1981**, *94*, 141206.

6. Hermann, K. H. Basic dyes. Ger. Offen. DE 2853822, 1980; *Chem. Abstr.* **1980**, *93*, 187749.
7. Chalkley, L. Photochemical preparation of stable dyes. U.S. Patent 2441561, 1948; *Chem. Abstr.* **1948**, *42*, 26594.
8. Augulis, V.; Sigg, E. B. Supravital staining and fixation of brain and spinal cord by intravascular perfusion. *Stain Technol.* **1971**, *46*, 183–190.
9. Zlotnick, G. W.; Gottlieb, M. A sensitive staining technique for the detection of phosphohydrolase activities after polyacrylamide gel electrophoresis. *Anal. Biochem.* **1986**, *153*, 121–125.
10. Booher, J. Apparatus containing polymer membranes and pH indicators for detecting bacterial growth beneath a wound dressing. PCT Int. Appl. WO 2006133430, 2006; *Chem. Abstr.* **2006**, *146*, 68867.
11. Braidwood, J. C. Use of bronopol for the treatment of diseases in fish. PCT Int. Appl. WO 9805311, 1998; *Chem. Abstr.* **1998**, *128*, 162867.
12. Kutushov, M. V. Use of malachite green as drug in treatment of malignant neoplasm. Russ. RU 2322973, 2008; *Chem. Abstr.* **2008**, *148*, 487155.
13. Nohr, R. S.; MacDonald, J. G. Light stable ink for print cartridges for ink jet printers and jet printing. U. S. Patent 5681380, 1997; *Chem. Abstr.* **1997**, *127*, 347733.
14. Nair, M.; Yang, X.; Jones, T. K.; Morgan, J. Electrophotographic toner porous particles containing hydrocolloids. U.S. Pat. Appl. Publ. US 2008176157, 2008; *Chem. Abstr.* **2008**, *149*, 211687.
15. Sato, M. Dry-film photoresists formulated based on expected swelling of their patterns in developers or in strippers. Jpn. Kokai Tokkyo Koho JP 2006292889, 2006; *Chem. Abstr.* **2006**, *145*, 446293.
16. Fujikura, S.; Iwasaki, M.; Maeda, M.; Wada, M. Photopolymerizable compositions for production of printing plates and photoresists. Ger. Offen. DE 3926667, 1990; *Chem. Abstr.* **1990**, *113*, 106446.
17. Munekazu, T.; Tsuchihashi, M.; Goto, S. Antistatic agents for photosensitive resins and photosensitive resin compositions containing the agents and basic dyes. Jpn. Kokai Tokkyo Koho JP 2009051991, 2009; *Chem. Abstr.* **2009**, *150*, 317752.
18. Diener, G.; Keil, D.; Kapler, R.; Opfermann, J.; Hoerhold, H. H.; Klee, J. Transparent electrophotographic recording material with improved photosensitivity. Ger. (East) DD 229231, 1985; *Chem. Abstr.* **1986**, *105*, 143500.
19. Iwasaki, M. Manufacture of printed circuit boards. Jpn. Kokai Tokkyo Koho JP 2005191284, 2005; *Chem. Abstr.* **2005**, *143*, 124465.
20. Chae, W. S.; Kang, T. S.; Kim, K. J. Photocurrent in a dye-sensitized nanoparticle TiO_2 solar cell. *J. Korean Electrochem. Soc.* **1999**, *2*, 232–236.
21. Bugner, D. E. Photoelectrographic imaging with dyes or pigments to effect a color density or hue shift. PCT Int. Appl. WO 9218905, 1992; *Chem. Abstr.* **1993**, *118*, 70111.
22. Capozzi, C. A.; Pye, L. D. Physical and optical properties of organically modified silicates doped with laser and NLO dyes. *Proc. SPIE-Int. Soc. Opt. Eng.* **1991**, *1513*, 320–329.
23. Imakomi, H.; Kondo, M.; Nitta, T. Dyes for solution dyeing of acrylic fibers. Jpn. Kokai Tokkyo Koho JP 01022971, 1989; *Chem. Abstr.* **1989**, *111*, 8892.
24. Cho, M. H.; Niles, A.; Huang, R.; Inglese, J.; Austin, C. P.; Riss, T.; Xia, M. A bioluminescent cytotoxicity assay for assessment of membrane integrity using a proteolytic biomarker. *Toxicol. In Vitro* **2008**, *22*, 1099–1106.
25. Huang, R.; Southall, N.; Cho, M. H.; Xia, M.; Inglese, J.; Austin, C. P. Characterization of diversity in toxicity mechanism using *in vitro* cytotoxicity assays in quantitative high throughput screening. *Chem. Res. Toxicol.* **2008**, *21*, 659–667.
26. Meyer, F. P.; Jorgenson, T. A. Teratological and other effects of malachite green on development of rainbow trout and rabbits. *Trans. Am. Fish. Soc.* **1983**, *112*, 818–824.
27. Jiao, B.; Yeung, E. K. C.; Chan, C. B.; Cheng, C. H. K. Establishment of a transgenic yeast screening system for estrogenicity and identification of the anti-estrogenic activity of malachite green. *J. Cell. Biochem.* **2008**, *105*, 1399–1409.

MEQ

CAS Registry Number 34373-76-7

Chemical Structure

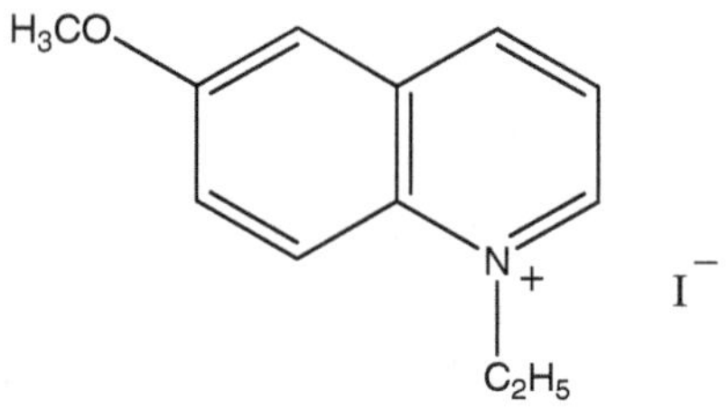

CA Index Name Quinolinium, 1-ethyl-6-methoxy-, iodide (1:1)

Other Names Quinolinium, 1-ethyl-6-methoxy-, iodide; MEQ; 6-Methoxy-*N*-ethylquinolinium iodide; 6-Methoxyquinoline ethiodide

Merck Index Number Not listed

Chemical/Dye Class Quinoline

Molecular Formula $C_{12}H_{14}INO$

Molecular Weight 315.15

Physical Form Solid

Solubility Soluble in water, dimethyl sulfoxide

Melting Point 182–183 °C

Absorption (λ_{max}) 344 nm

Emission (λ_{max}) 442 nm

Synthesis Synthetic methods[1–3]

Staining Applications Chloride ions[4–11]

Biological Applications Chloride indicator[4–11]

Industrial Applications Not reported

Safety/Toxicity No data available

REFERENCES

1. Peet, N. P.; Karrick, G. L.; Barbuch, R. J. Thermal dealkylations of quaternized methoxyquinolinium and methoxyisoquinolinium iodides in the mass spectrometer. *J. Heterocycl. Chem.* **1987**, *24*, 715–719.
2. Ardashev, B. I.; Mishchenko, V. F.; Olekhnovich, E. P. Cyanine dyes based on vinyl derivatives of pyridine and quinoline. *Khim. Geterotsikl. Soedin.* **1971**, *7*, 1679–1680; *Chem. Abstr.* **1972**, *76*, 128779.
3. Ishino, S.; Asakawa, T.; Ohta, A.; Miyagishi, S. Fluorescence quenching of quinoline derivatives in a micelle system. *J. Oleo Sci.* **2003**, *52*, 535–540.
4. Zhang, L. L.; Pathak, H. R.; Coulter, D. A.; Freed, M. A.; Vardi, N. Shift of intracellular chloride concentration in ganglion and amacrine cells of developing mouse retina. *J. Neurophysiol.* **2006**, *95*, 2404–2416.
5. Chub, N.; Mentis, G. Z.; O'Donovan, M. J. Chloride-sensitive MEQ fluorescence in chick embryo motoneurons following manipulations of chloride and during spontaneous network activity. *J. Neurophysiol.* **2006**, *95*, 323–330.
6. Ozawa, T.; Asakawa, T.; Ohta, A.; Miyagishi, S. Halide-sensitive fluorescent probes for self-assembly of novel cationic fluoro surfactants. *J. Oleo Sci.* **2005**, *54*, 135–142.
7. Mahlangu, D. A. F. D.; Dix, J. A. Halide fluxes in epithelial cells measured with an automated cell plate reader. *Anal. Biochem.* **2004**, *325*, 28–34.
8. Castle, N.; Ford, J. Ion channel permeability. PCT Int. Appl. WO 2001033219, 2001; *Chem. Abstr.* **2001**, *134*, 350285.
9. Inglefield, J. R.; Schwartz-Bloom, R. D. Using confocal microscopy and the fluorescent indicator, 6-methoxy-*N*-ethylquinolinium iodide, to measure changes in intracellular chloride. *Methods Enzymol.* **1999**, *307*, 469–481.
10. Fukuda, A.; Tanaka, M.; Yamada, Y.; Muramatsu, K.; Shimano, Y.; Nishino, H. Simultaneous optical imaging of intracellular Cl^- in neurons in different layers of rat neocortical slices: advantages and limitations. *Neurosci. Res.* **1998**, *32*, 363–371.
11. Woell, E.; Gschwentner, M.; Fuerst, J.; Hofer, S.; Buemberger, G.; Jungwirth, A.; Frick, J.; Deetjen, P.; Paulmichl, M. Fluorescence-optical measurements of chloride movements in cells using the membrane-permeable dye diH-MEQ. *Pfluegers Arch.* **1996**, *432*, 486–493.

METHYLENE BLUE

CAS Registry Number 61-73-4

Chemical Structure

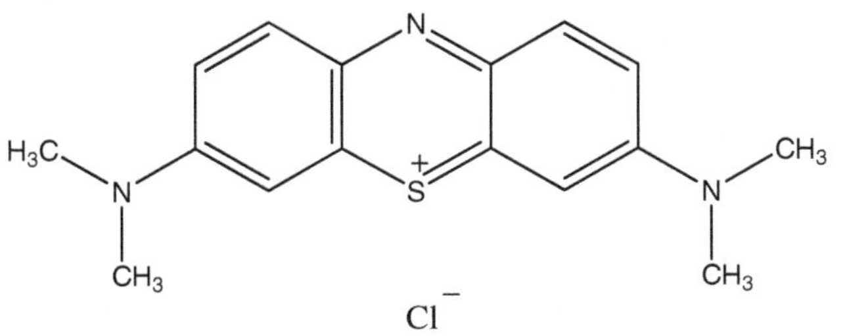

CA Index Name Phenothiazin-5-ium, 3,7-bis(dimethylamino)-, chloride (1:1)

Other Names C.I. Basic Blue 9; Phenothiazin-5-ium, 3,7-bis(dimethylamino)-, chloride; Aizen Methylene Blue BH; Aizen Methylene Blue FZ; Basic Blue 9; Basic Lake Blue; C.I. 52015; C.I. Solvent Blue 8; Calcozine Blue ZF; Chromosmon; Duasyn Basic Blue IAD; Ext D and C Blue No. 1; External Blue 1; Hidaco Methylene Blue Salt Free; Izit Crystal Dye; Leather Pure Blue HB; Maxilon Blue SG; Methylene Blue 2B; Methylene Blue 2BF; Methylene Blue 2BN; Methylene Blue 2BP; Methylene Blue A; Methylene Blue B; Methylene Blue BB; Methylene Blue BBA; Methylene Blue BD; Methylene Blue BP; Methylene Blue BPC; Methylene Blue BX; Methylene Blue BZ; Methylene Blue D; Methylene Blue FZ; Methylene Blue G; Methylene Blue GZ; Methylene Blue HGG; Methylene Blue IAD; Methylene Blue JFA; Methylene Blue MB; Methylene Blue N; Methylene Blue NF; Methylene Blue NZ; Methylene Blue SG; Methylene Blue SP; Methylene Blue ZF; Methylene Blue ZX; Methylene Blue Zinc Free; Methylene Blue chloride; Methylene Blue; Methylene Blue polychrome; Methylenium ceruleum; Methylthionine chloride; Methylthioninium chloride; Mitsui Methylene Blue; NSC 617593; Sandocryl Blue BRL; Schultz 1038; Solvent Blue 8; Swiss blue; Tetramethylthionine; Tetramethylthionine chloride; Urolene Blue; Yamamoto Methylene Blue B; Yamamoto Methylene Blue ZF

Merck Index Number 6060

Chemical/Dye Class Phenothiazine

Molecular Formula $C_{16}H_{18}ClN_3S$

Molecular Weight 319.85

Physical Form Green powder

Solubility Soluble in water, ethanol, ethylene glycol, methyl cellosolve

Melting Point 100–110 °C (decompose)

pK_a 2.6, 11.2

Absorption (λ_{max}) 661 nm

Synthesis Synthetic methods[1–8]

Staining Applications Blood;[9] bone marrow;[9] eye lens;[10] sentinel lymph nodes;[11] mammary tissues;[12] nucleic acids[13]

Biological Applications Detecting microorganisms;[14] treating diabetic retinopathy,[15] macular degeneration,[15] malignant uveal melanomas,[15] erysipelas,[16] hidradenitis suppurativa,[17] inflammation,[18] skin diseases[19]

Industrial Applications Thin films;[20] inks;[21,22] packaging material[23]

Safety/Toxicity Effluent toxicity;[24] genotoxicity;[25,26] hematotoxicity;[27] microbial toxicity;[28] mutagenicity;[29] neurotoxicity;[30] nucleic acid damage;[31,32] photodynamic toxicity;[33] reproductive toxicity;[34] teratogenicity[35]

REFERENCES

1. Fierz-David, H. E.; Blangey, L.; Vittum, P. W. *Fundamental Processes of Dye Chemistry*; Interscience Publishers, Inc.: New York, **1949**; p 311.
2. Wischik, C. M.; Storey, J. M. D.; Marshall, C.; Sinclair, J. P.; Baddeley, T. C. Preparation and purification of diaminophenothiazinium compounds for therapeutic use. PCT Int. Appl. WO 2008007074, 2008; *Chem. Abstr.* **2008**, *148*, 168729.
3. Aggarwal, P. Process for manufacturing zinc-free methylene blue. Indian Pat. Appl. IN 2005DE01921, 2006; *Chem. Abstr.* **2007**, *148*, 80622.
4. Wischik, C. M.; Rickard, J. E.; Harrington, C. R.; Horsley, D.; Storey, J. M. D.; Marshall, C.; Sinclair, J. P.; Tan, H. W. Process for preparation of diaminophenothiazinium compounds with high purity for therapeutic use. U.S. Pat. Appl. Publ. US 2006287523, 2006; *Chem. Abstr.* **2006**, *146*, 81874.
5. Storey, J. M. D.; Sinclair, J. P.; Marshall, C.; Tan, H. W.; Wischik, C. M. Synthesis and purification of diaminophenothiazinium compounds including methylthioninium chloride (MTC) via oxidative coupling and ring closure. PCT Int. Appl. WO 2006032879, 2006; *Chem. Abstr.* **2006**, *144*, 350699.

6. Mazur, Y.; Acher, A.; Shragina, L.; Avramoff, M. Preparation of phenothiazinium salts and their use for disinfecting aqueous effluents. U.S. Patent 5220009, 1993; *Chem. Abstr.* **1993**, *119*, 180802.
7. Kuban, V.; Jancarova, I. Preparation of highly pure methylene blue compositions by an extraction procedure. *Chem. Listy* **1985**, *79*, 1200–1205; *Chem. Abstr.* **1986**, *104*, 90471.
8. Konzelman, L. M.; Randvere, F. V. Methylene blue. U.S. Patent 4212971, 1980; *Chem. Abstr.* **1980**, *93*, 151662.
9. Dagiral, R. L. G. Cell staining method for blood and bone marrow. PCT Int. Appl. WO 2008043963, 2008; *Chem. Abstr.* **2008**, *148*, 466507.
10. Korsakova, N. V.; Sergeeva, V. E. Cytochemical analysis of cellular structures in eye lens of laboratory animals including vital–supravital staining by methylene blue. Russ. RU 2319132, 2008; *Chem. Abstr.* **2008**, *148*, 302857.
11. Maloney, T. J.; Camp, A. E.; Hernandez, J. J. Radiolabeled lymphatic staining agents and methods for making. U.S. Pat. Appl. Publ. US 2006002850, 2006; *Chem. Abstr.* **2006**, *144*, 114420.
12. Buehring, G. C.; Jensen, H. M. Lack of toxicity of methylene blue chloride to supravitally stained human mammary tissues. *Cancer Res.* **1983**, *43*, 6039–6044.
13. Wain, A. J.; Zhou, F. Scanning electrochemical microscopy imaging of DNA microarrays using methylene blue as a redox-active intercalator. *Langmuir* **2008**, *24*, 5155–5160.
14. Martin, S. M.; MacDonald, J. G.; Lye, J.; Sayre, C.; Thompson, K. Array for rapid detection of a microorganism based on dyes. PCT Int. Appl. WO 2008026104, 2008; *Chem. Abstr.* **2008**, *148*, 302867.
15. Chen, J. C. Photodynamic therapy comprising administration of a targeted photosensitizing agent, for treatment for eye disease such as diabetic retinopathy, macular degeneration, and malignant uveal melanomas. U.S. Patent 7320786, 2008; *Chem. Abstr.* **2008**, *148*, 152083.
16. Sufiyarov, I. F.; Khasanov, A. G.; Ibragimov, R. K.; Shaibakov, D. G. Method for the treatment of erysipelas. Russ. RU 2317085, 2008; *Chem. Abstr.* **2008**, *148*, 254182.
17. Tao, J. S.; Hunt, D. W. C. Photodynamic therapy for the treatment of hidradenitis suppurativa. PCT Int. Appl. WO 2008052350, 2008; *Chem. Abstr.* **2008**, *148*, 512019.
18. Blazhko, V. A.; Shebasheva, E. G.; Plotnikov, L. N.; Kalivradzhiyan, E. S.; Leshcheva, E. A.; Rudakov, O. B. A composition for treatment of inflammation in periodontium. Russ. RU 2323711, 2008; *Chem. Abstr.* **2008**, *148*, 503261.
19. Gilson, D. J.; Gilson, J. A.; DeBellis, J. A. Therapeutic device for delivering controlled air stream flow and/or photodynamic therapy of skin diseases. U.S. Pat. Appl. Publ. US 2008031924, 2008; *Chem. Abstr.* **2008**, *148*, 198849.
20. Murakami, A.; Yamaguchi, T.; Hirano, S. I.; Kikuta, K. Synthesis of porous titania thin films using carbonatation reaction and its hydrophilic property. *Thin Solid Films* **2008**, *516*, 3888–3892.
21. Sano, T. Clathrate compound-containing inks with suppressed bleed out of pigments for oxygen indicators, packaging materials using them, and packages comprising the packaging materials. Jpn. Kokai Tokkyo Koho JP 2008088201, 2008; *Chem. Abstr.* **2008**, *148*, 451332.
22. Udagawa, R. Inks for ink jet printers for light- and water-resistant images. Jpn. Kokai Tokkyo Koho JP 2003292858, 2003; *Chem. Abstr.* **2003**, *139*, 308991.
23. Kato, T. Oxygen indicator and packaging material. Jpn. Kokai Tokkyo Koho JP 2008102074, 2008; *Chem. Abstr.* **2008**, *148*, 528530.
24. Bailey, H. C.; Krassoi, R.; Elphick, J. R.; Mulhall, A. M.; Hunt, P.; Tedmanson, L.; Lovell, A. Whole effluent toxicity of sewage treatment plants in the Hawkesbury-Nepean watershed, New South Wales, Australia, to *Ceriodaphnia dubia* and *Selenastrum capricornutum*. *Environ. Toxicol. Chem.* **2000**, *19*, 72–81.
25. Choudhary, E.; Capalash, N.; Sharma, P. Genotoxicity of degradation products of textile dyes evaluated with rec-assay after photo Fenton and ligninase treatment. *J. Environ. Pathol., Toxicol. Oncol.* **2004**, *23*, 279–285.
26. Wagner, S. J.; Cifone, M. A.; Murli, H.; Dodd, R. Y.; Myhr, B.; Jerome, H. Mammalian genotoxicity assessment of methylene blue in plasma: implications for virus inactivation. *Transfusion* **1995**, *35*, 407–413.
27. Hejtmancik, M. R.; Ryan, M. J.; Toft, J. D.; Persing, R. L.; Kurtz, P. J.; Chhabra, R. S. Hematological effects in F344 rats and B6C3F1 mice during the 13-week gavage toxicity study of methylene blue trihydrate. *Toxicol. Sci.* **2002**, *65*, 126–134.
28. Oros, G.; Cserhati, T.; Forgacs, E. Separation of the strength and selectivity of the microbiological effect of synthetic dyes by spectral mapping technique. *Chemosphere* **2003**, *52*, 185–193.

29. Klopman, G.; Frierson, M. R.; Rosenkranz, H. S. The structural basis of the mutagenicity of chemicals in *Salmonella typhimurium*: the Gene-Tox data base. *Mutat. Res.* **1990**, *228*, 1–50.

30. Roginsky, V. A.; Bruchelt, G.; Bartuli, O. Ubiquinone-0 (2,3-dimethoxy-5-methyl-1,4-benzoquinone) as effective catalyzer of ascorbate and epinephrine oxidation and damager of neuroblastoma cells. *Biochem. Pharmacol.* **1998**, *55*, 85–91.

31. Kassam, S. N.; Rainbow, A. J. Deficient base excision repair of oxidative DNA damage induced by methylene blue plus visible light in xeroderma pigmentosum group C fibroblasts. *Biochem. Biophys. Res. Commun.* **2007**, *359*, 1004–1009.

32. Lazarova, M.; Labaj, J.; Eckl, P.; Slamenova, D. Comparative evaluation of DNA damage by genotoxicants in primary rat cells applying the comet assay. *Toxicol. Lett.* **2006**, *164*, 54–62.

33. Herkovits, J.; Perez-Coll, C. S.; Stockert, J. C.; Blazquez, A. The screening of photodynamic toxicity of dyes by means of a bioassay using amphibian embryos. *Res. J. Chem. Environ.* **2007**, *11*, 86–91.

34. Mahadevan, M. M.; Weitzman, G. A.; Hogan, S.; Breckinridge, S.; Miller, M. M. Methylene blue but not indigo carmine is toxic to human luteal cells *in vitro*. *Reprod. Toxicol.* **1993**, *7*, 631–633.

35. Cragan, J. D. Teratogen update: methylene blue. *Teratology* **1999**, *60*, 42–48.

METHYLENE BLUE TRIHYDRATE

CAS Registry Number 7220-79-3

Chemical Structure

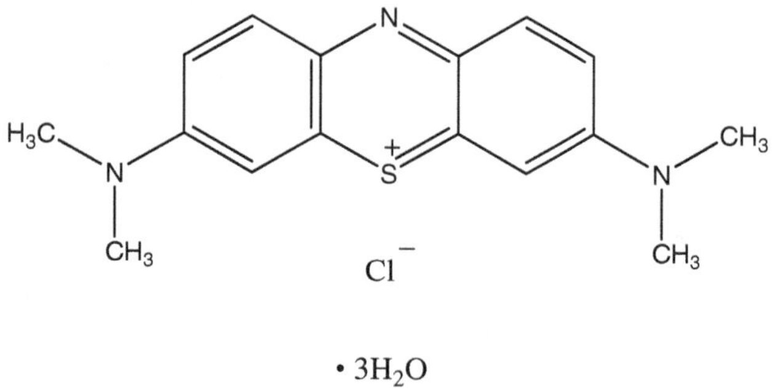

CA Index Name Phenothiazin-5-ium, 3,7-bis(dimethylamino)-, chloride, hydrate (1:1:3)

Other Names C.I. Basic Blue 9, trihydrate; C.I. 52015; Phenothiazin-5-ium, 3,7-bis(dimethylamino)-, chloride, trihydrate; 3,7-Bis(dimethylamino)phenazathionium chloride trihydrate; Methylene blue trihydrate

Merck Index Number 6060

Chemical/Dye Class Phenothiazine

Molecular Formula $C_{16}H_{18}ClN_3S{\cdot}3H_2O$

Molecular Weight 373.90

Physical Form Dark green needles or crystals or powder

Solubility Soluble in water, chloroform; sparingly soluble in ethanol

Melting Point 190 °C

pK_a 2.6, 11.2

Absorption (λ_{max}) 668 nm, 609 nm

Synthesis Synthetic methods[1–3]

Staining Applications Bacteria[4]

Biological Applications Diagnosis of tauopathy,[2] Alzheimer's disease (AD),[2] Pick's disease;[2] treating prophylaxis;[2] dental materials[5]

Industrial Applications Thin films;[6] data storage media;[7,8] photothermographic materials;[9] crayons[10]

Safety/Toxicity Acute toxicity;[11] cytotoxicity[12]

Certification/Approval Certified by Biological Stain Commission (BSC)

REFERENCES

1. Feraud, M.; Sayah, B. Process for preparing diaminophenothiazinium compounds with improved purity. PCT Int. Appl. WO 2008006979, 2008; *Chem. Abstr.* **2008**, *148*, 146491.
2. Wischik, C. M.; Storey, J. M. D.; Marshall, C.; Sinclair, J. P.; Baddeley, T. C. Preparation and purification of diaminophenothiazinium compounds for therapeutic use. PCT Int. Appl. WO 2008007074, 2008; *Chem. Abstr.* **2008**, *148*, 168729.
3. Shih, I. K.; Teare, F. W. Dye–drug salt formations and possible applications. I. Organic salt formation between a cationic dye (methylene blue) and certain organic acid ions. *Can. J. Pharm. Sci.* **1966**, *1*, 20–22.
4. Romanova, N. A.; Brovko, L. Y.; Moore, L.; Pometun, E.; Savitsky, A. P.; Ugarova, N. N.; Griffiths, M. W. Assessment of photodynamic destruction of *Escherichia coli* O157:H7 and *Listeria monocytogenes* by using ATP bioluminescence. *Appl. Environ. Microbiol.* **2003**, *69*, 6393–6398.
5. Fuchigami, S.; Kazama, H.; Iijima, K. Photopolymerizable dental compositions containing photofading dyes. Jpn. Kokai Tokkyo Koho JP 11139920, 1999; *Chem. Abstr.* **1999**, *131*, 35909.
6. El-Deeb, A. F.; El-Nahass, M. M.; Kamel, M. A.; Huthaily, S. Y. Influence of heat treatment and γ-irradiation on the optical properties of methylene blue trihydrate thin films. *Eur. Phys. J.* **2008**, *41*, 33–40.
7. Ezbiansky, K. A.; Olson, D. R.; Wisnudel, M. B.; Lindholm, E. P. Coating formulation for limited play data storage media. U.S. Pat. Appl. Publ. US 2007048664, 2007; *Chem. Abstr.* **2007**, *146*, 283995.
8. Wisnudel, M. B.; Olson, D. R.; Simon, D. A.; Longley, K. L.; Lindholm, E. P. Limited play data storage media, adhesive and coating formulations. U. S. Pat. Appl. Publ. US 2005051053, 2005; *Chem. Abstr.* 2005, *142*, 263184.
9. Mader, R. A.; Weigel, D. C.; Leichter, L. M.; Hoff, K. A. Photothermographic emulsions having stable color-forming developers. Eur. Pat. Appl. EP 273587, 1988; *Chem. Abstr.* **1989**, *110*, 31468.
10. Lytton, R. N. Water soluble water washable crayon composition. U.S. Patent 5380357, 1995; *Chem. Abstr.* **1995**, *123*, 12052.

11. Niculescu, S. P.; Atkinson, A.; Hammond, G.; Lewis, M. Using fragment chemistry data mining and probabilistic neural networks in screening chemicals for acute toxicity to the fathead minnow. *SAR QSAR Environ. Res.* **2004**, *15*, 293–309.

12. Huang, R.; Southall, N.; Cho, M. H.; Xia, M.; Inglese, J.; Austin, C. P. Characterization of diversity in toxicity mechanism using *in vitro* cytotoxicity assays in quantitative high throughput screening. *Chem. Res. Toxicol.* **2008**, *21*, 659–667.

METHYLENE VIOLET (METHYLENE VIOLET BERNTHSEN)

CAS Registry Number 2516-05-4

Chemical Structure

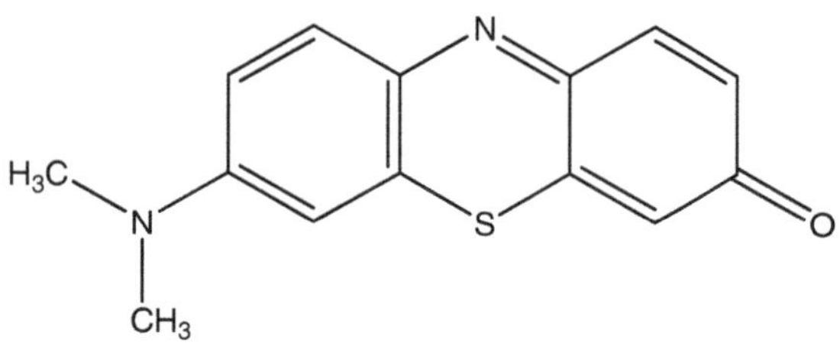

CA Index Name 3*H*-Phenothiazin-3-one, 7-(dimethylamino)-

Other Names C.I. 52041; Dimethylthionoline; Methylene Violet; Methylene Violet (biological stain); Methylene Violet Bernthsen; NSC 187694

Merck Index Number Not listed

Chemical/Dye Class Phenothiazine

Molecular Formula $C_{14}H_{12}N_2OS$

Molecular Weight 256.32

Physical Form Dark green powder

Solubility Insoluble in water, soluble in ethanol

Melting Point 216 °C (decompose)

Boiling Point (Calcd.) 429.6 ± 45.0 °C, pressure: 760 Torr

pK_a (Calcd.) 5.08 ± 0.20, most basic, temperature: 25 °C

Absorption (λ_{max}) 580 nm

Synthesis Synthetic method[1–7]

Staining Applications Bacteria;[8] chromosomes;[9] neurons;[10,11] nuclei;[12] nucleic acids;[13,14] skin;[15] yeast[16,17]

Biological Applications Detecting microorganisms;[18] medical devices[19]

Industrial Applications Adhesives;[20,21] liquid crystal displays;[22,23] recording materials;[24] toners;[25] photographic films[26]

Safety/Toxicity No data available

Certification/Approval Certified by Biological Stain Commission (BSC)

REFERENCES

1. McLean, M. R.; Badr, M.; Dalton, L. R.; Devine, R. L. S.; Steier, W. H. Fused, three-ring donor–acceptor molecules as potential materials for efficient second harmonic generation. *Mater. Res. Soc. Symp. Proc.* **1990**, *173*, 563–566.
2. Nikolaev, L. A.; Novozhilov, Yu. A.; Murzina, L. P. Interconversions of thiazine dyes. *Zh. Fiz. Khim.* **1976**, *50*, 2106–2108; *Chem. Abstr.* **1976**, *85*, 178955.
3. Marshall, P. N. The composition of stains produced by the oxidation of Methylene Blue. *Histochem. J.* **1976**, *8*, 431–442.
4. Lillie, R. D.; Pizzolato, P. Brilliant cresyl blue in acid alcohol as stain for gastric and other mucins. *Acta Histochem., Suppl.* **1971**, 669–676.
5. Wanzlick, H. W.; Pankow, B.; Daneke, J. Syntheses with quinones *in situ*. VII. Preparation of 7-dimethylamino-2-hydroxyphenothiazin-3-one. *Justus Liebigs Ann. Chem.* **1969**, *727*, 106–109.
6. Tzung, C. Methylene Violet (Bernthsen), by zinc-alkali-chlorate hydrolysis of methylene blue. *Stain Technol.* **1964**, *39*, 351–358.
7. Gurr, E. pH of ionic dye solutions. *Nature* **1964**, *202*, 920.
8. Oka, A. Dyeability discriminant method for identifying Gram negative bacteria. Jpn. Kokai Tokkyo Koho JP 2004208526, 2004; *Chem. Abstr.* **2004**, *141*, 119787.
9. Curtis, D.; Horobin, R. W. Chromosome banding: specification of structural features of dyes giving rise to G-banding. *Histochem. J.* **1982**, *14*, 911–928.
10. Smith, T. M.; Squires, R. F. Differential inhibition of brain specific [^{3}H]flunitrazepam binding by several types of dyes. *Neurochem. Res.* **1983**, *8*, 1177–1183.
11. Arthur, R. P.; Shelley, W. B. The technology of *in vitro* staining of nerves in human skin with thiazine dyes. *J. Invest. Dermatol.* **1959**, *33*, 121–133.
12. Dutt, M. K. *In situ* absorption spectra of PAS and feulgen-positive substances stained with methylene violet. *Indian J. Exp. Biol.* **1978**, *16*, 253–254.
13. Gupta, A.; Will, S. G. Light emission modifiers and their uses in nucleic acid detection, amplification and analysis. Eur. Pat. Appl. EP 1739190, 2007; *Chem. Abstr.* **2007**, *146*, 115979.

14. Dutt, M. K. Schiff-type dyes in the cytochemical staining of DNA. *Indian J. Exp. Biol.* **1975**, *13*, 142–146.
15. Hurley, H. J.; Witkowski, J. Dye clearance and eccrine sweat secretion in human skin. *J. Invest. Dermatol.* **1961**, *36*, 259–272.
16. Van Zandycke, S. M.; Simal, O.; Gualdoni, S.; Smart, K. A. Determination of yeast viability using fluorophores. *J. Am. Soc. Brewing Chem.* **2003**, *61*, 15–22.
17. Smart, K. A.; Chambers, K. M.; Lambert, I.; Jenkins, C. Use of methylene violet staining procedures to determine yeast viability and vitality. *J. Am. Soc. Brewing Chem.* **1999**, *57*, 18–23.
18. Martin, S. M.; MacDonald, J. G.; Lye, J.; Sayre, C.; Thompson, K. Rapid assessment of upper respiratory conditions and detection of microorganisms using test strip comprising at least one broad spectrum indicator. PCT Int. Appl. WO 2009027855, 2009; *Chem. Abstr.* **2009**, *150*, 278103.
19. Ash, S. R.; Steczko, J. Antimicrobial medical devices. PCT Int. Appl. WO 2008140753, 2008; *Chem. Abstr.* **2008**, *149*, 541832.
20. Cole, B. M.; Dower, W. V.; Oxman, J. D. Optical fiber connector using colored photocurable adhesive. PCT Int. Appl. WO 2000022060, 2000; *Chem. Abstr.* **2000**, *132*, 294806.
21. Bauer, H.; Piestert, G.; Gilch, H. G. Adhesive compositions. U.S. Patent 4395301, 1983; *Chem. Abstr.* **1983**, *99*, 123839.
22. Kiyohara, K.; Takeda, A. Polarizing films with excellent wet–heat resistance, their manufacture, polarizers, and liquid crystal displays therewith. Jpn. Kokai Tokkyo Koho JP 2008052170, 2008; *Chem. Abstr.* **2008**, *148*, 295623.
23. Shepherd, J. P.; Shen, S. S.; Marr, B. B.; Charbonneau, L. F. Polarizer films comprising wholly aromatic liquid crystalline polymers and dichroic dyes for liquid-crystal display devices. U. S. Patent 5738803, 1998; *Chem. Abstr.* **1998**, *128*, 302179.
24. Matsumoto, H.; Takashima, M.; Washisu, S.; Kawamura, K.; Sorori, T. Photopolymerizable compositions and their recording materials. Jpn. Kokai Tokkyo Koho JP 2001209176, 2001; *Chem. Abstr.* **2001**, *135*, 160144.
25. Ikeda, T.; Iizawa, T.; Honda, N. Electrophotographic toners. Jpn. Kokai Tokkyo Koho JP 51098031, 1976; *Chem. Abstr.* **1977**, *86*, 113702.
26. Landow, B. E. Photographic film having enhanced Herschel effect. U.S. Patent 2857273, 1958; *Chem. Abstr.* **1959**, *53*, 93402.

METHYL GREEN

CAS Registry Number 7114-03-6

Chemical Structure

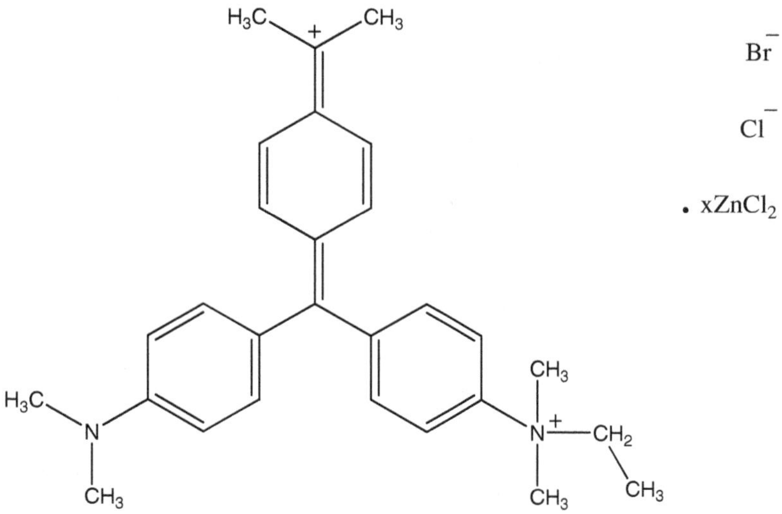

Chemical/Dye Class Triphenylmethane

Molecular Formula $C_{27}H_{35}BrClN_3 \cdot xZnCl_2$

Molecular Weight 653.24

Physical Form Red-brown powder

CA Index Name Benzenaminium, 4-[[4-(dimethylamino)phenyl][4-(dimethyliminio)-2,5-cyclohexadien-1-ylidene]methyl]-*N*-ethyl-*N*,*N*-dimethyl-, bromide chloride, compd. with zinc chloride ($ZnCl_2$) (1:1:1:?)

Other Names Ammonium, [α-[*p*-(dimethylamino)phenyl]-α-[*p*-(dimethyliminio)-2,5-cyclohexadien-1-ylidene]-*p*-tolyl]ethyldimethyl-, bromide chloride, compd. with zinc chloride; Benzenaminium, 4-[[4-(dimethylamino)phenyl][4-(dimethyliminio)-2,5-clohexadien-1-ylidene]methyl]-*N*-ethyl-*N*,*N*-dimethyl-, bromide chloride, compd. with zinc chloride; Benzenaminium, 4-[[4-(dimethylamino)phenyl][4-(dimethyliminio)-2,5-cyclohexadien-1-ylidene]methyl]-*N*-ethyl-*N*,*N*-dimethyl-, bromide chloride, compd. with zinc chloride ($ZnCl_2$); Zinc chloride ($ZnCl_2$), compd. with 4-[[4-(dimethylamino)phenyl][4-(dimethyliminio)-2,5-cyclohexadien-1-ylidene] methyl]-*N*-ethyl-*N*,*N*-dimethylbenzenaminium bromide chloride; Ethyl green; Methyl green; Sigma Ethyl green; Sigma Methyl green; C. I. 42590

Merck Index Number Not listed

Solubility Soluble in water, ethanol; insoluble in xylene

Melting Point >300 °C

pH Range 0.1–2.3

Color Change at pH Yellow (0.1) to greenish-blue (2.3)

Absorption (λ_{max}) 629 nm, 423 nm

Synthesis Synthetic method[1]

Staining Applications Brain;[2] spinal cord;[2] bacteria;[3] chromosomes;[4] nicotinic acetylcholine receptor;[5] nucleic acids;[6–8] phosphates;[9] polyphosphates;[9] tocopherol;[10] tocopherol acetate[10]

Biological Applications Detecting phosphates,[9] polyphosphates,[9] tocopherol,[10] tocopherol acetate[10]

Industrial Applications Spatial light modulators;[11] security applications[12]

Safety/Toxicity No data available

Certification/Approval Certified by Biological Stain Commission (BSC)

REFERENCES

1. Tikhonov, E. A.; Shpak, M. T. Second-harmonic generation of a neodymium laser by crystalline powders of organic compounds. *Ukr. Fiz. Zh. (Russ. Ed.)* **1972**, *17*, 190–202; *Chem. Abstr.* **1972**, *76*, 119684.
2. Augulis, V.; Sigg, E. B. Supravital staining and fixation of brain and spinal cord by intravascular perfusion. *Stain Technol.* **1971**, *46*, 183–190.
3. Cohen, E.; Rozen, H.; Joseph, T.; Braun, S.; Margulies, L. Photoprotection of *Bacillus*

thuringiensis kurstaki from ultraviolet irradiation. *J. Invertebr. Pathol.* **1991**, *57*, 343–351.

4. Saitoh, Y.; Laemmli, U. K. Metaphase chromosome structure: bands arise from a differential folding path of the highly AT-rich scaffold. *Cell* **1994**, *76*, 609–622.
5. Lurtz, M. M.; Pedersen, S. E. Aminotriarylmethane dyes are high-affinity noncompetitive antagonists of the nicotinic acetylcholine receptor. *Mol. Pharmacol.* **1999**, *55*, 159–167.
6. Telford, C. B. Nucleic acids and methyl green—pyronine: fixation, pyronine purification, and theory of staining. *Can. J. Med. Technol.* **1974**, *36*, 6, 11–18, 20–24.
7. Selander, R. K. Binding of quinacrine mustard to nucleic acids. *Acta Chem. Scand., Ser. B* **1974**, *28*, 45–55.
8. Dutt, M. K. Aqueous solutions of some basic dyes in the specific staining of acid hydrolyzed DNA. *Acta Histochem.* **1974**, *49*, 128–131.
9. Omelon, S.; Grynpas, M. A nonradioactive method for detecting phosphates and polyphosphates separated by PAGE. *Electrophoresis* **2007**, *28*, 2808–2811.
10. Pyka, A. Analytical evaluation of visualizing reagents used to detect tocopherol and tocopherol acetate on thin layer. *J. Liq. Chromatogr. Relat. Technol.* **2009**, *32*, 312–330.
11. Powles, C. M. J. Thin light-blocking layers for spatial light modulators. Eur. Pat. Appl. EP 329427, 1989; *Chem. Abstr.* **1990**, *112*, 14039.
12. Peters, A. Je.; Wisnudel, M. B.; Patel, B. P.; Misner, M. J.; Krishnan, K.; Nagarkar, K. R. Optical article having an electrically responsive layer as an anti-theft feature and a system and method for inhibiting theft. U.S. Pat. Appl. Publ. US 2008019258, 2008; *Chem. Abstr.* **2008**, *148*, 202188.

METHYL ORANGE (ORANGE III)

CAS Registry Number 547-58-0

Chemical Structure

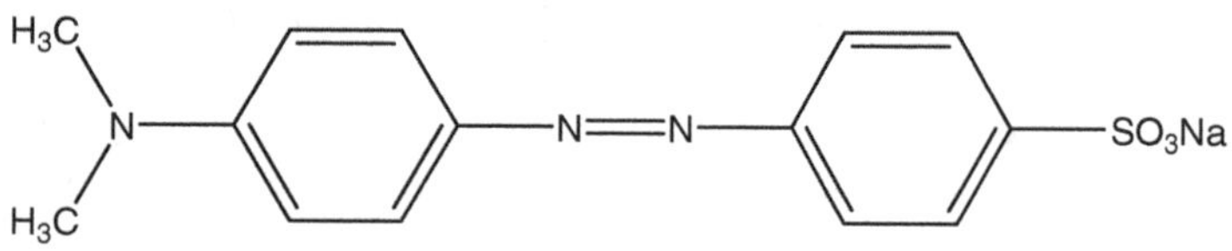

CA Index Name Benzenesulfonic acid, 4-[2-[4-(dimethylamino)phenyl]diazenyl]-, sodium salt (1:1)

Other Names Benzenesulfonic acid, 4-[[4-(dimethylamino)phenyl]azo]-, sodium salt; Benzenesulfonic acid, *p*-[[*p*-(dimethylamino)phenyl]azo]-, sodium salt; Orange III; 4-Dimethylaminoazobenzene-4′-sulfonic acid sodium salt; Acid Orange 52; Albion Methyl Orange; C.I. 13025; C.I. Acid Orange 52; Diazoben; Eniamethyl Orange; Gold Orange; Helianthine; Helianthine B; KCA Methyl Orange; Methyl Orange B; Methyl Orange; Orange 3; Sodium 4′-(dimethylamino)azobenzene-4-sulfonate; Sodium 4-(dimethylamino)azobenzene-4′-sulfonate; Sodium 4-[4-(dimethylamino)phenylazo]benzenesulfonate; Sodium *p*-[[*p*-(dimethylamino)phenyl]azo]benzenesulfonate; Sodium *p*-dimethylaminoazobenzenesulfonate; Tropaeolin D

Merck Index Number 6105

Chemical/Dye Class Azo

Molecular Formula $C_{14}H_{14}N_3NaO_3S$

Molecular Weight 327.33

Physical Form Orange-yellow powder or crystals

Solubility Slightly soluble in water, more soluble in hot water; practically insoluble in ethanol

Melting Point >300 °C

pH Range 3.0–4.4

Color Change at pH Red (3.0) to yellow (4.4)

p*Ka* 3.76, 3.40

Absorption (λ_{max}) 507 nm

Synthesis Synthetic methods[1–11]

Staining Applications Albumin;[12] leukocytes;[13] nucleic acids;[1,14,15] protein;[16] nails;[17] skin;[18] hairs[19]

Biological Applications Detecting microorganisms;[20] treating dermatological diseases,[21] vaginal affections;[22] dental materials;[1,23] wound dressing materials[1,24,25]

Industrial Applications Thin films;[1,26,27] nanoparticles;[28] liquid crystals;[1,29–31] sol–gel matrix;[1,32] waveguides;[1,33] paints;[1,34] glass[1,35]

Safety/Toxicity Carcinogenicity;[1,36,37] genotoxicity;[1,31,38,39] mutagenicity[1,40–42]

Certification/Approval Certified by Biological Stain Commission (BSC)

REFERENCES

1. Sabnis, R. W. *Handbook of Acid–Base Indicators*; CRC Press: Boca Raton, **2008**; pp 228–229.
2. Welcher, F. J. *Organic Analytical Reagents*; D. Van Nostrand Company: New York, **1948**; Vol. 4, pp 531–536.
3. Dabbagh, H. A.; Teimouri, A.; Chermahini, A. N. Green and efficient diazotization and diazo coupling reactions on clays. *Dyes Pigments* **2007**, *73*, 239–244.
4. Astolfi, D. L.; Mayville, F. C. Synthesis of methyl orange using ionic liquids. *Tetrahedron Lett.* **2003**, *44*, 9223–9224.
5. Wang, R.; Wang, H.; Wang, L.; Liu, X.; Wang, J. Experimental improvement of one-step method for the synthesis of methyl orange. *Yunnan Minzu Daxue Xuebao, Ziran Kexueban* **2004**, *13*, 325–327; *Chem. Abstr.* **2005**, *142*, 239740.
6. Huddersman, K.; Patruno, V.; Blake, G. J.; Dahm, R. H. Azo dyes encapsulated within aluminosilicate microporous materials. *J. Soc. Dyers Colourists* **1998**, *114*, 155–159.
7. Liu, J.; Sun, D. Synthesis of methyl orange in one step at room temperature. *Huaxue Shiji* **1997**, *19*, 374; *Chem. Abstr.* **1998**, *128*, 155471.
8. Kunzelmann, M. Highly concentrated azo compounds. Ger. Offen. DE 3234335, 1984; *Chem. Abstr.* **1984**, *101*, 56481.

9. Becker, G.; Wolf, A. Methyl orange. *Praxis Naturwissenschaften, Chem.* **1973**, *22*, 201–207; *Chem. Abstr.* **1974**, *80*, 69767.
10. Thiel, A.; Peter, O. Indicators. XII. Azo-indicators. *Z. Anorg. Allgem. Chem.* **1928**, *173*, 169–198; *Chem. Abstr.* **1929**, *23*, 26849.
11. Hunter, G. Color standards for use in the determination of imidazoles. *Biochem. J.* **1925**, *19*, 42–46.
12. Carter, D. C.; Ho, J.; Wang, Z. Albumin binding sites for evaluating drug interactions and methods of evaluating or designing drugs based on their albumin binding properties. PCT Int. Appl. WO 2005041895, 2005; *Chem. Abstr.* **2005**, *142*, 441828.
13. Takarada, K.; Kouzuki, C.; Hyousa, Y.; Sakata, T.; Akai, Y. A method for classifying and counting leukocytes. Eur. Pat. Appl. EP 695936, 1996; *Chem. Abstr.* **1996**, *124*, 283704.
14. Hwang, S. Y.; Jin, L. T.; Yoo, G. S.; Choi, J. K. Counterion-dye staining for DNA in electrophoresed gels using indoine blue and methyl orange. *Electrophoresis* **2006**, *27*, 1739–1743.
15. Nakagawa, K.; Ishikawa, M.; Yamazaki, H. Method for measuring target nucleic acid, and sensor. Jpn. Kokai Tokkyo Koho JP 2005034031, 2005; *Chem. Abstr.* **2005**, *142*, 213325.
16. Liu, J.; Yang, L.; Li, Y.; Pan, D.; Hopfinger, A. J. Constructing plasma protein binding model based on a combination of cluster analysis and 4D-fingerprint molecular similarity analyses. *Bioorg. Med. Chem.* **2006**, *14*, 611–621.
17. Borsakian, B.; Faraci, J. Color changing nail polish. PCT Int. Appl. WO 2003007675, 2003; *Chem. Abstr.* **2003**, *138*, 142188.
18. Yerazunis, W. S.; Lau, K. T.; Shepherd, R. L.; Crowley, K. H. Semi-permanent skin adhering device for detecting biological conditions. U.S. Pat. Appl. Publ. US 2008069727, 2008; *Chem. Abstr.* **2008**, *148*, 350489.
19. Hoeffkes, H.; Mucha, T. Preparation for simultaneous dyeing and lightening of hair containing a direct dye and a carbonate or carbonate precursor. Ger. Offen. DE 102005013488, 2006; *Chem. Abstr.* **2006**, *145*, 362835.
20. Martin, S. M.; MacDonald, J. G.; Lye, J.; Sayre, C.; Thompson, K. Array for rapid detection of a microorganism based on dyes. PCT Int. Appl. WO 2008026104, 2008; *Chem. Abstr.* **2008**, *148*, 302867.
21. Pellet, A. Exfoliant composition containing a cellulosic polymer and its applications in cosmetic or topical dermatological treatments. Fr. Demande FR 2902322, 2007; *Chem. Abstr.* **2007**, *148*, 61286.
22. De Simone, C. Gloves bear a reporter substance for identifying vaginal affections and method for the treatment thereof. PCT Int. Appl. WO 2006080035, 2006; *Chem. Abstr.* **2006**, *145*, 195910.
23. Takahashi, M.; Saka, K.; Sakakibara, R.; Nagao, A.; Otsuka, K. Dental compositions and kits containing color markers indicating progress of curing, manufacture of cured dental materials, and their application. Jpn. Kokai Tokkyo Koho JP 2004051555, 2004; *Chem. Abstr.* **2004**, *140*, 187453.
24. Addison, D.; Greenhalgh, D.; Cullen, B. M. Antioxidant and antimicrobial wound dressing materials. PCT Int. Appl. WO 2005049101, 2005; *Chem. Abstr.* **2005**, *143*, 13454.
25. Cullen, B. M.; Addison, D.; Greenhalgh, D. Bioabsorbable wound dressing containing an antioxidant dye. Brit. UK Pat. Appl. GB 2402882, 2004; *Chem. Abstr.* **2004**, *142*, 62794.
26. Verma, A.; Goyal, A.; Sharma, R. K. Microstructural, photocatalysis and electrochemical investigations on $CeTi_2O_6$ thin films. *Thin Solid Films* **2008**, *516*, 4925–4933.
27. Ge, L.; Xu, M. X.; Sun, M. Synthesis and characterization of TiO_2 photocatalytic thin films prepared from refluxed PTA sols. *Mater. Lett.* **2006**, *60*, 287–290.
28. Hong, R. Y.; Chen, L. L.; Li, J. H.; Li, H. Z.; Zheng, Y.; Ding, J. Preparation and application of polystyrene-grafted ZnO nanoparticles. *Polym. Adv. Technol.* **2007**, *18*, 901–909.
29. Motoc, C.; Iacobescu, G. Birefringence of azo-dye doped nematic liquid crystals. *J. Optoelectron. Adv. Mater.* **2006**, *8*, 295–298.
30. Wu, W. Y.; Fuh, A. Y. G. Rewritable liquid crystal gratings fabricated using photoalignment effect in dye-doped poly(vinyl alcohol) film. *Jpn. J. Appl. Phys., Part 1* **2007**, *46*, 6761–6766.
31. Lavretovich, O. D.; Schneider, T.; Golovin, A. B.; Lee, J. C. A design and fabrication of an aligned lyotropic chromonic liquid crystal films with improved optical properties and alignment uniformity. U.S. Pat. Appl. Publ. US 2006040069, 2006; *Chem. Abstr.* **2006**, *144*, 243534.
32. Zaggout, F. R.; El-Ashgar, N. M.; Zourab, S. M.; El-Nahhal, I. M.; Motaweh, H. Encapsulation of methyl

orange pH-indicator into a sol–gel matrix. *Mater. Lett.* **2005**, *59*, 2928–2931.

33. Kumar, R.; Singh, A. P.; Kapoor, A.; Tripathi, K. N. Effect of dye doping in poly(vinyl alcohol) waveguides. *J. Mod. Opt.* **2005**, *52*, 1471–1483.
34. Simon, P. L.; Toreki, W.; Fischer, P. G. Temporary visual indicators for paint and other compositions, latex formulation, making custom latex paint, application latex, and application kit. U.S. Pat. Appl. Publ. US 2006008912, 2006; *Chem. Abstr.* **2006**, *144*, 109843.
35. Ibragimov, S. N.; Fatkhutdinov, R. K.; Zaripov, I. N.; Nikitaev, S. P. Antimisting and antiicing coating compositions for glass. Russ. RU 2260028, 2005; *Chem. Abstr.* **2005**, *143*, 268449.
36. Dorado, G.; Pueyo, C. L-Arabinose resistance test with *Salmonella typhimurium* as a primary tool for carcinogen screening. *Cancer Res.* **1988**, *48*, 907–912.
37. Mamber, S. W.; Bryson, V.; Katz, S. E. The *Escherichia coli* WP2/WP100 rec assay for detection of potential chemical carcinogens. *Mutat. Res. Lett.* **1983**, *119*, 135–144.
38. Ben Mansour, H.; Corroler, D.; Barillier, D.; Ghedira, K.; Chekir, L.; Mosrati, R. Evaluation of genotoxicity and pro-oxidant effect of the azo dyes: Acids yellow 17, violet 7 and orange 52, and of their degradation products by *Pseudomonas putida* mt-2. *Food Chem. Toxicol.* **2007**, *45*, 1670–1677.
39. Choudhary, E.; Capalash, N.; Sharma, P. Genotoxicity of degradation products of textile dyes evaluated with rec-assay after photo Fenton and ligninase treatment. *J. Environ. Pathol., Toxicol. Oncol.* **2004**, *23*, 279–285.
40. Klopman, G.; Frierson, M. R.; Rosenkranz, H. S. The structural basis of the mutagenicity of chemicals in *Salmonella typhimurium*: the Gene-Tox data base. *Mutat. Res.* **1990**, *228*, 1–50.
41. Rosenkranz, H. S.; Klopman, G. Structural basis of the mutagenicity of phenylazoaniline dyes. *Mutat. Res.* **1989**, *221*, 217–234.
42. Brown, J. P.; Dietrich, P. S. Mutagenicity of selected sulfonated azo dyes in the *Salmonella*/microsome assay: use of aerobic and anaerobic activation procedures. *Mutat. Res.* **1983**, *116*, 305–315.

METHYLTHIAZOLETETRAZOLIUM (MTT)

CAS Registry Number 298-93-1

Chemical Structure

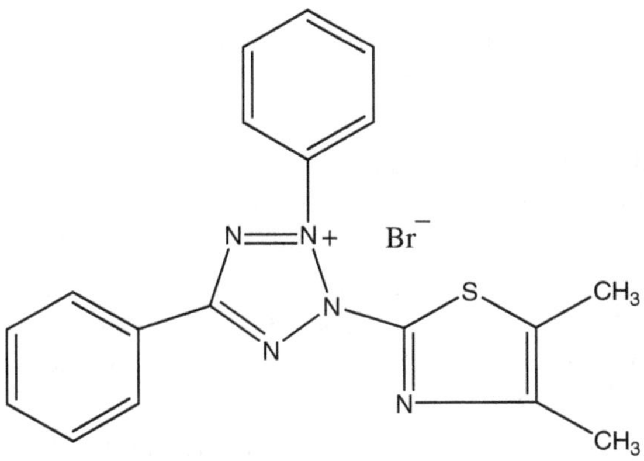

CA Index Name 2*H*-Tetrazolium, 2-(4,5-dimethyl-2-thiazolyl)-3,5-diphenyl-, bromide (1:1)

Other Names 2*H*-Tetrazolium, 2-(4,5-dimethyl-2-thiazolyl)-3,5-diphenyl, bromide; 3-(4,5-Dimethyl-2-thiazolyl)-2,5-diphenyl-2*H*-tetrazolium bromide; 2,5-Diphenyl-3-(4,5-dimethylthiazol-2-yl)-2*H*-tetrazolium bromide; 2,5-Diphenyl-3-(4,5-dimethylthiazol-2-yl)tetrazolium bromide; 2-(4,5-Dimethylthiazol-2-yl)-3,5-diphenyl-2*H*-tetrazolium bromide; 3-(4′,5′-Dimethyl-2-thiazolyl)-2,5-diphenyltetrazolium bromide; 3-(4,5-Dimethyl-2-thiazolyl)-2,5-diphenyltetrazolium bromide; 3-(4,5-Dimethylthiazolyl)-2,5-diphenyl-2*H*-tetrazolium bromide; 3-(4,5-Dimethylthiazolyl)-2,5-diphenyltetrazolium bromide; MMT Tetrazolium; MTT; MTT Tetrazolium; Methylthiazoletetrazolium; NSC 60102; Methylthiazolyldiphenyl Tetrazolium; Thiazolyl Blue Monotetrazolium; Thiazolyl blue; Thiazolyl blue (Sigma); Thiazolyl blue tetrazolium bromide

Merck Index Number Not listed

Chemical/Dye Class Tetrazolium salt

Molecular Formula $C_{18}H_{16}BrN_5S$

Molecular Weight 414.32

Physical Form Yellow to orange powder

Solubility Soluble in water, ethanol, methanol, dimethyl sulfoxide

Melting Point 195 °C (decompose)

Absorption (λ_{max}) 378 nm, 242 nm

Synthesis Synthetic methods[1–6]

Staining Applications Apoptotic and necrotic cells;[7] cells;[8–10] endospores;[11] glutathione peroxidase;[12] lyases;[13,14] neurons;[15] nucleic acids;[16] oospores;[17] serum cholinesterase;[18] skin;[19] sperms;[20,21] tissues[22]

Biological Applications Cell viability assay;[8–10] microbial growth assays;[23,24] DNA quantification assays;[25] tissue viability assays;[22] detecting enzymes;[26] measuring membrane potential;[27] treating Alzheimer's disease,[28] asthma,[29] cancer[30]

Industrial Applications Display devices;[31] photographic materials[32]

Safety/Toxicity Aquatic toxicity;[33] bacterial toxicity;[34] carcinogenicity;[35] cytotoxicity;[36–39] dental toxicity;[40,41] DNA damage;[42] embryotoxicity;[43,44] genotoxicity;[45] immunotoxicity;[46] mitochondrial toxicity;[47] nephrotoxicity;[48] neurotoxicity;[49–52] phototoxicity;[53] skin toxicity[54]

REFERENCES

1. Cui, X.; Vlahakis, J. Z.; Crandall, I. E.; Szarek, W. A. Anti-*Plasmodium* activity of tetrazolium salts. *Bioorg. Med. Chem.* **2008**, *16*, 1927–1947.
2. James, P. R. Tetrazolium salts as steam sterilization process indicators. PCT Int. Appl. WO 9837157, 1998; *Chem. Abstr.* **1998**, *129*, 218237.
3. Blatt, J. M.; Hatch, R. P. Use of specific counteranions to modify the solubility of tetrazolium salts used as indicators. U.S. Patent 5250695, 1993; *Chem. Abstr.* **1994**, *120*, 164187.
4. Barltrop, J. A.; Owen, T. C.; Cory, A. H.; Cory, J. G. 5-(3-Carboxymethoxyphenyl)-2-(4,5-dimethylthiazolyl)-3-(4-sulfophenyl)tetrazolium, inner salt (MTS) and related analogs of 3-(4,5-dimethylthiazolyl)-2,5-diphenyltetrazolium bromide (MTT) reducing to purple water-soluble formazans as cell-viability indicators. *Bioorg. Med. Chem. Lett.* **1991**, *1*, 611–614.
5. Johne, S.; Schaks, A.; Haertling, S.; Scharf, K. D.; Nover, L. Preparation of some new 2-[thiazolyl-(2)] tetrazolium derivatives and their testing on their suitability for the detection of oxidoreductase activity in plant tissues and tissue cultures. *Pharmazie* **1979**, *34*, 790–794.
6. Beyer, H.; Pyl, T. Thiazoles. XXIV. C,N-Diphenyl-*N*′-thiazolyl-2-formazans and their tetrazolium salts. *Chem. Ber.* **1954**, *87*, 1505–1511.

7. Yuan, Y. J.; Ge, Z. Q.; Li, J. C.; Wu, J. C.; Hu, Z. D. Differentiation of apoptotic and necrotic cells in suspension cultures of *Taxus cuspidata* by the combined use of fluorescent dyeing and histochemical staining methods. *Biotechnol. Lett.* **2002**, *24*, 71–76.
8. Nath, B. B.; Babrekar, A. A.; Parthasarathy, B. Assessment of cell viability in intact glandular tissue in *Chironomus ramosus* using dye-exclusion and colorimetric assays. *Cytotechnology* **2005**, *49*, 59–65.
9. Pozzolini, M.; Scarfi, S.; Benatti, U.; Giovine, M. Interference in MTT cell viability assay in activated macrophage cell line. *Anal. Biochem.* **2003**, *313*, 338–341.
10. Morgan, D. M. L. Tetrazolium (MTT) assay for cellular viability and activity. *Methods Mol. Biol.* **1998**, *79*, 179–183.
11. Arseculeratne, S. N.; Atapattu, D. N. The assessment of the viability of the endospores of *Rhinosporidium seeberi* with MTT (3-[4,5-dimethyl-2-thiazolyl]-2, 5-diphenyl-2*H*-tetrazolium bromide). *Mycol. Res.* **2004**, *108*, 1423–1430.
12. Lin, C. L.; Chen, H. J.; Hou, W. C. Activity staining of glutathione peroxidase after electrophoresis on native and sodium dodecyl sulfate polyacrylamide gels. *Electrophoresis* **2002**, *23*, 513–516.
13. Hou, W. C.; Chen, H. J.; Lin, Y. H.; Chen, Y. C.; Yang, L. L.; Lee, M. H. Activity staining of isocitrate lyase after electrophoresis on either native or sodium dodecyl sulfate polyacrylamide gels. *Electrophoresis* **2001**, *22*, 2653–2655.
14. Ukai, K.; Sekiya, J. A new staining method for lyases catalyzing cleavage of a C–S bond in sulfur-containing compounds after polyacrylamide gel electrophoresis. *Biosci., Biotechnol., Biochem.* **1997**, *61*, 124–126.
15. Takahashi, S.; Abe, T.; Gotoh, J.; Fukuuchi, Y. Substrate-dependence of reduction of MTT: a tetrazolium dye differs in cultured astroglia and neurons. *Neurochem. Int.* **2002**, *40*, 441–448.
16. Cheng, X.; Altosaar, I. Methods for high throughput screening and cloning of biocidal DNA molecules involving use of colony viability staining assay. U.S. Pat. Appl. Publ. US 2006068397, 2006; *Chem. Abstr.* **2006**, *144*, 325880.
17. Van Der Gaag, D. J. The effect of pH on staining of oospores of *Peronospora viciae* with tetrazolium bromide. *Mycologia* **1994**, *86*, 454–457.
18. Maekawa, M.; Sudo, K.; Kanno, T. A simple staining method to detect serum cholinesterase activity. *Clin. Chim. Acta* **1987**, *164*, 241–242.
19. Hood, H. L.; Bronaugh, R. L. A comparison of skin viability assays for *in vitro* skin absorption/metabolism studies. *In Vitro Mol. Toxicol.* **1999**, *12*, 3–9.
20. Aziz, D. M.; Ahlswede, L.; Enbergs, H. Application of MTT reduction assay to evaluate equine sperm viability. *Theriogenology* **2005**, *64*, 1350–1356.
21. Wishart, G. J. H. Method for assessing fowl sperm quality using a tetrazolium dye. PCT Int. Appl. WO 2001027616, 2001; *Chem. Abstr.* **2001**, *134*, 277611.
22. Alotto, D.; Ariotti, S.; Graziano, S.; Verrua, R.; Stella, M.; Magliacani, G.; Castagnoli, C. The role of quality control in a skin bank: tissue viability determination. *Cell Tissue Bank.* **2002**, *3*, 3–10.
23. Storey, D.; Field, T.; Ramage, G. Microbial growth assay and assay device. PCT Int. Appl. WO 2005098023, 2005; *Chem. Abstr.* **2005**, *143*, 402159.
24. Ushiyama, M. Tetrazolium salts and polyhydric alcohols for microbial growth and detection. Jpn. Kokai Tokkyo Koho JP 2005287452, 2005; *Chem. Abstr.* **2005**, *143*, 363250.
25. Pabbruwe, M. B.; Stewart, K.; Chaudhuri, J. B. A comparison of colorimetric and DNA quantification assays for the assessment of meniscal fibrochondrocyte proliferation in microcarrier culture. *Biotechnol. Lett.* **2005**, *27*, 1451–1455.
26. Self, C. H. Enzymatic detection method and use. Eur. Pat. Appl. EP 58539, 1982; *Chem. Abstr.* **1982**, *97*, 177838.
27. Reungpatthanaphong, P.; Dechsupa, S.; Meesungnoen, J.; Loetchutinat, C.; Mankhetkorn, S. Rhodamine B as a mitochondrial probe for measurement and monitoring of mitochondrial membrane potential in drug-sensitive and resistant cells. *J. Biochem. Biophys. Methods* **2003**, *57*, 1–16.
28. Bush, A. I.; Huang, X.; Atwood, C. S.; Tanzi, R. E. Method of screening for drugs useful in treating Alzheimer's disease. PCT Int. Appl. WO 2000066181, 2000; *Chem. Abstr.* **2000**, *133*, 344631.
29. Follettie, M.; Donaldson, D. D. Inhibitors of RegIII proteins as asthma therapeutics. PCT Int. Appl. WO 2005118615, 2005; *Chem. Abstr.* **2005**, *144*, 45487.
30. Kong, Q. Method and composition for treating cancer using cellular organelle crystallizing agents. U.S. Patent 6376525, 2002; *Chem. Abstr.* **2002**, *136*, 319375.

31. Manda, E.; Matsumoto, M.; Kawabata, K. Electrochromic electrodes for display devices. Jpn. Kokai Tokkyo Koho JP 01230026, 1989; *Chem. Abstr.* **1990**, *113*, 32036.

32. Wilson, R. B.; Williams, J. V. Color photographic images. Fr. FR 1551112, 1968; *Chem. Abstr.* **1970**, *72*, 95308.

33. Zvinavashe, E.; van den Berg, H.; Soffers, A. E M F.; Vervoort, J.; Freidig, A.; Murk, A. J.; Rietjens, I. M. C. M. QSAR models for predicting *in vivo* aquatic toxicity of chlorinated alkanes to fish. *Chem. Res. Toxicol.* **2008**, *21*, 739–745.

34. Sriram, D.; Yogeeswari, P.; Dinakaran, M.; Thirumurugan, R. Antimycobacterial activity of novel 1-(5-cyclobutyl-1,3-oxazol-2-yl)-3-(sub) phenyl/pyridylthiourea compounds endowed with high activity toward multidrug-resistant *Mycobacterium tuberculosis*. *J. Antimicrob. Chemother.* **2007**, *59*, 1194–1196.

35. Neychev, V. K.; Nikolova, E.; Zhelev, N.; Mitev, V. I. Saponins from *Tribulus terrestris* L. are less toxic for normal human fibroblasts than for many cancer lines: influence on apoptosis and proliferation. *Exp. Biol. Med.* **2007**, *232*, 126–133.

36. Froscio, S. M.; Fanok, S.; Humpage, A. R. Cytotoxicity screening for the cyanobacterial toxin cylindrospermopsin. *J. Toxicol. Environ. Health Part A* **2009**, *72*, 345–349.

37. Othmen, Z. O. B.; Golli, E. E.; Abid-Essefi, S.; Bacha, H. Cytotoxicity effects induced by Zearalenone metabolites, alpha Zearalenol and beta Zearalenol, on cultured Vero cells. *Toxicology* **2008**, *252*, 72–77.

38. Lee, I.; Yoo, J. K.; Na, M.; Min, B. S.; Lee, J.; Yun, B. S.; Jin, W.; Kim, H.; Youn, U.; Chen, Q. C.; Song, K. S.; Seong, Y. H.; Bae, K. Cytotoxicity of triterpenes isolated from *Aceriphyllum rossii*. *Chem. Pharm. Bull.* **2007**, *55*, 1376–1378.

39. Silva, A. M. M.; Silva, A. R.; Pinheiro, A. M.; Freitas, S. R. V. B.; Silva, V. D. A.; Souza, C. S.; Hughes, J. B.; El-Bacha, R. S.; Costa, M. F. D.; Velozo, E. S.; Tardy, M.; Costa, S. L. Alkaloids from *Prosopis juliflora* leaves induce glial activation, cytotoxicity and stimulate NO production. *Toxicon* **2007**, *49*, 601–614.

40. Lodiene, G.; Morisbak, E.; Bruzell, E.; Orstavik, D. Toxicity evaluation of root canal sealers *in vitro*. *Int. Endod. J.* **2008**, *41*, 72–77.

41. Kuo, T.; Lee, B.; Kang, S.; Lin, F.; Lin, C. Cytotoxicity of DP-bioglass paste used for treatment of dentin hypersensitivity. *J. Endod.* **2007**, *33*, 451–454.

42. Myllynen, P.; Kurttila, T.; Vaskivuo, L.; Vahakangas, K. DNA damage caused by benzo(*a*)pyrene in MCF-7 cells is increased by verapamil, probenecid and PSC833. *Toxicol. Lett.* **2007**, *169*, 3–12.

43. Stummann, T. C.; Hareng, L.; Bremer, S. Embryotoxicity hazard assessment of cadmium and arsenic compounds using embryonic stem cells. *Toxicology* **2008**, *252*, 118–122.

44. D'Alessandro, S.; Gelati, M.; Basilico, N.; Parati, E. A.; Haynes, R. K.; Taramelli, D. Differential effects on angiogenesis of two antimalarial compounds, dihydroartemisinin and artemisone: implications for embryotoxicity. *Toxicology* **2007**, *241*, 66–74.

45. Arbillaga, L.; Azqueta, A.; van Delft, J. H M.; Lopez de, C. A. *In vitro* gene expression data supporting a DNA non-reactive genotoxic mechanism for ochratoxin A. *Toxicol. Appl. Pharmacol.* **2007**, *220*, 216–224.

46. Wu, X.; Murphy, P.; Cunnick, J.; Hendrich, S. Synthesis and characterization of deoxynivalenol glucuronide: its comparative immunotoxicity with deoxynivalenol. *Food Chem. Toxicol.* **2007**, *45*, 1846–1855.

47. Zhang, H.; Chen, Q. Y.; Xiang, M. L.; Ma, C. Y.; Huang, Q.; Yang, S. Y. *In silico* prediction of mitochondrial toxicity by using GA-CG-SVM approach. *Toxicol. In Vitro* **2009**, *23*, 134–140.

48. Heinrich, M. C.; Kuhlmann, M. K.; Kohlbacher, S.; Scheer, M.; Grgic, A.; Heckmann, M. B.; Uder, M. Cytotoxicity of iodinated and gadolinium-based contrast agents in renal tubular cells at angiographic concentrations: *in vitro* study. *Radiology* **2007**, *242*, 425–434.

49. Hamann, K.; Nehrt, G.; Ouyang, H.; Duerstock, B.; Shi, R. Hydralazine inhibits compression and acrolein-mediated injuries in *ex vivo* spinal cord. *J. Neurochem.* **2008**, *104*, 708–718.

50. Kaur, P.; Schulz, K.; Aschner, M.; Syversen, T. Role of docosahexaenoic acid in modulating methylmercury-induced neurotoxicity. *Toxicol. Sci.* **2007**, *100*, 423–432.

51. Yan, Q.; Feng, M.; Yan, S. RNA interference-mediated inhibition of brain-derived neurotrophic factor expression increases cocaine's cytotoxicity in cultured cells. *Neurosci. Lett.* **2007**, *414*, 165–169.

52. Liu, Y.; Ritter, C.; Riek, R.; Schubert, D. The formation of bioactive amyloid species by prion

proteins *in vitro* and in cells. *Neurosci. Lett.* **2006**, *406*, 200–204.

53. Bernard, F. X.; Barrault, C.; Deguercy, A.; De Wever, B.; Rosdy, M. Development of a highly sensitive *in vitro* phototoxicity assay using the SkinEthic reconstructed human epidermis. *Cell Biol. Toxicol.* **2000**, *16*, 391–400.

54. Lim, Y.; Moon, S.; An, S; Lee, S.; Kim, S.; Chang, I.; Park, K.; Kim, H.; Heo, Y. Suitability of macrophage inflammatory protein-1beta production by THP-1 cells in differentiating skin sensitizers from irritant chemicals. *Contact Dermatitis* **2008**, *58*, 193–198.

METHYL VIOLET 2B (METHYL VIOLET)

CAS Registry Number 8004-87-3

Chemical Structure

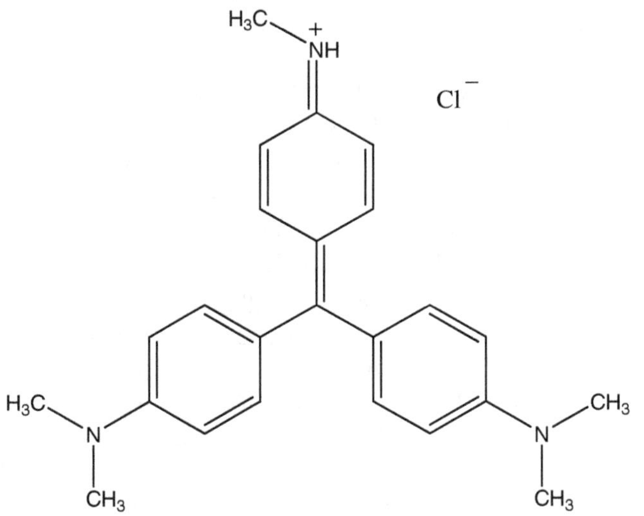

CA Index Name C.I. Basic Violet 1

Other Names Methyl Violet; Aizen Methyl Violet BB; Aizen Methyl Violet Pure Special; Basic Violet 1; Basic Violet 5BN; Basic Violet K; Basonyl Violet 600; C Ext. Violet 5; C.I. 42535; Gentian Violet B; Methyl Violet 2B; Methyl Violet 6B; Methyl Violet B; Methyl Violet BB; Methyl Violet FN; Methyl Violet N; Methyl Violet Pure Special; Methyl Violet Pure SP; Paris Violet R; Pyoktanin Blue; Pyoktaninum coeruleum; Violet Powder H 2503

Merck Index Number Not listed

Chemical/Dye Class Triphenylmethane

Molecular Formula Mixture (tetra, penta and hexa) of the hydrochloride salts of N-methylated forms of pararosanilines, predominantly the penta (chemical structure given above)

Molecular Weight Mixture (tetra, penta and hexa) of the hydrochloride salts of N-methylated forms of pararosanilines, predominantly the penta (chemical structure given above)

Physical Form Green to dark green crystalline powder

Solubility Soluble in water, ethanol; insoluble in xylene

Melting Point 137 °C (decompose)

pH Range 0.15–3.2

Color Change at pH Yellow (0.15) to violet (3.2)

Absorption (λ_{max}) 584 nm

Emission (λ_{max}) 590 nm

Synthesis Synthetic method[1–21]

Staining Applications Bacteria;[22] cells;[23] fish;[24] heparin;[25] nematicide granules;[26] nucleic acids;[1,27] polypeptides;[38] proteins;[28,29] skin;[30,31] carious tissue;[32] viscous tissue;[33] keratin fibers;[34] hairs[1,35,36]

Biological Applications Antimalarial agent;[37] detecting enzyme activity,[38] protein–protein interactions;[38] treating diabetes,[39] ringworm;[40] agrochemicals;[41] pesticides;[41] cosmetics;[1,42] wound dressing materials[1,43]

Industrial Applications Solar cells;[1,44] solar energy;[45] display devices;[1,46] photoresists;[1,47] inks;[1,48] toners;[49] highlighters;[1,50] adhesives;[51,52] detergent;[53] packaging materials;[1,54] lithium battery;[1,55] rubber;[1,56] cement;[57] stainless steel;[58] petroleum products;[1,59] leather;[60] paper;[61,62] textiles[63]

Safety/Toxicity Acute toxicity;[1,64] aquatic toxicity;[1,65] carcinogenicity;[1,66] microbial toxicity;[1,67,68] mutagenicity[1,69,70]

Certification/Approval Certified by Biological Stain Commission (BSC)

REFERENCES

1. Sabnis, R. W. *Handbook of Acid–Base Indicators*; CRC Press: Boca Raton, **2008**; pp 238–239.
2. Mukhortova, L. I.; Neboikov, P. L.; Devyatin, A. P.; Kazakov, V. V.; Andreev, O. K.; Gudkova, O. A. Composition for preparing liquid form of basic dye. Russ. RU 2222561, 2004; *Chem. Abstr.* **2004**, *141*, 158510.
3. Marshall, C. J., Jr. Composite colorants containing water-soluble dyes and mineral pigments and their manufacture and use. U.S. Patent 5106420, 1992; *Chem. Abstr.* **1992**, *117*, 152141.
4. Aizawa, T. Manufacture of triphenylmethane dye by electrochemical oxidation. Jpn. Kokai Tokkyo Koho JP 02194188, 1990; *Chem. Abstr.* **1991**, *114*, 52054.
5. Kotabe, S. Manufacture of methyl violet. Jpn. Kokai Tokkyo Koho JP 54040828, 1979; *Chem. Abstr.* **1979**, *91*, 58704.

6. Nakata, O. Methyl violet. Jpn. Kokai Tokkyo Koho JP 51017221, 1976; *Chem. Abstr.* **1976**, *85*, 7290.
7. Utamaru, A. Methyl violet. Jpn. Kokai Tokkyo Koho JP 50002728, 1975; *Chem. Abstr.* **1975**, *83*, 12213.
8. Feldman, M. L.; Gordon, J. E. Recovery and purification of basic triarylmethane dyes. U.S. Patent 3679713, 1972; *Chem. Abstr.* **1972**, *77*, 141488.
9. Feldman, M. L.; Gordon, J. E. Isolation of triphenylmethane dyes. Ger. Offen. DE 2102404, 1971; *Chem. Abstr.* **1971**, *75*, 89326.
10. Herrick, G. S.; Conger, J. C.; Savio, M. W. Methyl violet. U.S. Patent 2816900, 1957; *Chem. Abstr.* **1958**, *52*, 32641.
11. Yokozawa, S.; Watanabe, K. Methyl violet. JP 30002384, 1955; *Chem. Abstr.* **1957**, *51*, 85952.
12. Yokozawa, S.; Watanabe, K. Methyl violet. JP 30002385, 1955; *Chem. Abstr.* **1957**, *51*, 81748.
13. Yokoyama, K. Methyl violet. JP 27003087, 1952; *Chem. Abstr.* **1954**, *48*, 14990.
14. Frank, O. S. Methyl violet purification. U.S. Patent 2181369, 1939; *Chem. Abstr.* **1940**, *34*, 12895.
15. Ignat'ev, S. N.; Vasin, I. I. Methyl violet. RU 26391, 1929; *Chem. Abstr.* **1932**, *26*, 60046.
16. Kade, A. S.; Ignat'ev, S. N. Methyl violet. RU 23917, 1929; *Chem. Abstr.* **1932**, *26*, 60045.
17. Ignat'ev, S. N. Methyl violet. RU 23500, 1931; *Chem. Abstr.* **1932**, *26*, 15129.
18. Suichev, G. D. New synthesis of methyl violet. *Zh. Khim. Promysh.* **1928**, *5*, 72–73; *Chem. Abstr.* **1928**, *22*, 31131.
19. Sansone, R. Future possibilities in the production of methyl violets. I. *Color Trade J.* **1922**, *11*, 51–53.
20. Creighton, H. J. M. Method for making methyl violet. *Proc. Trans. Nova Scotian Inst. Sci.* **1919**, *15*, 57–61; *Chem. Abstr.* **1922**, *16*, 17386.
21. Miller, I. Dyestuff manufacture (methyl violet B). *Color Trade J.* **1918**, *2*, 18–23.
22. Yazdankhah, S. P.; Gogstad, G. O.; Sorum, R. H.; Larsen, H. J. S. Diagnostic analytical method and kit containing bacteria-retaining substrate and dye for detecting bacteria. PCT Int. Appl. WO 2001049872, 2001; *Chem. Abstr.* **2001**, *135*, 89497.
23. Gelvan, D. J.; Goltsman, L.; Chausovsky, A. Staining methods and compositions for identifying a cell phenotype. PCT Int. Appl. WO 2007102146, 2007; *Chem. Abstr.* **2007**, *147*, 339010.
24. Moen, V.; Nordtug, T.; Zahlsen, K. Mixture and procedure for marking fish. Norw. NO 317169, 2004; *Chem. Abstr.* **2007**, *146*, 141497.
25. Sun, W.; Jiao, K.; Ding, Y. Q. Electrochemical studies on the binding reaction of methyl violet with heparin and its analytical application. *Huaxue Xuebao* **2006**, *64*, 397–402; *Chem. Abstr.* **2006**, *144*, 440236.
26. Yu, J.; Song, G.; Li, R. Nematicide granule containing fosthiazate and abamectin. Faming Zhuanli Shenqing Gongkai Shuomingshu CN 101180969, 2008; *Chem. Abstr.* **2008**, *149*, 47034.
27. Zhang, W. J.; Xu, H. P.; Xue, C. X.; Chen, X. G.; De Hu, Z. Resonance light scattering for the determination of nucleic acids with methyl violet. *Anal. Lett.* **2001**, *34*, 553–568.
28. Choi, J. K. Detection method of proteins on polyacrylamide gels using a dye composition as a silver ion sensitizing agent. U.S. Pat. Appl. Publ. US 2004023401, 2004; *Chem. Abstr.* **2004**, *140*, 160128.
29. Levine, F. Test kit for the determination of proteins in body fluids, especially in urine. Ger. Offen. DE 19858949, 2000; *Chem. Abstr.* **2000**, *133*, 40205.
30. Yerazunis, W. S.; Lau, K. T.; Shepherd, R. L.; Crowley, K. H. Semi-permanent skin adhering device for detecting biological conditions. U.S. Pat. Appl. Publ. US 2008069727, 2008; *Chem. Abstr.* **2008**, *148*, 350489.
31. Zhang, L.; Liu, J. Manufacture of antiphlogistic and analgesic skin medicine for painless injection and cleaning wound. Faming Zhuanli Shenqing Gongkai Shuomingshu CN 1628848, 2005; *Chem. Abstr.* **2006**, *144*, 198760.
32. Guggenberger, R.; Luchterhandt, T.; Haeberlein, I.; Kappler, O.; Hansen, M. Dental composition for detection of carious tissue, detection method. PCT Int. Appl. WO 2006125650, 2006; *Chem. Abstr.* **2006**, *146*, 33158.
33. Sugitachi, A.; Wakabayashi, T. Viscous tissue markers containing chitosan for endoscopic surgery, surgery, or animal experiments. Jpn. Kokai Tokkyo Koho JP 2007262062, 2007; *Chem. Abstr.* **2007**, *147*, 422441.
34. Plos, G.; Guerin, F. Process for fast coloring and decoloring human keratinic fibers with direct dyes. Fr. Demande FR 2845906, 2004; *Chem. Abstr.* **2004**, *140*, 326616.
35. Trigg, D. L.; Jones, J. L. Multistep hair color revitalizing kit comprising hair dye and glossing agent. U.S. Pat. Appl. Publ. US 2008189876, 2008; *Chem. Abstr.* **2008**, *149*, 251967.

36. Hoeffkes, H.; Gross, W. Hair dyes containing silylcarbonates and/or sylilcarbamates for simultaneous dyeing and bleaching of hair. Ger. Offen. DE 102004052060, 2006; *Chem. Abstr.* **2006**, *144*, 419031.

37. Vennerstrom, J. L.; Makler, M. T.; Angerhofer, C. K.; Williams, J. A. Antimalarial dyes revisited: xanthenes, azines, oxazines, and thiazines. *Antimicrob. Agents Chemother.* **1995**, *39*, 2671–2677.

38. Lawrence, D. S.; Sharma, V.; Agnes, R. S. Deeply quenched enzyme sensors and protein–protein binding sensors. PCT Int. Appl. WO 2008070152, 2008; *Chem. Abstr.* **2008**, *149*, 26898.

39. Geier, A.; Kanety, H.; Karasik, A. Biologically active synthetic dye compounds mimicking the physiological activity of insulin or insulin-like growth factor, and their therapeutic use. U.S. Patent 6020374, 2000; *Chem. Abstr.* **2000**, *132*, 117549.

40. Guo, J. Medicinal preparation for treating ringworm. Faming Zhuanli Shenqing Gongkai Shuomingshu CN 1117381, 1996; *Chem. Abstr.* **1999**, *131*, 175053.

41. Melker, R. J. Materials and methods for creating customized agrochemical, pesticide or topical compositions having a temporary visual indicator. U.S. Pat. Appl. Publ. US 2008227856, 2008; *Chem. Abstr.* **2008**, *149*, 349309.

42. Roballey, T.; Feldman, N.; Schwibner, B. H. Cosmetic and reconstructive prosthesis containing a biologically compatible rupture indicator. U.S. Pat. Appl. Publ. US 2005149186, 2005; *Chem. Abstr.* **2005**, *143*, 103364.

43. Cullen, B. M.; Addison, D.; Greenhalgh, D. Bioabsorbable wound dressing containing an antioxidant dye. Brit. UK Pat. Appl. GB 2402882, 2004; *Chem. Abstr.* **2004**, *142*, 62794.

44. Senevirathna, M. K. I.; Pitigala, P. K. D. D. P.; Perera, V. P. S.; Tennakone, K. Molecular rectification: application in dye-sensitized solar cells. *Langmuir* **2005**, *21*, 2997–3001.

45. Ghani, A. J. A.; Al-Abbasi, M. A.; Ziada, B. A. The use of composite dye systems in solar energy applications. *J. Sol. Energy Res.* **1986**, *4*, 31–49.

46. Ogiwara, Y.; Setagawa, Y. Electrophoretic displays showing high contrast with metallic luster, their particles, and solutions therewith. Jpn. Kokai Tokkyo Koho JP 2008152211, 2008; *Chem. Abstr.* **2008**, *149*, 91650.

47. Abramov, V. N.; Aref'ev, N. M.; Kochukov, A. V.; Yakovlev, V. B. Film photoresist for stenciling and method for its manufacture. Russ. RU 2321037, 2008; *Chem. Abstr.* **2008**, *148*, 414240.

48. Vavru, J.; Lesna, V.; Duchackova, M.; Schlosserova, D. Ink exhibiting increased resistance to drying. Czech Rep. CZ 296589, 2006; *Chem. Abstr.* **2006**, *146*, 423667.

49. Yamaka, H.; Shimura, I.; Matsura, H.; Saito, H. Toner for developing electrostatic image. Jpn. Kokai Tokkyo Koho JP 10003183, 1998; *Chem. Abstr.* **1998**, *128*, 147470.

50. Davies-Smith, L.; Sum, V. W. Highlightable and highlighted mixtures, marking instruments, eradicator solution and kits, using the mixtures in highlighting. U.S. Pat. Appl. Publ. US 2005120919, 2005; *Chem. Abstr.* **2005**, *143*, 9331.

51. Zimmermann, D. Multicolor component adhesive system comprising polymerizable compositions, different coloring agents and a curing component. U.S. Pat. Appl. Publ. US 2008300321, 2008; *Chem. Abstr.* **2008**, *150*, 20872.

52. Tomokuni, M.; Adachi, M.; Kitai, M. Adhesive sheet for preparing membrane–electrode assembly for fuel cell. Jpn. Kokai Tokkyo Koho JP 2008243804, 2008; *Chem. Abstr.* **2008**, *149*, 451815.

53. Sadlowski, E. S.; Cummings, M. D. Laundry detergent compositions with efficient hueing dye. U.S. Pat. Appl. Publ. US 2006183658, 2006; *Chem. Abstr.* **2006**, *145*, 232298.

54. Oe, T.; Sugita, Y.; Hamada, M. Packaging materials containing detectors and packages using them. Jpn. Kokai Tokkyo Koho JP 2006273403, 2006; *Chem. Abstr.* **2006**, *145*, 420187.

55. Huang, W.; Xia, X.; Wang, Y.; Xu, C.; Zhu, W.; Qian, L. Preparation of ultrathin electrolytic copper foil for lithium battery. Faming Zhuanli Shenqing Gongkai Shuomingshu CN 1563502, 2005; *Chem. Abstr.* **2005**, *144*, 282223.

56. Talalaev, A. P.; Zinov'ev, V. M.; Sechin, S. N.; Zraichenko, L. I.; Luzina, M. A.; Kolchanova, T. A. Production of cast-molded urethane rubber articles. Russ. RU 2208019, 2003; *Chem. Abstr.* **2003**, *140*, 254837.

57. Oiwa, T. Coloring agent penetrating and reacting with cement product for forming weather- and wear-resistant colored coated film on cement. Jpn. Kokai Tokkyo Koho JP 11199349, 1999; *Chem. Abstr.* **1999**, *131*, 119735.

58. Zhang, L.; Fang, Z.; He, J.; Wu, Y.; Cao, B. Method for printing images on stainless steel. Faming Zhuanli Shenqing Gongkai Shuomingshu CN 1119993, 1996; *Chem. Abstr.* **1999**, *130*, 189430.

59. Smith, M. J.; Desai, B.; Frederico, J. J. Acid-developed water-insoluble azo dyes for marking of commercial petroleum products. U.S. Patent 6514917, 2003; *Chem. Abstr.* **2003**, *138*, 156109.

60. Feigel'son, B. R.; Alatyrtseva, I. N.; Popova, V. N.; Kobzev, Yu. P.; Kulagina, S. B.; Atamanova, M. I. Composition for dyeing natural leather black. U.S.S.R. SU 1174449, 1985; *Chem. Abstr.* **1986**, *104*, 20954.

61. Krmpotic, M. Process for coloring a paper fiber molded article. Aust. Pat. Appl. AU 2003204700, 2005; *Chem. Abstr.* **2006**, *144*, 469919.

62. Kielhorn-bayer, S.; Kuehn, I.; Tresch, R. Aqueous, dye-containing, polymer-based compositions for coloring of paper. Ger. Offen. DE 19937261, 2001; *Chem. Abstr.* **2001**, *134*, 164727.

63. Sugimura, M.; Ichikawa, M. Basic dye compositions for dyeing of textiles. Jpn. Kokai Tokkyo Koho JP 07033999, 1995; *Chem. Abstr.* **1995**, *123*, 172520.

64. Niculescu, S. P.; Atkinson, A.; Hammond, G.; Lewis, M. Using fragment chemistry data mining and probabilistic neural networks in screening chemicals for acute toxicity to the fathead minnow. *SAR QSAR Environ. Res.* **2004**, *15*, 293–309.

65. Nishiuchi, Y. Toxicity of agrochemicals to freshwater organisms. 2. Dyes. *Suisan Zoshoku* **1984**, *32*, 61–64; *Chem. Abstr.* **1986**, *105*, 166384.

66. Vachalkova, A.; Novotny, L.; Blesova, M. Polarographic reduction of some triphenylmethane dyes and their potential carcinogenic activity. *Neoplasma* **1996**, *43*, 113–117.

67. Oros, G.; Cserhati, T.; Forgacs, E. Separation of the strength and selectivity of the microbiological effect of synthetic dyes by spectral mapping technique. *Chemosphere* **2003**, *52*, 185–193.

68. Michaels, G. B.; Lewis, D. L. Sorption and toxicity of azo and triphenylmethane dyes to aquatic microbial populations. *Environ. Toxicol. Chem.* **1985**, *4*, 45–50.

69. Zimina, T. A.; Pavlenko, V. V. Toxic and mutagenic effects of some dyes. *Genetika (Moscow)* **1990**, *26*, 2246–2249; *Chem. Abstr.* **1991**, *115*, 2934.

70. Chung, K. T.; Fulk, G. E.; Andrews, A. W. Mutagenicity testing of some commonly used dyes. *Appl. Environ. Microbiol.* **1981**, *42*, 641–648.

MONOBROMOBIMANE (mBBr)

CAS Registry Number 71418-44-5

Chemical Structure

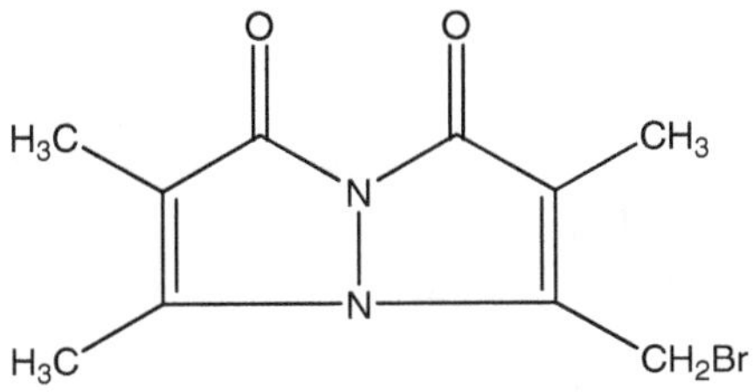

CA Index Name 1*H*,7*H*-Pyrazolo[1,2-*a*]pyrazole-1,7-dione, 3-(bromomethyl)-2,5,6-trimethyl-

Other Names Bromobimane; Monobromobimane; NSC 608544; Thiolyte MB

Merck Index Number Not listed

Chemical/Dye Class Pyrazole

Molecular Formula $C_{10}H_{11}BrN_2O_2$

Molecular Weight 271.11

Physical Form Yellow powder

Solubility Soluble in acetonitrile, dichloromethane, methanol, *N*,*N*-dimethyl formamide, dimethyl sulfoxide

Melting Point 160–161 °C, 152–154 °C

Boiling Point (Calcd.) 327.8 ± 44.0 °C, pressure: 760 Torr

pK_a (Calcd.) −3.57 ± 0.70, most basic, temperature: 25 °C

Absorption (λ_{max}) 398 nm

Emission (λ_{max}) 490 nm

Synthesis Synthetic method[1–11]

Staining Applications Cystine;[12,13] cysteine;[14–16] cysteine-17;[17] cysteine-111;[17] captopril;[18] erythrocytes;[19] mutants;[20] neurons;[21] nucleic acids;[22] proteins;[23–28] peptides;[28] sulfhydryls[29,30]

Biological Applications Glutathione S-transferase substrates;[31] detecting glutathione S-transferase,[32–38] thiols,[39–42] sulfite,[43] homocysteine,[44] mycothiol,[45] sulfur compounds;[45] thiol-reactive probes[11,39–42]

Industrial Applications Photographic materials[46]

Safety/Toxicity Hepatotoxicity;[47] mutagenicity;[48] reproductive toxicity[49]

REFERENCES

1. Chinn, P. C.; Pigiet, V.; Fahey, R. C. Determination of thiol proteins using monobromobimane labeling and high-performance liquid chromatographic analysis: application to *Escherichia coli* thioredoxin. *Anal. Biochem.* **1986**, *159*, 143–149.
2. Kobayashi, M.; Sugimoto, N. Labeling of polydeoxyribonucleotides or polyribonucleotides with fluorescent substances for use as analysis reagents. Jpn. Kokai Tokkyo Koho JP 61044352, 1986; *Chem. Abstr.* **1986**, *105*, 21305.
3. Kosower, E. M.; Pazhenchevsky, B.; Dodiuk, H.; Kanety, H.; Faust, D. Bimanes. 6. Reactive halogen derivatives of syn- and anti-1,5-diazabicyclo[3.3.0] octadienediones (9,10-dioxabimanes). *J. Org. Chem.* **1981**, *46*, 1666–1673.
4. Kosower, E. M.; Pazhenchevsky, B. Bimanes. 5. Synthesis and properties of syn- and anti-1,5-diazabicyclo[3.3.0]octadienediones (9,10-dioxabimanes). *J. Am. Chem. Soc.* **1980**, *102*, 4983–4993.
5. Perez-Rama, M.; Torres, V. E.; Abalde, A. J. Capillary zone electrophoresis for analysis of phytochelatins and other thiol peptides in complex biological samples derivatized with monobromobimane. *Electrophoresis* **2005**, *26*, 610–620.
6. Berggren, K. N.; Chernokalskaya, E.; Lopez, M. F.; Beechem, J. M.; Patton, W. F. Comparison of three different fluorescent visualization strategies for detecting *Escherichia coli* ATP synthase subunits after sodium dodecyl sulfate-polyacrylamide gel electrophoresis. *Proteomics* **2001**, *1*, 54–65.
7. Sneller, F. E.; van Heerwaarden, L. M.; Koevoets, P. L.; Vooijs, R.; Schat, H.; Verkleij, J. A. Derivatization of phytochelatins from *Silene vulgaris*, induced upon exposure to arsenate and cadmium: comparison of derivatization with Ellman's reagent and monobromobimane. *J. Agric. Food Chem.* **2000**, *48*, 4014–4019.
8. Hu, L.; Colman, R. F. Monobromobimane as an affinity label of the xenobiotic binding site of rat glutathione S-transferase 3-3. *J. Biol. Chem.* **1995**, *270*, 21875–21883.
9. Krauth-Siegel, R. L.; Jacoby, E. M.; Schirmer, R. H. Trypanothione and N1-glutathionylspermidine: isolation and determination. *Methods Enzymol.* **1995**, *251*, 287–294.
10. Vogel, F.; Lumper, L. Fluorescence labelling of NADPH-cytochrome P-450 reductase with the

monobromomethyl derivative of syn-9,10-dioxabimane. *Biochem. J.* **1983**, *215*, 159–166.

11. Newton, G. L.; Dorian, R.; Fahey, R. C. Analysis of biological thiols: derivatization with monobromobimane and separation by reverse-phase high-performance liquid chromatography. *Anal. Biochem.* **1981**, *114*, 383–387.
12. Pastore, A.; Lo Russo, A.; Greco, M.; Rizzoni, G.; Federici, G. Semiautomated method for determination of cystine concentration in polymorphonuclear leukocytes. *Clin. Chem.* **2000**, *46*, 574–576.
13. De Graff-Hess, A.; Trijbels, F.; Blom, H. New method for determining cystine in leukocytes and fibroblasts. *Clin. Chem.* **1999**, *45*, 2224–2228.
14. Petrotchenko, E. V.; Pasek, D.; Elms, P.; Dokholyan, N. V.; Meissner, G.; Borchers, C. H. Combining fluorescence detection and mass spectrometric analysis for comprehensive and quantitative analysis of redox-sensitive cysteines in native membrane proteins. *Anal. Chem.* **2006**, *78*, 7959–7966.
15. Kim, Y. J.; Pannell, L. K.; Sackett, D. L. Mass spectrometric measurement of differential reactivity of cysteine to localize protein–ligand binding sites. *Anal. Biochem.* **2004**, *332*, 376–383.
16. Catrina, S. B.; Coculescu, M.; Andersson, M. A chemical method to isolate hypothalamic nonapeptides by coupling cysteine with bimane. *J. Cell. Mol. Med.* **2001**, *5*, 195–197.
17. Hu, L.; Colman, R. F. Resonance energy transfer between sites in rat liver glutathione S-transferase, 1-1, selectively modified at cysteine-17 and cysteine-111. *Biochemistry* **1997**, *36*, 1635–1645.
18. Kok, R. J.; Visser, J.; Moolenaar, F.; de Zeeuw, D.; Meijer, D. K. F. Bioanalysis of captopril: two sensitive high-performance liquid chromatographic methods with pre- or postcolumn fluorescent labeling. *J. Chromatogr., B* **1997**, *693*, 181–189.
19. McNamara, K. P.; Yeung, E. S.; Rosenzweig, N.; Rosenzweig, Z. Dynamic analytical chemistry: a kinetic study of the labeling of normal and age fractionated human erythrocytes with monobromobimane. *Anal. Chim. Acta* **1997**, *356*, 75–83.
20. Mansoor, S. E.; Mchaourab, H. S.; Farrens, D. L. Determination of protein secondary structure and solvent accessibility using site-directed fluorescence labeling: studies of T4 lysozyme using the fluorescent probe monobromobimane. *Biochemistry* **1999**, *38*, 16383–16393.
21. Abe, T.; Takagi, N.; Nakano, M.; Takeo, S. The effects of monobromobimane on neuronal cell death in the hippocampus after transient global cerebral ischemia in rats. *Neurosci. Lett.* **2004**, *357*, 227–231.
22. Brush, C. K.; Gupta, V. Post-synthesis labeling of nucleic acids and uses thereof. PCT Int. Appl. WO 2003052115, 2003; *Chem. Abstr.* **2003**, *139*, 48134.
23. Alvarez, S.; Wilson, G. H.; Chen, S. Determination of *in vivo* disulfide-bonded proteins in *Arabidopsis*. *J. Chromatogr., B* **2009**, *877*, 101–104.
24. Momma, M.; Seki, T.; Hajika, M. Analysis of disulfide bonds in soybean proteins using fluorescence labeling by monobromobimane (mBBr). *Nippon Shokuhin Kagaku Kogaku Kaishi* **2004**, *51*, 406–412; *Chem. Abstr.* **2005**, *143*, 345619.
25. Herick, K.; Jackson, P.; Wersch, G.; Burkovski, A. Detection of fluorescence dye-labeled proteins in 2-D gels using an Arthur 1442 multi-wavelength fluoroimager. *BioTechniques* **2001**, *31*, 146–149.
26. Berggren, K.; Chernokalskaya, E.; Steinberg, T. H.; Kemper, C.; Lopez, M. F.; Diwu, Z.; Haugland, R. P.; Patton, W. F. Background-free, high sensitivity staining of proteins in one- and two-dimensional sodium dodecyl sulfate-polyacrylamide gels using a luminescent ruthenium complex. *Electrophoresis* **2000**, *21*, 2509–2521.
27. O'Keefe, D. O. Quantitative electrophoretic analysis of proteins labeled with monobromobimane. *Anal. Biochem.* **1994**, *222*, 86–94.
28. Nordheim, A.; Cahill, M. Labelling of peptides and proteins in electrophoresis. Eur. Pat. Appl. EP 1059531, 2000; *Chem. Abstr.* **2000**, *134*, 27289.
29. Wong, J. H.; Yano, H.; Lee, Y. M.; Cho, M. J.; Buchanan, B. B. Identification of thioredoxin-linked proteins by fluorescence labeling combined with isoelectric focusing/sodium dodecyl sulfate-polyacrylamide gel electrophoresis. *Methods Enzymol.* **2002**, *347*, 339–349.
30. Bennett, T. A.; Edwards, B. S.; Sklar, L. A.; Rogelj, S. Sulfhydryl regulation of L-selectin shedding: phenylarsine oxide promotes activation-independent L-selectin shedding from leukocytes. *J. Immunol.* **2000**, *164*, 4120–4129.
31. Fujikawa, Y.; Urano, Y.; Komatsu, T.; Hanaoka, K.; Kojima, H.; Terai, T.; Inoue, H.; Nagano, T. Design and synthesis of highly sensitive fluorogenic substrates for glutathione S-transferase and application for activity imaging in living cells. *J. Am. Chem. Soc.* **2008**, *130*, 14533–14543.
32. Stepigova, J.; Vrablikova, H.; Lang, J.; Vecerova, K.; Bartak, M. Glutathione and zeaxanthin formation

during high light stress in foliose lichens. *Plant, Soil Environ.* **2007**, *53*, 340–344.

33. Sebastia, J.; Cristofol, R.; Martin, M.; Rodriguez-Farre, E.; Sanfeliu, C. Evaluation of fluorescent dyes for measuring intracellular glutathione content in primary cultures of human neurons and neuroblastoma SH-SY5Y. *Cytometry* **2003**, *51A*, 16–25.
34. Anderson, M. T.; Trudell, J. R.; Voehringer, D. W.; Tjioe, I. M.; Herzenberg, L. A.; Herzenberg, L. A. An improved monobromobimane assay for glutathione utilizing tris-(2-carboxyethyl)phosphine as the reductant. *Anal. Biochem.* **1999**, *272*, 107–109.
35. Hogarth, L. A.; Rabello, C. M. A.; Hall, A. G. Measurement of reduced glutathione using high-pressure liquid chromatography. *Methods Mol. Med.* **1999**, *28*, 91–94.
36. Keij, J. F.; Bell-Prince, C.; Steinkamp, J. A. Simultaneous analysis of relative DNA and glutathione content in viable cells by phase-resolved flow cytometry. *Cytometry* **1999**, *35*, 48–54.
37. Hedley, D.; Chow, S. Glutathione and cellular resistance to anti-cancer drugs. *Methods Cell Biol.* **1994**, *42*, 31–44.
38. Hedley, D. W.; Chow, S. Evaluation of methods for measuring cellular glutathione content using flow cytometry. *Cytometry* **1994**, *15*, 349–358.
39. Patsoukis, N.; Georgiou, C. D. Fluorometric determination of thiol redox state. *Anal. Bioanal. Chem.* **2005**, *383*, 923–929.
40. Horan, A. D.; Chan, C. Y.; Pletcher, C. H.; Menon, C.; Evans, S. M.; Moore, J. S.; Koch, C. J. Analysis of tumor thiol concentrations: comparison of flow cytometric with chemical and biochemical techniques. *Cytometry* **1997**, *29*, 76–82.
41. Newton, G. L.; Fahey, R. C. Determination of biothiols by bromobimane labeling and high-performance liquid chromatography. *Methods Enzymol.* **1995**, *251*, 148–166.
42. Kosower, E. M.; Kosower, N. S. Bromobimane probes for thiols. *Methods Enzymol.* **1995**, *251*, 133–148.
43. Ji, A. J.; Savon, S. R.; Jacobsen, D. W. Determination of total serum sulfite by HPLC with fluorescence detection. *Clin. Chem.* **1995**, *41*, 897–903.
44. Melnikov, I. O.; Nazimov, I. V.; Stukacheva, E. A.; Glubokov, Yu. M. Determination of homocysteine and other low-molecular-weight amino thiols in blood plasma. *J. Anal. Chem.* **2006**, *61*, 1093–1099.
45. Rawat, M.; Uppal, M.; Newton, G.; Steffek, M.; Fahey, R. C.; Av-Gay, Y. Targeted mutagenesis of the *Mycobacterium smegmatis* mca gene, encoding a mycothiol-dependent detoxification protein. *J. Bacteriol.* **2004**, *186*, 6050–6058.
46. Kita, H.; Kaneko, Y.; Ishige, O. Silver halide color photographic material. Eur. Pat. Appl. EP 448241, 1991; *Chem. Abstr.* **1992**, *116*, 245153.
47. Gupta, S.; Rogers, L. K.; Taylor, S. K.; Smith, C. V. Inhibition of carbamyl phosphate synthetase-I and glutamine synthetase by hepatotoxic doses of acetaminophen in mice. *Toxicol. Appl. Pharmacol.* **1997**, *146*, 317–327.
48. Olive, P. L. Patterns of mutagen binding and penetration in multicell spheroids. *Environ. Mutagen.* **1986**, *8*, 705–715.
49. Czarny, N. A.; Mate, K. E.; Rodger, J. C. Acrosome stability in the spermatozoa of dasyurid marsupials. *Reprod., Fertil., Dev.* **2008**, *20*, 295–302.

MQAE

CAS Registry Number 162558-52-3

Chemical Structure

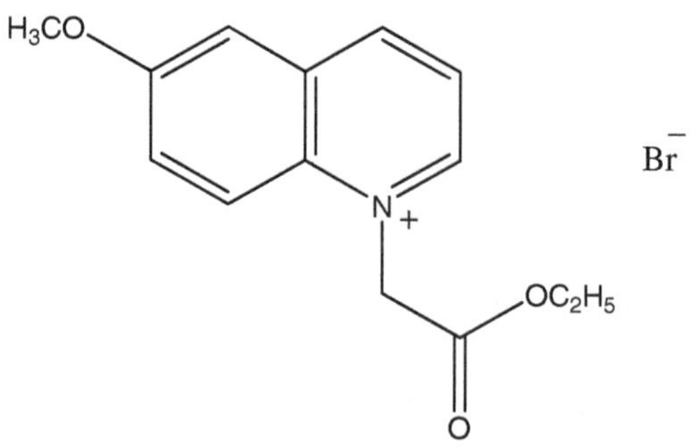

CA Index Name Quinolinium, 1-(2-ethoxy-2-oxoethyl)-6-methoxy-, bromide (1:1)

Other Names 1-(Ethoxycarbonylmethyl)-6-methoxyquinolinium bromide; (6-Methoxyquinolinio)acetic acid ethyl ester bromide; MQAE; Quinolinium, 1-(2-ethoxy-2-oxoethyl)-6-methoxy-, bromide

Merck Index Number Not listed

Chemical/Dye Class Quinoline

Molecular Formula $C_{14}H_{16}BrNO_3$

Molecular Weight 326.19

Physical Form White powder

Solubility Soluble in water, methanol, dimethyl sulfoxide

Melting Point 177–179 °C

Absorption (λ_{max}) 350 nm

Emission (λ_{max}) 460 nm

Synthesis Synthetic method[1]

Staining Applications Chloride ions[1–6]

Biological Applications Chloride indicator;[1–6] diagnosis of diseases caused by elemental imbalances;[7] detecting cancer cells,[8] spores,[8] stress biomarkers[9]

Industrial Applications Pulp fiber products[10]

Safety/Toxicity Reproductive toxicity[11]

REFERENCES

1. Verkman, A. S.; Sellers, M. C.; Chao, A. C.; Leung, T.; Ketcham, R. Synthesis and characterization of improved chloride-sensitive fluorescent indicators for biological applications. *Anal. Biochem.* **1989**, *178*, 355–361.
2. Miyazaki, H.; Shiozaki, A.; Niisato, N.; Marunaka, Y. Physiological significance of hypotonicity-induced regulatory volume decrease: reduction in intracellular Cl^- concentration acting as an intracellular signaling. *Am. J. Physiol.* **2007**, *292*, F1411–F1417.
3. Andersson, C.; Roomans, G. M. Determination of chloride efflux by X-ray microanalysis versus MQAE-fluorescence. *Microsc. Res. Tech.* **2002**, *59*, 531–535.
4. Kaneko, H.; Nakamura, T.; Lindemann, B. Noninvasive measurement of chloride concentration in rat olfactory receptor cells with use of a fluorescent dye. *Am. J. Physiol.* **2001**, *280*, C1387–C1393.
5. Yagodin, S.; Hoyland, J.; Mason, W. T.; Miyake, T.; Sattelle, D. B. Imaging of intracellular Ca^{2+}, pH and Cl^- transients in *Drosophila* cell lines. *Bioimages* **1997**, *5*, 111–118.
6. Koncz, C.; Daugirdas, J. T. Use of MQAE for measurement of intracellular [Cl^-] in cultured aortic smooth muscle cells. *Am. J. Physiol.* **1994**, *267*, H2114–H2123.
7. Rupp, M. E. Determination of elements in body fluids and test kit including the necessary reagents for diagnosis of diseases caused by elemental imbalances. U.S. Pat. Appl. Publ. US 2003203495, 2003; *Chem. Abstr.* **2003**, *139*, 347692.
8. Levy, M.; Ellington, A. D.; Pai, S. Proximity ligation assays with peptide conjugate 'burrs' and aptamers for the sensitive detection of spores and cancer cells. PCT Int. Appl. WO 2007027796, 2007; *Chem. Abstr.* **2007**, *146*, 312239.
9. Cullum, M. E.; Duplessis, C. A.; Crepeau, L. J. Method for the detection of stress biomarkers including cortisol by fluorescence polarization. U. S. Pat. Appl. Publ. US 2006105397, 2006; *Chem. Abstr.* **2006**, *144*, 463810.
10. Mathews, J.; Gustafson, R.; Hodgson, K. A method to determine the charge demand of single pulp fibers. *Nordic Pulp Paper Res. J.* **2004**, *19*, 453–459.
11. Phillips, K. P.; Zhou, W. L.; Baltz, J. M. Fluorophore toxicity in mouse eggs and zygotes. *Zygote* **1998**, *6*, 113–123.

NBD C_6-CERAMIDE

CAS Registry Number 86701-10-2

Chemical Structure

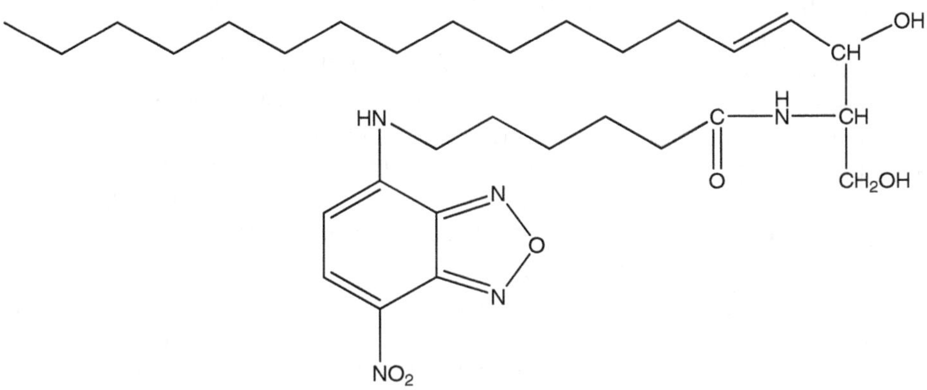

Solubility Soluble in chloroform, methanol, dimethyl sulfoxide

Melting Point >200 °C

pK_a (Calcd.) 13.54 ± 0.20, most acidic, temperature: 25 °C; −0.78 ± 0.70, most basic, temperature: 25 °C

CA Index Name Hexanamide, *N*-[2-hydroxy-1-(hydroxymethyl)-3-heptadecen-1-yl]-6-[(7-nitro-2,1,3-benzoxadiazol-4-yl)amino]-

Other Names Hexanamide, *N*-[2-hydroxy-1-(hydroxymethyl)-3-heptadecenyl]-6-[(7-nitro-2,1,3-benzoxadiazol-4-yl)amino]-; Hexanamide, *N*-[2-hydroxy-1-(hydroxymethyl)-3-heptadecenyl]-6-[(7-nitro-4-benzofurazanyl)amino]-; 2,1,3-Benzoxadiazole, hexanamide derivative; *N*-(NBD-aminohexanoyl)sphingosine; NBD C_6-ceramide

Merck Index Number Not listed

Chemical/Dye Class Benzoxadiazole

Molecular Formula $C_{30}H_{49}N_5O_6$

Molecular Weight 575.74

Physical Form Orange powder

Absorption (λ_{max}) 466 nm

Emission (λ_{max}) 536 nm

Synthesis Synthetic methods[1–12]

Staining Applications Golgi apparatus;[9–11,13–27] lipids;[8,28–32] lipopolysaccharides;[33] lipoproteins;[34–36] sphingolipids;[37–39] vacuoles[40]

Biological Applications Analyzing lipoproteins;[34–36] lipid metabolism;[8,28–32] sphingolipid metabolism;[37] sphingolipid transport;[38] drug screening assay;[41] inositol phosphatidyl ceramide (IPC) synthase assay;[42] sphingomyelinase activity assay;[43] treating fungal infections[44]

Industrial Applications Semiconductors[45]

Safety/Toxicity Fungal toxicity;[44] tumor necrosis[5]

REFERENCES

1. Allan, D. Lipid metabolic changes caused by short-chain ceramides and the connection with apoptosis. *Biochem. J.* **2000**, *345*, 603–610.
2. Fischl, A. S.; Liu, Y.; Browdy, A.; Cremesti, A. E. Inositolphosphoryl ceramide synthase from yeast. *Methods Enzymol.* **2000**, *311*, 123–130.
3. Nikolova-Karakashian, M. Assays for the biosynthesis of sphingomyelin and ceramide phosphoethanolamine. *Methods Enzymol.* **2000**, *311*, 31–42.
4. Ledesma, M. D.; Brugger, B.; Bunning, C.; Wieland, F. T.; Dotti, C. G. Maturation of the axonal plasma membrane requires upregulation of sphingomyelin synthesis and formation of protein–lipid complexes. *EMBO J.* **1999**, *18*, 1761–1771.
5. Bourteele, S.; Hausser, A.; Doppler, H.; Horn-Muller, J.; Ropke, C.; Schwarzmann, G.; Pfizenmaier, K.; Muller, G. Tumor necrosis factor induces ceramide oscillations and negatively controls sphingolipid

synthases by caspases in apoptotic Kym-1 cells. *J. Biol. Chem.* **1998**, *273*, 31245–31251.

6. Boldin, S.; Futerman, A. H. Glucosylceramide synthesis is required for basic fibroblast growth factor and laminin to stimulate axonal growth. *J. Neurochem.* **1997**, *68*, 882–885.
7. Sando, G. N.; Howard, E. J.; Madison, K. C. Induction of ceramide glucosyltransferase activity in cultured human keratinocytes: correlation with culture differentiation. *J. Biol. Chem.* **1996**, *271*, 22044–22051.
8. Hackstadt, T.; Scidmore, M. A.; Rockey, D. D. Lipid metabolism in *Chlamydia trachomatis*-infected cells: directed trafficking of Golgi-derived sphingolipids to the chlamydial inclusion. *Proc. Natl. Acad. Sci. U.S.A* **1995**, *92*, 4877–4881.
9. Pagano, R. E. The Golgi apparatus: insights from lipid biochemistry. *Biochem. Soc. Trans.* **1990**, *18*, 361–366.
10. Pagano, R. E. A fluorescent derivative of ceramide: physical properties and use in studying the Golgi apparatus of animal cells. *Methods Cell Biol.* **1989**, *29*, 75–85.
11. Lipsky, N. G.; Pagano, R. E. A vital stain for the Golgi apparatus. *Science* **1985**, *228*, 745–747.
12. Lipsky, N. G.; Pagano, R. E. Sphingolipid metabolism in cultured fibroblasts: microscopic and biochemical studies employing a fluorescent ceramide analogue. *Proc. Natl. Acad. Sci. U.S.A* **1983**, *80*, 2608–2612.
13. Kaufmann, A. M.; Toro-Ramos, A. J.; Krise, J. P. Assessment of Golgi apparatus versus plasma membrane-localized multi-drug resistance-associated protein 1. *Mol. Pharm.* **2008**, *5*, 787–794.
14. Maier, O.; Hoekstra, D. Trans-Golgi network and subapical compartment of HepG2 cells display different properties in sorting and exiting of sphingolipids. *J. Biol. Chem.* **2003**, *278*, 164–173.
15. Sandoval, R.; Leiser, J.; Molitoris, B. A. Aminoglycoside antibiotics traffic to the Golgi complex in LLC-PK1 cells. *J. Am. Soc. Nephrol.* **1998**, *9*, 167–174.
16. Hackstadt, T.; Rockey, D. D.; Heinzen, R. A.; Scidmore, M. A. *Chlamydia trachomatis* interrupts an exocytic pathway to acquire endogenously synthesized sphingomyelin in transit from the Golgi apparatus to the plasma membrane. *EMBO J.* **1996**, *15*, 964–977.
17. Seksek, O.; Biwersi, J.; Verkman, A. S. Direct measurement of trans-Golgi pH in living cells and regulation by second messengers. *J. Biol. Chem.* **1995**, *270*, 4967–4970.
18. Lujan, H. D.; Marotta, A.; Mowatt, M. R.; Sciaky, N.; Lippincott-Schwartz, J.; Nash, T. E. Developmental induction of Golgi structure and function in the primitive eukaryote *Giardia lamblia*. *J. Biol. Chem.* **1995**, *270*, 4612–4618.
19. Wu, Y. N.; Gadina, M.; Tao-Cheng, J. H.; Youle, R. J. Retinoic acid disrupts the Golgi apparatus and increases the cytosolic routing of specific protein toxins. *J. Cell Biol.* **1994**, *125*, 743–753.
20. de Vries, H.; Schrage, C.; Hoekstra, K.; Kok, J. W.; van der Haar, M. E.; Kalicharan, D.; Liem, R. S.; Copray, J. C.; Hoekstra, D. Outstations of the Golgi complex are present in the processes of cultured rat oligodendrocytes. *J. Neurosci. Res.* **1993**, *36*, 336–343.
21. Martin, O. C.; Comly, M. E.; Blanchette-Mackie, E. J.; Pentchev, P. G.; Pagano, R. E. Cholesterol deprivation affects the fluorescence properties of a ceramide analog at the Golgi apparatus of living cells. *Proc. Natl. Acad. Sci. U.S.A* **1993**, *90*, 2661–2665.
22. Johnson, K. J.; Boekelheide, K. Visualization of Golgi complexes and spermatogonial cohorts of viable, intact seminiferous tubules. *J. Histochem. Cytochem.* **1993**, *41*, 299–306.
23. Robenek, H.; Schmitz, G. Abnormal processing of Golgi elements and lysosomes in Tangier disease. *Arterioscler. Thromb.* **1991**, *11*, 1007–1020.
24. Chandra, S.; Kable, E. P.; Morrison, G. H.; Webb, W. W. Calcium sequestration in the Golgi apparatus of cultured mammalian cells revealed by laser scanning confocal microscopy and ion microscopy. *J. Cell Sci.* **1991**, *100*, 747–752.
25. Pagano, R. E.; Martin, O. C.; Kang, H. C.; Haugland, R. P. A novel fluorescent ceramide analogue for studying membrane traffic in animal cells: accumulation at the Golgi apparatus results in altered spectral properties of the sphingolipid precursor. *J. Cell Biol.* **1991**, *113*, 1267–1279.
26. Lippincott-Schwartz, J.; Glickman, J.; Donaldson, J. G.; Robbins, J.; Kreis, T. E.; Seamon, K. B.; Sheetz, M. P.; Klausner, R. D. Forskolin inhibits and reverses the effects of brefeldin A on Golgi morphology by a cAMP-independent mechanism. *J. Cell Biol.* **1991**, *112*, 567–577.
27. Ho, W. C.; Allan, V. J.; van Meer, G.; Berger, E. G.; Kreis, T. E. Reclustering of scattered Golgi elements occurs along microtubules. *Eur. J. Cell Biol.* **1989**, *48*, 250–263.
28. Alzhanov, D. T.; Suchland, R. J.; Bakke, A. C.; Stamm, W. E.; Rockey, D. D. Clonal isolation of chlamydia-infected cells using flow cytometry. *J. Microbiol. Methods* **2007**, *68*, 201–208.

29. Togo, T. Disruption of the plasma membrane stimulates rearrangement of microtubules and lipid traffic toward the wound site. *J. Cell Sci.* **2006**, *119*, 2780–2786.

30. Makino, A.; Ishii, K.; Murate, M.; Hayakawa, T.; Suzuki, Y.; Suzuki, M.; Ito, K.; Fujisawa, T.; Matsuo, H.; Ishitsuka, R.; Kobayashi, T. D-threo-1-phenyl-2-decanoylamino-3-morpholino-1-propanol alters cellular cholesterol homeostasis by modulating the endosome lipid domains. *Biochemistry* **2006**, *45*, 4530–4541.

31. Kuerschner, L.; Ejsing, C. S.; Ekroos, K.; Shevchenko, A.; Anderson, K. I.; Thiele, C. Polyene-lipids: a new tool to image lipids. *Nat. Methods* **2005**, *2*, 39–45.

32. Moffat, D.; Kusel, J. R. Fluorescent lipid uptake and transport in adult *Schistosoma mansoni. Parasitology* **1992**, *105*, 81–89.

33. Zimmermann, C.; Ginis, I.; Furuya, K.; Klimanis, D.; Ruetzler, C.; Spatz, M.; Hallenbeck, J. M. Lipopolysaccharide-induced ischemic tolerance is associated with increased levels of ceramide in brain and in plasma. *Brain Res.* **2001**, *895*, 59–65.

34. Troup, J. M. Method for analyzing blood for lipoprotein components. U.S. Pat. Appl. Publ. US 2008038762, 2008; *Chem. Abstr.* **2008**, *148*, 209664.

35. Troup, J. M. Method for analyzing blood for lipoprotein components. U.S. Pat. Appl. Publ. US 2008038763, 2008; *Chem. Abstr.* **2008**, *148*, 209663.

36. Boyanovsky, B.; Karakashian, A.; King, K.; Giltiay, N.; Nikolova-Karakashian, M. Uptake and metabolism of low density lipoproteins with elevated ceramide content by human microvascular endothelial cells: implications for the regulation of apoptosis. *J. Biol. Chem.* **2003**, *278*, 26992–26999.

37. Milis, D. G.; Moore, M. K.; Atshaves, B. P.; Schroeder, F.; Jefferson, J. R. Sterol carrier protein-2 expression alters sphingolipid metabolism in transfected mouse L-cell fibroblasts. *Mol. Cell. Biochem.* **2006**, *283*, 57–66.

38. Zegers, M. M.; Hoekstra, D. Sphingolipid transport to the apical plasma membrane domain in human hepatoma cells is controlled by PKC and PKA activity: a correlation with cell polarity in HepG2 cells. *J. Cell Biol.* **1997**, *138*, 307–321.

39. Babia, T.; Kok, J. W.; Hulstaert, C.; de Weerd, H.; Hoekstra, D. Differential metabolism and trafficking of sphingolipids in differentiated versus undifferentiated HT29 cells. *Int. J. Cancer* **1993**, *54*, 839–845.

40. Voronina, S. G.; Sherwood, M. W.; Gerasimenko, O. V.; Petersen, O. H.; Tepikin, A. V. Visualizing formation and dynamics of vacuoles in living cells using contrasting dextran-bound indicator: endocytic and nonendocytic vacuoles. *Am. J. Physiol.* **2007**, *293*, G1333-G1338.

41. Dykens, J. A.; Velicelebi, G.; Ghosh, S. S. Compositions and methods for assaying subcellular conditions and processes using energy transfer for drug screening. PCT Int. Appl. WO 2000079274, 2000; *Chem. Abstr.* **2000**, *134*, 66121.

42. Aeed, P. A.; Sperry, A. E.; Young, C. L.; Nagiec, M. M.; Elhammer, A. P. Effect of membrane perturbants on the activity and phase distribution of inositol phosphorylceramide synthase; development of a novel assay. *Biochemistry* **2004**, *43*, 8483–8493.

43. Loidl, A.; Claus, R.; Deigner, H. P.; Hermetter, A. High-precision fluorescence assay for sphingomyelinase activity of isolated enzymes and cell lysates. *J. Lipid Res.* **2002**, *43*, 815–823.

44. Del Poeta, M.; Luberto, C.; Kechichian, T. Methods for the diagnosis and treatment of fungal infections caused by microorganisms producing glucosylceramide. U.S. Pat. Appl. Publ. US 2008014192, 2008; *Chem. Abstr.* **2008**, *148*, 160094.

45. Alivisatos, A. P.; Larabell, C. A.; Parak, W. J.; Le Gros, M.; Boudreau, R. Semiconductor nanocrystal-based cellular imaging. PCT Int. Appl. WO 2003038401, 2003; *Chem. Abstr.* **2003**, *138*, 334028.

NBD METHYLHYDRAZINE

CAS Registry Number 214147-22-5

Chemical Structure

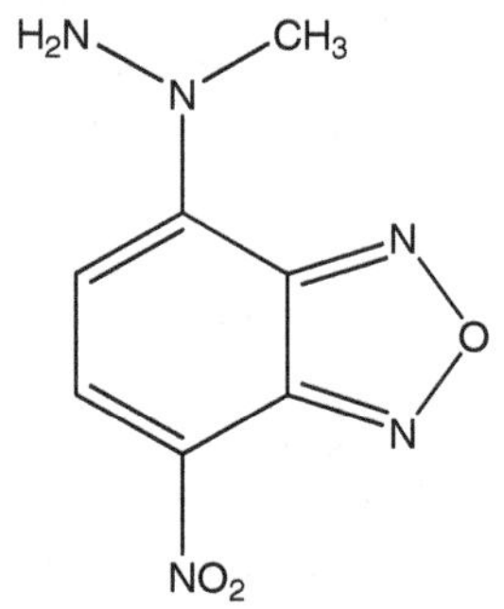

CA Index Name 2,1,3-Benzoxadiazole, 4-(1-methylhydrazinyl)-7-nitro-

Other Names 2,1,3-Benzoxadiazole, 4-(1-methylhydrazino)-7-nitro-; MNBDH; NBD methylhydrazine; 4-(1-Methylhydrazino)-7-nitrobenzofurazan; *N*-Methyl-4-hydrazino-7-nitrobenzofurazan; 4-(1-Methylhydrazino)-7-nitro-benzooxadiazole; 4-(*N*-Methylhydrazino)-7-nitro-1,2,3-benzooxadiazole

Merck Index Number Not listed

Chemical/Dye Class Benzoxadiazole

Molecular Formula $C_7H_7N_5O_3$

Molecular Weight 209.16

Physical Form Dark brown powder

Solubility Soluble in acetonitrile, methanol

Melting Point 160 °C

Boiling Point (Calcd.) 425.7 ± 55.0 °C, pressure: 760 Torr

pK_a (Calcd.) 2.07 ± 0.30, most basic, temperature: 25 °C

Absorption (λ_{max}) 487 nm

Synthesis Synthetic methods[1,2]

Staining Applications Nitrite ions[3]

Biological Applications Nitrite indicator;[3] detecting aldehydes and/or ketones (carbonyl compounds),[1,2,4–9] nitroaromatic compounds,[10] creatinine in body fluids,[11] telmisartan,[12] hydrogen peroxide,[13] peroxides,[13] as a peroxidase substrate[14]

Industrial Applications Not reported

Safety/Toxicity No data available

REFERENCES

1. Bueldt, A.; Karst, U. *N*-Methyl-4-hydrazino-7-nitrobenzofurazan as a new reagent for air monitoring of aldehydes and ketones. *Anal. Chem.* **1999**, *71*, 1893–1898.
2. Bueldt, A.; Karst, U. New benzoxadiazoles. Ger. Offen. DE 19800537, 1998; *Chem. Abstr.* **1998**, *129*, 290139.
3. Bueldt, A.; Karst, U. Determination of nitrite in waters by microplate fluorescence spectroscopy and HPLC with fluorescence detection. *Anal. Chem.* **1999**, *71*, 3003–3007.
4. Olsen, R.; Thorud, S.; Hersson, M.; Ovrebo, S.; Lundanes, E.; Greibrokk, T.; Ellingsen, D. G.; Thomassen, Y.; Molander, P. Determination of the dialdehyde glyoxal in workroom air development of personal sampling methodology. *J. Environ. Monit.* **2007**, *9*, 687–694.
5. Paolacci, H.; Tran, T. T. H. Nanoporous material for direct optical sensing of aldehydes. Fr. Demande FR 2890745, 2007; *Chem. Abstr.* **2007**, *146*, 350197.
6. Possanzini, M.; Di Palo, V.; Cecinato, A. Field evaluation of *n*-methyl-4-hydrazino-7-nitrobenzofurazan (MNBDH) coated silica gel cartridges for the measurement of lower carbonyls in air. *Chromatographia* **2004**, *60*, 715–719.
7. Schulte-Ladbeck, R.; Lindahl, R.; Levin, J. O.; Karst, U. Characterization of chemical interferences in the determination of unsaturated aldehydes using aromatic hydrazine reagents and liquid chromatography. *J. Environ. Monit.* **2001**, *3*, 306–310.
8. Zurek, G.; Buldt, A.; Karst, U. Determination of acetaldehyde in tobacco smoke using *N*-methyl-4-hydrazino-7-nitrobenzofurazan and liquid chromatography/mass spectrometry. *Fresenius' J. Anal. Chem.* **2000**, *366*, 396–399.
9. Buldt, A.; Lindahl, R.; Levin, J. O.; Karst, U. A diffusive sampling device for the determination of formaldehyde in air using *N*-methyl-4-hydrazino-7-nitrobenzofurazan (MNBDH) as reagent. *J. Environ. Monit.* **1999**, *1*, 39–43.
10. Hayen, H.; Jachmann, N.; Vogel, M.; Karst, U. LC-electron capture APCI-MS for the determination of nitroaromatic compounds. *Analyst* **2002**, *127*, 1027–1030.

11. Albarella, J. P.; Hatch, R. P. Fluorescent creatinine assay. U.S. Pat. Appl. Publ. US 2004132200, 2004; *Chem. Abstr.* **2004**, *141*, 85142.
12. Hempen, C.; Glaesle-Schwarz, L.; Kunz, U.; Karst, U. Determination of telmisartan in human blood plasma. *Anal. Chim. Acta* **2006**, *560*, 35–40.
13. Meyer, J.; Jachmann, N.; Bueldt, A.; Karst, U. Method for the determination of hydrogen peroxide and peroxides using peroxidase and *N*-methyl-4-hydrazino-7-nitrobenzofurazan (MNDBH). Ger. Offen. DE 19932380, 2000; *Chem. Abstr.* **2000**, *132*, 134354.
14. Meyer, J.; Buldt, A.; Vogel, M.; Karst, U. 4-(*N*-Methylhydrazino)-7-nitro-1,2,3-benzooxadiazole (MNBDH): a novel fluorogenic peroxidase substrate. *Angew. Chem. Int. Ed.* **2000**, *39*, 1453–1455.

NEOTETRAZOLIUM (NT)

CAS Registry Number 298-95-3

Chemical Structure

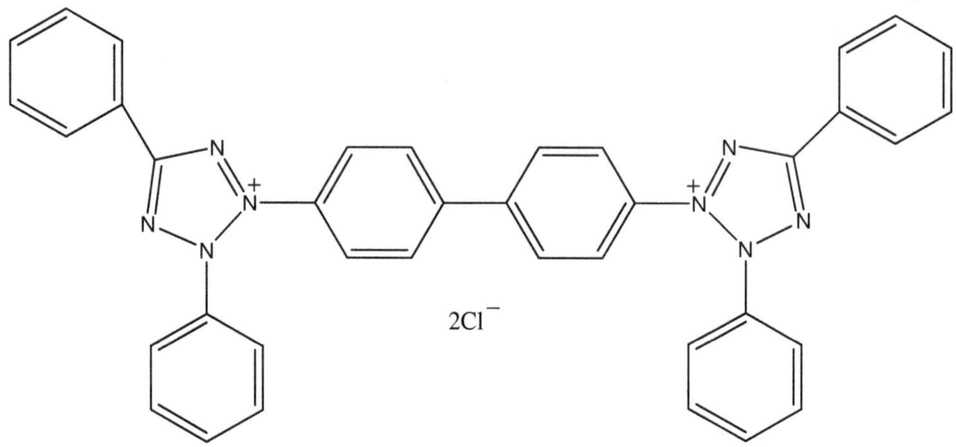

CA Index Name 2*H*-Tetrazolium, 3,3′-[1,1′-biphenyl]-4,4′-diylbis[2,5-diphenyl]-,chloride (1:2)

Other Names 2*H*-Tetrazolium, 3,3′-(4,4′-biphenylylene)bis[2,5-diphenyl]-, dichloride; 2*H*-Tetrazolium, 3,3′-[1,1′-biphenyl]-4,4′-diylbis[2,5-diphenyl]-, dichloride; 3,3′-(4,4′-Biphenylylene)bis[2,5-diphenyl-2*H*-tetrazolium chloride]; 2,2′-(*p*-Diphenylene)bis(3,5-diphenyl) ditetrazolium chloride; 3,3′-(4,4′-Biphenylene)bis[2,5-diphenyltetrazolium chloride]; NSC 27621; NTC; Neo-T; Neotetrazolium; Neotetrazolium blue; Neotetrazolium chloride; TP

Merck Index Number 6466

Chemical/Dye Class Tetrazolium salt

Molecular Formula $C_{38}H_{28}Cl_2N_8$

Molecular Weight 667.59

Physical Form Light yellow or tan powder

Solubility Soluble in water, ethanol, *N*,*N*-dimethyl formamide; insoluble in ether

Melting Point 297 °C (decompose)

Absorption (λ_{max}) 248 nm

Synthesis Synthetic methods[1–7]

Staining Applications Lipoproteins;[8] skin[9]

Biological Applications Antifungal agent;[10] glucose-6-phosphate dehydrogenase activity assays;[11] microbial growth assays;[12] detecting bacteria,[13] cancer cells,[14] gamma-hydroxybutyric acid (GHB),[15] microorganisms;[16] treating cancer[17]

Industrial Applications Recording materials;[18,19] toner[20,21]

Safety/Toxicity No data available

REFERENCES

1. Cui, X.; Vlahakis, J. Z.; Crandall, I. E.; Szarek, W. A. Anti-*Plasmodium* activity of tetrazolium salts. *Bioorg. Med. Chem.* **2008**, *16*, 1927–1947.
2. James, P. R. Tetrazolium salts as steam sterilization process indicators. PCT Int. Appl. WO 9837157, 1998; *Chem. Abstr.* **1998**, *129*, 218237.
3. Jones, G. R. N. Purification and some properties of neotetrazolium chloride and its chief monotetrazolium salt contaminant. *Histochem. J.* **1968**, *1*, 59–67.
4. Ostrovskaya, V. M.; Pryanishnikov, A. A. Neotetrazolium chloride. 2,2′,5,5′-Tetraphenyl-3,-3′-(4,4′-biphenylene)-di(tetrazolium chloride). *Metody Poluch. Khim. Reakt. Prep.* **1964**, 14–16; *Chem. Abstr.* **1967**, *66*, 37539.
5. Glantz, M. D.; Fried, F. Microseparation of tetrazolium salts by paper electrophoresis. *Microchem. J.* **1959**, *3*, 214–216.
6. Pannone, L. J.; Rust, J. B. Tetrazolium salts. U.S. Patent 2713581, 1955; *Chem. Abstr.* **1956**, *50*, 6481l.
7. Seiler, H.; Schmid, H. The synthesis of some bisformazan and bistetrazolium salts. I. *Helv. Chim. Acta* **1954**, *37*, 1–6.
8. Busbee, D. L.; Payne, D. M.; Jasheway, D. W.; Carlisle, S.; Lacko, A. G. Separation and detection of lipoproteins in human serum by use of

size-exclusion liquid chromatography: a preliminary report. *Clin. Chem.* **1981**, *27*, 2052–2058.

9. Kalopissis, G.; Bugaut, A. Tetrazolium skin dyes. Ger. Offen. DE 1912897, 1969; *Chem. Abstr.* **1970**, *72*, 24519.
10. Pastor, L.; Garcia-Domenech, R.; Galvez, J.; Wolski, S.; Garcia, M. D. New antifungals selected by molecular topology. *Bioorg. Med. Chem. Lett.* **1998**, *8*, 2577–2582.
11. De Jong, J. S.; Frederiks, W. M.; Van Noorden, C. J. Oxygen insensitivity of the histochemical assay of glucose-6-phosphate dehydrogenase activity for the detection of (pre)neoplasm in rat liver. *J. Histochem. Cytochem.* **2001**, *49*, 565–572.
12. Storey, D.; Field, T.; Ramage, G. Microbial growth assay and assay device. PCT Int. Appl. WO 2005098023, 2005; *Chem. Abstr.* **2005**, *143*, 402159.
13. Ushiyama, M. Film culture medium for detecting bacteria in water-system environment. Jpn. Kokai Tokkyo Koho JP 2007020434, 2007; *Chem. Abstr.* **2007**, *146*, 180320.
14. Griffini, P.; Vigorelli, E.; Jonges, G. N.; Van Noorden, C. J. F. The histochemical G6PDH reaction but not the LDH reaction with neotetrazolium is suitable for the oxygen sensitivity test to detect cancer cells. *J. Histochem. Cytochem.* **1994**, *42*, 1355–1363.
15. Parsons, S. M.; Harris, D. O.; Bravo, D. T. Methods, compositions and apparatuses for detection of gamma-hydroxybutyric acid (GHB). U.S. Pat. Appl. Publ. US 2003175846, 2003; *Chem. Abstr.* **2003**, *139*, 256601.
16. Ushiyama, M. Oxidizable-reducible chromogen pair-containing culture medium for microorganism colorimetric detection for use in food or environmental analysis. U.S. Pat. Appl. Publ. US 2006008867, 2006; *Chem. Abstr.* **2006**, *144*, 124523.
17. Kong, Q. Method and composition for treating cancer using cellular organelle crystallizing agents. U.S. Pat. Appl. Publ. US 2006286170, 2006; *Chem. Abstr.* **2006**, *146*, 55482.
18. Miyakawa, N.; Kokado, H.; Inoue, E. Electrically conducting material for electrically sensitive recording materials. Ger. Offen. DE 2905976, 1979; *Chem. Abstr.* **1980**, *92*, 85946.
19. Takatori, Y.; Haruta, M.; Shimozawa, A.; Nishide, K. Image reproduction or recording material. Ger. Offen. DE 2429729, 1975; *Chem. Abstr.* **1975**, *83*, 155788.
20. Yamamoto, A.; Takagi, S.; Fukumoto, H.; Tanaka, K. Toner for development of electrostatic images. Jpn. Kokai Tokkyo Koho JP 61137162, 1986; *Chem. Abstr.* **1987**, *106*, 58907.
21. Suzuki, T. Toner for development of latent electrostatic images. Ger. Offen. DE 3527480, 1986; *Chem. Abstr.* **1986**, *104*, 234268.

NEUTRAL RED

CAS Registry Number 553-24-2

Chemical Structure

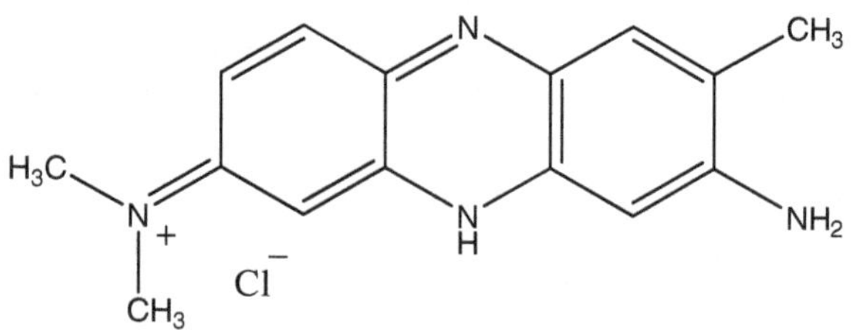

CA Index Name 2,8-Phenazinediamine, N8,N8,3-trimethyl-, hydrochloride

Other Names 3-Amino-7-dimethylamino-2-methylphenazine hydrochloride; Aminodimethylaminotoluaminozine hydrochloride; Cerven neutralni; Cerven toluylenova; C.I. 50040; C.I. Basic Red 5; C.I. Basic Red 5, monohydrochloride; Kernechrot; Michrome No. 226; Neutral Red; Neutral Red W; Nuclear Fast Red (basic dye); 2,8-Phenazinediamine, N8,N8,3-trimethyl-, monohydrochloride; Toluylene red

Merck Index Number 6488

Chemical/Dye Class Phenazine

Molecular Formula $C_{15}H_{17}ClN_4$

Molecular Weight 288.78

Physical Form Dark green or brownish-black powder

Solubility Soluble in water, ethanol, ethylene glycol; practically insoluble in xylene

Melting Point 290 °C (decompose)

pH Range 6.8–8.0

Color Change at pH Red (6.8) to yellow (8.0)

pK_a 6.7, 7.4

Absorption (λ_{max}) 540 nm, 533 nm

Emission (λ_{max}) 640 nm

Synthesis Synthetic methods[1–3]

Staining Applications Cells;[4] lysosomes;[5,6] nuclei;[7] nucleic acids;[8,9] retina[10]

Biological Applications Detecting pathogens,[11] bacterial infections;[1,12] treating age-related macular degeneration,[1,13] burns,[1,14] cancer,[1,15] diabetes,[1,16] obesity,[1,16] fungal infections,[17,18] viral diseases[1,19]

Industrial Applications Liquid crystal displays;[1,20] fuel cells;[1,21] photovoltaic cells;[22] solar cells;[1,22] photochromic materials;[23] sensors;[1,24] thermochromic materials;[1,25] detergents;[1,26] wood[1,27]

Safety/Toxicity Combustion toxicity;[1,28] cytotoxicity;[1,29–31] DNA damage;[32,33] lysosomal stability;[32,33] genotoxicity;[1,34–36] microbial toxicity;[37] mutagenicity;[1,38,39] nephrotoxicity;[1,40] neurotoxicity;[32] phototoxocity;[1,41–43] soil toxicity[1,44]

Certification/Approval Certified by Biological Stain Commission (BSC)

REFERENCES

1. Sabnis, R. W. *Handbook of Acid–Base Indicators*; CRC Press: Boca Raton, **2008**; pp 260–261.
2. Welcher, F. J. *Organic Analytical Reagents*; D. Van Nostrand Company: New York, **1948**; Vol. 4, pp 592–593.
3. Phillips, M.; Cohen, B. The preparation of vital neutral red. *Stain Technol.* **1927**, *2*, 74–79.
4. Nath, B. B.; Babrekar, A. A.; Parthasarathy, B. Assessment of cell viability in intact glandular tissue in *Chironomus ramosus* using dye-exclusion and colorimetric assays. *Cytotechnology* **2005**, *49*, 59–65.
5. Ivanova, L.; Uhlig, S. A bioassay for the simultaneous measurement of metabolic activity, membrane integrity, and lysosomal activity in cell cultures. *Anal. Biochem.* **2008**, *379*, 16–19.
6. Repetto, G.; del Peso, A.; Zurita, J. L. Neutral red uptake assay for the estimation of cell viability/cytotoxicity. *Nat. Protoc.* **2008**, *3*, 1125–1131.
7. Horobin, R. W.; Stockert, J. C.; Rashid-Doubell, F. Fluorescent cationic probes for nuclei of living cells: why are they selective? A quantitative structure–activity relations analysis. *Histochem. Cell Biol.* **2006**, *126*, 165–175.
8. Ni, Y.; Du, S.; Kokot, S. Interaction between quercetin–copper(II) complex and DNA with the use of the neutral red dye fluorophor probe. *Anal. Chim. Acta* **2007**, *584*, 19–27.
9. Jia, X.; Jia, L. Isolation of nucleic acids from *Escherichia coli* using colored buffers. U.S. Pat. Appl. Publ. US 2007015169, 2007; *Chem. Abstr.* **2007**, *146*, 136368.
10. Jackson, T. L.; Griffin, L.; Vote, B.; Hillenkamp, J.; Marshall, J. An experimental method for testing novel retinal vital stains. *Exp. Eye Res.* **2005**, *81*, 446–454.
11. Brook, M. A.; Brennan, J. D.; Pelton, R.; Voss, R.; Ye, L. Biomolecule compatible silica particles and use for

the detection of bacterial pathogens. PCT Int. Appl. WO 2008061363, 2008; *Chem. Abstr.* **2008**, *148*, 579895.

12. Stickler, D. J.; Waters, M. G. J. pH sensor for urinary catheters to detect bacterial infections. PCT Int. Appl. WO 2006000764, 2006; *Chem. Abstr.* **2006**, *144*, 94481.
13. Hofmann, R. F. Use of targeted oxidative therapeutic formulation in treatment of age-related macular degeneration. U.S. Pat. Appl. Publ. US 2005250756, 2005; *Chem. Abstr.* **2005**, *143*, 416314.
14. Hofmann, R. F. Use of targeted oxidative therapeutic formulation in treatment of burns. PCT Int. Appl. WO 2006002302, 2006; *Chem. Abstr.* **2006**, *144*, 114433.
15. Hofmann, R. F. Use of targeted oxidative therapeutic formulation in treatment of cancer. U.S. Pat. Appl. Publ. US 2005250757, 2005; *Chem. Abstr.* **2005**, *143*, 416223.
16. Hofmann, R. F. Use of targeted oxidative therapeutic formulation in treatment of diabetes and obesity. PCT Int. Appl. WO 2005107728, 2005; *Chem. Abstr.* **2005**, *143*, 466228.
17. Simon, M. M.; Pardo, J. Gliotoxin inhibition for the treatment of fungal infections. PCT Int. Appl. WO 2007128555, 2007; *Chem. Abstr.* **2007**, *147*, 515032.
18. Yamamoto, K. Method for evaluation of antifungal activity. Jpn. Kokai Tokkyo Koho JP 2007292481, 2007; *Chem. Abstr.* **2007**, *147*, 496293.
19. Hofmann, R. F. Targeted oxidative therapeutic formulation in treatment of viral diseases. U.S. Pat. Appl. Publ. US 2005192267, 2005; *Chem. Abstr.* **2005**, *143*, 272504.
20. Klein, S.; Geisow, A. D. Liquid crystal display device. Brit. UK Pat. Appl. GB 2415703, 2006; *Chem. Abstr.* **2006**, *144*, 97812.
21. Yamamoto, M.; Harada, T. Water purification apparatus and its use in fuel cell power generation system. Jpn. Kokai Tokkyo Koho JP 2004321934, 2004; *Chem. Abstr.* **2004**, *141*, 427669.
22. Chane, C. J. Y.; Corma, A.; Garcia, H. Multivalent metal cation-doped cerium oxide materials for photovoltaic cells and solar cells. Fr. Demande FR 2857783, 2005; *Chem. Abstr.* **2005**, *142*, 159487.
23. Ito, Y. Photochromic particles and their production. Jpn. Kokai Tokkyo Koho JP 2008037974, 2008; *Chem. Abstr.* **2008**, *148*, 264385.
24. Kopelman, R. A. Reusable pH sensor device and related methods. U.S. Pat. Appl. Publ. US 2005265895, 2005; *Chem. Abstr.* **2005**, *143*, 487933.
25. Zhang, X. Manufacture of reversible organic thermochromic materials from neutral red. Faming Zhuanli Shenqing Gongkai Shuomingshu CN 1546603, 2004; *Chem. Abstr.* **2005**, *143*, 356252.
26. Macdonald, J. G.; Huang, Y.; Yang, K.; Kim, J.; Wei, N. Single phase color change agents for liquid detergents for cleaning surfaces. U.S. Pat. Appl. Publ. US 2005049157, 2005; *Chem. Abstr.* **2005**, *142*, 263578.
27. Leach, R. M.; Zhang, J. Composition and process for coloring and preserving wood. PCT Int. Appl. WO 2005056257, 2005; *Chem. Abstr.* **2005**, *143*, 61604.
28. Lestari, F.; Markovic, B.; Green, A. R.; Chattopadhyay, G.; Hayes, A. J. Comparative assessment of three *in vitro* exposure methods for combustion toxicity. *J. Appl. Toxicol.* **2006**, *26*, 99–114.
29. Canal-Raffin, M.; l'Azou, B.; Jorly, J.; Hurtier, A.; Cambar, J.; Brochard, P. Cytotoxicity of folpet fungicide on human bronchial epithelial cells. *Toxicology* **2008**, *249*, 160–166.
30. Burlando, B.; Parodi, Al.; Volante, A.; Bassi, A. M. Comparison of the irritation potentials of *Boswellia serrata* gum resin and of acetyl-11-keto-beta-boswellic acid by *in vitro* cytotoxicity tests on human skin-derived cell lines. *Toxicol. Lett.* **2008**, *177*, 144–149.
31. Lu, B.; Kerepesi, L.; Wisse, L.; Hitchman, K.; Meng, Q. R. Cytotoxicity and gene expression profiles in cell cultures exposed to whole smoke from three types of cigarettes. *Toxicol. Sci.* **2007**, *98*, 469–478.
32. Rank, J.; Lehtonen, K. K.; Strand, J.; Laursen, M. DNA damage, acetylcholinesterase activity and lysosomal stability in native and transplanted mussels (*Mytilus edulis*) in areas close to coastal chemical dumping sites in Denmark. *Aquat. Toxicol.* **2007**, *84*, 50–61.
33. Regoli, F.; Frenzilli, G.; Bocchetti, R.; Annarumma, F.; Scarcelli, V.; Fattorini, D.; Nigro, M. Time-course variations of oxyradical metabolism, DNA integrity and lysosomal stability in mussels, *Mytilus galloprovincialis*, during a field translocation experiment. *Aquat. Toxicol.* **2004**, *68*, 167–178.
34. Knight, A. W.; Billinton, N.; Cahill, P. A.; Scott, A.; Harvey, J. S.; Roberts, K. J.; Tweats, D. J.; Keenan, P. O.; Walmsley, R. M. An analysis of results from 305 compounds tested with the yeast RAD54-GFP genotoxicity assay (GreenScreen GC) including relative predictivity of regulatory tests and rodent carcinogenesis and performance with autofluorescent and colored compounds. *Mutagenesis* **2007**, *22*, 409–416.

35. Sanchez-Fortun, S.; Llorente, M. T.; Castano, A. Genotoxic effects of selected biocides on RTG-2 fish cells by means of a modified Fast Micromethod Assay. *Aquat. Toxicol.* **2005**, *73*, 55–64.
36. Di Virgilio, A. L.; Iwami, K.; Watjen, W.; Kahl, R.; Degen, G. H. Genotoxicity of the isoflavones genistein, daidzein and equol in V79 cells. *Toxicol. Lett.* **2004**, *151*, 151–162.
37. Oros, G.; Cserhati, T.; Forgacs, E. Separation of the strength and selectivity of the microbiological effect of synthetic dyes by spectral mapping technique. *Chemosphere* **2003**, *52*, 185–193.
38. Klopman, G.; Frierson, M. R.; Rosenkranz, H. S. The structural basis of the mutagenicity of chemicals in *Salmonella typhimurium*: the Gene-Tox data base. *Mutat. Res.* **1990**, *228*, 1–50.
39. Chung, K. T.; Fulk, G. E.; Andrews, A. W. Mutagenicity testing of some commonly used dyes. *Appl. Environ. Microbiol.* **1981**, *42*, 641–648.
40. van der Harst, M. R.; Bull, S.; Laffont, C. M.; Klein, W. R. Gentamicin nephrotoxicity: a comparison of *in vitro* findings with *in vivo* experiments in equines. *Vet. Res. Commun.* **2005**, *29*, 247–261.
41. Onoue, S.; Igarashi, N.; Yamauchi, Y.; Murase, N.; Zhou, Y.; Kojima, T.; Yamada, S.; Tsuda, Y. *In vitro* phototoxicity of dihydropyridine derivatives: a photochemical and photobiological study. *Eur. J. Pharm. Sci.* **2008**, *33*, 262–270.
42. Kejlova, K.; Jirova, D.; Bendova, H.; Kandarova, H.; Weidenhoffer, Z.; Kolarova, H.; Liebsch, M. Phototoxicity of bergamot oil assessed by *in vitro* techniques in combination with human patch tests. *Toxicol. In Vitro* **2007**, *21*, 1298–1303.
43. Sharma, S.; Dub, A.; Bose, B.; Gupta, P. K. Pharmacokinetics and phototoxicity of purpurin-18 in human colon carcinoma cells using liposomes as delivery vehicles. *Cancer Chemother. Pharmacol.* **2006**, *57*, 500–506.
44. Robidoux, P. Y.; Dubois, C.; Hawari, J.; Sunahara, G. I. Assessment of soil toxicity from an antitank firing range using *Lumbricus terrestris* and *Eisenia andrei* in mesocosms and laboratory studies. *Ecotoxicology* **2004**, *13*, 603–614.

NEWPORT GREEN DCF

CAS Registry Number 288374-37-8

Chemical Structure

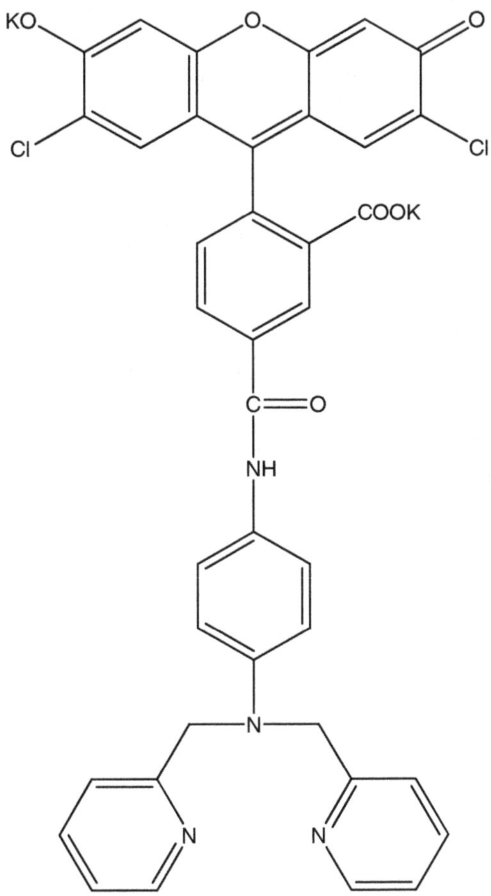

CA Index Name Spiro[isobenzofuran-1(3*H*),9′-[9*H*]xanthene]-5-carboxamide, *N*-[4-[bis(2-pyridinylmethyl)amino]phenyl]-2′,7′-dichloro-3′,6′-dihydroxy-3-oxo-, potassium salt (1:2)

Other Names Spiro[isobenzofuran-1(3*H*),9′-[9*H*]xanthene]-5-carboxamide, *N*-[4-[bis(2-pyridinylmethyl)amino]phenyl]-2′,7′-dichloro-3′,6′-dihydroxy-3-oxo-, dipotassium salt; Newport Green; Newport Green DCF dipotassium salt

Merck Index Number Not listed

Chemical/Dye Class Xanthene

Molecular Formula $C_{39}H_{24}Cl_2K_2N_4O_6$

Molecular Weight 793.74

Physical Form Solid

Solubility Soluble in water

Melting Point >200 °C

Absorption (λ_{max}) 506 nm

Emission (λ_{max}) 535 nm

Synthesis Synthetic method[1]

Staining Applications Zinc ions;[1–12,16] chromium;[2] manganese;[2] iron;[2] cobalt;[2,16] copper;[2,15] nickel;[2] cadmium[12–16]

Biological Applications Zinc indicator;[1–12,16] early diagnosis of prostate cancer;[3] treating herpes virus infection[10]

Industrial Applications Not reported

Safety/Toxicity No data available

REFERENCES

1. Gee, K. R.; Zhou, Z. L.; Ton-That, D.; Sensi, S. L.; Weiss, J. H. Measuring zinc in living cells: a new generation of sensitive and selective fluorescent probes. *Cell Calcium* **2002**, *31*, 245–251.
2. Zhao, J.; Bertoglio, B. A.; Devinney, M. J., Jr.; Dineley, K. E.; Kay, A. R. The interaction of biological and noxious transition metals with the zinc probes FluoZin-3 and Newport Green. *Anal. Biochem.* **2009**, *384*, 34–41.
3. Frederickson, C. J.; Costello, L. C.; Franklin, R. B. Zinc-based screening test and kit for early diagnosis of prostate cancer. U.S. Pat. Appl. Publ. US 2007207509, 2007; *Chem. Abstr.* **2007**, *147*, 317778.
4. Stork, C. J.; Li, Y. V. Measuring cell viability with membrane impermeable zinc fluorescent indicator. *J. Neurosci. Methods* **2006**, *155*, 180–186.
5. Reynolds, I. J. Fluorescence detection of redox-sensitive metals in neuronal culture: focus on iron and zinc. *Ann. N. Y. Acad. Sci.* **2004**, *1012*, 27–36.
6. Centre Hospitalier Regional et Universitaire de Lille Chru, Fr. Method for identification and purification of human pancreatic beta cells using a specific fluorescent zinc probe. Fr. Demande FR 2822954, 2002; *Chem. Abstr.* **2002**, *137*, 365962.
7. Dineley, K. E.; Malaiyandi, L. M.; Reynolds, I. J. A reevaluation of neuronal zinc measurements: artifacts

associated with high intracellular dye concentration. *Mol. Pharmacol.* **2002**, *62*, 618–627.

8. Thompson, R. B.; Peterson, D.; Mahoney, W.; Cramer, M.; Maliwal, B. P.; Suh, S. W.; Frederickson, C.; Fierke, C.; Herman, P. Fluorescent zinc indicators for neurobiology. *J. Neurosci. Methods* **2002**, *118*, 63–75.
9. Sumner, J. P.; Aylott, J. W.; Monson, E.; Kopelman, R. A fluorescent PEBBLE nanosensor for intracellular free zinc. *Analyst* **2002**, *127*, 11–16.
10. Clements, J. B.; MacLean, A. R. Zinc finger motif sequences from herpes simplex virus protein IE63 and uses thereof in drug screening for treating herpes virus infection. PCT Int. Appl. WO 2002004492, 2002; *Chem. Abstr.* **2002**, *136*, 112623.
11. Lukowiak, B.; Vandewalle, B.; Riachy, R.; Kerr-Conte, J.; Gmyr, V.; Belaich, S.; Lefebvre, J.; Pattou, F. Identification and purification of functional human β-cells by a new specific zinc-fluorescent probe. *J. Histochem. Cytochem.* **2001**, *49*, 519–527.
12. Zhang, J.; Niessner, R. Simultaneous determination of trace Ni(II) and Zn(II) in water by using fluorescence-based flow injection analysis. *Fenxi Shiyanshi* **2002**, *21*, 1–4; *Chem. Abstr.* **2002**, *137*, 237307.
13. Thierse, H. J.; Helm, S.; Pink, M.; Weltzien, H. U. Novel fluorescence assay for tracking molecular and cellular allergen–protein interactions. *J. Immunol. Methods* **2007**, *328*, 14–20.
14. Debus, N. P.; Sydow, S.; Hofmann, B.; Briel, A.; Roessling, G. L-Selectin contrast agents for depicting changes in lymph nodes. PCT Int. Appl. WO 2001017566, 2001; *Chem. Abstr.* **2001**, *134*, 227402.
15. Prestel, H.; Gahr, A.; Niessner, R. Detection of heavy metals in water by fluorescence spectroscopy: on the way to a suitable sensor system. *Fresenius' J. Anal. Chem.* **2000**, *368*, 182–191.
16. Wuertz, S.; Muller, E.; Spaeth, R.; Pfleiderer, P.; Flemming, H. C. Detection of heavy metals in bacterial biofilms and microbial flocs with the fluorescent complexing agent Newport Green. *J. Ind. Microbiol. Biotechnol.* **2000**, *24*, 116–123.

NEWPORT GREEN PDX

CAS Registry Number 612502-05-3

Chemical Structure

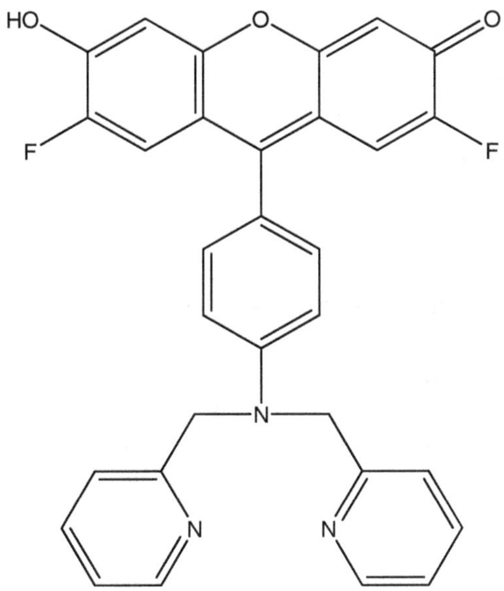

CA Index Name 3*H*-Xanthen-3-one, 9-[4-[bis(2-pyridinylmethyl)amino]phenyl]-2,7-difluoro-6-hydroxy-

Other Names Newport Green PDX

Merck Index Number Not listed

Chemical/Dye Class Xanthene

Molecular Formula $C_{31}H_{21}F_2N_3O_3$

Molecular Weight 521.51

Physical Form Solid

Solubility Soluble in water

Melting Point >200 °C

Boiling Point (Calcd.) 750.1 ± 60.0 °C, pressure: 760 Torr

pK_a (Calcd.) 7.93 ± 0.60, most acidic, temperature: 25 °C; 4.28 ± 0.12, most basic, temperature: 25 °C

Absorption (λ_{max}) 490 nm

Emission (λ_{max}) 518 nm

Synthesis Synthetic method[1]

Staining Applications Zinc ions;[1] mitochondria[2]

Biological Applications Zinc indicator[1]

Industrial Applications Not reported

Safety/Toxicity No data available

REFERENCES

1. Gee, K. R.; Zhou, Z. L.; Ton-That, D.; Sensi, S. L.; Weiss, J. H. Measuring zinc in living cells: a new generation of sensitive and selective fluorescent probes. *Cell Calcium* **2002**, *31*, 245–251.
2. Ichii, H.; Ricordi, C. Assessment of pancreatic islet cellular composition and fractional viability for prediction of post-transplant outcomes. PCT Int. Appl. WO 2007146811, 2007; *Chem. Abstr.* **2007**, *148*, 49116.

NIGROSIN

CAS Registry Number 8005-03-6

Chemical Structure No structure diagram available as it is a mixture

CA Index Name C.I. Acid Black 2

Other Names Nigrosine; ADC Nigrosine Black B; Acid Black 2; Acid Black WR; Acid Nigrosine; C.I. 50420; Calco Nigrosine O 2P; Conacid Black B; Cramity 81; Dinacid Nigrosine WS; Dyacid Nigrosine NBS Xtals; Dycosacid Nigrosine NBL (Bluish); Dycosacid Nigrosine NBL (Reddish); Lurazol Deep Blue EB; Nigrosin; Nigrosine (Crystals)NBL(Blush); Nigrosine (Crystals)NBL (Reddish); Nigrosine B; Nigrosine CBRS; Nigrosine Crystals; Nigrosine MS; Nigrosine NB; Nigrosine NB conc; Nigrosine NBSS Water Soluble; Nigrosine NN Water Soluble; Nigrosine WL Water Soluble; Nigrosine WLAH; Nigrosine WLF; Nigrosine WLF Uncut; Nigrosine WLL; Nigrosine WSB; Nigrosine Water Soluble; Nigrosine black; Nigtrosine WS; Nubian Black PA 9801; Nubian Black PA9803; Oil Black FS Special; Oil Black S; Orient Nigrosine BR; Orient Nigrosine OZ; Orient Nubian Black PA 9801; Orient Nubian Black PA 9803; Orient Water Black R 455; Orient Water Black R 456; Orient Water Black R 500; Orient Water Black R 510; Pacid Black 2; Pacid Leather Black; Sandopel Basic Black BHLN; Triacid Nigrosine WS; Vicoacid Nigrosine Black; Victacid Nigrosine Black Crystals; Water Black 179128; Water Black R 455; Water Black R 500; Water Black R 510

Merck Index Number Not listed

Chemical/Dye Class Azine

Molecular Formula Mixture, unspecified

Molecular Weight Mixture

Physical Form Black crystals or powder

Solubility Soluble in water; slightly soluble in ethanol

Melting Point Mixture

Absorption (λ_{max}) 570 nm

Synthesis Synthetic method[1–11]

Staining Applications Cells;[12] gel;[13] olives;[14] proteins;[15,16] soy sauce;[17] sperms;[18,19] lips;[20] eyelids;[20] eyebrows;[20] keratin fibers;[21] hairs[22]

Biological Applications Diagnostic assay;[23] rodenticide[24]

Industrial Applications Semiconductor devices;[25] fuel cells;[26] color filters;[27] liquid crystal displays;[27] electrophoretic display;[28] photoresists;[29,30] inks;[31–33] toners;[34,35] photography;[36] thermoplastic materials;[37,38] furniture and building materials;[39] paints;[40] leather;[41,42] plastics;[43] waxes;[43] cleaners;[43] steel;[44] wood[45]

Safety/Toxicity Genotoxicity;[46] mutagenicity;[47] percutaneous toxicity[48]

Certification/Approval Certified by Biological Stain Commission (BSC)

REFERENCES

1. Yamazaki, T. Electrophotographic dry toner containing nigrosine dye charge-controlling agent and its manufacture. Jpn. Kokai Tokkyo Koho JP 2006171501, 2006; *Chem. Abstr.* **2006**, *145*, 73329.
2. Dumitrache, V. C.; Dumitrache, L.; Athanasiu, A. A.; Cojocaru, M.; Bonte, T. R.; Murzea, S. Process for preparation of nigrosins. Rom. RO 120265, 2005; *Chem. Abstr.* **2006**, *144*, 313697.
3. Yamazaki, T. Electrophotographic magnetic toner using nigrosine dye. Jpn. Kokai Tokkyo Koho JP 2003035970, 2003; *Chem. Abstr.* **2003**, *138*, 128992.
4. Maekawa, Y. Nigrosine dyes and manufacture thereof. Jpn. Kokai Tokkyo Koho JP 09087535, 1997; *Chem. Abstr.* **1997**, *127*, 35962.
5. Mauer, W.; Zimmer, J. Aniline condensation dyes. Ger. Offen. DE 3000529, 1981; *Chem. Abstr.* **1981**, *95*, 99340.
6. Maekawa, Y. Semicontinuously producing aniline condensed dyes. U.S. Patent 4056530, 1977; *Chem. Abstr.* **1978**, *88*, 24256.
7. Kleinke, H. Induline and nigrosine. Ger. (East) DD 74068, 1970; *Chem. Abstr.* **1971**, *74*, 65581.
8. Mann, A.; Kaeppler, E.; Fuerst, H. Nigrosine and induline. Ger. (East) DD 64761, 1968; *Chem. Abstr.* **1969**, *71*, 40224.
9. Belic, I.; Gabrijelcic, Z.; Pogacnik, M.; Premru, V.; Stojkovic, I.; Zisic, M.; Zupan, G.; Zupan, V. The synthesis of organic dyes. *Kem. Zbornik* **1951**, 128–136; *Chem. Abstr.* **1954**, *48*, 34247.
10. Krasnovskii, A. A. Nigrosine. *Promysh. Organ. Khim.* **1937**, *4*, 14–19; *Chem. Abstr.* **1937**, *31*, 64971.
11. Koons, L. O. The manufacture of nigrosine. *Color Trade J.* **1920**, *7*, 57–58.
12. Nath, B. B.; Babrekar, A. A.; Parthasarathy, B. Assessment of cell viability in intact glandular

tissue in *Chironomus ramosus* using dye-exclusion and colorimetric assays. *Cytotechnology* **2005**, *49*, 59–65.

13. Wada, A. Method for staining and destaining gel, electrophoresis destaining device for gel, and kit for staining and destaining gel. U.S. Pat. Appl. Publ. US 2006275917, 2006; *Chem. Abstr.* **2006**, *146*, 41434.
14. Ziena, H. M. S.; Youssef, M. M.; Aman, M. E. Quality attributes of black olives as affected by different darkening methods. *Food Chem.* **1997**, *60*, 501–508.
15. Kutter, D.; Kremer, A.; Bousser, F.; Kambris, Z. A simple and inexpensive screening test for low protein levels in urine. *Clin. Chim. Acta* **1997**, *258*, 231–239.
16. Kohn, J. Nigrosine the forgotten protein stain. *Clin. Chim. Acta* **1987**, *166*, 335–336.
17. Mori, S.; Iida, M. Manufacture of standard color for soy sauce. Jpn. Kokai Tokkyo Koho JP 06194305, 1994; *Chem. Abstr.* **1994**, *121*, 299679.
18. Siemieniuch, M.; Dubiel, A. Preservation of tomcat (*Felis catus*) semen in variable temperatures. *Anim. Reprod. Sci.* **2007**, *99*, 135–144.
19. Bjoerndahl, L.; Soederlund, I.; Kvist, U. Evaluation of the one-step eosin–nigrosin staining technique for human sperm vitality assessment. *Hum. Reprod.* **2003**, *18*, 813–816.
20. Odenthal, M.; Paulig, H. Liquid permanent skin dyes containing glycols and/or benzotriazole. Ger. Offen. DE 19526624, 1997; *Chem. Abstr.* **1997**, *126*, 162005.
21. Bore, P.; De Labbey, A. Evaluation of the state of change of keratin fibers. Brit. UK Pat. Appl. GB 2027195, 1980; *Chem. Abstr.* **1980**, *93*, 101362.
22. Sallwey, A.; Schmitt, M.; Dieckow, A. Dyestuff for hair and fibers. Eur. Pat. Appl. EP 976385, 2000; *Chem. Abstr.* **2000**, *132*, 127457.
23. Wong, S. T.; Hiltibran, R. G.; Huang, T. M.; Calfin, B. B.; Pope, M. R.; Schapira, T. G.; Shain, E. B.; Young, D. F. Diagnostic assay and multilayer assay apparatus for a sample of biological fluid. U.S. Patent 6312888, 2001; *Chem. Abstr.* **2001**, *135*, 341131.
24. Shutova, M. I.; Shustikova, S. A. Preparation of rodenticidal formulation containing diphenacin, a gel-forming component, a dye, and an attractant. Russ. RU 2110916, 1998; *Chem. Abstr.* **2000**, *133*, 85628.
25. Takahashi, H. Uniformly color thermosetting resin laminates, their manufacture, substrates for semiconductor packaging, and semiconductor devices. Jpn. Kokai Tokkyo Koho JP 2008221624, 2008; *Chem. Abstr.* **2008**, *149*, 379947.
26. Joon, D. H.; Peck, D. H.; Shin, D. R.; Hwan, S. D.; Park, J. S. Powder molding materials with good dispersibility for fuel cell separators. Jpn. Kokai Tokkyo Koho JP 2004352986, 2004; *Chem. Abstr.* **2004**, *142*, 25943.
27. Enokimoto, K.; Izumi, Y.; Han, L. Color filters, manufacture thereof, and liquid crystal display using the same. Jpn. Kokai Tokkyo Koho JP 09318812, 1997; *Chem. Abstr.* **1997**, *128*, 103493.
28. Mitsumoto, Y.; Nagataki, Y.; Kono, K.; Asami, A.; Takao, N. Electrophoretic particle composition containing charge-controlling agent for electrophoretic display and method for manufacture thereof. Jpn. Kokai Tokkyo Koho JP 2004279732, 2004; *Chem. Abstr.* **2004**, *141*, 340521.
29. Urano, T.; Takahama, K.; Yamaoka, T. Photoresist for thermal-laser imaging. *J. Photopolym. Sci. Technol.* **2000**, *13*, 89–96.
30. Urano, T.; Takahama, K.; Yamaoka, T. Azide photoresist for thermal-laser imaging. *Imaging Sci. J.* **1999**, *47*, 133–140.
31. Onuki, I.; Ito, H.; Kurosawa, Y. Water-based ball-point pen ink compositions with good blotting and ball-holder wear resistance containing viscosity modifiers and lubricants. Jpn. Kokai Tokkyo Koho JP 2008115210, 2008; *Chem. Abstr.* **2008**, *148*, 540077.
32. Ogasawara, H. Aqueous inks with efficient permeation through cores for fountain pens. Jpn. Kokai Tokkyo Koho JP 2007297519, 2007; *Chem. Abstr.* **2007**, *147*, 523978.
33. Shibata, Y. Oil-based black writing ink compositions and writing instruments using them. Jpn. Kokai Tokkyo Koho JP 2007211198, 2007; *Chem. Abstr.* **2007**, *147*, 259107.
34. Yamazaki, T. Charge-controlling agent, its pulverization, and electrophotographic toner using it. Jpn. Kokai Tokkyo Koho JP 2006139061, 2006; *Chem. Abstr.* **2006**, *144*, 498242.
35. Yamazaki, T. Magnetic toner for electrostatic development from nigrosine base and xylene resin composition. Jpn. Kokai Tokkyo Koho JP 2005157168, 2005; *Chem. Abstr.* **2005**, *143*, 35126.
36. Kato, E.; Ishii, K.; Ishibashi, H.; Dan, S.; Sera, H. Liquid developer for electrostatic photography. Eur. Pat. Appl. EP 155788, 1985; *Chem. Abstr.* **1986**, *104*, 216433.
37. Gepraegs, M. Flame-retardant black thermoplastic molding materials. Ger. Offen. DE 10218902, 2003; *Chem. Abstr.* **2003**, *139*, 351362.

38. Krishnan, S.; Pyles, R. A.; Archey, R. L.; Johnson, J. B. Tinted plastic articles and thermoplastic composition for its preparation. Eur. Pat. Appl. EP 1213320, 2002; *Chem. Abstr.* **2002**, *137*, 21539.
39. Eguchi, Y.; Isetani, A. Colored polyamide compositions and their articles for furniture and building materials with good surface gloss. Jpn. Kokai Tokkyo Koho JP 2003231807, 2003; *Chem. Abstr.* **2003**, *139*, 181138.
40. Ishii, K. Color-loaded polymer microparticles for compounding in paints. Can. CA 1327419, 1994; *Chem. Abstr.* **1995**, *122*, 108780.
41. Ivanov, A. M.; Ivanov, I. A. Method of production of black dyeing composition for fatliquoring substances used in leather manufacturing. Russ. RU 2263713, 2005; *Chem. Abstr.* **2005**, *143*, 442411.
42. Ivanov, A. M.; Ivanov, I. A. Method for preparing black dyeing composition for leather compatible with fatliquoring components. Russ. RU 2243264, 2004; *Chem. Abstr.* **2004**, *142*, 76536.
43. Zhao, X. E.; Suddeth, K. C.; Moore, P. D.; Batlaw, R. Homogeneous liquid complexes of anionic organic dyes and quaternary ammonium compounds for coloring plastics, inks, waxes, and aqueous cleaners. U.S. Patent 5948152, 1999; *Chem. Abstr.* **1999**, *131*, 186356.
44. Yasnogorodskaya, E. I.; Pisanenko, E. P. Salt solution for chemical marking of steel articles. U.S.S.R. SU 1733503, 1992; *Chem. Abstr.* **1993**, *119*, 77265.
45. Yasutake, A. Multicolored wood and method for coloring the wood. Jpn. Kokai Tokkyo Koho JP 08244006, 1996; *Chem. Abstr.* **1996**, *126*, 9428.
46. Rajaguru, P.; Kalpana, R.; Hema, A.; Suba, S.; Baskarasethupathi, B.; Kumar, P. A.; Kalaiselvi, K. Genotoxicity of some sulfur dyes on tadpoles (*Rana hexadactyla*) measured using the comet assay. *Environ. Mol. Mutagen.* **2001**, *38*, 316–322.
47. Chung, K. T.; Fulk, G. E.; Andrews, A. W. Mutagenicity testing of some commonly used dyes. *Appl. Environ. Microbiol.* **1981**, *42*, 641–648.
48. Burnett, C.; Goldenthal, E. I.; Harris, S. B.; Wazeter, F. X.; Strausburg, J.; Kapp, R.; Voelker, R. Teratology and percutaneous toxicity studies on hair dyes. *J. Toxicol. Environ. Health* **1976**, *1*, 1027–1040.

NILE BLUE A

CAS Registry Number 3625-57-8

Chemical Structure

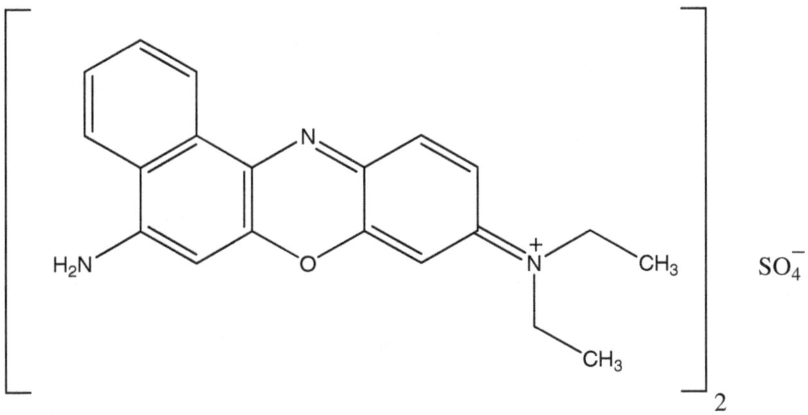

CA Index Name Benzo[*a*]phenoxazin-7-ium, 5-amino-9-(diethylamino)-, sulfate (2:1)

Other Names Bis[5-amino-9-(diethylamino)benzo[*a*]phenazoxonium] sulfate; C.I. 51180; C.I. Basic Blue 12; Nile blue sulfate; 5-Amino-9-(diethylamino)benzo[*a*]phenoxazinium sulfate; Nile Blue A; Nile Blue A sulfate

Merck Index Number Not listed

Chemical/Dye Class Phenoxazine

Molecular Formula $C_{40}H_{40}N_6O_6S$

Molecular Weight 732.85

Physical Form Dark green powder

Solubility Soluble in water, ethanol

Melting Point >300 °C (decompose)

Absorption (λ_{max}) 633 nm

Synthesis Synthetic method[1–5]

Staining Applications Bacteria;[6] erythrocytes;[7] leukocytes;[8] erythroblasts;[8] fungi;[9] granulated cells;[10] intima;[11] lesions;[11] neurons;[12] nucleic acids;[13–16] phospholipids;[17] proteins;[18,19] lipids;[19–21] tumor cells[22]

Biological Applications Detecting microorganisms;[23] treating virus infectious diseases;[24] photodynamic therapy[25]

Industrial Applications Dye laser;[3] electrochromic display device;[26] light-emitting diode;[27] semiconductors;[28] thin film materials;[29] laundry detergent;[30] paper;[31] textiles[32]

Safety/Toxicity Acute toxicity;[33] cytotoxicity;[34] microbial toxicity[35]

Certification/Approval Certified by Biological Stain Commission (BSC)

REFERENCES

1. Ali, M.; Sarkar, A.; Pandey, M. D.; Pandey, S. Efficient precipitation of dyes from dilute aqueous solutions of ionic liquids. *Anal. Sci.* **2006**, *22*, 1051–1053.
2. Presentey, B.; Jerushalmy, Z. On the specificity of various methods used to identify phospholipids in tissues. *Acta Histochem. Cytochem.* **1979**, *12*, 330–336.
3. Drexhage, K. H. Phenoxazine dye laser. Fr. Demande FR 2178158, 1973; *Chem. Abstr.* **1974**, *80*, 114727.
4. Dunnigan, M. G. Chromatographic separation and photometric analysis of the components of Nile blue sulfate. *Stain Technol.* **1968**, *43*, 243–248.
5. Sen, A. K.; Okuda, T.; Price, C. C.; Rutman, R. Preparation of Nile blue mustards. *Acta Unio Int. Contra Cancrum* **1960**, *26*, 774–779.
6. Betscheider, D.; Jose, J. Nile blue A for staining *Escherichia coli* in flow cytometer experiments. *Anal. Biochem.* **2009**, *384*, 194–196.
7. Tsuji, T.; Itose, H.; Konishi, A. Method for automatically analyzing nucleated bone marrow

cell. Jpn. Kokai Tokkyo Koho JP 2003329668, 2003; *Chem. Abstr.* **2003**, *139*, 393149.

8. Tsuji, T.; Sakata, T.; Ikeuchi, Y.; Oguni, S. Erythroblast diagnostic flow-cytometry method and reagents. Eur. Pat. Appl. EP 1004880, 2000; *Chem. Abstr.* **2000**, *133*, 14302.
9. Peruanskii, Y. V.; Peruanskaya, O. N. Ability of the rust fungi uredospores in bind Nile blue. *Tsitologiya* **1967**, *9*, 480–483; *Chem. Abstr.* **1967**, *67*, 30051.
10. Szokoly, V.; Gomba, S.; Soltesz, M. B. Vital staining of the granulated juxtaglomerular cells in the mouse kidney. *Nature* **1965**, *208*, 1331–1332.
11. Slater, R. S.; Smith, E. B. Microdissection of large atherosclerotic plaques to give morphologically and topographically defined fractions for analysis. 2. "Nile blue" cells. *Atherosclerosis* **1972**, *15*, 57–69.
12. Fisher, J. A. N.; Salzberg, B. M.; Yodh, A. G. Near infrared two-photon excitation cross-sections of voltage-sensitive dyes. *J. Neurosci. Methods* **2005**, *148*, 94–102.
13. Zhang, Z.; Huang, W.; Wang, E.; Dong, S. Investigation of the influence on conformational transition of DNA induced by cationic lipid vesicles. *Spectrochim. Acta* **2003**, *59A*, 255–263.
14. Huang, C. Z.; Li, Y. F.; Pu, Q. H.; Lai, L. J. Interactions of Nile blue sulfate with nucleic acids as studied by resonance light-scattering measurements and determination of nucleic acids at nanogram levels. *Anal. Lett.* **1999**, *32*, 2395–2415.
15. Huang, C. Z.; Li, Y. F.; Hu, X. L.; Li, N. B. Three-dimensional spectra of the long-range assembly of Nile blue sulfate on the molecular surface of DNA and determination of DNA by light-scattering. *Anal. Chim. Acta* **1999**, *395*, 187–197.
16. Iogannsen, M. G.; Dolgopyatova, M. A. Interaction of nucleic acids isolated from tumor cells with dyes. *Biokhimiya (Moscow)* **1972**, *37*, 1082–1087; *Chem. Abstr.* **1973**, *78*, 54308.
17. Dunnigan, M. G. Use of Nile blue sulfate in histochemical identification of phospholipids. *Stain Technol.* **1968**, *43*, 249–256.
18. Salomi, B. S. B.; Mitra, C. K.; Gorton, L. Electrochemistry of labeled proteins for biosensors. *Asian J. Phys.* **2005**, *14*, 161–165.
19. Kado, T. A method for detecting stains on a contact lens using an aqueous composition containing a dye. Jpn. Kokai Tokkyo Koho JP 2000065840, 2000; *Chem. Abstr.* **2000**, *132*, 199093.
20. Canete, M.; Hazen, M. J.; Stockert, J. C. Nile blue sulfate staining for demonstration of lipids in fluorescence microscopy. *Acta Histochem. Cytochem.* **1983**, *16*, 286–288.
21. Sire, M. F.; Vernier, J. M. Lipid staining of semithin sections with Sudan black B or Nile blue sulfate: application to intestinal fat absorption. *Acta Histochem. Cytochem.* **1980**, *13*, 193–201.
22. Bastos, A. L.; Marques, D. Nile blue inducible fluorescence of tumor cells. *Z. Naturforsch. B* **1972**, *27*, 1395–1398.
23. Martin, S. M.; MacDonald, J. G.; Lye, J.; Sayre, C.; Thompson, K. Rapid assessment of upper respiratory conditions and detection of microorganisms using test strip comprising at least one broad spectrum indicator. PCT Int. Appl. WO 2009027855, 2009; *Chem. Abstr.* **2009**, *150*, 278103.
24. Yoshinaka, Y.; Kato, I.; Igawa, Y.; Adachi, S. Virus protease inhibitors for treatment of virus infectious diseases. Jpn. Kokai Tokkyo Koho JP 03291227, 1991; *Chem. Abstr.* **1992**, *117*, 20488.
25. Zacal, N.; Rainbow, A. J. Photodynamic therapy resistant human colon carcinoma HT29 cells show cross-resistance to UVA but not UVC light. *Photochem. Photobiol.* **2007**, *83*, 730–737.
26. Vincent, J. B.; Flick, D. W. Electrochromic display device. U.S. Pat. Appl. Publ. US 2003179432, 2003; *Chem. Abstr.* **2003**, *139*, 283537.
27. Leu, C.; Yu, T. C.; Chen, G. L. Light-emitting diode and backlight system using the same. U.S. Pat. Appl. Publ. US 2005117320, 2005; *Chem. Abstr.* **2005**, *142*, 490207.
28. Chatterjee, D.; Dasgupta, S.; Rao, N. N. Visible light assisted photodegradation of halocarbons on the dye modified TiO_2 surface using visible light. *Sol. Energy Mater. Sol. Cells* **2006**, *90*, 1013–1020.
29. El-Nadi, L.; Al-Houty, L.; Omar, M. M.; Ragab, M. Organic thin film materials producing novel blue laser. *Chem. Phys. Lett.* **1998**, *286*, 9–14.
30. Sadlowski, E. S.; Cummings, M. D. Laundry detergent compositions containing hueing dye. U.S. Pat. Appl. Publ. US 2005288207, 2005; *Chem. Abstr.* **2005**, *144*, 71505.
31. Iftime, G.; Norsten, T. B.; Kazmaier, P. M. Inkless printing paper and method. U.S. Pat. Appl. Publ. US 2008311517, 2008; *Chem. Abstr.* **2008**, *150*, 67204.
32. Depoot, K. J. M. Liquid fabric treatment composition comprising deposition aids and fabric hueing dyes. Eur. Pat. Appl. EP 1975226, 2008; *Chem. Abstr.* **2008**, *149*, 404250.

33. Lutty, G. A. The acute intravenous toxicity of biological stains, dyes, and other fluorescent substances. *Toxicol. Appl. Pharmacol.* **1978**, *44*, 225–249.
34. Tong, Z.; Singh, G.; Rainbow, A. J. Extreme dark cytotoxicity of Nile blue A in normal human fibroblasts. *Photochem. Photobiol.* **2001**, *74*, 707–711.
35. Oros, G.; Cserhati, T.; Forgacs, E. Separation of the strength and selectivity of the microbiological effect of synthetic dyes by spectral mapping technique. *Chemosphere* **2003**, *52*, 185–193.

NITRO BLUE TETRAZOLIUM (NBT)

CAS Registry Number 298-83-9

Chemical Structure

Merck Index Number Not listed

Chemical/Dye Class Tetrazolium salt

Molecular Formula $C_{40}H_{30}Cl_2N_{10}O_6$

Molecular Weight 817.64

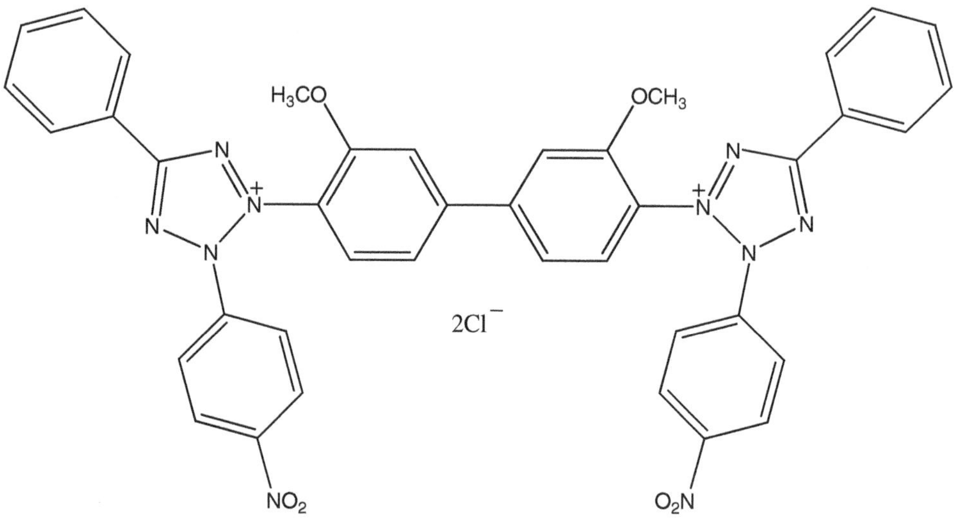

CA Index Name 2*H*-Tetrazolium, 2,2′-(3,3′-dimethoxy [1,1′-biphenyl]-4,4′-diyl)bis[3-(4-nitrophenyl)-5-phenyl]-, chloride (1:2)

Other Names 2*H*-Tetrazolium, 2,2′-(3,3′-dimethoxy [1,1′-biphenyl]-4,4′-diyl)bis[3-(4-nitrophenyl)-5-phenyl]-, dichloride; 2*H*-Tetrazolium, 3,3′-(3,3′-dimethoxy-4,4′-biphenylylene)bis[2-(*p*-nitrophenyl)-5-phenyl]-, dichloride; Tetrazolium, 3,3′-(3,3′-dimethoxy-4,4′-biphenylylene)bis[2-(*p*-nitrophenyl)-5-phenyl-2*H*-chloride]; 2, 2′-Bis(*p*-nitrophenyl)-5,5′-diphenyl-3,3′-(3,3′-dimethoxy-4,4′-biphenylylene)ditetrazolium chloride; 2,2′-Bis(*p*-nitrophenyl)-5,5′-diphenyl-3,3′-(3,3′-dimethoxy-4,4′-diphenylene)ditetrazolium chloride; 2,2′-Di-*p*-nitrophenyl-5,5′-diphenyl-3,3′-bis(3,3′-dimethoxy-4,4′-biphenylene) ditetrazolium chloride; 2,2′-Dinitrophenyl-5,5′-diphenyl-3,3′-dimethoxy-4,4′-diphenylene)ditetrazolium chloride; 3,3′-(3,3′-Dimethoxy-4,4′-biphenylylene)bis[2-(*p*-nitrophenyl)-5-phenyl-2*H*-tetrazolium chloride]; 3,3′-(3,3′-Dimethoxy-4,4′-diphenylene)bis[2-(*p*-nitrophenyl)-5-phenyltetrazolium chloride]; NBT; NBT (dye); NSC 27622; NTB; Nitro BT; Nitro Blue Tetrazolium; Nitro Blue Tetrazolium chloride; Nitro Tetrazolium BT; Nitrotetrazolium Chloride Blue; Nitrotetrazolium blue; Tetrazolium Nitro BT; Tetrazolium nitro blue; *p*-NBT; *p*-Nitro blue tetrazolium; *p*-Nitro blue tetrazolium chloride; *p*-Nitrotetrazolium blue

Physical Form Yellow powder

Solubility Soluble in water, ethanol, methanol, dimethyl sulfoxide

Melting Point 189 °C

Absorption (λ_{max}) 256 nm

Synthesis Synthetic methods[1–6]

Staining Applications Antibody;[7] antigen;[8] cells;[9] dihydrolipoamide dehydrogenase (DLDH) diaphorase;[10] enzymes;[11,12] gene expression[13]

Biological Applications Diagnosis of Alzheimer's disease,[14] bacterial vaginosis,[15] behavioral disturbances in children,[16] Hirschsprung disease;[11] phosphatase assay;[17] phytase assay;[18] detecting bacteria,[19] microorganisms,[20] phosphoinositides,[21] yeast;[22] generating and detecting reactive oxygen species;[23–25] treating cancer,[26] neuropathic pain[27]

Industrial Applications Microchip[28]

Safety/Toxicity Carcinogenicity;[29] chronic toxicity;[30] cytotoxicity;[31–33] hepatotoxicity;[34] mitochondrial damage;[35] mutagenicity;[36] nephrotoxicity;[37] neurotoxicity;[38–40] ototoxicity;[41] phototoxicity;[42,43] pulmonary toxicity[44]

REFERENCES

1. Cui, X.; Vlahakis, J. Z.; Crandall, I. E.; Szarek, W. A. Anti-*Plasmodium* activity of tetrazolium salts. *Bioorg. Med. Chem.* **2008**, *16*, 1927–1947.
2. James, P. R. Tetrazolium salts as steam sterilization process indicators. PCT Int. Appl. WO 9837157, 1998; *Chem. Abstr.* **1998**, *129*, 218237.
3. Ostrovskaya, V. M.; Yakushina, V. I. Nitrotetrazolium chloride blue. [3,3′-(3,3′-Dimethoxy-4,4′-biphenylylene)bis[2-(*p*-nitrophenyl)-5-phenyl-2*H*-tetrazolium dichloride]]. *Metody Poluch. Khim. Reakt. Prep.* **1969**, 154–157; *Chem. Abstr.* **1972**, *76*, 128781.
4. Malev, M. N. Preparation of nitro blue tetrazolium. U. S.S.R. SU 189105, 1966; *Chem. Abstr.* **1967**, *67*, 74472.
5. Pryanishnikov, A. A.; Ostrovskaya, V. M.; Raikhlin, N. T. Synthesis, properties, and histochemical application of salts of tetrazole with electron-accepting substituents. *Tr. Vses. Nauchno-Issled. Inst. Khim. Reakt.* **1963**, 139–148; *Chem. Abstr.* **1964**, *61*, 4214.
6. Tsou, K.; Cheng, C.; Nachlas, M. M.; Seligman, A. M. Syntheses of some *p*-nitrophenyl-substituted tetrazolium salts as electron acceptors for the demonstration of dehydrogenases. *J. Am. Chem. Soc.* **1956**, *78*, 6139–6144.
7. Wang, S.; Indrawati, L.; Wooters, M.; Caro-Aguilar, I.; Field, J.; Kaufhold, R.; Payne, A.; Caulfield, M. J.; Smith, J. G.; Heinrichs, J. H. A novel automated method for enumeration of *Chlamydia trachomatis* inclusion forming units. *J. Immunol. Methods* **2007**, *324*, 84–91.
8. Moulin, P.; Guiot, Y.; Rahier, J. Method and kit for determining antigen content in a sample using double immunohistochemical detection. PCT Int. Appl. WO 2008023055, 2008; *Chem. Abstr.* **2008**, *148*, 302883.
9. Berridge, M. V.; Herst, P. M.; Tan, A. S. Tetrazolium dyes as tools in cell biology: new insights into their cellular reduction. *Biotechnol. Annu. Rev.* **2005**, *11*, 127–152.
10. Yan, L. J.; Yang, S. H.; Shu, H.; Prokai, L.; Forster, M. J. Histochemical staining and quantification of dihydrolipoamide dehydrogenase diaphorase activity using blue native PAGE. *Electrophoresis* **2007**, *28*, 1036–1045.
11. Martucciello, G.; Favre, A. Diagnostic kit with pre-measured, stored components for histochemical determination of enzymic markers of Hirschsprung disease (congenital megacolon). Eur. Pat. Appl. EP 1724358, 2006; *Chem. Abstr.* **2006**, *145*, 501385.
12. Baud, S.; Graham, I. A. A spatiotemporal analysis of enzymatic activities associated with carbon metabolism in wild-type and mutant embryos of *Arabidopsis* using *in situ* histochemistry. *Plant J.* **2006**, *46*, 155–169.
13. Trinh, L. A.; McCutchen, M. D.; Bonner-Fraser, M.; Fraser, S. E.; Bumm, L. A.; McCauley, D. W. Fluorescent *in situ* hybridization employing the conventional NBT/BCIP chromogenic stain. *BioTechniques* **2007**, *42*, 756–759.
14. Mendez, E. Method for the *in vitro* diagnosis of Alzheimer's disease using a monoclonal antibody. PCT Int. Appl. WO 2006095041, 2006; *Chem. Abstr.* **2006**, *145*, 331220.
15. Dwir, O. S.; Keren, T.; Fish, F. Dry format sialidase colorimetric assay for bacterial vaginosis diagnosis. PCT Int. Appl. WO 2008018074, 2008; *Chem. Abstr.* **2008**, *148*, 232647.
16. Pokrovskii, D. G. Early diagnosis of affective and behavioral disturbances in children based on anxiety level, leukocyte formula, spontaneous NBT test and catamnesis. Russ. RU 2305566, 2007; *Chem. Abstr.* **2007**, *147*, 341665.
17. Tsukatani, T.; Ide, S.; Ono, M.; Matsumoto, K. New tetrazolium method for phosphatase assay using ascorbic acid 2-phosphate as a substrate. *Talanta* **2007**, *73*, 471–475.
18. Kumar, V.; Miasnikov, A. Colorimetric assay for phytase using phosphorylated polyhydroxybenzene as the substrate, and feed analysis applications. PCT Int. Appl. WO 2006123254, 2006; *Chem. Abstr.* **2006**, *146*, 3106.
19. Ushiyama, M. Film culture medium for detecting bacteria in water-system environment. Jpn. Kokai Tokkyo Koho JP 2007020434, 2007; *Chem. Abstr.* **2007**, *146*, 180320.
20. Martin, S. M.; MacDonald, J. G.; Lye, J.; Sayre, C.; Thompson, K. Array for rapid detection of a microorganism based on dyes. PCT Int. Appl. WO 2008026104, 2008; *Chem. Abstr.* **2008**, *148*, 302867.
21. Furutani, M.; Itoh, T.; Ijuin, T.; Tsujita, K.; Takenawa, T. Thin layer chromatography-blotting, a novel method for the detection of phosphoinositides. *J. Biochem.* **2006**, *139*, 663–670.
22. O'Neill, M.; Lebrun, S. Rapid, low-cost assay for detecting Brettanomyces and other spoilage yeast in

wine. U.S. Pat. Appl. Publ. US 2007196877, 2007; *Chem. Abstr.* **2007**, *147*, 276295.

23. Tai, Y. S. Detection of reactive oxygen species can be used to distinguish ToxA-induced cell death from the hypersensitive response. *Res. J. Bot.* **2007**, *2*, 1–12.
24. Armstrong, J. S.; Whiteman, M. Measurement of reactive oxygen species in cells and mitochondria. *Methods Cell Biol.* **2007**, *80*, 355–377.
25. Hoffmann, A.; Hammes, E.; Plieth, C.; Desel, C.; Sattelmacher, B.; Hansen, U. P. Effect of CO_2 supply on formation of reactive oxygen species in *Arabidopsis thaliana*. *Protoplasma* **2005**, *227*, 3–9.
26. Kong, Q. Method and composition for treating cancer using cellular organelle crystallizing agents. U.S. Pat. Appl. Publ. US 2006286170, 2006; *Chem. Abstr.* **2006**, *146*, 55482.
27. Tanabe, T. Remedy for neuropathic pain. PCT Int. Appl. WO 2006049312, 2006; *Chem. Abstr.* **2006**, *144*, 445378.
28. Yang, B.; Sudo, Y.; Sakaino, Y.; Karaki, H.; Wakabayashi, A. Microchip and analysis method using the same. Eur. Pat. Appl. EP 1717585, 2006; *Chem. Abstr.* **2006**, *145*, 467668.
29. Ohkawa, Y.; Iwata, K.; Shibuya, H.; Fujiki, H.; Inui, N. A rapid, simple screening method for skin-tumor promoters using mouse peritoneal macrophages *in vitro*. *Cancer Lett.* **1984**, *21*, 253–260.
30. Garg, U. K.; Pal, A. K.; Jha, G. J.; Jadhao, S. B. Haemato-biochemical and immuno-pathophysiological effects of chronic toxicity with synthetic pyrethroid, organophosphate and chlorinated pesticides in broiler chicks. *Int. Immunopharmacol.* **2004**, *4*, 1709–1722.
31. Anazetti, M. C.; Melo, P. S.; Duran, N.; Haun, M. Comparative cytotoxicity of dimethylamide-crotonin in the promyelocytic leukemia cell line (HL60) and human peripheral blood mononuclear cells. *Toxicology* **2003**, *188*, 261–274.
32. Ermolli, M.; Menne, C.; Pozzi, G.; Serra, M. A.; Clerici, L. A. Nickel, cobalt and chromium-induced cytotoxicity and intracellular accumulation in human hacat keratinocytes. *Toxicology* **2001**, *159*, 23–31.
33. Shimada, H.; Hirai, K.; Simamura, E.; Pan, J. Mitochondrial NADH-quinone oxidoreductase of the outer membrane is responsible for paraquat cytotoxicity in rat livers. *Arch. Biochem. Biophys.* **1998**, *351*, 75–81.
34. Mochida, S.; Ohta, Y.; Ogata, I.; Fujiwara, K. Gut-derived substances in activation of hepatic macrophages after partial hepatectomy in rats. *J. Hepatol.* **1992**, *16*, 266–272.
35. Maharaj, H.; Maharaj, D. S.; Daya, S. Acetylsalicylic acid and acetaminophen protect against MPP^+-induced mitochondrial damage and superoxide anion generation. *Life Sci.* **2006**, *78*, 2438–2443.
36. Venitt, S.; Crofton-Sleigh, C. Bacterial mutagenicity tests of phenazine methosulfate and three tetrazolium salts. *Mutat. Res.* **1979**, *68*, 107–116.
37. Romero, M.; Mosquera, J.; Rodriguez-Iturbe, B. A simple method to identify NBT-positive cells in isolated glomeruli. *Nephrol. Dial. Transplant.* **1997**, *12*, 174–179.
38. Maharaj, D. S.; Saravanan, K. S.; Maharaj, H.; Mohanakumar, K. P.; Daya, S. Acetaminophen and aspirin inhibit superoxide anion generation and lipid peroxidation, and protect against 1-methyl-4-phenyl pyridinium-induced dopaminergic neurotoxicity in rats. *Neurochem. Int.* **2004**, *44*, 355–360.
39. Chromy, B. A.; Nowak, R. J.; Lambert, M. P.; Viola, K. L.; Chang, L.; Velasco, P. T.; Jones, B. W.; Fernandez, S. J.; Lacor, P. N.; Horowitz, P.; Finch, C. E.; Krafft, G. A.; Klein, W. L. Self-assembly of Abeta(1–42) into globular neurotoxins. *Biochemistry* **2003**, *42*, 12749–12760.
40. Romero, I. A.; Lister, T.; Richards, H. K.; Seville, M. P.; Wylie, S. P.; Ray, D. E. Early metabolic changes during *m*-Dinitrobenzene neurotoxicity and the possible role of oxidative stress. *Free Radical Biol. Med.* **1995**, *18*, 311–319.
41. Dehne, N.; Lautermann, J.; Petrat, F.; Rauen, U.; de Groot, H. Cisplatin ototoxicity: involvement of iron and enhanced formation of superoxide anion radicals. *Toxicol. Appl. Pharmacol.* **2001**, *174*, 27–34.
42. Vargas, F.; Carbonell, K.; Camacho, M. Photochemistry and *in vitro* phototoxicity studies of levomepromazine (methotrimeprazine), a phototoxic neuroleptic drug. *Pharmazie* **2003**, *58*, 315–319.
43. Vargas, F.; Mendez, H.; Fuentes, A.; Sequera, J.; Fraile, G.; Velasquez, M.; Caceres, G.; Cuello, K. Photosensitizing activity of thiocolchicoside: photochemical and *in vitro* phototoxicity studies. *Pharmazie* **2001**, *56*, 83–88.
44. Nelson, S.; Summer, W. R.; Terry, P. B.; Warr, G. A.; Jakab, G. J. Erythromycin-induced suppression of pulmonary antibacterial defenses: a potential mechanism of superinfection in the lung. *Am. Rev. Respir. Dis.* **1987**, *136*, 1207–1212.

NONYL-ACRIDINE ORANGE (NAO)

CAS Registry Number 75168-11-5

Chemical Structure

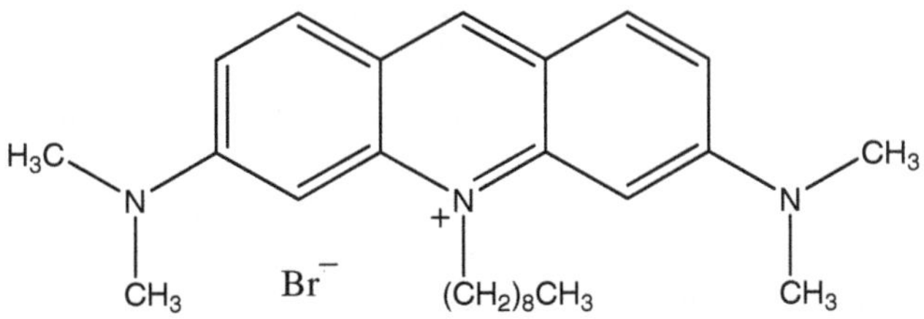

CA Index Name Acridinium, 3,6-bis(dimethylamino)-10-nonyl-, bromide (1:1)

Other Names Acridinium, 3,6-bis(dimethylamino)-10-nonyl-, bromide; 10-nonyl acridine orange; A 1372; Acridine Orange 10-nonyl bromide; Nonylacridine orange; NAO

Merck Index Number Not listed

Chemical/Dye Class Acridine

Molecular Formula $C_{26}H_{38}BrN_3$

Molecular Weight 472.50

Physical Form Orange-red powder

Solubility Soluble in ethanol, methanol, dimethyl sulfoxide

Melting Point >250 °C

Absorption (λ_{max}) 495 nm

Emission (λ_{max}) 519 nm

Synthesis Synthetic method[1]

Staining Applications Mitochondria;[1–11,21,23,25,28] liposomes;[12] nerve terminals;[13] cardiolipin;[14–18] blood cells[27]

Biological Applications Detecting cardiolipin,[14–18] mitochondrial membrane potential,[19,20] risk of type 2 diabetes,[21] cardiotoxicity,[22] prostate cancer;[23] apoptosis assay;[24,25] drug screening assay;[26] hematotoxicity assay;[27] treating arthritic disorders;[28] in photodynamic therapy;[29] ophthalmic applications[30]

Industrial Applications Not reported

Safety/Toxicity No data available

REFERENCES

1. Septinus, M.; Seiffert, W.; Zimmermann, H. W. Hydrophobic acridine dyes for fluorescence staining of mitochondria in living cells. 1. Thermodynamic and spectroscopic properties of 10-*n*-alkylacridine orange chlorides. *Histochemistry* **1983**, *79*, 443–456.
2. Hattori, F.; Fukuda, K. Method for selecting myocardial cells using intracellular mitochondria labeled with fluorescent indicator. PCT Int. Appl. WO 2006022377, 2006; *Chem. Abstr.* **2006**, *144*, 270175.
3. Ahmadzadeh, H.; Thompson, L. V.; Arriaga, E. A. On-column labeling for capillary electrophoretic analysis of individual mitochondria directly sampled from tissue cross sections. *Anal. Bioanal. Chem.* **2006**, *384*, 169–174.
4. Ahmadzadeh, H.; Johnson, R. D.; Thompson, L.; Arriaga, E. A. Direct sampling from muscle cross sections for electrophoretic analysis of individual mitochondria. *Anal. Chem.* **2004**, *76*, 315–321.
5. Benel, L.; Ronot, X.; Mounolou, J. C.; Gaudemer, F.; Adolphe, M. Compared flow cytometric analysis of mitochondria using 10-*N*-nonyl acridine orange and rhodamine 123. *Basic Appl. Histochem.* **1989**, *33*, 71–80.
6. Maftah, A.; Petit, J. M.; Ratinaud, M. H.; Julien, R. 10-*N*-Nonyl-acridine orange: a fluorescent probe which stains mitochondria independently of their energetic state. *Biochem. Biophys. Res. Commun.* **1989**, *164*, 185–190.
7. Ratinaud, M. H.; Leprat, P.; Julien, R. *In situ* flow cytometric analysis of nonyl acridine orange-stained mitochondria from splenocytes. *Cytometry* **1988**, *9*, 206–212.
8. Septinus, M.; Berthold, T.; Naujok, A.; Zimmermann, H. W. Hydrophobic acridine dyes for fluorescent staining of mitochondria in living cells. 3. Specific accumulation of the fluorescent dye NAO on the mitochondrial membranes in HeLa cells by hydrophobic interaction. Depression of respiratory activity, changes in the ultrastructure of mitochondria due to NAO. Increase of fluorescence in vital stained mitochondria *in situ* by irradiation. *Histochemistry* **1985**, *82*, 51–66.
9. Erbrich, U.; Septinus, M.; Naujok, A.; Zimmermann, H. W. Hydrophobic acridine dyes for fluorescence

staining of mitochondria in living cells. 2. Comparison of staining of living and fixed HeLa-cells with NAO and DPPAO. *Histochemistry* **1984**, *80*, 385–388.

10. Jacobson, J.; Duchen, M. R.; Heales, S. J. R. Intracellular distribution of the fluorescent dye nonyl acridine orange responds to the mitochondrial membrane potential: implications for assays of cardioliponyl acridine orangein and mitochondrial mass. *J. Neurochem.* **2002**, *82*, 224–233.
11. Keij, J. F.; Bell-Prince, C.; Steinkamp, J. A. Staining of mitochondrial membranes with 10-nonylacridine orange, MitoFluor Green, and MitoTracker Green is affected by mitochondrial membrane potential altering drugs. *Cytometry* **2000**, *39*, 203–210.
12. Agafonov, A. V.; Gritsenko, E. N.; Shlyapnikova, E. A.; Kharakoz, D. P.; Belosludtseva, N. V.; Lezhnev, E. I.; Saris, N. E. L.; Mironova, G. D. Ca^{2+}-induced phase separation in the membrane of palmitate-containing liposomes and its possible relation to membrane permeabilization. *J. Membr. Biol.* **2007**, *215*, 57–68.
13. Herrera, A. A.; Banner, L. R. The use and effects of vital fluorescent dyes: observation of motor nerve terminals and satellite cells in living frog muscles. *J. Neurocytol.* **1990**, *19*, 67–83.
14. Gohil, V. M.; Gvozdenovic-Jeremic, J.; Schlame, M.; Greenberg, M. L. Binding of 10-*N*-nonylacridine orange to cardiolipin-deficient yeast cells: implications for assay of cardiolipin. *Anal. Biochem.* **2005**, *343*, 350–352.
15. Mileykovskaya, E.; Dowhan, W.; Birke, R. L.; Zheng, D.; Lutterodt, L.; Haines, T. H. Cardiolipin binds nonyl acridine orange by aggregating the dye at exposed hydrophobic domains on bilayer surfaces. *FEBS Lett.* **2001**, *507*, 187–190.
16. Kaewsuya, P.; Danielson, N. D.; Ekhterae, D. Fluorescent determination of cardiolipin using 10-*N*-nonyl acridine orange. *Anal. Bioanal. Chem.* **2007**, *387*, 2775–2782.
17. Garcia Fernandez, M. I.; Ceccarelli, D.; Muscatello, U. Use of the fluorescent dye 10-*N*-nonyl acridine orange in quantitative and location assays of cardiolipin: a study on different experimental models. *Anal. Biochem.* **2004**, *328*, 174–180.
18. Fuller, K. M.; Duffy, C. F.; Arriaga, E. A. Determination of the cardiolipin content of individual mitochondria by capillary electrophoresis with laser-induced fluorescence detection. *Electrophoresis* **2002**, *23*, 1571–1576.
19. Dykens, J. A.; Fleck, B.; Ghosh, S.; Lewis, M.; Velicelebi, G.; Ward, M. W. High-throughput assessment of mitochondrial membrane potential *in situ* using fluorescence resonance energy transfer. *Mitochondrion* **2002**, *1*, 461–473.
20. Fujii, H.; Cody, S. H.; Seydel, U.; Papadimitriou, J. M.; Wood, D. J.; Zheng, M. H. Recording of mitochondrial transmembrane potential and volume in cultured rat osteoclasts by confocal laser scanning microscopy. *Histochem. J.* **1997**, *29*, 571–581.
21. Anderson, C. M.; Davis, R. E. Indicators of altered mitochondrial function in predictive methods for determining risk of type 2 diabetes mellitus. U.S. Patent 6140067, 2000; *Chem. Abstr.* **2000**, *133*, 319305.
22. Crane, P. D.; Orlandi, C. Labeled ether-substituted isonitriles for detecting the cardiotoxicity of compounds. U.S. Patent 5604112, 1997; *Chem. Abstr.* **1997**, *126*, 207495.
23. Dickman, D. Methods of detecting prostate cancer. PCT Int. Appl. WO 2006054296, 2006; *Chem. Abstr.* **2006**, *145*, 3346.
24. Ferlini, C.; Scambia, G. Assay for apoptosis using the mitochondrial probes, Rhodamine123 and 10-*N*-nonyl acridine orange. *Nat. Protoc.* **2007**, *2*, 3111–3114.
25. King, M. A.; Eddaoudi, A.; Davies, D. C. A comparison of three flow cytometry method for evaluating mitochondrial damage during staurosporine-induced apoptosis in Jurkat cells. *Cytometry* **2007**, *71A*, 668–674.
26. Dykens, J. A.; Velicelebi, G.; Ghosh, S. S. Compositions and methods for assaying subcellular conditions and processes using energy transfer for drug screening. PCT Int. Appl. WO 2000079274, 2000; *Chem. Abstr.* **2000**, *134*, 66121.
27. Dertinger, S. D.; Bemis, J. C.; Bryce, S. M. Method for measuring *in vivo* hematotoxicity with an emphasis on radiation exposure assessment. U.S. Pat. Appl. Publ. US 2008311586, 2008; *Chem. Abstr.* **2008**, *150*, 50134.
28. Murphy, A. N.; Dykens, J. A.; Ghosh, S. S.; Davis, R. E.; Granston, A. E., Jr.; Terkeltaub, R. Altered mitochondrial function indicator-based methods and compositions for diagnosing and treating arthritic disorders and regulating bone mass. PCT

Int. Appl. WO 2001020018, 2001; *Chem. Abstr.* **2001**, *134*, 231892.

29. Woo, T.; Miller, G. G.; Madiyalakan, R. Method and device for photodynamic therapy comprising administering and activating a photosensitizer with a chemiluminescent and/or a light-emitting diode light source in dermal contact with the patient. PCT Int. Appl. WO 2008011707, 2008; *Chem. Abstr.* **2008**, *148*, 222105.

30. Mentak, K. Ultra violet, violet, and blue light filtering polymers for ophthalmic applications. U.S. Pat. Appl. Publ. US 2006252844, 2006; *Chem. Abstr.* **2006**, *145*, 495756.

NUCLEAR YELLOW

CAS Registry Number 74681-68-8

Chemical Structure

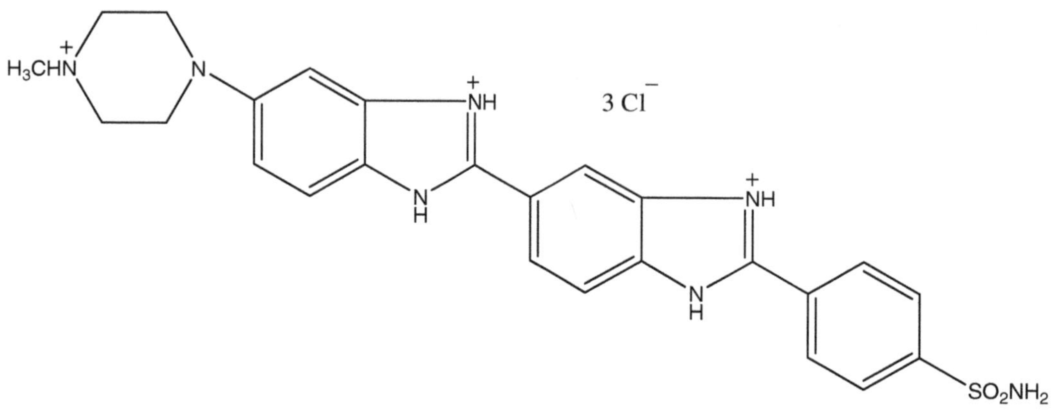

CA Index Name Benzenesulfonamide, 4-[5-(4-methyl-1-piperazinyl)[2,5′-bi-1*H*-benzimidazol]-2′-yl]-, hydrochloride (1:3)

Other Names Benzenesulfonamide, 4-[5-(4-methyl-1-piperazinyl)[2,5′-bi-1*H*-benzimidazol]-2′-yl]-, trihydrochloride; 4-[5-(4-Methyl-1-piperazinyl)[2,5′-bi-1*H*-benzimidazol]-2′-yl]-benzenesulfonamide; Hoechst S769121; Hoechst S769121 trihydrochloride; Nuclear yellow

Merck Index Number Not listed

Chemical/Dye Class Benzimidazole

Molecular Formula $C_{25}H_{28}Cl_3N_7O_2S$

Molecular Weight 596.96

Physical Form Yellow to green powder

Solubility Soluble in dimethyl sulfoxide, water

Melting Point >300 °C

Absorption (λ_{max}) 355 nm

Emission (λ_{max}) 495 nm

Synthesis Synthetic method[1]

Staining Applications Nucleic acids;[2–5] cells;[6] chromosomes;[7] lysosomes;[8] mitochondria;[9] neurons;[1,10–13] spinal cord[14]

Biological Applications Detecting nucleic acid,[2–5,15] proteins,[15] carbohydrates,[15] lipids;[15] carrying out polymerase chain reaction (PCR);[16] drug-targeting agent;[17] nerve mapping[18]

Industrial Applications Not reported

Safety/Toxicity No data available

REFERENCES

1. Bentivoglio, M.; Kuypers, H. G. J. M.; Catsman-Berrevoets, C. E.; Loewe, H.; Dann, O. Two new fluorescent retrograde neuronal tracers which are transported over long distances. *Neurosci. Lett.* **1980**, *18*, 25–30.
2. Weber, J.; Brendler, A.; Bley, T. Method and kit for the analysis of the DNA content in cells, especially plant cells using barium hydroxide treatment prior staining. Ger. Offen. DE 102006046079, 2008; *Chem. Abstr.* **2008**, *148*, 373652.
3. Kobayashi, M.; Kaji, S.; Omi, M.; Tamiya, E. Electrochemical DNA quantification based on aggregation induced by phosphate group-binding substance. Jpn. Kokai Tokkyo Koho JP 2006145342, 2006; *Chem. Abstr.* **2006**, *145*, 2140.
4. Stockert, J. C.; Pinna-Senn, E.; Bella, J. L.; Lisanti, J. A. DNA-binding fluorochromes: correlation between C-banding of mouse metaphase chromosomes and hydrogen bonding to adenine-thymine base pairs. *Acta Histochem.* **2005**, *106*, 413–420.
5. Curtis, S. K.; Cowden, R. R. Evaluation of five basic fluorochromes of potential use in microfluorometric studies of nucleic acids. *Histochemistry* **1983**, *78*, 503–511.
6. Anderson, A. L.; Knutson, C. R.; Mueth, D.; Plewa, J.; Tanner, E. Methods for staining cells for identification

and sorting. U.S. Pat. Appl. Publ. US 2006172315, 2006; *Chem. Abstr.* **2006**, *145*, 183714.

7. Pinna-Senn, E.; Lisanti, J. A.; Ortiz, M. I.; Dalmasso, G.; Bella, J. L.; Gosalvez, J.; Stockert, J. C. Specific heterochromatic banding of metaphase chromosomes using nuclear yellow. *Biotech. Histochem.* **2000**, *75*, 132–140.
8. Rashid, F.; Horobin, R. W.; Williams, M. A. Predicting the behavior and selectivity of fluorescent probes for lysosomes and related structures by means of structure–activity models. *Histochem. J.* **1991**, *23*, 450–459.
9. Rashid, F.; Horobin, R. W. Interaction of molecular probes with living cells and tissues. Part 2. A structure–activity analysis of mitochondrial staining by cationic probes, and a discussion of the synergistic nature of image-based and biochemical approaches. *Histochemistry* **1990**, *94*, 303–308.
10. Ishihara, A.; Taguchi, S.; Araki, H.; Nishihira, Y. Retrograde neuronal labeling of motoneurons in the rat by fluorescent tracers, and quantitative analysis of oxidative enzyme activity in labeled neurons. *Neurosci. Lett.* **1991**, *124*, 141–143.
11. So, K. F.; Xiao, Y. M.; Diao, Y. C. Effects on the growth of damaged ganglion cell axons after peripheral nerve transplantation in adult hamsters. *Brain Res.* **1986**, *377*, 168–172.
12. Van der Krans, A.; Hoogland, P. V. Labeling of neurons following intravenous injections of fluorescent tracers in mice. *J. Neurosci. Methods* **1983**, *9*, 95–103.
13. Katan, S.; Gottschall, J.; Neuhuber, W. Simultaneous visualization of horseradish peroxidase and nuclear yellow in tissue sections for neuronal double labeling. *Neurosci. Lett.* **1982**, *28*, 121–126.
14. Boulenguez, P.; Gestreau, C.; Vinit, S.; Stamegna, J. C.; Kastner, A.; Gauthier, P. Specific and artifactual labeling in the rat spinal cord and medulla after injection of monosynaptic retrograde tracers into the diaphragm. *Neurosci. Lett.* **2007**, *417*, 206–211.
15. Nakayama, M. Electrochemical chip or biosensor and electrochemically active substances for detection of biological molecule such as nucleic acid, protein, carbohydrate and lipid in medical, food or environmental field. PCT Int. Appl. WO 2008096757, 2008; *Chem. Abstr.* **2008**, *149*, 262178.
16. Gauer, C.; Mann, W.; Alunni-Fabbroni, M. Methods for carrying out enzymic reactions, such as PCR, in single cells. PCT Int. Appl. WO 2008064730, 2008; *Chem. Abstr.* **2008**, *148*, 578429.
17. Kuchiiwa, S.; Kuchiiwa, T. Agent for targeting drug to cerebral neuron. PCT Int. Appl. WO 2007086587, 2007; *Chem. Abstr.* **2007**, *147*, 220081.
18. Mangat, G.; Brzozowski, L. Intra-operative head and neck nerve mapping. PCT Int. Appl. WO 2007016790, 2007; *Chem. Abstr.* **2007**, *146*, 223533.

OIL RED O

CAS Registry Number 1320-06-5

Chemical Structure

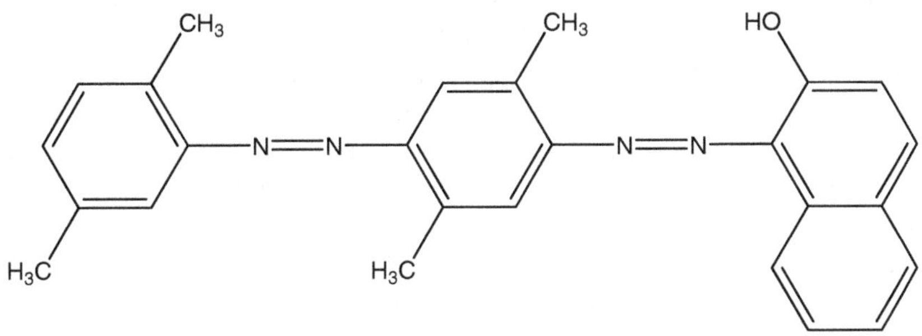

CA Index Name 2-Naphthalenol, 1-[2-[4-[2-(dimethylphenyl)diazenyl]dimethylphenyl]diazenyl]-

Other Names 2-Naphthalenol, 1-[[4-[(dimethylphenyl) azo]dimethylphenyl]azo]-; C.I. Solvent Red 27; Oil Red O; Aizen SOT Red 2; C.I. 26125; D and C Red No. 18; Fat Red 5B; Fat Red 5B02; Oil Red 5303; Oil Red 5B; Oil Red 6B; Oil Red OS; Orient Oil Red 5B; Red OS; SOT Red 2; Solvent Red 27; 1-([4-(Xylylazo)xylyl]azo)-2-naphthol

Merck Index Number Not listed

Chemical/Dye Class Azo

Molecular Formula $C_{26}H_{24}N_4O$

Molecular Weight 408.49

Physical Form Red to dark greenish-brown powder

Solubility Slightly soluble in water, ethanol; soluble in acetone, benzene, ethyl acetate, toluene, xylene, ethylene glycol

Melting Point 120 °C (decompose)

Absorption (λ_{max}) 518 nm, 359 nm

Synthesis Synthetic method[1–4]

Staining Applications Aortic valves;[5] cornea verticillata;[6] human adipose-derived stem cells;[7] intramyocellular lipid;[8] lipids;[9,10] lipid droplets;[11–13] mitochondria;[13] liposomes;[14] lipoproteins;[15] apolipoproteins;[15] neutral lipids;[16] nematodes;[17] natural fibers;[18] plant cuticle;[18] pancreatic fibroblastoid/stellate cells;[19] plant;[20] sirolimus;[21] spinal fluid;[22] wheat flour;[23] fats/oils;[23] hairs[24]

Biological Applications Detecting early stages of atherosclerotic lesion formation;[25] treating obesity,[26] hyperlipidemia,[26] fatty liver,[26] leptin production disorders;[27] dental impression material;[28] measuring phagocytosis[29]

Industrial Applications Color filters;[30] liquid crystals;[31,32] photoresists;[33] liquid crystal displays;[34,35] inks;[36,37] toners;[38] adhesive paper;[39] electrophotography;[40] optical devices;[41] air freshners;[42] furniture polish;[43] pesticide;[44] plastic lenses;[45] steel materials;[46] textiles;[47] petroleum products;[48,49] perfumes;[50] wood products[51]

Safety/Toxicity Cytotoxicity;[52,53] deformation of lipid droplets;[12] cardiovascular toxicity;[54] nephrotoxicity[55]

Certification/Approval Certified by Biological Stain Commission (BSC)

REFERENCES

1. Zhang, L.; Chen, R. Preparation of antioxidants for fuel oils. Faming Zhuanli Shenqing Gongkai Shuomingshu CN 1211611, 1999; *Chem. Abstr.* **2000**, *132*, 295986.
2. Proctor, G. B.; Horobin, R. W. Purification of Oil Red O using preparative paper chromatography. *Stain Technol.* **1985**, *60*, 247–248.
3. Reiners, W. Isolation and thin-layer chromatography of fat-soluble dyes. *Fresenius' Z. Anal. Chem.* **1967**, *229*, 406–409.
4. French, R. W. Fat stains. *Stain Technol.* **1926**, *1*, 79.
5. Olsson, M.; Thyberg, J.; Nilsson, J. Presence of oxidized low density lipoprotein in nonrheumatic

Handbook of Biological Dyes and Stains By R. W. Sabnis

stenotic aortic valves. *Arterioscler. Thromb. Vasc. Biol.* **1999**, *19*, 1218–1222.

6. Hirano, K.; Murata, K.; Miyagawa, A.; Terasaki, H.; Saigusa, J.; Nagasaka, T.; Kobayashi, M. Histopathologic findings of cornea verticillata in a woman heterozygous for Fabry's disease. *Cornea* **2001**, *20*, 233–236.
7. Wolbank, S.; Peterbauer, A.; Wassermann, E.; Hennerbichler, S.; Voglauer, R.; van Griensven, M.; Duba, H. C.; Gabriel, C.; Redl, H. Labelling of human adipose-derived stem cells for non-invasive *in vivo* cell tracking. *Cell Tissue Banking* **2007**, *8*, 163–177.
8. De Bock, K.; Dresselaers, T.; Kiens, B.; Richter, E. A.; Van Hecke, P.; Hespel, P. Evaluation of intramyocellular lipid breakdown during exercise by biochemical assay, NMR spectroscopy, and Oil Red O staining. *Am. J. Physiol.* **2007**, *293*, E428–E434.
9. Rizk, T.; Montero-Menei, C.; Jollivet, C.; Benoit, J. P.; Menei, P. Pitfalls in the detection of lipid vectors in neural cell culture and in brain tissue. *J. Biomed. Mater. Res.* **2004**, *68A*, 360–364.
10. Churukian, C. J. Lillie's oil red O method for neutral lipids. *J. Histotechnol.* **1999**, *22*, 309–311.
11. Wang, H.; Wang, H.; Xiong, W.; Chen, Y.; Ma, Q.; Ma, J.; Ge, Y.; Han, D. Evaluation on the phagocytosis of apoptotic spermatogenic cells by Sertoli cells *in vitro* through detecting lipid droplet formation by Oil Red O staining. *Reproduction* **2006**, *132*, 485–492.
12. Fukumoto, S.; Fujimoto, T. Deformation of lipid droplets in fixed samples. *Histochem. Cell Biol.* **2002**, *118*, 423–428.
13. Shaw, C. S.; Jones, D. A.; Wagenmakers, A. J. M. Network distribution of mitochondria and lipid droplets in human muscle fibers. *Histochem. Cell Biol.* **2008**, *129*, 65–72.
14. Loukas, Y. L. A computer-based expert system designs and analyzes a 2(k-p) fractional factorial design for the formulation optimization of novel multicomponent liposomes. *J. Pharm. Biomed. Anal.* **1998**, *17*, 133–140.
15. Koren, E.; Koscec, M. Antibodies to lipoproteins and apolipoproteins and methods of use thereof. PCT Int. Appl. WO 9600903, 1996; *Chem. Abstr.* **1996**, *124*, 197747.
16. Patel, M. N.; Stolinski, M.; Wright, D. J. Neutral lipids and the assessment of infectivity in entomopathogenic nematodes: observations on four *Steinernema* species. *Parasitology* **1997**, *114*, 489–496.
17. Fitters, P. F. L.; Meijer, E. M. J.; Wright, D. J.; Griffin, C. T. Estimation of lipid reserves in unstained living and dead nematodes by image analysis. *J. Nematol.* **1997**, *29*, 160–167.
18. Akin, D. E.; Rigsby, L. L.; Morrison, W. H. Oil Red as a histochemical stain for natural fibers and plant cuticle. *Ind. Crops Prod.* **2004**, *19*, 119–124.
19. Kruse, M. L.; Hildebrand, P. B.; Timke, C.; Folsch, U. R.; Schafer, H.; Schmidt, W. E. Isolation, long-term culture, and characterization of rat pancreatic fibroblastoid/stellate cells. *Pancreas* **2001**, *23*, 49–54.
20. Bamba, T.; Fukusaki, E. I.; Nakazawa, Y.; Kobayashi, A. *In-situ* chemical analyses of *trans*-polyisoprene by histochemical staining and Fourier transform infrared microspectroscopy in a rubber-producing plant, *Eucommia ulmoides* Oliver. *Planta* **2002**, *215*, 934–939.
21. Ma, K. L.; Ruan, X. Z.; Powis, S. H.; Moorhead, J. F.; Varghese, Z. Anti-atherosclerotic effects of sirolimus on human vascular smooth muscle cells. *Am. J. Physiol.* **2007**, *292*, H2721–H2728.
22. Masuoka, S.; Yoneda, K.; Oka, M. Apparatus and method for analyzing spinal fluid by staining and imaging analysis. Jpn. Kokai Tokkyo Koho JP 2004132787, 2004; *Chem. Abstr.* **2004**, *140*, 371466.
23. Ashida, Y. Examination of food containing wheat flour and fats/oils by double staining and staining solution kits therefor. Jpn. Kokai Tokkyo Koho JP 2008122401, 2008; *Chem. Abstr.* **2008**, *148*, 584596.
24. L'Oreal, Fr. Hair dye composition comprising an electrophilic monomer, a hydrophobic direct dye, and a liquid organic solvent. Fr. Demande FR 2899816, 2007; *Chem. Abstr.* **2007**, *147*, 474290.
25. Rogers, K. A.; Karnovsky, M. J. A rapid method for the detection of early stages of atherosclerotic lesion formation. *Am. J. Pathol.* **1988**, *133*, 451–455.
26. Huh, T. L.; Koh, H. J.; Choi, M. S.; Jung, U. J. cDNA encoding cytoplasmic isocitrate dehydrogenase and its inhibitors for use in the treatment of obesity, hyperlipidemia, and fatty liver. PCT Int. Appl. WO 2002033063, 2002; *Chem. Abstr.* **2002**, *136*, 321290.
27. Gerhardt, C.; Romero-Romero, I. A.; Strosberg, A. D. Medicines useful for treating disorders of regulation of body fat and diseases related to disorders of leptin production. PCT Int. Appl. WO 2000066149, 2000; *Chem. Abstr.* **2000**, *133*, 346766.
28. Oxman, J. D.; Konings, M. S.; Tiers, G. V. D.; Vogel, K. M.; Vogel, D. E. Dental impression material with cure-indicating dye. PCT Int. Appl. WO 9600560, 1996; *Chem. Abstr.* **1996**, *124*, 242386.

29. Kalmar, J. R. Measurement of opsonic phagocytosis by human polymorphonuclear neutrophils. *Methods Enzymol.* **1994**, *236*, 108–119.

30. Deno, T.; Eichenberger, T.; Voigt, J. Use of pyrimido [5,4-*g*]pteridines as shading component in color filter colorant compositions. PCT Int. Appl. WO 2006037728, 2006; *Chem. Abstr.* **2006**, *144*, 400791.

31. Lukishova, S. G.; Lepeshkin, N.; Boyd, R. W.; Marshall, K. L. Far-field patterns from dye-doped planar-aligned nematic liquid crystals under nanosecond laser irradiation. *Mol. Cryst. Liq. Cryst.* **2006**, *453*, 393–401.

32. Marshall, K. L.; Adelsberger, K.; Myhre, G.; Griffin, D. W. The LCPDI: a compact and robust phase-shifting point-diffraction interferometer based on dye-doped LC technology. *Mol. Cryst. Liq. Cryst.* **2006**, *454*, 425–447.

33. Pampalone, T. R.; Kuyan, F. A. Contrast enhancing additives for positive photoresist. *J. Electrochem. Soc.* **1988**, *135*, 471–476.

34. Hughes, C. J.; Coates, D.; Alkins, A. O. Cholesteric liquid crystal display device. PCT Int. Appl. WO 2006054050, 2006; *Chem. Abstr.* **2006**, *145*, 17884.

35. Motoi, F.; Kitahora, T.; Kobayashi, N.; Ueda, H. Reflective liquid crystal display having chiral nematic liquid crystal layer selectively reflecting specific wavelength light. Jpn. Kokai Tokkyo Koho JP 2002023198, 2002; *Chem. Abstr.* **2002**, *136*, 126651.

36. Bui, L.; Babu, R.; Desai, S.; Brucker, B.; Ranganathan, N. K. Ink composition for drop-on-demand printing and method of use. PCT Int. Appl. WO 2008136795, 2008; *Chem. Abstr.* **2008**, *149*, 558420.

37. Nakamura, K. Oil-based black ink compositions for marking pens giving water-resistant writing. Jpn. Kokai Tokkyo Koho JP 2006056946, 2006; *Chem. Abstr.* **2006**, *144*, 234791.

38. Otaki, K.; Suguro, Y. Electrostatographic developer toners containing polymer particles coated with charge-controlling agent. Jpn. Kokai Tokkyo Koho JP 04039673, 1992; *Chem. Abstr.* **1992**, *117*, 80023.

39. Ikeda, T.; Nozaki, A.; Ebe, K. Adhesive papers for copying with edge face covered with urethane acrylate polymer. Eur. Pat. Appl. EP 255396, 1988; *Chem. Abstr.* **1988**, *109*, 139238.

40. Sugiyama, T.; Watanabe, Y.; Otani, S.; Saito, T.; Yamashita, H. Method and apparatus for forming full-color sharp images by electrophotography without cleaning failure nor toner fusion. Jpn. Kokai Tokkyo Koho JP 2005070188, 2005; *Chem. Abstr.* **2005**, *142*, 306398.

41. Kosc, T. Z.; Marshall, K. L.; Jacobs, S. D. Electrically addressable optical devices using a system of composite layered flakes suspended in a fluid host to obtain angularly dependent optical effects. U.S. Pat. Appl. Publ. US 2004233508, 2004; *Chem. Abstr.* **2004**, *141*, 417702.

42. Harris, M.; Hannington, P.; Ewin, C. Air fresheners. PCT Int. Appl. WO 2002066084, 2002; *Chem. Abstr.* **2002**, *137*, 190411.

43. Moodycliffe, T. I.; Werkowski, L. M. Wax-free furniture polish composition containing petroleum distillates. U.S. Pat. Appl. Publ. US 2003075073, 2003; *Chem. Abstr.* **2003**, *138*, 339807.

44. Ian Malcolm, S.; Wright, T.; Perrin, B.; Mulqueen, P.; Waller, A.; Pierce, A. Capsule formulation containing pesticide and photoprotectant. PCT Int. Appl. WO 2006089747, 2006; *Chem. Abstr.* **2006**, *145*, 286603.

45. Parker, T. L.; Soane, D. S. Quantitative tinting and primer compositions for plastic lenses. PCT Int. Appl. WO 9615472, 1996; *Chem. Abstr.* **1996**, *125*, 89300.

46. Fujiwara, K. Water-washable marking inks for use in flaw detection of steel materials. Jpn. Kokai Tokkyo Koho JP 01139659, 1989; *Chem. Abstr.* **1990**, *112*, 79667.

47. Himmelsbach, D. S.; Akin, D. E.; Kim, J.; Hardin, I. R. Chemical structural investigation of the cotton fiber base and associated seed coat: Fourier-transform infrared mapping and histochemistry. *Text. Res. J.* **2003**, *73*, 281–288.

48. Smith, M. J.; Desai, B.; Frederico, J. J. Molecular tags for organic solvent systems, especially petroleum products. PCT Int. Appl. WO 2003022942, 2003; *Chem. Abstr.* **2003**, *138*, 257641.

49. Niizawa, A.; Yamaguchi, M. Standard dye solutions for calibration of instruments for measurement of ASTM color of petroleum products. Ger. Offen. DE 4310597, 1993; *Chem. Abstr.* **1994**, *120*, 34368.

50. Shah, S. A. A. Coloured anhydrous gel element for perfuming or deodorising air. PCT Int. Appl. WO 2001078794, 2001; *Chem. Abstr.* **2001**, *135*, 322183.

51. Barlow, I. J. Aqueous scratch-covering compositions for finished wooden articles. PCT Int. Appl. WO 9410237, 1994; *Chem. Abstr.* **1995**, *122*, 242479.

52. Wei, T.; Geijer, S.; Lindberg, M.; Berne, B.; Torma, H. Detergents with different chemical properties induce variable degree of cytotoxicity and mRNA expression of lipid-metabolizing enzymes and differentiation markers in cultured keratinocytes. *Toxicol. In Vitro* **2006**, *20*, 1387–1394.

53. Speidel, M. T.; Booyse, F. M.; Abrams, A.; Moore, M. A.; Chung, B. H. Lipolyzed hypertriglyceridemic serum and triglyceride-rich lipoprotein cause lipid accumulation in and are cytotoxic to cultured human endothelial cells: high density lipoproteins inhibit this cytotoxicity. *Thromb. Res.* **1990**, *58*, 251–264.

54. Van, S. E. W.; Simmons, J. E.; Sloane, R. A.; Moorman, M. P.; Adkins, B., Jr.; Cockrell, B. Y. Failure of carbon disulfide and levothyroxine to modify the cardiovascular response of rabbits to a high-cholesterol diet. *Toxicology* **1986**, *40*, 45–58.

55. Bach, P. H.; Scholey, D. J.; Delacruz, L.; Moret, M.; Nichol, S. Renal and urinary lipid changes associated with an acutely induced renal papillary necrosis in rats. *Food Chem. Toxicol.* **1991**, *29*, 211–219.

ORANGE II (TROPAEOLIN OOO)

CAS Registry Number 633-96-5

Chemical Structure

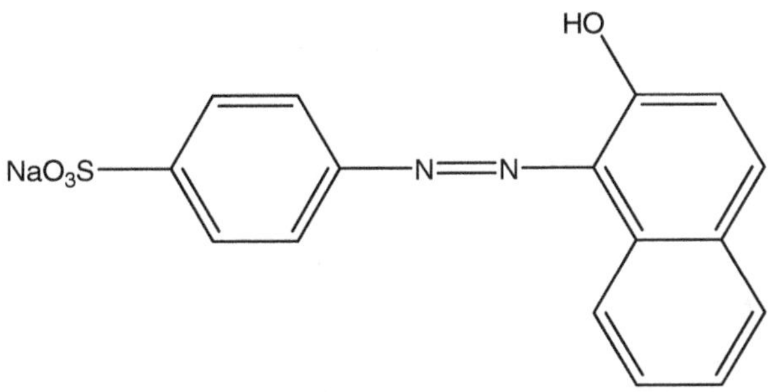

CA Index Name Benzenesulfonic acid, 4-[2-(2-hydroxy-1-naphthalenyl)diazenyl]-, sodium salt (1:1)

Other Names Acid Orange; Acid Orange A; Benzenesulfonic acid, 4-[(2-hydroxy-1-naphthalenyl)azo]-, monosodium salt; C.I. Acid Orange 7, monosodium salt; Benzenesulfonic acid, *p*-(2-hydroxy-1-naphthylazo)-, sodium salt; β-Naphthol Orange; β-Naphthyl orange; 11550 Orange; 2-Naphthol Orange II; AO 7; Acid Leather Orange PRW; Acid Leather Orange extra; Acid Orange 7; Acid Orange II; Acilan Orange II; Airedale Orange II; Amacid Orange Y; Amacid Orange Y Conc; Apollo Acid Orange II; Atul Acid Orange II; BTK Orange II; Basacid Orange 280; Basacid Orange 282; Benzenesulfonic acid, 4-[2-(2-hydroxy-1-naphthalenyl)diazenyl]-, monosodium salt; Betanaphthol orange; Borunil Orange A 2R; Brasilan Orange A; Bucacid Orange A; C Ext. Orange 8; C.I. 15510; C.I. Acid Orange 7; Calcocid Orange Y; Certiqual Orange II; Colacid Orange; Colocid Orange II; Colorosacid Orange; Conacid Orange L; Concorde Acid Orange II; Concorde Leather Orange MT; Covalene Orange II; Curol Orange; D and C Orange 4 Aluminum Lake; D and C Orange No. 4; D&C Orange #4; D&C Orange No. 4; D&C Orange No. 4-307005; D&C Orange No. 4-90121; Daedo Acid Orange 2G; Derma Fur Orange R 125; Dermacid Orange 2R; Diacid Orange II; Dinacid Orange II; Duasyn Acid Orange P; Dyacid Orange II; Dycosacid Orange Yellow II; Dynacid Orange Y; Egacid Orange II; Eniacid Orange II; Erio Orange II; Everacid Orange II; Fenazo Orange; Hidacid Orange II; Hispacid Orange AF; Ichoacid Orange II; Indacid Orange II; Japan Orange 205; Japan Orange No. 205; Java Orange II; Kemacid Orange II; Keyacid Orange II; Kiton Orange II; Kromon Lake Orange Toner; Lake Orange A; Lake Orange II YS; Leather Orange Extra; Libacid Orange LII; Lurazol Orange E; Lurazol Orange EBR; Lutetia Orange 3JR; Mandarin G; Multacid Orange II; Multicuer Orange II; Naphthalene Lake Orange G; Naphthalene Orange G; Naphthol Orange; Naphthol Red J; Naphtocard Orange II; Neelicol Orange II; Neklacid Orange II; No. 177 Orange Lake; Nubilon Orange R; Orange 2; Orange 2 sodium salt; Orange Extra N; Orange Extra P; Orange II; Orange II 307004; Orange II 90121; Orange II Special For Lacquer; Orange II for Lakes; Orange IIC; Orange IIP; Orange IIS; Orange IISM; Orange No. 205; Orange Toner GRT; Orange Y; Orange YA; Orange YZ; Pacid Orange II; Peeracid Orange II; Persian Orange; Persian Orange Lake; Persian Orange X; Pure Orange II S; Rifa Acid Orange 2G; Rifa Acid Orange II; Rifa Leather Orange BR; Ritacid Orange II; Romexal Orange; Rybacel Orange A; Sandal Acid Orange II; Sanyo Gum Orange A; Sodium 4-(2-hydroxy-1-naphthylazo)benzenesulfonate; Solar Orange; Special Orange GR; Special Orange H; Sulfacide Orange JR; Symuler Orange Lake 43; Symulon Acid Orange II; Takaoka Acid Orange II; Tangarine Lake X 917; Tertracid Orange II; Triacid Orange II; Tropaeolin OOO; Tropaeolin OOO 2; Tropeolin OOO; Vibra Color Orange AOR 7; Vicoacid Orange II; Vondacid Orange II; Vopsider Orange A; Water Orange 060506; Wool Orange A; *p*-(2-Hydroxy-1-naphthylazo)benzenesulfonic acid sodium salt

Merck Index Number 6858

Chemical/Dye Class Azo

Molecular Formula $C_{16}H_{11}N_2NaO_4S$

Molecular Weight 350.32

Physical Form Orange-brown powder

Solubility Very soluble in water; very slightly soluble in ethanol

Melting Point 164 °C

pH Range 7.4–8.6; 10.2–11.8

Color Change at pH Amber (7.4) to orange (8.6); orange (10.2) to red (11.8)

pK_a 8.26, 11.4

Absorption (λ_{max}) 483 nm

Synthesis Synthetic methods[1–9]

Staining Applications Drinks;[10] candies;[10] Heinz bodies in erythrocytes;[11] Horny cells;[12] eye shadow;[13] lips;[14,15] skin;[14–17] tooth;[18] hairs;[1,19–22] keratin fibers[23,24]

Biological Applications Cosmetics;[1,15,25] wound dressing materials[1,26]

Industrial Applications Organic light emitting devices;[1,27] inks;[1,28] cleansing products;[29] textiles[1,30]

Safety/Toxicity Acute toxicity;[1,31] carcinogenicity;[1,32] environmental toxicity;[33] fish toxicity;[1,34] genotoxicity;[1,35] mutagenicity[1,36,37]

REFERENCES

1. Sabnis, R. W. *Handbook of Acid–Base Indicators*; CRC Press: Boca Raton, **2008**; pp 283–285.
2. Dabbagh, H. A.; Teimouri, A.; Chermahini, A. N. Green and efficient diazotization and diazo coupling reactions on clays. *Dyes Pigm.* **2007**, *73*, 239–244.
3. Vesely, M.; Kubac, L.; Polackova, E.; Bohac, Z. Three-step diazo coupling procedure for the manufacture of orange dyes. Czech. CS 277137, 1992; *Chem. Abstr.* **1994**, *121*, 282309.
4. Arsac, A.; Frank, P. Azo dyes. Fr. Demande FR 2439222, 1980; *Chem. Abstr.* **1981**, *94*, 4948.
5. Badzynski, M.; Sokol, W.; Wielunski, J. Studies of the continuous synthesis of Acid Orange. *Przem. Chem.* **1978**, *57*, 134–136; *Chem. Abstr.* **1978**, *88*, 171768.
6. Kawasaki, S.; Kawashita, H.; Ishihara, M.; Miki, A. Manufacture of sulfo group-containing azo dyes. Jpn. Kokai Tokkyo Koho JP 52019725, 1977; *Chem. Abstr.* **1977**, *86*, 191321.
7. Blatt, A. H. *Organic Synthesis Collective Volume*; John Wiley & Sons: New York, **1943**; Vol. 2, p 36.
8. Vesely, M.; Bohac, Z. Automatic manufacture of azo dyes. Ger. Offen. DE 2224643, 1972; *Chem. Abstr.* **1973**, *78*, 73625.
9. Kindler, H.; Schuler, D. Continuous diazotization process. FR 1419092, 1965; *Chem. Abstr.* **1966**, *65*, 91168.
10. Yoshioka, N.; Ichihashi, K. Determination of 40 synthetic food colors in drinks and candies by high-performance liquid chromatography using a short column with photodiode array detection. *Talanta* **2008**, *74*, 1408–1413.
11. Rofe, P. Azo dyes and Heinz bodies. *Br. J. Ind. Med.* **1957**, *14*, 275–280.
12. Yamazaki, K.; Fujiwara, N.; Seto, T.; Yang, P. Horny cell-staining liquid, and horny cell-staining method. PCT Int. Appl. WO 2007029437, 2007; *Chem. Abstr.* **2007**, *146*, 312248.
13. Bouchard, F. Eye shadow comprising least 50% colorants. Eur. Pat. Appl. EP 1757262, 2007; *Chem. Abstr.* **2007**, *146*, 280374.
14. Blin, X. Range of compositions for application to the skin, lips, and/or superficial body growths. Fr. Demande FR 2868697, 2005; *Chem. Abstr.* **2005**, *143*, 392523.
15. Dumousseaux, C. Cosmetic composition for application to the skin, lips, and/or body surface. Fr. Demande FR 2868696, 2005; *Chem. Abstr.* **2005**, *143*, 392522.
16. Teboul, K. Method for coloring dark skins. PCT Int. Appl. WO 2008087308, 2008; *Chem. Abstr.* **2008**, *149*, 160219.
17. Arnaud, P.; Feltin, C. Method for coloring of the skin comprising direct dye. Fr. Demande FR 2909873, 2008; *Chem. Abstr.* **2008**, *149*, 61572.
18. Lee, B. S.; Huang, S. H.; Chiang, Y. C.; Chien, Y. S.; Mou, C. Y.; Lin, C. P. Development of *in vitro* tooth staining model and usage of catalysts to elevate the effectiveness of tooth bleaching. *Dent. Mater.* **2008**, *24*, 57–66.
19. Inoue, M. Hair dye emulsions containing water-soluble direct dyes. Jpn. Kokai Tokkyo Koho JP 2008074775, 2008; *Chem. Abstr.* **2008**, *148*, 386508.
20. Kawai, S.; Oikawa, S. Hair dye compositions containing viscosity enhancing agents. Jpn. Kokai Tokkyo Koho JP 2008063312, 2008; *Chem. Abstr.* **2008**, *148*, 362851.
21. Hoeffkes, H.; Fuhr, D. Hair lightening and/or coloring agent containing heterocycles for hydrogen peroxide activation. Ger. Offen. DE 102006041292, 2008; *Chem. Abstr.* **2008**, *148*, 314415.
22. Tsuge, S. Storage-stable acidic hair dye compositions containing hydrophilic organic solvents, lipophilic penetrants, and acryloyldimethyltaurine salt copolymer thickeners. Jpn. Kokai Tokkyo Koho JP 2008019203, 2008; *Chem. Abstr.* **2008**, *148*, 198115.
23. Brun, G.; Gourlaouen, L. Composition including a compound X and a compound Y of which one at least is silicone, and a hydrophilic direct dye. Fr. Demande FR 2910297, 2008; *Chem. Abstr.* **2008**, *149*, 86948.
24. Wood, J.; Noecker, B.; Wilz, R. Compositions for coloring of keratin fibers. Eur. Pat. Appl. EP 1598049, 2005; *Chem. Abstr.* **2005**, *143*, 482621.
25. Sato, Y.; Susaki, S.; Morimoto, H.; Miyamoto, K. Light-resistant colored cosmetics. Jpn. Kokai Tokkyo Koho JP 2006169144, 2006; *Chem. Abstr.* **2006**, *145*, 109757.
26. Addison, D.; Greenhalgh, D.; Cullen, B. M. Antioxidant and antimicrobial wound dressing materials. PCT Int. Appl. WO 2005049101, 2005; *Chem. Abstr.* **2005**, *143*, 13454.
27. Klubek, K. P.; Jarikov, V. V.; Liao, L. S.; Brown, C. T.; Tang, C. W. OLEDs with improved operational lifetime employing a host with multiple components. U.S. Pat. Appl. Publ. US 2006040131, 2006; *Chem. Abstr.* **2006**, *144*, 243090.

28. Takuhara, H.; Tomioka, H.; Sugaya, T.; Tochihara, S. Ink-jet printing forming lightfast images with expanded color reproduction range and inks therefor. Jpn. Kokai Tokkyo Koho JP 2006169295, 2006; *Chem. Abstr.* **2006**, *145*, 64660.

29. Krzysik, D. G.; Utschig, J. M.; Cole, D. B. Color changing liquid cleansing products containing surfactants, electrolytes and coloring agents. U.S. Pat. Appl. Publ. US 2005148490, 2005; *Chem. Abstr.* **2005**, *143*, 120038.

30. Iyota, S.; Asada, T.; Kanayama, S. Agent and method for fiber treatment. Jpn. Kokai Tokkyo Koho JP 2007230900, 2007; *Chem. Abstr.* **2007**, *147*, 307508.

31. Singh, R. L.; Khanna, S. K.; Singh, G. B. Acute and short-term toxicity of a popular blend of metanil yellow and orange II in albino rats. *Indian J. Exp. Biol.* **1988**, *26*, 105–111.

32. Mamber, S. W.; Bryson, V.; Katz, S. E. Evaluation of the *Escherichia coli* K12 inductest for detection of potential chemical carcinogens. *Mutat. Res.* **1984**, *130*, 141–151.

33. Tonogai, Y.; Ogawa, S.; Ito, Y.; Iwaida, M. Actual survey on TLm (median tolerance limit) values of environmental pollutants, especially on amines, nitriles, aromatic nitrogen compounds and artificial dyes. *J. Toxicol. Sci.* **1982**, *7*, 193–203.

34. Tonogai, Y.; Ito, Y.; Iwaida, M.; Tati, M.; Ose, Y.; Hori, M. Studies on the toxicity of coal-tar dyes III. Reason of acute toxicity to fish caused by coal-tar dyes and their industrial effluents. *J. Toxicol. Sci.* **1980**, *5*, 23–33.

35. Wever, J.; Muenzner, R.; Renner, H. W. Testing of Sunset Yellow and Orange II for genotoxicity in different laboratory animal species. *Environ. Mol. Mutagen.* **1989**, *13*, 271–276.

36. Rafii, F.; Hall, J. D.; Cerniglia, C. E. Mutagenicity of azo dyes used in foods, drugs and cosmetics before and after reduction by *Clostridium* species from the human intestinal tract. *Food Chem. Toxicol.* **1997**, *35*, 897–901.

37. Chung, K. T.; Fulk, G. E.; Andrews, A. W. Mutagenicity testing of some commonly used dyes. *Appl. Environ. Microbiol.* **1981**, *42*, 641–648.

ORANGE G

CAS Registry Number 1936-15-8

Chemical Structure

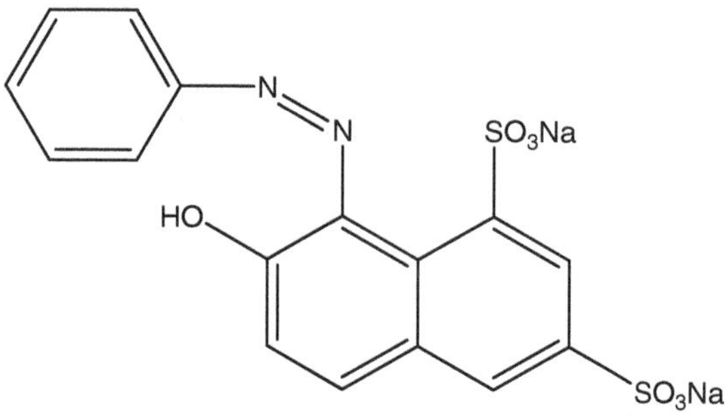

CA Index Name 1,3-Naphthalenedisulfonic acid, 7-hydroxy-8-(2-phenyldiazenyl)-, sodium salt (1:2)

Other Names 1,3-Naphthalenedisulfonic acid, 7-hydroxy-8-(phenylazo)-, disodium salt; 2-Naphthol-6,8-disulfonic acid, 1-phenylazo-, disodium salt; C.I. Acid Orange 10; C.I. Acid Orange 10, disodium salt; 1370 Orange; Acid Crystal Orange; Acid Fast Orange EGG; Acid Fast Orange G; Acid Fast Orange GG; Acid Leather Orange G; Acid Leather Orange KG; Acid Leather Orange PGW; Acid Lightfast Orange; Acid Orange 10; Acid Orange 2G; Acid Orange 2GL; Acid Orange EGG; Acid Orange G; Acid Orange G Crystal; Acid Orange GG; Acid Orange Lightfast; Acidal Fast Orange; Acilan Orange GX; Amacid Crystal Orange; Amacid Light Orange G; Anadurm Orange A-G; Apocid Orange 2G; Atul Acid Crystal Orange G; Brasilan Orange 2G; Bucacid Fast Orange G; C.I. 16230; C.I. Food Orange 4; Calcocid Fast Light Orange 2G; Certicol Orange GS; Cetil Light Orange GG; Colacid Orange G; Colocid Orange 2G; Colorosacid Orange 10; Concorde Acid Orange L-GX; Crystal Orange 2G; Crystal Orange G; D and C Orange No. 3; Dinacid Crystal Orange G; Dolkwal Orange G; Duasyn Acid Orange GG; Dyacid Orange G; Dynacid Orange 2G; Egacid Orange GG; Eniacid Light Orange G; Erio Fast Orange AS; Fabracid Orange FL; Fast Acid Orange G; Fast Light Orange G; Fast Light Orange GA-CF; Fenazo Light Orange 2G; Food Orange 4; Food Orange GG; Hexacol Orange G; Hexacol Orange GG Crystals; Hidacid Fast Orange G; Hispacid Fast Orange 2G; Indacid Crystal Orange; Kemacid Crystal Orange G; Keyacid Orange 2G; Kiton Fast Orange G; Light Orange G; Multacid Orange 2G; Naphthalene Fast Orange 2G; Naphthalene Fast Orange 2GS; Naphthazine Orange 2G; Neelicol Orange G; Neklacid Fast Orange GG; Orange 2G; Orange BPC; Orange G; Orange G dye; Orange GBPC; Orange GG; Pacid Orange G 10; Sandolan Fast Orange E-GL; Sandolan Orange E-GL; Solar Light Orange GX; Straight Orange G; Tetracid Light Orange G; Triacid Light Orange 2G; Vicoacid Orange IJ; Victacid Orange IJ; Vondacid Light Orange 2G; Water Orange 145924; Wool Orange 2G; Wool Orange G; Xylene Fast Orange G

Merck Index Number Not listed

Chemical/Dye Class Azo

Molecular Formula $C_{16}H_{10}N_2Na_2O_7S_2$

Molecular Weight 452.37

Physical Form Orange to red powder

Solubility Soluble in water, methyl cellosolve; slightly soluble in ethanol; insoluble in xylene

Melting Point 141 °C

pH Range 11.5–14.0

Color Change at pH Yellow (11.5) to pink (14.0)

pK_a 12.8

Absorption (λ_{max}) 475 nm

Synthesis Synthetic method[1–6]

Staining Applications Adenohypophyseal cells;[7] clavanins;[8] cells;[9] eosinophil leukocyte granules;[10] fungi;[11] food;[12] drinks;[13] candies;[13] lymph nodes;[14] pap smears;[15] proteins;[16,17] keratin fibers;[18] hairs;[19] skin;[20,21] sputum;[22] glial tumors;[23] zein membranes;[24] urinary sediments[25]

Biological Applications Detecting lung cancer metastasis;[14] measuring glycated proteins;[26] ophthalmic devices[27]

Industrial Applications Inks;[28,29] highlighters;[30] photography;[31] plastics;[32] paper;[33] textiles;[34–36] wood[37]

Safety/Toxicity Acute oral toxicity;[38] bacterial toxicity;[40] carcinogenicity;[39,41–46] chromosomal aberration;[46–48] freshwater shrimp toxicity;[49] genotoxicity;[50,51] hematotoxicity;[52] mutagenicity;[41,42,53–57] serological toxicity;[52] short-term toxicity[58,59]

Certification/Approval Certified by Biological Stain Commission (BSC)

REFERENCES

1. Hrehorowicz, E.; Gmaj, J.; Balinski, J. Manufacture of disodium 7-hydroxy-8-phenylazonaphthalene-1,3-disulfonate dye. Pol. PL 136238, 1987; *Chem. Abstr.* **1990**, *113*, 233338.
2. Pla-Delfina, J. M. Systematic identification of food, drug, and cosmetic azo dyes. *J. Soc. Cosmet. Chem.* **1962**, *13*, 214–244.

3. Rofe, P. Azo dyes and Heinz bodies. *Br. J. Ind. Med.* **1957**, *14*, 275–280.

4. Kressman, T. R. E. Ion-exchange separations based on ionic size. *J. Phys. Chem.* **1952**, *56*, 118–123.

5. Rais, J. Orange GG. *Chem. Obzor* **1950**, *25*, 97–101.

6. Proctor, G. B.; Horobin, R. W. Purification of oil red O using preparative paper chromatography. *Stain Technol.* **1985**, *60*, 247–248.

7. Kikui, Y.; Miki, A. A differential staining method for adenohypophyseal cells. *Arch. Histol. Cytol.* **1995**, *58*, 375–378.

8. Menzel, L. P.; Lee, I. H.; Sjostrand, B.; Lehrer, R. I. Immunolocalization of clavanins in *Styela clava* hemocytes. *Dev. Comp. Immunol.* **2002**, *26*, 505–515.

9. Cairns, H. Staining physiological samples. Brit. UK Pat. Appl. GB 2372811, 2002; *Chem. Abstr.* **2002**, *138*, 35727.

10. Isabel, T. C.; Del, C. P.; Carlos, S. J. Influence of inorganic salts on the staining reaction of eosinophil leukocyte granules by anionic dyes. *Acta Histochem.* **1992**, *93*, 313–318.

11. Eichlerova, I.; Homolka, L.; Lisa, L.; Nerud, F. Orange G and Remazol Brilliant Blue R decolorization by white rot fungi *Dichomitus squalens*, *Ischnoderma resinosum* and *Pleurotus calyptratus*. *Chemosphere* **2005**, *60*, 398–404.

12. Jones, A. V.; Thomas, J. D. R. The influence of certain metal ions on the visible spectra of food dyes. *J. Food Technol.* **1968**, *3*, 1–14.

13. Yoshioka, N.; Ichihashi, K. Determination of 40 synthetic food colors in drinks and candies by high-performance liquid chromatography using a short column with photodiode array detection. *Talanta* **2008**, *74*, 1408–1413.

14. Okubo, K.; Kato, T.; Hara, A.; Yoshimi, N.; Takeda, K.; Iwao, F. Imprint cytology for detecting metastasis of lung cancer in mediastinal lymph nodes. *Ann. Thorac. Surg.* **2004**, *78*, 1190–1193.

15. Akinremi, T. O.; Nazeer, S.; Totsch, M. Reduced alcohol use in the staining of Pap smears: a satisfactory low-cost protocol for cervical cancer screening. *Acta Cytol.* **2005**, *49*, 169–172.

16. Sun, W.; Han, J.; Ren, Y.; Jiao, K. Voltammetric studies on the interaction of orange G with proteins: analytical applications. *J. Braz. Chem. Soc.* **2006**, *17*, 510–517.

17. Molnar-Perl, I.; Pinter-Szakacs, M.; Kovago, A.; Petroczy, I.; Kralovanszky, U. P.; Matyas, J. Dye-binding stoichiometry of AO 12, AB 10B and OG with etalon proteins, feed and feedingstuffs and its application for reactive lysine determination. *Food Chem.* **1986**, *20*, 21–38.

18. Kravtchenko, S. Agent for simultaneously bleaching and coloring of keratin fibers comprising an anionic or nonionic dye and an inert organic liquid. Eur. Pat. Appl. EP 1738801, 2007; *Chem. Abstr.* **2007**, *146*, 128004.

19. Plos, G. Hair dye composition containing an oxidase alcohol and a direct dye having an acid function or its salt for dyeing keratin fibers. Fr. Demande FR 2865396, 2005; *Chem. Abstr.* **2005**, *143*, 179094.

20. Lee, H.; Soistier, N.; Simon, J. C. Cosmetic composition of foundation type for making up dark skins. U.S. Pat. Appl. Publ. US 2005019285, 2005; *Chem. Abstr.* **2005**, *142*, 162067.

21. Lee, H.; Soistier, N.; Simon, J. C. Cosmetic makeup foundations for darkened skins. Fr. Demande FR 2848822, 2004; *Chem. Abstr.* **2004**, *141*, 76373.

22. Cavaliere, M. J.; Maeda, M. Y.; Longatto, F. A.; Shirata, N. K.; Santos, R. T.; Kitamura, C.; Ueki, S. Y.; Martins, M. C. Frequency of *Candida* sp infection in tuberculous patients with acquired immunodeficiency syndrome: morphological and immunocytochemical study in sputum. *Pathologica* **1994**, *86*, 409–411.

23. Szydlowska, H. Orange-staining structures in glial tumors stained with polychromes. *Neuropatol. Polska* **1966**, *4*, 187–202; *Chem. Abstr.* **1968**, *68*, 67202.

24. Oh, Y. K.; Flanagan, D. R. Swelling and permeability characteristics of zein membranes. *J. Pharm. Sci. Technol.* **2003**, *57*, 208–217.

25. Matsuoka, M.; Fukada, S.; Shimada, M.; Furuhata, T.; Shimidzu, M.; Itabashi, A.; Okada, S. Investigation of a new urinary staining as an alternative to Sternheimer staining. *Igaku Kensa* **2003**, *52*, 187–193; *Chem. Abstr.* **2003**, *138*, 381508.

26. Ueda, S.; Matsuoka, T.; Suga, A. Stabilization of leuco dyes with reducing agents, liquid compositions containing them, method for reduction of nonspecific color reactions of leuco dyes, method and reagents for determination of hydrogen pero-xide, and reagents for measurement of glycated proteins. Jpn. Kokai Tokkyo Koho JP 2008201968, 2008; *Chem. Abstr.* **2008**, *149*, 326772.

27. Mentak, K. Ultra violet, violet, and blue light filtering polymers for ophthalmic applications. U. S. Pat. Appl. Publ. US 2006252844, 2006; *Chem. Abstr.* **2006**, *145*, 495756.

28. Godbout, D. A.; Vincent Kwan, W. S. Water-based, resin-free and solvent-free eradicable and re-write

ball-pen inks or tinted fluid. U.S. Pat. Appl. Publ. US 2006032398, 2006; *Chem. Abstr.* **2006**, *144*, 214528.

29. Inoue, S.; Fujita, T. Storage-stable anticlogging water-thinned black inks. Jpn. Kokai Tokkyo Koho JP 11012525, 1999; *Chem. Abstr.* **1999**, *130*, 126413.
30. Davies-Smith, L.; Sum, V. W. Highlightable and highlighted mixtures, marking instruments, eradicator solution and kits, using the mixtures in highlighting. U.S. Pat. Appl. Publ. US 2005120919, 2005; *Chem. Abstr.* **2005**, *143*, 9331.
31. Mihara, Y.; Ishigaki, K. Silver halide photographic photosensitive materials. Jpn. Kokai Tokkyo Koho JP 61174540, 1986; *Chem. Abstr.* **1987**, *106*, 111265.
32. Bodi, J. A.; Phillips, T. L. Daylight/nightglow colored phosphorescent plastic compositions and articles. PCT Int. Appl. WO 2000027908, 2000; *Chem. Abstr.* **2000**, *132*, 335413.
33. Astbury, J. H.; Cooper, H. M. Dyeing of paper. Brit. GB 1193294, 1970; *Chem. Abstr.* **1970**, *73*, 57410.
34. Sun, G.; Kim, Y. H. Antimicrobial treatment of polymers or dyed textiles. PCT Int. Appl. WO 2000015897, 2000; *Chem. Abstr.* **2000**, *132*, 223814.
35. Gooding, J. J.; Compton, R. G.; Brennan, C. M.; Atherton, J. H. The dyeing of nylon and cotton cloth with azo dyes: kinetics and mechanism. *J. Colloid Interface Sci.* **1996**, *180*, 605–613.
36. Wlochowicz, A.; Wojciechowska-Bujok, E. Effect of dyeing on sorption of water vapor by wool. *Text. Res. J.* **1987**, *57*, 503–508.
37. Kitamura, Y. Deep ebony stains for wood. Jpn. Kokai Tokkyo Koho JP 03266601, 1991; *Chem. Abstr.* **1992**, *116*, 154081.
38. Schafer, E. W., Jr.; Bowles, W. A., Jr. Acute oral toxicity and repellency of 933 chemicals to house and deer mice. *Arch. Environ. Contam. Toxicol.* **1985**, *14*, 111–129.
39. Rosenkranz, H. S.; Karol, M. H. Chemical carcinogenicity: can it be predicted from knowledge of mutagenicity and allergic contact dermatitis? *Mutat. Res.* **1999**, *431*, 81–91.
40. Oros, G.; Cserhati, T.; Forgacs, E. Separation of the strength and selectivity of the microbiological effect of synthetic dyes by spectral mapping technique. *Chemosphere* **2003**, *52*, 185–193.
41. Benigni, R. Mouse bone marrow micronucleus assay: relationships with *in vitro* mutagenicity and rodent carcinogenicity. *J. Toxicol. Environ. Health* **1995**, *45*, 337–347.
42. Ashby, J.; Tennant, R. W. Definitive relationships among che-mical structure, carcinogenicity and mutagenicity for 301 che-micals tested by the U.S. NTP. *Mutat. Res.* **1991**, *257*, 229–306.
43. Haseman, J. K.; Clark, A. M. Carcinogenicity results for 114 laboratory animal studies used to assess the predictivity of four in vitro genetic toxicity assays for rodent carcinogenicity. *Environ. Mol. Mutagen.* **1990**, *16*, 15–31.
44. Zeiger, E. Carcinogenicity of mutagens: predictive capability of the *Salmonella* mutagenesis assay for rodent carcinogenicity. *Cancer Res.* **1987**, *47*, 1287–1296.
45. Haseman, J. K. Issues in carcinogenicity testing: dose selection. *Fundam. Appl. Toxicol.* **1985**, *5*, 66–78.
46. Rosenkranz, H. S.; Ennever, F. K.; Klopman, G. Relationship between carcinogenicity in rodents and the induction of sister chromatid exchanges and chromosomal aberrations in Chinese hamster ovary cells. *Mutagenesis* **1990**, *5*, 559–571.
47. Shelby, M. D.; Witt, K. L. Comparison of results from mouse bone marrow chromosome aberration and micronucleus tests. *Environ. Mol. Mutagen.* **1995**, *25*, 302–313.
48. Rosenkranz, H. S.; Ennever, F. K.; Dimayuga, M.; Klopman, G. Significant differences in the structural basis of the induction of sister chromatid exchanges and chromosomal aberrations in Chinese hamster ovary cells. *Environ. Mol. Mutagen.* **1990**, *16*, 149–177.
49. Ogugbue, C. J.; Oranusi, N. A. Quantal response of freshwater shrimp (*Desmocaris trispinosa*) to toxicity of azo dyes. *Global J. Environ. Sci.* **2005**, *4*, 41–45.
50. Kirkland, D.; Aardema, M.; Henderson, L.; Mueller, L. Evaluation of the ability of a battery of three *in vitro* genotoxicity tests to discriminate rodent carcinogens and non-carcinogens. I. Sensitivity, specificity and relative predictivity. *Mutat. Res.* **2005**, *584*, 1–256.
51. Haveland-Smith, R. B.; Combes, R. D. Screening of food dyes for genotoxic activity. *Food Cosmet. Toxicol.* **1980**, *18*, 215–221.
52. Chakravarty, G.; Goyal, R. P.; Sharma, S.; Sharma, A. Hematological and serological toxicity of Orange G in Swiss albino mice, *Mus musculus. Nat., Environ. Pollut. Technol.* **2006**, *5*, 95–99.
53. Serafimova, R.; Todorov, M.; Pavlov, T.; Kotov, S.; Jacob, E.; Aptula, A.; Mekenyan, O. Identification of the structural requirements for mutagenicity, by incorporating molecular flexibility and metabolic activation of chemicals. II. General Ames mutagenicity model. *Chem. Res. Toxicol.* **2007**, *20*, 662–676.

54. Benigni, R.; Andreoli, C.; Giuliani, A. Relationships among *in vitro* mutagenicity assays: quantitative vs. qualitative test results. *Environ. Mol. Mutagen.* **1995**, *26*, 155–162.

55. Rosenkranz, H. S.; Klopman, G. The structural basis of the mutagenicity of chemicals in *Salmonella typhimurium*: the National toxicology program data base. *Mutat. Res.* **1990**, *228*, 51–80.

56. Zeiger, E.; Anderson, B.; Haworth, S.; Lawlor, T.; Mortelmans, K. *Salmonella* mutagenicity tests. IV. Results from the testing of 300 chemicals. *Environ. Mol. Mutagen.* **1988**, *11*, 1–157.

57. Cameron, T. P.; Hughes, T. J.; Kirby, P. E.; Fung, V. A.; Dunkel, V. C. Mutagenic activity of 27 dyes and related chemicals in the *Salmonella*/microsome and mouse lymphoma $TK^{+/-}$ assays. *Mutat. Res.* **1987**, *189*, 223–261.

58. Gaunt, I. F.; Kiss, I. S.; Grasso, P.; Gangolli, S. D. Short-term toxicity of orange G in pigs. *Food Cosmet. Toxicol.* **1973**, *11*, 367–374.

59. Gaunt, I. F.; Wright, M.; Grasso, P.; Gangolli, S. D. Short-term toxicity of Orange G in rats. *Food Cosmet. Toxicol.* **1971**, *9*, 329–342.

ORCEIN

CAS Registry Number 1400-62-0

Chemical Structure Note: Orcein is a mixture of 14 dyes. The major components of orcein are given below:

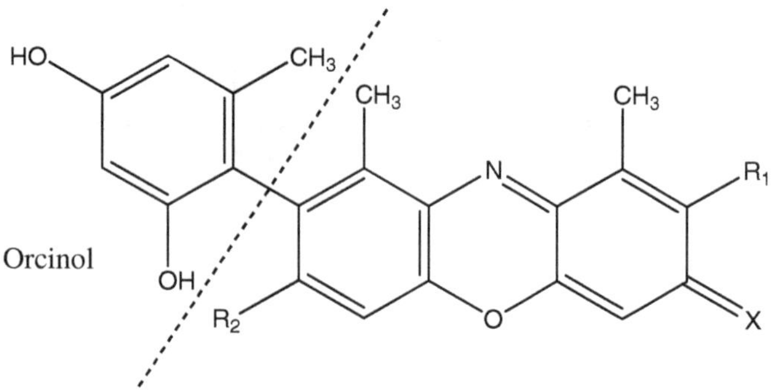

Orcein	R_1	R_2	X
α-Aminoorcein	H	NH_2	O
α-Hydroxyorcein	H	OH	O
β- and γ-Aminoorcein	Orcinol	NH_2	O
β- and γ-Hydroxyorcein	Orcinol	OH	O
β- and γ-Aminoorceimine	Orcinol	NH_2	NH

CA Index Name Orcein

Other Names C.I. 1242; C.I. Natural Red 28

Merck Index Number 6863

Chemical/Dye Class Phenoxazine

Molecular Formula Unspecified (it is a mixture)

Molecular Weight Unspecified (it is a mixture)

Physical Form Brownish-red or brownish-black powder or crystals

Solubility Insoluble in water, benzene, chloroform, ether; soluble in ethanol, acetone, acetic acid; soluble in aqueous alkali

Melting Point Unspecified (it is a mixture)

Boiling Point Unspecified (it is a mixture)

pK_a 4.0, 6.9, 13.4

Absorption (λ_{max}) 575 nm

Emission (λ_{max}) 585–590 nm

Synthesis Synthetic methods[1–17]

Staining Applications Bronchial elastic fibers;[18] chromosomes;[19] eosinophil granules;[20] elastic fibers;[21,22] embryos;[23] fibronectin;[24] vitronectin;[24] hepatocytes;[25] hepatitis B virus;[26] hepatitis;[27] hepatocellular carcinoma;[27–31] hepatitis B surface antigen (HBsAg);[32–34] histamine;[35] Kupffer cells;[36] leukocytes;[37] lipids;[38] mucins;[39] lipoproteins;[40] plasma proteins;[40] nucleic acids;[41] oocytes;[42] hairs[43,44]

Biological Applications Diagnosis of liver biopsy;[45] detecting *Candida*,[46] hepatitis B surface antigen in fixed tissues,[30] histamines,[35] HBsAg in liver cell,[32] microorganisms[47]

Industrial Applications Textiles[48]

Safety/Toxicity Carcinogenicity[49]

Certification/Approval Certified by Biological Stain Commission (BSC)

REFERENCES

1. Beecken, H.; Gottschalk, E. M.; Gizycki, U.; Kramer, H.; Maassen, D.; Matthies, H. G.; Musso, H.; Rathjen, C.; Zdhorszky, U. Orcein and litmus. *Biotech. Histochem.* **2003**, *78*, 289–302.
2. Brown, T. M.; Cooksey, C. J.; Dronsfield, A. T. Orchil and litmus—the first synthetic dyes? *Educ. Chem.* **2001**, *38*, 151–153.
3. Musso, H. Orcein pigments. XXI. Synthesis of α-aminoorcein. *Chem. Ber.* **1963**, *96*, 1936–1944.
4. Musso, H.; Zahorszky, U. I. Orcein pigments. XX. The autoxidation products of 2,5-dimethylresorcinol in ammonia and potassium hydroxide. *Chem. Ber.* **1963**, *96*, 1593–1609.
5. Musso, H.; Zahorszky, U. I.; Maassen, D.; Seeger, I. Orcein pigments. XVIII. The autoxidation products of resorcinol and 2-methylresorcinol in ammonia. *Chem. Ber.* **1963**, *96*, 1579–1587.
6. Musso, H.; Maassen, D.; Bormann, D. Orcein pigments. XVII. The autoxidation of 4,5,6-trimethylresorcinol. *Chem. Ber.* **1962**, *95*, 2837–2843.
7. Musso, H.; Maassen, D. Orcein pigments. XVI. The autoxidation of resorcinol derivatives, particularly of 4,5-dimethylresorcinol. *Chem. Ber.* **1962**, *95*, 2831–2836.
8. Beecken, H.; Gottschalk, E. M.; Gizycki, U.; Kraemer, H.; Maassen, D.; Matthies, H. G.; Musso,

H.; Rathjen, C.; Zahorszky, U. I. Orcein and litmus. *Angew. Chem.* **1961**, *73*, 665–673.

9. Musso, H.; Mannsfeld, S. P. Orcein pigments. XIII. Synthesis and rate of acetylation of α-aminoisoorcein. *Chem. Ber.* **1961**, *94*, 2585–2589.
10. Musso, H. Orcein and litmus pigments: constitutional elucidation and constitutional proof by synthesis. *Planta Med.* **1960**, *8*, 432–446.
11. Musso, H.; Kramer, H. Orcein dyes. VIII. The constitution of the hydroxyorceins and their oxidation products. *Chem. Ber.* **1958**, *91*, 2001–2016.
12. Musso, H. Orcein dyes. VII. Synthesis, constitution, and light absorption of Henrich's quinone. *Chem. Ber.* **1958**, *91*, 349–363.
13. Musso, H. Orcein dyes. II. The separation of orcein into its components. *Chem. Ber.* **1956**, *89*, 1659–1673.
14. Pavolini, T.; Gambarin, F.; Fiorenzato, B. Synthesis of *N*-homo.ovrddot.orceins. *Ann. Chim.* **1950**, *40*, 234–236.
15. Wolff, J. A new function of the catalyzer called peroxidase, and the biochemical transformation of orcinol into orcein. *Compt. Rend.* **1913**, *155*, 1031-1033.
16. Wolff, J. A new function of the catalyzer called peroxidase, and the biochemical transformation of orcinol into orcein. *Biochem. Bull.* **1913**, *2*, 53–54.
17. Zulkowski, K.; Peters, K. Orcein. *Monatsh. Chem.* **1890**, *11*, 227–245.
18. Bousquet, J.; Lacoste, J. Y.; Chanez, P.; Vic, P.; Godard, P.; Michel, F. B. Bronchial elastic fibers in normal subjects and asthmatic patients. *Am. J. Respir. Crit. Care Med.* **1996**, *153*, 1648–1654.
19. Lavania, U. C.; Sharma, A. K. Trypsin-orcein banding in plant chromosomes. *Stain Technol.* **1979**, *54*, 261–263.
20. Goldstein, D. J. Selective staining of eosinophil granules in sections by alkaline orcein in a concentrated urea solution. *Stain Technol.* **1963**, *38*, 49–51.
21. Ono, H.; Moriyasu, T.; Ohmori, Y.; Nishimura, T.; Saiga, T. Improved method of elastic fiber staining for evaluation of vascular invasion by cancer cells. *Igaku Kensa* **1995**, *44*, 1410–1415; *Chem. Abstr.* **1995**, *123*, 334055.
22. Dawson, J. F.; Brochier, J.; Schmitt, D.; Saeland, S.; Thivolet, J. Elastic fibres: histological correlation with orcein and a new monoclonal antibody, HB8. *Br. J. Dermatol.* **1984**, *110*, 539–546.
23. El-Maghraby, M. A. H. A.; Gardner, D. L. Synthetic orcein as a stain for chick embryo cartilage matrix. *Stain Technol.* **1969**, *44*, 127–129.
24. Loridon-Rosa, B.; Vielh, P.; Cuadrado, C.; Burtin, P. Comparative distribution of fibronectin and vitronectin in human breast and colon carcinomas: an immunofluorescence study. *Am. J. Clin. Pathol.* **1988**, *90*, 7–16.
25. Uehara, M.; Takezawa, J.; Yamada, S.; Sakurai, S.; Takagai, H.; Ojima, T.; Kuwabara, H.; Kobayashi, S.; Ito, H.; Joshita, H. A case of liver injury complicated with progressive muscular dystrophy (limb-girdle type)—a histological investigation of orcein negative ground glass inclusions in many of the hepatocytes. *Jpn. J. Gastroenterol.* **1989**, *86*, 95–99.
26. Mori, J.; Omata, M.; Yokosuka, O.; Imazeki, F.; Ito, Y.; Uchiumi, K.; Matsuyama, Y.; Ye, W. F.; Okuda, K. Liver orcein stain and viral DNA in duck hepatitis B virus infection in Chinese ducks and experimentally infected Japanese ducklings. *Hepatology* **1984**, *4*, 1124–1128.
27. Une, Y.; Tatara, S.; Nomura, Y.; Takahashi, R.; Saito, Y. Hepatitis and hepatocellular carcinoma in two prairie dogs (*Cynomys ludovicianus*). *J. Vet. Med. Sci.* **1996**, *58*, 933–935.
28. Panicker, J. N.; Shenoy, K. T.; Augustine, J. A cytological study of hepatitis B surface antigen localization using orcein staining in hepatocellular carcinoma. *Indian J. Med. Res.* **1996**, *104*, 374–376.
29. Kitagawa, K.; Matsui, O.; Kadoya, M.; Takashima, T.; Kawamori, Y.; Yamahana, T.; Kidani, H.; Hirano, M.; Masuda, S.; Nakanuma, Y. Hepatocellular carcinomas with excessive copper accumulation: CT and MR findings. *Radiology* **1991**, *180*, 623–628.
30. Perez-Barrios, A.; Colina-Ruizdelgado, F.; Gallego, I.; Martinez-Tello, F. J. Hepatocellular carcinoma: a study of 50 autopsy cases with detection of hepatitis B surface antigen in fixed tissues. *Pathol. Res. Pract.* **1983**, *176*, 131–144.
31. Nakashima, T.; Kojiro, M.; Kawano, Y.; Shirai, F.; Takemoto, N.; Tomimatsu, Y.; Kawasaki, H.; Okuda, K. Histologic growth pattern of hepatocellular carcinoma: relationship to orcein (hepatitis B surface antigen)-positive cells in cancer tissue. *Hum. Pathol.* **1982**, *13*, 563–568.
32. Murthy, D. P.; SenGupta, S. K.; Kelkar, S. S. Orcein staining for hepatitis B surface antigen (HBsAg) in liver diseases in Papua New Guinea. *Papua New Guinea Med. J.* **1988**, *31*, 179–183.
33. Kerr, R.; Hall, P. An evaluation of orcein methods for demonstrating hepatitis B surface antigen and

copper-associated protein in human liver. *Stain Technol.* **1986**, *61*, 243–247.

34. Volmer, J.; Popp, A.; Luders, C. J.; Henning, H. The value of various stains for the determination of HBsAg in the liver cell: comparative quantitative histomorphometric studies. *Pathologe* **1984**, *5*, 21–26.
35. Kimoto, H.; Oda, T. Detection of histamine in rat mast cell granules by orcein-water blue stain. *Acta Histochem. Cytochem.* **1979**, *12*, 292–300.
36. Uehara, C.; Sakurai, M.; Ishizu, H.; Mitsutani, N.; Koizumi, T. Orcein-positive material in Kupffer cells from the liver. *Hepato-gastroenterology* **1985**, *32*, 232–234.
37. Shousha, S.; Mitchell, T. R. Orcein staining of leukocytes. *Histochem. J.* **1983**, *15*, 563–569.
38. McKinney, B.; Riley, M. Orcein-Oil Red O stain for concomitant demonstration of elastic tissue and lipid. *Stain Technol.* **1967**, *42*, 245–248.
39. Dev, G.; Saxena, R. Orcein—a stain for mucins in colorectal carcinomas. *Indian J. Cancer* **1984**, *21*, 68–70.
40. Muckle, T. J.; Bradford, J. Dye induced alterations in lipoproteins and other plasma proteins. *Acta Histochem.* **1980**, *67*, 32–39.
41. Ginsberg, S. D.; Che, S. Amplification of RNA through a cDNA intermediate with bacteriophage promoters incorporated at the 3′-end of the cDNA. PCT Int. Appl. WO 2002065093, 2002; *Chem. Abstr.* **2002**, *137*, 180748.
42. Fajer, A. B.; Tsafriri, A.; Reese, A. H. Long-lasting preparations of mammalian oocytes stained by the acetoorcein method. *Stain Technol.* **1978**, *53*, 244–246.
43. De Boni, M. Use of natural dyes for dyeing human hair. PCT Int. Appl. WO 2008047055, 2008; *Chem. Abstr.* **2008**, *148*, 479061.
44. Greaves, E.; Greaves, J. T. Hair coloring compositions containing plant dyes and metal salts or mineral salts. U.S. Pat. Appl. Publ. US 2007251024, 2007; *Chem. Abstr.* **2007**, *147*, 508055.
45. Guarascio, P.; Yentis, F.; Cevikbas, U.; Portmann, B.; Williams, R. Value of copper-associated protein in diagnostic assessment of liver biopsy. *J. Clin. Pathol.* **1983**, *36*, 18–23.
46. Martin, S. M.; Long, A. M.; MacDonald, J. G.; Lye, J. Method for detecting *Candida* on skin using dyes. PCT Int. Appl. WO 2008026105, 2008; *Chem. Abstr.* **2008**, *148*, 302866.
47. Martin, S. M.; MacDonald, J. G.; Lye, J.; Sayre, C.; Thompson, K. Array for rapid detection of a microorganism based on dyes. PCT Int. Appl. WO 2008026104, 2008; *Chem. Abstr.* **2008**, *148*, 302867.
48. Joosten, I.; van Bommel, M. R.; Hofmann-de Keijzer, R.; Reschreiter, H. Micro analysis of Hallstatt textiles: color and condition. *Microchim. Acta* **2006**, *155*, 169–174.
49. Vecchio, F. M.; Dina, M. A. HBs Ag and hepatocellular carcinoma associated with cirrhosis. A retrospective study based on autopsy. *Hepato-gastroenterology* **1980**, *27*, 365–368.

OXONOL V

CAS Registry Number 61389-30-8

Chemical Structure

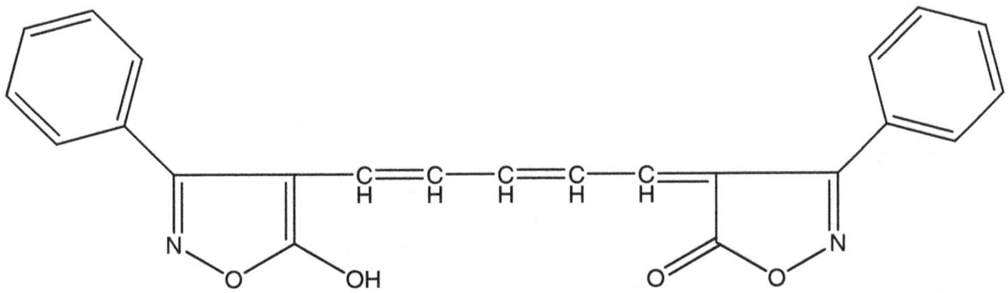

CA Index Name 5(4*H*)-Isoxazolone, 4-[5-(5-hydroxy-3-phenyl-4-isoxazolyl)-2,4-pentadien-1-ylidene]-3-phenyl-

Other Names Bis[3-phenyl-5-oxoisoxazol-4-yl]pentamethineoxonol; 4-[5-(5-Hydroxy-3-phenyl-4-isoxazolyl)-2,4-pentadienylidene]-3-phenyl-5(4*H*)-isoxazolone; 5(4*H*)-Isoxazolone, 4-[5-(5-hydroxy-3-phenyl-4-isoxazolyl)-2,4-pentadienylidene]-3-phenyl-; 1,5-Penta-1,3-dien-1-yl-5-ylidene-bis(5-oxo-3-phenyl-4-isoxazole); OX-V; Oxonol V

Merck Index Number Not listed

Chemical/Dye Class Oxonol; Isoxazole

Molecular Formula $C_{23}H_{16}N_2O_4$

Molecular Weight 384.38

Physical Form Dark green powder

Solubility Soluble in ethanol, methanol, dimethyl sulfoxide

Melting Point 200.5–201.5 °C (decompose)

Boiling Point (Calcd.) 634.1 ± 65.0 °C, pressure: 760 Torr

pK_a (Calcd.) 5.06 ± 0.50, most acidic, temperature: 25 °C; −3.45 ± 0.50, most basic, temperature: 25 °C

Absorption (λ_{max}) 610 nm

Emission (λ_{max}) 639 nm

Synthesis Synthetic methods[1,2]

Staining Applications Cerebral cortex;[3] chromaffin granule;[4] lipid vesicles;[5] liposomes;[6] proteoliposomes;[7] submitochondrial particles[8]

Biological Applications Measuring membrane potential;[4,7,9–14] potassium channel openers;[15] BK channel activator;[16] pyrophosphate assay[17]

Industrial Applications Photographic materials[18]

Safety/Toxicity No data available

REFERENCES

1. Stolle, R.; Bach, G. Trimethinoxonols from pentamethinoxonols. Ger. (East) DD 123465, 1976; *Chem. Abstr.* **1977**, *87*, 86339.
2. Smith, J. C.; Russ, P.; Cooperman, B. S.; Chance, B. Synthesis, structure determination, spectral properties, and energy-linked spectral responses of the extrinsic probe oxonol V in membranes. *Biochemistry* **1976**, *15*, 5094–5105.
3. Evans, D.; Smith, J. C. Seizure activity and cortical spreading depression monitored by an extrinsic potential-sensitive molecular probe. *Brain Res.* **1987**, *409*, 350–357.
4. Scherman, D.; Henry, J. P. Oxonol-V as a probe of chromaffin granule membrane potentials. *Biochim. Biophys. Acta, Biomembr.* **1980**, *599*, 150–166.
5. Clarke, R. J.; Apell, H. J. A stopped-flow kinetic study of the interaction of potential-sensitive oxonol dyes with lipid vesicles. *Biophys. Chem.* **1989**, *34*, 225–237.
6. Cooper, C. E.; Wrigglesworth, J. M.; Nicholls, P. The mechanism of potassium movement across the liposomal membrane. *Biochem. Biophys. Res. Commun.* **1990**, *173*, 1008–1012.

7. Cooper, C. E.; Bruce, D.; Nicholls, P. Use of oxonol V as a probe of membrane potential in proteoliposomes containing cytochrome oxidase in the submitochondrial orientation. *Biochemistry* **1990**, *29*, 3859–3865.
8. Smith, J. C.; Hallidy, L.; Topp, M. R. The behavior of the fluorescence lifetime and polarization of oxonol potential-sensitive extrinsic probes in solution and in beef heart submitochondrial particles. *J. Membr. Biol.* **1981**, *60*, 173–185.
9. Deutsch, M.; Namer, Y. A method and system for measuring membrane potential based on fluorescence polarization. PCT Int. Appl. WO 2007122602, 2007; *Chem. Abstr.* **2007**, *147*, 498288.
10. Holoubek, A.; Vecer, J.; Opekarova, M.; Sigler, K. Ratiometric fluorescence measurements of membrane potential generated by yeast plasma membrane H^+-ATPase reconstituted into vesicles. *Biochim. Biophys. Acta, Biomembr.* **2003**, *1609*, 71–79.
11. Okun, I.; Okun, A.; Kaler, G. Identifying compounds that alter membrane biological potentials with fluorescent ionic dyes. U.S. Patent 6287758, 2001; *Chem. Abstr.* **2001**, *135*, 223771.
12. Murakami, T. Cytotoxicity test method by measuring membrane electric potential. Jpn. Kokai Tokkyo Koho JP 2000300290, 2000; *Chem. Abstr.* **2000**, *133*, 330852.
13. Ozer, N. K.; Bashford, C. L.; Carter, N. D.; Pasternak, C. A. Plasma membrane potential of lymphocytes from ataxia telangiectasia patients. *Clin. Biochem.* **1989**, *22*, 469–473.
14. Bashford, C. L.; Alder, G. M.; Gray, M. A.; Micklem, K. J.; Taylor, C. C.; Turek, P. J.; Pasternak, C. A. Oxonol dyes as monitors of membrane potential: the effect of viruses and toxins on the plasma membrane potential of animal cells in monolayer culture and in suspension. *J. Cell. Physiol.* **1985**, *123*, 326–336.
15. Imaizumi, Y.; Oya, S.; Morimoto, T. Compounds with oligo methine chain containing oxooxoanion stable stereostructure as potassium channel openers. Jpn. Kokai Tokkyo Koho JP 2008266161, 2008; *Chem. Abstr.* **2008**, *149*, 525486.
16. Morimoto, T.; Sakamoto, K.; Sade, H.; Ohya, S.; Muraki, K.; Imaizumi, Y. Voltage-sensitive oxonol dyes are novel large-conductance Ca^{2+}-activated K^+ channel activators selective for β 1 and β 4 but not for β 2 subunits. *Mol. Pharmacol.* **2007**, *71*, 1075–1088.
17. Yaku, H.; Maeshima, M.; Nakanishi, Y.; Hirono, M.; Yukimasa, T.; Oka, H. Pyrophosphate assay with H^+-pyrophosphatase and application to primer extension detection. PCT Int. Appl. WO 2005093088, 2005; *Chem. Abstr.* **2005**, *143*, 362871.
18. Ohno, S. Silver halide photographic material. Eur. Pat. Appl. EP 552646, 1993; *Chem. Abstr.* **1994**, *120*, 311308.

OXONOL VI

CAS Registry Number 64724-75-0

Chemical Structure

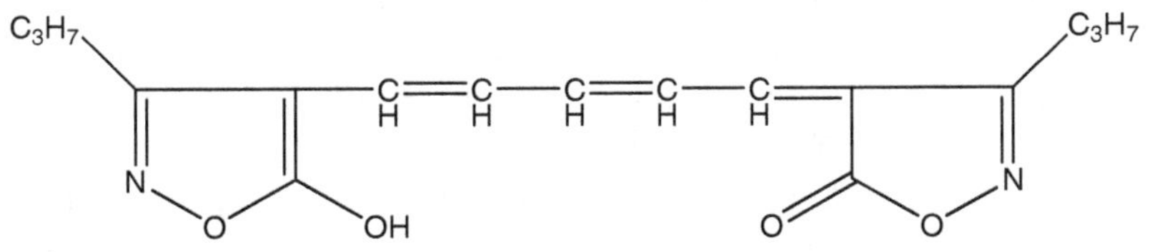

CA Index Name 5(4*H*)-Isoxazolone, 4-[5-(5-hydroxy-3-propyl-4-isoxazolyl)-2,4-pentadien-1-ylidene]-3-propyl-

Other Names 1,5-Bis(5-oxo-3-propylisoxazol-4-yl) pentamethine oxonol; Bis[3-porpyl-5-oxoisoxazol-4-yl] pentamethineoxonol; 5(4*H*)-Isoxazolone, 4-[5-(5-hydroxy-3-propyl-4-isoxazolyl)-2,4-pentadienylidene]-3-propyl-; OX-VI; Oxonol VI

Merck Index Number Not listed

Chemical/Dye Class Oxonol; Isoxazole

Molecular Formula $C_{17}H_{20}N_2O_4$

Molecular Weight 316.35

Physical Form Dark brown powder

Solubility Soluble in ethanol, methanol, dimethyl sulfoxide

Boiling Point (Calcd.) 499.4 ± 55.0 °C, pressure: 760 Torr

pK_a (Calcd.) 5.40 ± 0.50, most acidic, temperature: 25 °C; −2.35 ± 0.50, most basic, temperature: 25 °C

Absorption (λ_{max}) 599 nm

Emission (λ_{max}) 634 nm

Synthesis Synthetic methods[1,2]

Staining Applications Bacteria;[3] liposomes;[4] proteoliposomes;[5,6] lipid particles;[7] lipid vesicles;[8,9] proteins;[10] sarcoplasmic reticulum;[11] submitochondrial particles[12]

Biological Applications Measuring membrane potential;[3,4,6,9,11,13–20] treating Alzheimer's disease;[21] potassium channel openers[22]

Industrial Applications Not reported

Safety/Toxicity No data available

REFERENCES

1. Bashford, C. L.; Thayer, W. S. Thermodynamics of the electrochemical proton gradient in bovine heart submitochondrial particles. *J. Biol. Chem.* **1977**, *252*, 8459–8463.
2. Smith, J. C.; Russ, P.; Cooperman, B. S.; Chance, B. Synthesis, structure determination, spectral properties, and energy-linked spectral responses of the extrinsic probe oxonol V in membranes. *Biochemistry* **1976**, *15*, 5094–5105.
3. Bashford, C. L.; Chance, B.; Prince, R. C. Oxonol dyes as monitors of membrane potential: their behavior in photosynthetic bacteria. *Biochim. Biophys. Acta, Bioenerg.* **1979**, *545*, 46–57.
4. Venema, K.; Gibrat, R.; Grouzis, J. P.; Grignon, C. Quantitative measurement of cationic fluxes, selectivity and membrane potential using liposomes multilabeled with fluorescent probes. *Biochim. Biophys. Acta, Biomembr.* **1993**, *1146*, 87–96.
5. Goldshleger, R.; Shahak, Y.; Karlish, S. J. Electrogenic and electroneutral transport modes of renal Na/K ATPase reconstituted into proteoliposomes. *J. Membr. Biol.* **1990**, *113*, 139–154.
6. Van Walraven, H. S.; Krab, K.; Hagendoorn, M. J. M.; Kraayenhof, R. The use of carotenoids and oxonol VI as probes for membrane potential in proteoliposomes. *FEBS Lett.* **1985**, *184*, 96–99.
7. Doranz, B. J. Lipid particles containing ion channels and membrane potential-sensitive dyes and their use in screening for effectors of ion channels. PCT Int. Appl. WO 2007089582, 2007; *Chem. Abstr.* **2007**, *147*, 250560.
8. Clarke, R. J.; Apell, H. J. A stopped-flow kinetic study of the interaction of potential-sensitive oxonol dyes with lipid vesicles. *Biophys. Chem.* **1989**, *34*, 225–237.

9. Apell, H. J.; Bersch, B. Oxonol VI as an optical indicator for membrane potentials in lipid vesicles. *Biochim. Biophys. Acta, Biomembr.* **1987**, *903*, 480–494.
10. Patti, J. T.; Montemagno, C. D. Fluorometric functional assay for ion channel proteins in lipid nanovesicle membranes. *Nanotechnology* **2007**, *18*, 325103/1–325103/5.
11. Beeler, T. J.; Farmen, R. H.; Martonosi, A. N. The mechanism of voltage-sensitive dye responses on sarcoplasmic reticulum. *J. Membr. Biol.* **1981**, *62*, 113–137.
12. Smith, J. C.; Hallidy, L.; Topp, M. R. The behavior of the fluorescence lifetime and polarization of oxonol potential-sensitive extrinsic probes in solution and in beef heart submitochondrial particles. *J. Membr. Biol.* **1981**, *60*, 173–185.
13. Deutsch, M.; Namer, Y. A method and system for measuring membrane potential based on fluorescence polarization. PCT Int. Appl. WO 2007122602, 2007; *Chem. Abstr.* **2007**, *147*, 498288.
14. Richter, P. R.; Schuster, M.; Meyer, I.; Lebert, M.; Hader, D. P. Indications for acceleration-dependent changes of membrane potential in the flagellate *Euglena gracilis*. *Protoplasma* **2006**, *229*, 101–108.
15. Richter, P.; Lebert, M.; Korn, R.; Hader, D. P. Possible involvement of the membrane potential in the gravitactic orientation of *Euglena gracilis*. *J. Plant Physiol.* **2001**, *158*, 35–39.
16. Okun, I.; Okun, A.; Kaler, G. Identifying compounds that alter membrane biological potentials with fluorescent ionic dyes. U.S. Patent 6287758, 2001; *Chem. Abstr.* **2001**, *135*, 223771.
17. Murakami, T. Cytotoxicity test method by measuring membrane electric potential. Jpn. Kokai Tokkyo Koho JP 2000300290, 2000; *Chem. Abstr.* **2000**, *133*, 330852.
18. Ahmed, I.; Krishnamoorthy, G. Anomalous response of oxonol-V to membrane potential in mitochondrial proton pumps. *Biochim. Biophys. Acta, Bioenerg.* **1994**, *1188*, 131–138.
19. Armitage, J. P.; Evans, M. C. W. Comparison of the carotenoid bandshift and oxanol dyes to measure membrane potential changes during chemotactic stimulation of *Rhodopseudomonas sphaeroides* and *Escherichia coli*. *FEBS Lett.* **1981**, *126*, 98–102.
20. Schuurmans, J. J.; Casey, R. P.; Kraayenhof, R. Transmembrane electrical potential formation in spinach chloroplasts: investigation using a rapidly-responding extrinsic probe. *FEBS Lett.* **1978**, *94*, 405–409.
21. Kudo, Y.; Arai, H.; Okamura, N.; Maruyama, M.; Furumoto, S.; Doh-Ura, K. Long-wavelength fluorescent substance for conformation disease diagnosis or treatment, and its use. PCT Int. Appl. WO 2007063950, 2007; *Chem. Abstr.* **2007**, *147*, 26487.
22. Imaizumi, Y.; Oya, S.; Morimoto, T. Compounds with oligo methine chain containing oxooxoanion stable stereostructure as potassium channel openers. Jpn. Kokai Tokkyo Koho JP 2008266161, 2008; *Chem. Abstr.* **2008**, *149*, 525486.

PARAROSANILINE HYDROCHLORIDE

CAS Registry Number 569-61-9

Chemical Structure

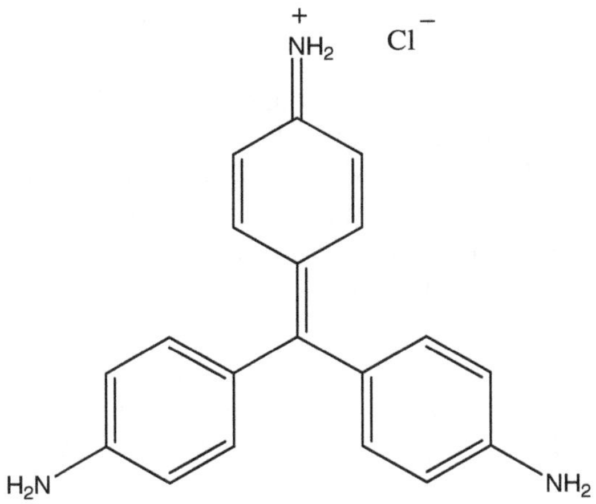

CA Index Name Benzenamine, 4,4′-[(4-imino-2,5-cyclohexadien-1-ylidene)methylene]bis-, hydrochloride (1:1)

Other Names Benzenamine, 4,4′-[(4-imino-2,5-cyclohexadien-1-ylidene)methylene]bis-, monohydrochloride; Benzenamine, 4-[(4-aminophenyl)(4-imino-2,5-cyclohexadien-1-ylidene)methyl]-, monohydrochloride; C.I. Basic Red 9, monohydrochloride; Para Magenta; Basic Parafuchsine; Basic Red 9; C.I. 42500; C.I. Basic Red 9; Calcozine Magenta N; Fuchsine DR 001; Fuchsine SP; Fuchsine SPC; Orient Para Magenta Base; Parafuchsin; Parafuchsine; Pararosaniline; Pararosaniline chloride; Pararosaniline hydrochloride; *p*-Fuchsin; *p*-Fuchsine; *p*-Rosaniline hydrochloride

Merck Index Number Not listed

Chemical/Dye Class Triphenylmethane

Molecular Formula $C_{19}H_{18}ClN_3$

Molecular Weight 323.82

Physical Form Dark green crystals

Solubility Soluble in water, ethanol; insoluble in ether

Melting Point 268–270 °C (decompose)

Absorption (λ_{max}) 545 nm

Synthesis Synthetic methods[1–11]

Staining Applications Cytokines;[12] antigen-specific antibody;[12] leukocytes;[13] nuclei;[14] nucleic acids;[15,16] parasites;[17] precancerous cells;[18] prions;[19] hairs;[20] keratine fibers[21]

Biological Applications Detecting breast cancer;[22] treating pathogens[23]

Industrial Applications Optical waveguides;[24] semiconductor materials;[25] nanoparticles;[26] inks;[27–29] highlighters;[30] toner;[31] textiles;[32] wood[33]

Safety/Toxicity Carcinogenicity;[34–41] chemical toxicity;[42] chromosomal aberration;[39,43,44] genotoxicity;[45] mutagenicity[37,42,46–48]

Certification/Approval Certified by Biological Stain Commission (BSC)

REFERENCES

1. Ellis, G. D.; Dimarcello, B. J.; Bradshaw, D. J. Preparation of a dye for coloring protein-based fibers and cellulose-based materials from the oxidation byproducts of the manufacture of a triphenylmethane dye. Eur. Pat. Appl. EP 909794, 1999; *Chem. Abstr.* **1999**, *130*, 313189.
2. Zhang, Y. Selection of the synthetic methods of tri (aminophenyl)methanol and approach of the new technological process. *Ranliao Gongye* **1998**, *35*, 28–32; *Chem. Abstr.* **1999**, *131*, 59993.
3. Turrill, F. H.; Squire, L. E.; Towle, J. L. Separation of pararosaniline from aniline solutions. Ger. Offen. DE 2355735, 1974; *Chem. Abstr.* **1974**, *81*, 137596.
4. Van Duijn, P.; Riddersma, S. H. Purification of pararosaniline and atebrine by chromatography on lipophilic Sephadex LH-20. *Histochem. J.* **1973**, *5*, 169–172.
5. Gabler, W. Purification of parafuchsin for the preparation of Schiff's reagent: paper chromatographic separation of the products of the reaction between aldehydes and Schiff's reagent. *Acta Histochem.* **1965**, *21*, 387–392.
6. Fry, B. A. Basic triphenylmethane dyes and the inhibition of glutamine synthesis by *Staphylococcus aureus* (*Micrococcus pyogenes* var. *aureus*). *J. Gen. Microbiol.* **1957**, *16*, 341–349.

Handbook of Biological Dyes and Stains By R. W. Sabnis

7. Yarbo, C. L.; Miller, B.; Anderson, C. E. Purifying pararosaniline for use in the colorless Schiff reagent. *Stain Technol.* **1954**, *29*, 299–300.
8. Loukomsky, S. A.; Herrick, G. S. Polyaryl methane dyes of the rosaniline type. U.S. Patent 2542544, 1951; *Chem. Abstr.* **1951**, *45*, 43085.
9. Goldacre, R. J.; Phillips, J. N. The ionization of basic triphenylmethane dyes. *J. Chem. Soc.* **1949**, 1724–1732.
10. Ignatyev, S. N.; Vasin, I. I. Parafuchsin. RU 59307, 1929; *Chem. Abstr.* **1931**, *25*, 23167.
11. Zoller, H. F. Oxalyl chloride in the synthesis of the triphenylmethane dyes. *Science* **1920**, *52*, 207.
12. Schrijver, I. A.; Melief, M. J.; Van Meurs, M.; Companjen, A. R.; Laman, J. D. Pararosaniline fixation for detection of co-stimulatory molecules, cytokines, and specific antibody. *J. Histochem. Cytochem.* **2000**, *48*, 95–103.
13. Takarada, K.; Kouzuki, C.; Hyousa, Y.; Sakata, T.; Akai, Y. A method for classifying and counting leukocytes. Eur. Pat. Appl. EP 695936, 1996; *Chem. Abstr.* **1996**, *124*, 283704.
14. Horobin, R. W.; Stockert, J. C.; Rashid-Doubell, F. Fluorescent cationic probes for nuclei of living cells: why are they selective? A quantitative structure–activity relations analysis. *Histochem. Cell Biol.* **2006**, *126*, 165–175.
15. Garner, D. M.; Todorovic, C.; Lee, W. E. Cytological stain composition for cytological analysis of cellular DNA. U.S. Pat. Appl. Publ. US 2006199243, 2006; *Chem. Abstr.* **2006**, *145*, 288084.
16. Kamme, F. C.; Meurers, B. H.; Talantov, D.; Yu, J. Methods for preservation of RNA in a biological samples for quantitation in immunostaining and other assays. PCT Int. Appl. WO 2004096984, 2004; *Chem. Abstr.* **2004**, *141*, 391521.
17. Peribanez, M. A.; Lucientes, J.; Arce, S.; Morales, M.; Castillo, J. A.; Gracia, M. J. Histochemical differentiation of *Dirofilaria immitis*, *Dirofilaria repens* and *Acanthocheilonema dracunculoides* microfilariae by staining with a commercial kit, Leucognost-SP. *Vet. Parasitol.* **2001**, *102*, 173–175.
18. Cottington, K. Detecting cancer by histochemical staining. Brit. UK Pat. Appl. GB 2355790, 2001; *Chem. Abstr.* **2001**, *135*, 177699.
19. Orser, C. S.; Pan, T.; Sethi, J. Detecting and removing pathogenic misfolded proteins, such as prions, using conformational capture peptide probes and double-labeled detection-amplification peptides. PCT Int. Appl. WO 2006088823, 2006; *Chem. Abstr.* **2006**, *145*, 265232.
20. Moeller, H.; Hoeffkes, H.; Oberkobusch, D. Oxidative hair dyes composed of primary amino group-containing chromophores and reactive carbonyl compounds. Ger. Offen. DE 10260881, 2004; *Chem. Abstr.* **2004**, *141*, 93976.
21. Plos, G.; Guerin, F. Process for fast coloring and decoloring human keratinic fibers with direct dyes. Fr. Demande FR 2845906, 2004; *Chem. Abstr.* **2004**, *140*, 326616.
22. Krepinsky, J. J.; Furrer, R.; Yeung, K. S. Methods for detection of breast cancer. PCT Int. Appl. WO 2004034059, 2004; *Chem. Abstr.* **2004**, *140*, 335240.
23. Lurie, R. Phototherapy-based method and composition for treating pathogens. PCT Int. Appl. WO 9904628, 1999; *Chem. Abstr.* **1999**, *130*, 150435.
24. Tomomatsu, I.; Iwase, M. Polymer optical waveguides containing colorants. Jpn. Kokai Tokkyo Koho JP 08075938, 1996; *Chem. Abstr.* **1996**, *124*, 355889.
25. Werner, A.; Pfeiffer, M.; Fritz, T.; Leo, K. Doped organic semiconductor material and method for production thereof. PCT Int. Appl. WO 2003070822, 2003; *Chem. Abstr.* **2003**, *139*, 189329.
26. Jungmann, N.; Schmidt, M.; Ebenhoch, J.; Weis, J.; Maskos, M. Dye loading of amphiphilic poly (organosiloxane) nanoparticles. *Angew. Chem. Int. Ed.* **2003**, *42*, 1714–1717.
27. Ryu, S. M.; Jung, Y. K. Inkjet ink composition and multi-color inkjet ink set having the same. U.S. Pat. Appl. Publ. US 2007081061, 2007; *Chem. Abstr.* **2007**, *146*, 403849.
28. Hiromichi, N. Ink-jet inks containing photopolymerization initiators and recording method. Eur. Pat. Appl. EP 953613, 1999; *Chem. Abstr.* **1999**, *131*, 323974.
29. Yamazaki, H. Anticlogging jet-printing inks with low sodium content. Jpn. Kokai Tokkyo Koho JP 08333542, 1996; *Chem. Abstr.* **1997**, *126*, 173173.
30. Davies-Smith, L.; Sum, V. W. Highlightable and highlighted mixtures, marking instruments, eradicator solution and kits, using the mixtures in highlighting. U.S. Pat. Appl. Publ. US 2005120919, 2005; *Chem. Abstr.* **2005**, *143*, 9331.
31. Yamaka, H.; Shimura, I.; Matsura, H.; Saito, H. Toner for developing electrostatic image. Jpn. Kokai Tokkyo Koho JP 10003183, 1998; *Chem. Abstr.* **1998**, *128*, 147470.
32. Sun, G.; Kim, Y. H. Antimicrobial polymers, their manufacture and textiles. PCT Int. Appl. WO 2002022923, 2002; *Chem. Abstr.* **2002**, *136*, 248989.

33. Leach, R. M.; Zhang, J. Composition and process for coloring and preserving wood. PCT Int. Appl. WO 2005056257, 2005; *Chem. Abstr.* **2005**, *143*, 61604.

34. Sanner, T.; Dybing, E.; Willems, M. I.; Kroese, E. D. A simple method for quantitative risk assessment of non-threshold carcinogens based on the dose descriptor T25. *Pharmacol. Toxicol.* **2001**, *88*, 331–341.

35. Kowalski, L. A.; Laitinen, A. M.; Martazavi-Asl, B.; Wee, R. K. H.; Erb, H. E.; Assi, K. P.; Madden, Z. *In vitro* determination of carcinogenicity of sixty-four compounds using a bovine papillomavirus DNA-carrying C3H/10T1/2 cell line. *Environ. Mol. Mutagen.* **2000**, *35*, 300–311.

36. Albert, R. E. Allergic contact sensitizing chemicals as environmental carcinogens. *Environ. Health Perspect.* **1997**, *105*, 940–948.

37. Dybing, E.; Sanner, T.; Roelfzema, H.; Kroese, D.; Tennant, R. W. T25: a simplified carcinogenic potency index: description of the system and study of correlations between carcinogenic potency and species/site specificity and mutagenicity. *Pharmacol. Toxicol.* **1997**, *80*, 272–279.

38. Haseman, J. K.; Lockhart, A. The relationship between use of the maximum tolerated dose and study sensitivity for detecting rodent carcinogenicity. *Fundam. Appl. Toxicol.* **1994**, *22*, 382–391.

39. Rosenkranz, H. S.; Ennever, F. K.; Klopman, G. Relationship between carcinogenicity in rodents and the induction of sister chromatid exchanges and chromosomal aberrations in Chinese hamster ovary cells. *Mutagenesis* **1990**, *5*, 559–571.

40. Zeiger, E. Carcinogenicity of mutagens: predictive capability of the *Salmonella* mutagenesis assay for rodent carcinogenicity. *Cancer Res.* **1987**, *47*, 1287–1296.

41. Traul, K. A.; Takayama, K.; Kachevsky, V.; Hink, R. J.; Wolff, J. S. Rapid *in vitro* assay for carcinogenicity of chemical substances in mammalian cells utilizing an attachment-independence endpoint. 2. Assay validation. *J. Appl. Toxicol.* **1981**, *1*, 190–195.

42. Hayes, S.; Gordon, A.; Sadowski, I.; Hayes, C. RK bacterial test for independently measuring chemical toxicity and mutagenicity: short-term forward selection assay. *Mutat. Res.* **1984**, *130*, 97–106.

43. Yamada, T.; Odawara, K.; Kaneko, H. Concurrent detection of gene mutations and chromosome aberrations induced by five chemicals in a CHL/IU cell line incorporating a gpt shuttle vector. *Mutat. Res.* **2000**, *471*, 29–36.

44. Rosenkranz, H. S.; Ennever, F. K.; Dimayuga, M.; Klopman, G. Significant differences in the structural basis of the induction of sister chromatid exchanges and chromosomal aberrations in Chinese hamster ovary cells. *Environ. Mol. Mutagen.* **1990**, *16*, 149–177.

45. Kirkland, D.; Aardema, M.; Henderson, L.; Mueller, L. Evaluation of the ability of a battery of three *in vitro* genotoxicity tests to discriminate rodent carcinogens and non-carcinogens. I. Sensitivity, specificity and relative predictivity. *Mutat. Res.* **2005**, *584*, 1–256.

46. Serafimova, R.; Todorov, M.; Pavlov, T.; Kotov, S.; Jacob, E.; Aptula, A.; Mekenyan, O. Identification of the structural requirements for mutagenicity, by incorporating molecular flexibility and metabolic activation of chemicals. II. General Ames mutagenicity model. *Chem. Res. Toxicol.* **2007**, *20*, 662–676.

47. Benigni, R.; Andreoli, C.; Giuliani, A. Relationships among *in vitro* mutagenicity assays: quantitative vs. qualitative test results. *Environ. Mol. Mutagen.* **1995**, *26*, 155–162.

48. Zeiger, E. Mutagenicity of 42 chemicals in *Salmonella*. Rodent tumor profiles, *Salmonella* mutagenicity and risk assessment. *Environ. Mol. Mutagen.* **1990**, *16*, 32–54.

PBFI

CAS Registry Number 124549-11-7

Chemical Structure

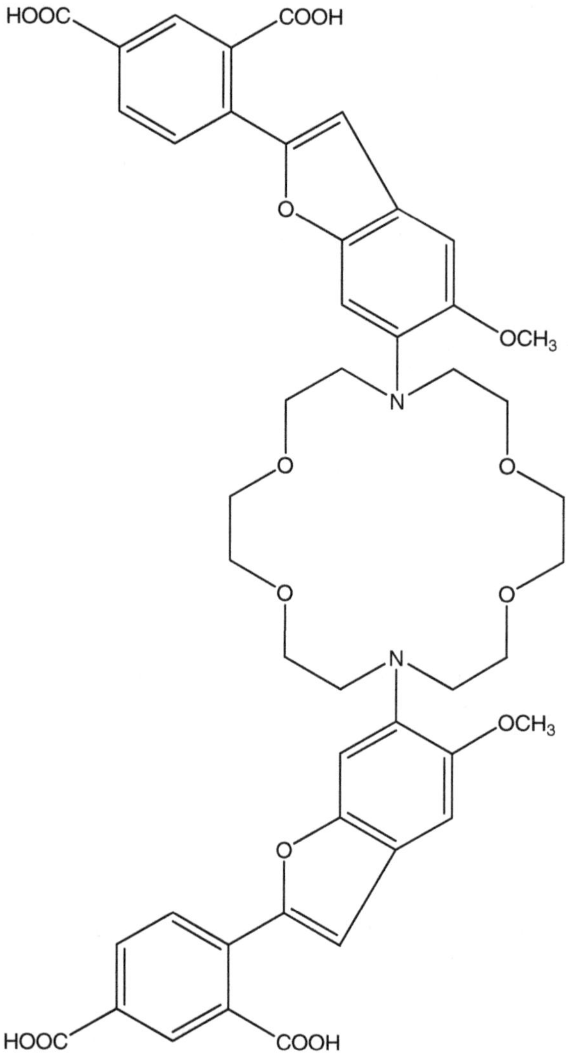

CA Index Name 1,3-Benzenedicarboxylic acid, 4,4′-[1,4,10,13-tetraoxa-7,16-diazacyclooctadecane-7,16-diylbis(5-methoxy-6,2-benzofurandiyl)]bis-

Other Names 1,4,10,13-Tetraoxa-7,16-diazacyclooctadecane, 1,3-benzenedicarboxylic acid derivative; PBFI

Merck Index Number Not listed

Chemical/Dye Class Benzofuran

Molecular Formula $C_{46}H_{46}N_2O_{16}$

Molecular Weight 882.86

Physical Form Brownish-yellow powder

Solubility Soluble in methanol, dimethyl sulfoxide

Melting Point >200 °C

Boiling Point (Calcd.) 1058.8 ± 65.0 °C, pressure: 760 Torr

pK_a (Calcd.) 2.96 ± 0.36, most acidic, temperature: 25 °C; 4.06 ± 0.40, most basic, temperature: 25 °C

Absorption (λ_{max}) 336 nm

Emission (λ_{max}) 557 nm

Synthesis Synthetic methods[1–3]

Staining Applications Potassium ions;[1–15] cells;[16] liposomes[4]

Biological Applications Potassium indicator;[1–15] monitoring cellular processes[16]

Industrial Applications Not reported

Safety/Toxicity Carcinogenicity[17]

REFERENCES

1. Crossley, R.; Goolamali, Z.; Gosper, J. J.; Sammes, P. G. Synthesis and spectral properties of new fluorescent probes for potassium. *J. Chem. Soc., Perkin Trans. 2* **1994**, 513–520.
2. Tsien, R. Y.; Minta, A. Preparation and use of selective fluorescent indicator dyes for alkali metal cation analysis. Eur. Pat. Appl. EP 369733, 1990; *Chem. Abstr.* **1991**, *115*, 227789.
3. Minta, A.; Tsien, R. Y. Fluorescent indicators for cytosolic sodium. *J. Biol. Chem.* **1989**, *264*, 19449–19457.
4. Venema, K.; Gibrat, R.; Grouzis, J. P.; Grignon, C. Quantitative measurement of cationic fluxes, selectivity and membrane potential using liposomes multilabeled with fluorescent probes. *Biochim. Biophys. Acta, Biomembr.* **1993**, *1146*, 87–96.

5. Zeilinger, C. Process and apparatus for ion channel characterization. PCT Int. Appl. WO 2008022658, 2008; *Chem. Abstr.* **2008**, *148*, 302860.
6. Brown, J. Q.; McShane, M. J. Core-referenced ratiometric fluorescent potassium ion sensors using self-assembled ultrathin films on europium nanoparticles. *IEEE Sens. J.* **2005**, *5*, 1197–1205.
7. Rehman, S.; Yoo, N. H.; Park, M. R.; Yun, S. J. Confocal potassium imaging: giving new insight into potassium concentrated at the aperture area of barley (*Hordeum vulgare* L.) pollen. *Plant Sci.* **2005**, *169*, 457–459.
8. Halperin, S. J.; Lynch, J. P. Effects of salinity on cytosolic Na^+ and K^+ in root hairs of *Arabidopsis thaliana*: *in vivo* measurements using the fluorescent dyes SBFI and PBFI. *J. Exp. Bot.* **2003**, *54*, 2035–2043.
9. Muhling, K. H.; Lauchli, A. Light-induced pH and K^+ changes in the apoplast of intact leaves. *Planta* **2000**, *212*, 9–15.
10. Denda, M.; Hosoi, J.; Asida, Y. Visual imaging of ion distribution in human epidermis. *Biochem. Biophys. Res. Commun.* **2000**, *272*, 134–137.
11. Szmacinski, H.; Lakowicz, J. R. Potassium and sodium measurements at clinical concentrations using phase-modulation fluorometry. *Sens. Actuators, B* **1999**, *B60*, 8–18.
12. Kowalczyk, A.; Boens, N.; Meuwis, K.; Ameloot, M. Potential misevaluation of the ground-state dissociation constant from fluorimetric titrations: application to the ion indicators SBFI, PBFI, and Fura-2. *Anal. Biochem.* **1997**, *245*, 28–37.
13. Freisleben, H. J.; Zwicker, K.; Jezek, P.; John, G.; Bettin-Bogutzki, A.; Ring, K.; Nawroth, T. Reconstitution of bacteriorhodopsin and ATP synthase from *Micrococcus luteus* into liposomes of the purified main tetraether lipid from *Thermoplasma acidophilum*: proton conductance and light-driven ATP synthesis. *Chem. Phys. Lipids* **1995**, *78*, 137–147.
14. Kasner, S. E.; Ganz, M. B. Regulation of intracellular potassium in mesangial cells: a fluorescence analysis using the dye, PBFI. *Am. J. Physiol.* **1992**, *262*, F462–F467.
15. Jezek, P.; Mahdi, F.; Garlid, K. D. Reconstitution of the beef heart and rat liver mitochondrial potassium/hydrogen ion (sodium/hydrogen ion) antiporter: quantitation of potassium transport with the novel fluorescent probe, PBFI. *J. Biol. Chem.* **1990**, *265*, 10522–10526.
16. Cousin, M. A.; Evans, G. J. O. Assay method. PCT Int. Appl. WO 2006134319, 2006; *Chem. Abstr.* **2006**, *146*, 77602.
17. Andersson, B.; Janson, V.; Behnam-Motlagh, P.; Henriksson, R.; Grankvist, K. Induction of apoptosis by intracellular potassium ion depletion: using the fluorescent dye PBFI in a 96-well plate method in cultured lung cancer cells. *Toxicol. In Vitro* **2006**, *20*, 986–994.

PBFI AM

CAS Registry Number 124549-23-1

Chemical Structure

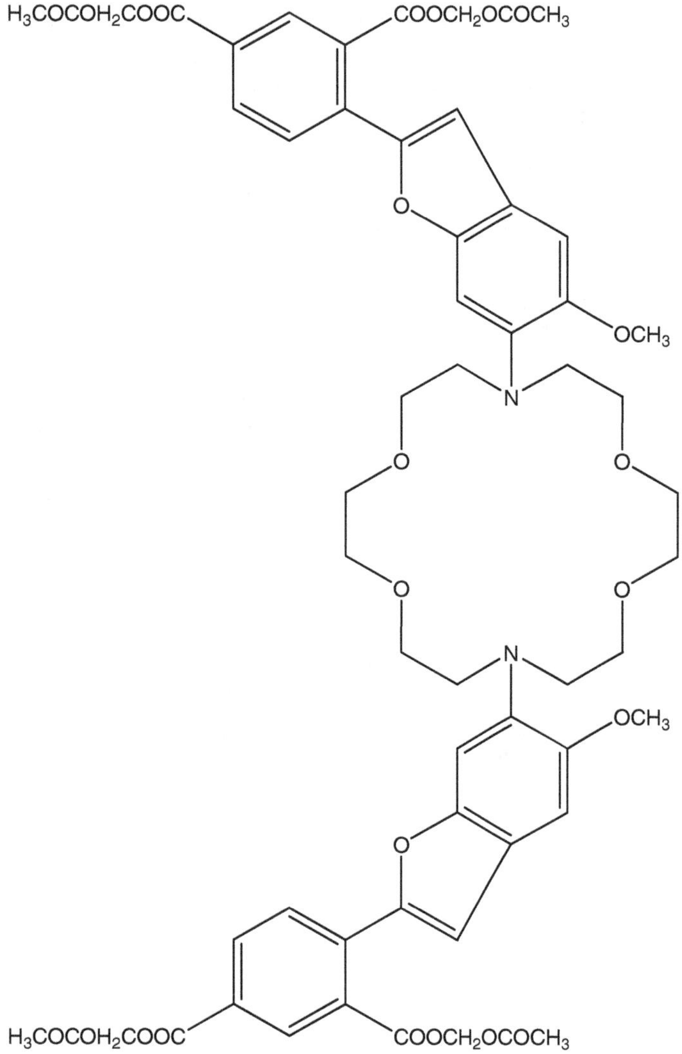

CA Index Name 1,3-Benzenedicarboxylic acid, 4,4′-[1,4,10,13-tetraoxa-7,16-diazacyclooctadecane-7,16-diylbis(5-methoxy-6,2-benzofurandiyl)]bis-, tetrakis [(acetyloxy)methyl] ester

Other Names 1,4,10,13-Tetraoxa-7,16-diazacyclooctadecane, 1,3-benzenedicarboxylic acid derivative; PBFI AM

Merck Index Number Not listed

Chemical/Dye Class Benzofuran

Molecular Formula $C_{58}H_{62}N_2O_{24}$

Molecular Weight 1171.11

Physical Form Yellowish-brown powder

Solubility Soluble in methanol, dimethyl sulfoxide

Melting Point >200 °C

Boiling Point (Calcd.) 1136.6 ± 65.0 °C, pressure: 760 Torr

pK_a (Calcd.) 3.79 ± 0.40, most basic, temperature: 25 °C

Absorption (λ_{max}) 369 nm

Emission (λ_{max}) Fluorescence is very weak

Synthesis Synthetic methods[1,2]

Staining Applications Potassium ions[1,2]

Biological Applications Potassium indicator[1,2]

Industrial Applications Not reported

Safety/Toxicity No data available

REFERENCES

1. Tsien, R. Y.; Minta, A. Preparation and use of selective fluorescent indicator dyes for alkali metal cation analysis. Eur. Pat. Appl. EP 369733, 1990; *Chem. Abstr*. **1991**, *115*, 227789.
2. Minta, A.; Tsien, R. Y. Fluorescent indicators for cytosolic sodium. *J. Biol. Chem.* **1989**, *264*, 19449–19457.

PHEN GREEN SK

CAS Registry Number 234075-34-4

Chemical Structure

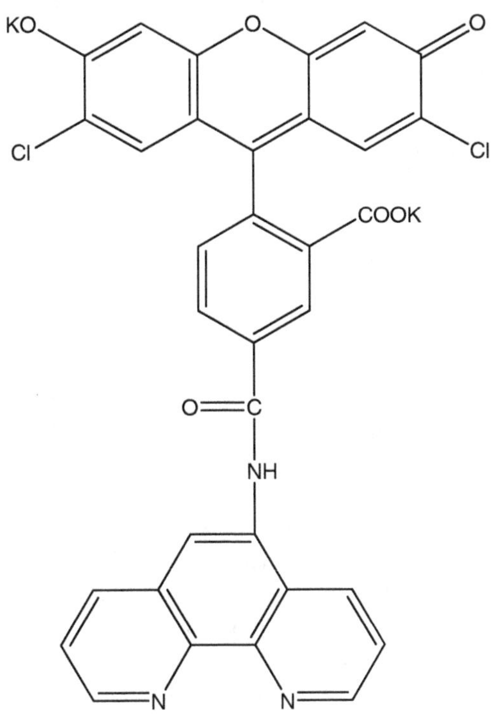

CA Index Name Spiro[isobenzofuran-1(3*H*),9′-[9*H*]xanthene]-5-carboxamide, 2′,7′-dichloro-3′,6′-dihydroxy-3-oxo-*N*-1,10-phenanthrolin-5-yl-, potassium salt (1:2)

Other Names Spiro[isobenzofuran-1(3*H*),9′-[9*H*]xanthene]-5-carboxamide, 2′,7′-dichloro-3′,6′-dihydroxy-3-oxo-*N*-1,10-phenanthrolin-5-yl-, dipotassium salt; Phen Green SK; Phen Green dipotassium salt

Merck Index Number Not listed

Chemical/Dye Class Xanthene

Molecular Formula $C_{33}H_{15}Cl_2K_2N_3O_6$

Molecular Weight 698.60

Physical Form Solid

Solubility Soluble in water

Melting Point >200 °C

Absorption (λ_{max}) 507 nm

Emission (λ_{max}) 532 nm

Synthesis Synthetic methods[1,2]

Staining Applications Copper ions;[1,2,5] iron ions;[6–8] cells;[3] neurons[4]

Biological Applications Copper indicators;[1,2,5] iron ions[6–8]

Industrial Applications Not reported

Safety/Toxicity No data available

REFERENCES

1. Kuhn, M. A.; Haugland, R. P.; Hoyland, B. M. Methods of sensing with fluorescent conjugates of metal-chelating nitrogen heterocycles. U.S. Patent 5648270, 1997; *Chem. Abstr.* **1997**, *127*, 199374.
2. Kuhn, M. A.; Hoyland, B.; Carter, S.; Zhang, C.; Haugland, R. P. Fluorescent ion indicators for detecting heavy metals. *Proc. SPIE-Int. Soc. Opt. Eng.* **1995**, *2388*, 238–244.
3. Petrat, F.; de Groot, H.; Rauen, U. Determination of the chelatable iron pool of single intact cells by laser scanning microscopy. *Arch. Biochem. Biophys.* **2000**, *376*, 74–81.
4. Reynolds, I. J. Fluorescence detection of redox-sensitive metals in neuronal culture: focus on iron and zinc. *Ann. N.Y. Acad. Sci.* **2004**, *1012*, 27–36.
5. Kim, H. S.; Ahner, B. A. Calibration of Phen Green for use as a Cu(I)-selective fluorescent indicator. *Anal. Chim. Acta* **2006**, *575*, 223–229.
6. Shingles, R.; North, M.; McCarty, R. E. Ferrous ion transport across chloroplast inner envelope membranes. *Plant Physiol.* **2002**, *128*, 1022–1030.
7. Shingles, R.; North, M.; McCarty, R. E. Direct measurement of ferrous ion transport across membranes using a sensitive fluorometric assay. *Anal. Biochem.* **2001**, *296*, 106–113.
8. Petrat, F.; Rauen, U.; De Groot, H. Determination of the chelatable iron pool of isolated rat hepatocytes by digital fluorescence microscopy using the fluorescent probe, Phen Green SK. *Hepatology* **1999**, *29*, 1171–1179.

PHLOXINE B

CAS Registry Number 18472-87-2

Chemical Structure

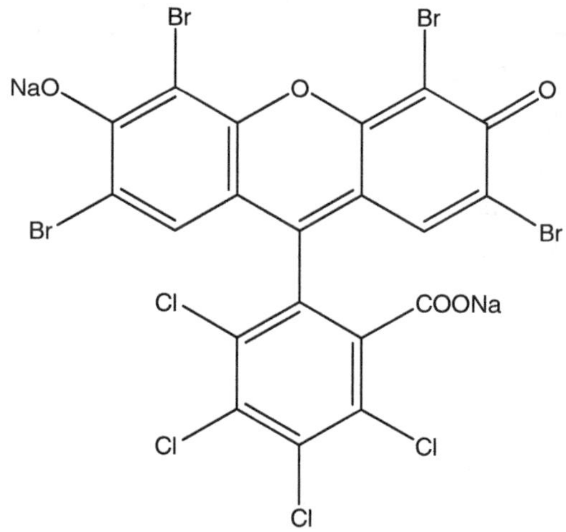

CA Index Name Spiro[isobenzofuran-1(3*H*),9′-[9*H*]xanthen]-3-one, 2′,4′,5′,7′-tetrabromo-4,5,6,7-tetrachloro-3′,6′-dihydroxy-, sodium salt (1:2)

Other Names Fluorescein, 2′,4′,5′,7′-tetrabromo-4,5,6,7-tetrachloro-, disodium salt; Phloxin B; Spiro[isobenzofuran-1(3*H*),9′-[9*H*]xanthen]-3-one, 2′,4′,5′,7′-tetrabromo-4,5,6,7-tetrachloro-3′,6′-dihydroxy-, disodium salt; 11969 Red; 3427 Veri Pur Pink; Acid Red 92; Aizen Acid Phloxine PB; C.I. 45410; C.I. Acid Red 92; Cyanosin; Cyanosin (acid dye); Cyanosin B; Cyanosine; D and C Red No. 28; D&C Red 28; D&C Red 28-308267; D&C Red No. 28; D&C Red No. 28-15347; D&C Red No. 28-38015; Daiwa Red 104WB; Daiwa Red 45; Disodium 9-(3′,4′,5′,6′-tetrachloro-*o*-carboxyphenyl)-6-hydroxy-6-hydroxy-2,4,5,7-tetrabromo-3-isoxanthone; Eosin blue; Eosin bluish; Eosine blue; Eosine bluish; Food Red 104; Food Red No. 104; Japan Red 104; Japan Red 104-1; Japan Red No. 104-1; Orient Water Pink 2; Phloxine B; Phloxine B 38015; Phloxine P; Phyloxine B; Red 104; Red No. 104; Red No. 104-1; Triacid Floxine 2G; Water Pink 2

Merck Index Number Not listed

Chemical/Dye Class Xanthene

Molecular Formula $C_{20}H_2Br_4Cl_4Na_2O_5$

Molecular Weight 829.66

Physical Form Red to brown crystals

Solubility Soluble in water, ethylene glycol; slightly soluble in ethanol, methanol

Melting Point >250 °C

pH Range 1.1–3.3; 3.4–5.0

Color Change at pH Colorless (1.1) to purple (3.3); colorless (3.4) to yellow (5.0)

Absorption (λ_{max}) 548 nm, 510 nm

Synthesis Synthetic methods[1–7]

Staining Applications Albumin;[8,9] aluminum;[10] bacteria;[11] candies;[12] drinks;[12] dental plaque;[1,13,14] lymph node;[15] proteins;[16,17] peptides;[17] carious tissue;[18] teeth;[19] sunscreen;[20,21] eyebrows;[29] eyelashes;[29] eye shadow;[22] fingerprint powders;[23] lipsticks;[24] lips;[1,25–27,29,30] nails;[30] skin;[1,25–34] hairs;[1,28,29,35–39] keratin fibers[40]

Biological Applications Detecting proteins;[41] treating microbial infection,[42] parasitic infection,[42] skin,[43,44] mouth,[43,44] digestive tract,[43,44] urinary tract,[43,44] reproductive tract,[43,44] respiratory tract,[43,44] circulatory system,[43,44] head,[43,44] neck,[43,44] endocrine system,[43,44] lymphoreticular system,[43,44] cystic fibrosis,[45] disseminated bronchectasis,[45] pulmonary infections,[45] chronic pancreatitis,[45] male infertility,[45] long QT syndrome,[45] protozoan infections,[46] yeast infection,[47] fungal infections;[47] antimicrobial agent;[48] insecticides;[49] herbicides[50]

Industrial Applications Color filter;[1,51] liquid-crystal displays;[51] solar cells;[52] optical waveguides;[1,53] electrochromic displays;[1,54] photoresists;[55] recording materials;[56] photographic materials;[57] inks;[1,58–61] toner;[62] highlighters;[1,63] paints;[64] adhesives;[65] petroleum markers;[66] cleansing products[67,68]

Safety/Toxicity Acute toxicity;[1,69] chronic toxicity;[1,69] cytotoxicity;[1,70–74] genotoxicity;[75] insect toxicity;[1,76] microbial toxicity;[77] mutagenicity;[1,78–80] oral toxicity;[1,81] photodynamic toxicity;[82] phototoxicity;[1,83,84] teratogenicity;[1,85] yeast toxicity[86]

Certification/Approval Certified by Biological Stain Commission (BSC)

REFERENCES

1. Sabnis, R. W. *Handbook of Acid–Base Indicators*; CRC Press: Boca Raton, **2008**; pp 318–320.
2. Welcher, F. J. *Organic Analytical Reagents*; D. Van Nostrand Company: New York, **1948**; Vol. 4, pp 385–386.
3. Rogers, R.; Horwitz, E. P.; Bond, A. H. Process for separating and recovering an anionic dye from an aqueous solution. U.S. Patent 5707525, 1998; *Chem. Abstr*. **1998**, *128*, 142665.

4. Oinuma, M.; Hashimoto, S.; Shimoyama, S.; Noda, H. Xanthene pigments fixed on supports, their preparation, and makeup cosmetics containing them. Jpn. Kokai Tokkyo Koho JP 10316877, 1998; *Chem. Abstr.* **1998**, *130*, 17113.
5. Rogers, R.; Horwitz, E. P.; Bond, A. H. Separating anionic dyes from aqueous solution. PCT Int. Appl. WO 9640397, 1996; *Chem. Abstr.* **1997**, *126*, 132452.
6. Weisz, A.; Andrzejewski, D.; Ito, Y. Preparative separation of components of the color additive D and C Red No. 28 (phloxine B) by pH-zone-refining counter-current chromatography. *J. Chromatogr. A* **1994**, *678*, 77–84.
7. Weisz, A.; Langowski, A. J.; Meyers, M. B.; Thieken, M. A.; Ito, Y. Preparative purification of tetrabromotetrachlorofluorescein and phloxine B by centrifugal counter-current chromatography. *J. Chromatogr.* **1991**, *538*, 157–164.
8. Kosaka, H.; Sakamoto, H.; Sotozono, E.; Sasano, Z.; Osawa, S. Liquid reagent for measuring albumin in liquid sample. Jpn. Kokai Tokkyo Koho JP 2008046052, 2008; *Chem. Abstr.* **2008**, *148*, 302905.
9. Kosaka, H.; Sakamoto, H. Liquid reagent for measuring albumin in liquid sample. Jpn. Kokai Tokkyo Koho JP 2008046018, 2008; *Chem. Abstr.* **2008**, *148*, 302904.
10. Walton, J. R. A bright field/fluorescent stain for aluminum: its specificity, validation, and staining characteristics. *Biotech. Histochem.* **2004**, *79*, 169–176.
11. Rasooly, R. Phloxine B, a versatile bacterial stain. *FEMS Immunol. Med. Microbiol.* **2007**, *49*, 261–265.
12. Yoshioka, N.; Ichihashi, K. Determination of 40 synthetic food colors in drinks and candies by high-performance liquid chromatography using a short column with photodiode array detection. *Talanta* **2008**, *74*, 1408–1413.
13. Walsh, L. J.; Ota, R.; Nagao, K.; Sato, T. Dental plaque-disclosing/dyeing compositions for caries activity evaluation. Jpn. Kokai Tokkyo Koho JP 2008189575, 2008; *Chem. Abstr.* **2008**, *149*, 252011.
14. Kadobayashi, I.; Yamane, H. Natural pigments for staining dental plaque. Jpn. Kokai Tokkyo Koho JP 2006028152, 2006; *Chem. Abstr.* **2006**, *144*, 177589.
15. Morton, D. L. Methods for lymph node identification. U.S. Patent 6815170, 2004; *Chem. Abstr.* **2004**, *141*, 391557.
16. Kosaka, H.; Sakamoto, H.; Sotozono, E.; Sasano, Z.; Osawa, S. Liquid reagent for measuring total protein quantity in liquid sample. Jpn. Kokai Tokkyo Koho JP 2008046053, 2008; *Chem. Abstr.* **2008**, *148*, 302906.
17. Orser, C. S.; Pan, T.; Sethi, J. Detecting and removing pathogenic misfolded proteins, such as prions, using conformational capture peptide probes and double-labeled detection-amplification peptides. PCT Int. Appl. WO 2006088823, 2006; *Chem. Abstr.* **2006**, *145*, 265232.
18. Guggenberger, R.; Luchterhandt, T.; Haeberlein, I.; Kappler, O.; Hansen, M. Dental composition for detection of carious tissue, detection method. PCT Int. Appl. WO 2006125650, 2006; *Chem. Abstr.* **2006**, *146*, 33158.
19. Horiuchi, T.; Takusagawa, H. Dye compositions for tooth stain. Jpn. Kokai Tokkyo Koho JP 08143477, 1996; *Chem. Abstr.* **1996**, *125*, 151144.
20. Lanzendoerfer, G.; Schulz, J.; Knueppel, A.; Kovacevic, M.; Bormann, A.; Grundt, W.; Schleyer, K. Cosmetic and dermatological sunscreening compositions containing UV-absorbers, inorganic pigments and organic dyes. Eur. Pat. Appl. EP 1371359, 2003; *Chem. Abstr.* **2003**, *140*, 31181.
21. Ross, J. S.; Morefield, E. M. Sunscreen with disappearing color indicator. U.S. Patent 6290936, 2001; *Chem. Abstr.* **2001**, *135*, 215789.
22. Bouchard, F. Eye shadow comprising least 50% colorants. Eur. Pat. Appl. EP 1757262, 2007; *Chem. Abstr.* **2007**, *146*, 280374.
23. Sodhi, G. S.; Kaur, J. Organic fingerprint powders based on fluorescent phloxine B dye. *Defence Sci. J.* **2000**, *50*, 213–215.
24. Scalia, S.; Simeoni, S. Assay of xanthene dyes in lipsticks by inverse supercritical fluid extraction and HPLC. *Chromatographia* **2001**, *53*, 490–494.
25. Blin, X. Range of compositions for application to the skin, lips, and/or superficial body growths. Fr. Demande FR 2868697, 2005; *Chem. Abstr.* **2005**, *143*, 392523.
26. Dumousseaux, C. Cosmetic composition for application to the skin, lips, and/or body surface. Fr. Demande FR 2868696, 2005; *Chem. Abstr.* **2005**, *143*, 392522.
27. Dumousseaux, C. Cosmetic composition for application to the skin, the lips, and/or superficial body growths. Fr. Demande FR 2868695, 2005; *Chem. Abstr.* **2005**, *143*, 392521.
28. Morrissey, C. T.; Kalla, K. K. Cosmetic compositions comprising colorants with low free dye. U.S. Pat. Appl. Publ. US 2006171909, 2006; *Chem. Abstr.* **2006**, *145*, 173628.

29. Zheng, T.; Rothouse, J.; Fleissman, L. Two-part cosmetic product. U.S. Pat. Appl. Publ. US 2006140895, 2006; *Chem. Abstr.* **2006**, *145*, 109760.
30. Zheng, T.; Sandstrom, G. A. Long wear topical composition having improved glossy appearance. U.S. Pat. Appl. Publ. US 2006134035, 2006; *Chem. Abstr.* **2006**, *145*, 89391.
31. Arnaud, P.; Feltin, C. Method for coloring of the skin comprising direct dye. Fr. Demande FR 2909873, 2008; *Chem. Abstr.* **2008**, *149*, 61572.
32. Bazelaire, E.; Petit, C. B.; Arnaud, P.; Feltin, C. Process and composition for coloring of dark skin. Fr. Demande FR 2901474, 2007; *Chem. Abstr.* **2007**, *148*, 16895.
33. Arnaud, P.; Feltin, C.; Bazelaire, E. Process for coloring of dark skin using a direct dye. Fr. Demande FR 2901475, 2007; *Chem. Abstr.* **2007**, *148*, 39160.
34. Sakazaki, Y.; Suzuki, Y. Makeup compositions containing fibers with modified cross-sections for correction of skin color. Jpn. Kokai Tokkyo Koho JP 2007077098, 2007; *Chem. Abstr.* **2007**, *146*, 343745.
35. Trigg, D. L.; Jones, J. L. Multistep hair color revitalizing kit comprising hair dye and glossing agent. U.S. Pat. Appl. Publ. US 2008189876, 2008; *Chem. Abstr.* **2008**, *149*, 251967.
36. Shami, F.; Ghannad, A. D. Composition for simultaneously lightening and coloring hair. Can. Pat. Appl. CA 2573991, 2007; *Chem. Abstr.* **2007**, *147*, 172969.
37. Hercouet, L. Hair dye composition comprising a derivative of diamino-*N*,*N*-dihydropyrazolone, a coupler, and a heterocyclic direct dye. Fr. Demande FR 2886132, 2006; *Chem. Abstr.* **2006**, *146*, 49738.
38. Hoeffkes, H.; Gross, W. Hair dyes containing silylcarbonates and/or sylilcarbamates for simultaneous dyeing and bleaching of hair. Ger. Offen. DE 102004052060, 2006; *Chem. Abstr.* **2006**, *144*, 419031.
39. Shinkai, M. Oxidative hair dye first compositions. Jpn. Kokai Tokkyo Koho JP 2004059468, 2004; *Chem. Abstr.* **2004**, *140*, 186967.
40. Javet, M.; Mueller, C.; Weber, I. Bleaching keratin fibers with compositions containing an agent having a basic pH and sulfamate bleach booster. Eur. Pat. Appl. EP 1759684, 2007; *Chem. Abstr.* **2007**, *146*, 301740.
41. Jenkins, D. B.; Tucker, E. B.; Kozikoski, T. E. Protein detection system in mammalian urine using phloxine B and application to animal litter. U.S. Pat. Appl. Publ. US 2006270051, 2006; *Chem. Abstr.* **2006**, *146*, 23033.
42. Dees, H. C.; Scott, T. C.; Smolik, J.; Wachter, E. A.; Fisher, W. G. Topical halogenated xanthenes for photodynamic treatment. U.S. Pat. Appl. Publ. US 2008118578, 2008; *Chem. Abstr.* **2008**, *148*, 568986.
43. Dees, H. C.; Scott, T. C.; Wachter, E. A.; Fisher, W. G.; Smolik, J. Intracorporeal medicaments for high energy phototherapeutic treatment of disease. U.S. Pat. Appl. Publ. US 2007208076, 2007; *Chem. Abstr.* **2007**, *147*, 317211.
44. Scott, T. C.; Dees, H. C.; Wachter, E. A. Medicaments comprising halogenated xanthenes, such as Rose Bengal, for chemotherapeutic treatment of disease. U.S. Pat. Appl. Publ. US 2008118567, 2008; *Chem. Abstr.* **2008**, *148*, 568984.
45. Sheppard, D. N.; Gai, Z. Activation of the cystic fibrosis transmembrane conductance regulator chloride channel. PCT Int. Appl. WO 2002005793, 2002; *Chem. Abstr.* **2002**, *136*, 112661.
46. Blair, B. G. Method of treatment of protozoan infections in fish. U.S. Pat. Appl. Publ. US 2002037921, 2002; *Chem. Abstr.* **2002**, *136*, 260249.
47. Severin, F.; Hyman, A. Yeast pheromones for the treatment of infectious diseases, and method for identification of modulators of programmed cell death. PCT Int. Appl. WO 2003080657, 2003; *Chem. Abstr.* **2003**, *139*, 271021.
48. Rasooly, R. Phloxine B as an antimicrobial agent. U. S. Pat. Appl. Publ. US 2002165267, 2002; *Chem. Abstr.* **2002**, *137*, 307298.
49. Simchoni-Barak, M.; Schlein, Y.; Muller, G. C. Insecticidal compositions for control of sugar-feeding insects. U.S. Pat. Appl. Publ. US 2007269404, 2007; *Chem. Abstr.* **2007**, *147*, 516458.
50. Putsche, F. W., Jr. Fluorescent dyes as herbicides. U. S. Patent 5310725, 1994; *Chem. Abstr.* **1994**, *121*, 52396.
51. Nakazawa, K.; Shirota, K.; Miyazaki, T.; Kashiwazaki, A.; Hirose, M. Color filter for liquid-crystal display panel. Eur. Pat. Appl. EP 965874, 1999; *Chem. Abstr.* **1999**, *132*, 71457.
52. Okamoto, T.; Ito, S. Dye-sensitized solar cell and its manufacture by electrodepositing zinc oxide with template compound. Jpn. Kokai Tokkyo Koho JP 2008177099, 2008; *Chem. Abstr.* **2008**, *149*, 180291.
53. Wolfe, D. B.; Love, J. C.; Gates, B. D.; Whitesides, G. M.; Conroy, R. S.; Prentiss, M. Fabrication of planar optical waveguides by electrical microcontact printing. *Appl. Phys. Lett.* **2004**, *84*, 1623–1625.

54. Vincent, J. B.; Flick, D. W. Electrochromic display device. U.S. Pat. Appl. Publ. US 2003179432, 2003; *Chem. Abstr.* **2003**, *139*, 283537.

55. Rode, K.; Gerstorf, J.; Mohr, D.; Frass, W. Photopolymerizable compositions for photoresists and printing plate preparation. Eur. Pat. Appl. EP 321826, 1989; *Chem. Abstr.* **1989**, *111*, 222178.

56. Lawrence, J. R.; O'Neill, F. T.; Sheridan, J. T. Photopolymer holographic recording material. *Optik* **2001**, *112*, 449–463.

57. Tsuji, N.; Sakuma, H. Silver halide photographic material containing acidic dye. Jpn. Kokai Tokkyo Koho JP 2001075212, 2001; *Chem. Abstr.* **2001**, *134*, 245185.

58. Watanabe, K. Aqueous ink compositions with leakage prevention for ball-point pens. Jpn. Kokai Tokkyo Koho JP 2008274146, 2008; *Chem. Abstr.* **2008**, *149*, 536066.

59. Onuki, Y.; Takagishi, I. Saturated long fatty acid-containing aqueous inks for writing pens. Jpn. Kokai Tokkyo Koho JP 2008214405, 2008; *Chem. Abstr.* **2008**, *149*, 357653.

60. Takagi, M. Water-thinned ink compositions for ball-point pens, and ball-point pen refills and ball-point pens containing them. Jpn. Kokai Tokkyo Koho JP 2008111020, 2008; *Chem. Abstr.* **2008**, *148*, 519176.

61. Takasu, Y. Aqueous ink compositions with bubble prevention and writing instruments using them. Jpn. Kokai Tokkyo Koho JP 2006306907, 2006; *Chem. Abstr.* **2006**, *145*, 473212.

62. Katagiri, Y.; Ko, K.; Tateiwa, Y.; Date, H.; Kashiwagawa, T.; Saruwatari, N. Impact-resistant electrophotographic toner and its manufacture. Jpn. Kokai Tokkyo Koho JP 01169463, 1989; *Chem. Abstr.* **1990**, *112*, 88231.

63. Kwan, W. S. V.; Duong, C. Highlighting marking compositions, highlighting kits, highlighted complexes, and application of eradicator. U.S. Pat. Appl. Publ. US 2007017413, 2007; *Chem. Abstr.* **2007**, *146*, 164696.

64. Miller, R. E.; Dereamus, R. C. Color changing systems using pan paint compositions and markers. U.S. Patent 5498282, 1996; *Chem. Abstr.* **1996**, *124*, 346248.

65. Nikutowski, E. A.; James, D. S.; Oxman, J. D. Adhesive for use in the oral environment having color-changing capabilities. PCT Int. Appl. WO 2002030363, 2002; *Chem. Abstr.* **2002**, *136*, 330628.

66. Smith, M. J. Fluorescent petroleum markers. U.S. Patent 5498808, 1996; *Chem. Abstr.* **1996**, *124*, 294047.

67. Lant, N. J. Enzyme and photobleach containing cleaning compositions. PCT Int. Appl. WO 2007087259, 2007; *Chem. Abstr.* **2007**, *147*, 214039.

68. Krzysik, D. G.; Utschig, J. M.; Cole, D. B. Color changing liquid cleansing products containing surfactants, electrolytes and coloring agents. U.S. Pat. Appl. Publ. US 2005148490, 2005; *Chem. Abstr.* **2005**, *143*, 120038.

69. Walthall, W. K.; Stark, J. D. The acute and chronic toxicity of two xanthene dyes, fluorescein sodium salt and phloxine B, to *Daphnia pulex*. *Environ. Pollut.* **1999**, *104*, 207–215.

70. Takahashi, Y.; Koike, M.; Honda, H.; Ito, Y.; Sakaguchi, H.; Suzuki, H.; Nishiyama, N. Development of the short time exposure (STE) test: an *in vitro* eye irritation test using SIRC cells. *Toxicol. In Vitro* **2008**, *22*, 760–770.

71. Okumura, H.; Arashima, M.; Ohuchi, J.; Kasai, Y.; Tsukumo, K.; Kakishima, H.; Kotani, M.; Kojima, H.; Kurishita, A.; Hayashi, M.; Miyajima, A.; Sunouchi, M.; Ohno, Y. Interlaboratory validation of the *in vitro* eye irritation tests for cosmetic ingredients. 10. Evaluation of cytotoxicity test on CHL cells. *Toxicol. In Vitro* **1999**, *13*, 199–208.

72. Chiba, K.; Makino, I.; Ohuchi, J.; Kasai, Y.; Kakishima, H.; Tsukumo, K.; Uchiyama, T.; Miyai, E.; Akiyama, J.; Okamoto, Y.; Kojima, H.; Okumura, H.; Tsurumi, Y.; Usami, M.; Katoh, K.; Sugiura, S.; Kurishita, A.; Sunouchi, M.; Miyajima, A.; Hayashi, M.; Ohno, Y. Interlaboratory validation of the *in vitro* eye irritation tests for cosmetic ingredients. 9. Evaluation of cytotoxicity test on HeLa cells. *Toxicol. In Vitro* **1999**, *13*, 189–198.

73. Tani, N.; Kinoshita, S.; Okamoto, Y.; Kotani, M.; Itagaki, H.; Murakami, N.; Sugiura, S.; Usami, M.; Kato, K.; Kojima, H.; Ohno, T.; Saijo, K.; Kato, M.; Hayashi, M.; Ohno, Y. Interlaboratory validation of the *in vitro* eye irritation tests for cosmetic ingredients. 8. Evaluation of cytotoxicity tests on SIRC cells. *Toxicol. In Vitro* **1999**, *13*, 175–187.

74. Sako, F.; Kobayashi, N.; Watabe, H.; Taniguchi, N. Cytotoxicity of food dyes on cultured fetal rat hepatocytes. *Toxicol. Appl. Pharmacol.* **1980**, *54*, 285–292.

75. Sasaki, Y. F.; Kawaguchi, S.; Kamaya, A.; Ohshita, M.; Kabasawa, K.; Iwama, K.; Taniguchi, K.; Tsuda, S. The comet assay with 8 mouse organs: results with 39 currently used food additives. *Mutat. Res.* **2002**, *519*, 103–119.

76. Yoho, T. P.; Butler, L.; Weaver, J. E. Photodynamic killing of house flies fed food, drug, and cosmetic dye additives. *Environ. Entomol.* **1976**, *5*, 203–204.

77. Oros, G.; Cserhati, T.; Forgacs, E. Separation of the strength and selectivity of the microbiological effect of synthetic dyes by spectral mapping technique. *Chemosphere* **2003**, *52*, 185–193.

78. Aoki, H.; Ogawa, Y.; Yukawa, C.; Nakamura, M.; Nakazawa, H. Lack of mutagenicity of a red food colour impurity (1-carboxyl-5,7-dibromo-6-hydroxy-2,3,4-trichloroxanthone) in phloxine B. *Food Addit. Contam.* **2002**, *19*, 350–356.

79. Klopman, G.; Frierson, M. R.; Rosenkranz, H. S. The structural basis of the mutagenicity of chemicals in *Salmonella typhimurium*: the Gene-Tox data base. *Mutat. Res.* **1990**, *228*, 1–50.

80. Ishidate, M., Jr.; Sofuni, T.; Yoshikawa, K.; Hayashi, M.; Nohmi, T.; Sawada, M.; Matsuoka, A. Primary mutagenicity screening of food additives currently used in Japan. *Food Chem. Toxicol.* **1984**, *22*, 623–636.

81. Hagiwara, A.; Sano, M.; Ichihara, T.; Yoshino, H.; Miyata, E.; Tamano, S.; Aoki, H.; Yukawa, C.; Koda, T.; Nakamura, M.; Shirai, T. Four-week oral toxicity study of 1-carboxy-5,7-dibromo-6-hydroxy-2,3,4-trichloroxanthone (HXCA), an impurity of phloxine B, in F344 rats. *J. Toxicol. Sci.* **2003**, *28*, 445–453.

82. Herkovits, J.; Perez-Coll, C. S.; Stockert, J. C.; Blazquez, A. The screening of photodynamic toxicity of dyes by means of a bioassay using amphibian embryos. *Res. J. Chem. Environ.* **2007**, *11*, 86–91.

83. Wang, H.; Lu, L.; Zhu, S.; Li, Y.; Cai, W. The phototoxicity of xanthene derivatives against *Escherichia coli*, *Staphylococcus aureus*, and *Saccharomyces cerevisiae*. *Curr. Microbiol.* **2006**, *52*, 1–5.

84. Inbaraj, J. J.; Kukielczak, B. M.; Chignell, C. F. Phloxine B phototoxicity: a mechanistic study using HaCaT keratinocytes. *Photochem. Photobiol.* **2005**, *81*, 81–88.

85. Seno, M.; Fukuda, S.; Umisa, H. A teratogenicity study of phloxine B in ICR mice. *Food Chem. Toxicol.* **1984**, *22*, 55–60.

86. Mutoh, N.; Kawabata, M.; Nakagawa, C. W.; Kitajima, S. Pro-oxidant action of phloxine B on fission yeast *Schizosaccharomyces pombe*. *Yeast* **2005**, *22*, 91–97.

PONCEAU SX

CAS Registry Number 4548-53-2

Chemical Structure

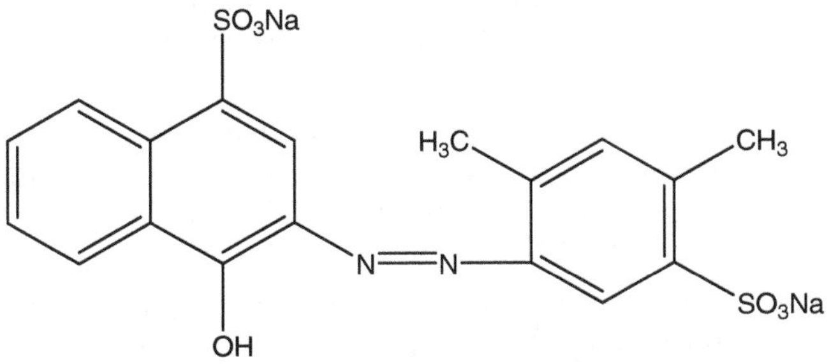

CA Index Name 1-Naphthalenesulfonic acid, 3-[2-(2,4-dimethyl-5-sulfophenyl)diazenyl]-4-hydroxy-, sodium salt (1:2)

Other Names 1-Naphthalenesulfonic acid, 3-[(2,4-dimethyl-5-sulfophenyl)azo]-4-hydroxy-, disodium salt; C. I. Food Red 1; C.I. Food Red 1, disodium salt; Ponceau SX; 12101 Red; 1306 Red; C.I. 14700; Certicol Ponceau SXS; D and C Red No. 4; Edicol Supra Ponceau SX; FD and C Red No. 4; FD&C Red No. 4; FDC Red 4; Food Red 1; Food Red 4; Hexacol Ponceau SX; Japan Red 4; Japan Red 504; Japan Red No. 4; Japan Red No. 504; Red 4; Red No. 1; Red No. 504; Usacert Red No. 4

Merck Index Number 7591

Chemical/Dye Class Azo

Molecular Formula $C_{18}H_{14}N_2Na_2O_7S_2$

Molecular Weight 480.42

Physical Form Red crystals

Solubility Soluble in water, ethanol

Melting Point >300 °C

Absorption (λ_{max}) 500 nm

Synthesis Synthetic methods[1–4]

Staining Applications Animal feed;[5] drinks;[6] candies;[6] frozen food;[7] meat;[8] microorganisms;[9] sausage casings;[10] lotion;[11] eye shadows;[12] lips;[13,14] skin;[14–17] hairs;[18–21] keratin fibers[22,23]

Biological Applications Detecting proteins;[24] treating acquired resistance to GABAergic (ARG) agents;[25] Shampoos;[26] soaps[27]

Industrial Applications Inks;[28,29] cleansing products;[30] fabric softener;[31,32] herbicides;[33] perfumes[34]

Safety/Toxicity Acute toxicity;[35] carcinogenicity;[36,37] chronic toxicity;[38] Genotoxicity;[39,40] mutagenicity[41–44]

REFERENCES

1. Guillon, M.; Mondet, J.; Papantoniou, C.; Vandenbossche, C. Polymer salts with dyes for cosmetic preparations. Ger. Offen. DE 3238166, 1983; *Chem. Abstr.* **1983**, *99*, 43322.
2. Guillon, M.; Mondet, J.; Papantoniou, C.; Vandenbossche, C. Make-up compositions containing acid dye salts and copolymers containing tertiary amine functions. Fr. Demande FR 2480770, 1981; *Chem. Abstr.* **1982**, *96*, 91493.
3. Wenninger, J. A.; Jones, J. H.; Dolinsky, M. Coal-tar colors. XXIV. FD & C Red No. 4. *J. Assoc. Off. Agric. Chem.* **1960**, *43*, 805–809.
4. Willheim, R.; Ivy, A. C.; Janecek, H. M. Decrease in resistance to benzopyrene skin tumors in mice due to the feeding of certified food dyes. *Exp. Med. Surg.* **1956**, *14*, 354–362.
5. MacMillan, M. J.; Ely, C. M. Visual method for detecting fat-soluble vitamins in animal feed. U.S. Patent 3438781, 1969; *Chem. Abstr.* **1969**, *71*, 69464.
6. Yoshioka, N.; Ichihashi, K. Determination of 40 synthetic food colors in drinks and candies by high-performance liquid chromatography using a short column with photodiode array detection. *Talanta* **2008**, *74*, 1408–1413.
7. Del Valle, F. High protein frozen food product. U.S. Patent 6558731, 2003; *Chem. Abstr.* **2003**, *138*, 336930.
8. Orcutt, M. W.; Mueller, I.; Mertle, T. J.; Sandoval, A.; Altemueller, P. A.; Downey, J. Meat compositions comprising colored structured protein products. PCT Int. Appl. WO 2008124629, 2008; *Chem. Abstr.* **2008**, *149*, 447004.
9. Noda, N.; Mizutani, T. Microorganism-measuring method using multiple staining. Jpn. Kokai Tokkyo Koho JP 2006340684, 2006; *Chem. Abstr.* **2006**, *146*, 77563.
10. Winkler, B. Edible dyed sausage casings. Ger. Offen. DE 2212399, 1973; *Chem. Abstr.* **1974**, *80*, 13817.
11. Sato, Y.; Susaki, S.; Morimoto, H.; Miyamoto, K. Light-resistant colored cosmetics. Jpn. Kokai Tokkyo Koho JP 2006169144, 2006; *Chem. Abstr.* **2006**, *145*, 109757.

12. Bouchard, F. Eye shadow comprising least 50% colorants. Eur. Pat. Appl. EP 1757262, 2007; *Chem. Abstr.* **2007**, *146*, 280374.
13. Gross, P. Lip coloring agent. Ger. Offen. DE 19627931, 1998; *Chem. Abstr.* **1998**, *128*, 132267.
14. Dumousseaux, C. Cosmetic composition for application to the skin and lips containing pigments. Fr. Demande FR 2845277, 2004; *Chem. Abstr.* **2004**, *140*, 326628.
15. Teboul, K. Method for coloring dark skins. PCT Int. Appl. WO 2008087308, 2008; *Chem. Abstr.* **2008**, *149*, 160219.
16. Bazelaire, E.; Petit, C. B.; Arnaud, P.; Feltin, C. Process and composition for coloring of dark skin. Fr. Demande FR 2901474, 2007; *Chem. Abstr.* **2007**, *148*, 16895.
17. Lee, H.; Soistier, N.; Simon, J. C. Cosmetic makeup foundations for darkened skins. Fr. Demande FR 2848822, 2004; *Chem. Abstr.* **2004**, *141*, 76373.
18. Kawai, S.; Oikawa, S. Hair dye compositions containing viscosity enhancing agents. Jpn. Kokai Tokkyo Koho JP 2008063312, 2008; *Chem. Abstr.* **2008**, *148*, 362851.
19. Hoeffkes, H.; Gross, W. Hair dyes containing silylcarbonates and/or sylilcarbamates for simultaneous dying and bleaching of hair. Ger. Offen. DE 102004052060, 2006; *Chem. Abstr.* **2006**, *144*, 419031.
20. Pollack, G. Hair dye compositions containing polymers. U.S. Pat. Appl. Publ. US 2004025264, 2004; *Chem. Abstr.* **2004**, *140*, 169236.
21. Tsuge, T.; Mori, K.; Mita, S.; Kato, S.; Arai, A. Hair dye compositions containing acidic dyes. Jpn. Kokai Tokkyo Koho JP 2003073238, 2003; *Chem. Abstr.* **2003**, *138*, 209917.
22. Kravtchenko, S. A composition for simultaneously bleaching and coloring of keratin fibers comprising an anionic or nonionic dye and a polymer. Eur. Pat. Appl. EP 1738800, 2007; *Chem. Abstr.* **2007**, *146*, 128003.
23. Plos, G. Hair dye composition containing an oxidase alcohol and a direct dye having an acid function or its salt for dyeing keratin fibers. Fr. Demande FR 2865396, 2005; *Chem. Abstr.* **2005**, *143*, 179094.
24. Toyonaga, M.; Watanabe, M. Composition for the detection of protein. Jpn. Kokai Tokkyo Koho JP 61164158, 1986; *Chem. Abstr.* **1987**, *106*, 192294.
25. Johnson, J. Methods and compositions for treating ARG. PCT Int. Appl. WO 2006124609, 2006; *Chem. Abstr.* **2006**, *145*, 500136.
26. Ogino, S.; Kumagai, S.; Hirota, H. Shampoos containing amphoteric and anionic surfactants, fatty acid esters, and dyes. Jpn. Kokai Tokkyo Koho JP 01149709, 1989; *Chem. Abstr.* **1990**, *112*, 42231.
27. Yamazaki, T. Decoloring-resistant soap compositions with bright color tone for framed soaps. Jpn. Kokai Tokkyo Koho JP 2005002254, 2005; *Chem. Abstr.* **2005**, *142*, 96340.
28. Ortalano, D. M.; Vissing, C. J. Dye based aqueous pigment dispersions and ink composition therewith. PCT Int. Appl. WO 9961534, 1999; *Chem. Abstr.* **1999**, *132*, 23888.
29. Nakayama, T.; Kurokawa, F. Screen-printing ink compositions. Jpn. Kokai Tokkyo Koho JP 63056579, 1988; *Chem. Abstr.* **1988**, *109*, 75452.
30. Krzysik, D. G.; Utschig, J. M.; Cole, D. B. Color changing liquid cleansing products containing surfactants, electrolytes and coloring agents. U.S. Pat. Appl. Publ. US 2005148490, 2005; *Chem. Abstr.* **2005**, *143*, 120038.
31. Nihei, S.; Oota, S. Colored liquid fabric softener compositions with improved storage stability. Jpn. Kokai Tokkyo Koho JP 07018573, 1995; *Chem. Abstr.* **1995**, *123*, 290476.
32. Wahl, E. H.; Vimba, A. I. Dyes for imparting pink color to aqueous fabric softener compositions. Eur. Pat. Appl. EP 304410, 1989; *Chem. Abstr.* **1989**, *111*, 176797.
33. Yamaji, T.; Shinozaki, T.; Ishikawa, M. Warning color stabilization in bipyridinium herbicides. Jpn. Kokai Tokkyo Koho JP 62148405, 1987; *Chem. Abstr.* **1987**, *107*, 193041.
34. Masuda, M. Stable colored perfume compositions. Jpn. Kokai Tokkyo Koho JP 2005146075, 2005; *Chem. Abstr.* **2005**, *143*, 13029.
35. Tonogai, Y.; Iwaida, M.; Tati, M.; Ose, Y.; Sato, T. Biochemical decomposition of coal-tar dyes. II. Acute toxicity of coal-tar dyes and their decomposed products. *J. Toxicol. Sci.* **1978**, *3*, 205–214.
36. Matthews, E. J.; Contrera, J. F. A new highly specific method for predicting the carcinogenic potential of pharmaceuticals in rodents using enhanced MCASE QSAR-ES software. *Regul. Toxicol. Pharmacol.* **1998**, *28*, 242–264.
37. Ashby, J.; Paton, D. The Influence of chemical structure on the extent and sites of carcinogenesis for 522 rodent carcinogens and 55 different human carcinogen exposures. *Mutat. Res.* **1993**, *286*, 3–74.

38. Davis, K. J.; Nelson, A. A.; Zwickey, R. E.; Hansen, W. H.; Fitzhugh, O. G. Chronic toxicity of Ponceau SX to rats, mice, and dogs. *Toxicol. Appl. Pharmacol.* **1966**, *8*, 306–317.

39. Kirkland, D.; Aardema, M.; Henderson, L.; Mueller, L. Evaluation of the ability of a battery of three *in vitro* genotoxicity tests to discriminate rodent carcinogens and non-carcinogens. I. Sensitivity, specificity and relative predictivity. *Mutat. Res.* **2005**, *584*, 1–256.

40. Haveland-Smith, R. B.; Combes, R. D. Screening of food dyes for genotoxic activity. *Food Cosmet. Toxicol.* **1980**, *18*, 215–221.

41. Serafimova, R.; Todorov, M.; Pavlov, T.; Kotov, S.; Jacob, E.; Aptula, A.; Mekenyan, O. Identification of the structural requirements for mutagenicity by incorporating molecular flexibility and metabolic activation of chemicals. II. General Ames mutagenicity model. *Chem. Res. Toxicol.* **2007**, *20*, 662–676.

42. Rafii, F.; Hall, J. D.; Cerniglia, C. E. Mutagenicity of azo dyes used in foods, drugs and cosmetics before and after reduction by *Clostridium* species from the human intestinal tract. *Food Chem. Toxicol.* **1997**, *35*, 897–901.

43. Zeiger, E.; Anderson, B.; Haworth, S.; Lawlor, T.; Mortelmans, K. *Salmonella* mutagenicity tests: V. Results from the testing of 311 chemicals. *Environ. Mol. Mutagen.* **1992**, *19*, 2–141.

44. Chung, K. T.; Fulk, G. E.; Andrews, A. W. Mutagenicity testing of some commonly used dyes. *Appl. Environ. Microbiol.* **1981**, *42*, 641–648.

POPO 1

CAS Registry Number 169454-15-3

Chemical Structure

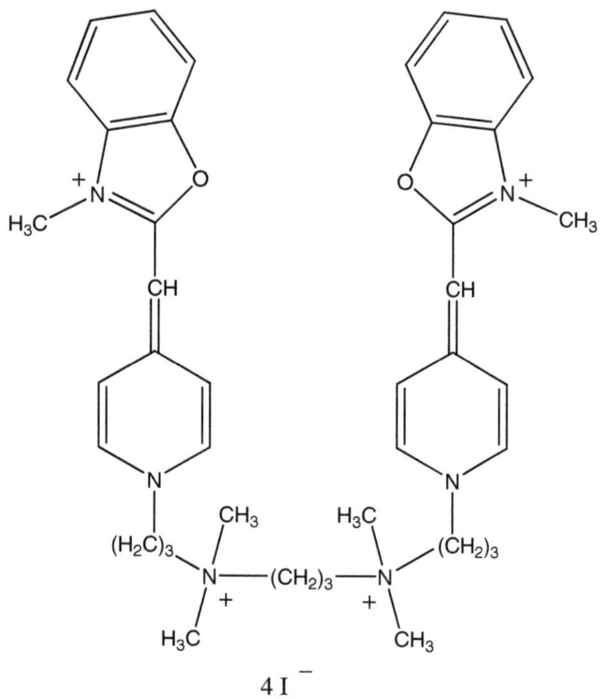

CA Index Name Benzoxazolium, 2,2′-[1,3-propanediylbis[(dimethyliminio)-3,1-propanediyl-1(4*H*)-pyridinyl-4-ylidenemethylidyne]]bis[3-methyl-, iodide (1:4)

Other Names Benzoxazolium, 2,2′-[1,3-propanediylbis[(dimethyliminio)-3,1-propanediyl-1(4*H*)-pyridinyl-4-ylidenemethylidyne]]bis[3-methyl-, tetraiodide; POPO 1; POPO 1 iodide

Merck Index Number Not listed

Chemical/Dye Class Cyanine

Molecular Formula $C_{41}H_{54}I_4N_6O_2$

Molecular Weight 1170.53

Physical Form Yellow-brown powder

Solubility Soluble in dimethyl sulfoxide

Melting Point >250 °C

Absorption (λ_{max}) 434 nm

Emission (λ_{max}) 456 nm

Synthesis Synthetic method[1]

Staining Applications Nucleic acids;[1–5] cells;[6,7] leukocytes;[16] nuclei;[16] megakaryocytes;[8] microorganisms;[9] sperms;[10] hairs[11]

Biological Applications Nucleic acid hybridization;[12,13] detecting nucleic acids,[1–5,14] cells,[6,7] pathogens;[15] counting embryoblasts[16]

Industrial Applications Not reported

Safety/Toxicity No data available

REFERENCES

1. Yue, S. T.; Haugland, R. P. Dimers of unsymmetrical cyanine dyes containing pyridinium moieties. U.S. Patent 5410030, 1995; *Chem. Abstr.* **1995**, *123*, 259753.
2. Exner, M.; Rogers, A. Methods for identifying nucleic acids and determining melting temperature using multiple fluorophores and FRET. U.S. Pat. Appl. Publ. US 2007172836, 2007; *Chem. Abstr.* **2007**, *147*, 182489.
3. Wittwer, C. T.; Dujols, V. E.; Reed, G.; Zhou, L. Genotyping by thermal denaturation analysis using loss of fluorescence of double-stranded DNA-specific dyes as a reporter to monitor denaturation. PCT Int. Appl. WO 2004038038, 2004; *Chem. Abstr.* **2004**, *140*, 386988.
4. Morrison, T. Improved selective ligation and nucleic acid amplification assay. PCT Int. Appl. WO 2005059178, 2005; *Chem. Abstr.* **2005**, *143*, 72747.
5. Winter, S.; Loeber, G. DNA-binding and fluorescence properties of the DNA bis-intercalating purple oxazole dimer POPO-1. *J. Biomed. Opt.* **1997**, *2*, 125–130.
6. Anderson, A. L.; Knutson, C. R.; Mueth, D.; Plewa, J.; Tanner, E. Methods for staining cells for identification and sorting. U.S. Pat. Appl. Publ. US 2006172315, 2006; *Chem. Abstr.* **2006**, *145*, 183714.
7. Millard, P. J.; Roth, B. L.; Yue, S. T.; Haugland, R. P. Fluorescent viability assay using cyclic-substituted unsymmetrical cyanine dyes. U.S. Patent 5534416, 1996; *Chem. Abstr.* **1996**, *125*, 162751.
8. Minakami, T.; Mori, Y.; Tsuji, T.; Ikeuchi, Y. Megakaryocyte classification/counting method by double fluorescent staining and flow cytometry. Jpn. Kokai Tokkyo Koho JP 2006275985, 2006; *Chem. Abstr.* **2006**, *145*, 413679.
9. Noda, N.; Mizutani, T. Microorganism-measuring method using multiple staining. Jpn. Kokai Tokkyo

Koho JP 2006340684, 2006; *Chem. Abstr.* **2006**, *146*, 77563.

10. Matsumoto, T.; Okada, H.; Hamaguchi, Y. Method and reagent for counting sperm by flow cytometry. Jpn. Kokai Tokkyo Koho JP 2001242168, 2001; *Chem. Abstr.* **2001**, *135*, 207869.
11. Lagrange, A. Hair dye compositions containing a polycationic direct dye. Fr. Demande FR 2848840, 2004; *Chem. Abstr.* **2004**, *141*, 76344.
12. Hahn, J. H.; Park, N. Electrochemical device and methods for detection method of nucleic acid hybridization. PCT Int. Appl. WO 2003010338, 2003; *Chem. Abstr.* **2003**, *138*, 149875.
13. Erikson, G. H.; Daksis, J. I. Improving the signal/noise ratio of nucleic acid hybridization assays by preincubation of primer and target with nucleic acid binding agents. U.S. Pat. Appl. Publ. US 2004180345, 2004; *Chem. Abstr.* **2004**, *141*, 255469.
14. Erikson, G. H.; Daksis, J. I.; Kandic, I.; Picard, P. Conditions for formation of three- and four-stranded nucleic acid complexes and their detection and analytical use. PCT Int. Appl. WO 2002103051, 2002; *Chem. Abstr.* **2002**, *138*, 50812.
15. Vannier, E. Methods for detection of pathogens in red blood cells. PCT Int. Appl. WO 2006031544, 2006; *Chem. Abstr.* **2006**, *144*, 307966.
16. Heuven, B.; Wong, F. S.; Tsuji, T.; Sakata, T.; Hamaguchi, I. Method for classifying and counting erythroblasts by flow cytometry. Jpn. Kokai Tokkyo Koho JP 11326323, 1999; *Chem. Abstr.* **1999**, *132*, 1804.

POPO 3

CAS Registry Number 154757-99-0

Chemical Structure

Solubility Soluble in *N,N*-dimethylformamide

Melting Point >250 °C

Absorption (λ_{max}) 534 nm

Emission (λ_{max}) 570 nm

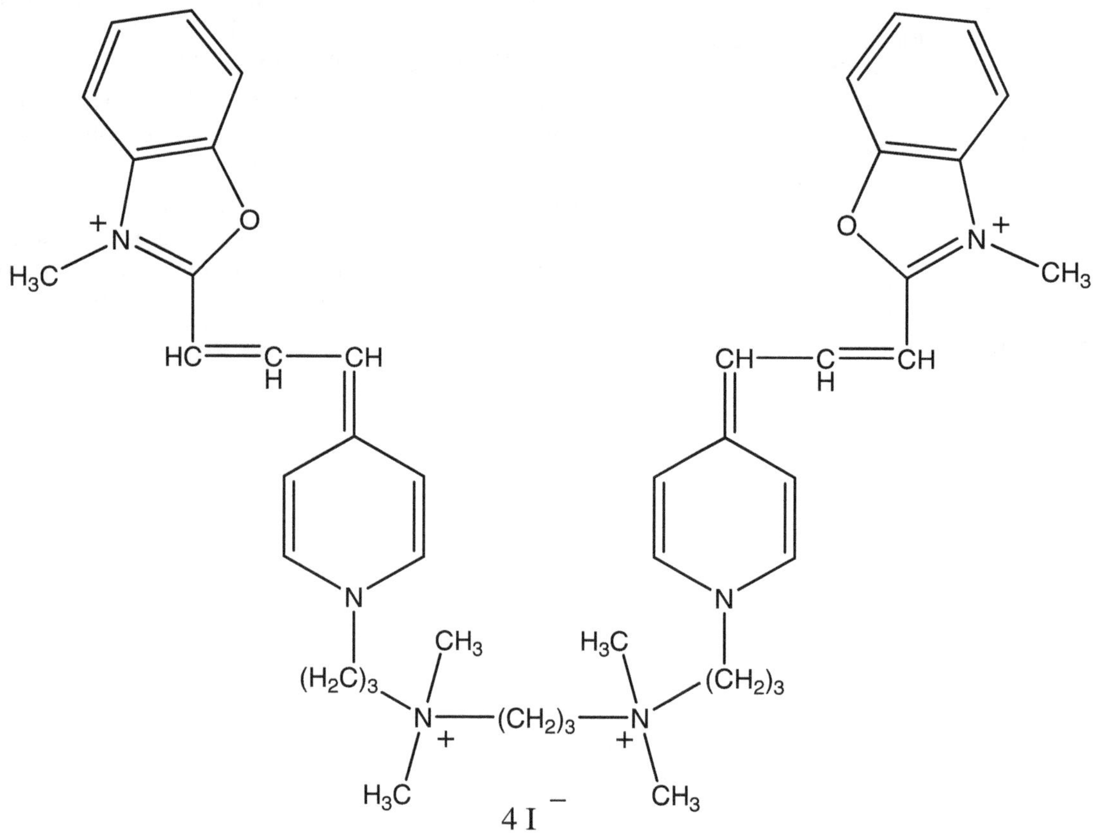

CA Index Name Benzoxazolium, 2,2′-[1,3-propanediylbis[(dimethyliminio)-3,1-propanediyl-1(4*H*)-pyridinyl-4-ylidene-1-propen-1-yl-3-ylidene]]bis[3-methyl, tetraiodide

Other Names POPO 3; POPO 3 iodide

Merck Index Number Not listed

Chemical/Dye Class Cyanine

Molecular Formula $C_{45}H_{58}I_4N_6O_2$

Molecular Weight 1222.61

Physical Form Yellow-brown powder

Synthesis Synthetic method[1]

Staining Applications Nucleic acids;[2–7] cells;[8] leukocytes;[16] nuclei;[16] megakaryocyte;[9] microorganisms;[10] sperms[11]

Biological Applications Nucleic acid hybridization;[12,13] detecting nucleic acids,[2–7] cells,[8] hepatitis B virus,[14] pathogens;[15] counting embryoblasts;[16] biochips;[17] nucleic acid fragment sizing[6]

Industrial Applications Not reported

Safety/Toxicity No data available

REFERENCES

1. Yue, S. T.; Haugland, R. P. Dimers of unsymmetrical cyanine dyes containing pyridinium moieties. U.S. Patent 5410030, 1995; *Chem. Abstr.* **1995**, *123*, 259753.

2. Exner, M.; Rogers, A. Methods for identifying nucleic acids and determining melting temperature using multiple fluorophores and FRET. U.S. Pat. Appl. Publ. US 2007172836, 2007; *Chem. Abstr.* **2007**, *147*, 182489.
3. Wittwer, C. T.; Dujols, V. E.; Reed, G.; Zhou, L. Genotyping by thermal denaturation analysis using loss of fluorescence of double-stranded DNA-specific dyes as a reporter to monitor denaturation. PCT Int. Appl. WO 2004038038, 2004; *Chem. Abstr.* **2004**, *140*, 386988.
4. Erikson, G. H.; Daksis, J. I.; Kandic, I.; Picard, P. Conditions for formation of three- and four-stranded nucleic acid complexes and their detection and analytical use. PCT Int. Appl. WO 2002103051, 2002; *Chem. Abstr.* **2002**, *138*, 50812.
5. Deka, C.; Gordon, K. M.; Gupta, R.; Horton, A. Methods and compositions for rapid staining of nucleic acids in whole cells. U.S. Patent 6271035, 2001; *Chem. Abstr.* **2001**, *135*, 149591.
6. Yan, X.; Grace, W. K.; Yoshida, T. M.; Habbersett, R. C.; Velappan, N.; Jett, J. H.; Keller, R. A.; Marrone, B. L. Characteristics of different nucleic acid staining dyes for DNA fragment sizing by flow cytometry. *Anal. Chem.* **1999**, *71*, 5470–5480.
7. Figeys, D.; Arriaga, E.; Renborg, A.; Dovichi, N. J. Use of the fluorescent intercalating dyes POPO-3, YOYO-3 and YOYO-1 for ultrasensitive detection of double-stranded DNA separated by capillary electrophoresis with hydroxypropylmethyl cellulose and non-crosslinked polyacrylamide. *J. Chromatogr., A* **1994**, *669*, 205–216.
8. Anderson, A. L.; Knutson, C. R.; Mueth, D.; Plewa, J.; Tanner, E. Methods for staining cells for identification and sorting. U.S. Pat. Appl. Publ. US 2006172315, 2006; *Chem. Abstr.* **2006**, *145*, 183714.
9. Minakami, T.; Mori, Y.; Tsuji, T.; Ikeuchi, Y. Megakaryocyte classification/counting method by double fluorescent staining and flow cytometry. Jpn. Kokai Tokkyo Koho JP 2006275985, 2006; *Chem. Abstr.* **2006**, *145*, 413679.
10. Tobin, K. J.; Onstott, T. C.; DeFlaun, M. F.; Colwell, F. S.; Fredrickson, J. *In situ* imaging of microorganisms in geologic material. *J. Microbiol. Methods* **1999**, *37*, 201–213.
11. Matsumoto, T.; Okada, H.; Hamaguchi, Y. Method and reagent for counting sperm by flow cytometry. Jpn. Kokai Tokkyo Koho JP 2001242168, 2001; *Chem. Abstr.* **2001**, *135*, 207869.
12. Hahn, J. H.; Park, N. Electrochemical device and methods for detection method of nucleic acid hybridization. PCT Int. Appl. WO 2003010338, 2003; *Chem. Abstr.* **2003**, *138*, 149875.
13. Erikson, G. H.; Daksis, J. I. Improving the signal/noise ratio of nucleic acid hybridization assays by preincubation of primer and target with nucleic acid binding agents. U.S. Pat. Appl. Publ. US 2004180345, 2004; *Chem. Abstr.* **2004**, *141*, 255469.
14. Tan, W. G.; Tyrrell, D. L. J.; Dovichi, N. J. Detection of duck hepatitis B virus DNA fragments using on-column intercalating dye labeling with capillary electrophoresis-laser-induced fluorescence. *J. Chromatogr., A* **1999**, *853*, 309–319.
15. Vannier, E. Methods for detection of pathogens in red blood cells. PCT Int. Appl. WO 2006031544, 2006; *Chem. Abstr.* **2006**, *144*, 307966.
16. Heuven, B.; Wong, F. S.; Tsuji, T.; Sakata, T.; Hamaguchi, I. Method for classifying and counting erythroblasts by flow cytometry. Jpn. Kokai Tokkyo Koho JP 11326323, 1999; *Chem. Abstr.* **1999**, *132*, 1804.
17. Yeung, E. S.; Shortreed, M.; Ma, Y. Biochip apparatus and methods for high-throughput screening individual molecules in samples comprising multiple molecules. PCT Int. Appl. WO 2001090748, 2001; *Chem. Abstr.* **2001**, *136*, 2456.

PO-PRO 1

CAS Registry Number 157199-56-9

Chemical Structure

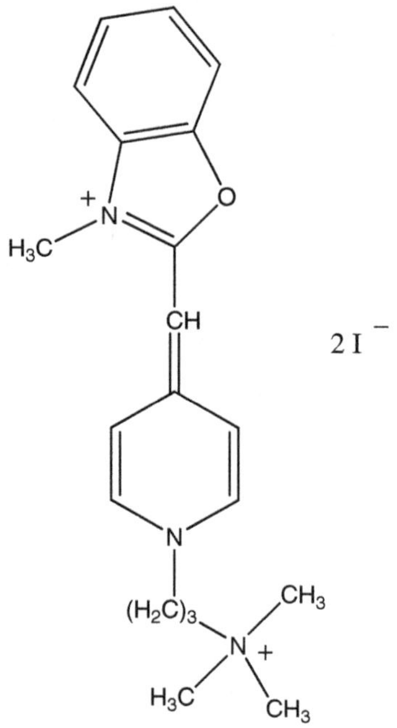

CA Index Name Benzoxazolium, 3-methyl-2-[[1-[3-(trimethylammonio)propyl]-4(1*H*)-pyridinylidene]methyl]-, iodide (1:2)

Other Names Benzoxazolium, 3-methyl-2-[[1-[3-(trimethylammonio)propyl]-4(1*H*)-pyridinylidene]methyl]-, diiodide; PO-PRO 1; PO-PRO 1 iodide

Merck Index Number Not listed

Chemical/Dye Class Cyanine

Molecular Formula $C_{20}H_{27}I_2N_3O$

Molecular Weight 579.26

Physical Form Yellow-brown powder

Solubility Soluble in dimethyl sulfoxide

Melting Point >250 °C

Absorption (λ_{max}) 435 nm

Emission (λ_{max}) 455 nm

Synthesis Synthetic methods[1,2]

Staining Applications Nucleic acids;[3–7] cells;[1,8,9] microorganisms[10]

Biological Applications Nucleic acid hybridization;[11,12] detecting nucleic acids;[3–7] cells;[1,8,9] nucleic acid sequencing[7]

Industrial Applications Not reported

Safety/Toxicity No data available

REFERENCES

1. Millard, P. J.; Roth, B. L.; Yue, S. T.; Haugland, R. P. Fluorescent viability assay using cyclic-substituted unsymmetrical cyanine dyes. U.S. Patent 5534416, 1996; *Chem. Abstr.* **1996**, *125*, 162751.
2. Yue, S. T.; Johnson, I. D.; Huang, Z.; Haugland, R. P. Unsymmetrical cyanine dyes with a cationic side chain for enhanced fluorescence on binding with nucleic acids. U.S. Patent 5321130, 1994; *Chem. Abstr.* **1994**, *121*, 129393.
3. Exner, M.; Rogers, A. Methods for identifying nucleic acids and determining melting temperature using multiple fluorophores and FRET. U.S. Pat. Appl. Publ. US 2007172836, 2007; *Chem. Abstr.* **2007**, *147*, 182489.
4. Wittwer, C. T.; Dujols, V. E.; Reed, G.; Zhou, L. Genotyping by thermal denaturation analysis using loss of fluorescence of double-stranded DNA-specific dyes as a reporter to monitor denaturation. PCT Int. Appl. WO 2004038038, 2004; *Chem. Abstr.* **2004**, *140*, 386988.
5. Erikson, G. H.; Daksis, J. I.; Kandic, I.; Picard, P. Conditions for formation of three- and four-stranded nucleic acid complexes and their detection and analytical use. PCT Int. Appl. WO 2002103051, 2002; *Chem. Abstr.* **2002**, *138*, 50812.
6. Sutherland, J. W.; Patterson, D. R. Homogeneous method for assay of double-stranded nucleic acid products in nucleic acid amplification using fluorescent dyes. Eur. Pat. Appl. EP 684316, 1995; *Chem. Abstr.* **1996**, *124*, 137781.
7. Hoser, M. J. Methods for nucleic acid sequencing. PCT Int. Appl. WO 2004074503, 2004; *Chem. Abstr.* **2004**, *141*, 237698.
8. Anderson, A. L.; Knutson, C. R.; Mueth, D.; Plewa, J.; Tanner, E. Methods for staining cells for identification and sorting. U.S. Pat. Appl. Publ. US 2006172315, 2006; *Chem. Abstr.* **2006**, *145*, 183714.
9. Hoshi, H.; O'Brien, J.; Mills, S. L. A novel fluorescent tracer for visualizing coupled cells in neural circuits

of living tissue. *J. Histochem. Cytochem.* **2006**, *54*, 1169–1176.

10. Sunamura, T.; Maruyama, A.; Kurane, R. Method for detecting and counting microorganism. Jpn. Kokai Tokkyo Koho JP 2002291499, 2002; *Chem. Abstr.* **2002**, *137*, 275332.
11. Hahn, J. H.; Park, N. Electrochemical device and methods for detection method of nucleic acid hybridization. PCT Int. Appl. WO 2003010338, 2003; *Chem. Abstr.* **2003**, *138*, 149875.
12. Erikson, G. H.; Daksis, J. I. Improving the signal/noise ratio of nucleic acid hybridization assays by preincubation of primer and target with nucleic acid binding agents. U.S. Pat. Appl. Publ. US 2004180345, 2004; *Chem. Abstr.* **2004**, *141*, 255469.

PO-PRO 3

CAS Registry Number 161016-55-3

Chemical Structure

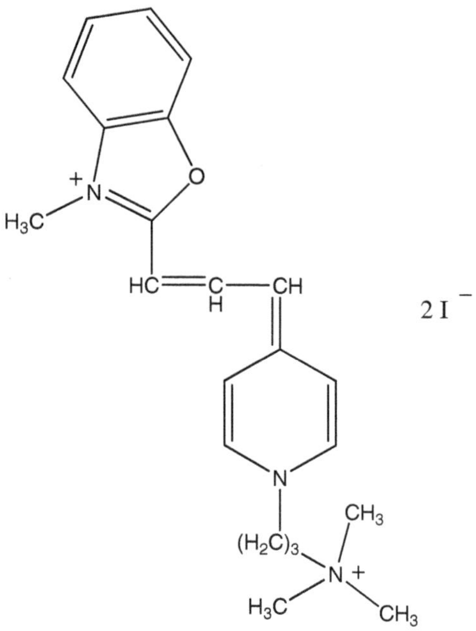

CA Index Name Benzoxazolium, 3-methyl-2-[3-[1-[3-(trimethylammonio)propyl]-4(1*H*)-pyridinylidene]-1-propenyl]-, diiodide

Other Names PO-PRO 3; PO-PRO 3 iodide

Merck Index Number Not listed

Chemical/Dye Class Cyanine

Molecular Formula $C_{22}H_{29}I_2N_3O$

Molecular Weight 605.30

Physical Form Yellow-brown powder

Solubility Soluble in dimethyl sulfoxide

Melting Point >250 °C

Absorption (λ_{max}) 539 nm

Emission (λ_{max}) 567 nm

Synthesis Synthetic method[1]

Staining Applications Nucleic acids;[2–8] cells[1,9]

Biological Applications Nucleic acid hybridization;[10,11] detecting nucleic acids,[2–8] cells;[1,9] nucleic acid fingerprinting;[4] nucleic acid fragment sizing;[6] nucleic acid amplification[7]

Industrial Applications Not reported

Safety/Toxicity No data available

REFERENCES

1. Millard, P. J.; Roth, B. L.; Yue, S. T.; Haugland, R. P. Fluorescent viability assay using cyclic-substituted unsymmetrical cyanine dyes. U.S. Patent 5534416, 1996; *Chem. Abstr.* **1996**, *125*, 162751.
2. Exner, M.; Rogers, A. Methods for identifying nucleic acids and determining melting temperature using multiple fluorophores and FRET. U.S. Pat. Appl. Publ. US 2007172836, 2007; *Chem. Abstr.* **2007**, *147*, 182489.
3. Erikson, G. H.; Daksis, J. I.; Kandic, I.; Picard, P. Conditions for formation of three- and four-stranded nucleic acid complexes and their detection and analytical use. PCT Int. Appl. WO 2002103051, 2002; *Chem. Abstr.* **2002**, *138*, 50812.
4. Pena, S. D. J. Multiplex PCR analysis of human microsatellite DNA and use in DNA fingerprinting. Braz. Pedido PI BR 2099004359, 2001; *Chem. Abstr.* **2002**, *136*, 242894.
5. Deka, C.; Gordon, K. M.; Gupta, R.; Horton, A. Methods and compositions for rapid staining of nucleic acids in whole cells. U.S. Patent 6271035, 2001; *Chem. Abstr.* **2001**, *135*, 149591.
6. Yan, X.; Grace, W. K.; Yoshida, T. M.; Habbersett, R. C.; Velappan, N.; Jett, J. H.; Keller, R. A.; Marrone, B. L. Characteristics of different nucleic acid staining dyes for DNA fragment sizing by flow cytometry. *Anal. Chem.* **1999**, *71*, 5470–5480.
7. Sutherland, J. W.; Patterson, D. R. Homogeneous method for assay of double-stranded nucleic acid products in nucleic acid amplification using fluorescent dyes. Eur. Pat. Appl. EP 684316, 1995; *Chem. Abstr.* **1996**, *124*, 137781.
8. Sherwood, C. S.; Haynes, C. A.; Turner, R. F. B. Nanogram-level micro-volume DNA assay based on the monomeric cyanine dye PO-PRO-3 iodide. *BioTechniques* **1995**, *18*, 136–141.
9. Anderson, A. L.; Knutson, C. R.; Mueth, D.; Plewa, J.; Tanner, E. Methods for staining cells for identification and sorting. U.S. Pat. Appl. Publ.

US 2006172315, 2006; *Chem. Abstr.* **2006**, *145*, 183714.

10. Hahn, J. H.; Park, N. Electrochemical device and methods for detection method of nucleic acid hybridization. PCT Int. Appl. WO 2003010338, 2003; *Chem. Abstr.* **2003**, *138*, 149875.

11. Erikson, G. H.; Daksis, J. I. Improving the signal/noise ratio of nucleic acid hybridization assays by preincubation of primer and target with nucleic acid binding agents. U.S. Pat. Appl. Publ. US 2004180345, 2004; *Chem. Abstr.* **2004**, *141*, 255469.

PROPIDIUM IODIDE

CAS Registry Number 25535-16-4

Chemical Structure

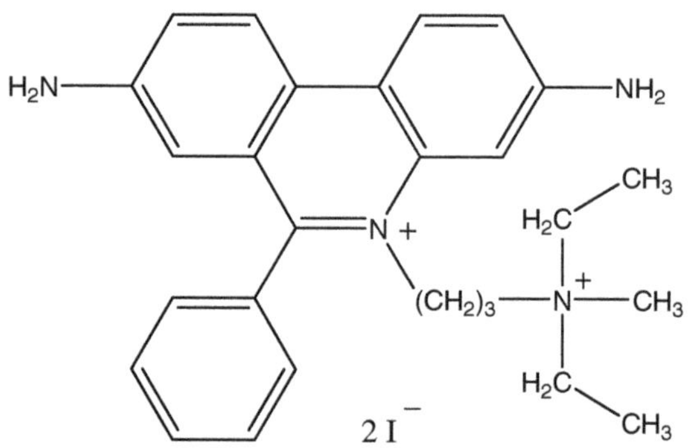

CA Index Name Phenanthridinium, 3,8-diamino-5-[3-(diethylmethylammonio)propyl]-6-phenyl-, iodide (1:2)

Other Names 3,8-Diamino-5-(3-diethylaminopropyl)-6-phenylphenanthridinium iodide methiodide; Phenanthridinium, 3,8-diamino-5-[3-(diethylmethylammonio) propyl]-6-phenyl-, diiodide; Ammonium, [3-(3,8-diamino-6-phenyl-5-phenanthridinio)propyl]diethylmethyl-, diiodide; 3,8-Diamino-5-(diethylmethylaminopropyl)-6-phenylphenanthridinium diiodide; PI; Propidium; Propidium diiodide; Propidium iodide

Merck Index Number Not listed

Chemical/Dye Class Phenanthridine

Molecular Formula $C_{27}H_{34}I_2N_4$

Molecular Weight 668.39

Physical Form Red powder

Solubility Soluble in water, *N,N*-dimethylformamide, dimethyl sulfoxide

Melting Point 220–225 °C (decompose)

Absorption (λ_{max}) 493 nm, 535 nm

Emission (λ_{max}) 636 nm, 617 nm

Synthesis Synthetic methods[1–5]

Staining Applications Nucleic acids;[5–7] cells;[8,9] HeLa cells;[10] metastatic cancer cells;[11] stem cells;[12] yeast cells;[13] leukocytes;[14] lymph nodes;[15] lymphoma cell lines;[26] megakaryocyte;[16] microorganisms;[17–19] parasites;[20] proteins;[5] sperms[21]

Biological Applications Detecting nucleic acids,[5–7,22] cancer cells,[23] spores,[23] activity of hydrolase,[24] microorganisms;[17–19] apoptosis assay;[25,26] measuring beta-amyloid in Alzheimer's disease (AD);[32] treating diseases associated with androgens,[27] cancer[28]

Industrial Applications Not reported

Safety/Toxicity Cytotoxicity;[29] embryotoxicity;[30] mutagenicity;[31] neurotoxicity[32]

REFERENCES

1. Hahn, K. M.; Toutchkine, A.; Muthyala, R.; Kraynov, V.; Bark, S. J.; Burton, D. R.; Chamberlain, C. Labeled peptides, proteins and antibodies and processes and intermediates useful for their preparation. U.S. Pat. Appl. Publ. US 2002055133, 2002; *Chem. Abstr.* **2002**, *136*, 366139.
2. Hahn, K. M.; Toutchkine, A.; Muthyala, R.; Kraynov, V.; Bark, S. J.; Burton, D. R.; Chamberlain, C. Labeled peptides, proteins and antibodies and processes and intermediates useful for their preparation. PCT Int. Appl. WO 2002008245, 2002; *Chem. Abstr.* **2002**, *136*, 163716.
3. Church, J. G.; Stapleton, E. A.; Reilly, B. D. Isolation of high quality mRNA from a discrete cell cycle population identified using a nonvital dye and fluorescence activated sorting. *Cytometry* **1993**, *14*, 271–275.
4. Patel, D. J.; Canuel, L. L. Biphasic helix–coil transition of the ethidium bromide.poly(dA-dT) and the propidium diiodide.poly(dA-dT) complexes: stabilization of base-pair regions centered about the intercalation site. *Biopolymers* **1977**, *16*, 857–873.
5. Crissman, H. A.; Oka, M. S.; Steinkamp, J. A. Rapid staining methods for analysis of deoxyribonucleic acid and protein in mammalian cells. *J. Histochem. Cytochem.* **1976**, *24*, 64–71.
6. Maniotis, A. J.; Folberg, R.; Valyi-Nagy, K.; Valyi-Nagy, T. Methods for determining the pathogenicity of viral strains based on exposure of nuclease sites on host cell's nuclear DNA. PCT Int. Appl. WO 2007030815, 2007; *Chem. Abstr.* **2007**, *146*, 312258.
7. Swerts, K.; Van Roy, N.; Benoit, Y.; Laureys, G.; Philippe, J. DRAQ5: improved flow cytometric DNA content analysis and minimal residual disease detection in childhood malignancies. *Clin. Chim. Acta* **2007**, *379*, 154–157.
8. Lacombe, F.; Belloc, F.; Veriac, S.; Lefevre, D. Method of discrimination of at least two cellular populations by immunofluorescence flow cytometry and application to oncopathology. Fr.

Demande FR 2895087, 2007; *Chem. Abstr.* **2007**, *147*, 93934.

9. Buchegger, F.; Dupertuis, Y. M.; Perillo-Adamer, F. A pitfall of propidium iodide staining in fluorescence-activated cell sorting cell cycle analysis? *Cancer Res.* **2007**, *67*, 5576–5577.
10. He, H.; Chang, D. C.; Lee, Y. K. Using a micro electroporation chip to determine the optimal physical parameters in the uptake of biomolecules in HeLa cells. *Bioelectrochemistry* **2007**, *70*, 363–368.
11. Su, J.; Sun, A.; Tian, J. Immunohistochemical reagent and method for rapidly detecting lymphatic metastasis in breast cancer. Faming Zhuanli Shenqing Gongkai Shuomingshu CN 1945333, 2007; *Chem. Abstr.* **2007**, *146*, 417863.
12. Sidhu, K. S.; Tuch, B. E. Human embryonic stem cell clones. U.S. Pat. Appl. Publ. US 2007160974, 2007; *Chem. Abstr.* **2007**, *147*, 137096.
13. Thomas, R. A. Flow transducers and flow cytometers for making simultaneous electronic and optical measurements on particles for identifying viable cells and budding yeast. PCT Int. Appl. WO 2007044617, 2007; *Chem. Abstr.* **2007**, *146*, 397768.
14. Baskic, D.; Popovic, S.; Ristic, P.; Arsenijevic, N. N. Analysis of cycloheximide-induced apoptosis in human leukocytes: fluorescence microscopy using annexin V/propidium iodide *versus* acridine orange/ethidium bromide. *Cell Biol. Int.* **2006**, *30*, 924–932.
15. Mangat, G.; Brzozowski, L. Intra-operative head and neck nerve mapping. PCT Int. Appl. WO 2007016790, 2007; *Chem. Abstr.* **2007**, *146*, 223533.
16. Minakami, T.; Mori, Y.; Tsuji, T.; Ikeuchi, Y. Megakaryocyte classification/counting method by double fluorescent staining and flow cytometry. Jpn. Kokai Tokkyo Koho JP 2006275985, 2006; *Chem. Abstr.* **2006**, *145*, 413679.
17. Kinbara, K.; Shimomura, Y. Microorganism bioactivity evaluation method, and kit. Jpn. Kokai Tokkyo Koho JP 2007097532, 2007; *Chem. Abstr.* **2007**, *146*, 397795.
18. Novak, J. S.; Yuan, J. T. C. High throughput screening method for antimicrobial formulations. U.S. Pat. Appl. Publ. US 2007042453, 2007; *Chem. Abstr.* **2007**, *146*, 247365.
19. Noda, N.; Mizutani, T. Microorganism-measuring method using multiple staining. Jpn. Kokai Tokkyo Koho JP 2006340684, 2006; *Chem. Abstr.* **2006**, *146*, 77563.
20. Escaron, C. J.; Lees, D. M.; Tewari, R.; Smith, D. F.; Caron, E. A simple, robust and versatile method to characterize intracellular parasitism. *Mol. Biochem. Parasitol.* **2007**, *153*, 72–76.
21. Siemieniuch, M.; Dubiel, A. Preservation of tomcat (*Felis catus*) semen in variable temperatures. *Anim. Reprod. Sci.* **2007**, *99*, 135–144.
22. Lin, Y.; Zheng, F. Flow cytometer-microcarrier gene chip for high-throughput analysis of nucleic acid. Faming Zhuanli Shenqing Gongkai Shuomingshu CN 1970789, 2007; *Chem. Abstr.* **2007**, *147*, 65555.
23. Levy, M.; Ellington, A. D.; Pai, S. Proximity ligation assays with peptide conjugate 'burrs' and aptamers for the sensitive detection of spores and cancer cells. PCT Int. Appl. WO 2007027796, 2007; *Chem. Abstr.* **2007**, *146*, 312239.
24. Karuso, P. H.; Choi, H. Y. Determination of the activity of a hydrolase or other hydrolytic agent and real-time monitoring of biomolecule digestion using a fluorescent reporter dye. PCT Int. Appl. WO 2007051257, 2007; *Chem. Abstr.* **2007**, *146*, 477766.
25. Buenz, E. J.; Limburg, P. J.; Howe, C. L. A high-throughput 3-parameter flow cytometry-based cell death assay. *Cytometry* **2007**, *71A*, 170–173.
26. Mattes, M. J. Apoptosis assays with lymphoma cell lines: problems and pitfalls. *Br. J. Cancer* **2007**, *96*, 928–936.
27. Singh, M.; Gatson, J. W. Membrane androgen receptor as a therapeutic target for the prevention/promotion of cell death. U.S. Pat. Appl. Publ. US 2007141581, 2007; *Chem. Abstr.* **2007**, *147*, 88350.
28. D'Andrea, A. D.; Taniguchi, T. Compositions and methods for the treatment of cancer. U.S. Pat. Appl. Publ. US 2007105130, 2007; *Chem. Abstr.* **2007**, *146*, 475663.
29. Merrick, P.; Nieminen, A. L.; Harper, R. A.; Herman, B.; Lemasters, J. J. Cytotoxicity screening of surfactant-based shampoos using a multiwell fluorescence scanner: correlation with Draize eye scores. *Toxicol. In Vitro* **1996**, *10*, 101.
30. Kohler, M.; Kundig, A.; Reist, H. W.; Michel, C. Modification of *in vitro* mouse embryogenesis by X-rays and fluorochromes. *Radiat. Environ. Biophys.* **1994**, *33*, 341–351.
31. Fukunaga, M.; Yielding, L. W. Structure–function characterization of phenanthridinium compounds as mutagens in *Salmonella*. *Mutat. Res.* **1983**, *121*, 89–94.
32. Patel, D.; Good, T. A rapid method to measure beta-amyloid induced neurotoxicity *in vitro*. *J. Neurosci. Methods* **2007**, *161*, 1–10.

PYRONIN B

CAS Registry Number 2150-48-3

Chemical Structure

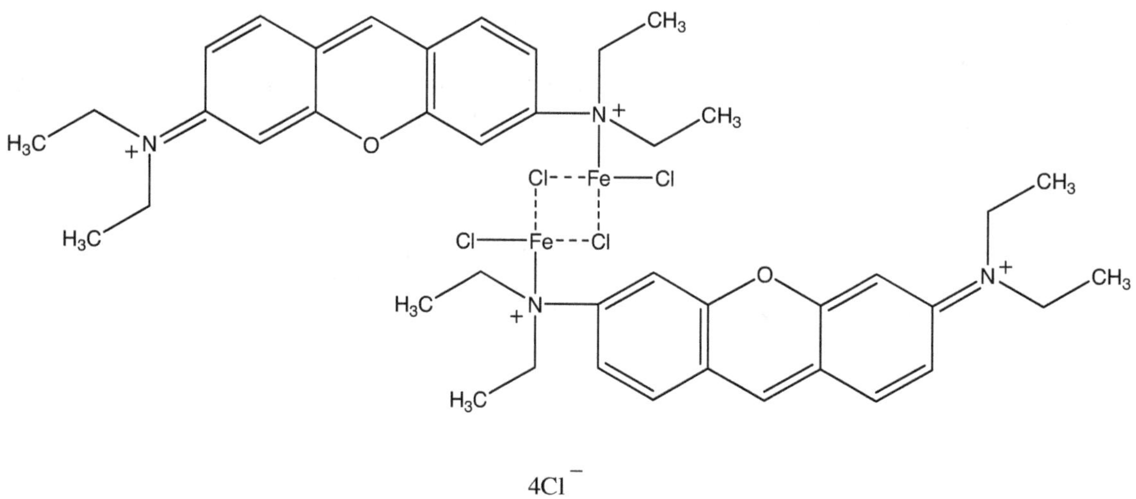

Melting Point 176–178 °C

Absorption (λ_{max}) 553 nm

Synthesis Synthetic methods[1–3]

CA Index Name Xanthylium, 3,6-bis(diethylamino)-, chloride (1:1), ferric chloride complex

Other Names Ammonium, [6-(diethylamino)-3*H*-xanthen-3-ylidene]diethyl-, chloride; Ethanaminium, *N*-[6-(diethylamino)-3*H*-xanthen-3-ylidene]-*N*-ethyl-, chloride; Pyronine B; Xanthylium, 3,6-bis(diethylamino)-, chloride; 6-(Diethylamino)-3*H*-xanthen-3-ylidene]diethylammonium chloride; C.I. 45010; Pyronin B; Pyronine B (By)

Merck Index Number 8006

Chemical/Dye Class Xanthene

Molecular Formula $C_{42}H_{54}Cl_8Fe_2N_4O_2$

Molecular Weight 1042.28

Physical Form Green needles or crystals

Solubility Soluble in water, ethyl acetate, methyl acetate, ethylene glycol; slightly soluble in ethanol, methanol, methyl cellosolve

Staining Applications Cytoplasm;[4] nucleic acids;[5–7]plant cells;[8] urine sediments[9–11]

Biological Applications Detecting apoptosis in live cells;[12] determination of DNA,[13] hydrogen peroxide,[14] glucose;[14] diagnosis of diseases related to amyloid accumulation;[15] urinary tract infection;[16] treating protozoan infections in fish[17]

Industrial Applications Organic light emitting diodes;[18] photovoltaic devices;[19] transistors;[20] lithographic printing plates;[21] organic thin films;[22] semiconductor materials;[23] recording materials;[24] electrophotography;[25] determination of gold,[26] ruthenium,[27] thallium;[28] herbicides[29]

Safety/Toxicity No data available

Certification/Approval Certified by Biological Stain Commission (BSC)

REFERENCES

1. Horobin, R. W.; Murgatroyd, L. B. Identification and purification of pyronin and rhodamine dyes. *Stain Technol.* **1969**, *44*, 297–302.
2. Chamberlin, E. M.; Powell, B. F.; Williams, D. E.; Conn, J. The structure of commercial pyronine B. *J. Org. Chem.* **1962**, *27*, 2263–2264.
3. Albert, A. Acridine syntheses and reactions. III. Synthesis of aminoacridines from formic acid and amines. *J. Chem. Soc.* **1947**, 244–250.
4. Diaz, G.; Diana, A.; Setzu, M. D.; Sogos, V.; Zucca, A. Nuclear dyes and cytoplasmic staining. *Cytometry* **1997**, *29*, 375–376.

5. Dutt, M. K. Schiff-type dyes in the cytochemical staining of DNA. *Indian J. Exp. Biol.* **1975**, *13*, 142–146.
6. Iogannsen, M. G.; Dolgopyatova, M. A. Interaction of nucleic acids isolated from tumor cells with dyes. *Biokhimiya* **1972**, *37*, 1082–1087; *Chem. Abstr.* **1973**, *78*, 54308.
7. Paolillo, D. J., Jr. The selectivity of certain pyronines for ribonucleic acid in shoot tips of Ephedra. *Acta Histochem.* **1964**, *18*, 283–294.
8. Michel, E. An analysis of vital staining in the presence of humic acid. I. Vital fluorescence staining of plant cells. *Protoplasma* **1963**, *56*, 466–490.
9. Matsuoka, M.; Fukada, S.; Shimada, M.; Furuhata, T.; Shimidzu, M.; Itabashi, A.; Okada, S. Investigation of a new urinary staining as an alternative to Sternheimer staining. *Igaku Kensa* **2003**, *52*, 187–193; *Chem. Abstr.* **2003**, *138*, 381508.
10. Sternheimer, R. Urinary sediment stain. U.S. Patent 3961039, 1976; *Chem. Abstr.* **1976**, *85*, 43435.
11. Sternheimer, R. Staining composition for non-fixed wet urine sediments and cell elements of serous discharges. Ger. Offen. DE 2451174, 1975; *Chem. Abstr.* **1975**, *83*, 110866.
12. Frey, T. Nucleic acid dyes for detection of apoptosis in live cells. *Cytometry* **1995**, *21*, 265–274.
13. Sun, W.; You, J.; Hu, X.; Jiao, K. Utilization of pyronine B as voltammetric probe for the determination of DNA. *Anal. Lett.* **2006**, *39*, 33–45.
14. Chen, L. H.; Liu, L. Z.; Shen, H. X. Studies on mimic peroxidase behaviors of molybdenum(VI)–sodium dodecyl sulphate complex for the determination of hydrogen peroxide and glucose. *Anal. Lett.* **2004**, *37*, 561–573.
15. Kudo, K.; Suemoto, T.; Suzuki, M.; Tojo, H.; Shimazu, H. Pyronine B analogs as imaging agents and probes for diagnosis of diseases related to amyloid accumulation. Jpn. Kokai Tokkyo Koho JP 2000344684, 2000; *Chem. Abstr.* **2000**, *134*, 27006.
16. Inoue, J.; Nishizaki, M. Analytical method for formative components in urine. Jpn. Kokai Tokkyo Koho JP 11083849, 1999; *Chem. Abstr.* **1999**, *130*, 293613.
17. Blair, B. G. Method of treatment of protozoan infections in fish. U.S. Pat. Appl. Publ. US 2002037921, 2002; *Chem. Abstr.* **2002**, *136*, 260249.
18. Kim, T. S.; Song, O. K.; Jeong, H. I.; Koo, Y. M. Organic light emitting diode and organic light emitting display having the same. U.S. Pat. Appl. Publ. US 2008118775, 2008; *Chem. Abstr.* **2008**, *148*, 572638.
19. Jung, B. J.; Lee, M. J.; Kim, D. W.; Lee, C. H.; Kim, J. Y. Photovoltaic device. U.S. Pat. Appl. Publ. US 2007144579, 2007; *Chem. Abstr.* **2007**, *147*, 121863.
20. Kawakami, H.; Kato, H.; Maeda, T.; Sekine, N. Transistor. Eur. Pat. Appl. EP 1684365, 2006; *Chem. Abstr.* **2006**, *145*, 178689.
21. Goto, T.; Murota, Y. Light-sensitive lithographic printing plate precursors and method for manufacture thereof. Jpn. Kokai Tokkyo Koho JP 2005165012, 2005; *Chem. Abstr.* **2005**, *143*, 86722.
22. Werner, A.; Li, F.; Harada, K.; Pfeiffer, M.; Fritz, T.; Leo, K.; Machill, S. n-Type doping of organic thin films using cationic dyes. *Adv. Funct. Mater.* **2004**, *14*, 255–260.
23. Werner, A.; Pfeiffer, M.; Fritz, T.; Leo, K. Doped organic semiconductor material and method for production thereof. PCT Int. Appl. WO 2003070822, 2003; *Chem. Abstr.* **2003**, *139*, 189329.
24. Murofushi, K.; Hosoda, K. Photodecolorizing recording material. Jpn. Kokai Tokkyo Koho JP 06236000, 1994; *Chem. Abstr.* **1995**, *123*, 127664.
25. Mori, Y.; Nagata, N.; Morishita, H.; Shito, N. Photoconductive material for color electrophotography. Jpn. Kokai Tokkyo Koho JP 49009242, 1974; *Chem. Abstr.* **1974**, *80*, 151159.
26. Geokchyan, N. O.; Khachatryan, A. G. Extraction–absorptiometric determination of gold by pyronine B. *Khim. Zh. Armenii* **1996**, *49*, 111–112; *Chem. Abstr.* **1998**, *128*, 316620.
27. Razaei, B.; Majidi, N. Simple and selective flow injection catalytic determination of ruthenium with spectrophotometric detection. *J. Korean Chem. Soc.* **2007**, *51*, 147–153.
28. Mikaelyan, D. A.; Artsruni, V. Zh.; Khachatryan, A. G. Extraction-fluorometric determination of thallium traces with pyronin B. *J. Anal. Chem.* **1995**, *50*, 149.
29. Kitaguchi, N.; Shimizu, T. 3,6-Diaminoxanthylium herbicides. Jpn. Kokai Tokkyo Koho JP 61165307, 1986; *Chem. Abstr.* **1987**, *106*, 115253.

PYRONIN Y

CAS Registry Number 92-32-0

Chemical Structure

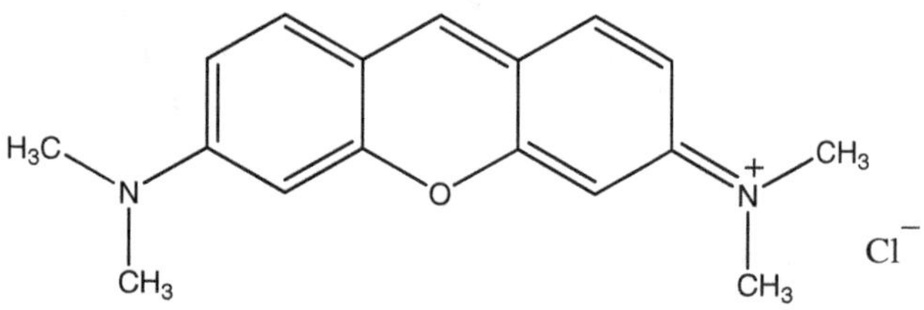

CA Index Name Xanthylium, 3,6-bis(dimethylamino)-, chloride (1:1)

Other Names Ammonium, [6-(dimethylamino)-3*H*-xanthen-3-ylidene]dimethyl-, chloride; Methanaminium, *N*-[6-(dimethylamino)-3*H*-xanthen-3-ylidene]-*N*-methyl-, chloride; Pyronine G; Xanthylium, 3,6-bis(dimethylamino), chloride; 3,6-Bis(dimethylamino)xanthylium chloride; C.I. 45005; Pyronin G; Pyronin GS; Pyronin J; Pyronin Y; Pyronin Yellow; Pyronine; Pyronine GS; Pyronine GY; Pyronine Y; Schultz no. 853

Merck Index Number 8007

Chemical/Dye Class Xanthene

Molecular Formula $C_{17}H_{19}ClN_2O$

Molecular Weight 302.80

Physical Form Green powder or crystals

Solubility Soluble in water; sparingly soluble in ethanol, ethylene glycol, methyl cellosolve

Melting Point 250–260 °C

Absorption (λ_{max}) 548 nm

Emission (λ_{max}) 570 nm

Synthesis Synthetic method[1–3]

Staining Applications Cells;[4,5] erythrocytes;[6] hemopoietic stem cells;[7] nucleic acids;[2,8–12] paraffin tissues;[13] precancerous and cancerous cells;[14] proteins;[12,15] human skin explant cultures;[16] spinal fluid;[17] tattoos;[18] urinary sediments;[19] hairs;[20,21] keratin fibers[22,23]

Biological Applications Antimalarial agent;[24] antiviral agent;[25] nucleic acid sequencing;[26] treating protozoan infections;[27] herbicides[28]

Industrial Applications Photovoltaic cells;[29] solar cells;[29] thin films;[30] lithographic printing plates[31]

Safety/Toxicity Bacterial toxicity;[32] chromosomal aberrations;[33] mutagenicity;[34] phototoxicity;[35,36] skin toxicity[16]

Certification/Approval Certified by Biological Stain Commission (BSC)

REFERENCES

1. Jakobsen, P.; Andersen, A. P.; Lyon, H.; Treppendahl, S. Preparation and characterization of pyronin Y. *Microsc. Acta* **1983**, *87*, 41–47.
2. Dutt, M. K. Purification of methyl green-pyronin and its use in cytochemical detection of DNA and RNA. *Nucleus* **1978**, *21*, 201–205.
3. Biehringer, J. Two new methods of preparation of dyes of the pyronine group. *Justus Liebigs Ann. Chem.* **1912**, *391*, 308–325.
4. Gelvan, D. J.; Goltsman, L.; Chausovsky, A. Staining methods and compositions for identifying a cell phenotype. PCT Int. Appl. WO 2007102146, 2007; *Chem. Abstr.* **2007**, *147*, 339010.
5. Lacombe, F.; Belloc, F.; Veriac, S.; Lefevre, D. Method of discrimination of at least two cellular populations by immunofluorescence flow cytometry and application to oncopathology. Fr. Demande FR 2895087, 2007; *Chem. Abstr.* **2007**, *147*, 93934.
6. Dertinger, S. D.; Torous, D. K.; Tometsko, K. R. Method for the enumeration of micronucleated erythrocyte populations with a single laser flow cytometer. U.S. Patent 6100038, 2000; *Chem. Abstr.* **2000**, *133*, 132120.
7. Huttmann, A.; Liu, S. L.; Boyd, A. W.; Li, C. L. Functional heterogeneity within rhodamine123lo Hoechst 33342lo/sp primitive hemopoietic stem cells revealed by pyronin Y. *Exp. Hematol.* **2001**, *29*, 1109–1116.
8. Wei, Q.; Zhang, H.; Duan, C.; Li, Y.; Du, B. High sensitive fluorophotometric determination of nucleic acids with pyronine G sensitized by *N,N*-dimethylformamide. *Ann. Chim.* **2006**, *96*, 273–284.
9. Prento, P.; Lyon, H. O. Methyl green-pyronin Y staining of nucleic acids: studies on the effects of staining time, dye composition and diffusion rates. *Biotech. Histochem.* **2003**, *78*, 27–33.
10. Zhao, W. P.; Liu, X. G.; Wang, W.; Gao, X. M. Large quantity of ribosomal RNA exists extracellularly in mouse spleen. *Histochem. J.* **2002**, *34*, 391–396.

11. Deka, C.; Gordon, K. M.; Gupta, R.; Horton, A. Methods and compositions for rapid staining of nucleic acids in whole cells. U.S. Patent 6271035, 2001; *Chem. Abstr.* **2001**, *135*, 149591.

12. Pigarevskii, P. V.; Nagornev, V. A.; Zagol'skaya, V. N. Method for simultaneous determination of lysosomal cationic proteins in granulocytes and of DNA and RNA in lymphocytes in a single tissue preparation. Russ. RU 2180104, 2002; *Chem. Abstr.* **2002**, *137*, 348817.

13. Li, B.; Wu, Y.; Gao, X. M. Pyronin Y as a fluorescent stain for paraffin sections. *Histochem. J.* **2002**, *34*, 299–303.

14. Burkett, D. D. Improved diagnostic method for detecting dysplastic epithelial tissue. PCT Int. Appl. WO 2002007693, 2002; *Chem. Abstr.* **2002**, *136*, 131224.

15. Feng, S.; Pan, Z.; Fan, J. Determination of trace proteins with pyronine Y and SDS by resonance light scattering. *Anal. Bioanal. Chem.* **2005**, *383*, 255–260.

16. Jacobs, J. J. L.; Lehe, C.; Cammans, K. D. A.; Das, P. K.; Elliott, G. R. An *in vitro* model for detecting skin irritants: methyl green-pyronine staining of human skin explant cultures. *Toxicol. In Vitro* **2002**, *16*, 581–588.

17. Masuoka, S.; Yoneda, K.; Oka, M. Apparatus and method for analyzing spinal fluid by staining and imaging analysis. Jpn. Kokai Tokkyo Koho JP 2004132787, 2004; *Chem. Abstr.* **2004**, *140*, 371466.

18. Anderson, R. R.; Mlynarczyk-Evans, S. K.; Drill, C. A. Variable appearance tissue markings such as tattoos. PCT Int. Appl. WO 2005046620, 2005; *Chem. Abstr.* **2005**, *142*, 487174.

19. Hashimoto, H.; Kaneda, T.; Imai, T. A new preservative method for urinary sediment by Alcian blue pyronine stain. *Igaku Kensa* **1998**, *47*, 856–859; *Chem. Abstr.* **1998**, *129*, 2279.

20. Lagrange, A. Hair dye composition containing a fluorindine compound and a coloring composition. Fr. Demande FR 2864782, 2005; *Chem. Abstr.* **2005**, *143*, 120052.

21. Kravtchenko, S.; Lagrange, A.; Guerin, F. Hair dye composition containing an azo-pyridinio-pyridone direct dye and a synthetic different dye, and a method for coloring of keratinic fibers. Fr. Demande FR 2864781, 2005; *Chem. Abstr.* **2005**, *143*, 120051.

22. Greaves, A.; David, H.; Samain, H. Composition containing mixed dyes based on azo or tri(hetero) aryl chromophores. Fr. Demande FR 2866806, 2005; *Chem. Abstr.* **2005**, *143*, 271925.

23. Moeckli, P. Cationic dyes for keratin-containing fibers. PCT Int. Appl. WO 9515144, 1995; *Chem. Abstr.* **1995**, *123*, 179094.

24. Vennerstrom, J. L.; Makler, M. T.; Angerhofer, C. K.; Williams, J. A. Antimalarial dyes revisited: xanthenes, azines, oxazines, and thiazines. *Antimicrob. Agents Chemother.* **1995**, *39*, 2671–2677.

25. Jamison, J. M.; Krabill, K.; Hatwalkar, A.; Jamison, E.; Tsai, C. C. Potentiation of the antiviral activity of poly r(A-U) by xanthene dyes. *Cell Biol. Int. Rep.* **1990**, *14*, 1075–1084.

26. Seidel, C.; Fritz, H. J.; Griesinger, C.; Gaiko, N. N.; Berger, S.; Fries, J. Method and apparatus for nucleic acid sequencing by phosphorolytic cleavage of nucleic acids, fluorescently-labeled phosphates, permeability-mediated spatial resolution of degradation products, and detection by fluorometry. PCT Int. Appl. WO 2003080861, 2003; *Chem. Abstr.* **2003**, *139*, 287263.

27. Blair, B. G. Method of treatment of protozoan infections in fish. U.S. Pat. Appl. Publ. US 2002037921, 2002; *Chem. Abstr.* **2002**, *136*, 260249.

28. Kitaguchi, N.; Shimizu, T. 3,6-Diaminoxanthylium herbicides. Jpn. Kokai Tokkyo Koho JP 61165307, 1986; *Chem. Abstr.* **1987**, *106*, 115253.

29. Chane, C. J. Y.; Corma, A.; Garcia, H. Multivalent metal cation-doped cerium oxide materials for photovoltaic cells and solar cells. Fr. Demande FR 2857783, 2005; *Chem. Abstr.* **2005**, *142*, 159487.

30. Sharma, G. D.; Sharma, S. K.; Roy, M. S. Photovoltaic properties of Schottky device based on dye sensitized poly (3-phenylazomethine thiophene) thin film. *Thin Solid Films* **2004**, *468*, 208–215.

31. Endo, A. Lithographic printing master plate containing dye compound and phenolic resin. Jpn. Kokai Tokkyo Koho JP 2004061947, 2004; *Chem. Abstr.* **2004**, *140*, 225820.

32. Kazama, H.; Hamashima, H.; Sasatsu, M.; Arai, T. Distribution of the antiseptic-resistance gene qacEΔ 1 in Gram-positive bacteria. *FEMS Microbiol. Lett.* **1998**, *165*, 295–299.

33. Cortes, F.; Hazen, M. J. Chromosomal aberrations and SCEs in *Allium cepa* root-tip cells treated with caffeine and pyronine Y. *Mutat. Res.* **1984**, *127*, 119–122.

34. Ferguson, L. R.; Baguley, B. C. Verapamil as a co-mutagen in the *Salmonella*/mammalian microsome

mutagenicity test. *Mutat. Res. Lett.* **1988**, *209*, 57–62.

35. Hussain, S.; Harris, F.; Phoenix, D. A. The phototoxicity of phenothiazinium-based photosensitizers to bacterial membranes. *FEMS Immunol. Med. Microbiol.* **2006**, *46*, 124–130.

36. Phoenix, D. A.; Sayed, Z.; Hussain, S.; Harris, F.; Wainwright, M. The phototoxicity of phenothiazinium derivatives against *Escherichia coli* and *Staphylococcus aureus*. *FEMS Immunol. Med. Microbiol.* **2003**, *39*, 17–22.

QSY 7 CARBOXYLIC ACID, SUCCINIMIDYL ESTER

CAS Registry Number 304014-12-8

Chemical Structure

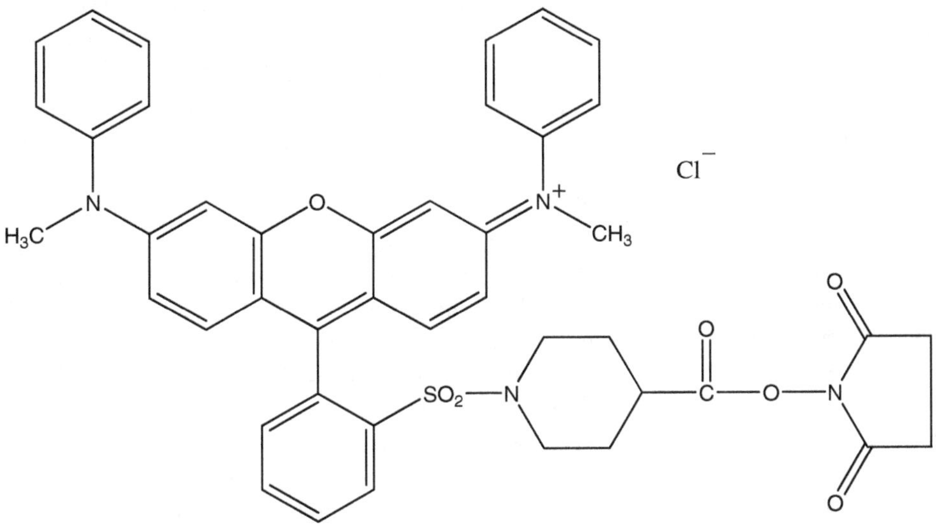

CA Index Name Xanthylium, 9-[2-[[4-[[(2,5-dioxo-1-pyrrolidinyl)oxy]carbonyl]-1-piperidinyl]sulfonyl]phenyl]-3,6-bis(methylphenylamino)-, chloride (1:1)

Other Names Xanthylium, 9-[2-[[4-[[(2,5-dioxo-1-pyrrolidinyl)oxy]carbonyl]-1-piperidinyl]sulfonyl]phenyl]-3,6-bis(methylphenylamino)-, chloride; QSY 7 carboxylic acid succinimidyl ester; QSY 7SE; QSY-7 succinimidyl ester

Merck Index Number Not listed

Chemical/Dye Class Xanthene

Molecular Formula $C_{43}H_{39}ClN_4O_7S$

Molecular Weight 791.32

Physical Form Solid

Solubility Soluble in dimethyl sulfoxide, methanol

Melting Point >200 °C

Absorption (λ_{max}) 560 nm

Synthesis Synthetic methods[1,2]

Staining Applications Nucleic acids;[13–15] oligonucleotides;[1] bacteria;[3] peptides;[4] tumors[5]

Biological Applications Amplification assays;[1] hybridization assays;[1,6,7] litigation assays;[1] FRET assays;[3,4,8] hepatitis C virus (HCV) NS3 protease;[9] FRET probes;[10–12] nucleic acid hybridization;[13,14] detecting nucleic acids,[13–15] human immunodeficiency virus type 1 DNA,[16] polymorphisms,[17,18] RNase enzyme,[19] protease enzyme;[20] nucleic acid quantitation;[21] oligonucleotide probes;[22,23] quenchers[5,24,25]

Industrial Applications Light-emitting device[26]

Safety/Toxicity No data available

REFERENCES

1. Singer, V. L.; Haugland, R. P. Oligonucleotides labeled with energy transfer acceptors for use in amplification, hybridization, and ligation assays employing fluorescent nucleic acid stains. U.S. Patent 6323337, 2001; *Chem. Abstr.* **2001**, *136*, 1576.
2. Haugland, R. P.; Singer, V. L.; Yue, S. T. Xanthene dyes and their application as luminescence quenching compounds. PCT Int. Appl. WO 2000064988, 2000; *Chem. Abstr.* **2000**, *133*, 336549.
3. Bruno, J. G.; Ulvick, S. J.; Uzzell, G. L.; Tabb, J. S.; Valdes, E. R.; Batt, C. A. Novel immuno-FRET assay method for bacillus spores and *Escherichia coli* O157: H7. *Biochem. Biophys. Res. Commun.* **2001**, *287*, 875–880.

4. Steward, L. E.; Fernandez-Salas, E.; Aoki, K. R. Fluorophore-labeled peptides and FRET assays for clostridial toxins. U.S. Pat. Appl. Publ. US 2003143651, 2003; *Chem. Abstr.* **2003**, *139*, 129339.
5. Weissleder, R.; Tung, C. H.; Mahmood, U.; Josephson, L.; Bogdanov, A. Intramolecularly-quenched near infrared fluorescent probes for tumor imaging. U.S. Patent 6592847, 2003; *Chem. Abstr.* **2003**, *139*, 97659.
6. Papkovsky, D. B.; O'Sullivan, P.; Burke, M.; O'Riordan, T. Phosphorescent metalloporphyrin labeled probes for use in separation-free hybridization assays. PCT Int. Appl. WO 2005080596, 2005; *Chem. Abstr.* **2005**, *143*, 263101.
7. Burke, M.; O'Sullivan, P. J.; Soini, A. E.; Berney, H.; Papkovsky, D. B. Evaluation of the phosphorescent palladium(II)-coproporphyrin labels in separation-free hybridization assays. *Anal. Biochem.* **2003**, *320*, 273–280.
8. Kokko, T.; Kokko, L.; Soukka, T.; Loevgren, T. Homogeneous non-competitive bioaffinity assay based on fluorescence resonance energy transfer. *Anal. Chim. Acta* **2007**, *585*, 120–125.
9. Mao, S. S.; Lewis, S. D. Time-resolved fluorescence assay for HCV NS3 protease using europium-labeled substrate containing quenching group, and drug screening use. PCT Int. Appl. WO 2006102087, 2006; *Chem. Abstr.* **2006**, *145*, 371745.
10. Mandecki, W. FRET-based single molecule fluorescence assay to sequence nucleic acids using ribosomal translation component labeled with fluorophore-quencher pairs. U.S. Pat. Appl. Publ. US 2005282173, 2005; *Chem. Abstr.* **2005**, *144*, 66376.
11. Song, N. W.; Kim, H. H. Methods and FRET probes for detection of herbicide-resistant EPSPS gene-containing soybean plants using laser induced fluorescence. PCT Int. Appl. WO 2006121277, 2006; *Chem. Abstr.* **2006**, *145*, 501856.
12. Burroughs-Tencza, S. FRET-based peptide biosensors for detecting anthrax lethal factor protease and *Bacillus anthracis*. PCT Int. Appl. WO 2001059149, 2001; *Chem. Abstr.* **2001**, *135*, 177260.
13. Schmitt-John, T.; Palmisano, R.; Plessow, R.; Brockhinke, A.; Weidner, J. Detection of a nucleic acid by hybridization with pairs of probes labeled with dyes that interact by FRET. Ger. Offen. DE 10133308, 2003; *Chem. Abstr.* **2003**, *138*, 118424.
14. Kushon, S. A.; Ley, K. D.; Bradford, K.; Jones, R. M.; McBranch, D.; Whitten, D. Detection of DNA hybridization via fluorescent polymer superquenching. *Langmuir* **2002**, *18*, 7245–7249.
15. Tao, S.; Cheng, J.; Max, X.; Zhou, Y. Lab-on-a-chip system for analyzing nucleic acid. PCT Int. Appl. WO 2004079002, 2004; *Chem. Abstr.* **2004**, *141*, 256931.
16. Luo, W.; Yang, H.; Rathbun, K.; Pau, C. P.; Ou, C. Y. Detection of human immunodeficiency virus type 1 DNA in dried blood spots by a duplex real-time PCR assay. *J. Clin. Microbiol.* **2005**, *43*, 1851–1857.
17. Lee, M. A. Methods and kits for detecting polymorphisms in nucleic acids. PCT Int. Appl. WO 2002097132, 2002; *Chem. Abstr.* **2002**, *138*, 20433.
18. Knapp, M. R.; Kofp-Sill, A. R.; Simeonov, A.; Peponnet, C. Analytical systems and methods for high throughput screening and detection of genetic polymorphisms by nucleic acid hybridization. PCT Int. Appl. WO 2002083952, 2002; *Chem. Abstr.* **2002**, *137*, 333999.
19. Walder, J. A.; Behlke, M. A.; Devor, E. J.; Huang, L. Visual detection assays for RNase using nucleic acid substrates with RNase-cleavable domain flanked by a fluorescence reporter group and a dark fluorescence quencher. U.S. Pat. Appl. Publ. US 2004137479, 2004; *Chem. Abstr.* **2004**, *141*, 119302.
20. Kumaraswamy, S.; Bergstedt, T.; Shi, X.; Rininsland, F.; Kushon, S.; Xia, W.; Ley, K.; Achyuthan, K.; McBranch, D.; Whitten, D. Fluorescent-conjugated polymer superquenching facilitates highly sensitive detection of proteases. *Proc. Natl. Acad. Sci. U.S.A.* **2004**, *101*, 7511–7515.
21. Nicklas, J. A.; Buel, E. Development of an Alu-based, QSY 7-labeled primer PCR method for quantitation of human DNA in forensic samples. *J. Forensic Sci.* **2003**, *48*, 282–291.
22. Iwaki, Y.; Shinoki, H.; Seshimoto, O.; Nakamura, K. Methods for reducing background fluorescence of oligonucleotide probes by quenching agents for improved sensitivity of microarrays. U.S. Pat. Appl. Publ. US 2003044830, 2003; *Chem. Abstr.* **2003**, *138*, 216455.
23. Lyles, F. E. Oligonucleotide probes containing two targeting sequences and proximity-modulated signal generating system for nucleic acid detection and disease diagnosis. PCT Int. Appl. WO 2003095666, 2003; *Chem. Abstr.* **2003**, *139*, 376191.
24. Moreira, B. G.; You, Y.; Behlke, M. A.; Owczarzy, R. Effects of fluorescent dyes, quenchers, and dangling ends on DNA duplex stability. *Biochem. Biophys. Res. Commun.* **2005**, *327*, 473–484.

25. Adamczyk, M.; Moore, J. A.; Shreder, K. Quenching of biotinylated aequorin bioluminescence by dye-labeled avidin conjugates: application to homogeneous bioluminescence resonance energy transfer assays. *Org. Lett.* **2001**, *3*, 1797–1800.

26. Shinar, R.; Choudhury, B.; Zhou, Z.; Wu, H. S.; Tabatabai, L. B.; Shinar, J. Structurally integrated organic light-emitting device-based sensors for oxygen, glucose, hydrazine, and anthrax. *Proc. SPIE-Int. Soc. Opt. Eng.* **2004**, *5588*, 59–69.

QSY 9 CARBOXYLIC ACID, SUCCINIMIDYL ESTER

CAS Registry Number 700834-40-8

Chemical Structure

Molecular Weight 951.43

Physical Form Solid

Solubility Soluble in water, dimethyl sulfoxide, methanol

Melting Point >200 °C

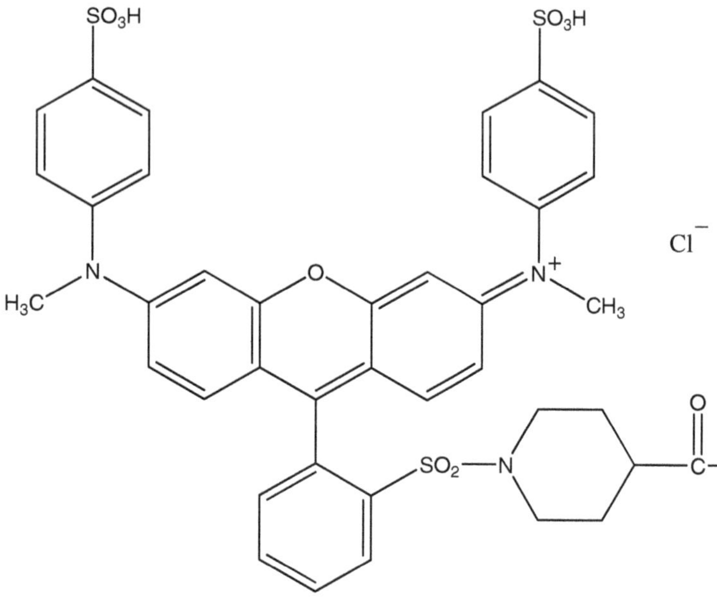

CA Index Name Xanthylium, 9-[2-[[4-[[(2,5-dioxo-1-pyrrolidinyl)oxy]carbonyl]-1-piperidinyl]sulfonyl]phenyl]-3,6-bis[methyl(4-sulfophenyl)amino]-, chloride (1:1)

Other Names Xanthylium, 9-[2-[[4-[[(2,5-dioxo-1-pyrrolidinyl)oxy]carbonyl]-1-piperidinyl]sulfonyl]phenyl]-3,6-bis[methyl(4-sulfophenyl)amino]-, chloride; QSY 9; QSY 9 carboxylic acid succinimidyl ester

Merck Index Number Not listed

Chemical/Dye Class Xanthene

Molecular Formula $C_{43}H_{39}ClN_4O_{13}S_3$

Absorption (λ_{max}) 562 nm

Synthesis Synthetic methods[1,2]

Staining Applications Nucleic acids;[3] oligonucleotides;[1] antibody;[4] G-protein[5]

Biological Applications Amplification assays;[1] hybridization assays;[1] litigation assays;[1] ligand assays;[5,7] FRET assays;[6] detecting analyte,[8] antigen,[4] phosphoinositide kinase and phosphatase activity[9]

Industrial Applications Not reported

Safety/Toxicity No data available

REFERENCES

1. Singer, V. L.; Haugland, R. P. Oligonucleotides labeled with energy transfer acceptors for use in amplification, hybridization, and ligation assays employing fluorescent nucleic acid stains. U.S. Patent 6323337, 2001; *Chem. Abstr.* **2001**, *136*, 1576.
2. Haugland, R. P.; Singer, V. L.; Yue, S. T. Xanthene dyes and their application as luminescence quenching compounds. PCT Int. Appl. WO 2000064988, 2000; *Chem. Abstr.* **2000**, *133*, 336549.
3. Chiuman, W.; Li, Y. Efficient signaling platforms built from a small catalytic DNA and doubly labeled fluorogenic substrates. *Nucleic Acids Res.* **2007**, *35*, 401–405.
4. Beechem, J.; Hagen, D.; Johnson, I. Antibody-label complexes and methods for antigen or ligand immunolabeling or detection, diagnosis and therapy. U.S. Pat. Appl. Publ. US 2007269902, 2007; *Chem. Abstr.* **2007**, *148*, 9415.

5. McMurchie, E. J.; Leifert, W. R. Cell free G-protein coupled receptor and ligand assays involving G-protein subunits labeled with an energy donor or acceptor and optionally a GPCR. PCT Int. Appl. WO 2005121755, 2005; *Chem. Abstr.* **2005**, *144*, 45719.
6. Kokko, T.; Kokko, L.; Soukka, T.; Loevgren, T. Homogeneous non-competitive bioaffinity assay based on fluorescence resonance energy transfer. *Anal. Chim. Acta* **2007**, *585*, 120–125.
7. Beechem, J.; Gee, K.; Hagen, D.; Johnson, I.; Kang, H. C.; Pastula, C. Competitive immunoassay using a ligand analog covalently bonded to a fluorescent reporter molecule. PCT Int. Appl. WO 2005050206, 2005; *Chem. Abstr.* **2005**, *143*, 22621.
8. Volland, H.; Creminon, C.; Neuburger, L. M.; Grassi, J. Apparatus and process for the continuous detection of an analyte using a trifunctional detection reagent. Fr. Demande FR 2847984, 2004; *Chem. Abstr.* **2004**, *141*, 33109.
9. Drees, B. E.; Neilsen, P. O.; Branch, A. M.; Weipert, A.; Hudson, H. A.; Feng, L.; Prestwich, G. Assays for detection of phosphoinositide kinase and phosphatase activity. U.S. Pat. Appl. Publ. US 2005009124, 2005; *Chem. Abstr.* **2005**, *142*, 129786.

QSY 21 CARBOXYLIC ACID, SUCCINIMIDYL ESTER

CAS Registry Number 304014-13-9

Chemical Structure

Molecular Weight 815.34

Physical Form Solid

Solubility Soluble in dimethyl sulfoxide, methanol

Melting Point >200 °C

Absorption (λ_{max}) 661 nm

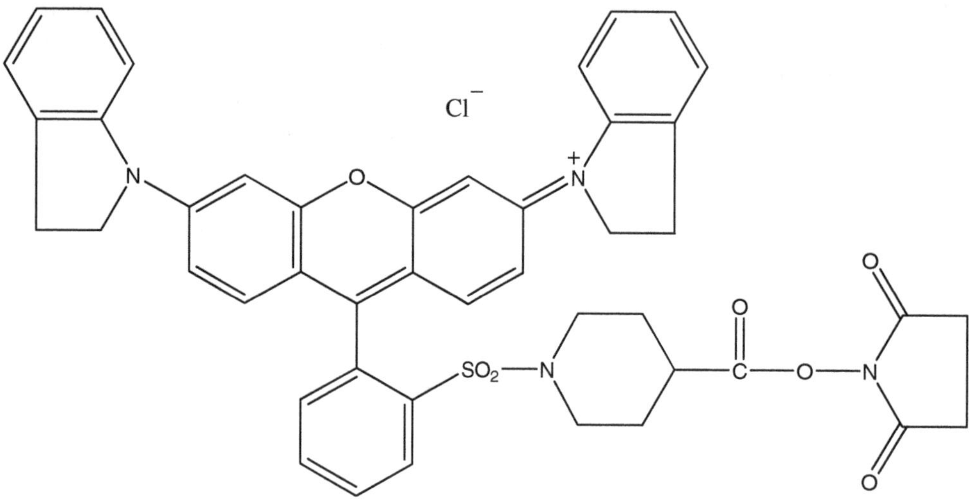

CA Index Name Xanthylium, 3,6-bis(2,3-dihydro-1*H*-indol-1-yl)-9-[2-[[4-[[(2,5-dioxo-1-pyrrolidinyl)oxy]carbonyl]-1-piperidinyl]sulfonyl]phenyl]-, chloride (1:1)

Other Names Xanthylium, 3,6-bis(2,3-dihydro-1*H*-indol-1-yl)-9-[2-[[4-[[(2,5-dioxo-1-pyrrolidinyl)oxy]carbonyl]-1-piperidinyl]sulfonyl]phenyl]-, chloride; QSY 21; QSY 21 carboxylic acid succinimidyl ester; QSY 21NHS

Merck Index Number Not listed

Chemical/Dye Class Xanthene

Molecular Formula $C_{45}H_{39}ClN_4O_7S$

Synthesis Synthetic methods[1,2]

Staining Applications Nucleic acids;[3] cells;[4] oligonucleotides;[1] G-proteins;[5] cysteine protease;[6] tumors[7]

Biological Applications Amplification assays;[1] hybridization assays;[1] litigation assays;[1] ligand assays;[5] detecting analytes,[8] enzymes,[9] optical microcavities,[10] protein–DNA interactions;[11] imaging apoptosis,[12] β-lactamase activity;[13] FRET probes[14]

Industrial Applications Not reported

Safety/Toxicity No data available

REFERENCES

1. Singer, V. L.; Haugland, R. P. Oligonucleotides labeled with energy transfer acceptors for use in amplification, hybridization, and ligation assays employing fluorescent nucleic acid stains. U.S. Patent 6323337, 2001; *Chem. Abstr.* **2001**, *136*, 1576.
2. Haugland, R. P.; Singer, V. L.; Yue, S. T. Xanthene dyes and their application as luminescence quenching compounds. PCT Int. Appl. WO 2000064988, 2000; *Chem. Abstr.* **2000**, *133*, 336549.
3. Chiuman, W.; Li, Y. Efficient signaling platforms built from a small catalytic DNA and doubly labeled fluorogenic substrates. *Nucleic Acids Res.* **2007**, *35*, 401–405.
4. Texier-Nogues, I.; Coll, J. L.; Dumy, P.; Boturyn, D.; Favrot, M. Molecular constructs providing an imaging function activatable in intracellular environment. PCT Int. Appl. WO 2007010128, 2007; *Chem. Abstr.* **2007**, *146*, 190484.
5. McMurchie, E. J.; Leifert, W. R. Cell free G-protein coupled receptor and ligand assays involving G-protein subunits labeled with an energy donor or acceptor and optionally a GPCR. PCT Int. Appl.

WO 2005121755, 2005; *Chem. Abstr.* **2005**, *144*, 45719.

6. Blum, G.; von Degenfeld, G.; Merchant, M. J.; Blau, H. M.; Bogyo, M. Noninvasive optical imaging of cysteine protease activity using fluorescently quenched activity-based probes. *Nat. Chem. Biol.* **2007**, *3*, 668–677.
7. Razkin, J.; Josserand, V.; Boturyn, D.; Jin, Z. H.; Dumy, P.; Favrot, M.; Coll, J. L.; Texier, I. Activatable fluorescent probes for tumour-targeting imaging in live mice. *ChemMedChem* **2006**, *1*, 1069–1072.
8. Volland, H.; Creminon, C.; Neuburger, L. M.; Grassi, J. Apparatus and process for the continuous detection of an analyte using a trifunctional detection reagent. Fr. Demande FR 2847984, 2004; *Chem. Abstr.* **2004**, *141*, 33109.
9. Texier-Nogues, I.; Robert, V.; Coll, J. L.; Imberty, A. Fluorescent saccharide enzyme substrates for enzyme determination and *in vivo* imaging. PCT Int. Appl. WO 2007010145, 2007; *Chem. Abstr.* **2007**, *146*, 137595.
10. Armani, A. M.; Kulkarni, R. P.; Fraser, S. E.; Flagan, R. C.; Vahala, K. J. Label-free, single-molecule detection with optical microcavities. *Science* **2007**, *317*, 783–787.
11. Bonham, A. J.; Braun, G.; Pavel, I.; Moskovits, M.; Reich, N. O. Detection of sequence-specific protein–DNA interactions via surface enhanced resonance Raman scattering. *J. Am. Chem. Soc.* **2007**, *129*, 14572–14573.
12. Bullok, K.; Piwnica-Worms, D. Synthesis and characterization of a small, membrane-permeant, caspase-activatable far-red fluorescent peptide for imaging apoptosis. *J. Med. Chem.* **2005**, *48*, 5404–5407.
13. Xing, B.; Khanamiryan, A.; Rao, J. Cell-permeable near-infrared fluorogenic substrates for imaging β-lactamase activity. *J. Am. Chem. Soc.* **2005**, *127*, 4158–4159.
14. Texier, I.; Heinrich, E. Inhibition of fluorescent dyes for the design of efficient activatable probes dedicated to non-invasive small animal imaging. *Proc. SPIE-Int. Soc. Opt. Eng.* **2007**, *6449*, 64490I/1–64490I/11.

QUIN 2

CAS Registry Number 73630-23-6

Chemical Structure

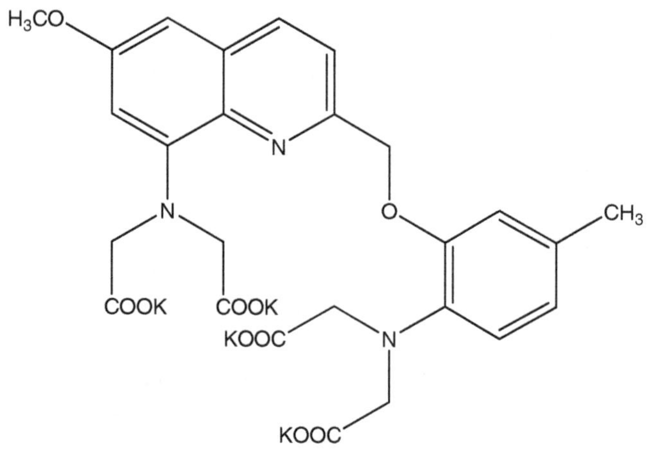

CA Index Name Glycine, *N*-[2-[[8-[bis(carboxymethyl)amino]-6-methoxy-2-quinolinyl]methoxy]-4-methylphenyl]-*N*-(carboxymethyl)-, potassium salt (1:4)

Other Names 2-[(2-Amino-5-methylphenoxy)methyl]-6-methoxy-8-aminoquinoline-*N*,*N*,*N'*,*N'*-tetraacetic acid tetrapotassium salt; 2-{[2-Bis-(carboxymethyl)amino-5-methylphenoxy]-methyl}-6-methoxy-8-bis-(carboxymethyl)aminoquinoline tetrapotassium salt; Glycine, *N*-[2-[[8-[bis(carboxymethyl)amino]-6-methoxy-2-quinolinyl]methoxy]-4-methylphenyl]-*N*-(carboxymethyl)-, tetrapotassium salt; Quin 2; Quin 2 tetrapotassium salt

Merck Index Number 8042

Chemical/Dye Class Quinoline

Molecular Formula $C_{26}H_{23}K_4N_3O_{10}$

Molecular Weight 693.87

Physical Form Light yellow powder

Solubility Soluble in water

Melting Point >250 °C

Absorption (λ_{max}) 353 nm, 333 nm

Emission (λ_{max}) 495 nm

Synthesis Synthetic method[1]

Staining Applications Calcium ions;[3–18] cadmium ions;[19] iron ions;[20,21] lanthanum ions;[22] zinc ions;[23] cells[2]

Biological Applications Calcium indicator;[3–18] cadmium indicator;[19] iron indicator;[20,21] lanthanum indicator;[22] zinc indicator;[23] identifying taste modulators[24]

Industrial Applications Not reported

Safety/Toxicity DNA damage;[25] hepatotoxicity[26,27]

REFERENCES

1. Tsien, R. Y. New calcium indicators and buffers with high selectivity against magnesium and protons: design, synthesis, and properties of prototype structures. *Biochemistry* **1980**, *19*, 2396–2404.
2. Chinn, K. S.; Jim, K. T. C.; Yen, B.; Wuh, H. Apparatus and method for detecting activity of living cells. PCT Int. Appl. WO 2007041308, 2007; *Chem. Abstr.* **2007**, *146*, 397842.
3. Knoop, S.; de Groot, H.; Rauen, U. Little evidence for a major role of Ca^{2+} in cold-induced injury of liver cells. *Cryobiology* **2008**, *56*, 103–113.
4. Ooi, A.; Soematsu, H. Insight into nucleotide and Ca^{2+} binding to carp α-actin. *Fish. Sci.* **2007**, *73*, 684–693.
5. Johnson, J. D.; Tikunova, S. B. Fluorescence methods for measuring calcium affinity and calcium exchange with proteins. *Methods Mol. Biol.* **2002**, *173*, 89–102.
6. White, D. M.; Cousins, M. J. Effect of subcutaneous administration of calcium channel blockers on nerve injury-induced hyperalgesia. *Brain Res.* **1998**, *801*, 50–58.
7. Valant, P. A.; Adjei, P. N.; Haynes, D. H. Rapid calcium extrusion via the sodium/calcium exchanger of the human platelet. *J. Membr. Biol.* **1992**, *130*, 63–82.
8. Lakowicz, J. R.; Szmacinski, H.; Nowaczyk, K.; Johnson, M. L. Fluorescence lifetime imaging of calcium using Quin-2. *Cell Calcium* **1992**, *13*, 131–147.
9. Martinez-Serrano, A.; Blanco, P.; Satrustegui, J. Calcium binding to the cytosol and calcium extrusion mechanisms in intact synaptosomes and their alterations with aging. *J. Biol. Chem.* **1992**, *267*, 4672–4679.
10. Wabnitz, H. Determination of calcium ion concentration in cells by scanning fluorescence microscopy. Ger. (East) DD 292084, 1991; *Chem. Abstr.* **1991**, *115*, 251639.
11. Maxfield, F. R. Measurement of vacuolar pH and cytoplasmic calcium in living cells using fluorescence microscopy. *Methods Enzymol.* **1989**, *173*, 745–771.

12. Mazorow, D. L.; Millar, D. B. Quin-2 and Fura-2 measure calcium differently. *Anal. Biochem.* **1990**, *186*, 28–30.

13. Crofts, J. N.; Barritt, G. J. The measurement of calcium inflow across the liver cell plasma membrane by using quin2 and studies of the roles of sodium and extracellular calcium in the mechanism of calcium inflow. *Biochem. J.* **1989**, *264*, 61–70.

14. Gilroy, S.; Hughes, W. A.; Trewavas, A. J. A comparison between Quin-2 and aequorin as indicators of cytoplasmic calcium levels in higher plant cell protoplasts. *Plant Physiol.* **1989**, *90*, 482–491.

15. Verhage, M.; Besselsen, E.; Lopes da Silva, F. H.; Ghijsen, W. E. J. M. Evaluation of the calcium concentration in purified nerve terminals: relationship between calcium homeostasis and synaptosomal preparation. *J. Neurochem.* **1988**, *51*, 1667–1674.

16. Jacob, R.; Murphy, E.; Lieberman, M. Free calcium in isolated chick embryo heart cells measured using quin2 and fura-2. *Am. J. Physiol.* **1987**, *253*, C337–C342.

17. Hart, G. R.; Ray, K. P.; Wallis, M. Use of quin 2 to measure calcium concentrations in ovine anterior pituitary cells and the effects of quin 2 on secretion of growth hormone and prolactin. *FEBS Lett.* **1986**, *203*, 77–81.

18. Wolniak, S. M.; Bart, K. M. The buffering of calcium with quin2 reversibly forestalls anaphase onset in stamen hair cells of *Tradescantia*. *Eur. J. Cell Biol.* **1985**, *39*, 33–40.

19. Hinkle, P. M.; Shanshala, E. D., II; Nelson, E. J. Measurement of intracellular cadmium with fluorescent dyes. Further evidence for the role of calcium channels in cadmium uptake. *J. Biol. Chem.* **1992**, *267*, 25553–25559.

20. Sandstrom, B. E.; Svoboda, P.; Granstrom, M.; Harms-Ringdahl, M.; Candeias, L. P. H_2O_2-driven reduction of the Fe^{3+}-quin2 chelate and the subsequent formation of oxidizing species. *Free Radical Biol. Med.* **1997**, *23*, 744–753.

21. Sandstroem, B. E.; Granstroem, M.; Vezin, H.; Bienvenu, P.; Marklund, S. L. A comparison of four assays detecting oxidizing species: correlated reactivity of Fe(III)-quin2, but not Fe(III)-EDTA, with hydrogen peroxide. *Biol. Trace Elem. Res.* **1995**, *47*, 29–36.

22. Garmo, O. A.; Lehto, N. J.; Zhang, H.; Davison, W.; Royset, O.; Steinnes, E. Dynamic aspects of DGT as demonstrated by experiments with lanthanide complexes of a multidentate Ligand. *Environ. Sci. Technol.* **2006**, *40*, 4754–4760.

23. Jefferson, J. R.; Hunt, J. B.; Ginsburg, A. Characterization of indo-1 and quin-2 as spectroscopic probes for $zinc^{(2+)}$–protein interactions. *Anal. Biochem.* **1990**, *187*, 328–336.

24. Servant, G.; Kamdar, P.; Rivadeneyra, A. Novel cell-based assays for identifying enhancers or inhibitors of T1R taste receptors (T1R2/T1R3 sweet) and umami (T1R1/T1R3 umami) taste receptors. U.S. Pat. Appl. Publ. US 2008305500, 2008; *Chem. Abstr.* **2008**, *150*, 31118.

25. Sandstroem, B. E. Effects of quin2 acetoxymethyl ester on H_2O_2-induced DNA single-strand breakage in mammalian cells: H_2O_2-concentration-dependent inhibition of damage and additive protective effect with the hydroxyl-radical scavenger dimethyl sulfoxide. *Biochem. J.* **1995**, *305*, 181–185.

26. Carpenter-Deyo, L.; Reed, D. J. Involvement of calcium and iron in Quin 2 toxicity to isolated hepatocytes. II. *J. Pharmacol. Exp. Ther.* **1991**, *258*, 747–752.

27. Carpenter-Deyo, L.; Duimstra, J. R.; Hedstrom, O.; Reed, D. J. Toxicity to isolated hepatocytes caused by the intracellular calcium indicator, Quin 2. I. *J. Pharmacol. Exp. Ther.* **1991**, *258*, 739–746.

QUIN 2 AM

CAS Registry Number 83104-85-2

Chemical Structure

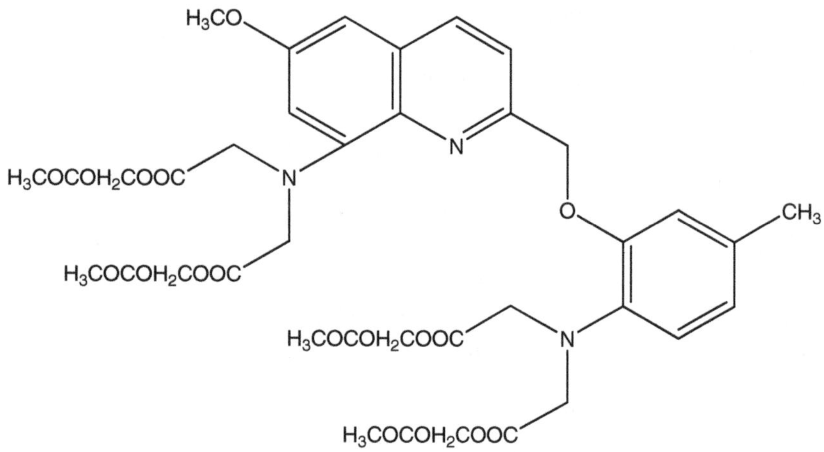

CA Index Name Glycine, *N*-[2-[(acetyloxy)methoxy]-2-oxoethyl]-*N*-[2-[[8-[bis[2-[(acetyloxy)methoxy]-2-oxoethyl]amino]-6-methoxy-2-quinolinyl]methoxy]-4-methylphenyl]-, (acetyloxy)methyl ester

Other Names 2-[(2-Amino-5-methylphenoxy)methyl]-6-methoxy-8-aminoquinoline-*N*,*N*,*N'*,*N'*-tetraacetic acid tetrakis(acetoxymethyl ester); 2-{[2-Bis(carboxymethyl)amino-5-methylphenoxy]-methyl}-6-methoxy-8-bis(carboxymethyl)aminoquinoline tetrakis(acetoxymethyl) ester; Quin 2 acetoxymethyl ester; Quin 2 AM; Quin 2 AM ester

Merck Index Number 8042

Chemical/Dye Class Quinoline

Molecular Formula $C_{38}H_{43}N_3O_{18}$

Molecular Weight 829.76

Physical Form Yellow powder

Solubility Insoluble in water; soluble in dimethyl sulfoxide, methanol

Melting Point 113 °C

Boiling Point (Calcd.) 865.6 ± 65.0 °C, pressure: 760 Torr

pK$_a$ (Calcd.) 2.33 ± 0.50, most basic, temperature: 25 °C

Absorption (λ_{max}) 348 nm

Emission (λ_{max}) 446 nm

Synthesis Synthetic method[1]

Staining Applications Calcium ions;[1,3–21] leukocyte tumor cells[2]

Biological Applications Calcium indicator[1,3–21]

Industrial Applications Not reported

Safety/Toxicity Carcinogenicity;[22] cytotoxicity;[23,24] DNA damage;[25] hepatotoxicity[26–28]

REFERENCES

1. Tsien, R. Y.; Pozzan, T.; Rink, T. J. Calcium homeostasis in intact lymphocytes: cytoplasmic free calcium monitored with a new, intracellularly trapped fluorescent indicator. *J. Cell Biol.* **1982**, *94*, 325–334.
2. Minakami, T.; Tsuji, T.; Oguni, S.; Hamaguchi, Y.; Tsuruta, K.; Kamihiro, T. Method for detecting leukocyte tumor cells. Jpn. Kokai Tokkyo Koho JP 2002207036, 2002; *Chem. Abstr.* **2002**, *137*, 106038.
3. Knoop, S.; de Groot, H.; Rauen, U. Little evidence for a major role of Ca^{2+} in cold-induced injury of liver cells. *Cryobiology* **2008**, *56*, 103–113.
4. Byrne, A. M.; Lemasters, J. J.; Nieminen, A. L. Contribution of increased mitochondrial free Ca^{2+} to the mitochondrial permeability transition induced by *tert*-butylhydroperoxide in rat hepatocytes. *Hepatology* **1999**, *29*, 1523–1531.
5. Kushnareva, Y. E.; Haley, L. M.; Sokolove, P. M. The role of low (< or = 1 mM) phosphate concentrations in regulation of mitochondrial permeability: modulation of matrix free Ca^{2+} concentration. *Arch. Biochem. Biophys.* **1999**, *363*, 155–162.
6. Weinstock, R. S.; Saville, C. M.; Messina, J. L. Role of cytosolic calcium in regulation of cytoskeletal gene

expression by insulin. *Am. J. Physiol.* **1993**, *264*, E519–E525.

7. Hoggard, N.; Davis, J. R.; Berwaer, M.; Monget, P.; Peers, B.; Belayew, A.; Martial, J. A. Pit-1 binding sequences permit calcium regulation of human prolactin gene expression. *Mol. Endocrinol.* **1991**, *5*, 1748–1754.
8. Tucker, R. W.; Fay, F. S. Distribution of intracellular free calcium in quiescent BALB/c 3T3 cells stimulated by platelet-derived growth factor. *Eur. J. Cell Biol.* **1990**, *51*, 120–127.
9. Yatomi, Y.; Higashihara, M.; Ozaki, Y.; Kume, S.; Kurokawa, K. Intracellular ionized calcium mobilization in CD9 monoclonal antibody-activated human platelets. *Biochem. Biophys. Res. Commun.* **1990**, *171*, 109–115.
10. Rao, A. K.; Kowalska, M. A.; Disa, J. Impaired cytoplasmic ionized calcium mobilization in inherited platelet secretion defects. *Blood* **1989**, *74*, 664–672.
11. Kudo, Y.; Takeda, K.; Hicks, T. P.; Ogura, A.; Kawasaki, Y. A new device for monitoring concentrations of intracellular calcium in CNS preparations and its application to the frog's spinal cord. *J. Neurosci. Methods* **1989**, *30*, 161–168.
12. McConkey, D. J.; Hartzell, P.; Duddy, S. K.; Hakansson, H.; Orrenius, S. 2,3,7,8-Tetrachlorodibenzo-*p*-dioxin kills immature thymocytes by Ca^{2+}-mediated endonuclease activation. *Science* **1988**, *242*, 256–259.
13. Miyoshi, N.; Hara, K.; Fukuda, M. Spectroscopic study of fluorescence probe (Quin-2AM) for calcium and its application. *Photomed. Photobiol.* **1988**, *10*, 209–213.
14. De Chaffoy de Courcelles, D.; Roevens, P.; Verheyen, F.; Van Belle, H.; De Clerck, F. The effect of the intracellular calcium chelator Quin-2 on the platelet phosphoinositide metabolism, protein phosphorylation and morphology. *Thromb. Haemostasis* **1987**, *58*, 927–931.
15. Csermely, P.; Somogyi, J. The possible pitfalls of the measurement of intracellular calcium concentration of lymphocytes with the fluorescent indicator quin2. *Immunobiology* **1987**, *174*, 380–394.
16. Pershadsingh, H. A.; Gale, R. D.; McDonald, J. M. Chelation of intracellular calcium prevents stimulation of glucose transport by insulin and insulinomimetic agents in the adipocyte: evidence for a common mechanism. *Endocrinology* **1987**, *121*, 1727–1732.
17. Simpson, A. M.; White, I. G. Measurement and manipulation of cytoplasmic free calcium of ram and boar spermatozoa using quin 2. *Cell Calcium* **1988**, *9*, 45–56.
18. Lukacs, G. L.; Kapus, A. Measurement of the matrix free calcium concentration in heart mitochondria by entrapped fura 2 and quin 2. *Biochem. J.* **1987**, *248*, 609–613.
19. Avdonin, P. V.; Cheglakov, I. B.; Boogry, E. M.; Svitina-Ulitina, I. V.; Mazaev, A. V.; Tkachuk, V. A. Evidence for the receptor-operated calcium channels in human platelet plasma membrane. *Thromb. Res.* **1987**, *46*, 29–37.
20. Cork, R. J. Problems with the application of quin-2-AM to measuring cytoplasmic free calcium in plant cells. *Plant, Cell Environ.* **1986**, *9*, 157–161.
21. Hatayama, K.; Kambayashi, J.; Nakamura, K.; Ohshiro, T.; Mori, T. Fluorescent calcium indicator quin 2 as an intracellular calcium antagonist in platelet reaction. *Thromb. Res.* **1985**, *38*, 505–512.
22. Lazenby, C. M.; Thompson, M. G.; Hickman, J. A. Elevation of leukemic cell intracellular calcium by the ether lipid SRI 62-834. *Cancer Res.* **1990**, *50*, 3327–3330.
23. Nakagawa, Y. Cytotoxic effects of β-thujaplicin on rat thymocytes and prevention by the compound in tributyltin-induced thymocyte damage. *Bull. Environ. Contam. Toxicol.* **2001**, *67*, 12–19.
24. McConkey, D. J.; Chow, S. C.; Orrenius, S.; Jondal, M. NK cell-induced cytotoxicity is dependent on a Ca^{2+} increase in the target. *FASEB J.* **1990**, *4*, 2661–2664.
25. Sandstroem, B. E. Effects of quin2 acetoxymethyl ester on H_2O_2-induced DNA single-strand breakage in mammalian cells: H_2O_2-concentration-dependent inhibition of damage and additive protective effect with the hydroxyl-radical scavenger dimethyl sulfoxide. *Biochem. J.* **1995**, *305*, 181–185.
26. Schmitz, G.; Lepper, H.; Estler, C. J. Failure of calcium antagonistic agents to prevent hepatotoxicity induced by diclofenac. *Pharmacol. Toxicol.* **1995**, *77*, 32–35.
27. Carpenter-Deyo, L.; Reed, D. J. Involvement of calcium and iron in Quin 2 toxicity to isolated hepatocytes. II. *J. Pharmacol. Exp. Ther.* **1991**, *258*, 747–752.
28. Carpenter-Deyo, L.; Duimstra, J. R.; Hedstrom, O.; Reed, D. J. Toxicity to isolated hepatocytes caused by the intracellular calcium indicator, Quin 2. I. *J. Pharmacol. Exp. Ther.* **1991**, *258*, 739–746.

REDOXSENSOR RED CC-1

CAS Registry Number 296277-09-3

Chemical Structure

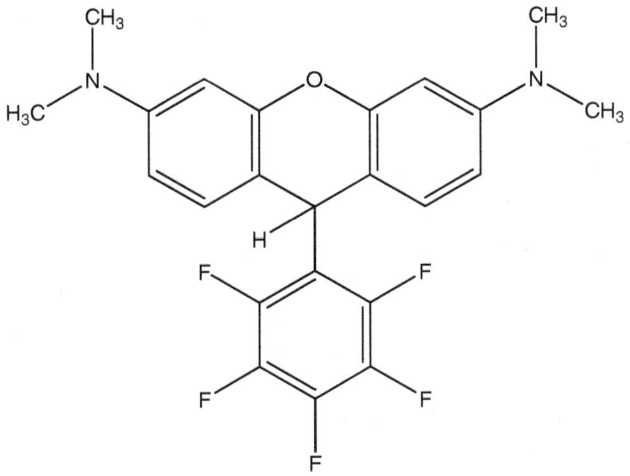

CA Index Name 9*H*-Xanthene-3,6-diamine, *N*,*N*,*N*′,*N*′-tetramethyl-9-(pentafluorophenyl)-

Other Names R 14060; RedoxSensor; RedoxSensor CC 1; RedoxSensor Red; RedoxSensor Red CC 1

Merck Index Number Not listed

Chemical/Dye Class Xanthene

Molecular Formula $C_{23}H_{19}F_5N_2O$

Molecular Weight 434.40

Physical Form Off-white powder

Solubility Soluble in methanol, dimethyl sulfoxide

Melting Point >200 °C

Boiling Point (Calcd.) 445.1 ± 45.0 °C, pressure: 760 Torr

pK_a (Calcd.) 4.83 ± 0.40, most basic, temperature: 25 °C

Absorption (λ_{max}) 239 nm

Synthesis Synthetic method[1]

Staining Applications Mitochondria;[1,2,4] lysosomes;[1] cells[3]

Biological Applications Detecting prostate cancer[4]

Industrial Applications Assaying reactive oxidants on smoke[5]

Safety/Toxicity No data available

REFERENCES

1. Chen, C. S.; Gee, K. R. Redox-dependent trafficking of 2,3,4,5,6-pentafluorodihydrotetramethylrosamine, a novel fluorogenic indicator of cellular oxidative activity. *Free Radical Biol. Med.* **2000**, *28*, 1266–1278.
2. Hattori, F.; Fukuda, K. Method for selecting myocardial cells using intracellular mitochondria labeled with fluorescent indicator. PCT Int. Appl. WO 2006022377, 2006; *Chem. Abstr.* **2006**, *144*, 270175.
3. Nadeau, J. L.; Cohen, N. Use of quantum dots for biological labels and sensors. PCT Int. Appl. WO 2006037226, 2006; *Chem. Abstr.* **2006**, *144*, 385738.
4. Dickman, D. Methods of detecting prostate cancer. PCT Int. Appl. WO 2006054296, 2006; *Chem. Abstr.* **2006**, *145*, 3346.
5. Huang, D.; Ou, B. Method for assaying reactive oxidants in smoke. U.S. Pat. Appl. Publ. US 2004126891, 2004; *Chem. Abstr.* **2004**, *141*, 66653.

Handbook of Biological Dyes and Stains By R. W. Sabnis

RESAZURIN SODIUM SALT

CAS Registry Number 62758-13-8

Chemical Structure

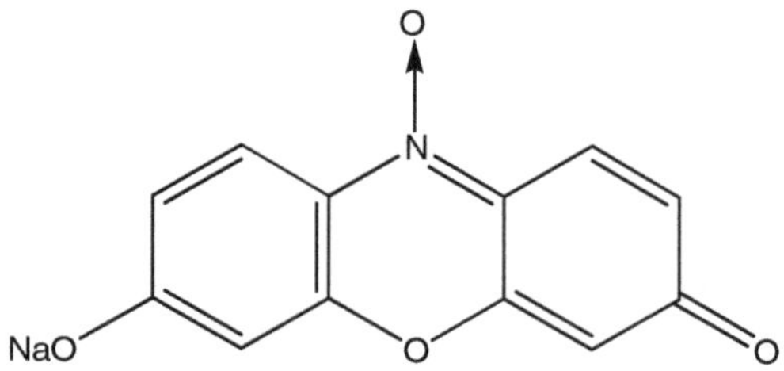

CA Index Name 3*H*-Phenoxazin-3-one, 7-hydroxy-, 10-oxide, sodium salt (1:1)

Other Names 3*H*-Phenoxazin-3-one, 7-hydroxy-, 10-oxide, sodium salt; Resazurin sodium salt; Sodium resazurin

Merck Index Number Not listed

Chemical/Dye Class Phenoxazine

Molecular Formula $C_{12}H_6NNaO_4$

Molecular Weight 251.17

Physical Form Dark green to black powder

Solubility Soluble in water; slightly soluble in ethanol, acetic acid; insoluble in ether

Melting Point >250 °C

pH Range 3.8–6.5

Color Change at pH Orange (3.8) to dark violet (6.5)

pK_a 6.71

Absorption (λ_{max}) 598 nm, 380 nm

Synthesis Synthetic methods[1–3]

Staining Applications D-Arabinitol;[4] bacteria;[5] hypoxic cells;[6] microorganisms;[7–9] hairs[10]

Biological Applications Detecting bile acids,[11] glycosidase activity;[12] as hydrolase substrate;[13] in beer production;[14] treating diabetic complications;[15] enzymatic microplate assay;[16] β-glucosidase assay;[17] glutathione transferase assay[18]

Industrial Applications Electrochemichromic devices;[19] oxygen-barrier packaging materials[20]

Safety/Toxicity No data available

Certification/Approval Certified by Biological Stain Commission (BSC)

REFERENCES

1. Sabnis, R. W. *Handbook of Acid–Base Indicators*; CRC Press: Boca Raton, **2008**; pp 337–338.
2. Green, F. J. *The Sigma–Aldrich Handbook of Stains, Dyes and Indicators*; Aldrich Chemical Company Inc.: Milwaukee, **1991**; pp 622–623.
3. Welcher, F. J. *Organic Analytical Reagents*; D. Van Nostrand Company: New York, **1948**; Vol. 4, pp 548–549.
4. Anderson-Mauser, L. Diagnosis of *Candida vaginitis* by detecting D-arabinitol in vaginal fluid. Eur. Pat. Appl. EP 556725, 1993; *Chem. Abstr.* **1993**, *119*, 177140.
5. Toyama, K.; Fukuwatari, Y.; Yano, Y.; Kiyotaki, K.; Nakagawa, M.; Kano, K.; Sasaki, K. Method and kit for detecting Coliform bacteria. Jpn. Kokai Tokkyo Koho JP 2002360296, 2002; *Chem. Abstr.* **2002**, *138*, 21812.
6. Hodgkiss, R. J.; Begg, A. C.; Middleton, R. W.; Parrick, J.; Stratford, M. R. L.; Wardman, P.; Wilson, G. D. Fluorescent markers for hypoxic cells: a study of novel heterocyclic compounds that undergo bioreductive binding. *Biochem. Pharmacol.* **1991**, *41*, 533–541.
7. Cooper, A. J.; Enzien, M. V.; Hatch, S. R.; Ho, B. P.; Wu, M. M. Method for locating hidden microorganism contaminated surfaces in industrial water systems. U.S. Pat. Appl. Publ. US 2004029211, 2004; *Chem. Abstr.* **2004**, *140*, 151531.
8. Bochner, B. R.; Naleway, J. J. Gel matrix with redox purple for testing and characterizing microorganisms. U.S. Patent 5882882, 1999; *Chem. Abstr.* **1999**, *130*, 220169.
9. Bochner, B. R.; Naleway, J. J. Gel matrix with redox purple for growing and testing microorganisms. PCT Int. Appl. WO 9826270, 1998; *Chem. Abstr.* **1998**, *129*, 78824.
10. Kobayashi, S. Hair preparations containing direct dyes. Jpn. Kokai Tokkyo Koho JP 2006265158, 2006; *Chem. Abstr.* **2006**, *145*, 382935.
11. Nakasuga, A. Enzymic–chromatographic–spectrometric determination of bile acids. Jpn. Kokai Tokkyo Koho JP 61260896, 1986; *Chem. Abstr.* **1987**, *106*, 210560.
12. Klein, C.; Batz, H. G.; Sernetz, M.; Hofmann, J. Glycosides of resorufin derivatives, useful in determining the activity of glycosidases. Ger.

Offen. DE 3411574, 1985; *Chem. Abstr.* **1986**, *104*, 168764.

13. Guder, H. J.; Von der Eltz, H.; Eltz, H. V. New dihydroresorufin derivatives for use as hydrolase substrates. Ger. Offen. DE 3644401, 1988; *Chem. Abstr.* **1989**, *110*, 91267.
14. Nishikawa, N.; Kamata, K. Determination of the quality of malt for use in beer production. Jpn. Kokai Tokkyo Koho JP 62019099, 1987; *Chem. Abstr.* **1987**, *106*, 212577.
15. Inoue, J. Pharmaceutical compositions and methods using quinoline and oxophenoxazine derivatives for inhibition of Maillard's reaction. Can. Pat. Appl. CA 2029421, 1991; *Chem. Abstr.* **1992**, *116*, 51606.
16. Yamamoto, N.; Kawasaki, K.; Sato, T.; Hirose, Y.; Muroyama, K. A nonradioisotope, enzymatic microplate assay for *in vivo* evaluation of 2-deoxyglucose uptake in muscle tissue. *Anal. Biochem.* **2008**, *375*, 397–399.
17. Tokutake, S.; Kasai, K.; Tomikura, T.; Yamaji, N.; Kato, M. Glycosides having chromophores as substrates for sensitive enzyme analysis. II. Synthesis of phenolindophenyl-β-D-glucopyranosides having an electron-withdrawing substituent as substrates for β-glucosidase. *Chem. Pharm. Bull.* **1990**, *38*, 3466–3470.
18. Suvorov, A. A.; Stulovskij, A. V.; Vilyatser, A. Y.; Voznyj, I. V.; Rozengart, E. V.; Khovanskikh, A. E. A chromogenic substrate for glutathione transferase assay. U.S.S.R. SU 1759874, 1992; *Chem. Abstr.* **1993**, *118*, 229168.
19. Varaprasad, D. V.; Looman, S. D.; Zhao, M.; Habibi, H. R.; Lynam, N. R. Electrochemichromic solutions, processes for preparing and using the same, and devices manufactured with the same. U.S. Patent 5500760, 1996; *Chem. Abstr.* **1996**, *124*, 328070.
20. Kamiyama, M. Oxygen-barrier packaging materials. Jpn. Kokai Tokkyo Koho JP 09124076, 1997; *Chem. Abstr.* **1997**, *127*, 66976.

RH 237

CAS Registry Number 83668-91-1

Chemical Structure

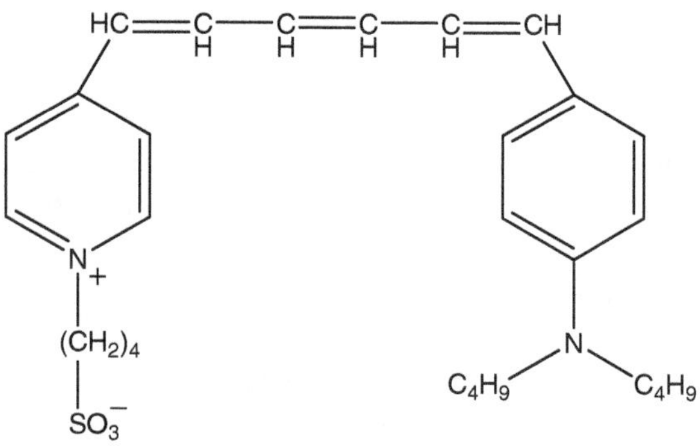

CA Index Name Pyridinium, 4-[6-[4-(dibutylamino) phenyl]-1,3,5-hexatrienyl]-1-(4-sulfobutyl)-, inner salt

Other Names 4-(*N,N*-Dibutylanilino)hexatrienyl-4′-pyridinium butanesulfonate; RH 237

Merck Index Number Not listed

Chemical/Dye Class Styryl

Molecular Formula $C_{29}H_{40}N_2O_3S$

Molecular Weight 496.71

Physical Form Solid

Solubility Soluble in ethanol, methanol, dimethyl sulfoxide

Melting Point >200 °C

Absorption (λ_{max}) 528 nm

Emission (λ_{max}) 782 nm

Synthesis Synthetic method[1]

Staining Applications Atria;[2] cardiac tissues;[3–5] cardiac myocytes;[6–8] neurons;[9,10] proteins;[11] Na^+/K^+-ATPase[12]

Biological Applications Measuring membrane potential;[7,8,13–16] detecting Na^+/K^+-ATPase;[12] examining activity of ion channels[17]

Industrial Applications Monitoring of polymerization[18]

Safety/Toxicity Cardiac toxicity[2]

REFERENCES

1. Grinvald, A.; Hildesheim, R.; Farber, I. C.; Anglister, L. Improved fluorescent probes for the measurement of rapid changes in membrane potential. *Biophys. J.* **1982**, *39*, 301–308.
2. Mueller, W.; Windisch, H.; Tritthart, H. A. Fluorescent styryl dyes applied as fast optical probes of cardiac action potential. *Eur. Biophys. J.* **1986**, *14*, 103–111.
3. Byars, J. L.; Smith, W. M.; Ideker, R. E.; Fast, V. G. Development of an optrode for intramural multisite optical recordings of Vm in the heart. *J. Cardiovasc. Electrophysiol.* **2003**, *14*, 1196–1202.
4. Fast, V. G.; Kleber, A. G. Anisotropic conduction in monolayers of neonatal rat heart cells cultured on collagen substrate. *Circ. Res.* **1994**, *75*, 591–595.
5. Fast, V. G.; Kleber, A. G. Microscopic conduction in cultured strands of neonatal rat heart cells measured with voltage-sensitive dyes. *Circ. Res.* **1993**, *73*, 914–925.
6. Thomas, S. P.; Bircher-Lehmann, L.; Thomas, S. A.; Zhuang, J.; Saffitz, J. E.; Kleber, A. G. Synthetic strands of neonatal mouse cardiac myocytes: structural and electrophysiological properties. *Circ. Res.* **2000**, *87*, 467–473.
7. Fast, V. G.; Ideker, R. E. Simultaneous optical mapping of transmembrane potential and intracellular calcium in myocyte cultures. *J. Cardiovasc. Electrophysiol.* **2000**, *11*, 547–556.
8. Windisch, H.; Muller, W.; Tritthart, H. A. Fluorescence monitoring of rapid changes in membrane potential in heart muscle. *Biophys. J.* **1985**, *48*, 877–884.
9. Savtchenko, L. P.; Gogan, P.; Korogod, S. M.; Tyc-Dumont, S. Imaging stochastic spatial variability of active channel clusters during excitation of single neurons. *Neurosci. Res.* **2001**, *39*, 431–446.
10. Kim, J. H.; Dunn, M. B.; Hua, Y.; Rydberg, J.; Yae, H.; Elias, S. A.; Ebner, T. J. Imaging of cerebellar surface activation *in vivo* using voltage sensitive dyes. *Neuroscience* **1989**, *31*, 613–623.
11. Haugland, R. P.; Singer, V. L.; Jones, L. J.; Steinberg, T. H. Merocyanine dye protein stains. PCT Int. Appl. WO 9636882, 1996; *Chem. Abstr.* **1997**, *126*, 44635.
12. Malkov, D. Y.; Pavlov, K. V.; Sokolov, V. S. Dipole potential drop due to RH-dye adsorption on the lipid bilayer and its influence on Na^+/K^+-ATPase activity. *Ann. N.Y. Acad. Sci.* **1997**, *834*, 357–360.

13. Murakami, T. Cytotoxicity test method by measuring membrane electric potential. Jpn. Kokai Tokkyo Koho JP 2000300290, 2000; *Chem. Abstr.* **2000**, *133*, 330852.

14. Passechnik, V. I.; Sokolov, V. S. Application of inner field compensation method to determine the immersion depth of electrochromic styryl dyes inside the membrane. *Biol. Membr.* **2003**, *20*, 433–442; *Chem. Abstr.* **2004**, *141*, 136566.

15. Malkov, D.; Sokolov, V. S. Dipole potential drop on the membrane/solution boundary due to adsorption of styryl dyes RH-421, RH-237 and RH-160. *Biol. Membr.* **1995**, *12*, 652–663; *Chem. Abstr.* **1996**, *124*, 283615.

16. Malkov, D. Y.; Sokolov, V. S. Fluorescent styryl dyes of the RH series affect a potential drop on the membrane/solution boundary. *Biochim. Biophys. Acta Biomembr.* **1996**, *1278*, 197–204.

17. Ebneth, A.; Netzer, R.; Hahn, U. Method for examining the activity of ion channels. PCT Int. Appl. WO 2005069008, 2005; *Chem. Abstr.* **2005**, *143*, 149507.

18. Wang, F. W.; Lowry, R. E. Wavelength-shift fluorescent probes for monitoring of polymerization. *J. Appl. Polym. Sci.* **2006**, *101*, 747–750.

RH 414

CAS Registry Number 161433-30-3

Chemical Structure

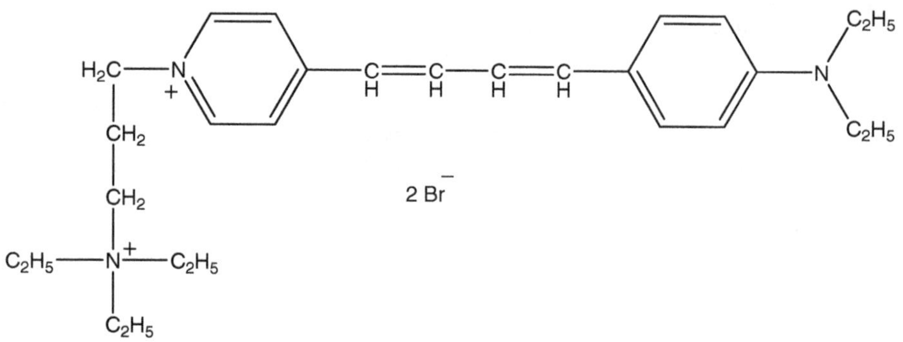

CA Index Name Pyridinium, 4-[4-[4-(diethylamino) phenyl]-1,3-butadien-1-yl]-1-[3-(triethylammonio)propyl]-, bromide (1:2)

Other Names Pyridinium, 4-[4-[4-(diethylamino)phenyl]-1,3-butadienyl]-1-[3-(triethylammonio)propyl]-, dibromide; RH 414

Merck Index Number Not listed

Chemical/Dye Class Styryl

Molecular Formula $C_{28}H_{43}Br_2N_3$

Molecular Weight 581.48

Physical Form Solid

Solubility Soluble in water, ethanol, methanol, dimethyl sulfoxide

Melting Point >200 °C

Absorption (λ_{max}) 532 nm

Emission (λ_{max}) 716 nm

Synthesis Synthetic method[1]

Staining Applications Clostridial toxin substrate;[2] membranes;[3] flavoproteins;[4] neurons[5,6]

Biological Applications Measuring membrane potential;[7,8] detecting Clostridia toxin activity;[2] probes for Na, K-ATPase reaction mechanism;[9] examining activity of ion channels[10]

Industrial Applications Not reported

Safety/Toxicity No data available

REFERENCES

1. Haugland, R. P. *The Handbook: A Guide to Fluorescent Probes and Labeling Technologies*; Molecular Probes Inc.: Eugene, **2005**; pp 976–978.
2. Fernandez-Salas, E.; Steward, L. E.; Aoki, K. R. Lipophilic dye-containing cells producing fluorophore-labeled Clostridial toxin substrate for FRET detection of toxin activity. PCT Int. Appl. WO 2006107921, 2006; *Chem. Abstr.* **2006**, *145*, 390749.
3. Krolenko, S. A.; Adamyan, S. Ya.; Belyaeva, T. N.; Mozhenok, T. P. Acridine orange bioaccumulation in acid organelles of normal and vacuolated frog skeletal muscle fibres. *Cell Biol. Int.* **2006**, *30*, 933–939.
4. Samhan-Arias, A. K.; Garcia-Bereguiain, M. A.; Martin-Romero, F. J.; Gutierrez-Merino, C. Regionalization of plasma membrane-bound flavoproteins of cerebellar granule neurons in culture by fluorescence energy transfer imaging. *J. Fluoresc.* **2006**, *16*, 393–401.
5. Fisher, J. A. N.; Salzberg, B. M.; Yodh, A. G. Near infrared two-photon excitation cross-sections of voltage-sensitive dyes. *J. Neurosci. Methods* **2005**, *148*, 94–102.
6. Cinelli, A. R. High-definition mapping of neural activity using voltage-sensitive dyes. *Methods* **2000**, *21*, 349–372.
7. Murakami, T. Cytotoxicity test method by measuring membrane electric potential. Jpn. Kokai Tokkyo Koho JP 2000300290, 2000; *Chem. Abstr.* **2000**, *133*, 330852.
8. Sinha, S. R.; Saggau, P. Simultaneous optical recording of membrane potential and intracellular calcium from brain slices. *Methods* **1999**, *18*, 204–214.

9. Fedosova, N. U.; Cornelius, F.; Klodos, I. Fluorescent styryl dyes as probes for Na,K-ATPase reaction mechanism: significance of the charge of the hydrophilic moiety of RH dyes. *Biochemistry* **1995**, *34*, 16806–16814.

10. Ebneth, A.; Netzer, R.; Hahn, U. Method for examining the activity of ion channels. PCT Int. Appl. WO 2005069008, 2005; *Chem. Abstr*. **2005**, *143*, 149507.

RH 421

CAS Registry Number 107610-19-5

Chemical Structure

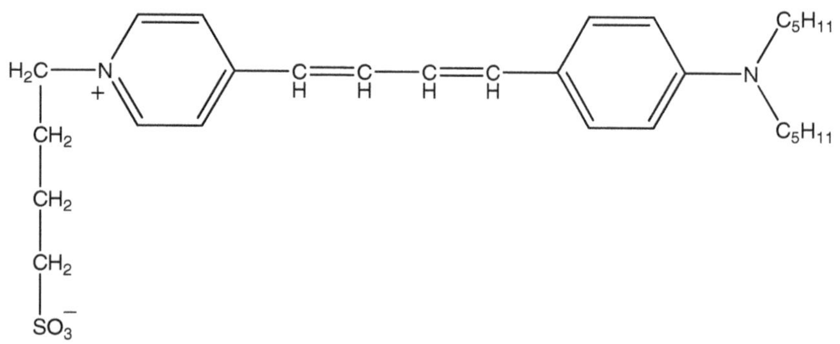

CA Index Name Pyridinium, 4-[4-[4-(dipentylamino) phenyl]-1,3-butadienyl]-1-(4-sulfobutyl)-, inner salt

Other Names 4-{4-[4-(Dipentylamino)phenyl]-1,3-butadienyl}-1-(4-sulfobutyl)pyridinium hydroxide; RH 421

Merck Index Number Not listed

Chemical/Dye Class Styryl

Molecular Formula $C_{29}H_{42}N_2O_3S$

Molecular Weight 498.72

Physical Form Red powder

Solubility Soluble in ethanol, methanol, dimethyl sulfoxide, *N,N*-dimethylformamide

Melting Point >200 °C

Absorption (λ_{max}) 515 nm

Emission (λ_{max}) 704 nm

Synthesis Synthetic method[1]

Staining Applications Cardiac tissues;[2,3] lipid membranes;[4–6,8] proteins;[7,8] neurons[9]

Biological Applications Measuring membrane potential[10–12]

Industrial Applications Photoresists[13]

Safety/Toxicity No data available

REFERENCES

1. Haugland, R. P. *The Handbook: A Guide to Fluorescent Probes and Labeling Technologies*; Molecular Probes Inc.: Eugene, **2005**; pp 976–978.
2. Salama, G.; Kanai, A.; Efimov, I. R. Subthreshold stimulation of Purkinje fibers interrupts ventricular tachycardia in intact hearts: experimental study with voltage-sensitive dyes and imaging techniques. *Circ. Res.* **1994**, *74*, 604–619.
3. Mueller, W.; Windisch, H.; Tritthart, H. A. Fluorescent styryl dyes applied as fast optical probes of cardiac action potential. *Eur. Biophys. J.* **1986**, *14*, 103–111.
4. Krishna, M. M. G.; Periasamy, N. Fluorescence of organic dyes in lipid membranes: site of solubilization and effects of viscosity and refractive index on lifetimes. *J. Fluoresc.* **1998**, *8*, 81–91.
5. Rokitskaya, T. I.; Antonenko, Y. N.; Kotova, E. A. Effect of the dipole potential of a bilayer lipid membrane on gramicidin channel dissociation kinetics. *Biophys. J.* **1997**, *73*, 850–854.
6. Visser, N. V.; van Hoek, A.; Visser, A. J. W. G.; Frank, J.; Apell, H. J.; Clarke, R. J. Time-resolved fluorescence investigations of the interaction of the voltage-sensitive probe RH421 with lipid membranes and proteins. *Biochemistry* **1995**, *34*, 11777–11784.
7. Haugland, R. P.; Singer, V. L.; Jones, L. J.; Steinberg, T. H. Merocyanine dye protein stains. PCT Int. Appl. WO 9636882, 1996; *Chem. Abstr.* **1997**, *126*, 44635.
8. Frank, J.; Zouni, A.; van Hoek, A.; Visser, A. J. W. G.; Clarke, R. J. Interaction of the fluorescent probe RH421 with ribulose-1,5-bisphosphate carboxylase/oxygenase and with Na^+, K^+-ATPase membrane fragments. *Biochim. Biophys. Acta, Biomembr.* **1996**, *1280*, 51–64.
9. Fromherz, P.; Muller, C. O. Cable properties of a straight neurite of a leech neuron probed by a voltage-

sensitive dye. *Proc. Natl. Acad. Sci. U.S.A.* **1994**, *91*, 4604–4608.

10. Murakami, T. Cytotoxicity test method by measuring membrane electric potential. Jpn. Kokai Tokkyo Koho JP 2000300290, 2000; *Chem. Abstr.* **2000**, *133*, 330852.
11. Clarke, R. J.; Kane, D. J. Optical detection of membrane dipole potential: avoidance of fluidity and dye-induced effects. *Biochim. Biophys. Acta, Biomembr.* **1997**, *1323*, 223–239.
12. Malkov, D. Y.; Sokolov, V. S. Fluorescent styryl dyes of the RH series affect a potential drop on the membrane/solution boundary. *Biochim. Biophys. Acta, Biomembr.* **1996**, *1278*, 197–204.
13. Farid, S. Y.; Haley, N. F.; Moody, R. E.; Specht, D. P. Negative working photoresists responsive to shorter wavelength visible light and novel coated articles. U. S. Patent 4743529, 1988; *Chem. Abstr.* **1988**, *109*, 160608.

RH 795

CAS Registry Number 172807-13-5

Chemical Structure

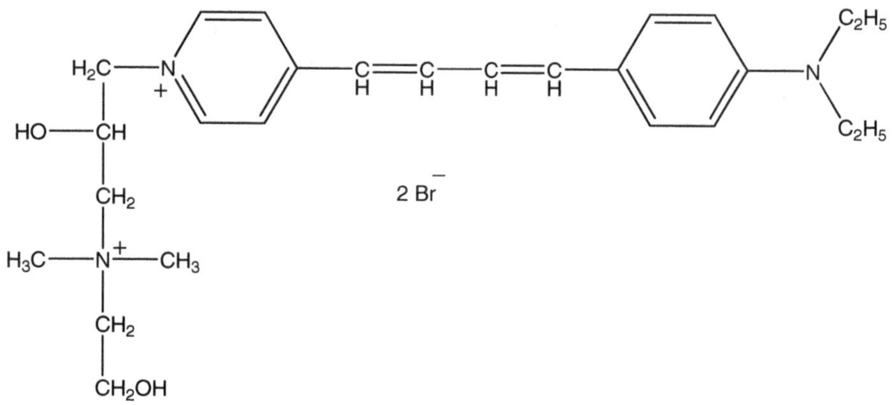

Solubility Soluble in ethanol, methanol, dimethyl sulfoxide

Melting Point >200 °C

Absorption (λ_{max}) 530 nm

CA Index Name Pyridinium, 4-[4-[4-(diethylamino) phenyl]-1,3-butadienyl]-1-[2-hydroxy-3-[(2-hydroxyethyl)dimethylammonio]propyl]-, dibromide

Other Names RH 795

Merck Index Number Not listed

Chemical/Dye Class Styryl

Molecular Formula $C_{26}H_{39}Br_2N_3O_2$

Molecular Weight 585.42

Physical Form Solid

Emission (λ_{max}) 712 nm

Synthesis Synthetic method[1]

Staining Applications Brainstem;[2] cortex;[3] neurons[4]

Biological Applications Measuring membrane potential;[4,5] probes for Na, K-ATPase reaction mechanism;[6] examining activity of ion channels[7]

Industrial Applications Not reported

Safety/Toxicity No data available

REFERENCES

1. Haugland, R. P. *The Handbook: A Guide to Fluorescent Probes and Labeling Technologies*; Molecular Probes Inc.: Eugene, **2005**; pp 976–978.
2. Okada, Y.; Chen, Z.; Eldridge, F. L. Central respiratory chemoreception in the developing rat brainstem: dynamic imaging analysis with fluorescent voltage-sensitive dye. *Keio Univ. Symp. Life Sci. Med.* **1999**, *2*, 410–415.
3. Tsytsarev, V.; Premachandra, K.; Takeshita, D.; Bahar, S. Imaging cortical electrical stimulation *in vivo*: fast intrinsic optical signal versus voltage-sensitive dyes. *Opt. Lett.* **2008**, *33*, 1032–1034.
4. Fisher, J. A. N.; Salzberg, B. M.; Yodh, A. G. Near infrared two-photon excitation cross-sections of voltage-sensitive dyes. *J. Neurosci. Methods* **2005**, *148*, 94–102.
5. Murakami, T. Cytotoxicity test method by measuring membrane electric potential. Jpn. Kokai Tokkyo Koho JP 2000300290, 2000; *Chem. Abstr.* **2000**, *133*, 330852.
6. Fedosova, N. U.; Cornelius, F.; Klodos, I. Fluorescent styryl dyes as probes for Na,K-ATPase reaction mechanism: significance of the charge of the hydrophilic moiety of RH dyes. *Biochemistry* **1995**, *34*, 16806–16814.
7. Ebneth, A.; Netzer, R.; Hahn, U. Method for examining the activity of ion channels. PCT Int. Appl. WO 2005069008, 2005; *Chem. Abstr.* **2005**, *143*, 149507.

RHODAMINE 6G

CAS Registry Number 989-38-8

Chemical Structure

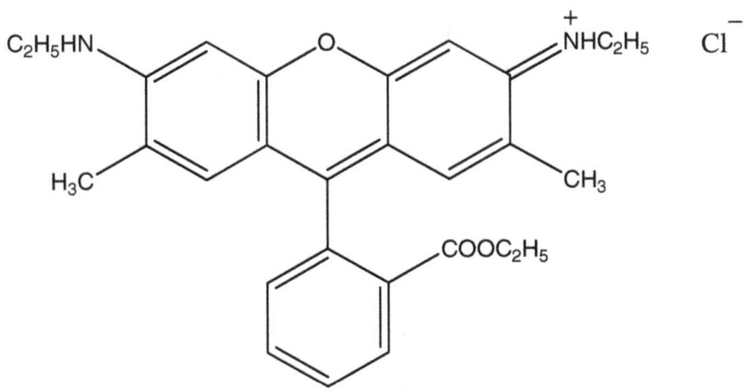

CA Index Name Xanthylium, 9-[2-(ethoxycarbonyl) phenyl]-3,6-bis(ethylamino)-2,7-dimethyl-, chloride (1:1)

Other Names Benzoic acid, *o*-[6-(ethylamino)-3-(ethylimino)-2,7-dimethyl-3*H*-xanthen-9-yl]-, ethyl ester, monohydrochloride; Rhodamine 6GCP; Xanthylium, 9-[2-(ethoxycarbonyl)phenyl]-3,6-bis(ethylamino)-2,7-dimethyl, chloride; Aizen Rhodamine 6GCP; Basic Red 1; Basonyl Red 482; Basonyl Red 483; C.I. 45160; C.I. Basic Red 1; Calcozine Red 6G; Calcozine Rhodamine 6GX; Eljon Pink Toner; Exciton 590; Fanal Pink B; Fanal Pink GFK; Fanal Red 25532; Flexo Red 482; Heliostable Brilliant Pink B extra; Mitsui Rhodamine 6GCP; NSC 36345; Nyco Liquid Red GF; R 634; R 6G; Rh 6G; Rhodamin 6G; Rhodamine 4GD; Rhodamine 4GH; Rhodamine 590 Chloride; Rhodamine 5GDN; Rhodamine 5GDN Extra; Rhodamine 5GL; Rhodamine 6G; Rhodamine 6G Extra; Rhodamine 6G Extra Base; Rhodamine 6G chloride; Rhodamine 6GB; Rhodamine 6GBN; Rhodamine 6GD; Rhodamine 6GDN; Rhodamine 6GDN Extra; Rhodamine 6GEx ethyl ester; Rhodamine 6GH; Rhodamine 6GO; Rhodamine 6GX; Rhodamine 6JH; Rhodamine 6JH-SA; Rhodamine 6JH-SA Extra 1150; Rhodamine 6Zh-DN; Rhodamine F 5G; Rhodamine F 5GL; Rhodamine GDN; Rhodamine GDN Extra; Rhodamine Y 20-7425; Rhodamine Zh; Silosuper Pink B; Vali Fast Red 1308

Merck Index Number Not listed

Chemical/Dye Class Xanthene

Molecular Formula $C_{28}H_{31}ClN_2O_3$

Molecular Weight 479.02

Physical Form Red-brown or maroon powder or crystals

Solubility Soluble in water, ethanol, methanol

Melting Point 290 °C

Absorption (λ_{max}) 528 nm

Emission (λ_{max}) 551 nm

Synthesis Synthetic methods[1–11]

Staining Applications Mitochondria;[12–14,23] internal limiting membrane;[15] vitreous;[15] lens capsule of eye;[15] serum albumin;[16] hairs[17]

Biological Applications Detecting nucleic acids,[18] prostate cancer,[19] stress biomarkers,[20] protease activity,[21] urogenital infection;[22] measuring membrane potential;[23] photodynamic therapy;[24] apoptosis assays;[25] stents[26]

Industrial Applications Paints;[27] inks;[28] color filters;[29] dye lasers;[30,31] electroluminescent displays;[32] laser devices;[33] light-emitting devices;[34] liquid crystals;[35] photovoltaic device;[36] thin films;[37] waveguides[38]

Safety/Toxicity Carcinogenicity;[39,40] DNA damage;[41] ecotoxicity;[42] genotoxicity;[43] mutagenicity[44,45]

REFERENCES

1. Romieu, A.; Brossard, D.; Hamon, M.; Outaabout, H.; Portal, C.; Renard, P. Postsynthetic derivatization of fluorophores with alpha-sulfo-beta-alanine dipeptide linker: application to the preparation of water-soluble cyanine and rhodamine dyes. *Bioconjugate Chem.* **2008**, *19*, 279–289.
2. Garcia-Moreno, I.; Costela, A.; Cuesta, A.; Garcia, O.; del, A. D.; Sastre, R. Synthesis, structure, and physical properties of hybrid nanocomposites for solid-state dye lasers. *J. Phys. Chem. B* **2005**, *109*, 21618–21626.
3. Venner, M. R.; Case, A. D.; Fulker, D. J.; Griffiths, J.; Mama, J. Lasing properties of novel near-infrared laser dyes. *Proc. SPIE-Int. Soc. Opt. Eng.* **2005**, *5707*, 227–236.
4. Xiao, G.; Zhu, D. Bulk-scale preparation of ethyl 2-[3,6-bis(*N*-ethylamino)-2,7-dimethyl-3-hydroxanth-9-yl]benzoate hydrochloride. Faming Zhuanli Shenqing Gongkai Shuomingshu CN 1083058, 1994; *Chem. Abstr.* **1995**, *122*, 290714.
5. Sekima, H. Manufacture of rhodamine dye alkyl esters. Jpn. Kokai Tokkyo Koho JP 05271559, 1993; *Chem. Abstr.* **1993**, *120*, 194038.

6. Beger, V. N.; Kolesnikova, T. D.; Sechkarev, A. V. Optical density reduction effect of activated porous glass due to formation of dye molecule fractal clusters. *Proc. SPIE-Int. Soc. Opt. Eng.* **1993**, *1983*, 775–776.
7. Aburada, K.; Akagi, M. Preparation of rhodamines. Eur. Pat. Appl. EP 468821, 1992; *Chem. Abstr.* **1992**, *116*, 237370.
8. Haghighat, S.; Castro, D. J.; Lufkin, R. B.; Fetterman, H. R.; Castro, D. J.; Soudant, J.; Ward, P. H.; Saxton, R. E. Laser dyes for experimental phototherapy of human cancer: comparison of three rhodamines. *Laryngoscope* **1992**, *102*, 81–87.
9. Marczenko, Z.; Kalinowski, K. Sensitive flotation–spectrophotometric determination of platinum based on systems with chlorostannate(II) and basic dyes. *Anal. Chim. Acta* **1983**, *153*, 219–227.
10. Gaunt, T. N. Fluorescing color pigments. Ger. Offen DE 961575, 1957; *Chem. Abstr.* **1960**, *54*, 47327.
11. Levshin, V. L.; Baranova, E. G. Nature of concentration effects in rhodamine solutions. *Bull. Acad. Sci. USSR, Phys. Ser.* **1956**, *20*, 387–393.
12. Hu, S.; Zhao, H.; Yin, X. J.; Ma, J. K. H. Role of mitochondria in silica-induced apoptosis of alveolar macrophages: inhibition of apoptosis by rhodamine 6G and *N*-acetyl-L-cysteine. *J. Toxicol. Environ. Health, Part A* **2007**, *70*, 1403–1415.
13. Hattori, F.; Fukuda, K. Method for selecting myocardial cells using intracellular mitochondria labeled with fluorescent indicator. PCT Int. Appl. WO 2006022377, 2006; *Chem. Abstr.* **2006**, *144*, 270175.
14. Rashid, F.; Horobin, R. W. Interaction of molecular probes with living cells and tissues. Part 2. A structure–activity analysis of mitochondrial staining by cationic probes, and a discussion of the synergistic nature of image-based and biochemical approaches. *Histochemistry* **1990**, *94*, 303–308.
15. Haritoglou, C. Method, dye and medicament for staining the internal limiting membrane, the vitreous and/or the capsule of an eye. PCT Int. Appl. WO 2006133903, 2006; *Chem. Abstr.* **2006**, *146*, 77565.
16. Nizomov, N.; Ismailov, Z. F.; Kurtaliev, E. N.; Nizamov, Sh. N.; Khodzhaev, G.; Patsenker, L. D. Luminescent spectral properties of rhodamine derivatives while binding to serum albumin. *J. Appl. Spectrosc.* **2006**, *73*, 432–436.
17. Hercouet, L. Hair dye composition comprising a derivative of diamino-*N,N*-dihydropyrazolone, a coupler, and a heterocyclic direct dye. Fr. Demande FR 2886132, 2006; *Chem. Abstr.* **2006**, *146*, 49738.
18. Seitz, O.; Grossmann, T. Detection of target nucleic acids using template-catalyzed transfer reactions. Eur. Pat. Appl. EP 1860197, 2007; *Chem. Abstr.* **2007**, *148*, 2012.
19. Dickman, D. Methods of detecting prostate cancer. PCT Int. Appl. WO 2006054296, 2006; *Chem. Abstr.* **2006**, *145*, 3346.
20. Cullum, M. E.; Duplessis, C. A.; Crepeau, L. J. Method for the detection of stress biomarkers including cortisol by fluorescence polarization. U. S. Pat. Appl. Publ. US 2006105397, 2006; *Chem. Abstr.* **2006**, *144*, 463810.
21. Liu, G. L.; Rosa-Bauza, Y. T.; Salisbury, C. M.; Craik, C.; Ellman, J. A.; Chen, F. F.; Lee, L. P. Peptide–nanoparticle hybrid SERS probes for optical detection of protease activity. *J. Nanosci. Nanotechnol.* **2007**, *7*, 2323–2330.
22. Granovsky, I. E.; Beletsky, I. P.; Shlyapnikova, E. A.; Gavryushkin, A. V.; Biryukov, S. V. Method for detecting urogenital infection using multiplex asymmetrical PCR, oligonucleotides, oligonucleotide combinations, DNA microchips, and a test kit based thereon. PCT Int. Appl. WO 2007136303, 2007; *Chem. Abstr.* **2007**, *148*, 25154.
23. Mandalà, M.; Serck-Hanssen, G.; Martino, G.; Helle, K. B. The fluorescent cationic dye rhodamine 6G as a probe for membrane potential in bovine aortic endothelial cells. *Anal. Biochem.* **1999**, *274*, 1–6.
24. Woo, T.; Miller, G. G.; Madiyalakan, R. Method and device for photodynamic therapy comprising administering and activating a photosensitizer with a chemiluminescent and/or a light-emitting diode light source in dermal contact with the patient. PCT Int. Appl. WO 2008011707, 2008; *Chem. Abstr.* **2008**, *148*, 222105.
25. Singh, M.; Gatson, J. W. Membrane androgen receptor as a therapeutic target for the prevention/promotion of cell death. U.S. Pat. Appl. Publ. US 2007141581, 2007; *Chem. Abstr.* **2007**, *147*, 88350.
26. Popowski, Y. Medical stent provided with inhibitors of ATP synthesis. PCT Int. Appl. WO 2006024488, 2006; *Chem. Abstr.* **2006**, *144*, 280677.
27. Sabnis, R. W.; Kehoe, T. D.; Balchunis, R. J. Paint compositions comprising color changing phthalein acid–base indicator. PCT Int. Appl. WO 2006105191, 2006; *Chem. Abstr.* **2006**, *145*, 403489.
28. Likavec, W. R.; Jatla, A.; Waters, J. F. Ink jet printing inks with alicyclic polyester dye carrier. PCT Int.

Appl. WO 2000050525, 2000; *Chem. Abstr.* **2000**, *133*, 194791.

29. Fukuhara, S.; Asano, M. Color filter plate for organic electroluminescent device. Jpn. Kokai Tokkyo Koho JP 2007250437, 2007; *Chem. Abstr.* **2007**, *147*, 394876.
30. Chen, F.; Gindre, D.; Nunzi, J. M. First order distributed feedback dye laser effect in reflection pumping geometry for nonlinear optical measurements. *Proc. SPIE-Int. Soc. Opt. Eng.* **2007**, *6653*, 665304/1–665304/5.
31. Shopova, S. I.; Zhou, H.; Fan, X.; Zhang, P. Optofluidic ring resonator based dye laser. *Appl. Phys. Lett.* **2007**, *90*, 221101/1–221101/3.
32. Ogino, S. Organic EL (electroluminescent) display having color conversion layer and its manufacture. Jpn. Kokai Tokkyo Koho JP 2008047493, 2008; *Chem. Abstr.* **2008**, *148*, 273265.
33. Nenchev, M.; Deneva, M. Laser device for generating fixed frequency radiation along a reference atomic absorption line. Bulg. Pat. Appl. BG 109300, 2007; *Chem. Abstr.* **2007**, *147*, 511120.
34. Hama, T. Methods for manufacture color conversion filters, their substrates, and multicolor light-emitting devices. Jpn. Kokai Tokkyo Koho JP 2007213993, 2007; *Chem. Abstr.* **2007**, *147*, 265503.
35. Ilchishin, I. P.; Maslov, P. Yu.; Tikhonov, E. A.; Lipnitsky, S. O.; Stepanov, A. A. Lasing in dye-doped nematic liquid crystals at a dynamic distributed feedback for two-scheme excitation. *Mol. Cryst. Liq. Cryst.* **2007**, *467*, 235–245.
36. Jung, B. J.; Lee, M. J.; Kim, D. W.; Lee, C. H.; Kim, J. Y. Photovoltaic device. U.S. Pat. Appl. Publ. US 2007144579, 2007; *Chem. Abstr.* **2007**, *147*, 121863.
37. Palomino-Merino, R.; Torres-Kauffman, J.; Lozada-Morales, R.; Portillo-Moreno, O.; Garcia-Rocha, M.; Zelaya-Angel, O. Photoluminescence of rhodamine 6G-doped amorphous TiO_2 thin films grown by sol–gel. *Vacuum* **2007**, *81*, 1480–1483.
38. Kuzyk, A.; Pettersson, M.; Toppari, J. J.; Hakala, T. K.; Tikkanen, H.; Kunttu, H.; Torma, P. Molecular coupling of light with plasmonic waveguides. *Opt. Express* **2007**, *15*, 9908–9917.
39. Johnson, F. M. Carcinogenic chemical-response "fingerprint" for male F344 rats exposed to a series of 195 chemicals: implications for predicting carcinogens with transgenic models. *Environ. Mol. Mutagen.* **1999**, *34*, 234–245.
40. Albert, R. E. Allergic contact sensitizing chemicals as environmental carcinogens. *Environ. Health Perspect.* **1997**, *105*, 940–948.
41. Nestmann, E. R.; Douglas, G. R.; Matula, T. I.; Grant, C. E.; Kowbel, D. J. Mutagenic activity of rhodamine dyes and their impurities as detected by mutation induction in *Salmonella* and DNA damage in Chinese hamster ovary cells. *Cancer Res.* **1979**, *39*, 4412–4417.
42. Benoit-Guyod, J. L.; Rochat, J.; Alary, J.; Andre, C.; Taillandier, G. Correlations between physico-chemical properties and ecotoxicity of fluorescent xanthenic water tracers. *Toxicol. Eur. Res.* **1979**, *2*, 241–246.
43. Kirkland, D.; Aardema, M.; Henderson, L.; Mueller, L. Evaluation of the ability of a battery of three *in vitro* genotoxicity tests to discriminate rodent carcinogens and non-carcinogens. I. Sensitivity, specificity and relative predictivity. *Mutat. Res.* **2005**, *584*, 1–256.
44. Serafimova, R.; Todorov, M.; Pavlov, T.; Kotov, S.; Jacob, E.; Aptula, A.; Mekenyan, O. Identification of the structural requirements for mutagenicity, by incorporating molecular flexibility and metabolic activation of chemicals. II. General Ames mutagenicity model. *Chem. Res. Toxicol.* **2007**, *20*, 662–676.
45. Zeiger, E.; Anderson, B.; Haworth, S.; Lawlor, T.; Mortelmans, K.; Speck, W. *Salmonella* mutagenicity tests. III. Results from the testing of 255 chemicals. *Environ. Mutagen.* **1987**, *9*, 1–109.

RHODAMINE 123

CAS Registry Number 62669-70-9

Chemical Structure

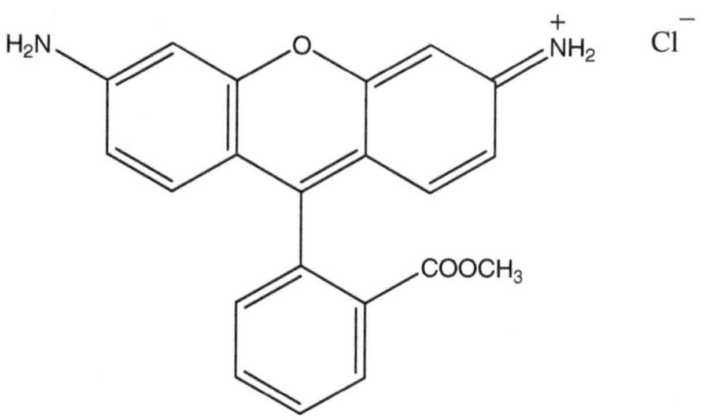

CA Index Name Xanthylium, 3,6-diamino-9-[2-(methoxycarbonyl)phenyl]-, chloride (1:1)

Other Names 2-(6-Amino-3-imino-3*H*-xanthen-9-yl) benzoic acid methyl ester; Benzoic acid, 2-(6-amino-3-imino-3*H*-xanthen-9-yl), methyl ester, monohydrochloride; 3,6-Diamino-9-[2-(methoxycarbonyl)phenyl]xanthylium chloride; Xanthylium, 3,6-diamino-9-[2-(methoxycarbonyl)phenyl]-, chloride; R 22420; R 302; RH 123; Rhodamine 123

Merck Index Number Not listed

Chemical/Dye Class Xanthene

Molecular Formula $C_{21}H_{17}ClN_2O_3$

Molecular Weight 380.83

Physical Form Brownish-red crystals or powder

Solubility Slightly soluble in water; soluble in methanol, ethanol, ether, *N,N*-dimethylformamide

Melting Point 235 °C

Absorption (λ_{max}) 501 nm, 507 nm

Emission (λ_{max}) 529 nm

Synthesis Synthetic methods[1–6]

Staining Applications Mitochondria;[7–9] cells;[10,11] P-glycoprotein;[12–14] sperms[15]

Biological Applications Measuring membrane potential;[16] detecting cancer cells,[17] spores,[17] prostate cancer,[18] stress biomarkers;[19] treating disc degenerative disease,[20] epilepsy,[21] erectile dysfunction;[22] apoptosis assays;[23] tumor cell multidrug resistance assay;[24] implantable medical devices[25]

Industrial Applications Colored bubbles;[26] paints[27]

Safety/Toxicity Acute toxicity;[28] carcinogenicity;[29,30] cytotoxocity;[31,32] metabolic toxicity;[33] mutagenicity;[34] neurotoxicity;[35] ocular toxicity[36]

REFERENCES

1. Ross, J. A.; Ross, B. P.; Rubinsztein-Dunlop, H.; McGeary, R. P. Facile synthesis of rhodamine esters using acetyl chloride in alcohol solution. *Synth. Commun.* **2006**, 36, 1745–1750.
2. Spassova, M.; Koid, P.; Watanabe, K. Rhodanol synthons for fluorescent labeling and drug conjugates. *Collect. Czech. Chem. Commun.* **1996**, *61*, S290–S293.
3. Vora, M. M.; Lambrecht, R. M. Optimized synthesis of radioiodinated rhodamine-123. *J. Labelled Compd. Radiopharm.* **1989**, *27*, 789–802.
4. Thakur, M. L.; Leeper, D. B.; Rowley, R.; Park, C. H. Radioiodinated rhodamine-123: preparation and preliminary evaluation as an agent for tumor scintigraphy. *Nucl. Med. Biol.* **1988**, *15*, 517–524.
5. Kinsey, B. M.; Kassis, A. I.; Fayad, F.; Layne, W. W.; Adelstein, S. J. Synthesis and biological studies of iodinated ($^{127/125}$I) derivatives of rhodamine 123. *J. Med. Chem.* **1987**, *30*, 1757–1761.
6. Moonen, P.; Gorree, G. C. M.; Hoekstra, A. Radioiodinated rhodamine-123: a potential cationic hepatobiliary imaging agent. *Nucl. Med. Commun.* **1987**, *8*, 99–104.
7. Swayne, T. C.; Gay, A. C.; Pon, L. A. Visualization of mitochondria in budding yeast. *Methods Cell Biol.* **2007**, *80*, 591–626.
8. Lemasters, J. J.; Ramsheh, V. K. Imaging of mitochondrial polarization of depolarization with cationic fluorophores. *Methods Cell Biol.* **2007**, *80*, 283–295.
9. Hattori, F.; Fukuda, K. Method for selecting myocardial cells using intracellular mitochondria labeled with fluorescent indicator. PCT Int. Appl. WO 2006022377, 2006; *Chem. Abstr.* **2006**, *144*, 270175.
10. Hatz, S.; Lambert, J. D. C.; Ogilby, P. R. Measuring the lifetime of singlet oxygen in a single cell: addressing the issue of cell viability. *Photochem. Photobiol. Sci.* **2007**, *6*, 1106–1116.

11. Lacombe, F.; Belloc, F.; Veriac, S.; Lefevre, D. Method of discrimination of at least two cellular populations by immunofluorescence flow cytometry and application to oncopathology. Fr. Demande FR 2895087, 2007; *Chem. Abstr.* **2007**, *147*, 93934.

12. Foeger, F.; Hoyer, H.; Kafedjiiski, K.; Thaurer, M.; Bernkop-Schnuerch, A. *In vivo* comparison of various polymeric and low molecular mass inhibitors of intestinal P-glycoprotein. *Biomaterials* **2006**, *27*, 5855–5860.

13. Chaoui, D.; Faussat, A. M.; Majdak, P.; Tang, R.; Perrot, J. Y.; Pasco, S.; Klein, C.; Marie, J. P.; Legrand, O. JC-1, a sensitive probe for a simultaneous detection of P-glycoprotein activity and apoptosis in leukemic cells. *Cytometry* **2006**, *70B*, 189–196.

14. Constable, P. A.; Lawrenson, J. G.; Dolman, D. E. M.; Arden, G. B.; Abbott, N. J. P-Glycoprotein expression in human retinal pigment epithelium cell lines. *Exp. Eye Res.* **2006**, *83*, 24–30.

15. Fraser, L.; Dziekonska, A.; Strzezek, R.; Strzezek, J. Dialysis of boar semen prior to freezing–thawing: its effects on post-thaw sperm characteristics. *Theriogenology* **2007**, *67*, 994–1003.

16. Deutsch, M.; Namer, Y. A method and system for measuring membrane potential based on fluorescence polarization. PCT Int. Appl. WO 2007122602, 2007; *Chem. Abstr.* **2007**, *147*, 498288.

17. Levy, M.; Ellington, A. D.; Pai, S. Proximity ligation assays with peptide conjugate 'burrs' and aptamers for the sensitive detection of spores and cancer cells. PCT Int. Appl. WO 2007027796, 2007; *Chem. Abstr.* **2007**, *146*, 312239.

18. Dickman, D. Methods of detecting prostate cancer. PCT Int. Appl. WO 2006054296, 2006; *Chem. Abstr.* **2006**, *145*, 3346.

19. Cullum, M. E.; Duplessis, C. A.; Crepeau, L. J. Method for the detection of stress biomarkers including cortisol by fluorescence polarization. U. S. Pat. Appl. Publ. US 2006105397, 2006; *Chem. Abstr.* **2006**, *144*, 463810.

20. Ichim, T. E. Treatment of disc degenerative disease using cells able to increase angiogenesis alone or in combination with growth factors or a matrix and compositions for same. PCT Int. Appl. WO 2007136673, 2007; *Chem. Abstr.* **2007**, *148*, 24846.

21. Nedergaard, M.; Tian, G. F. Method for treatment and prevention of epilepsy. PCT Int. Appl. WO 2006062683, 2006; *Chem. Abstr.* **2006**, *145*, 55958.

22. Ichim, T. E. Treatment of erectile dysfunction by stem cell therapy. PCT Int. Appl. WO 2007149548, 2007; *Chem. Abstr.* **2007**, *148*, 52294.

23. Ferlini, C.; Scambia, G. Assay for apoptosis using the mitochondrial probes, rhodamine123 and 10-*N*-nonyl acridine orange. *Nat. Protoc.* **2007**, *2*, 3111–3114.

24. Loken, M. R. Immunocytometric tumor cell multidrug resistance assay. U.S. Pat. Appl. Publ. US 2006263834, 2006; *Chem. Abstr.* **2006**, *145*, 500017.

25. Fischer, F. J.; Miller, J. W.; Andrews, M. O. Implantable medical device with anti-neoplastic drug. U.S. Pat. Appl. Publ. US 2006030826, 2006; *Chem. Abstr.* **2006**, *144*, 198874.

26. Sabnis, R. W.; Kehoe, T. D. Composition and method for producing colored bubbles. U.S. Pat. Appl. Publ. US 2006004110, 2006; *Chem. Abstr.* **2006**, *144*, 110164.

27. Sabnis, R. W.; Kehoe, T. D.; Balchunis, R. J. Paint compositions comprising color changing phthalein acid–base indicator. PCT Int. Appl. WO 2006105191, 2006; *Chem. Abstr.* **2006**, *145*, 403489.

28. Hartig, S.; Fries, S.; Balcarcel, R. R. Reduced mitochondrial membrane potential and metabolism correspond to acute chloroform toxicity of *in vitro* hepatocytes. *J. Appl. Toxicol.* **2005**, *25*, 310–317.

29. Vasconcelos, F. C.; Cavalcanti, G. B.; Silva, K. L.; de Meis, E.; Kwee, J. K.; Rumjanek, V. M.; Maia, R. C. Contrasting features of MDR phenotype in leukemias by using two fluorochromes: implications for clinical practice. *Leuk. Res.* **2007**, *31*, 445–454.

30. Arcadi, J. A. The effect of rhodamine-123 on 3 prostate tumors from the rat. *J. Urol.* **1998**, *160*, 2402–2406.

31. Pourahmad, J.; Rabiei, M.; Jokar, F.; O'brien, P. J. A comparison of hepatocyte cytotoxic mechanisms for chromate and arsenite. *Toxicology* **2005**, *206*, 449–460.

32. Rahn, C. A.; Bombick, D. W.; Doolittle, D. J. Assessment of mitochondrial membrane potential as an indicator of cytotoxicity. *Fundam. Appl. Toxicol.* **1991**, *16*, 435–448.

33. Downing, T. W.; Garner, D. L.; Ericsson, S. A.; Redelman, D. Metabolic toxicity of fluorescent stains on thawed cryopreserved bovine sperm cells. *J. Histochem. Cytochem.* **1991**, *39*, 485–489.

34. Ferguson, L. R.; Baguley, B. C. Verapamil as a co-mutagen in the *Salmonella*/mammalian microsome mutagenicity test. *Mutat. Res. Lett.* **1988**, *209*, 57–62.

35. Uchida, K.; Yamada, M.; Hayashi, T.; Mine, Y.; Kawase, T. Possible harmful effects on central nervous system cells in the use of physiological saline as an irrigant during neurosurgical procedures. *Surg. Neurol.* **2004**, *62*, 96–105.

36. Bantseev, V.; McCanna, D.; Banh, A.; Wong, W. W.; Moran, K. L.; Dixon, D. G.; Trevithick, J. R.; Sivak, J. G. Mechanisms of ocular toxicity using the *in vitro* bovine lens and sodium dodecyl sulfate as a chemical model. *Toxicol. Sci.* **2003**, *73*, 98–107.

RHODZIN 3

CAS Registry Number 677716-65-3

Chemical Structure

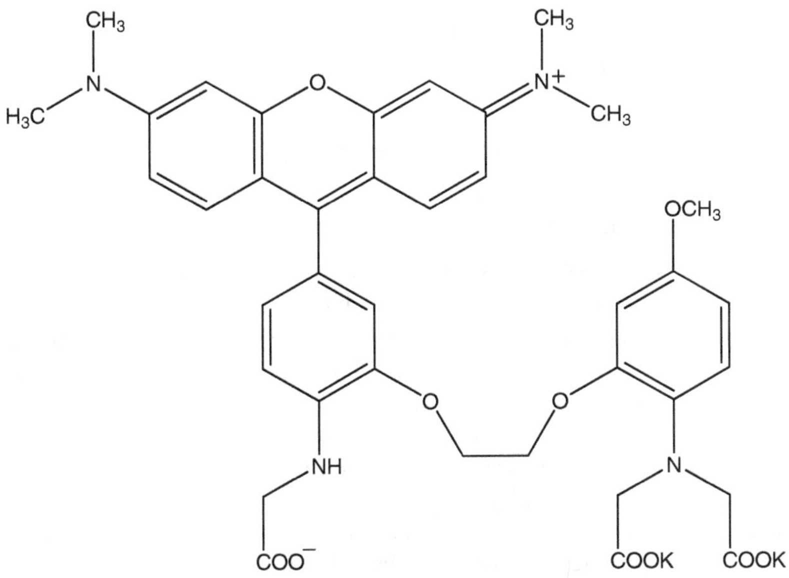

Molecular Formula $C_{38}H_{38}K_2N_4O_{10}$

Molecular Weight 788.94

Physical Form Solid

Solubility Soluble in water

CA Index Name Xanthylium, 9-[3-[2-[2-[bis(carboxymethyl)amino]-5-methoxyphenoxy]ethoxy]-4-[(carboxymethyl)amino]phenyl]-3,6-bis(dimethylamino)-, inner salt, potassium salt (1:2)

Other Names Xanthylium, 9-[3-[2-[2-[bis(carboxymethyl)amino]-5-methoxyphenoxy]ethoxy]-4-[(carboxymethyl)amino]phenyl]-3,6-bis(dimethylamino)-, inner salt, dipotassium salt; RhodZin 3; RhodZin 3 dipotassium salt

Merck Index Number Not listed

Chemical/Dye Class Xanthene

Melting Point >250 °C

Absorption (λ_{max}) 549 nm

Emission (λ_{max}) 576 nm

Synthesis Synthetic methods[1,2]

Staining Applications Zinc ions[1–3]

Biological Applications Zinc indicator[1–3]

Industrial Applications Not reported

Safety/Toxicity No data available

REFERENCES

1. Gee, K. R. Zinc binding compounds and their method of use. U.S. Pat. Appl. Publ. US 2005250214, 2005; *Chem. Abstr.* **2005**, *143*, 472091.
2. Sensi, S. L.; Ton-That, D.; Weiss, J. H.; Rothe, A.; Gee, K. R. A new mitochondrial fluorescent zinc sensor. *Cell Calcium* **2003**, *34*, 281–284.
3. Krezel, A.; Maret, W. Dual nanomolar and picomolar Zn(II) binding properties of metallothionein. *J. Am. Chem. Soc.* **2007**, *129*, 10911–10921.

ROSE BENGAL

CAS Registry Number 632-69-9

Chemical Structure

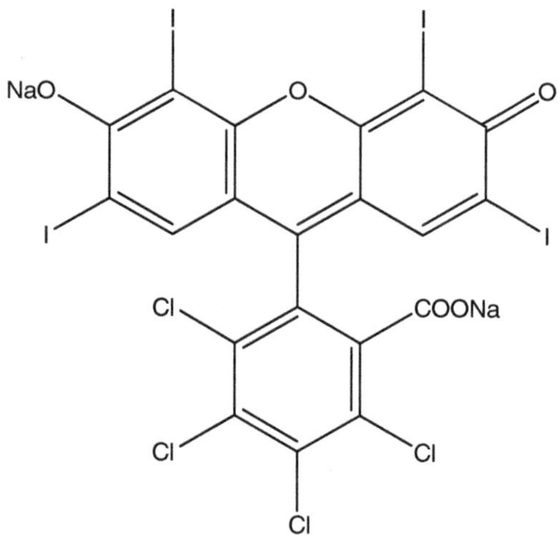

CA Index Name Spiro[isobenzofuran-1(3*H*),9′-[9*H*]xanthen]-3-one, 4,5,6,7-tetrachloro-3′,6′-dihydroxy-2′,4′,5′,7′-tetraiodo-, sodium salt (1:2)

Other Names C.I. 45440; C.I. Acid Red 94; Bengal Rose B sodium salt; Fluorescein, 4,5,6,7-tetrachloro-2′,4′,5′,7′-tetraiodo-, disodium salt; Spiro[isobenzofuran-1(3*H*),9′-[9*H*]xanthen]-3-one, 4,5,6,7-tetrachloro-3′,6′-dihydroxy-2′,4′,5′,7′-tetraiodo-, disodium salt; 2′,4′,5′,7′-Tetraiodo-3,4,5,6-tetrachlorofluorescein disodium salt; 3′,4′,5′,6′-Tetrachloro-2,4,5,7-tetraiodofluorescein disodium salt; Japan Red 105-1; Rose Bengal Extra; Rose Bengal disodium salt; Rose Bengal extra; Rose Bengal sodium; Rose Bengal sodium salt; Sodium tetraiodotetrachlorofluorescein

Merck Index Number 8262

Chemical/Dye Class Xanthene

Molecular Formula $C_{20}H_2Cl_4I_4Na_2O_5$

Molecular Weight 1017.64

Physical Form Red-brown powder

Solubility Soluble in water, ethanol

Melting Point >200 °C

pK_a 3.9, 4.7

Absorption (λ_{max}) 548 nm

Emission (λ_{max}) 567 nm

Synthesis Synthetic methods[1–4]

Staining Applications Bacteria;[5] apoptotic cells;[9] bacterial spores;[17] proteins;[6] hairs[7,8]

Biological Applications Apoptosis assay;[9,10] diagnosis of diseases related to amyloid accumulation;[11] controlling plant diseases;[12] identifying fungi;[12,13] treating skin,[14] mouth,[14] digestive tract,[14] urinary tract,[14] reproductive tract,[14] respiratory tract,[14] circulatory system,[14] head and neck,[14] endocrine system,[14] lymphoreticular system,[14] rhinitis,[15] asthma,[15] urticaria,[15] atopic dermatitis,[15] anaphylactic shock;[15] photodynamic therapy;[16,17] pesticides;[18] insecticides;[19] fungicides[20]

Industrial Applications Organic photodetectors;[21] memory devices;[22] solar cells;[23] semiconductor film;[24] imaging method;[25] optical filters;[26] antireflection films;[26] as polymerization photoinitiators;[27] recording materials;[28] thermoplastics;[29] cements;[30] photographic materials[31]

Safety/Toxicity Genotoxicity;[32] microbial toxicity[33]

Certification/Approval Certified by Biological Stain Commission (BSC)

REFERENCES

1. Sugita, N.; Kawabata, K.; Sasaki, K.; Sakata, I.; Umemura, S. Synthesis of amphiphilic derivatives of rose bengal and their tumor accumulation. *Bioconjugate Chem.* **2007**, *18*, 866–873.
2. Paczkowski, J.; Paczkowska, B.; Neckers, D. C. Photolysis products of Rose Bengal lactone diacetate. *J. Photochem. Photobiol., A Chem.* **1991**, *61*, 131–136.
3. Amt-Guerri, F.; Lopez-Gonzalez, M. M. C.; Martinez-Utrilla, R.; Sastra, R. Synthesis and spectroscopic properties of new Rose Bengal and Eosin Y derivatives. *Dyes Pigm.* **1990**, *12*, 249–272.
4. Johansson, S. Analysis and purification of Rose Bengal Sodium for use as reference substance and in pharmaceutical preparations. *Svensk Farm. Tidskrift* **1973**, *77*, 641–647; *Chem. Abstr.* **1974**, *80*, 19447.
5. Jemli, M.; Alouini, Z.; Sabbahi, S.; Gueddari, M. Destruction of fecal bacteria in wastewater by three photosensitizers. *J. Environ. Monit.* **2002**, *4*, 511–516.
6. Kosaka, H.; Sakamoto, H. Development of a colorimetric quantitation method of protein in solution by excluding the influence of creatinine.

PCT Int. Appl. WO 2005071420, 2005; *Chem. Abstr.* **2005**, *143*, 188838.

7. Shinkai, M. Oxidative hair dye first compositions. Jpn. Kokai Tokkyo Koho JP 2004059468, 2004; *Chem. Abstr.* **2004**, *140*, 186967.
8. Kanda, N. Acidic hair dye compositions containing specified acidic dyes, acids, and aromatic alcohols. Jpn. Kokai Tokkyo Koho JP 2004018505, 2004; *Chem. Abstr.* **2004**, *140*, 116929.
9. Meyer, M.; Essack, M.; Kanyanda, S.; Rees, J. G. A low-cost flow cytometric assay for the detection and quantification of apoptosis using an anionic halogenated fluorescein dye. *BioTechniques* **2008**, *45*, 317–318, 320.
10. Elliott, R. J. Apoptosis assay. Brit. UK Pat. Appl. GB 2356929, 2001; *Chem. Abstr.* **2001**, *135*, 254107.
11. Kudo, K.; Suemoto, T.; Suzuki, M.; Tojo, H.; Shimazu, H. Pyronine B analogs as imaging agents and probes for diagnosis of diseases related to amyloid accumulation. Jpn. Kokai Tokkyo Koho JP 2000344684, 2000; *Chem. Abstr.* **2000**, *134*, 27006.
12. Takahi, Y.; Kato, S.; Kondo, Y.; Takeshiba, H.; Kajino, H.; Ookochi, T. Safe microbicide compositions containing F-155 and Rose Bengal, coating of seeds with the compositions, control of plant diseases with them, and seeds coated with Rose Bengal. Jpn. Kokai Tokkyo Koho JP 07228505, 1995; *Chem. Abstr.* **1995**, *123*, 332734.
13. Li, H. Fungal culture media for identification, count, separation, and subculture of fungi. Faming Zhuanli Shenqing Gongkai Shuomingshu CN 1834223, 2006; *Chem. Abstr.* **2006**, *145*, 414006.
14. Dees, H. C.; Scott, T. C.; Wachter, E. A.; Fisher, W. G.; Smolik, J. Intracorporeal medicaments for high energy phototherapeutic treatment of disease. U.S. Pat. Appl. Publ. US 2007208076, 2007; *Chem. Abstr.* **2007**, *147*, 317211.
15. Cheng, Y. S. E.; Liu, Y.; Chu, J.; Kinet, J. P.; Jouvin, M. H.; Sudo, Y.; Qian, X. Inhibition of the binding of human IgE to its receptor by tetracyclic compounds for the alleviation of IgE-mediated immune response. PCT Int. Appl. WO 9740033, 1997; *Chem. Abstr.* **1997**, *128*, 368.
16. Dees, H. C.; Scott, T. C.; Smolik, J.; Wachter, E. A.; Fisher, W. G. Topical halogenated xanthenes for photodynamic treatment. U.S. Pat. Appl. Publ. US 2008118578, 2008; *Chem. Abstr.* **2008**, *148*, 568986.
17. Demidova, T. N.; Hamblin, M. R. Photodynamic inactivation of *Bacillus* spores, mediated by phenothiazinium dyes. *Appl. Environ. Microbiol.* **2005**, *71*, 6918–6925.
18. Ian Malcolm, S.; Wright, T.; Perrin, B.; Mulqueen, P.; Waller, A.; Pierce, A. Capsule formulation containing pesticide and photoprotectant. PCT Int. Appl. WO 2006089747, 2006; *Chem. Abstr.* **2006**, *145*, 286603.
19. Alder, A.; Rindlisbacher, A.; Streibert, H. P.; Baenninger, R. *N*-Phenylthiourea insecticides and photosensitizer enhancers. Eur. Pat. Appl. EP 390743, 1990; *Chem. Abstr.* **1991**, *115*, 177449.
20. Thielert, W.; Scheinpflug, H.; Lubosch, J. Triazole-containing synergistic fungicidal seed treatment agent. Eur. Pat. Appl. EP 297426, 1989; *Chem. Abstr.* **1989**, *110*, 168093.
21. Sahu, S.; Batabyal, S. K.; Pal, A. J. Organic photodetectors with electrically bistable electron acceptors and nanotubes. *Appl. Phys. Lett.* **2007**, *90*, 142112/1–142112/3.
22. Jakobsson, F. L. E.; Crispin, X.; Coelle, M.; Buechel, M.; de Leeuw, D. M.; Berggren, M. On the switching mechanism in Rose Bengal-based memory devices. *Org. Electron.* **2007**, *8*, 559–565.
23. Pradhan, B.; Batabyal, S. K.; Pal, A. J. Vertically aligned ZnO nanowire arrays in Rose Bengal-based dye-sensitized solar cells. *Sol. Energy Mater. Sol. Cells* **2007**, *91*, 769–773.
24. Takeshita, K.; Sasaki, Y.; Kobashi, M.; Tanaka, Y.; Maeda, S.; Yamakata, A.; Ishibashi, T.; Onishi, H. Photophysics and electron dynamics in dye-sensitized semiconductor film studied by time-resolved mid-IR spectroscopy. *J. Phys. Chem.* **2003**, *107*, 4156–4161.
25. Barr, R. K.; O'Connor, C. Imaging methods. U.S. Pat. Appl. Publ. US 2007117042, 2007; *Chem. Abstr.* **2007**, *146*, 531343.
26. Yabuki, Y. Optical filters and antireflection films. Jpn. Kokai Tokkyo Koho JP 2000352613, 2000; *Chem. Abstr.* **2000**, *134*, 78720.
27. Burget, D.; Fouassier, J. P.; Amat-Guerri, F.; Mallavia, R.; Sastre, R. Enhanced activity as polymerization photoinitiators of Rose Bengal and Eosin esters with an *O*-benzoyl-α-oxooxime group: the role of the excited state reactivity. *Acta Polym.* **1999**, *50*, 337–346.
28. Kanda, N.; Nakamura, M. Thermal recording materials colored with specific colorant. Jpn. Kokai Tokkyo Koho JP 06092027, 1994; *Chem. Abstr.* **1994**, *121*, 145456.
29. Saenger, D.; Barzynski, H. Light-degradable thermoplastics. Ger. Offen. DE 2209138, 1973; *Chem. Abstr.* **1974**, *80*, 27904.

30. Allam, P. C. Cements for colorless, transparent, photoactivatable hydrophobic sheet material. Brit. GB 1144547, 1969; *Chem. Abstr.* **1969**, *70*, 107093.

31. Tani, T.; Honda, K.; Kikuchi, S. Studies on spectral sensitization and desensitization in photography. XIII. Discussions on spectral sensitization of dyes from electronic energy levels, photographic half-wave potentials, and excitation energies. *Kogyo Kagaku Zasshi* **1968**, *71*, 42–47; *Chem. Abstr.* **1968**, *68*, 118488.

32. Sasaki, Y. F.; Kawaguchi, S.; Kamaya, A.; Ohshita, M.; Kabasawa, K.; Iwama, K.; Taniguchi, K.; Tsuda, S. The comet assay with 8 mouse organs: results with 39 currently used food additives. *Mutat. Res.* **2002**, *519*, 103–119.

33. Oros, G.; Cserhati, T.; Forgacs, E. Separation of the strength and selectivity of the microbiological effect of synthetic dyes by spectral mapping technique. *Chemosphere* **2003**, *52*, 185–193.

SAFRANIN O

CAS Registry Number 477-73-6

Chemical Structure

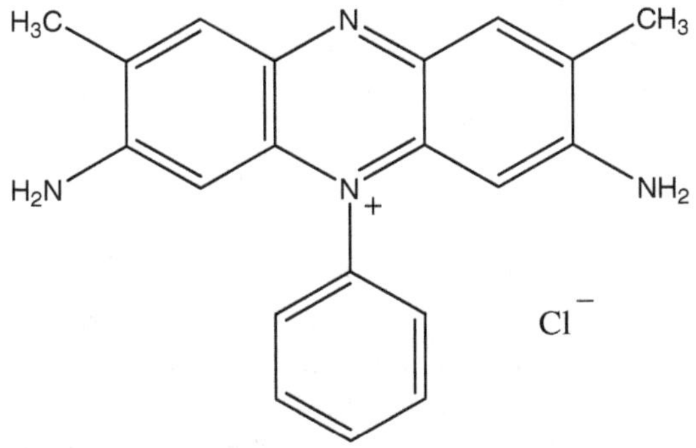

CA Index Name Phenazinium, 3,7-diamino-2,8-dimethyl-5-phenyl-, chloride (1 : 1)

Other Names C.I. Basic Red 2; Phenazinium, 3,7-diamino-2,8-dimethyl-5-phenyl-, chloride; Safranine O; 2,8-Dimethylphenosafranine; 3,7-Diamino-2,8-dimethyl-5-phenylphenazinium chloride; Basic Pink; Basic Red 2; Brilliant Safranine BR; Brilliant Safranine G; Brilliant Safranine GR; C.I. 50240; Calcozine Red Y; Cotton Red; Duasyn Basic Red TH; Gossypimine; Hidaco Safranine; Leather Red HT; Lowacryl Red 2; Lowacryl Red 2 Conc; Mitsui Safranine T; Nippon Kagaku Safranine GK; Nippon Kagaku Safranine T; Safranin; Safranin O; Safranin T; Safranine; Safranine A; Safranine B; Safranine G; Safranine GF; Safranine J; Safranine OK; Safranine OK 70:100; Safranine Superfine G; Safranine T; Safranine TH; Safranine TN; Safranine TS; Safranine Y; Safranine YN; Safranine Zh; Tolusafranine

Merck Index Number Not listed

Chemical/Dye Class Phenazine

Molecular Formula $C_{20}H_{19}ClN_4$

Molecular Weight 350.84

Physical Form Dark red to dark green powder

Solubility Soluble in water, ethanol, ethylene glycol, methyl cellosolve, pyridine

Melting Point >240 °C (decompose)

pK_a 6.4

Absorption (λ_{max}) 530 nm

Synthesis Synthetic methods[1–16]

Staining Applications Antigen;[17] bacteria;[18] brain;[19] cellulose;[20] lignin;[20] mitochondria;[21,22] nucleated and non-nucleated blood cells;[28] nucleic acids;[23,24] proteins;[25] spinal cord;[19] hairs[26,27]

Biological Applications Hematotoxicity assays;[28] measuring membrane potential;[21,22,29] detecting microorganisms;[30] treating diabetes-associated pain,[31] mechanical allodynia,[32] oncological diseases;[33] food packaging materials[34]

Industrial Applications Photogalvanic cells;[35] polymeric electro-mechanic devices;[36] recording material;[37] inks;[38,39] textiles[40]

Safety/Toxicity Bacterial toxicity;[41] carcinogenicity;[42–44] environmental toxicity;[45] microbial toxicity;[46] mitochondrial toxicity;[47] mutagenicity;[48] nucleic acid damage[49,50]

Certification/Approval Certified by Biological Stain Commission (BSC)

REFERENCES

1. Ali, M.; Sarkar, A.; Pandey, M. D.; Pandey, S. Efficient precipitation of dyes from dilute aqueous solutions of ionic liquids. *Anal. Sci.* **2006**, *22*, 1051–1053.
2. Li, T.; Pan, Q. Synthesis of safranine O. *Yingyong Huagong* **2005**, *34*, 437–439; *Chem. Abstr.* **2005**, *144*, 451990.
3. Chen, M.; Liao, W. Synthesis of safranine O. *Huaxue Shiji* **1986**, *8*, 164–167; *Chem. Abstr.* **1986**, *105*, 154658.
4. Shigehara, K.; Matsunaga, H.; Tsuchida, E. Synthesis and properties of polymeric azine and thiazine dyes. *J. Polym. Sci., Polym. Chem. Ed.* **1978**, *16*, 1853–1863.
5. Corbett, J. F. Mechanism of formation of phenosafranine dyes. *J. Soc. Dyers Colourists* **1972**, *88*, 438–443.
6. Mizuta, M. Separation and utilization of *o*- and *p*-nitrotoluene from the mononitrotoluenes prepared from the gasoline fraction of Syukkoko crude oil. III. Preparation of safranine and magenta: total

conclusions. *Kogyo Kagaku Zasshi* **1935**, *38*, 723–724; *Chem. Abstr.* **1936**, *30*, 15064.

7. Stiehler, R. D.; Chen, T.; Clark, W. M. Studies on oxidation–reduction. XVIII. Simple safranines. *J. Am. Chem. Soc.* **1933**, *55*, 891–908.
8. Cobenzl, A. Nitrosodialkylaniline, dyes therefrom, safranine and Meldola blue. *Chem.-Ztg.* **1926**, *50*, 494–495.
9. Cobenzl, A. Preparation of mauveine and various bases. *Oesterreich. Chem.-Ztg.* **1925**, *28*, 25–28; *Chem. Abstr.* **1925**, *19*, 19760.
10. Zoller, H. F. Oxalyl chloride in the synthesis of the triphenylmethane dyes. *Science* **1920**, *52*, 207.
11. Orlov, N. N. Synthesis of safranines. III. *J. Chem. Soc., Abstr.* **1912**, *100*, 89.
12. Orlov, N. N. Synthesis of safranines. III. *Zh. Russ. Fiz.-Khim. Obshch.* **1912**, *42*, 939–949; *Chem. Abstr.* **1912**, *6*, 14346.
13. Shaposhnikov, V.; Orlov, N. Synthesis of the simpler safranines. *Zh. Russ. Fiz.-Khim. Obshch.* **1911**, *42*, 512–522; *Chem. Abstr.* **1911**, *5*, 7768.
14. Shaposhnikov, V. Synthesis of safranine. *Zh. Russ. Fiz.-Khim. Obshch.* **1911**, *42*, 505–512; *Chem. Abstr.* **1911**, *5*, 7767.
15. Barbier, P.; Sisley, P. On a method of preparing the symmetrical safranines. *Bull. Soc. Chim.* **1907**, *35*, 1278–1282; *Chem. Abstr.* **1907**, *1*, 3480.
16. Salm, E. A study of indicators. *Z. Physik. Chem.* **1906**, *57*, 471–501; *Chem. Abstr.* **1907**, *1*, 2788.
17. Kaushik, K.; Khurana, S.; Wanchu, A.; Malla, N. Evaluation of staining techniques, antigen detection and nested PCR for the diagnosis of cryptosporidiosis in HIV seropositive and seronegative patients. *Acta Tropica* **2008**, *107*, 1–7.
18. Pandolfi, D.; Pons, M. N.; da Motta, M. Characterization of PHB storage in activated sludge extended filamentous bacteria by automated colour image analysis. *Biotechnol. Lett.* **2007**, *29*, 1263–1269.
19. Augulis, V.; Sigg, E. B. Supravital staining and fixation of brain and spinal cord by intravascular perfusion. *Stain Technol.* **1971**, *46*, 183–190.
20. De Micco, V.; Aronne, G. Combined histochemistry and autofluorescence for identifying lignin distribution in cell walls. *Biotech. Histochem.* **2007**, *82*, 209–216.
21. Bachurin, S. O.; Dubova, L. G.; Kireeva, E. G.; Shevtsova, E. F. Screening assay using safranin dye for determining the effect of cytoprotectants and cytostatics on mitochondrial permeability and membrane potential. Russ. RU 2315998, 2008; *Chem. Abstr.* **2008**, *148*, 206572.
22. Feldkamp, T.; Kribben, A.; Weinberg, J. M. Assessment of mitochondrial membrane potential in proximal tubules after hypoxia–reoxygenation. *Am. J. Physiol.* **2005**, *288*, F1092–F1102.
23. Sun, W.; Zhao, N.; Jiao, K. Determination of yeast RNA using safranine T with linear sweep voltammetry. *Acta Chim. Sloven.* **2007**, *54*, 778–783.
24. Sun, W.; You, J. Y.; Jiao, K. Electrochemical detection of DNA based on its interaction with safranine T. *J. Indian Chem. Soc.* **2007**, *84*, 171–175.
25. Keillor, J. W.; Michnick, S. W.; Girouard, S. Fluorescent labeling of specific protein targets *in vitro* and *in vivo*. U.S. Pat. Appl. Publ. US 2006147948, 2006; *Chem. Abstr.* **2006**, *145*, 120058.
26. Hoeffkes, H.; Bossmann, B. Use of coated alkali agents for oxidative hair bleaching and dyeing compositions to prevent foaming. Eur. Pat. Appl. EP 1752191, 2007; *Chem. Abstr.* **2007**, *146*, 212227.
27. Hercouet, L. Hair dye composition comprising a derivative of diamino-*N,N*-dihydropyrazolone, a coupler, and a heterocyclic direct dye. Fr. Demande FR 2886132, 2006; *Chem. Abstr.* **2006**, *146*, 49738.
28. Dertinger, S. D.; Bemis, J. C.; Bryce, S. M. Method for measuring *in vivo* hematotoxicity with an emphasis on radiation exposure assessment. U.S. Pat. Appl. Publ. US 2008311586, 2008; *Chem. Abstr.* **2008**, *150*, 50134.
29. Agarwal, A. Measuring $\Delta\Psi$ m in isolated tubules. *Am. J. Physiol.* **2005**, *288*, F1090–F1091.
30. Martin, S. M.; MacDonald, J. G.; Lye, J.; Sayre, C.; Thompson, K. Rapid assessment of upper respiratory conditions and detection of microorganisms using test strip comprising at least one broad spectrum indicator. PCT Int. Appl. WO 2009027855, 2009; *Chem. Abstr.* **2009**, *150*, 278103.
31. Buschman, H. H. Use of compounds binding to the sigma receptor for the treatment of diabetes-associated pain. Eur. Pat. Appl. EP 1787679, 2007; *Chem. Abstr.* **2007**, *146*, 514793.
32. Baeyens Cabrera, J. M. Use of compounds active on the sigma receptor for the treatment of mechanical allodynia. PCT Int. Appl. WO 2006010587, 2006; *Chem. Abstr.* **2006**, *144*, 184692.
33. Kutushov, M. V. Use of naphthalene derivatives in form of agents for treating oncological diseases. PCT Int. Appl. WO 2009028983, 2009; *Chem. Abstr.* **2009**, *150*, 290583.

34. Kawashima, M.; Kamisaka, K.; Tanaka, S.; Wada, T.; Endo, H. Color-changing oxygen detectors applied to freshness-keeping food packaging materials. Jpn. Kokai Tokkyo Koho JP 2007121007, 2007; *Chem. Abstr.* **2007**, *146*, 492118.

35. Genwa, K. R.; Khatri, N. C. Comparative study of photosensitizing dyes in photogalvanic cells for solar energy conversion and storage: Brij-35-diethylenetriamine pentaacetic acid (DTPA) system. *Energy Fuels* **2009**, *23*, 1024–1031.

36. Lira, L. M.; Cordoba de Torresi, S. I. Polymeric electro-mechanic devices applied to antibiotic-controlled release. *Sens. Actuators, B* **2008**, *B130*, 638–644.

37. Knocke, F. Photopolymerizable composition suitable for holographic recording material. PCT Int. Appl. WO 2005124456, 2005; *Chem. Abstr.* **2005**, *144*, 78022.

38. Kawashima, M.; Uesaka, K. Oxygen-detecting ink compositions, detection substrates coated with them, laminates with the substrates, and containers and bags comprising the laminates. Jpn. Kokai Tokkyo Koho JP 2008069278, 2008; *Chem. Abstr.* **2008**, *148*, 381565.

39. Yamaguchi, N.; Sano, K. Radical indicator ink compositions, indicators for radical disinfection containing them, and detection of disinfection completion by their color change. Jpn. Kokai Tokkyo Koho JP 2006328178, 2006; *Chem. Abstr.* **2006**, *146*, 38149.

40. Li, S.; Li, L.; Spoon, J. W. Color indicator for halamine treated fabric. U.S. Pat. Appl. Publ. US 2007218562, 2007; *Chem. Abstr.* **2007**, *147*, 387512.

41. Kazama, H.; Hamashima, H.; Sasatsu, M.; Arai, T. Distribution of the antiseptic-resistance gene qacEΔ 1 in Gram-positive bacteria. *FEMS Microbiol. Lett.* **1998**, *165*, 295–299.

42. Lampidis, T. J.; Hasin, Y.; Weiss, M. J.; Chen, L. B. Selective killing of carcinoma cells "*in vitro*" by lipophilic–cationic compounds: a cellular basis. *Biomed. Pharmacother.* **1985**, *39*, 220–226.

43. Parsons, P. G.; Musk, P. Uptake and toxicity of safranine and triphenylmethylphosphonium ion in human tumor cells. *Aust. J. Exp. Biol. Med. Sci.* **1983**, *61*, 117–125.

44. Piolatto, G.; Negri, E.; La Vecchia, C.; Pira, E.; Decarli, A.; Peto, J. Bladder cancer mortality of workers exposed to aromatic amines: an updated analysis. *Br. J. Cancer* **1991**, *63*, 457–459.

45. Tonogai, Y.; Ogawa, S.; Ito, Y.; Iwaida, M. Actual survey on TLm (median tolerance limit) values of environmental pollutants, especially on amines, nitriles, aromatic nitrogen compounds and artificial dyes. *J. Toxicol. Sci.* **1982**, *7*, 193–203.

46. Oros, G.; Cserhati, T.; Forgacs, E. Separation of the strength and selectivity of the microbiological effect of synthetic dyes by spectral mapping technique. *Chemosphere* **2003**, *52*, 185–193.

47. Zhang, H.; Chen, Q. Y.; Xiang, M. L.; Ma, C. Y.; Huang, Q.; Yang, S. Y. *In silico* prediction of mitochondrial toxicity by using GA-CG-SVM approach. *Toxicol. In Vitro* **2009**, *23*, 134–140.

48. Ferguson, L. R.; Baguley, B. C. Verapamil as a co-mutagen in the *Salmonella*/mammalian microsome mutagenicity test. *Mutat. Res.* **1988**, *209*, 57–62.

49. McCarroll, N. E.; Piper, C. E.; Keech, B. H. An *E. coli* microsuspension assay for the detection of DNA damage induced by direct-acting agents and promutagens. *Environ. Mutagen.* **1981**, *3*, 429–444.

50. Rhee, Y.; Valentine, M. R.; Termini, J. Oxidative base damage in RNA detected by reverse transcriptase. *Nucleic Acids Res.* **1995**, *23*, 3275–3282.

SBFI

CAS Registry Number 124549-08-2

Chemical Structure

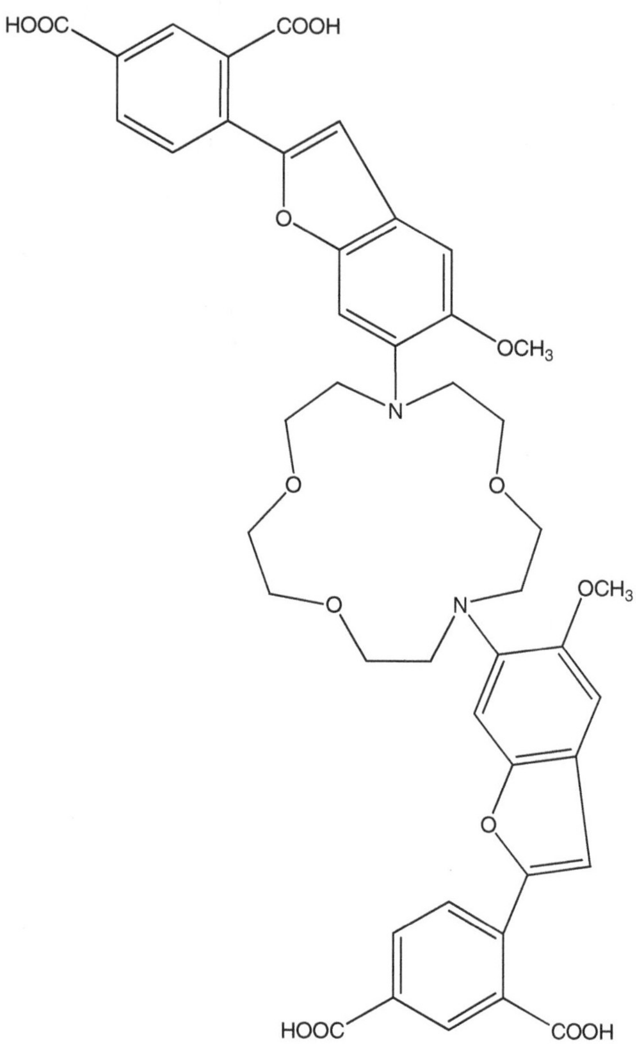

CA Index Name 1,3-Benzenedicarboxylic acid, 4,4′-[1,4,10-trioxa-7,13-diazacyclopentadecane-7,13-diylbis(5-methoxy-6,2-benzofurandiyl)]bis-

Other Names 1,4,10-Trioxa-7,13-diazacyclopentadecane, 1,3-benzenedicarboxylic acid derivative; SBFI

Merck Index Number Not listed

Chemical/Dye Class Benzofuran

Molecular Formula $C_{44}H_{42}N_2O_{15}$

Molecular Weight 838.81

Physical Form Solid

Solubility Soluble in methanol, dimethyl sulfoxide

Melting Point >200 °C

Boiling Point (Calcd.) 1034.1 ± 65.0 °C, pressure: 760 Torr

pK_a (Calcd.) 2.96 ± 0.36, most acidic, temperature: 25 °C; 4.09 ± 0.40, most basic, temperature: 25 °C

Absorption (λ_{max}) 339 nm

Emission (λ_{max}) 565 nm

Synthesis Synthetic methods[1,2]

Staining Applications Sodium ions;[1–32] cells[34]

Biological Applications Sodium indicator;[1–32] identifying genes;[33] monitoring cellular processes;[34] K^+-channel openers[35]

Industrial Applications Not reported

Safety/Toxicity Cardiotoxicity;[36] copper toxicity[37]

REFERENCES

1. Minta, A.; Tsien, R. Y. Fluorescent indicators for cytosolic sodium. *J. Biol. Chem.* **1989**, *264*, 19449–19457.
2. Tsien, R. Y.; Minta, A. Preparation and use of selective fluorescent indicator dyes for alkali metal cation analysis. Eur. Pat. Appl. EP 369733, 1990; *Chem. Abstr.* **1991**, *115*, 227789.
3. Kole, M. H. P.; Ilschner, S. U.; Kampa, B. M.; Williams, S. R.; Ruben, P. C.; Stuart, G. J. Action potential generation requires a high sodium channel density in the axon initial segment. *Nat. Neurosci.* **2008**, *11*, 178–186.
4. Meier, S. D.; Kovalchuk, Y.; Rose, C. R. Properties of the new fluorescent Na^+ indicator CoroNa Green: comparison with SBFI and confocal Na^+ imaging. *J. Neurosci. Methods* **2006**, *155*, 251–259.
5. Imanishi, T.; Matsushima, K.; Kawaguchi, A.; Wada, T.; Masuko, T.; Yoshida, S.; Ichida, S. Enhancement

of veratridine-induced sodium dynamics in NG108-15 cells during differentiation. *Biol. Pharm. Bull.* **2006**, *29*, 701–704.

6. Tashiro, M.; Tursun, P.; Konishi, M. Intracellular and extracellular concentrations of Na^+ modulate Mg^{2+} transport in rat ventricular myocytes. *Biophys. J.* **2005**, *89*, 3235–3247.
7. D'Onofrio, C.; Kader, A.; Lindberg, S. Uptake of sodium in quince, sugar beet, and wheat protoplasts determined by the fluorescent sodium-binding dye benzofuran isophthalate. *J. Plant Physiol.* **2005**, *162*, 421–428.
8. Kader, M. A.; Lindberg, S. Uptake of sodium in protoplasts of salt-sensitive and salt-tolerant cultivars of rice, *Oryza sativa* L. determined by the fluorescent dye SBFI. *J. Exp. Bot.* **2005**, *56*, 3149–3158.
9. Onizuka, S.; Kasaba, T.; Hamakawa, T.; Ibusuki, S.; Takasaki, M. Lidocaine increases intracellular sodium concentration through voltage-dependent sodium channels in an identified lymnaea neuron. *Anesthesiology* **2004**, *101*, 110–120.
10. Halperin, S. J.; Lynch, J. P. Effects of salinity on cytosolic Na^+ and K^+ in root hairs of *Arabidopsis thaliana*: *in vivo* measurements using the fluorescent dyes SBFI and PBFI. *J. Exp. Bot.* **2003**, *54*, 2035–2043.
11. Yeung, E. W.; Ballard, H. J.; Bourreau, J. P.; Allen, D. G. Intracellular sodium in mammalian muscle fibers after eccentric contractions. *J. Appl. Physiol.* **2003**, *94*, 2475–2482.
12. Diarra, A.; Sheldon, C.; Church, J. *In situ* calibration and $[H^+]$ sensitivity of the fluorescent Na^+ indicator SBFI. *Am. J. Physiol.* **2001**, *280*, C1623–C1633.
13. Chatton, J. Y.; Shimamoto, K.; Magistretti, P. J. Effects of glial glutamate transporter inhibitors on intracellular Na^+ in mouse astrocytes. *Brain Res.* **2001**, *893*, 46–52.
14. Despa, S.; Vecer, J.; Steels, P.; Ameloot, M. Fluorescence lifetime microscopy of the Na^+ indicator sodium green in HeLa cells. *Anal. Biochem.* **2000**, *281*, 159–175.
15. Despa, S.; Steels, P.; Ameloot, M. Fluorescence lifetime microscopy of the sodium indicator sodium-binding benzofuran isophthalate in HeLa cells. *Anal. Biochem.* **2000**, *280*, 227–241.
16. Baartscheer, A.; Schumacher, C. A.; Fiolet, J. W. T. Small changes of cytosolic sodium in rat ventricular myocytes measured with SBFI in emission ratio mode. *J. Mol. Cell. Cardiol.* **1997**, *29*, 3375–3383.
17. Lamont, C.; Burdyga, T. V.; Wray, S. Intracellular Na^+ measurements in smooth muscle using SBFI—changes in $[Na^+]$, Ca^{2+} and force in normal and Na^+-loaded ureter. *Pfluegers Arch.* **1998**, *435*, 523–527.
18. Mittmann, T.; Linton, S. M.; Schwindt, P.; Crill, W. Evidence for persistent Na^+ current in apical dendrites of rat neocortical neurons from imaging of Na^+-sensitive dye. *J. Neurophysiol.* **1997**, *78*, 1188–1192.
19. Rose, C. R.; Ransom, B. R. Regulation of intracellular sodium in cultured rat hippocampal neurones. *J. Physiol.* **1997**, *499*, 573–587.
20. Owen, C. S.; Wahl, M. L.; Leeper, D. B.; Perry, H. D.; Bobyock, S. B.; Russell, M.; Woodward, W. Accurate whole-spectrum measurements of intracellular pH and $[Na^+]$. *J. Fluoresc.* **1995**, *5*, 329–335.
21. Amorino, G. P.; Fox, M. H. Intracellular Na^+ measurements using sodium green tetraacetate with flow cytometry. *Cytometry* **1995**, *21*, 248–256.
22. Terada, H.; Hayashi, H.; Satoh, H.; Katoh, H.; Yamazaki, N. Simultaneous measurement of $[Na^+]_i$ and Ca^{2+} transients in an isolated myocyte: effects of strophanthidin. *Biochem. Biophys. Res. Commun.* **1994**, *203*, 1050–1056.
23. Turvey, S. E.; Allen, D. G. Changes in myoplasmic sodium concentration during exposure to lactate in perfused rat heart. *Cardiovasc. Res.* **1994**, *28*, 987–993.
24. Deri, Z.; Adam-Vizi, V. Detection of intracellular free Na^+ concentration of synaptosomes by a fluorescent indicator, Na(+)-binding benzofuran isophthalate: the effect of veratridine, ouabain, and alpha-latrotoxin. *J. Neurochem.* **1993**, *61*, 818–825.
25. van der Bent, V.; Demole, C.; Johnson, E. I.; Rossier, M. F.; Python, C. P.; Vallotton, M. B.; Capponi, A. M. Angiotensin-II induces changes in the cytosolic sodium concentration in bovine adrenal glomerulosa cells: involvement in the activation of aldosterone biosynthesis. *Endocrinology* **1993**, *133*, 1213–1220.
26. Donoso, P.; Mill, J. G.; O'Neill, S. C.; Eisner, D. A. Fluorescence measurements of cytoplasmic and mitochondrial sodium concentration in rat ventricular myocytes. *J. Physiol.* **1992**, *448*, 493–509.
27. Sage, S. O.; Rink, T. J.; Mahaut-Smith, M. P. Resting and ADP-evoked changes in cytosolic free sodium concentration in human platelets loaded with the indicator SBFI. *J. Physiol.* **1991**, *441*, 559–573.
28. Satoh, H.; Hayashi, H.; Noda, N.; Terada, H.; Kobayashi, A.; Yamashita, Y.; Kawai, T.; Hirano,

M.; Yamazaki, N. Quantification of intracellular free sodium ions by using a new fluorescent indicator, sodium-binding benzofuran isophthalate in guinea pig myocytes. *Biochem. Biophys. Res. Commun.* **1991**, *175*, 611–616.

29. Garlid, K. D.; Shariat-Madar, Z.; Nath, S.; Jezek, P. Reconstitution and partial purification of the sodium-selective sodium/hydrogen ion antiporter of beef heart mitochondria. *J. Biol. Chem.* **1991**, *266*, 6518–6523.

30. Naftalin, R. J.; Pedley, K. C. Video enhanced imaging of the fluorescent Na^+ probe SBFI indicates that colonic crypts absorb fluid by generating a hypertonic interstitial fluid. *FEBS Lett.* **1990**, *260*, 187–194.

31. Borin, M.; Siffert, W. Stimulation by thrombin increases the cytosolic free sodium concentration in human platelets: studies with the novel fluorescent cytosolic sodium indicator sodium-binding benzofuran isophthalate. *J. Biol. Chem.* **1990**, *265*, 19543–19550.

32. Harootunian, A. T.; Kao, J. P. Y.; Eckert, B. K.; Tsien, R. Y. Fluorescence ratio imaging of cytosolic free sodium in individual fibroblasts and lymphocytes. *J. Biol. Chem.* **1989**, *264*, 19458–19467.

33. Kim, E. J.; Barker, L.; Burnet, M.; Guse, J. H.; Luyten, K.; Tsotsou, G. Use of a yeast expression host to identify genes for transport proteins using fluorescent dye reporters. PCT Int. Appl. WO 2003038092, 2003; *Chem. Abstr.* **2003**, *138*, 363834.

34. Cousin, M. A.; Evans, G. J. O. Assay method. PCT Int. Appl. WO 2006134319, 2006; *Chem. Abstr.* **2006**, *146*, 77602.

35. Kobayashi, K.; Nishizawa, Y.; Sawada, K.; Ogura, H.; Miyabe, M. K^+-channel openers suppress epileptiform activities induced by 4-aminopyridine in cultured rat hippocampal neurons. *J. Pharmacol. Sci.* **2008**, *108*, 517–528.

36. Chen, M.; Xiao, C. Y.; Hashizume, H.; Abiko, Y. Differential effects of Ca^{2+} channel blockers on Ca^{2+} overload induced by lysophosphatidylcholine in cardiomyocytes. *Eur. J. Pharmacol.* **1997**, *333*, 261–268.

37. Benders, A. A.; Li, J.; Lock, R. A.; Bindels, R. J.; Bonga, S. E.; Veerkamp, J. H. Copper toxicity in cultured human skeletal muscle cells: the involvement of Na^+/K(+)-ATPase and the Na^+/Ca(2+)-exchanger. *Pfluegers Arch.* **1994**, *428*, 461–467.

SBFI AM

CAS Registry Number 129423-53-6
Chemical Structure

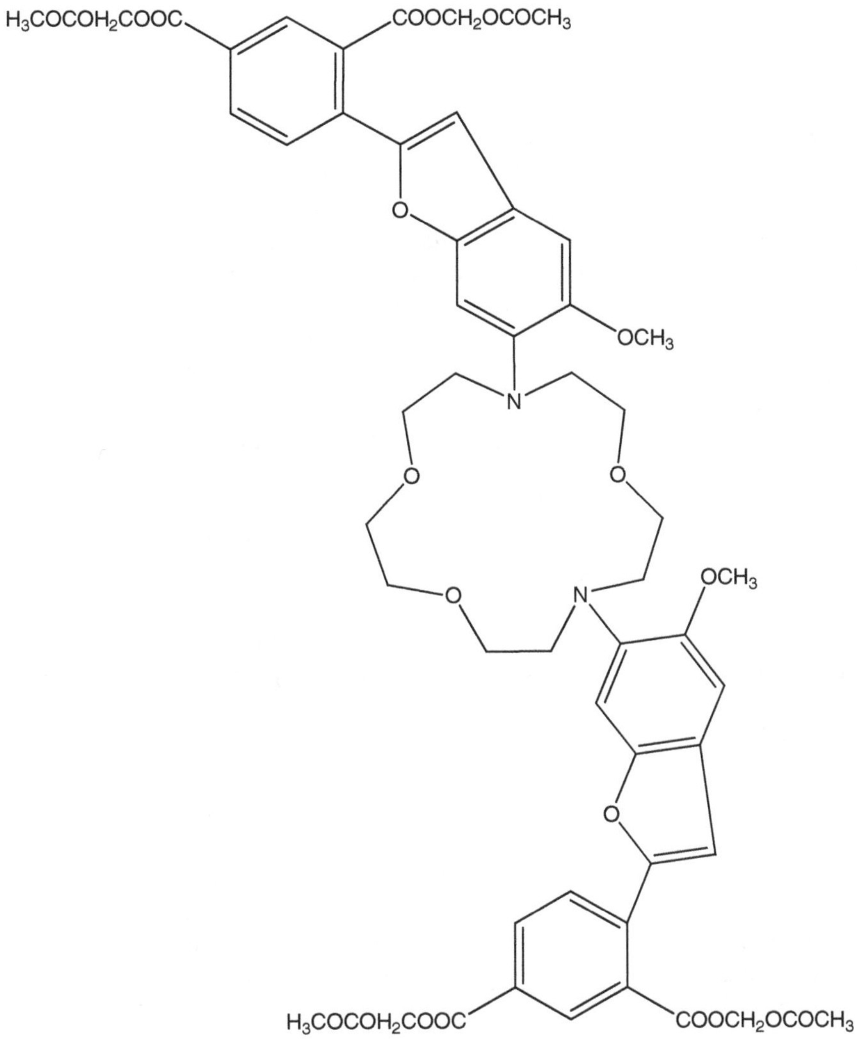

Merck Index Number Not listed
Chemical/Dye Class Benzofuran
Molecular Formula $C_{56}H_{58}N_2O_{23}$
Molecular Weight 1127.06

CA Index Name 1,3-Benzenedicarboxylic acid, 4,4′-[1,4,10-trioxa-7,13-diazacyclopentadecane-7,13-diylbis(5-methoxy-6,2-benzofurandiyl)]bis-, 1,1′,3,3′-tetrakis[(acetyloxy)methyl]ester

Other Names 1,3-Benzenedicarboxylic acid, 4,4′-[1,4,10-trioxa-7,13-diazacyclopentadecane-7,13-diylbis(5-methoxy-6,2-benzofurandiyl)]bis-,tetrakis[(acetyloxy)methyl]ester; 1,4,10-Trioxa-7,13-diazacyclopentadecane, 1,3-benzenedicarboxylic acid derivative; SBFI-AM, Sodium-binding benzofuran isophthalate-AM

Physical Form Orange powder
Solubility Soluble in methanol, dimethyl sulfoxide
Melting Point >200 °C
Boiling Point (Calcd.) 1110.3 ± 65.0 °C, pressure: 760 Torr
pK_a (Calcd.) 3.82 ± 0.40, most basic, temperature: 25 °C
Absorption (λ_{max}) 379 nm
Emission (λ_{max}) Fluorescence is very weak

Synthesis Synthetic methods[1,2]

Staining Applications Sodium ions[1–15]

Biological Applications Sodium indicator;[1–15] assays for identifying taste-specific genes;[16,17] treating depression[6]

Industrial Applications Not reported

Safety/Toxicity No data available

REFERENCES

1. Minta, A.; Tsien, R. Y. Fluorescent indicators for cytosolic sodium. *J. Biol. Chem.* **1989**, *264*, 19449–19457.
2. Tsien, R. Y.; Minta, A. Preparation and use of selective fluorescent indicator dyes for alkali metal cation analysis. Eur. Pat. Appl. EP 369733, 1990; *Chem. Abstr.* **1991**, *115*, 227789.
3. Moyer, B.; Zlotnik, A.; Hevezi, P.; Soto, H.; Lu, M.; Gao, N.; Servant, G.; Brust, P.; Williams, M.; Kalabat, D.; White, E. C. Identification of TRPML3 as a salty taste receptor and use in assays for identifying taste modulators and/or therapeutics that modulate sodium transport, absorption or excretion and/or aldosterone and/or vasopressin production or release. PCT Int. Appl. WO 2009008950, 2009; *Chem. Abstr.* **2009**, *150*, 164501.
4. Servant, G.; Brust, P.; Ray, S.; Hung, N. Improved cell-based fluorescent assays for identifying α and δ epithelial sodium channel modulators. PCT Int. Appl. WO 2008013969, 2008; *Chem. Abstr.* **2008**, *148*, 222369.
5. D'Onofrio, C.; Kader, A.; Lindberg, S. Uptake of sodium in quince, sugar beet, and wheat protoplasts determined by the fluorescent sodium-binding dye benzofuran isophthalate. *J. Plant Physiol.* **2005**, *162*, 421–428.
6. Singer, A.; Wonnemann, M. Treatment of depression by using compounds which elevate the intracellular sodium concentration. Eur. Pat. Appl. EP 1034782, 2000; *Chem. Abstr.* **2000**, *133*, 217714.
7. Despa, S.; Steels, P.; Ameloot, M. Fluorescence lifetime microscopy of the sodium indicator sodium-binding benzofuran isophthalate in HeLa cells. *Anal. Biochem.* **2000**, *280*, 227–241.
8. Ito, S.; Ohta, T.; Kadota, H.; Kitamura, N.; Nakazato, Y. Measurement of intracellular Na^+ concentration by a Na^+-sensitive fluorescent dye, sodium-binding benzofuran isophthalate, in porcine adrenal chromaffin cells—usage of palytoxin as a Na^+ ionophore. *J. Neurosci. Methods* **1997**, *75*, 21–27.
9. Cutaia, M.; Davis, R.; Parks, N.; Rounds, S. Effect of ATP-induced permeabilization on loading of the Na^+ probe SBFI into endothelial cells. *J. Appl. Physiol.* **1996**, *81*, 509–515.
10. Amorino, G. P.; Fox, M. H. Intracellular Na^+ measurements using sodium green tetraacetate with flow cytometry. *Cytometry* **1995**, *21*, 248–256.
11. Hayashi, H.; Satoh, H.; Noda, N.; Terada, H.; Hirano, M.; Yamashita, Y.; Kobayashi, A.; Yamazaki, N. Simultaneous measurement of intracellular Na^+ and Ca^{2+} during K^+-free perfusion in isolated myocytes. *Am. J. Physiol.* **1994**, *266*, C416–C422.
12. Deri, Z.; Adam-Vizi, V. Detection of intracellular free sodium concentration of synaptosomes by a fluorescent indicator, sodium-binding benzofuran isophthalate: the effect of veratridine, ouabain, and α-latrotoxin. *J. Neurochem.* **1993**, *61*, 818–825.
13. Borzak, S.; Reers, M.; Arruda, J.; Sharma, V. K.; Sheu, S. S.; Smith, T. W.; Marsh, J. D. Sodium efflux mechanisms in ventricular myocytes: measurement of $[Na^+]_i$ with sodium-binding benzofuran isophthalate. *Am. J. Physiol.* **1992**, *263*, H866–H874.
14. Jung, D. W.; Apel, L. M.; Brierley, G. P. Transmembrane gradients of free sodium in isolated heart mitochondria estimated using a fluorescent probe. *Am. J. Physiol.* **1992**, *262*, C1047–C1055.
15. Borin, M.; Siffert, W. Stimulation by thrombin increases the cytosolic free sodium concentration in human platelets: studies with the novel fluorescent cytosolic sodium indicator sodium-binding benzofuran isophthalate. *J. Biol. Chem.* **1990**, *265*, 19543–19550.
16. Moyer, B.; Zlotnik, A.; Hevezi, P.; Soto, H.; Kalabat, D.; Lu, M.; Gao, N.; White, E. Rationale, methods, and assays for identifying human and non-human primate taste-specific genes and use thereof in taste modulator and therapeutic screening assays. PCT Int. Appl. WO 2008153924, 2008; *Chem. Abstr.* **2008**, *150*, 49063.
17. Moyer, B.; Lu, M.; Echeverri, F.; Kalabat, D.; Gao, N.; Hevezi, P. Rationale, methods, and assays for identifying novel taste cell genes and salty taste receptor targets and assays using these identified genes or gene products. PCT Int. Appl. WO 2007146120, 2007; *Chem. Abstr.* **2007**, *148*, 49091.

SPQ

CAS Registry Number 83907-40-8

Chemical Structure

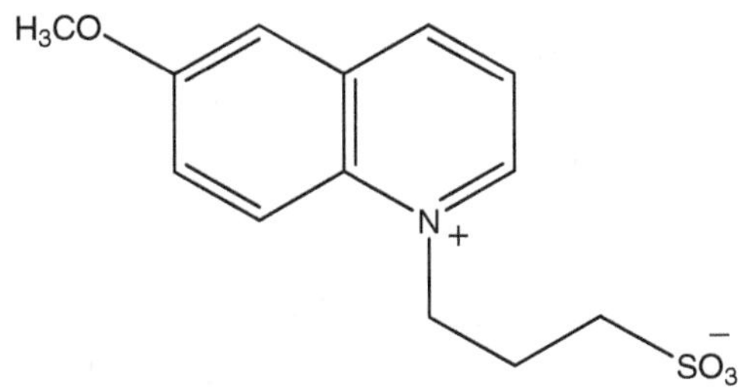

CA Index Name Quinolinium, 6-methoxy-1-(3-sulfopropyl)-, inner salt

Other Names 6-Methoxy-*N*-(3-sulfopropyl)quinolinium, inner salt; SPQ

Merck Index Number Not listed

Chemical/Dye Class Quinoline

Molecular Formula $C_{13}H_{15}NO_4S$

Molecular Weight 281.33

Physical Form White powder

Solubility Soluble in water, dimethyl sulfoxide

Melting Point >299 °C (decompose)

Absorption (λ_{max}) 344 nm

Emission (λ_{max}) 443 nm

Synthesis Synthetic methods[1,2]

Staining Applications Chloride ions;[3–31] nitrite ions[32]

Biological Applications Chloride indicator;[3–31] nitrite indicator;[32] diagnosis of diseases caused by elemental imbalances;[35] detecting cancer cells,[36] spores,[36] stress biomarkers;[37] identifying genes;[38] studying chloride–bicarbonate exchange[33,34]

Industrial Applications Not reported

Safety/Toxicity No data available

REFERENCES

1. Adamczyk, M.; Rege, S. Microwave assisted sulfopropylation of *N*-heterocycles using 1,3-propane sultone. *Tetrahedron Lett.* **1998**, *39*, 9587–9588.
2. Wolfbeis, O. S.; Urbano, E. Syntheses of fluorescent dyes. XIV. Standards for fluorescence measurements in the near neutral pH-range. *J. Heterocycl. Chem.* **1982**, *19*, 841–843.
3. Ozawa, T.; Asakawa, T.; Ohta, A.; Miyagishi, S. Halide-sensitive fluorescent probes for self-assembly of novel cationic fluoro surfactants. *J. Oleo Sci.* **2005**, *54*, 135–142.
4. Mahlangu, D. A. F. D.; Dix, J. A. Halide fluxes in epithelial cells measured with an automated cell plate reader. *Anal. Biochem.* **2004**, *325*, 28–34.
5. Lee, J. E.; Park, H. S.; Uhm, D. Y.; Kim, S. J. Effects of KCNQ1 channel blocker, 293B, on the acetylcholine-induced Cl^- secretion of rat pancreatic acini. *Pancreas* **2004**, *28*, 435–442.
6. Soldati, L.; Adamo, D.; Spaventa, R.; Bianchi, G.; Vezzoli, G. Chloride fluxes activated by parathyroid hormone in human erythrocytes. *Biochem. Biophys. Res. Commun.* **2000**, *269*, 470–473.
7. Garcia, N. H.; Plato, C. F.; Garvin, J. L. Fluorescent determination of chloride in nanoliter samples. *Kidney Int.* **1999**, *55*, 321–325.
8. Rubera, I.; Tauc, M.; Verheecke-Mauze, C.; Bidet, M.; Poujeol, C.; Touret, N.; Cuiller, B.; Poujeol, P. Regulation of cAMP-dependent chloride channels in DC1 immortalized rabbit distal tubule cells in culture. *Am. J. Physiol.* **1999**, *276*, F104–F121.
9. Srinivas, S. P.; Bonanno, J. A.; Hughes, B. A.; Morton, D. Assessment of swelling-activated Cl^- channels using the halide-sensitive fluorescent indicator 6-methoxy-*N*-(3-sulfopropyl)quinolinium. *Biophys. J.* **1998**, *75*, 115–123.
10. Dupuit, F.; Chinet, T.; Zahm, J. M.; Pierrot, D.; Hinnrasky, J.; Kaplan, H.; Bonnet, N.; Puchelle, E. Induction of a cAMP-stimulated chloride secretion in regenerating poorly differentiated airway epithelial cells by adenovirus-mediated CFTR gene transfer. *Hum. Gene Ther.* **1997**, *8*, 1439–1450.
11. Pilas, B.; Durack, G. A flow cytometric method for measurement of intracellular chloride concentration in lymphocytes using the halide-specific probe 6-methoxy-*N*-(3-sulfopropyl) quinolinium (SPQ). *Cytometry* **1997**, *28*, 316–322.
12. Garcia, L.; Rigoulet, M.; Georgescauld, D.; Dufy, B.; Sartor, P. Regulation of intracellular chloride concentration in rat lactotrophs: possible role of mitochondria. *FEBS Lett.* **1997**, *400*, 113–118.
13. Stern, M.; Munkonge, F. M.; Caplen, N. J.; Sorgi, F.; Huang, L.; Geddes, D. M.; Alton, E. W. Quantitative fluorescence measurements of chloride secretion in

native airway epithelium from CF and non-CF subjects. *Gene Ther*. **1995**, *2*, 766–774.

14. Fine, D. M.; Lo, C. F.; Aguillar, L.; Blackmon, D. L.; Montrose, M. H. Cellular chloride depletion inhibits cAMP-activated electrogenic chloride fluxes in HT29-18-C1 cells. *J. Membr. Biol.* **1995**, *145*, 129–141.
15. Nishikawa, K.; Ishihara, H.; Ozawa, K.; Tamura, K. Chloride transport mechanism in swine tracheal submucosal gland cells. *Respiration* **1995**, *62*, 274–279.
16. Huang, S. J.; Chan, H. C.; Wong, P. Y. D. Adrenaline-regulated Cl^- transport in cultured single rat epididymal cells measured by an entrapped Cl-sensitive fluorophore. *J. Physiol.* **1994**, *474*, 183–191.
17. Vasseur, M.; Frangne, R.; Alvarado, F. Buffer-dependent pH sensitivity of the fluorescent chloride-indicator dye SPQ. *Am. J. Physiol.* **1993**, *264*, C27–C31.
18. Orosz, D. E.; Garlid, K. D. A sensitive new fluorescence assay for measuring proton transport across liposomal membranes. *Anal. Biochem.* **1993**, *210*, 7–15.
19. Schummer, U.; Schiefer, H. G. Chloride fluxes across *Acholeplasma laidlawii* membranes. *FEMS Microbiol. Lett.* **1991**, *67*, 109–113.
20. Pope, A. J.; Leigh, R. A. Characterization of chloride transport at the tonoplast of higher plants using a chloride-sensitive fluorescent probe: effects of other anions, membrane potential, and transport inhibitors. *Planta* **1990**, *181*, 406–413.
21. Cabrini, G. Chloride sensitive fluorescent indicators for use in cytology. *Appl. Fluoresc. Technol.* **1989**, *1*, 11–12.
22. Verkman, A. S.; Takla, R.; Sefton, B.; Basbaum, C.; Widdicombe, J. H. Quantitative fluorescence measurement of chloride transport mechanisms in phospholipid vesicles. *Biochemistry* **1989**, *28*, 4240–4244.
23. Pope, A. J.; Leigh, R. A. The use of a chloride-sensitive fluorescent probe to measure chloride transport in isolated tonoplast vesicles. *Planta* **1988**, *176*, 451–460.
24. Dechecchi, M. C.; Cabrini, G. Chloride conductance in membrane vesicles from human placenta using a fluorescent probe: implications for cystic fibrosis. *Biochim. Biophys. Acta* **1988**, *945*, 113–120.
25. Krapf, R.; Berry, C. A.; Verkman, A. S. Estimation of intracellular chloride activity in isolated perfused rabbit proximal convoluted tubules using a fluorescent indicator. *Biophys. J.* **1988**, *53*, 955–962.
26. Krapf, R.; Illsley, N. P.; Tseng, H. C.; Verkman, A. S. Structure–activity relationships of chloride-sensitive fluorescent indicators for biological application. *Anal. Biochem.* **1988**, *169*, 142–150.
27. Chen, P. Y.; Verkman, A. S. Sodium-dependent chloride transport in basolateral membrane vesicles isolated from rabbit proximal tubule. *Biochemistry* **1988**, *27*, 655–660.
28. Chen, P. Y.; Illsley, N. P.; Verkman, A. S. Renal brush-border chloride transport mechanisms characterized using a fluorescent indicator. *Am. J. Physiol.* **1988**, *254*, F114–F120.
29. Illsley, N. P.; Verkman, A. S. Membrane chloride transport measured using a chloride-sensitive fluorescent probe. *Biochemistry* **1987**, *26*, 1215–1219.
30. Urbano, E.; Offenbacher, H.; Wolfbeis, O. S. Optical sensor for continuous determination of halides. *Anal. Chem.* **1984**, *56*, 427–429.
31. Wolfbeis, O. S.; Urbano, E. Fluorometric analysis. 3. Fluorometric, heavy-metal-free method for the determination of chlorine, bromine and iodine in organic materials. *Fresenius' Z. Anal. Chem.* **1983**, *314*, 577–581.
32. Shingles, R.; Roh, M. H.; McCarty, R. E. Direct measurement of nitrite transport across erythrocyte membrane vesicles using the fluorescent probe, 6-methoxy-*N*-(3-sulfopropyl) quinolinium. *J. Bioenerg. Biomembr.* **1997**, *29*, 611–616.
33. Greco, F. A.; Solomon, A. K. Kinetics of chloride–bicarbonate exchange across the human red blood cell membrane. *J. Membr. Biol.* **1997**, *159*, 197–208.
34. Calafut, T. M.; Dix, J. A. Chloride–bicarbonate exchange through the human red cell ghost membrane monitored by the fluorescent probe 6-methoxy-*N*-(3-sulfopropyl)quinolinium. *Anal. Biochem.* **1995**, *230*, 1–7.
35. Rupp, M. E. Determination of elements in body fluids and test kit including the necessary reagents for diagnosis of diseases caused by elemental imbalances. U.S. Pat. Appl. Publ. US 2003203495, 2003; *Chem. Abstr.* **2003**, *139*, 347692.
36. Levy, M.; Ellington, A. D.; Pai, S. Proximity ligation assays with peptide conjugate 'burrs' and aptamers for the sensitive detection of spores and cancer cells. PCT Int. Appl. WO 2007027796, 2007; *Chem. Abstr.* **2007**, *146*, 312239.

37. Cullum, M. E.; Duplessis, C. A.; Crepeau, L. J. Method for the detection of stress biomarkers including cortisol by fluorescence polarization. U. S. Pat. Appl. Publ. US 2006105397, 2006; *Chem. Abstr.* **2006**, *144*, 463810.

38. Kim, E. J.; Barker, L.; Burnet, M.; Guse, J. H.; Luyten, K.; Tsotsou, G. Use of a yeast expression host to identify genes for transport proteins using fluorescent dye reporters. PCT Int. Appl. WO 2003038092, 2003; *Chem. Abstr.* **2003**, *138*, 363834.

STAINS-ALL

CAS Registry Number 7423-31-6

Chemical Structure

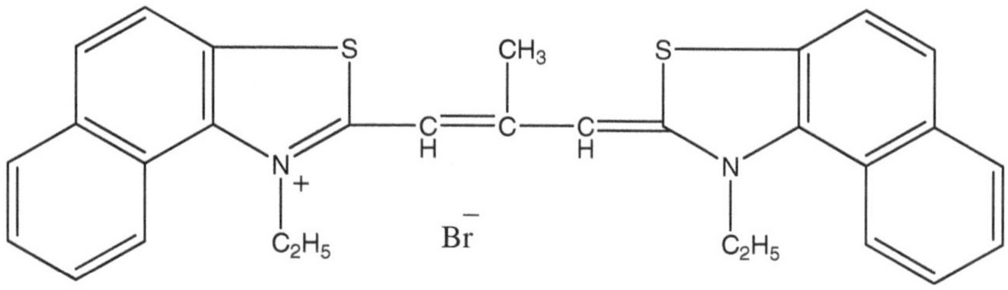

Chemical/Dye Class Naphthothiazole

Molecular Formula $C_{30}H_{27}BrN_2S_2$

Molecular Weight 559.58

Physical Form Dark green to dark grey powder

CA Index Name Naphtho[1,2-*d*]thiazolium, 1-ethyl-2-[3-(1-ethylnaphtho[1,2-*d*]thiazol-2(1*H*)-ylidene)-2-methyl-1-propen-1-yl]-, bromide (1 : 1)

Other Names 1-Ethyl-2-[3-(1-ethylnaphtho[1,2-*d*] thiazolin-2-ylidene)-2-methylpropenyl]naphtho[1,2-*d*] thiazolium bromide; Naphtho[1,2-*d*]thiazolium, 1-ethyl-2-[3-(1-ethylnaphtho[1,2-*d*]thiazol-2(1*H*)-ylidene)-2-methyl-1-propenyl]-, bromide; Naphtho[1,2-*d*]thiazolium, 1-ethyl-2-[3-(1-ethylnaphtho[1,2-*d*]thiazolin-2-ylidene)-2-methylpropenyl]-, bromide; 4,5,4′,5′-Dibenzothiacarbocyanine bromide, 3,3′-diethyl-9-methyl-; 3,3′-Diethyl-9-methyl-4,5,4′,5′-dibenzothiacarbocyanine bromide; 3,3′-Diethyl-9-methyl-4,5,4′,5′-naphthothiacarbocyanine bromide; 4,5,4′,5′-Dibenzo-3,3′-diethyl-9-methylthiacarbocyanine bromide; Carbocyanin DBTC; DBTC; Stains-all

Merck Index Number Not listed

Solubility Soluble in water, ethanol, chloroform

Melting Point >200 °C

Absorption (λ_{max}) 575 nm

Synthesis Synthetic methods[1–5]

Staining Applications Axonemal tubulins;[6] biomolecules;[7] elastin fibers;[8] eye lens proteins crystallins;[9] fungi;[10] protozoa;[10] glycosaminoglycans;[11] neurofilaments;[12] nucleic acids;[13–15,18,19] proteins;[16–19] conjugated proteins;[18] polar lipids;[18] phosphorpoteins[20]

Biological Applications Detecting nucleic acid hybridization;[21] identifying microorganisms[22]

Industrial Applications Lithographic printing plates;[23] optical discs;[24] optical recording materials;[25] photographic materials[26–28]

Safety/Toxicity Endotoxicity[29]

REFERENCES

1. Lu, M.; Guo, Q.; Seeman, N. C.; Kallenbach, N. R. Drug binding by branched DNA: selective interaction of the dye Stains-All with an immobile junction. *Biochemistry* **1990**, *29*, 3407–3412.
2. Homma, S.; Kobayashi, S.; Ono, H.; Takahashi, T. Meso-alkylcarbocyanine dyes. JP 39002626, 1964; *Chem. Abstr.* **1964**, *61*, 18788.
3. Carroll, B. H.; Jones, J. E. Supersensitization of photographic emulsions. U.S. Patent 2688545, 1954; *Chem. Abstr.* **1955**, *49*, 601.
4. Mizuno, Y.; Uemura, T. Syntheses of cyanine dyes. VIII. Syntheses of antibacterial thiocarbocyanines. *Yakugaku Zasshi* **1950**, *70*, 543–545; *Chem. Abstr.* **1951**, *45*, 41458.
5. Keyes, G. H. Meso-substituted carbocyanine dyes and thio ketones. U.S. Patent 2486173, 1949; *Chem. Abstr.* **1950**, *44*, 32934.
6. Nakamura, K.; Masuyama, E.; Wada, S.; Okuno, M. Application of Stains-all staining to the analysis of axonemal tubulins: identification of β-tubulin and β-isotubulins. *J. Biochem. Biophys. Methods* **1990**, *21*, 237–245.
7. Chirikjian, J. G. Articles of manufacture and methods for staining and labeling biomolecules. PCT Int. Appl. WO 9942620, 1999; *Chem. Abstr.* **1999**, *131*, 167357.
8. Horobin, R. W.; Flemming, L. Structure–staining relationships in histochemistry and biological staining. II. Mechanistic and practical aspects of the staining of elastic fibers. *J. Microsc.* **1980**, *119*, 357–372.
9. Sharma, Y.; Rao, C. M.; Rao, S. Binding site conformation dictates the color of the dye Stains-all. A study of the binding of this dye to the eye lens

proteins crystallins. *J. Biol. Chem.* **1989**, *264*, 20923–20927.

10. Dao, M. L. Method for staining fungi and protozoa. U.S. Patent 7157242, 2007; *Chem. Abstr.* **2007**, *146*, 96260.
11. Volpi, N.; Maccari, F. Detection of submicrogram quantities of glycosaminoglycans on agarose gels by sequential staining with toluidine blue and Stains-All. *Electrophoresis* **2002**, *23*, 4060–4066.
12. Ksiezak-Reding, H.; Yen, S. H. Phosphatase and carbocyanine dye binding define different types of phosphate groups in mammalian neurofilaments. *J. Neurosci.* **1987**, *7*, 3554–3560.
13. Matselyukh, B. P.; Yarmoluk, S. M.; Matselyukh, A. B.; Kovalska, V. B.; Kocheshev, I. O.; Kryvorotenko, D. V.; Lukashov, S. S. Interaction of cyanine dyes with nucleic acids: XXXI. Using of polymethine cyanine dyes for the visualization of DNA in agarose gels. *J. Biochem. Biophys. Methods* **2003**, *57*, 35–43.
14. Kovalska, V. B.; Valyukh, I. V.; Lukashov, S. S.; Slominskii, Yu. L.; Yarmoluk, S. M. An investigation of tricarbocyanines "Stains-All" and "*iso*-Stains-All" as fluorescent nucleic acids probes. *J. Fluoresc.* **2002**, *12*, 209–212.
15. Rueger, A. M.; Rueger, W. Differential staining of the subunits of DNA-dependent RNA polymerase. *FEBS Lett.* **1980**, *120*, 233–235.
16. Urban, P. F.; Ancel, R.; Virmaux, N. A soluble protein in rod outer segments stained blue by Stains-all. *Biochem. Soc. Trans.* **1988**, *16*, 299.
17. Campbell, K. P.; MacLennan, D. H.; Jorgensen, A. O. Staining of the calcium-binding proteins, calsequestrin, calmodulin, troponin C, and S-100, with the cationic carbocyanine dye "Stains-all". *J. Biol. Chem.* **1983**, *258*, 11267–11273.
18. Green, M. R. Simultaneous differential staining of nucleic acids, proteins, conjugated proteins, and polar lipids by a cationic carbocyanine dye. *J. Histochem. Cytochem.* **1975**, *23*, 411–423.
19. Green, M. R.; Pastewka, J. V. Simultaneous differential staining by a cationic carbocyanine dye of nucleic acids, proteins, and conjugated proteins. II. Carbohydrate and sulfated carbohydrate-containing proteins. *J. Histochem. Cytochem.* **1974**, *22*, 774–781.
20. Green, M. R.; Pastewka, J. V.; Peacock, A. C. Differential staining of phosphoproteins on polyacrylamide gels with a cationic carbocyanine dye. *Anal. Biochem.* **1973**, *56*, 43–51.
21. Bupp, C. R., II; Choi, K. Y.; Holmes-Davis, R. A.; Izmailov, A.; Koshinsky, H.; Nulf, C. J.; Urdea, M.; Wang, M.; Warner, B. D.; Zwick, M. Carbocyanine dye dimers linked by a conjugated alkenyl chain for use in detection of nucleic acid hybridization. PCT Int. Appl. WO 2007100711, 2007; *Chem. Abstr.* **2007**, *147*, 358263.
22. Fluit, A. C.; Widjojoatmodjo, M. N. A method for identifying microorganisms based on nucleic acid amplification and gel electrophoresis. PCT Int. Appl. WO 9513396, 1995; *Chem. Abstr.* **1995**, *123*, 220262.
23. Terashima, E.; Sumi, S. Lithographic printing plates using silver halide emulsion layers. Jpn. Kokai Tokkyo Koho JP 2006195201, 2006; *Chem. Abstr.* **2006**, *145*, 177299.
24. Larson, L. C.; Merry, J. B.; Cunningham, M. P.; Owens, J. C. Limited play optical discs. PCT Int. Appl. WO 2006130496, 2006; *Chem. Abstr.* **2006**, *146*, 52451.
25. Tanaka, H. Wavelength-multiple optical recording material. Jpn. Kokai Tokkyo Koho JP 03248147, 1991; *Chem. Abstr.* **1992**, *117*, 201998.
26. Mihara, J.; Myake, K. Photographic image formation. Jpn. Kokai Tokkyo Koho JP 09054409, 1997; *Chem. Abstr.* **1997**, *126*, 285245.
27. Yoshida, K.; Nishio, S. Spectrally sensitized silver halide photographic material containing polyhydroxybenzene. Jpn. Kokai Tokkyo Koho JP 03025432, 1991; *Chem. Abstr.* **1991**, *115*, 102751.
28. Yoshida, K. Spectrally sensitized silver halide photographic material containing cyanine dyes. Jpn. Kokai Tokkyo Koho JP 03001132, 1991; *Chem. Abstr.* **1991**, *115*, 38578.
29. Ogawa, Y.; Kanoh, S. Enhancement of endotoxicity and reactivity with carbocyanine dye by sonication of lipopolysaccharide. *Microbiol. Immunol.* **1984**, *28*, 1313–1323.

STILBENE ISOTHIOCYANATE SULFONIC ACID (SITS)

CAS Registry Number 51023-76-8

Chemical Structure

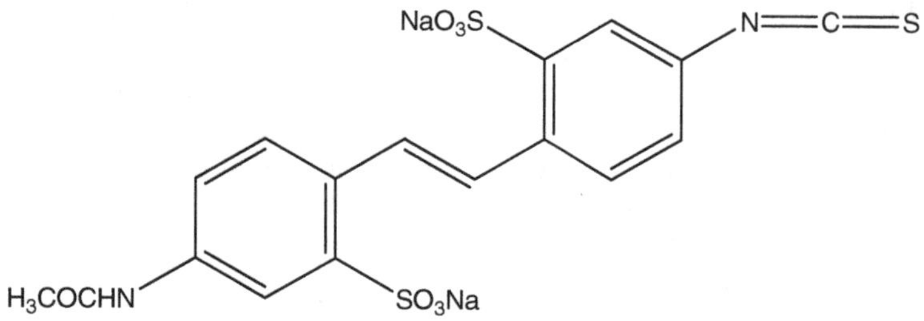

CA Index Name Benzenesulfonic acid, 5-(acetylamino)-2-[2-(4-isothiocyanato-2-sulfophenyl)ethenyl]-, sodium salt (1 : 2)

Other Names 4-Acetamido-4′-isothiocyanato-2,2′-stilbenedisulfonic acid disodium salt; Benzenesulfonic acid, 5-(acetylamino)-2-[2-(4-isothiocyanato-2-sulfophenyl) ethenyl]-, disodium salt; Disodium 4-acetamido-4′-isothiocyanatostilbene-2,2′-disulfonate; SITS

Merck Index Number Not listed

Chemical/Dye Class Stilbene

Molecular Formula $C_{17}H_{12}N_2Na_2O_7S_3$

Molecular Weight 498.46

Physical Form Yellow or tan-yellow powder

Solubility Soluble in water, methanol, dimethyl sulfoxide

Melting Point >200 °C

Absorption (λ_{max}) 336 nm

Emission (λ_{max}) 436 nm

Synthesis Synthetic methods[1–4]

Staining Applications Neuronal axons;[3] proteins;[3,4] photochromes[5]

Biological Applications Chloride channel blockers;[6–17] anion transport inhibitors;[18–20,25] treating cancer,[21] glaucoma,[22] heart diseases,[23] viral diseases[24]

Industrial Applications Not reported

Safety/Toxicity Effect on hemolysis;[25] lysosomal sulfate transport[26]

REFERENCES

1. Jozefiak, T. H.; Bastos, C. M.; Papoulis, A. T.; Holmes-Farley, S. R. Preparation of biaryl phosphate transport inhibitors. PCT Int. Appl. WO 2003057225, 2003; *Chem. Abstr.* **2003**, *139*, 117209.
2. Tobey, N. A.; Orlando, R. C. Cadherin-binding assay for identifying compounds which may protect stratified squamous epithelium against damage by noxious substances. PCT Int. Appl. WO 2002010767, 2002; *Chem. Abstr.* **2002**, *136*, 145195.
3. Horobin, R. W.; Payne, J. N.; Jakobsen, P. Histochemical implications of the chemical and biological properties of SITS and some related compounds. *J. Microsc.* **1987**, *146*, 87–96.
4. Szewczyk, B.; Bienkowska-Szewczyk, K.; Kozloff, L. M. Use of different fluorochromes for monitoring protein elution and transfer. *Electrophoresis* **1987**, *8*, 25–28.
5. Likhtenshtein, G. I.; Khudjakov, D. V.; Vogel, V. R. Photochrome-labeling method in study of dynamics of biological systems. *J. Biochem. Biophys. Methods* **1992**, *25*, 219–229.
6. Poletto Chaves, L. A.; Varanda, W. A. Volume-activated chloride channels in mice Leydig cells. *Pfluegers Arch.* **2008**, *457*, 493–504.
7. Liantonio, A.; Pusch, M.; Picollo, A.; Guida, P.; De Luca, A.; Pierno, S.; Fracchiolla, G.; Loiodice, F.; Tortorella, P.; Camerino, D. C. Investigations of pharmacologic properties of the renal CLC-K1 chloride channel co-expressed with barttin by the use of 2-(*p*-chlorophenoxy)propionic acid derivatives and other structurally unrelated chloride channels blockers. *J. Am. Soc. Nephrol.* **2004**, *15*, 13–20.
8. Eskandari, S.; Zampighi, G. A.; Leung, D. W.; Wright, E. M.; Loo, D. D. F. Inhibition of gap junction hemichannels by chloride channel blockers. *J. Membr. Biol.* **2002**, *185*, 93–102.
9. Plath, K. E. S.; Grabbe, J.; Strenzke, N.; Wolff, H. H.; Gibbs, B. F. Effect of chloride channel blockers on

anti-IgE-stimulated histamine and IL-4/IL-13 release from human basophils. *Inflamm. Res.* **2001**, *50*, S51–S52.

10. Eder, C.; Klee, R.; Heinemann, U. Involvement of stretch-activated Cl^- channels in ramification of *Murine microglia. J. Neurosci.* **1998**, *18*, 7127–7137.
11. Tanaka, H.; Matsui, S.; Kawanishi, T.; Shigenobu, K. Use of chloride blockers: a novel approach for cardioprotection against ischemia–reperfusion damage. *J. Pharmacol. Exp. Ther.* **1996**, *278*, 854–861.
12. Wilkinson, R. E.; Duncan, R. R.; Berry, C. Plant mineral contents of root tips from four sorghum cultivars after exposure to the anion channel blocker, SITS. *J. Plant Nutr.* **1994**, *17*, 2189–2192.
13. Lewis, R. S.; Ross, P. E.; Cahalan, M. D. Chloride channels activated by osmotic stress in T lymphocytes. *J. Gen. Physiol.* **1993**, *101*, 801–826.
14. Weaver, J. L.; Szabo, G., Jr.; Pine, P. S.; Gottesman, M. M.; Goldenberg, S.; Aszalos, A. The effect of ion channel blockers, immunosuppressive agents, and other drugs on the activity of the multi-drug transporter. *Int. J. Cancer* **1993**, *54*, 456–461.
15. Hidaka, J.; Ide, T.; Kawasaki, T.; Taguchi, T.; Kasai, M. Characterization of a chloride-channel from rabbit transverse tubules in the planar lipid bilayer system. *Biochem. Biophys. Res. Commun.* **1993**, *191*, 977–982.
16. Kokubun, S.; Saigusa, A.; Tamura, T. Blockade of chlorine channels by organic and inorganic blockers in vascular smooth muscle cells. *Pfluegers Arch.* **1991**, *418*, 204–213.
17. Yarbrough, J. W.; Merryman, J. I.; Barnhill, M. A.; Hahn, K. A. Inhibitors of intracellular chloride regulation induce cisplatin resistance in canine osteosarcoma cells. *In Vivo* **1999**, *13*, 375–384.
18. Beavis, A. D.; Davatol-Hag, H. The mitochondrial inner membrane anion channel is inhibited by DIDS. *J. Bioenerg. Biomembr.* **1996**, *28*, 207–214.
19. Luo, H.; Fanestil, D. D. Inhibition of binding of [^{3}H] metolazone to rat kidney membrane by stilbene disulfonates. *Biochem. Pharmacol.* **1991**, *41*, 1745–1750.
20. Klein-Nulend, J.; Raisz, L. G. Effects of two inhibitors of anion transport on bone resorption in organ culture. *Endocrinology* **1989**, *125*, 1019–1024.
21. Sato, N.; Suzuki, N.; Yamaguchi, M.; Yamaguchi, N.; Okuma, K. Proteolipid subunits of vacuolar H^+-ATPase (ATP6F) as tumor antigens, application to cancer therapy, and use of proton pump inhibitor as anticancer agent. Jpn. Kokai Tokkyo Koho JP 2001286284, 2001; *Chem. Abstr.* **2001**, *135*, 314438.
22. Laties, A.; Mitchell, C. P2X7 receptor-, A3 receptor-, and ATP-based methods for neuroprotection of retinal ganglion cells and treatment of glaucoma. PCT Int. Appl. WO 2007002139, 2007; *Chem. Abstr.* **2007**, *146*, 93596.
23. Okada, Y.; Tanabe, S. Drug screening for treatment of heart diseases. PCT Int. Appl. WO 2003014727, 2003; *Chem. Abstr.* **2003**, *138*, 131069.
24. Cardin, A. D.; Tyms, S. A. Sulfonic stilbene derivatives in the treatment of viral diseases. Eur. Pat. Appl. EP 498095, 1992; *Chem. Abstr.* **1992**, *117*, 208840.
25. Kurbanazarova, R. Sh.; Krasil'nikov, O. V.; Kragoe, E. D.; Sabirov, R. Z. Effect of anion transport inhibitors on hemolysis induced by melittin. *Dokl. Akad. Nauk Resp. Uzb.* **1996**, 47–49; *Chem. Abstr.* **1997**, *126*, 262222.
26. Koetters, P. J.; Chou, H.; Jonas, A. J. Lysosomal sulfate transport: inhibitor studies. *Biochim. Biophys. Acta, Biomembr.* **1995**, *1235*, 79–84.

SUDAN BLACK B

CAS Registry Number 4197-25-5

Chemical Structure

Physical Form Dark brown powder

Solubility Insoluble in water; soluble in ethanol, acetone, benzene, toluene, xylene, ethylene glycol

Melting Point 180–186 °C (decompose)

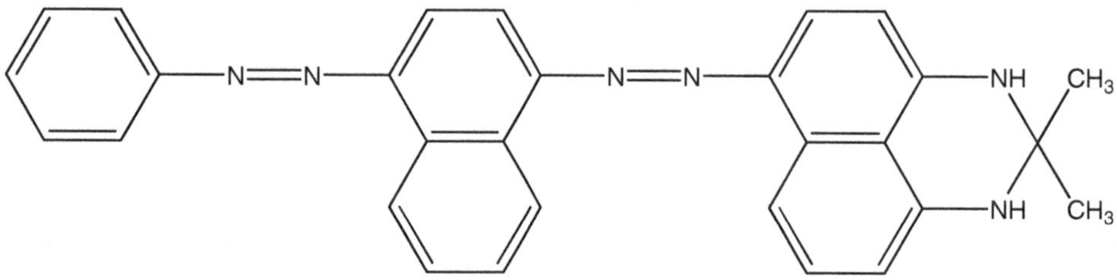

CA Index Name 1*H*-Perimidine, 2,3-dihydro-2,2-dimethyl-6-[2-[4-(2-phenyldiazenyl)-1-naphthalenyl] diazenyl]-

Other Names 1*H*-Perimidine, 2,3-dihydro-2,2-dimethyl-6-[[4-(phenylazo)-1-naphthalenyl]azo]-; C.I. Solvent Black 3; Sudan Black B; 2,3-Dihydro-2,2-dimethyl-6-[(4-phenylazo-1-naphthalenyl)-azo]-1*H*-perimidine; 2,3-Dihydro-2,2-dimethyl-6-[[4-(phenylazo)-1-naphthyl] azo]perimidine; 6-(1-Phenylazo-4-naphthylazo)-2,3-dihydro-2,2-dimethylperimidine; Acetylated Sudan Black B; Aizen SOT Black 6; C.I. 26150; Ceres Black BN; Chuo Sudan Black 141; Fast Black HBN; Fat Black HB; Fat Black HB 01; Hexatype Black B; Lacquer Black S; Lacquer Black VB; NSC 11239; Neptune Black X 60; Nubian Black BT; Oil Black 860; Oil Black BT; Oil Black HBB; Oil Black HZ; Orient Oil Black 860; Orient Oil Black HBB; SOT-Black 6; Solvent Black 3; Sudan Black 141; Sudan Black X 60; Sudan Deep Black BB; Sudan Deep Black BN; Typogen Black

Merck Index Number 8885

Chemical/Dye Class Azo

Molecular Formula $C_{29}H_{24}N_6$

Molecular Weight 456.54

Boiling Point (Calcd.) 726.0 ± 60.0 °C, pressure: 760 Torr

pK_a (Calcd.) 3.57 ± 0.40, most basic, temperature: 25 °C

Absorption (λ_{max}) 598 nm, 415 nm

Synthesis Synthetic methods[1–5]

Staining Applications Glycol methacrylate embedded tissue sections;[6] blast cells;[7–9] lipids;[10–12] lipid antigens;[13] lipoproteins;[14,15] apolipoproteins;[15] neuron-specific nuclear protein NeuN;[16] liposomes;[17] myelin;[18] mutants;[19] skin;[20] tissues;[21] keratin fibers;[22] hair[23]

Biological Applications Diagnosis of acute myeloid leukemia (AML);[7–9] detecting neuron-specific nuclear protein NeuN;[16] drug screening[24]

Industrial Applications Plasma display panels;[25] inks;[26–29] toners;[30] photographic materials;[31] adhesives;[32] pesticides;[33] leather;[34] textiles[34–36]

Safety/Toxicity Acute oral toxicity,[37] carcinogenicity[7,38]

Certification/Approval Certified by Biological Stain Commission (BSC)

REFERENCES

1. Senshu, H.; Otsuka, M. Purification of C.I. Solvent Black 3. Eur. Pat. Appl. EP 340792, 1989; *Chem. Abstr.* **1990**, *112*, 22344.
2. Patsch, M.; Ruske, M. Dyes containing imidazolylmethyl groups and their use. Eur. Pat. Appl. EP 34725, 1981; *Chem. Abstr.* **1982**, *96*, 36853.
3. Reiners, W. Isolation and thin-layer chromatography of fat-soluble dyes. *Fresenius' Z. Anal. Chem.* **1967**, *229*, 406–409; *Chem. Abstr.* **1967**, *67*, 87516.
4. Ikegami, H.; Izume, T.; Hiyama, H. Synthesis of Typophor Schwarz F. B. *Kagaku Kogyo* **1953**, *27*, 59–61; *Chem. Abstr.* **1955**, *49*, 67153.

5. Gurr, E. *Encyclopedia of Microscopic Stains*; Leonard Hill Books Ltd.: London, 1960; pp 376–377.
6. Cerri, P. S.; Sasso-Cerri, E. Staining methods applied to glycol methacrylate embedded tissue sections. *Micron* **2003**, *34*, 365–372.
7. Klobusicka, M.; Kusenda, J.; Babusikova, O. Myeloid enzymes profile related to the immunophenotypic characteristics of blast cells from patients with acute myeloid leukemia (AML) at diagnosis. *Neoplasma* **2005**, *52*, 211–218.
8. Ngan, M.; Chien, K.; Lee, S. Sudan black B positivity in acute lymphoblastic leukemia. *Mod. Pathol.* **1992**, *5*, 68–70.
9. Hoyle, C. F.; Gray, R. G.; Wheatley, K.; Swirsky, D.; de Bastos, M.; Sherrington, P.; Rees, J. K.; Hayhoe, F. G. Prognostic importance of Sudan Black positivity: a study of bone marrow slides from 1,386 patients with de novo acute myeloid leukaemia. *Br. J. Haematol.* **1991**, *79*, 398–407.
10. Seli, E.; Kayisli, U. A.; Selam, B.; Seli, M.; Arici, A. Estradiol suppresses vascular monocyte chemotactic protein-1 expression during early atherogenesis. *Am. J. Obstetrics Gynecol.* **2002**, *187*, 1544–1549.
11. Nozu, K. Lipid staining method. Jpn. Kokai Tokkyo Koho JP 11044673, 1999; *Chem. Abstr.* **1999**, *130*, 220184.
12. Haimovici, R.; Gantz, D. L.; Rumelt, S.; Freddo, T. F.; Small, D. M. The lipid composition of drusen, Bruch's membrane, and sclera by hot stage polarizing light microscopy. *Invest. Ophthalmol. Vis. Sci.* **2001**, *42*, 1592–1599.
13. Matsumoto, Y.; Kono, R.; Anraku, H. Lipophilic dye for immobilized lipid antigen for immunoassay of anti-phospholipid antibody. Jpn. Kokai Tokkyo Koho JP 08105891, 1996; *Chem. Abstr.* **1996**, *125*, 81277.
14. Machattie, C. T.; Berger, G. M. B. A comprehensive method for characterizing plasma lipoprotein and subfractions using gradient gel electrophoresis and immunoblotting. *Eur. J. Lab. Med.* **1995**, *3*, 87–94.
15. Koren, E.; Koscec, M. Antibodies to lipoproteins and apolipoproteins and methods of use thereof. PCT Int. Appl. WO 9600903, 1996; *Chem. Abstr.* **1996**, *124*, 197747.
16. Gill, S. K.; Ishak, M.; Rylett, R. J. Exposure of nuclear antigens in formalin-fixed, paraffin-embedded necropsy human spinal cord tissue: detection of NeuN. *J. Neurosci. Methods* **2005**, *148*, 26–35.
17. Yoshikawa, Y.; Miyazaki, M.; Houjou, T.; Komuta, Y.; Yoshikawa, H.; Takada, K. Biodistribution of cyclosporin encapsulated in liposomes modified with bioadhesive polymer. *J. Pharm. Pharmacol.* **1997**, *49*, 661–668.
18. Gerrits, P. O.; Brekelmans-Bartels, M.; Mast, L.; Gravenmade, E. J.; Horobin, R. W.; Holstege, G. Staining myelin and myelin-like degradation products in the spinal cords of chronic experimental allergic encephalomyelitis (Cr-EAE) rats using Sudan black B staining of glycol methacrylate-embedded material. *J. Neurosci. Methods* **1992**, *45*, 99–105.
19. Liu, M.; Gonzpalez, J. E.; Willis, L. B.; Walker, G. C. A novel screening method for isolating exopolysaccharide-deficient mutants. *Appl. Environ. Microbiol.* **1998**, *64*, 4600–4602.
20. Fujisawa, T. Ink compositions for skin marking. Jpn. Kokai Tokkyo Koho JP 07011183, 1995; *Chem. Abstr.* **1995**, *123*, 172955.
21. Baschong, W.; Suetterlin, R.; Laeng, R. H. Control of autofluorescence of archival formaldehyde-fixed, paraffin-embedded tissue in confocal laser scanning microscopy (CLSM). *J. Histochem. Cytochem.* **2001**, *49*, 1565–1571.
22. Brun, G. Composition for dyeing keratin fibers, comprising an electrophilic monomer, a hydrophobic direct dye and a liquid organic solvent. PCT Int. Appl. WO 2007118812, 2007; *Chem. Abstr.* **2007**, *147*, 491195.
23. Trigg, D. L.; Jones, J. L. Multistep hair color revitalizing kit comprising hair dye and glossing agent. U.S. Pat. Appl. Publ. US 2008189876, 2008; *Chem. Abstr.* **2008**, *149*, 251967.
24. Austen, M. Use of zebrafish or *Oryzias latipes* larvae as metabolic disease models and for drug screening. PCT Int. Appl. WO 2004044241, 2004; *Chem. Abstr.* **2004**, *140*, 405029.
25. Muroi, S.; Suwa, T. Neon light-cutting adhesive compositions and sheets, optical filters, and plasma display panels. Jpn. Kokai Tokkyo Koho JP 2008266439, 2008; *Chem. Abstr.* **2008**, *149*, 521435.
26. Otsubo, A.; Takagishi, I. Oil-based inks with smooth writing properties for ball-point pens. Jpn. Kokai Tokkyo Koho JP 2008274144, 2008; *Chem. Abstr.* **2008**, *149*, 536065.
27. Ogawa, H. Inks with good storage stability for textile printing and coloring agents for them. Jpn. Kokai Tokkyo Koho JP 2008069203, 2008; *Chem. Abstr.* **2008**, *148*, 404719.
28. Hiroshima, S. Ink jet recording method and apparatus using two kinds of inks. Jpn. Kokai Tokkyo Koho JP 2008000909, 2008; *Chem. Abstr.* **2008**, *148*, 109104.

29. Okabe, E. Oil-based black writing inks. Jpn. Kokai Tokkyo Koho JP 2006036951, 2006; *Chem. Abstr.* **2006**, *144*, 214506.

30. Maeda, S.; Hotsuta, Y.; Juchi, M.; Yoneda, S.; Yamada, Y. Electrophotographic color toners containing metal complex salts. Jpn. Kokai Tokkyo Koho JP 07043932, 1995; *Chem. Abstr.* **1995**, *123*, 70289.

31. Honda, N.; Ikeda, T.; Aizawa, T. Dry developer for electrostatic photography. Ger. Offen. DE 2828890, 1979; *Chem. Abstr.* **1979**, *91*, 47313.

32. Hoffmann, G.; Richter, P. Adhesives for relief printing plates. Eur. Pat. Appl. EP 59385, 1982; *Chem. Abstr.* **1983**, *98*, 225306.

33. Ian Malcolm, S.; Wright, T.; Perrin, B.; Mulqueen, P.; Waller, A.; Pierce, A. Capsule formulation containing pesticide and photoprotectant. PCT Int. Appl. WO 2006089747, 2006; *Chem. Abstr.* **2006**, *145*, 286603.

34. Peter, H. J. Protective impregnation spray for leather and textiles. Ger. Offen. DE 3438645, 1986; *Chem. Abstr.* **1986**, *105*, 193313.

35. Nelson, R. T. Liquid agent for treating textiles. Ger. Offen. DE 2732985, 1978; *Chem. Abstr.* **1078**, *88*, 154294.

36. Li, D.; Sun, G. Kinetics of thermo-fixation of solvent dyes and pigments in polyester fibres. *Color. Technol.* **2006**, *122*, 194–200.

37. Schafer, E. W., Jr.; Bowles, W. A., Jr.; Hurlbut, J. The acute oral toxicity, repellency, and hazard potential of 998 chemicals to one or more species of wild and domestic birds. *Arch. Environ. Contam. Toxicol.* **1983**, *12*, 355–382.

38. Shin, S. Y.; Koo, S. H.; Kwon, K. C.; Park, J. W.; Ko, C. S.; Jo, D. Y. Monosomy 7 as the sole abnormality of an acute basophilic leukemia. *Cancer Genet. Cytogenet.* **2007**, *172*, 168–171.

SUDAN III

CAS Registry Number 85-86-9

Chemical Structure

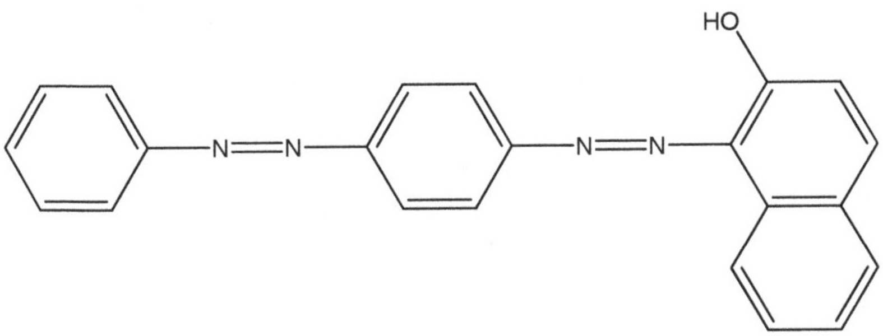

CA Index Name 2-Naphthalenol, 1-[2-[4-(2-phenyldiazenyl)phenyl]diazenyl]-

Other Names 2-Naphthalenol, 1-[[4-(phenylazo)phenyl]azo]-; 2-Naphthol, 1-(*p*-phenylazophenylazo)-; C.I. Solvent Red 23; 1-(*p*-Phenylazophenylazo)-2-naphthol; 111440 Red; Brasilazina Oil Scarlet; C.I. 26100; Certiqual Oil Red; D and C Red No. 17; D&C Red No. 17; FD and C Red No. 17; Fast Oil Scarlet III; Fat Red Bluish; Fat Red HRR; Fat Red R; Fat Red RS; Fat Scarlet LB; Fat Soluble Red Zh; Grasal Brilliant Red G; Grasan Brilliant Red G; Japan Red 225; Japan Red No. 225; NSC 65825; NSC 8995; Oil Red 3G; Oil Red AS; Oil Red DR 126; Oil Red Extra; Oil Scarlet G; Organol Red BS; Organol Scarlet; Red No. 225; Red Zh; Silotras Scarlet TB; Solvent Red 23; Somalia Red III; Stearix Scarlet; Sudan 3; Sudan III; Sudan P III; Sudan Red III; Sudan Red BK; Tetrazobenzene-β-naphthol; Toney Red; Tony Red

Merck Index Number 8884

Chemical/Dye Class Azo

Molecular Formula $C_{22}H_{16}N_4O$

Molecular Weight 352.39

Physical Form Brown-red powder

Solubility Insoluble in water; soluble in benzene, chloroform; moderately soluble in acetone, ether, petroleum ether; slightly soluble in ethanol, xylene

Melting Point 195 °C

Boiling Point (Calcd.) 584.6 ± 35.0 °C, pressure: 760 Torr

pK_a (Calcd.) 13.44 ± 0.50, most acidic, temperature: 25 °C

Absorption (λ_{max}) 507 nm, 354 nm

Synthesis Synthetic methods[1–6]

Staining Applications Lipids;[7–11] fats;[12] skin;[13–20] lips;[20] lipsticks;[21] contact lens;[22] eye shadow;[23] keratin fibers;[24] hairs;[25] latex particles;[26,27] nail enamel;[28] sunscreen;[29,30] spinal fluid[31]

Biological Applications Antiseptic;[32] preventing prostate cancer;[33] medical devices;[34] dental impression materials[35]

Industrial Applications Plasma display panels;[36] liquid crystal displays;[37] dielectric materials;[38] photoresists;[39,40] inks;[41] toners;[42] pesticides;[43] cleansing products;[44] textiles;[45] fabric softener;[46] oils;[47] fuel cells;[48] petroleum products[49]

Safety/Toxicity Acute cytogenetic effect;[50] carcinogenicity;[33,51,52] chromosomal aberration;[53] genotoxicity;[54,55] mutagenicity;[56] neurotoxicity;[57] skin toxicity[58]

Certification/Approval Certified by Biological Stain Commission (BSC)

REFERENCES

1. Bouchoul, A.; Paillous, P.; Bouaoud, S. E. Solvatochromic effect on azo-quinonehydrazone equilibrium of azonaphthol dyes. *J. Soc. Algerienne Chim.* **2003**, *13*, 49–62.
2. Walba, D. M.; Homan, T. C.; Richards, R. M.; Haltiwanger, R. C. Topological stereochemistry. 9. Synthesis and cutting “in half” of a molecular Moebius strip. *New J. Chem.* **1993**, *17*, 661–681.
3. Kuzukawa, M.; Okada, S.; Kanbe, R.; Ikeda, Y.; Yoshimura, K. Preparation of Red 225. Jpn. Kokai Tokkyo Koho JP 04028770, 1992; *Chem. Abstr.* **1992**, *117*, 9815.
4. Longo, M. L.; Torre, M. Dyeing properties and substantivity of disperse dyes derived from substituted 4-aminoazobenzenes on hydrophobic fibers. *Ann. Chim.* **1980**, *70*, 173–181.

5. Reiners, W. Isolation and thin-layer chromatography of fat-soluble dyes. *Fresenius' Z. Anal. Chem.* **1967**, *229,* 406–409.
6. Christman, J. F.; Cunningham, G. L. Chromatography and biological stains. V. Isolation and spectral analysis of the orange fat-staining component of Sudan III. *Stain Technol.* **1953**, *28*, 275–277.
7. Ershov, V. A.; Sidorin, V. S.; Ryllo, A. G. Method for differential diagnosis of primary and secondary liver cancer involving cytological and cytochemical analysis of tumor trepanobiopsy specimens stained with azure-eosine, alcian blue, or Sudan III to determine mucin or lipid inclusions. Russ. RU 2327422, 2008; *Chem. Abstr.* **2008**, *149*, 76596.
8. Zhang, M.; Wu, H. Q.; Hu, S. W.; Wen, J. M. Pathologic differential diagnosis of lipid storage myopathy from glycogen storage myopathy by histochemistry and immunohistochemistry. *Jiepouxue Yanjiu* **2007**, *29*, 361–363; *Chem. Abstr.* **2008**, *148*, 556557.
9. Ohira, T.; Ishikawa, K.; Masuda, I.; Yokoyama, M.; Honda, I. Histologic localization of lipid in the articular tissues in calcium pyrophosphate dihydrate crystal deposition disease. *Arthritis Rheum.* **1988**, *31*, 1057–1062.
10. Fukumoto, S.; Fujimoto, T. Deformation of lipid droplets in fixed samples. *Histochem. Cell Biol.* **2002**, *118*, 423–428.
11. Matsumoto, Y.; Kono, R.; Anraku, H. Lipophilic dye for immobilized lipid antigen for immunoassay of anti-phospholipid antibody. Jpn. Kokai Tokkyo Koho JP 08105891, 1996; *Chem. Abstr.* **1996**, *125*, 81277.
12. Fine, K. D.; Ogunji, F. A new method of quantitative fecal fat microscopy and its correlation with chemically measured fecal fat output. *Am. J. Clin. Pathol.* **2000**, *113*, 528–534.
13. Arnaud, P.; Feltin, C. Method for coloring of the skin comprising direct dye. Fr. Demande FR 2909873, 2008; *Chem. Abstr.* **2008**, *149*, 61572.
14. Thevenet, L.; Chevalier, G.; Girier-Dufournier, F. Method of making-up dark skin by using cosmetic compositions comprising red interference pigment capable of generating highlights when applied to skin. U.S. Pat. Appl. Publ. US 2008019933, 2008; *Chem. Abstr.* **2008**, *148*, 174945.
15. Arnaud, P.; Feltin, C. Method for coloring the skin of men. PCT Int. Appl. WO 2008074654, 2008; *Chem. Abstr.* **2008**, *149*, 86992.
16. Bazelaire, E.; Petit, C. B.; Arnaud, P.; Feltin, C. Process and composition for coloring of dark skin. Fr. Demande FR 2901474, 2007; *Chem. Abstr.* **2007**, *148*, 16895.
17. Haynes, C. T.; Bronaugh, R. L.; Yourick, J. J. *In vitro* dermal absorption and metabolism of D&C RED No.17. *J. Cosmet. Sci.* **2005**, *56*, 376–377.
18. Arnaud, P.; Feltin, C.; Bazelaire, E. Process for coloring of dark skin using a direct dye. Fr. Demande FR 2901475, 2007; *Chem. Abstr.* **2007**, *148*, 39160.
19. Yourick, J. J.; Sasik, C. T.; Bronaugh, R. L. *In vitro* dermal absorption and metabolism of D&C red no. 17 in human and porcine skin. *J. Cosmet. Sci.* **2007**, *58*, 255–266.
20. Dumousseaux, C. Cosmetic composition for application to the skin and lips containing pigments. Fr. Demande FR 2845277, 2004; *Chem. Abstr.* **2004**, *140*, 326628.
21. Lee, S. M.; Sung, G. H. Lipsticks comprising dextrin esters and oils. Repub. Korean Kongkae Taeho Kongbo KR 2008079033 2008; *Chem. Abstr.* **2008**, *149*, 499396.
22. Wu, W. Manufacture of color contact lens with three-dimensional effects. Faming Zhuanli Shenqing Gongkai Shuomingshu CN 101135782, 2008; *Chem. Abstr.* **2008**, *148*, 387371.
23. Bouchard, F. Eye shadow comprising least 50% colorants. Eur. Pat. Appl. EP 1757262, 2007; *Chem. Abstr.* **2007**, *146*, 280374.
24. Brun, G. Composition for dyeing keratin fibers, comprising an electrophilic monomer, a hydrophobic direct dye and a liquid organic solvent. PCT Int. Appl. WO 2007118812, 2007; *Chem. Abstr.* **2007**, *147*, 491195.
25. L'Oreal, Fr. Hair dye composition comprising an electrophilic monomer, a hydrophobic direct dye, and a liquid organic solvent. Fr. Demande FR 2899816, 2007; *Chem. Abstr.* **2007**, *147*, 474290.
26. Ooishi, K.; Obana, S. Manufacture of colored latex particles. Jpn. Kokai Tokkyo Koho JP 06306108, 1994; *Chem. Abstr.* **1995**, *123*, 56987.
27. Ooishi, K.; Obana, S. Preparation of colored chlorinated latex particles. Jpn. Kokai Tokkyo Koho JP 07062019, 1995; *Chem. Abstr.* **1995**, *123*, 171463.
28. Socci, R. L. Nail enamel composition comprising dye having a decorative color effect. U.S. Pat. Appl. Publ. US 2007231281, 2007; *Chem. Abstr.* **2007**, *147*, 412719.
29. Fowler, K. C. Sunscreen having disappearing color indicator. U.S. Patent 6042813, 2000; *Chem. Abstr.* **2000**, *132*, 241698.

30. Simon, J. C. Makeup products combining a photochromic pigment and a sunscreen. Fr. Demande FR 2780275, 1999; *Chem. Abstr.* **2000**, *132*, 185252.
31. Masuoka, S.; Yoneda, K.; Oka, M. Apparatus and method for analyzing spinal fluid by staining and imaging analysis. Jpn. Kokai Tokkyo Koho JP 2004132787, 2004; *Chem. Abstr.* **2004**, *140*, 371466.
32. Raad, I.; Hanna, H. A.; Nabulsi, N. Antiseptic composition with broad spectrum antimicrobial activity containing an active compound bound to a dye. PCT Int. Appl. WO 2002082907, 2002; *Chem. Abstr.* **2002**, *137*, 304750.
33. Brooks, J. D.; Goldberg, M. F.; Nelson, L. A.; Wu, D.; Nelson, W. G. Identification of potential prostate cancer preventive agents through induction of quinone reductase *in vitro*. *Cancer Epidemiol. Biomarkers Prev.* **2002**, *11*, 868–875.
34. Raad, I. Coating medical device surfaces with antimicrobial agents. PCT Int. Appl. WO 2007062306, 2007; *Chem. Abstr.* **2007**, *147*, 39321.
35. Oxman, J. D.; Konings, M. S.; Tiers, G. V. D.; Vogel, K. M.; Vogel, D. E. Dental impression material with cure-indicating dye. PCT Int. Appl. WO 9600560, 1996; *Chem. Abstr.* **1996**, *124*, 242386.
36. Abura, T.; Uchida, T.; Iguchi, Y. Method for manufacturing non-phosphor layers for plasma display panels. Jpn. Kokai Tokkyo Koho JP 2006225500, 2006; *Chem. Abstr.* **2006**, *145*, 281263.
37. Hughes, C. J.; Coates, D.; Alkins, A. O. Cholesteric liquid crystal display device. PCT Int. Appl. WO 2006054050, 2006; *Chem. Abstr.* **2006**, *145*, 17884.
38. Fujiwara, T.; Watanabe, H.; Inagaki, T. Ultraviolet-curable polymer dielectric material and film capacitor using it. Jpn. Kokai Tokkyo Koho JP 10321458, 1998; *Chem. Abstr.* **1998**, *130*, 74914.
39. Kohara, H.; Miyabe, M.; Tokutake, N.; Nakayama, T.; Asaumi, S.; Tanaka, H.; Arai, Y. Positive-working resist compositions containing azo dyes. Ger. Offen. DE 3720017, 1987; *Chem. Abstr.* **1988**, *109*, 83507.
40. Pampalone, T. R.; Kuyan, F. A. Contrast enhancing additives for positive photoresist. *J. Electrochem. Soc.* **1988**, *135*, 471–476.
41. Mehta, R.; Yeager, S. L.; Webendorfer, S. D. Chemically reactive security ink, use of security ink, and security documents incorporating security ink. U.S. Pat. Appl. Publ. US 2007281139, 2007; *Chem. Abstr.* **2007**, *148*, 35215.
42. Watanabe, H.; Shoji, H.; Kawakado, K.; Seto, N. Photoconductive toners. Jpn. Kokai Tokkyo Koho JP 60153054, 1985; *Chem. Abstr.* **1986**, *104*, 43162.
43. Ian Malcolm, S.; Wright, T.; Perrin, B.; Mulqueen, P.; Waller, A.; Pierce, A. Capsule formulation containing pesticide and photoprotectant. PCT Int. Appl. WO 2006089747, 2006; *Chem. Abstr.* **2006**, *145*, 286603.
44. Krzysik, D. G.; Utschig, J. M.; Cole, D. B. Color changing liquid cleansing products containing surfactants, electrolytes and coloring agents. U.S. Pat. Appl. Publ. US 2005148490, 2005; *Chem. Abstr.* **2005**, *143*, 120038.
45. Schmidt, A.; Bach, E.; Schollmeyer, E. Supercritical fluid dyeing of cotton modified with 2,4,6-trichloro-1,3,5-triazine. *Color. Technol.* **2003**, *119*, 31–36.
46. Wahl, E. H.; Vimba, A. I. Liquid fabric softener compositions containing dyes to provide a pink color. U.S. Patent 4897208, 1990; *Chem. Abstr.* **1990**, *112*, 201172.
47. Ishihara, M. Lubricating oil composition for power steering of automobiles. Jpn. Kokai Tokkyo Koho JP 02265998, 1990; *Chem. Abstr.* **1991**, *114*, 167634.
48. Abe, S.; Egawa, H.; Takemoto, S.; Ito, N. Fuel cell stack coolant composition. Eur. Pat. Appl. EP 1457542, 2004; *Chem. Abstr.* **2004**, *141*, 263417.
49. Smith, M. J.; Desai, B.; Frederico, J. J. Molecular tags for organic solvent systems, especially petroleum products. PCT Int. Appl. WO 2003022942, 2003; *Chem. Abstr.* **2003**, *138*, 257641.
50. Fujie, K.; Ito, Y.; Maeda, S. Acute cytogenetic effect of benzene on rat bone marrow cells *in vivo* and the effect of inducers or inhibitors of drug-metabolizing enzymes. *Mutat. Res.* **1992**, *298*, 81–90.
51. Mamber, S. W.; Bryson, V.; Katz, S. E. Evaluation of the *Escherichia coli* K12 inductest for detection of potential chemical carcinogens. *Mutat. Res.* **1984**, *130*, 141–151.
52. Mamber, S. W.; Bryson, V.; Katz, S. E. The *Escherichia coli* WP2/WP100 rec assay for detection of potential chemical carcinogens. *Mutat. Res.* **1983**, 119, 135–144.
53. Ito, Y.; Maeda, S.; Souno, K.; Ueda, N.; Sugiyama, T. Induction of hepatic glutathione transferase and suppression of 7,12-dimethylbenz[*a*]anthracene-induced chromosome aberrations in rat bone marrow cells by Sudan III and related azo dyes. *J. Natl. Cancer Inst.* **1984**, *73*, 177–183.
54. Hatakeyama, S.; Hayasaki, Y.; Masuda, M.; Kazusaka, A.; Fujita, S. Mechanism for mouse strain differences in the protective effect of Sudan III against the *in vivo* genotoxicity of 7,12-dimethylbenz[*a*]anthracene. *Toxicol. Lett.* **1996**, *89*, 231–239.

55. Hatakeyama, S.; Hayasaki, Y.; Masuda, M.; Kazusaka, A.; Fujita, S. Paradoxical effect of Sudan III on the *in vivo* and *in vitro* genotoxicity elicited by 7,12-dimethylbenz(*a*)anthracene. *J. Biochem. Toxicol.* **1995**, *10*, 143–149.

56. Muzzall, J. M.; Cook, W. L. Mutagenicity test of dyes used in cosmetics with the *Salmonella*/mammalian-microsome test. *Mutat. Res.* **1979**, *67*, 1–8.

57. Romero, F. J.; Roma, J.; Bosch-Morell, F.; Romero, B.; Segura-Aguilar, J.; Llombart-Bosch, A.; Ernster, L. Reduction of brain antioxidant defense upon treatment with butylated hydroxyanisole (BHA) and Sudan III in Syrian golden hamster. *Neurochem. Res.* **2000**, *25*, 389–393.

58. Xie, Z.; Hayakawa, R.; Sugiura, M.; Kojima, H.; Konishi, H.; Ichihara, G.; Takeuchi, Y. Experimental study on skin sensitization potencies and cross-reactivities of hair-dye-related chemicals in guinea pigs. *Contact Dermatitis* **2000**, *42*, 270–275.

SUDAN IV

CAS Registry Number 85-83-6

Chemical Structure

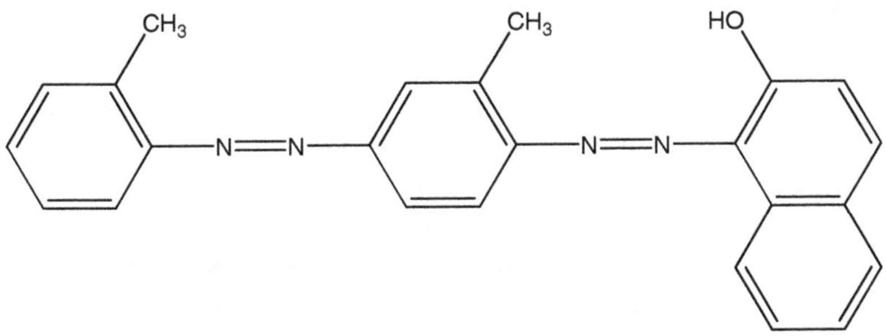

CA Index Name 2-Naphthalenol, 1-[2-[2-methyl-4-[2-(2-methylphenyl)diazenyl]phenyl]diazenyl]-

Other Names 2-Naphthalenol, 1-[[2-methyl-4-[(2-methylphenyl)azo]phenyl]azo]-; 2-Naphthol, 1-(4-*o*-tolylazo-*o*-tolylazo)-; C.I. Solvent Red 24; 1-(2′-Methyl-4′-(2″-methylphenylazo)phenylazo)-2-naphthol; 1-[4-(*o*-Tolylazo)-*o*-tolylazo]-2-naphthol; 2′,3-Dimethyl-4-(2-hydroxynaphthylazo)azobenzene; Aizen SOT Red 1; Atlasol Red 4B; Biebrich Scarlet BPC; Biebrich Scarlet Red; Brasilazina Oil Red B; C.I. 26105; Candle Scarlet 2B; Candle Scarlet B; Candle Scarlet G; Ceres Red BB; Dispersol Red PP; Enial Red IV; Fast Oil Red B; Fast Red BB; Fat Ponceau R; Fat Red 2B; Fat Red B; Fat Red BB; Fat Red BS; Fat Red TS; Fat Soluble Dark Red; Grasal Brilliant Red B; Grasan Brilliant Red B; Hidaco Oil Red; Japan Red 501; Lacquer Red V; Lacquer Red VS; NSC 10472; Oil Red; Oil Red 282; Oil Red 2B; Oil Red 3; Oil Red 3B; Oil Red 47; Oil Red 7; Oil Red A; Oil Red APT; Oil Red B; Oil Red BB; Oil Red BS; Oil Red D; Oil Red ED; Oil Red F; Oil Red GO; Oil Red RC; Oil Red RR; Oil Red S; Oil Red SST Extra; Oil Red TAX; Oil Red ZD; Oil Scarlet; Oil Scarlet 48; Oleal Red BB; Oleosol Red BB; Organol Red B; Orient Oil Red RR; Plastoresin Red F; Red 3R Soluble in Grease; Resinol Red 2B; SOT Red 1; Scarlet R; Scarlet R (solvent dye); Scarlet oil; Scarlet red; Silotras Red T 3B; Solvent Red 24; Somalia Red IV; Stearix Red 4B; Stearix Red 4S; Sudan (IV) dye; Sudan 4; Sudan IV; Sudan P; Sudan Red 380; Sudan Red 4BA; Sudan Red BB; Sudan Red BBA; Sudan Red IV; TU 6-14-8780; Tertrogras Red N; Toyo Oil Red BB; Waxoline Red O; Waxoline Red OM; Waxoline Red OS; *o*-Tolueneazo-*o*-toluene-β-naphthol; *o*-Tolylazo-*o*-tolylazo-β-naphthol; *o*-Tolylazo-*o*-tolylazo-2-naphthol

Merck Index Number 8393

Chemical/Dye Class Azo

Molecular Formula $C_{24}H_{20}N_4O$

Molecular Weight 380.44

Physical Form Dark red-brown powder

Solubility Insoluble in water; soluble in benzene, methanol, acetone, isopropanol; slightly soluble in ethanol

Melting Point 184–186 °C (decompose)

Boiling Point (Calcd.) 618.8 ± 55.0 °C, pressure: 760 Torr

pK_a (Calcd.) 13.52 ± 0.50, most acidic, temperature: 25 °C

Absorption (λ_{max}) 520 nm, 357 nm

Synthesis Synthetic methods[1–4]

Staining Applications Fat globules;[5] fats;[6] lipids;[6–9,13] fish oil;[10] lipoproteins;[11] CYP1A1 proteins;[12] proteins;[13] starch;[13] nucleic acids;[12,14,15] tissues;[16,17] waxes[18]

Biological Applications Detecting atherosclerosis;[10,11] diapers;[19] skin care products;[20] shampoos;[21] hair colors;[22] stents;[23] dental impression materials[24]

Industrial Applications Plasma display panels;[25] liquid crystal display device;[26] color filters;[27] semiconductor devices;[28] photoresists;[29,30] recording materials;[31] inks;[32,33] toners;[34] lenses;[35] pesticides;[36] laundry detergents;[37] textiles;[38,39] clay;[40] gasoline;[41,42] diesel fuels;[42] petroleum products[43]

Safety/Toxicity Mutagenicity[44–46]

Certification/Approval Certified by Biological Stain Commission (BSC)

REFERENCES

1. Pawlowski, N. E.; Norton, K. A. Water-soluble dyes with aminoalkyl groups and jet-printing inks containing them. Eur. Pat. Appl. EP 369643, 1990; *Chem. Abstr.* **1990**, *113*, 117062.
2. Proctor, G. B.; Horobin, R. W. Purification of Oil Red O using preparative paper chromatography. *Stain Technol.* **1985**, *60*, 247–248.
3. Shirado, Y. Soluble azo dyes. Jpn. Kokai Tokkyo Koho JP 49003923, 1974; *Chem. Abstr.* **1974**, *81*, 27193.
4. Komatsu, K.; Kuroki, N.; Konishi, K. Dyes for polypropylene fibers. *Bull. Univ. Osaka Pref., Ser. A* **1963**, *12*, 77–90.
5. Cherkasov, V. A.; Litvinenko, S. G.; Rudakov, A. G. Method for diagnosis of traumatic cerebral fat embolism involving microscopic determination of fat globules stained with Sudan IV in cerebrospinal fluid. Russ. RU 2176798, 2001; *Chem. Abstr.* **2002**, 137, 246069.
6. Mochizuki, S.; Moriya, T.; Naganuma, H.; Sasano, H.; Saito, T. Fat staining on Epon-embedded specimens: significance of serial sectioning for electron and light microscopy. *Byori Rinsho* **2000**, *18*, 1255–1259; *Chem. Abstr.* **2000**, *133*, 331621.
7. Richardson, M.; De Reske, M.; Delaney, K.; Fletch, A.; Wilcox, L. H.; Kinlough-Rathbone, R. L. Respiratory infection in lipid-fed rabbits enhances sudanophilia and the expression of VCAM-1. *Am. J. Pathol.* **1997**, *151*, 1009–1017.
8. Mourelle, M.; Perez, V. M.; Rojkind, M. Lipid quantitation in formalin-fixed liver sections. *J. Histochem. Cytochem.* **1988**, *36*, 1471–1474.
9. Bowyer, D. E. Recovery of lipids from aortas stained with Sudan IV. Comments. *Atherosclerosis* **1977**, *26*, 387–388.
10. Barbeau, M. L.; Klemp, K. F.; Guyton, J. R.; Rogers, K. A. Dietary fish oil. Influence on lesion regression in the porcine model of atherosclerosis. *Arterioscler. Thromb. Vasc. Biol.* **1997**, *17*, 688–694.
11. Xiao, W.; Scott, T. M.; Feng, L.; Yu, Z.; Wang, L.; Hughes, J. A.; Liu, H. Acetylated low-density lipoprotein-encapsulated cholesteryl 1,3-diiopanoate glyceryl ether for the detection of atherosclerosis in rabbits. *J. Nucl. Med.* **2003**, *44*, 770–773.
12. Refat, N. A. G. A.; Ibrahim, Z. S.; Moustafa, G. G.; Sakamoto, K. Q.; Ishizuka, M.; Fujita, S. The induction of cytochrome P450 1A1 by Sudan dyes. *J. Biochem. Mol. Toxicol.* **2008**, *22*, 77–84.
13. Ubach, M. C.; Jurado, L. A.; Vinjamoori, D. V.; Das, P.; Krohn, B.; Modiano, S. H. Method of predicting a trait of interest in plants by measuring the degree of starch–protein association. U.S. Pat. Appl. Publ. US 2007240241, 2007; *Chem. Abstr.* **2007**, *147*, 422488.
14. Ji, Y. B.; Ji, C. F.; Gao, S. Y.; Lang, L. A study of the effect of sudan I, III, and IV on the DNA/RNA ratio and 3D structure of HepG-2 using LCM. *J. Harbin Inst. Technol.* **2006**, *13*, 173–177.
15. Yamaguchi, T. Adsorption of carcinogenic and/or mutagenic pigments on DNA-binding Sepharose. *Agric. Biol. Chem.* **1988**, *52*, 845–847.
16. Li, L.; Hu, Z.; Xu, J.; Hu, W.; Chen, R.; Li, R. Method for staining tissue section for medical use at low cost and high efficiency. Faming Zhuanli Shenqing Gongkai Shuomingshu CN 101251449, 2008; *Chem. Abstr.* **2008**, *149*, 373328.
17. Juarranz, A.; Horobin, R. W.; Proctor, G. B. Prediction of *in situ* fluorescence of histochemical reagents using a structure–staining correlation procedure. *Histochemistry* **1986**, *84*, 426–431.
18. Russell, S. E.; Webster, J. R. Solid, shaped and colored wax articles. PCT Int. Appl. WO 2000022037, 2000; *Chem. Abstr.* **2000**, *132*, 295356.
19. Kataoka, T. Polyurethane elastic fiber roll containing hydrophobic dyes for disposable diapers. Jpn. Kokai Tokkyo Koho JP 2005154909, 2005; *Chem. Abstr.* **2005**, *143*, 65549.
20. Homola, A. M.; Dunton, R. K.; Pitts, G. Skin care and protective compositions containing transfer agents and barrier materials. PCT Int. Appl. WO 2000038617, 2000; *Chem. Abstr.* **2000**, *133*, 94281.
21. Castilla C., Carlos, A. Fast-acting antifungal and antibacterial shampoo. Span. ES 2061407, 1994; *Chem. Abstr.* **1995**, *122*, 248020.
22. Trigg, D. L.; Jones, J. L. Multistep hair color revitalizing kit comprising hair dye and glossing agent. U.S. Pat. Appl. Publ. US 2008189876, 2008; *Chem. Abstr.* **2008**, *149*, 251967.
23. Cao, C.; Lv, R.; Zhang, C.; Zou, Y.; Song, G.; Zhang, C. Properties and structure character of new type collagen stent materials. *Beijing Keji Daxue Xuebao* **2007**, *29*, 193–197; *Chem. Abstr.* **2007**, *147*, 101747.
24. Oxman, J. D.; Konings, M. S.; Tiers, G. V. D.; Vogel, K. M.; Vogel, D. E. Dental impression material with cure-indicating dye. PCT Int. Appl. WO 9600560, 1996; *Chem. Abstr.* **1996**, *124*, 242386.

25. Yamamoto, Y.; Kawabe, K. Photosensitive paste for manufacturing plasma display panels. Jpn. Kokai Tokkyo Koho JP 2008224940, 2008; *Chem. Abstr.* **2008**, *149*, 390347.

26. Hughes, C. J.; Coates, D.; Alkins, A. O. Cholesteric liquid crystal display device. PCT Int. Appl. WO 2006054050, 2006; *Chem. Abstr.* **2006**, *145*, 17884.

27. Matsumoto, Y. Method for manufacturing color filters for optical imaging devices. Jpn. Kokai Tokkyo Koho JP 2004020897, 2004; *Chem. Abstr.* **2004**, *140*, 119948.

28. Tanaka, T.; Masaki, T.; Ogata, T. Ceramic green sheets suitable for semiconductor devices. Jpn. Kokai Tokkyo Koho JP 10330167, 1998; *Chem. Abstr.* **1998**, *130*, 16613.

29. Uchigawa, K.; Komano, H.; Aoyama, T.; Oota, K. Photoresist composition for forming light-shielding film. Jpn. Kokai Tokkyo Koho JP 06051499, 1994; *Chem. Abstr.* **1994**, *121*, 289664.

30. Pampalone, T. R.; Kuyan, F. A. Contrast enhancing additives for positive photoresist. *J. Electrochem. Soc.* **1988**, *135*, 471–476.

31. Nogiwa, T.; Hotsuta, Y.; Konagaya, Y.; Morohoshi, K.; Kawaguchi, M.; Suzuki, A. Reversible thermographic recording materials providing visible images. Jpn. Kokai Tokkyo Koho JP 04303681, 1992; *Chem. Abstr.* **1993**, *119*, 105972.

32. Hiroshima, S. Jet printing inks giving images with excellent lightfastness and wide color reproduction. Jpn. Kokai Tokkyo Koho JP 2008024853, 2008; *Chem. Abstr.* **2008**, *148*, 193765.

33. Watarai, Y. Ink-jet inks applicable to mimeographic printing and ink-recorded articles. Jpn. Kokai Tokkyo Koho JP 2006117889, 2006; *Chem. Abstr.* **2006**, *144*, 452094.

34. Kaneko, G. Electrostatic latent image development magenta toner. Jpn. Kokai Tokkyo Koho JP 08137135, 1996; *Chem. Abstr.* **1996**, *125*, 181193.

35. Jethmalani, J. M.; Abdelsadek, G.; Cetin, E.; Mccarty, S.; Ge, J. Tinted lenses containing multiple dyes for high order aberrations. PCT Int. Appl. WO 2007120427, 2007; *Chem. Abstr.* **2007**, *147*, 491746.

36. Ian Malcolm, S.; Wright, T.; Perrin, B.; Mulqueen, P.; Waller, A.; Pierce, A. Capsule formulation containing pesticide and photoprotectant. PCT Int. Appl. WO 2006089747, 2006; *Chem. Abstr.* **2006**, *145*, 286603.

37. Li, H.; Wang, Y.; Hu, Y. Spectrometric determination of detergency of laundry detergents using sudan red VI as label. *Tenside Surfact. Deterg.* **2006**, *43*, 295–298.

38. Li, D.; Sun, G. Kinetics of thermo-fixation of solvent dyes and pigments in polyester fibres. *Color. Technol.* **2006**, *122*, 194–200.

39. Batchelor, S. N.; Bird, J. M. Method for laundry treating textile garment. PCT Int. Appl. WO 2006045375, 2006; *Chem. Abstr.* **2006**, *144*, 452334.

40. Kimura, S.; Yuyama, N.; Tanaka, S. Colorant for clay and water-based setting type colored synthetic clay therewith. Jpn. Kokai Tokkyo Koho JP 2005272766, 2005; *Chem. Abstr.* **2005**, *143*, 368072.

41. Gau, Y. Coloring agent for gasoline. *Petrotech* **2007**, *30*, 220–221; *Chem. Abstr.* **2007**, *148*, 541128.

42. He, Y.; Yu, D.; Tang, C.; Liu, B. Manufacture of mixed fuel containing methanol and gasoline or diesel fuel. Faming Zhuanli Shenqing Gongkai Shuomingshu CN 1912074, 2007; *Chem. Abstr.* **2007**, *146*, 254996.

43. Smith, M. J.; Desai, B.; Frederico, J. J. Molecular tags for organic solvent systems, especially petroleum products. PCT Int. Appl. WO 2003022942, 2003; *Chem. Abstr.* **2003**, *138*, 257641.

44. Klopman, G.; Frierson, M. R.; Rosenkranz, H. S. The structural basis of the mutagenicity of chemicals in *Salmonella typhimurium*: the Gene-Tox data base. *Mutat. Res.* **1990**, *228*, 1–50.

45. Rosenkranz, H. S.; Klopman, G. Structural basis of the mutagenicity of phenylazoaniline dyes. *Mutat. Res.* **1989**, *221*, 217–234.

46. Brown, J. P.; Roehm, G. W.; Brown, R. J. Mutagenicity testing of certified food colors and related azo, xanthene and triphenylmethane dyes with the *Salmonella*/microsome system. *Mutat. Res.* **1978**, *56*, 249–271.

SUNSET YELLOW FCF

CAS Registry Number 2783-94-0

Chemical Structure

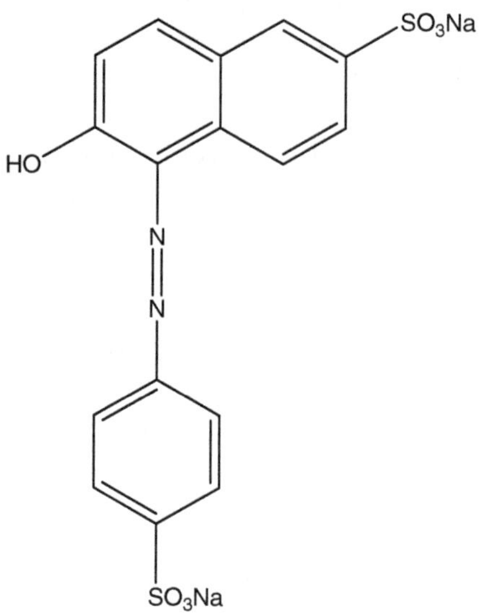

CA Index Name 2-Naphthalenesulfonic acid, 6-hydroxy-5-[2-(4-sulfophenyl)diazenyl]-, sodium salt (1 : 2)

Other Names 2-Naphthalenesulfonic acid, 6-hydroxy-5-[(4-sulfophenyl)azo]-, disodium salt; C.I. Food Yellow 3; C.I. Food Yellow 3, disodium salt; Sunset Yellow FCF; 1-*p*-Sulfophenylazo-2-hydroxynaphthalene-6-sulfonate disodium salt; 1351 Yellow; 1899 Yellow; Acid Food Yellow 3; Acid Yellow TRA; Aizen Food Yellow 5; Alabaster No. 3; Atul Sunset Yellow FCF; C Orange 10; C.I. 15985; Canacert Sunset Yellow FCF; Certicol Sunset Yellow CFS; Cilefa Orange S; Cogilor Orange 213.11; D and C Yellow No. 6; Dolkwal Sunset Yellow; E 110; E 110 (dye); Edicol Supra Yellow FC; Eniacid Sunset Yellow; Eurocert Orange FCF; FD & C Yellow 6; FD and C Yellow 6; FD and C Yellow No. 6; FD&C Yellow No. 6; Food Yellow 3; Food Yellow 5; Food Yellow No. 5; HD Sunset Yellow FCF; HD Sunset Yellow FCF Supra; Hexacol Sunset Yellow FCF; Hexacol Sunset Yellow FCF Supra; INS 110; Japan Food Yellow No. 5; Japan Yellow 5; Japan Yellow No. 5; KCA Foodcol Sunset Yellow FCF; L Orange Z 2010; Maple Sunset Yellow FCF; Orange G; Orange II R; Orange PAL; Orange Yellow 85; Orange Yellow S; Orange Yellow SFQ; Orient Water Yellow 2; Para Orange; Sun Orange A Geigy; Sun Yellow; Sun Yellow Extra Conc. A Export; Sun Yellow Extra Pure A; Sun Yellow FCF; Sunlight Yellow FCF; Sunset Yellow; Sunset Yellow 6; Twilight yellow; Usacert Yellow No. 6; Vitasyn Orange RGL 90; Water Yellow 2; Yellow 6; Yellow No. 5; Yellow No. 6; Yellow Orange S; Yellow S; Yellow SY; Yellow Sun

Merck Index Number 9001

Chemical/Dye Class Azo

Molecular Formula $C_{16}H_{10}N_2Na_2O_7S_2$

Molecular Weight 452.37

Physical Form Orange-red crystals

Solubility Soluble in water, ethanol

Melting Point 390 °C (decompose)

Absorption (λ_{max}) 480 nm

Synthesis Synthetic methods[1–13]

Staining Applications Cells;[14] animal feed;[15] bakery products;[16] candies;[20] beverages;[17–19] bread;[21] cereal;[21] confectionery products;[22] cough syrup;[23] dairy products;[24] dietary supplement;[25,29] drinks;[20,26,27] frozen food product;[28] grape extract;[29] meat products;[30] soft drinks;[31] soup;[32] sweeteners;[33,34] syrup;[19] capsule;[35] tablets;[22,36–38] dental bleaching gel;[39] eyebrows;[40] eye shadows;[41] lips;[42] skin;[42,43] tattoos;[44] tooth;[45] hairs;[46–48] keratin fibers[49]

Biological Applications Medical devices;[50] treating bone metabolic diseases,[51] dermatological disorders,[52] mitochondrial diseases,[53] respiratory illness[54]

Industrial Applications Inks;[55–59] paints;[60] thermoplastics;[61] batteries;[62] cleansing products;[63] colored bubbles;[64] textiles;[65] toys[66]

Safety/Toxicity Acute toxicity;[67] carcinogenicity;[68–73] childhood behavior effects;[74] chromosomal aberration;[70,75,76] cytotoxicity;[77] genotoxicity;[78–81] mutagenicity;[82–86] neurotoxicity;[87] reproductive toxicity[87]

Certification/Approval Approved by Food & Drugs Administration (FDA)

REFERENCES

1. Holtzinger, G.; Marie, G. Pigments based on clay-bound water-soluble dyes. Fr. Demande FR 2879097, 2006; *Chem. Abstr*. **2006**, *145*, 69332.
2. Yamada, M.; Morimoto, T.; Nakamura, M.; Goda, Y.; Nakazawa, H. The generative condition of the trisodium salt of 6-hydroxy-7-(4-sulfophenyl)-5-(4-

sulfophenyl-azo)-2-naphthalenesulfonic acid, a subsidiary color formed in the production of food yellow No. 5 (Sunset Yellow FCF). *Nippon Shokuhin Kagaku Gakkaishi* **2001**, *8*, 73–77; *Chem. Abstr.* **2001**, *136*, 308791.

3. Peiperl, M. D.; Prival, M. J.; Bell, S. J. Determination of combined benzidine in FD&C Yellow No. 6 (Sunset Yellow FCF). *Food Chem. Toxicol.* **1995**, *33*, 829–839.
4. Kimura, M.; Umemoto, M.; Tsuji, S.; Shibata, T.; Ito, Y. Studies on "sunset yellow FCF standard" for the dye standard of the National Institute of Health Sciences. *Bull. Natl. Inst. Hyg. Sci.* **1993**, 137–138.
5. Kabasawa, Y.; Tanimura, T.; Nakazawa, H.; Shinomiya, K. Application of counter alternative current chromatography to purification of food mono-azo dyes. *Anal. Sci.* **1992**, *8*, 351–353.
6. Sokolowska-Gajda, J.; Freeman, H. S. The synthesis of disperse and cationic dyes from acid dye structures. *Dyes Pigm.* **1990**, *14*, 35–48.
7. Rebhahn, R. W. J.; Cook, W. L. Ultrafiltration purification of food dyes. U.S. Patent 4560746, 1985; *Chem. Abstr.* **1986**, *105*, 116539.
8. Guillon, M.; Mondet, J.; Papantoniou, C.; Vandenbossche, C. Make-up compositions containing acid dye salts and copolymers containing tertiary amine functions. Fr. Demande FR 2480770, 1981; *Chem. Abstr.* **1982**, *96*, 91493.
9. Cooper, A. R.; Booth, R. G. Purification of polymeric dyes by ultrafiltration. *J. Appl. Polym. Sci.* **1979**, *23*, 1373–1384.
10. Demler, W. R.; Christ, P. G. Soluble azo dyes for coloring sugar. Fr. Demande FR 1546168, 1968; *Chem. Abstr.* **1970**, *72*, 80324.
11. Pla-Delfina, J. M. Systematic identification of food, drug, and cosmetic azo dyes. *J. Soc. Cosmet. Chem.* **1962**, *13*, 214–244.
12. Jablonski, C. F. Report on [determination of] coloring matters in foods. *J. Assoc. Off. Agric. Chem.* **1950**, *33*, 517–521.
13. Stein, C. Studies on coal-tar colors, FD&C Yellow No. 6 and C. I. No. 26. *J. Assoc. Off. Agric. Chem.* **1949**, *32*, 672–679.
14. Yamazaki, K.; Fujiwara, N.; Seto, T.; Yang, P. Horny cell-staining liquid, and horny cell-staining method. PCT Int. Appl. WO 2007029437, 2007; *Chem. Abstr.* **2007**, *146*, 312248.
15. Bayer, M. Animal feed for dogs and cats. Ger. Offen. DE 19958692, 2001; *Chem. Abstr.* **2001**, *135*, 4863.
16. Waldock, C. S. Method and apparatus for marking bakery products. PCT Int. Appl. WO 2005013698, 2005; *Chem. Abstr.* **2005**, *142*, 217857.
17. Smith, L. C. Carbonated beverage national school lunch meal. U.S. Pat. Appl. Publ. US 2007281059, 2007; *Chem. Abstr.* **2007**, *148*, 10213.
18. Pearce, T. M. Method for a consumer to create his own tasty beverage by flavor addition. U.S. Pat. Appl. Publ. US 2005100639, 2005; *Chem. Abstr.* **2005**, *142*, 462768.
19. Chang, P. K. Color-stable syrup and beverage compositions fortified with vitamin C. U.S. Patent 5336510, 1994; *Chem. Abstr.* **1994**, *121*, 254387.
20. Yoshioka, N.; Ichihashi, K. Determination of 40 synthetic food colors in drinks and candies by high-performance liquid chromatography using a short column with photodiode array detection. *Talanta* **2008**, *74*, 1408–1413.
21. Pearce, T. M. Flavor-sweetener additive for enhancing the desirability of a breakfast cereal or "bread". U.S. Pat. Appl. Publ. US 2005100651, 2005; *Chem. Abstr.* **2005**, *142*, 462727.
22. Grillo, S. M.; Saraceni, K.; Kelley, J. A. Coating of confectionery or tablets by using sugar/syrup or sugarless solutions containing dry color concentrate. U.S. Patent 5882707, 1999; *Chem. Abstr.* **1999**, *130*, 222390.
23. Neelwarne, B.; Rudrappa, T.; Narayan, M. S.; Chidambaramurthy, K. N.; Ravishankar, G. A. An improved process for the preparation of natural color stabilized cough syrup mixture. Indian Pat. Appl. IN 2003DE00395, 2007; *Chem. Abstr.* **2007**, *147*, 455620.
24. Harrison, P. A. Stabilized sugar-containing composition adapted to flavor or color dairy products. N.Z. Patent NZ 521401, 2005; *Chem. Abstr.* **2005**, *143*, 152353.
25. Dahanukar, D. S. Air tight dietary supplement sachet containing bovine colostrums, herbal extracts and variants thereof. Indian Pat. Appl. IN 2005MU00171, 2006; *Chem. Abstr.* **2007**, *147*, 528006.
26. Sherwood, S.; Jenkins, D. High energy carbonated protein drink and method of making. U.S. Pat. Appl. Publ. US 2006147602, 2006; *Chem. Abstr.* **2006**, *145*, 102723.
27. Bratanova, Z. V.; Spirichev, V. B.; Zhukov, S. P.; Bespalov, V. V.; Chistyakov, V. Y.; Drozdov, Y. A. Dry mixture and syrup for preparation of a drink and nonalcoholic drink. Russ. RU 2161423, 2001; *Chem. Abstr.* **2002**, *136*, 150308.

28. Del Valle, F. High protein frozen food product. U.S. Patent 6558731, 2003; *Chem. Abstr.* **2003**, *138*, 336930.

29. Shrikhande, A. J.; Wang, H.; Kupina, S. A. Grape extract, dietary supplement thereof, and processes therefor. U.S. Pat. Appl. Publ. US 2007071871, 2007; *Chem. Abstr.* **2007**, *146*, 357778.

30. Qiao, L.; Xu, H.; Sun, J. Determination of artificial pigment from meat products. *Huagong Shikan* **2007**, *21*, 52–54; *Chem. Abstr.* **2007**, *148*, 9806.

31. Chou, S. S.; Lin, Y. H.; Cheng, C. C.; Hwang, D. F. Determination of synthetic colors in soft drinks and confectioneries by micellar electrokinetic capillary chromatography. *J. Food Sci.* **2002**, *67*, 1314–1318.

32. Li, Z. Method for producing concentrated composite soup seasoning for beef noodle. Faming Zhuanli Shenqing Gongkai Shuomingshu CN 101095535, 2008; *Chem. Abstr.* **2008**, *148*, 120643.

33. Benedict, S. R.; Martling, S.; Moraru, G. High-intensity sweeteners and coloring agent compositions. PCT Int. Appl. WO 2007033227, 2007; *Chem. Abstr.* **2007**, *146*, 336926.

34. Catani, S. J.; Liao, S.; Loades, M. Self-mixing tabletop sweetener. U.S. Pat. Appl. Publ. US 2007059418, 2007; *Chem. Abstr.* **2007**, *146*, 315560.

35. Yu, S.; Liu, M. Method for manufacturing soft capsule. Faming Zhuanli Shenqing Gongkai Shuomingshu CN 101069678, 2007; *Chem. Abstr.* **2007**, *148*, 39571.

36. Flanagan, J.; Smith, T. L.; Barkley, A.; Nicholson, R. E.; Callahan, T. P. Gellan gum-based tablet coatings. U.S. Patent 6485747, 2002; *Chem. Abstr.* **2002**, *137*, 389149.

37. Chaudhari, S. P.; Chaudhari, P. D.; Mistry, C. J.; Patil, M. J.; Barhate, N. S. The effect of core and coating composition on drug release: from directly compressed time-controlled release tablets. *Pharm. Technol.* **2007**, *31*, 132, 134, 136, 138, 140, 142, 144.

38. Dahanukar, D. S. A chewable tablet dietary supplement containing spray dried powder of bovine colostrums, with herbal ingredients/and variants" and the process of making thereof. Indian Pat. Appl. IN 2005MU00172, 2006; *Chem. Abstr.* **2007**, *147*, 528007.

39. Banerjee, A.; Friedman, J. Dental bleaching gel composition, activator system and method for activating a dental bleaching gel. Eur. Pat. Appl. EP 1224925, 2002; *Chem. Abstr.* **2002**, *137*, 114276.

40. Akaishi, T. Eyebrow cosmetics containing polyether-modified silicones, and their applicators. Jpn. Kokai Tokkyo Koho JP 2005008547, 2005; *Chem. Abstr.* **2005**, *142*, 100017.

41. Bouchard, F. Eye shadow comprising least 50% colorants. Eur. Pat. Appl. EP 1757262, 2007; *Chem. Abstr.* **2007**, *146*, 280374.

42. Dumousseaux, C. Cosmetic composition for application to the skin, lips, and/or body surface. Fr. Demande FR 2868696, 2005; *Chem. Abstr.* **2005**, *143*, 392522.

43. Lee, H.; Soistier, N.; Simon, J. C. Cosmetic makeup foundations for darkened skins. Fr. Demande FR 2848822, 2004; *Chem. Abstr.* **2004**, *141*, 76373.

44. Kundakovic, L.; Klitzman, B.; Koger, K. E.; Mathiowitz, E.; Drill, C. A. Cellular or organelle-entrapped nanoparticles containing dyes for tattoos. PCT Int. Appl. WO 2007024429, 2007; *Chem. Abstr.* **2007**, *146*, 280376.

45. Hoic, D. A.; Curtis, J. P. Tooth coating compositions with resin, colorant, and pearlescent particles. U.S. Pat. Appl. Publ. US 2005175552, 2005; *Chem. Abstr.* **2005**, *143*, 216750.

46. Kawai, S.; Oikawa, S. Hair dye compositions containing viscosity enhancing agents. Jpn. Kokai Tokkyo Koho JP 2008063312, 2008; *Chem. Abstr.* **2008**, *148*, 362851.

47. Plos, G. Hair dye composition containing an oxidase alcohol and a direct dye having an acid function or its salt for dyeing keratin fibers. Fr. Demande FR 2865396, 2005; *Chem. Abstr.* **2005**, *143*, 179094.

48. Sallwey, A.; Schmitt, M.; Dieckow, A. Dyestuff for hair and fibers. Eur. Pat. Appl. EP 976385, 2000; *Chem. Abstr.* **2000**, *132*, 127457.

49. Kravtchenko, S. Agent for simultaneously bleaching and coloring of keratin fibers comprising an anionic or nonionic dye and an inert organic liquid. Eur. Pat. Appl. EP 1738801, 2007; *Chem. Abstr.* **2007**, *146*, 128004.

50. Raad, I. Coating medical device surfaces with antimicrobial agents. PCT Int. Appl. WO 2007062306, 2007; *Chem. Abstr.* **2007**, *147*, 39321.

51. Diaz, L. E. Liquid composition comprising alendronic acid for prevention and/or treatment of different bone metabolic diseases, uses thereof, and preparation process therefore. Eur. Pat. Appl. EP 1894557, 2008; *Chem. Abstr.* **2008**, *148*, 292170.

52. Murad, H. Methods for treating dermatological and other health-related conditions in a patient. PCT Int. Appl. WO 2005095959, 2005; *Chem. Abstr.* **2005**, *143*, 392984.

53. Ghosal, S. Mitochondria-targeted antioxidants comprising coenzyme Q10, reduced CoQ10, oxygenated dibenzo-α-pyrone, or amino acyl ester, for treatment of mitochondrial diseases. PCT Int. Appl. WO 2008016554, 2008; *Chem. Abstr.* **2008**, *148*, 222003.

54. Martin, K. L.; Khanolkar, J. E.; Gledhill, D. W.; Criss, S. E.; Ramji, N.; Costeines, E. R.; Huetter, T. E.; Iyer, R. R.; Anness, D. K. Compositions comprising phenylephrine and kits useful for treatment of respiratory illness. U.S. Pat. Appl. Publ. US 2007249727, 2007; *Chem. Abstr.* **2007**, *147*, 474743.

55. Ogasawara, H. Aqueous inks with efficient permeation through cores for fountain pens. Jpn. Kokai Tokkyo Koho JP 2007297519, 2007; *Chem. Abstr.* **2007**, *147*, 523978.

56. Sugai, A. Colored ink compositions with metallic gloss for writing tools. Jpn. Kokai Tokkyo Koho JP 2005239908, 2005; *Chem. Abstr.* **2005**, *143*, 268426.

57. Aoyama, T. Aqueous ink for ink-jet printing. Jpn. Kokai Tokkyo Koho JP 11315229, 1999; *Chem. Abstr.* **1999**, *131*, 338432.

58. Masugi, T. Water-based ink compositions. Jpn. Kokai Tokkyo Koho JP 07216281, 1995; *Chem. Abstr.* **1995**, *123*, 343776.

59. Kito, M. Aqueous inks suitable for children. Jpn. Kokai Tokkyo Koho JP 04298577, 1992; *Chem. Abstr.* **1993**, *118*, 236096.

60. Sabnis, R. W.; Kehoe, T. D.; Balchunis, R. J. Paint compositions comprising color changing phthalein acid–base indicator. PCT Int. Appl. WO 2006105191, 2006; *Chem. Abstr.* **2006**, *145*, 403489.

61. Van Den Oever, M. J. A.; Boeriu, C. G.; Blaauw, R.; Van Haveren, J. Colorants based on renewable resources and food-grade colorants for application in thermoplastics. *J. Appl. Polym. Sci.* **2004**, *92*, 2961–2969.

62. Gauthier, M.; Michot, C. Composite organic–inorganic cathodes for secondary lithium batteries prepared from inorganic oxides and functionalized organic compounds. PCT Int. Appl. WO 2006120332, 2006; *Chem. Abstr.* **2006**, *145*, 508528.

63. Krzysik, D. G.; Utschig, J. M.; Cole, D. B. Color changing liquid cleansing products containing surfactants, electrolytes and coloring agents. U.S. Pat. Appl. Publ. US 2005148490, 2005; *Chem. Abstr.* **2005**, *143*, 120038.

64. Sabnis, R. W.; Kehoe, T. D. Composition and method for producing colored bubbles. U.S. Pat. Appl. Publ. US 2006004110, 2006; *Chem. Abstr.* **2006**, *144*, 110164.

65. Yoshimura, T. Apparatus and method for drawing draft on fabric for handcraft and the fabric. Jpn. Kokai Tokkyo Koho JP 2001322316, 2001; *Chem. Abstr.* **2001**, *135*, 364564.

66. Ono, Y. Water-erasable markers for toys and figure toy sets therewith. Jpn. Kokai Tokkyo Koho JP 2002129085, 2002; *Chem. Abstr.* **2002**, *136*, 371209.

67. Tonogai, Y.; Iwaida, M.; Tati, M.; Ose, Y.; Sato, T. Biochemical decomposition of coal-tar dyes. II. Acute toxicity of coal-tar dyes and their decomposed products. *J. Toxicol. Sci.* **1978**, *3*, 205–214.

68. Helguera, A. M.; Cabrera Perez, M. A.; Gonzalez, M. P.; Ruiz, R. M.; Gonzalez, D. H. A topological substructural approach applied to the computational prediction of rodent carcinogenicity. *Bioorg. Med. Chem.* **2005**, *13*, 2477–2488.

69. Haseman, J. K.; Clark, A. M. Carcinogenicity results for 114 laboratory animal studies used to assess the predictivity of four *in vitro* genetic toxicity assays for rodent carcinogenicity. *Environ. Mol. Mutagen.* **1990**, *16*, 15–31.

70. Rosenkranz, H. S.; Ennever, F. K.; Klopman, G. Relationship between carcinogenicity in rodents and the induction of sister chromatid exchanges and chromosomal aberrations in Chinese hamster ovary cells. *Mutagenesis* **1990**, *5*, 559–571.

71. Bakale, G.; McCreary, R. D. A physicochemical screening test for chemical carcinogens: the ke test. *Carcinogenesis* **1987**, *8*, 253–264.

72. Haseman, J. K.; Winbush, J. S.; O'Donnell, M. W., Jr. Use of dual control groups to estimate false positive rates in laboratory animal carcinogenicity studies. *Fundam. Appl. Toxicol.* **1986**, *7*, 573–584.

73. Haseman, J. K. Issues in carcinogenicity testing: dose selection. *Fundam. Appl. Toxicol.* **1985**, *5*, 66–78.

74. McCann, D.; Barrett, A.; Cooper, A.; Crumpler, D.; Dalen, L.; Grimshaw, K.; Kitchin, E.; Lok, K.; Porteous, L.; Prince, E.; Sonuga-Barke, E.; Warner, J. O.; Stevenson, J. Food additives and hyperactive behaviour in 3-year-old and 8/9-year-old children in the community: a randomized, double-blinded placebo-controlled trial. *Lancet* **2007**, *370*, 1560–1567.

75. Serra, J. R.; Thompson, E. D.; Jurs, P. C. Development of binary classification of structural chromosome aberrations for a diverse set of organic compounds from molecular structure. *Chem. Res. Toxicol.* **2003**, *16*, 153–163.

76. Rosenkranz, H. S.; Ennever, F. K.; Dimayuga, M.; Klopman, G. Significant differences in the structural basis of the induction of sister chromatid exchanges and chromosomal aberrations in Chinese hamster ovary cells. *Environ. Mol. Mutagen.* **1990**, *16*, 149–177.
77. Yamashoji, S.; Isshiki, K. Rapid detection of cytotoxicity of food additives and contaminants by a novel cytotoxicity test, menadione-catalyzed H_2O_2 production assay. *Cytotechnology* **2002**, *37*, 171–178.
78. Knight, A. W.; Billinton, N.; Cahill, P. A.; Scott, A.; Harvey, J. S.; Roberts, K. J.; Tweats, D. J.; Keenan, P. O.; Walmsley, R. M. An analysis of results from 305 compounds tested with the yeast RAD54-GFP genotoxicity assay (GreenScreen GC)—including relative predictivity of regulatory tests and rodent carcinogenesis and performance with autofluorescent and colored compounds. *Mutagenesis* **2007**, *22*, 409–416.
79. Kirkland, D.; Aardema, M.; Henderson, L.; Mueller, L. Evaluation of the ability of a battery of three *in vitro* genotoxicity tests to discriminate rodent carcinogens and non-carcinogens. I. Sensitivity, specificity and relative predictivity. *Mutat. Res.* **2005**, *584*, 1–256.
80. Rosenkranz, H. S.; Klopman, G. Structural alerts to genotoxicity: the interaction of human and artificial intelligence. *Mutagenesis* **1990**, *5*, 333–361.
81. Haveland-Smith, R. B.; Combes, R. D. Screening of food dyes for genotoxic activity. *Food Cosmet. Toxicol.* **1980**, *18*, 215–221.
82. Serafimova, R.; Todorov, M.; Pavlov, T.; Kotov, S.; Jacob, E.; Aptula, A.; Mekenyan, O. Identification of the structural requirements for mutagenicity, by incorporating molecular flexibility and metabolic activation of chemicals. II. General Ames mutagenicity model. *Chem. Res. Toxicol.* **2007**, *20*, 662–676.
83. Rafii, F.; Hall, J. D.; Cerniglia, C. E. Mutagenicity of azo dyes used in foods, drugs and cosmetics before and after reduction by *Clostridium* species from the human intestinal tract. *Food Chem. Toxicol.* **1997**, *35*, 897–901.
84. Rosenkranz, H. S.; Klopman, G. Structural basis of the mutagenicity of 1-amino-2-naphthol-based azo dyes. *Mutagenesis* **1990**, *5*, 137–146.
85. Ishidate, M., Jr.; Sofuni, T.; Yoshikawa, K.; Hayashi, M.; Nohmi, T.; Sawada, M.; Matsuoka, A. Primary mutagenicity screening of food additives currently used in Japan. *Food Chem. Toxicol.* **1984**, *22*, 623–636.
86. Chung, K. T.; Fulk, G. E.; Andrews, A. W. Mutagenicity testing of some commonly used dyes. *Appl. Environ. Microbiol.* **1981**, *42*, 641–648.
87. Tanaka, T. Reproductive and neurobehavioral effects of Sunset Yellow FCF administered to mice in the diet. *Toxicol. Ind. Health* **1996**, *12*, 69–79.

TARTRAZINE

CAS Registry Number 1934-21-0

Chemical Structure

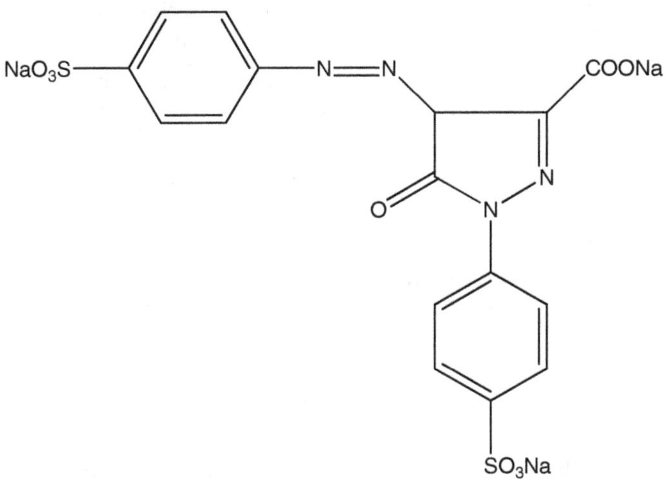

CA Index Name 1*H*-Pyrazole-3-carboxylic acid, 4,5-dihydro-5-oxo-1-(4-sulfophenyl)-4-[2-(4-sulfophenyl) diazenyl]-, sodium salt (1 : 3)

Other Names 1*H*-Pyrazole-3-carboxylic acid, 4,5-dihydro-5-oxo-1-(4-sulfophenyl)-4-[(4-sulfophenyl)azo]-, trisodium salt; C.I. Acid Yellow 23; C.I. Acid Yellow 23, trisodium salt; Tartrazine; 1310 Yellow; 1409 Yellow; 3-Carboxy-5-hydroxy-1-*p*-sulfophenyl-4-*p*-sulfophenylazopyrazole trisodium salt; A.F. Yellow No. 4; AY 23; Acid Leather Yellow T; Acid Yellow 23; Acid Yellow N; Acid Yellow T; Acid Yellow XX-SF; Acilan Yellow GG; Airedale Yellow T; Aizen Tartrazine; Amacid Yellow T; Amacid Yellow T-EX; Apollo Acid Tartrazine; Atul Tartrazine; B 3014; Basovit Yellow 133E; Bosovit Yellow 133E; Bucacid Tartrazine; C.I. 19140; C.I. Food Yellow 4; C.I. Solvent Yellow 57; Calcocid Yellow MCG; Calcocid Yellow XX; Canacert Tartrazine; Certicol Tartrazol Yellow S; Cilefa Yellow T; Cogilor Yellow 113.11; Cogilor Yellow 113.11-90003; Concorde Acid Tartrazine; Concorde Leather Yellow TTZ; Curon Fast Yellow 5G; D and C Yellow No. 5; Dinacid Tartrazine Yellow PFD; Dolkwal Tartrazine; Duasyn Acid Yellow XX; Duasyn Acid Yellow XX-SF; Duasyn Acid Yellow XX-SF-LP 413; Durkee Yellow Food Color; Dyacid Yellow N; Dye Yellow Lake; Dynacid Yellow K; E 102; E 102 (dye); Edicol Supra Tartrazine N; Egacid Yellow T; Egg Yellow A; Eniacid Tartrazine J; Erio Tartrazine; Erio Yellow T Supra; Eurocert Tartrazine; Eurocert Tartrazine 311840; FD & C Yellow No. 5-307008; FD and C Yellow No. 5; FD&C Yellow 5; FD&C Yellow No. 5; FD&C Yellow No. 5-37000; FD&C Yellow No. 5-37009; FD&C Yellow No. 5-90123; Fenazo Yellow T; Food Dye Yellow 4; Food Yellow 4; Food Yellow No. 4; Food dye yellow number 4; HD Tartrazine; HD Tartrazine Supra; Hexacert Yellow No. 5; Hexacol Tartrazine; Hidazid Tartrazine; Hispacid Fast Yellow T; Hydrazine yellow; Hydroxine Yellow L; INS 102; Japan Food Yellow No. 4; Japan Yellow 4; Japan Yellow No. 4; KCA Foodcol Tartrazine PF; KCA Tartrazine PF; Kako Tartrazine; Kayaku Food Colour Yellow No. 4; Kayaku Tartrazine; Kemacid Tartrazine NS; Kiton Yellow T; L Yellow Z 1020; Lake Yellow; Lemon Yellow; Lemon Yellow (Chinese food dye); Lemon Yellow A; Lemon Yellow A Geigy; Maple Tartrazol Yellow; Mitsui Tartrazine; Multacid Yellow T; NSC 4760; Naphtocard Yellow O; Necol Tartrazine; Neelicol Tartrazine; Neklacid Yellow T; Orient Water Yellow 1; Oxanal Yellow T; Pacid Tartrazin; Pro-Jet Yellow OAM; PuriColor Yellow AYE 23; Ritacid Yellow NS; Romexal Yellow T; San-ei Tartrazine; Sandolan Yellow E-TZ; Sicovit Tartrazin Lake E 102; Sicovit Tartrazine Lake E 102; Solvent Yellow 57; Sugai Tartrazine; Tartar Yellow FS; Tartar Yellow N; Tartar Yellow PF; Tartar Yellow S; Tartran Yellow; Tartraphenine; Tartrazin; Tartrazine 307043; Tartrazine 36008; Tartrazine 37000; Tartrazine 37009; Tartrazine 90123; Tartrazine A; Tartrazine A Export; Tartrazine B; Tartrazine B.P.C.; Tartrazine C; Tartrazine E 102; Tartrazine Extra Pure A; Tartrazine FQ; Tartrazine G; Tartrazine GR; Tartrazine Lake; Tartrazine Lake Yellow N; Tartrazine M; Tartrazine MCGL; Tartrazine N; Tartrazine NS; Tartrazine O; Tartrazine O Specially Pure; Tartrazine T; Tartrazine XX; Tartrazine XX Specially Pure; Tartrazine XXX; Tartrazine Yellow; Tartrazine Yellow 5; Tartrazine Yellow BF; Tartrazol BPC; Tartrazol Yellow; Tartrine Yellow O; Triacid Tartrazine Yellow O; Triacid Yellow T; Trisodium 3-carboxy-5-hydroxy-1-*p*-sulfophenyl-4-*p*-sulfophenylazopyrazole; Unitertracid Yellow TE; Usacert FD&C Yellow No. 5-310122; Usacert Yellow No. 5; Vitasyn Tartrazine X 90; Vitasyn Tartrazine XX 90; Vondacid Tartrazine; Water Yellow 1; Water Yellow 176552; Water Yellow 176571; Water Yellow 178758; Water Yellow 178819; Wool Yellow; Xylene Fast Yellow GT; Yellow 5; Yellow HO 203; Yellow Lake 69; Yellow No. 5

Merck Index Number 9072

Chemical/Dye Class Pyrazolone; Azo

Molecular Formula $C_{16}H_9N_4Na_3O_9S_2$

Molecular Weight 534.36

Physical Form Bright orange-yellow powder

Solubility Freely soluble in water; soluble in ethanol, cellosolve; insoluble in xylene

Melting Point 300 °C

Absorption (λ_{max}) 425 nm

Synthesis Synthetic methods[1–25]

Staining Applications Enzymes;[26] lymph nodes;[27] alcohol;[28] bakery products;[29] beverages;[30–33] candy;[34,36] cucumber pickle products;[35] drinks;[36,37] egg noodle;[38] fast food products;[38] fish;[39] ice cream;[40] meat products;[41] potato products;[42] seafood;[43] sweeteners;[44] syrup;[33] yoghurts;[45] capsules;[46] tablets;[47] mouthwash;[48] toothpastes;[49] sunscreen;[50] eyebrows;[51] eyelashes;[52] lips;[53–56] skin;[55–57] tattoos;[58] tooth;[59] hairs[60]

Biological Applications Treating hepatitis,[44,47] periodontal disease,[61] psoriasis[62]

Industrial Applications Color filters;[63,64] liquid crystal displays;[63,64] organic thin films;[65] inks;[66–68] highlighters;[69] paints;[70,71] adhesives;[72] photographic materials;[73] thermoplastics;[74] cleansing products;[75] detergents;[76] colored bubbles;[77] papers,[78] fabrics,[79] wood[80,81]

Safety/Toxicity Acute oral toxicity;[82] allergic contact dermatitis;[83] carcinogenicity;[84–88] chromosomal aberration;[89] chronic toxicity;[86,87] cytotoxicity;[90,91] dermal toxicity;[92] genotoxicity;[93–96] hyperactive behavior in children;[97] mutagenicity;[98–106] neurotoxicity;[107] pseudoallergic reactions;[108] reproductive toxicity[107]

Certification/Approval Approved by Food & Drugs Administration (FDA)

REFERENCES

1. Ding, Q.; Wang, L. Lemon Yellow and manufacturing methods therefor. Faming Zhuanli Shenqing Gongkai Shuomingshu CN 1952015, 2007; *Chem. Abstr.* **2007**, *146*, 520608.
2. Miyata, H.; Yamabi, S.; Ogawa, M. Colorants, their manufacture, compositions, inks, color filters, and electrophoretic particles containing the colorants, and electrophoretic display devices using them. Jpn. Kokai Tokkyo Koho JP 2006083345, 2006; *Chem. Abstr.* **2006**, *144*, 333092.
3. Huddleston, J. G.; Willauer, H. D.; Boaz, K. R.; Rogers, R. D. Separation and recovery of food coloring dyes using aqueous biphasic extraction chromatographic resins. *J. Chromatogr., B* **1998**, *711*, 237–244.
4. Rogers, R.; Horwitz, E. P.; Bond, A. H. Process for separating and recovering an anionic dye from an aqueous solution. U.S. Patent 5707525, 1998; *Chem. Abstr.* **1998**, *128*, 142665.
5. Pieschel, F.; Specht, W.; Hepp, W.; Raschke, D.; Lange, E.; Lueckert, D.; Schnabel, S.; Riedel, H. J.; Weise, H. Preparation of mono-, dis-, tetrakis-, and pentakisazo dyes. Ger. (East) DD 294394, 1991; *Chem. Abstr.* **1992**, *116*, 61570.
6. Vesely, M.; Polackova, E.; Havel, M.; Janecek, M. Tartrazine dye crystal production by salting-out with sodium chloride. Czech. CS 255665, 1988; *Chem. Abstr.* **1989**, *111*, 235047.
7. Ruzcka, K. One-step preparation of tartrazine. Czech. CS 246340, 1987; *Chem. Abstr.* **1988**, *109*, 112061.
8. Shirota, K.; Sakaeda, T.; Suga, Y. Purification of dyes for jet-printing inks. Jpn. Kokai Tokkyo Koho JP 62197462, 1987; *Chem. Abstr.* **1988**, *108*, 77352.
9. Ito, S.; Suzuki, M.; Hashimoto, K.; Tanaka, M. Synthesis of organic coloring materials on surfaces of polystyrene particles. *Zairyo Gijutsu* **1987**, *5*, 108–110; *Chem. Abstr.* **1987**, *107*, 116942.
10. Apostolache, S. V. Tartrazine. Rom. RO 66729, 1979; *Chem. Abstr.* **1981**, *95*, 63677.
11. Arsac, A.; Frank, P. Azo dyes. Fr. Demande FR 2439222, 1980; *Chem. Abstr.* **1981**, *94*, 4948.
12. Konishi, J.; Shiomitsu, T. Manufacture of tartrazine. Jpn. Kokai Tokkyo Koho JP 55012150, 1980; *Chem. Abstr.* **1980**, *93*, 73746.
13. Cooper, A. R.; Booth, R. G. Purification of polymeric dyes by ultrafiltration. *J. Appl. Polym. Sci.* **1979**, *23*, 1373–1384.
14. Nakajima, K. Tartrazine. Jpn. Kokai Tokkyo Koho JP 50100118, 1975; *Chem. Abstr.* **1975**, *83*, 195223.
15. Sachs, G.; Dunn, A. K.; Skuce, W. F. Edible water-soluble azo. U.S. Patent 3748745, 1973; *Chem. Abstr.* **1973**, *79*, 124900.
16. Fridman, S. A.; Antonova, E. G. Purification of tartrazine. U.S.S.R. SU 376421, 1973; *Chem. Abstr.* **1973**, *79*, 67810.
17. Koch, L. Oil-soluble dyes. Ger. Offen. DE 1954923, 1970; *Chem. Abstr.* **1970**, *73*, 110873.
18. Sato, N.; Nakamura, T.; Miyamoto, M. Pigment coating. I. Manufacture and properties of Food Yellow No. 4 (Tartrazine) aluminum Lake. 1.

Yakuzaigaku **1968**, *28*, 42–47; *Chem. Abstr.* **1968**, *69*, 97684.

19. McKeown, G. G. Separation of amaranth and tartrazine. *J. Assoc. Off. Agric. Chem.* **1954**, *37*, 527–529.
20. Sawdey, G. W.; Ruoff, M. K.; Vittum, P. W. The formation of azomethine dyes from 4-anisylidene- and 4,4′-anisylidenebispyrazolones. *J. Am. Chem. Soc.* **1950**, *72*, 4947–4949.
21. Freeman, K. A.; Jones, J. H.; Graichen, C. Coal-tar colors. VIII. FD&C yellow No. 5. *J. Assoc. Off. Agric. Chem.* **1950**, *33*, 937–942.
22. Kendall, J. D.; Fry, D. J. Pyrazolone azo dyes. U.S. Patent 2457823, 1949; *Chem. Abstr.* **1949**, *43*, 45298.
23. Kendall, J. D.; Fry, D. J. Azo dyes. GB 585781, 1947; *Chem. Abstr.* **1947**, *41*, 33602.
24. Norton, G. Tartrazine. *Z. Farben-Ind.* **1928**, *20*, 8–9; *Chem. Abstr.* **1928**, *22*, 22951.
25. Sansone, R. The manufacture of tartrazine. *Color Trade J.* **1923**, *12–13*, 50–53, 123–127; *Chem. Abstr.* **1924**, *18*, 5518.
26. Blum, R.; Kadow, S. Use of food dyes for staining of enzyme solutions. Ger. Offen. DE 10046034, 2002; *Chem. Abstr.* **2002**, *136*, 289893.
27. Morton, D. L. Methods for lymph node identification. U.S. Patent 6815170, 2004; *Chem. Abstr.* **2004**, *141*, 391557.
28. Nazarov, A. N.; Kocherov, M. B.; Tarasov, R. V. Denatured alcohol. Russ. RU 2229504, 2004; *Chem. Abstr.* **2004**, *141*, 259480.
29. Waldock, C. S. Method and apparatus for marking bakery products. PCT Int. Appl. WO 2005013698, 2005; *Chem. Abstr.* **2005**, *142*, 217857.
30. Smith, L. C. Carbonated beverage National School Lunch meal. U.S. Pat. Appl. Publ. US 2007281059, 2007; *Chem. Abstr.* **2007**, *148*, 10213.
31. Bezuglov, A. Y. Production of lemon-lime beverage with moderate alcohol content. Russ. RU 2245903, 2005; *Chem. Abstr.* **2005**, *142*, 197007.
32. Roy, G.; Berardi, R.; Chan, W.; Lee, T. Prevention of synthetic color fading in beverages using botanically derived color stabilizers such as phenylpropenoic carbonyl compounds. U.S. Pat. Appl. Publ. US 2004091589, 2004; *Chem. Abstr.* **2004**, *140*, 405940.
33. Chang, P. K. Color-stable syrup and beverage compositions fortified with vitamin C. U.S. Patent 5336510, 1994; *Chem. Abstr.* **1994**, *121*, 254387.
34. Ryzhikov, S. I.; Andreev, K. V.; Krasnyak, V. T.; Matveev, S. B.; Zubareva, N. A. Chewable candy containing fat, emulsifier, proteins, and sweeteners. Russ. RU 2169485, 2001; *Chem. Abstr.* **2002**, *137*, 46452.
35. Perez-Diaz, I. M.; Kelling, R. E.; Hale, S.; Breidt, F.; McFeeters, R. F. Lactobacilli and tartrazine as causative agents of red-color spoilage in cucumber pickle products. *J. Food Sci.* **2007**, *72*, M240–M245.
36. Yoshioka, N.; Ichihashi, K. Determination of 40 synthetic food colors in drinks and candies by high-performance liquid chromatography using a short column with photodiode array detection. *Talanta* **2008**, *74*, 1408–1413.
37. Donhowe, E. T. Product and process of making an alcohol containing sport drink. U.S. Pat. Appl. Publ. US 2003134007, 2003; *Chem. Abstr.* **2003**, *139*, 100270.
38. Nguen, C. K. Fast food product "Business menu", egg noodle for this product and method for its production. Russ. RU 2316229, 2008; *Chem. Abstr.* **2008**, *148*, 190827.
39. Mueller, I.; Altemueller, P. A.; Sandoval, A. Retorted fish compositions comprising structured plant protein products. PCT Int. Appl. WO 2008034063, 2008; *Chem. Abstr.* **2008**, *148*, 330364.
40. Emel'yanova, O. A.; Galashevskaya, O. A.; Lopatko, V. A. Method for production of glazed ice cream. Russ. RU 2268606, 2006; *Chem. Abstr.* **2006**, *144*, 149468.
41. Qiao, L.; Xu, H.; Sun, J. Determination of artificial pigment from meat products. *Huagong Shikan* **2007**, *21*, 52–54; *Chem. Abstr.* **2007**, *148*, 9806.
42. Nassar, A.; Nassar, W.; Nassar, M. H. Method of making French fry-style potato products. U.S. Patent 7273631, 2007; *Chem. Abstr.* **2007**, *147*, 384866.
43. Altemueller, A. G. Simulated seafood compositions comprising structured plant protein products and fatty acids. U.S. Pat. Appl. Publ. US 2008069927, 2008; *Chem. Abstr.* **2008**, *148*, 330362.
44. Wang, C. Granule suspensions containing entecavir and sweeteners for treating hepatitis. Faming Zhuanli Shenqing Gongkai Shuomingshu CN 101028273, 2007; *Chem. Abstr.* **2007**, *147*, 392313.
45. Calvo, C.; Salvador, A.; Fiszman, S. M. Influence of colour intensity on the perception of colour and sweetness in various fruit-flavoured yoghurts. *Eur. Food Res. Technol.* **2001**, *213*, 99–103.
46. Yu, S.; Liu, M. Method for manufacturing soft capsule. Faming Zhuanli Shenqing Gongkai

Shuomingshu CN 101069678, 2007; *Chem. Abstr.* **2007**, *148*, 39571.

47. Wen, L. Pharmaceutical effervescent tablets containing entecavir for treating hepatitis. Faming Zhuanli Shenqing Gongkai Shuomingshu CN 101057851, 2007; *Chem. Abstr.* **2007**, *147*, 508459.
48. Eduardo, M. F. A. de P. Mouthwash for infants. Braz. Pedido PI BR 2001005138, 2003; *Chem. Abstr.* **2005**, *142*, 360374.
49. Oniki, T.; Watanabe, T. Colored toothpastes with transparency. Jpn. Kokai Tokkyo Koho JP 2002080333, 2002; *Chem. Abstr.* **2002**, *136*, 236698.
50. Ross, J. S.; Morefield, E. M. Sunscreen with disappearing color indicator. U.S. Patent 6290936, 2001; *Chem. Abstr.* **2001**, *135*, 215789.
51. Kato, T.; Fukuoka, N. Quick-drying eyebrow coloring compositions with no color migration. Jpn. Kokai Tokkyo Koho JP 2002161024, 2002; *Chem. Abstr.* **2002**, *136*, 406610.
52. Okuyama, M. Eyelash cosmetic composition containing light interference color fibers. Jpn. Kokai Tokkyo Koho JP 2005314395, 2005; *Chem. Abstr.* **2005**, *143*, 446245.
53. Hagiwara, H.; Okuyama, M. Lip cosmetic compositions containing light interference color fiber. Jpn. Kokai Tokkyo Koho JP 2005314391, 2005; *Chem. Abstr.* **2005**, *143*, 446247.
54. Bobka, E.; Gibtner, T.; Mara, M. Water-free formulations for application to the lips. U.S. Pat. Appl. Publ. US 2005226832, 2005; *Chem. Abstr.* **2005**, *143*, 372863.
55. Blin, X. Range of compositions for application to the skin, lips, and/or superficial body growths. Fr. Demande FR 2868697, 2005; *Chem. Abstr.* **2005**, *143*, 392523.
56. Dumousseaux, C. Cosmetic composition for application to the skin, lips, and/or body surface. Fr. Demande FR 2868696, 2005; *Chem. Abstr.* **2005**, *143*, 392522.
57. Arnaud, P.; Feltin, C.; Bazelaire, E. Process for coloring of dark skin using a direct dye. Fr. Demande FR 2901475, 2007; *Chem. Abstr.* **2007**, *148*, 39160.
58. Yamaguchi, J. Skin-coloring compositions suitable for temporally tattoo. Jpn. Kokai Tokkyo Koho JP 2000178138, 2000; *Chem. Abstr.* **2000**, *133*, 63607.
59. Hoic, D. A.; Curtis, J. P. Tooth coating compositions with resin, colorant, and pearlescent particles. U.S. Pat. Appl. Publ. US 2005175552, 2005; *Chem. Abstr.* **2005**, *143*, 216750.
60. Sallwey, A.; Schmitt, M.; Dieckow, A. Dyestuff for hair and fibers. Eur. Pat. Appl. EP 976385, 2000; *Chem. Abstr.* **2000**, *132*, 127457.
61. Andersen, S. N.; Wilson, J. B. Dye composition and laser energy for treating periodontal disease. U.S. Pat. Appl. Publ. US 2003059379, 2003; *Chem. Abstr.* **2003**, *138*, 276252.
62. De Oliveira, M. Compositions for the treatment for psoriasis. Can. Pat. Appl. CA 2305062, 2001; *Chem. Abstr.* **2004**, *142*, 183350.
63. Iwata, K.; Sakamoto, J. Manufacture of color filter for liquid crystal display. Jpn. Kokai Tokkyo Koho JP 2003035812, 2003; *Chem. Abstr.* **2003**, *138*, 145179.
64. Shiba, S. Color filter panels for liquid-crystal display devices and ink-jet printing process for their manufacture. Jpn. Kokai Tokkyo Koho JP 2002196126, 2002; *Chem. Abstr.* **2002**, *137*, 80398.
65. Ito, Y.; Ohiso, S.; Ishidzuki, N. Organic thin film. Jpn. Kokai Tokkyo Koho JP 11021538, 1999; *Chem. Abstr.* **1999**, *130*, 145947.
66. Katsuragi, T.; Notoya, Y. Liquid compositions, ink-jet recording method, ink sets, recording units, ink cartridges, ink-jet recording apparatus, and inks for ink-jet recording. Jpn. Kokai Tokkyo Koho JP 2008037911, 2008; *Chem. Abstr.* **2008**, *148*, 264507.
67. Kobayashi, Y.; Maru, T. Water-thinned ink compositions giving color-changeable writing. Jpn. Kokai Tokkyo Koho JP 2004182770, 2004; *Chem. Abstr.* **2004**, *141*, 55900.
68. Kitamura, K.; Fukumoto, H.; Kanetani, M.; Kato, S. Dark yellow ink compositions and ink sets therewith. Jpn. Kokai Tokkyo Koho JP 2002060664, 2002; *Chem. Abstr.* **2002**, *136*, 201977.
69. Davies-Smith, L.; Sum, V. W. Highlightable and highlighted mixtures, marking instruments, eradicator solution and kits, using the mixtures in highlighting. U.S. Pat. Appl. Publ. US 2005120919, 2005; *Chem. Abstr.* **2005**, *143*, 9331.
70. Miller, R. E.; Dereamus, R. C. Color changing systems using pan paint compositions and markers. U.S. Patent 5498282, 1996; *Chem. Abstr.* **1996**, *124*, 346248.
71. Godly, G. Stain-free tempera paints. U.S. Patent 4826535, 1989; *Chem. Abstr.* **1989**, *111*, 136099.

72. Fukushima, Y. Nylon adhesives. Jpn. Kokai Tokkyo Koho JP 49121836, 1974; *Chem. Abstr.* **1975**, *83*, 207040.

73. Kawai, K.; Hioki, T. Silver halide color photographic materials and color image formation. Jpn. Kokai Tokkyo Koho JP 08015832, 1996; *Chem. Abstr.* **1996**, *124*, 302426.

74. Van Den Oever, M. J. A.; Boeriu, C. G.; Blaauw, R.; Van Haveren, J. Colorants based on renewable resources and food-grade colorants for application in thermoplastics. *J. Appl. Polym. Sci.* **2004**, *92*, 2961–2969.

75. Krzysik, D. G.; Utschig, J. M.; Cole, D. B. Color changing liquid cleansing products containing surfactants, electrolytes and coloring agents. U.S. Pat. Appl. Publ. US 2005148490, 2005; *Chem. Abstr.* **2005**, *143*, 120038.

76. Janke, H.; Harer, J.; Holderbaum, T. Colored detergent for dishwashing. PCT Int. Appl. WO 9960088, 1999; *Chem. Abstr.* **1999**, *132*, 4281.

77. Sabnis, R. W.; Kehoe, T. D. Composition and method for producing colored bubbles. U.S. Pat. Appl. Publ. US 2006004110, 2006; *Chem. Abstr.* **2006**, *144*, 110164.

78. Astbury, J. H.; Cooper, H. M. Dyeing of paper. Brit. GB 1193294, 1970; *Chem. Abstr.* **1970**, *73*, 57410.

79. Yoshioka, S. Colored fabrics for manufacture of dyed garments without defects. Jpn. Kokai Tokkyo Koho JP 06057646, 1994; *Chem. Abstr.* **1994**, *121*, 37584.

80. Seidelmann, K. M.; Schmid, A. Black staining compositions for wood. Ger. Offen. DE 19702190, 1998; *Chem. Abstr.* **1998**, *129*, 190512.

81. Geiger, W. Mordants for black dyeing of wood. Ger. Offen. DE 4237063, 1994; *Chem. Abstr.* **1994**, *121*, 282497.

82. Schafer, E. W., Jr.; Bowles, W. A., Jr. Acute oral toxicity and repellency of 933 chemicals to house and deer mice. *Arch. Environ. Contam. Toxicol.* **1985**, *14*, 111–129.

83. DiNardo, J.; Draelos, Z. D. An animal model assessment of common dye-induced allergic contact dermatitis. *J. Cosmet. Sci.* **2007**, *58*, 209–214.

84. Morales, A. H.; Perez, M. A. C.; Gonzalez, M. P. A radial distribution–function approach for predicting rodent carcinogenicity. *J. Mol. Model.* **2006**, *12*, 769–780.

85. Helguera, A. M.; Cabrera Perez, M. A.; Gonzalez, M. P.; Ruiz, R. M.; Gonzalez Diaz, H. A topological substructural approach applied to the computational prediction of rodent carcinogenicity. *Bioorg. Med. Chem.* **2005**, *13*, 2477–2488.

86. Borzelleca, J. F.; Hallagan, J. B. A chronic toxicity/carcinogenicity study of FD & C Yellow No. 5 (tartrazine) in mice. *Food Chem. Toxicol.* **1988**, *26*, 189–194.

87. Borzelleca, J. F.; Hallagan, J. B. Chronic toxicity/carcinogenicity studies of FD & C Yellow No. 5 (tartrazine) in rats. *Food Chem. Toxicol.* **1988**, *26*, 179–187.

88. Haseman, J. K.; Winbush, J. S.; O'Donnell, M. W., Jr. Use of dual control groups to estimate false positive rates in laboratory animal carcinogenicity studies. *Fundam. Appl. Toxicol.* **1986**, *7*, 573–584.

89. Serra, J. R.; Thompson, E. D.; Jurs, P. C. Development of binary classification of structural chromosome aberrations for a diverse set of organic compounds from molecular structure. *Chem. Res. Toxicol.* **2003**, *16*, 153–163.

90. Yadav, S.; Bhalla, M. Cytotoxic evaluation of Tartrazine on root tips of *Allium sativum* (L.). *Pollut. Res.* **2005**, *24*, 279–282.

91. Yamashoji, S.; Isshiki, K. Rapid detection of cytotoxicity of food additives and contaminants by a novel cytotoxicity test, menadione-catalyzed H_2O_2 production assay. *Cytotechnology* **2002**, *37*, 171–178.

92. Carson, S. Skin painting studies in mice with 14 FD & C and D & C colors: FD & C Blue No. 1, Red No. 3, and Yellow No. 5, D & C Red No. 7, Red No. 9, Red No. 10, Red No. 19, Red No. 21, Red No. 27, Red No. 31, Red No. 36, Orange No. 5, Orange No. 10, and Orange No. 17. *J. Toxicol., Cutan. Ocul. Toxicol.* **1984**, *3*, 357–370.

93. Kirkland, D.; Aardema, M.; Henderson, L.; Mueller, L. Evaluation of the ability of a battery of three *in vitro* genotoxicity tests to discriminate rodent carcinogens and non-carcinogens. I. Sensitivity, specificity and relative predictivity. *Mutat. Res.* **2005**, *584*, 1–256.

94. Das, A.; Mukherjee, A. Genotoxicity testing of the food colours amaranth and tartrazine. *Int. J. Hum. Genet.* **2004**, *4*, 277–280.

95. Tripathy, N. K.; Patnaik, K. K.; Nabi, M. J. Genotoxicity of tartrazine studied in two somatic assays of *Drosophila melanogaster*. *Mutat. Res.* **1989**, *224*, 479–483.

96. Haveland-Smith, R. B.; Combes, R. D. Screening of food dyes for genotoxic activity. *Food Cosmet. Toxicol.* **1980**, *18*, 215–221.

97. McCann, D.; Barrett, A.; Cooper, A.; Crumpler, D.; Dalen, L.; Grimshaw, K.; Kitchin, E.; Lok, K.; Porteous, L.; Prince, E.; Sonuga-Barke, E.; Warner, J. O.; Stevenson, J. Food additives and hyperactive behaviour in 3-year-old and 8/9-year-old children in the community: a randomized, double-blinded placebo-controlled trial. *Lancet* **2007**, *370*, 1560–1567.

98. Serafimova, R.; Todorov, M.; Pavlov, T.; Kotov, S.; Jacob, E.; Aptula, A.; Mekenyan, O. Identification of the structural requirements for mutagenicity, by incorporating molecular flexibility and metabolic activation of chemicals. II. General Ames mutagenicity model. *Chem. Res. Toxicol.* **2007**, *20*, 662–676.

99. Varella, S. D.; Pozetti, G. L.; Vilegas, W.; Varanda, E. A. Mutagenic activity of sweepings and pigments from a household-wax factory assayed with *Salmonella typhimurium*. *Food Chem. Toxicol.* **2004**, *42*, 2029–2035.

100. Rafii, F.; Hall, J. D.; Cerniglia, C. E. Mutagenicity of azo dyes used in foods, drugs and cosmetics before and after reduction by *Clostridium* species from the human intestinal tract. *Food Chem. Toxicol.* **1997**, *35*, 897–901.

101. Henschler, D.; Wild, D. Mutagenicity of tartrazine revisited. *Arch. Toxicol.* **1986**, *59*, 69–70.

102. Combes, R. D. On the mutagenicity of tartrazine (FD & C yellow No 5). *Arch. Toxicol.* **1986**, *59*, 67–68.

103. De France, B. F.; Carter, M. H.; Josephy, P. D. Comparative metabolism and mutagenicity of azo and hydrazone dyes in the Ames test. *Food Chem. Toxicol.* **1986**, *24*, 165–169.

104. Ishidate, M., Jr.; Sofuni, T.; Yoshikawa, K.; Hayashi, M.; Nohmi, T.; Sawada, M.; Matsuoka, A. Primary mutagenicity screening of food additives currently used in Japan. *Food Chem. Toxicol.* **1984**, *22*, 623–636.

105. Chung, K. T.; Fulk, G. E.; Andrews, A. W. Mutagenicity testing of some commonly used dyes. *Appl. Environ. Microbiol.* **1981**, *42*, 641–648.

106. Green, M. R.; Pastewka, J. V. Mutagenicity of some lipsticks and their dyes. *J. Natl. Cancer Inst.* **1980**, *64*, 665–669.

107. Tanaka, T. Reproductive and neurobehavioral toxicity study of tartrazine administered to mice in the diet. *Food Chem. Toxicol.* **2006**, *44*, 179–187.

108. Zuberbier, T.; Pfrommer, C.; Specht, K.; Vieths, S.; Bastl-Borrmann, R.; Worm, M.; Henz, B. M. Aromatic components of food as novel eliciting factors of pseudoallergic reactions in chronic urticaria. *J. Allergy Clin. Immunol.* **2002**, *109*, 343–348.

TETRANITRO BLUE TETRAZOLIUM (TNBT)

CAS Registry Number 1184-43-6

Chemical Structure

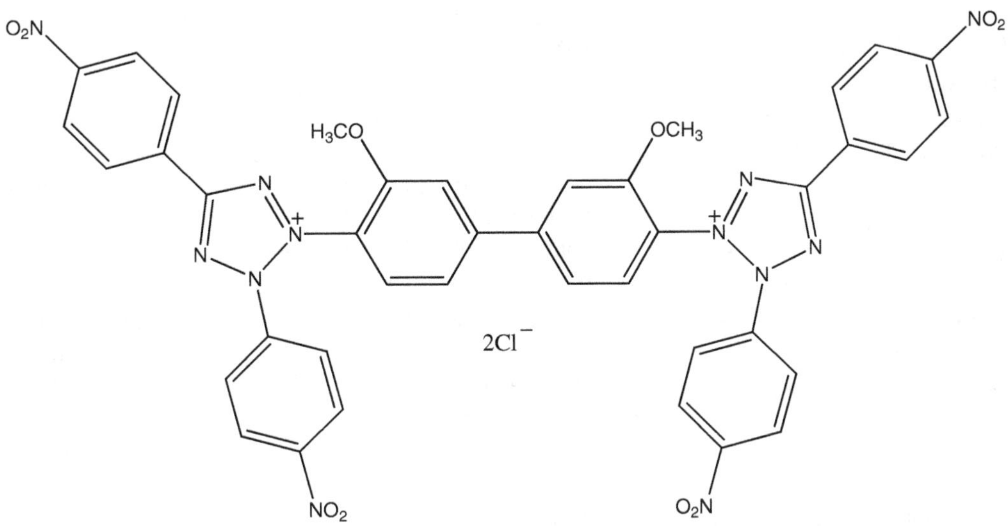

Chemical/Dye Class Tetrazolium salt

Molecular Formula $C_{40}H_{28}Cl_2N_{12}O_{10}$

Molecular Weight 907.63

Physical Form Yellow crystals or powder

CA Index Name 2*H*-Tetrazolium, 2,2′-(3,3′-dimethoxy [1,1′-biphenyl]-4,4′-diyl)bis[3,5-bis(4-nitrophenyl)-, chloride (1 : 2)

Other Names 2*H*-Tetrazolium, 2,2′-(3,3′-dimethoxy [1,1′-biphenyl]-4,4′-diyl)bis[3,5-bis(4-nitrophenyl)-, dichloride; 2*H*-Tetrazolium, 3,3′-(3,3′-dimethoxy-4,4′-biphenylylene)bis[2,5-bis(*p*-nitrophenyl)-, dichloride; [3,3′-(3,3′-Dimethoxy-4,4′-biphenylylene)bis[2,5-bis(*p*-nitrophenyl)-2*H*-tetrazolium chloride]]; 2,2′,5,5′-Tetra-*p*-nitrophenyl-3,3′-(3,3′-dimethoxy-4,4′-biphenylene)ditetrazolium chloride; 2,2′,5,5′-Tetrakis(*p*-nitrophenyl)-3,3′-(3,3′-dimethoxy-4,4′-diphenylene)ditetrazolium chloride; 2,2′,5,5′-Tetrakis-*p*-(nitrophenyl)-3,3′-bis(3,3′-dimethoxy-4,4′-diphenylene)ditetrazolium chloride; Blue *p*-nitrotetrazolium chloride; NSC 121208; Nitroblue monotetrazolium chloride; TNBT; Tetranitro Blue Tetrazolium; Tetranitro-BT; Tetranitroblue tetrazolium chloride; Tetranitrotetrazolium blue

Merck Index Number Not listed

Solubility Soluble in water, ethanol, methanol, *N,N*-dimethylformamide

Melting Point 170 °C (decompose)

Absorption (λ_{max}) 279 nm

Synthesis Synthetic methods[1–4]

Staining Applications Cancer tissues;[5] enzymes;[6] β-galactosidase;[7] glucose-6-phosphate dehydrogenase;[8,9] motor nerve terminals[10]

Biological Applications Diagnosis of bacterial vaginosis;[11] detecting alkaline phosphatase,[12] gamma-hydroxybutyric acid (GHB),[13] succinate dehydrogenase activity;[14] generating and detecting reactive oxygen species;[15,16] treating cancer[17]

Industrial Applications Recording materials;[18] steel products;[19] toner[20]

Safety/Toxicity Bacterial toxicity[21]

REFERENCES

1. Cui, X.; Vlahakis, J. Z.; Crandall, I. E.; Szarek, W. A. Anti-*Plasmodium* activity of tetrazolium salts. *Bioorg. Med. Chem.* **2008**, *16*, 1927–1947.
2. James, P. R. Tetrazolium salts as steam sterilization process indicators. PCT Int. Appl. WO 9837157, 1998; *Chem. Abstr.* **1998**, *129*, 218237.
3. Vecerek, B. Tetrazolium salts. Czech. CS 147890, 1973; *Chem. Abstr.* **1973**, *79*, 53332.
4. Reissenweber, N. J. Succinic dehydrogenase activity in rat spermatozoa as demonstrated with various tetrazolium salts. *Naturwissenschaften* **1966**, *53*, 133–134.

5. Cottington, K. Detecting cancer by histochemical staining. Brit. UK Pat. Appl. GB 2355790, 2001; *Chem. Abstr.* **2001**, *135*, 177699.
6. Moriwaki, Y.; Yamamoto, T.; Yamakita, J. I.; Takahashi, S.; Higashino, K. Enzymic histochemical localization of aldehyde oxidase in rat liver by the tetrazolium method. *Acta Histochem. Cytochem.* **1997**, *30*, 113–115.
7. Gugliotta, P.; Pacchioni, D.; Bussolati, G. Staining reaction for β-galactosidase in immunocytochemistry and *in situ* hybridization. *Eur. J. Histochem.* **1992**, *36*, 143–148.
8. Van Noorden, C. J. F.; Vogels, I. M. C. A sensitive cytochemical staining method for glucose-6-phosphate dehydrogenase activity in individual erythrocytes. II. Further improvements of the staining procedure and some observations with glucose-6-phosphate dehydrogenase deficiency. *Br. J. Haematol.* **1985**, *60*, 57–63.
9. Van Noorden, C. J. F.; Vogels, I. M. C.; James, J.; Tas, J. A sensitive cytochemical staining method for glucose-6-phosphate dehydrogenase activity in individual erythrocytes. I. Optimalization of the staining procedure. *Histochemistry* **1982**, *74*, 493–506.
10. Gunther, J. S.; Letinsky, M. S. Structure of motor nerve terminals in chickens with hereditary muscular dystrophy. *Muscle Nerve* **1985**, *8*, 568–575.
11. Dwir, O. S.; Keren, T.; Fish, F. Dry format sialidase colorimetric assay for bacterial vaginosis diagnosis. PCT Int. Appl. WO 2008018074, 2008; *Chem. Abstr.* **2008**, *148*, 232647.
12. Asan, E.; Kugler, P. Qualitative and quantitative detection of alkaline phosphatase coupled to an oligonucleotide probe for somatostatin mRNA after *in situ* hybridization using unfixed rat brain tissue. *Histochem. Cell Biol.* **1995**, *103*, 463–471.
13. Parsons, S. M.; Harris, D. O.; Bravo, D. T. Methods, compositions and apparatuses for detection of gamma-hydroxybutyric acid (GHB). U.S. Pat. Appl. Publ. US 2003175846, 2003; *Chem. Abstr.* **2003**, *139*, 256601.
14. Chikamori, K.; Shimizu, T.; Shinohara, T.; Fukui, K.; Araki, T.; Yamada, M. Cytophotometry for succinate dehydrogenase activity in relation to tissue protein content with Tetranitro Blue Tetrazolium (TNBT). *Acta Histochem. Cytochem.* **1986**, *19*, 241–251.
15. Freitas, I.; Griffini, P.; Bertone, V.; Bertone, R.; Fenoglio, C.; Milliery, R.; Vairetti, M. *In situ* detection of reactive oxygen species and nitric oxide production in normal and pathological tissues: improvement by differential interference contrast. *Exp. Gerontol.* **2002**, *37*, 591–602.
16. Peskin, A. V.; Khramtsov, A. V.; Morozov, I. A.; Zemskov, V. M.; Zbarskii, I. B. Visualization of reactive oxygen species formation by phagocytizing macrophages. *Exp. Cell Res.* **1984**, *151*, 247–251.
17. Kong, Q. Method and composition for treating cancer using cellular organelle crystallizing agents. U.S. Pat. Appl. Publ. US 2006286170, 2006; *Chem. Abstr.* **2006**, *146*, 55482.
18. Takatori, Y.; Haruta, M.; Shimozawa, A.; Nishide, K. Image reproduction or recording material. Ger. Offen. DE 2429729, 1975; *Chem. Abstr.* **1975**, *83*, 155788.
19. Ehrhardt, W. C.; Cheng, L.; Stasney, D.; Whitaker, K. A. Corrosion inhibitor mixtures with tetrazolium compounds for protection of steel and metal parts in aqueous systems. PCT Int. Appl. WO 2000066810, 2000; *Chem. Abstr.* **2000**, *133*, 353724.
20. Yamamoto, A.; Takagi, S.; Fukumoto, H.; Tanaka, K. Toner for development of electrostatic images. Jpn. Kokai Tokkyo Koho JP 61137162, 1986; *Chem. Abstr.* **1987**, *106*, 58907.
21. Mazul, M. M.; Danilov, V. S. Tetrazolium salt effect on the bioluminescence of *Escherichia coli* recombinant strain. *Biotekhnologiya* **2002**, 91–96; *Chem. Abstr.* **2002**, *138*, 68140.

TETRAZOLIUM VIOLET (TV)

CAS Registry Number 1719-71-7

Chemical Structure

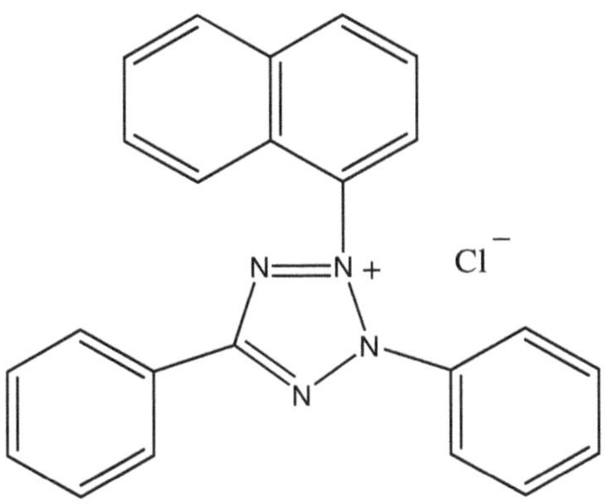

CA Index Name 2*H*-Tetrazolium, 3-(1-naphthalenyl)-2,5-diphenyl-, chloride (1 : 1)

Other Names 2*H*-Tetrazolium, 3-(1-naphthalenyl)-2,5-diphenyl-, chloride; 2*H*-Tetrazolium, 3-(1-naphthyl)-2,5-diphenyl-, chloride; 3-(1-Naphthyl)-2,5-diphenyl-2*H*-tetrazolium chloride; 2,5-Diphenyl-3-(1-naphthyl)-2*H*-tetrazolium chloride; 2,5-Diphenyl-3-(1-naphthyl)tetrazolium chloride; 2,5-Diphenyl-3-α-naphthyltetrazolium chloride; TV; Tetrazolium purple; Tetrazolium violet; Tetrazolium violet chloride; VT; Violet Tetrazolium

Merck Index Number Not listed

Chemical/Dye Class Tetrazolium salt

Molecular Formula $C_{23}H_{17}ClN_4$

Molecular Weight 384.86

Physical Form Yellow brown or tan powder

Solubility Soluble in water, ethanol, methanol

Melting Point 245–250 °C (decompose)

Absorption (λ_{max}) 244 nm

Synthesis Synthetic methods[1–3]

Staining Applications Bacteria;[4] fungi[5]

Biological Applications Analysis of microorganisms;[6] detecting anti-bacterial agent,[7] γ-hydroxybutyric acid (GHB);[8] treating cancer,[9] pain[10]

Industrial Applications Semiconductor powders;[11] photographic materials;[12] recording materials;[13] steel products[14]

Safety/Toxicity No data available

REFERENCES

1. Cui, X.; Vlahakis, J. Z.; Crandall, I. E.; Szarek, W. A. Anti-*Plasmodium* activity of tetrazolium salts. *Bioorg. Med. Chem.* **2008**, *16*, 1927–1947.
2. Gu, B.; Wang, M. Synthesis of 3-(1-naphthalenyl)-2,5-diphenyl-2*H*-tetrazolium chloride. *Zhongguo Yiyao Gongye Zazhi* **2006**, *37*, 10–11; *Chem. Abstr.* **2007**, *148*, 191893.
3. James, P. R. Tetrazolium salts as steam sterilization process indicators. PCT Int. Appl. WO 9837157, 1998; *Chem. Abstr.* **1998**, *129*, 218237.
4. Hayashi, S.; Kobayashi, T.; Honda, H. Simple and rapid cell growth assay using tetrazolium violet coloring method for screening of organic solvent tolerant bacteria. *J. Biosci. Bioeng.* **2003**, *96*, 360–363.
5. Komatsu, S. Coloration culture medium for identifying *Candida*. Jpn. Kokai Tokkyo Koho JP 2005080574, 2005; *Chem. Abstr.* **2005**, *142*, 351732.
6. Bochner, B.; Panomitros, E. Media and system for comparative phenotype analysis of microorganism. PCT Int. Appl. WO 9966066, 1999; *Chem. Abstr.* **1999**, *132*, 20799.
7. Ushiyama, M.; Aoyama, S. Sheet culture medium for detecting anti-bacterial agent, and kit for detecting anti-bacterial agent. Jpn. Kokai Tokkyo Koho JP 2003189896, 2003; *Chem. Abstr.* **2003**, *139*, 49506.
8. Parsons, S. M.; Harris, D. O.; Bravo, D. T. Methods, compositions and apparatuses for detection of gamma-hydroxybutyric acid (GHB). U.S. Pat. Appl. Publ. US 2003175846, 2003; *Chem. Abstr.* **2003**, *139*, 256601.
9. Kong, Q. Method and composition for treating cancer using cellular organelle crystallizing agents. U.S. Pat. Appl. Publ. US 2006286170, 2006; *Chem. Abstr.* **2006**, *146*, 55482.
10. Xia, Z. T. Combination products of a guanylate cyclase inhibitor and a local anesthetic for pain relief. PCT Int. Appl. WO 2000030630, 2000; *Chem. Abstr.* **2000**, *133*, 12755.
11. Kato, T.; Loo, B. H.; Yokomaku, M.; Butsugan, Y.; Sim, K. Y.; Fujishima, A. Photoelectrochemical reduction of tetrazolium salts to formazans on surfaces of semiconductor powders in alcohol solutions. *Spectrosc. Lett.* **1995**, *28*, 849–859.

12. Nishina, Y.; Matsuo, T.; Hase, T.; Takahashi, K.; Hanyu, T. Lith-type silver halide photographic materials and their processing. Jpn. Kokai Tokkyo Koho JP 53093015, 1978; *Chem. Abstr.* **1979**, *90*, 64397.

13. Fujiwara, M.; Matsuo, S.; Masukawa, T.; Kaneko, Y.; Kawasaki, M. Light-sensitive photographic silver halide recording material. Ger. Offen. DE 2734335, 1978; *Chem. Abstr.* **1978**, *88*, 161442.

14. Cheng, L. Corrosion inhibitors with tetrazolium salts for steel protection in aqueous systems. PCT Int. Appl. WO 2000066809, 2000; *Chem. Abstr.* **2000**, *133*, 353308.

THIONIN

Note: The literature is exclusively devoted to Thionin chloride; however, Thionin acetate is certified by Biological Stain Commission (BSC).

CAS Registry Number 78338-22-4

Chemical Structure

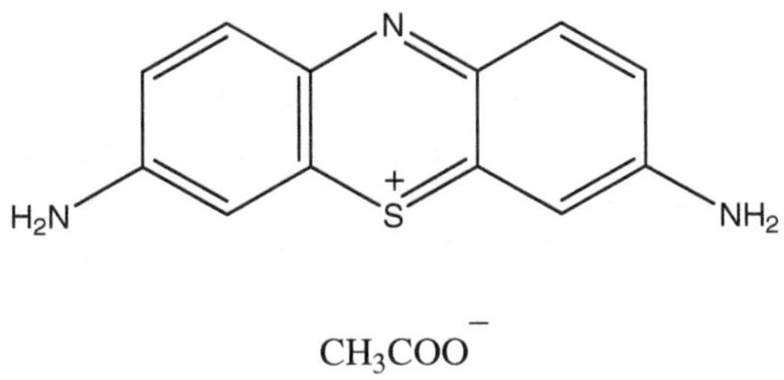

CA Index Name Phenothiazin-5-ium, 3,7-diamino-, acetate (1 : 1)

Other Names C.I. 52000; Lauth's violet; Phenothiazin-5-ium, 3,7-diamino-, acetate; 3,7-Diaminophenothiazin-5-ium acetate; Thionin; Thionin acetate; Thionin acetate salt; Thionine acetate

Merck Index Number Not listed (Chloride listed: 9346)

Chemical/Dye Class Phenothiazine

Molecular Formula $C_{14}H_{13}N_3O_2S$

Molecular Weight 287.34

Physical Form Dark green powder

Solubility Soluble in water, ethanol

Melting Point >200 °C

pK_a 2.5, 11.3

Absorption (λ_{max}) 598 nm

Emission (λ_{max}) 625 nm

Synthesis Synthetic methods[1,2]

Staining Applications Bacteria;[3] viruses;[4] nucleic acids;[4] glomerular deposits[5]

Biological Applications Biosensors;[6] diagnosis of diabetes;[7] detecting ascorbic acid,[8] uric acid,[8] glucose,[9] glomerular deposits[5]

Industrial Applications Batteries;[10] carbon nanotubes;[11,12] glass–polyester composite fibers;[13] magnetic composites;[14] electrochromic devices[15]

Safety/Toxicity No data available

Certification/Approval Certified by Biological Stain Commission (BSC)

REFERENCES

1. Natarajan, P.; Raja, C. Studies on the dynamics of poly(carboxylic acids) with covalently bound thionine and phenosafranine in dilute aqueous solutions. *Eur. Polym. J.* **2005**, *41*, 2496–2504.
2. Rodriguez-Castellon, E.; Jimenez-Lopez, A.; Olivera-Pastor, P.; Merida-Robles, J. M.; Perez-Reina, F. J.; Alcantara-Rodriguez, M.; Souto-Bachiller, F. A.; Rodriguez-Rodriguez, L. de los, A.; Siegel, G. G. Intercalation of thionine in colloidal α-zirconium phosphate. *Mater. Sci. Forum* **1994**, *152–153*, 379–382.
3. Lithgow, A. M.; Romero, L.; Sanchez, I. C.; Souto, F. A.; Vega, C. A. Interception of the electron transport chain in bacteria with hydrophilic redox mediators. Part 1. Selective improvement of the performance of biofuel cells with 2,6-disulfonated thionine as mediator. *J. Chem. Res., Synop.* **1986**, 178–179.
4. MacGregor, A. Inactivation of non-enveloped viruses with agents affecting viral capsid and nucleic acids. PCT Int. Appl. WO 2000034446, 2000; *Chem. Abstr.* **2000**, *133*, 34393.
5. Utsunomiya, Y. Detection of glomerular deposits in various renal diseases by light microscopy using periodic acid thionin [PATS]-chromotrope staining. *Jpn. J. Nephrol.* **1991**, *33*, 949–959.
6. Kirstein, D.; Kirstein, L.; Scheller, F.; Borcherding, H.; Ronnenberg, J.; Diekmann, S.; Steinrucke, P. Amperometric nitrate biosensors on the basis of *Pseudomonas stutzeri* nitrate reductase. *J. Electroanal. Chem.* **1999**, *474*, 43–51.
7. Umegae, Y.; Machida, R.; Takagi, H.; Irie, Y.; Yokoyama, T.; Tanabe, T. Test kits comprising redox enzyme, stabilizer and electrochemical electrode sensor chip for 1,5-anhydroglucitol determination in blood and diabetes diagnosis. PCT Int. Appl. WO 2008072702, 2008; *Chem. Abstr.* **2008**, *149*, 76607.
8. Shahrokhian, S.; Ghalkhani, M. Simultaneous voltammetric detection of ascorbic acid and uric acid at a carbon-paste modified electrode incorporating thionine-Nafion ion-pair as an electron mediator. *Electrochim. Acta* **2006**, *51*, 2599–2606.

9. Salimi, A.; Noorbakhsh, A.; Mamkhezri, H.; Ghavami, R. Electrocatalytic reduction of H_2O_2 and oxygen on the surface of thionin incorporated onto MWCNTs modified glassy carbon electrode: application to glucose detection. *Electroanalysis* **2007**, *19*, 1100–1108.
10. Brenneman, R. K.; Lichtin, N. N. Sunlight rechargeable storage cell. U.S. Patent 4262066, 1981; *Chem. Abstr.* **1981**, *95*, 46156.
11. Wang, Z.; Li, M.; Zhang, Y.; Yuan, J.; Shen, Y.; Niu, L.; Ivaska, A. Thionine-interlinked multi-walled carbon nanotube/gold nanoparticle composites. *Carbon* **2007**, *45*, 2111–2115.
12. Salimi, A.; Noorbakhash, A.; Karonian, F. S. Amperometric detection of nitrite, iodate and periodate on glassy carbon electrode modified with thionin and multi-wall carbon nanotubes. *Int. J. Electrochem. Sci.* **2006**, *1*, 435–446.
13. Medenouvo, F.; Fouassier, J. P. Photochemical manufacture of glass–polyester composite fibers. *Angew. Makromol. Chem.* **1995**, *229*, 29–62.
14. Safarik, I.; Safarikova, M.; Buricova, V. Sorption of water soluble organic dyes on magnetic poly(oxy-2,6-dimethyl-1,4-phenylene). *Collect. Czech. Chem. Commun.* **1995**, *60*, 1448–1456.
15. Varaprasad, D. V.; Habibi, H. R.; Looman, S. D.; Lynam, N. R.; Zhao, M. Electrochemichromic solutions, processes for preparing and using the same, and devices manufactured with the same. Eur. Pat. Appl. EP 531143, 1993; *Chem. Abstr.* **1994**, *120*, 229898.

TMRE

CAS Registry Number 115532-52-0

Chemical Structure

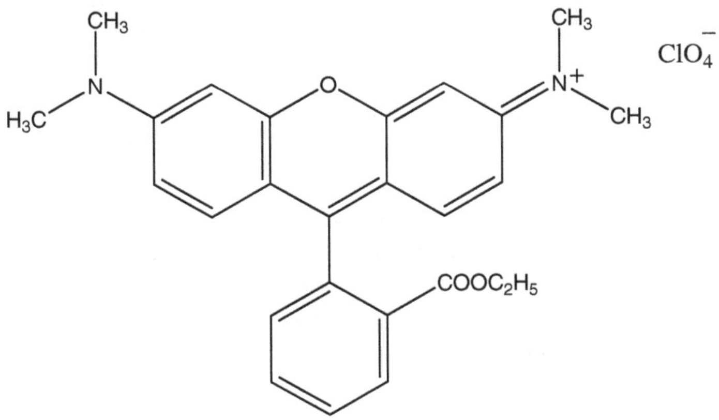

CA Index Name Xanthylium, 3,6-bis(dimethylamino)-9-[2-(ethoxycarbonyl)phenyl]-, perchlorate

Other Names T 669; TMRE; Tetramethylrhodamine ethyl ester perchlorate

Merck Index Number Not listed

Chemical/Dye Class Xanthene

Molecular Formula $C_{26}H_{27}ClN_2O_7$

Molecular Weight 514.95

Physical Form Dark green crystals

Solubility Soluble in ethanol, methanol, dimethyl sulfoxide

Melting Point 264–266 °C

Absorption (λ_{max}) 549 nm

Emission (λ_{max}) 574 nm

Synthesis Synthetic method[1]

Staining Applications Mitochondria;[2–10] cells[3]

Biological Applications Detecting mitochondrial membrane potential;[1,11,12] apoptosis assays;[13–15] multi-drug resistance assays[16]

Industrial Applications Semi-conducting polymer nanoparticles[17]

Safety/Toxicity No data available

REFERENCES

1. Ehrenberg, B.; Montana, V.; Wei, M. D.; Wuskell, J. P.; Loew, L. M. Membrane potential can be determined in individual cells from the Nernstian distribution of cationic dyes. *Biophys. J.* **1988**, *53*, 785–794.
2. Hattori, F.; Fukuda, K. Method for selecting myocardial cells using intracellular mitochondria labeled with fluorescent indicator. PCT Int. Appl. WO 2006022377, 2006; *Chem. Abstr.* **2006**, *144*, 270175.
3. Farkas, D. L.; Wei, M. D.; Febbroriello, P.; Carson, J. H.; Loew, L. M. Simultaneous imaging of cell and mitochondrial membrane potentials. *Biophys. J.* **1989**, *56*, 1053–1069.
4. Bernas, T.; Dobrucki, J. Mitochondrial and nonmitochondrial reduction of MTT: interaction of MTT with TMRE, JC-1, and NAO mitochondrial fluorescent probes. *Cytometry* **2002**, *47*, 236–242.
5. Yu, W. R.; Liu, T.; Fehlings, T. K.; Fehlings, M. G. Involvement of mitochondrial signaling pathways in the mechanism of Fas-mediated apoptosis after spinal cord injury. *Eur. J. Neurosci.* **2009**, *29*, 114–131.
6. Yin, Z.; Aschner, J. L.; dos Santos, A. P.; Aschner, M. Mitochondrial-dependent manganese neurotoxicity in rat primary astrocyte cultures. *Brain Res.* **2008**, *1203*, 1–11.
7. Mironov, S. L. ADP regulates movements of mitochondria in neurons. *Biophys. J.* **2007**, *92*, 2944–2952.
8. Spees, J. L.; Olson, S. D.; Whitney, M. J.; Prockop, D. J. Mitochondrial transfer between cells can rescue aerobic respiration. *Proc. Natl. Acad. Sci. U.S.A.* **2006**, *103*, 1283–1288.
9. Collins, T. J.; Bootman, M. D. Mitochondria are morphologically heterogeneous within cells. *J. Exp. Biol.* **2003**, *206*, 1993–2000.

10. Collins, T. J.; Berridge, M. J.; Lipp, P.; Bootman, M. D. Mitochondria are morphologically and functionally heterogeneous within cells. *EMBO J.* **2002**, *21*, 1616–1627.

11. Chalmers, S.; McCarron, J. G. The mitochondrial membrane potential and Ca^{2+} oscillations in smooth muscle. *J. Cell Sci.* **2008**, *121*, 75–85.

12. Nicholls, D. G.; Ward, M. W. Mitochondrial membrane potential and neuronal glutamate excitotoxicity: mortality and millivolts. *Trends Neurosci.* **2000**, *23*, 166–174.

13. Galluzzi, L.; Zamzami, N.; de La Motte Rouge, T.; Lemaire, C.; Brenner, C.; Kroemer, G. Methods for the assessment of mitochondrial membrane permeabilization in apoptosis. *Apoptosis* **2007**, *12*, 803–813.

14. Goldstein, J. C.; Muñoz-Pinedo, C.; Ricci, J. E.; Adams, S. R.; Kelekar, A.; Schuler, M.; Tsien, R. Y.; Green, D. R. Cytochrome *c* is released in a single step during apoptosis. *Cell Death Differ.* **2005**, *12*, 453–462.

15. MacKenzie, F.; Duriez, P.; Wong, F.; Noseda, M.; Karsan, A. Notch4 inhibits endothelial apoptosis via RBP-Jkappa-dependent and -independent pathways. *J. Biol. Chem.* **2004**, *279*, 11657–11663.

16. Eytan, G.; Assaraf, Y. Assay for multidrug resistance based on fluorescent dye exclusion. PCT Int. Appl. WO 9807034, 1998; *Chem. Abstr.* **1998**, *128*, 188616.

17. Grigalevicius, S.; Forster, M.; Ellinger, S.; Landfester, K.; Scherf, U. Excitation energy transfer from semiconducting polymer nanoparticles to surface-bound fluorescent dyes. *Macromol. Rapid Commun.* **2006**, *27*, 200–202.

TMRM

CAS Registry Number 115532-50-8

Chemical Structure

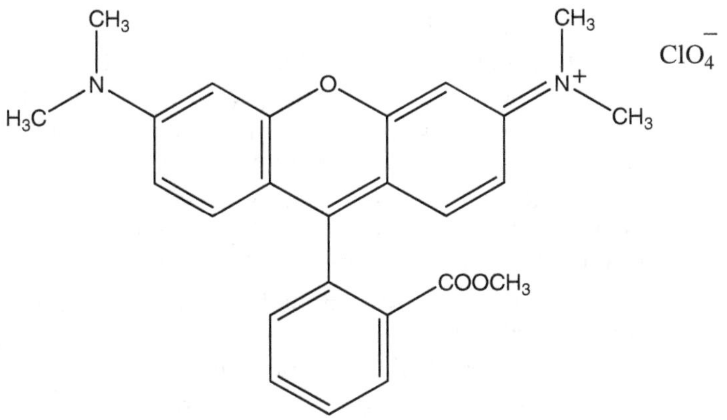

CA Index Name Xanthylium, 3,6-bis(dimethylamino)-9-[2-(methoxycarbonyl)phenyl]-, perchlorate

Other Names T 668; TMRM; Tetramethylrhodamine methyl ester perchlorate

Merck Index Number Not listed

Chemical/Dye Class Xanthene

Molecular Formula $C_{25}H_{25}ClN_2O_7$

Molecular Weight 500.93

Physical Form Dark green crystals

Solubility Soluble in ethanol, methanol, dimethyl sulfoxide

Melting Point 274–276 °C

Absorption (λ_{max}) 549 nm

Emission (λ_{max}) 573 nm

Synthesis Synthetic method[1]

Staining Applications Mitochondria;[2–4] cells[3]

Biological Applications Detecting mitochondrial membrane potential;[1,3,5,6] apoptosis assays;[7,8] multidrug resistance assays[9]

Industrial Applications Not reported

Safety/Toxicity No data available

REFERENCES

1. Ehrenberg, B.; Montana, V.; Wei, M. D.; Wuskell, J. P.; Loew, L. M. Membrane potential can be determined in individual cells from the Nernstian distribution of cationic dyes. *Biophys. J.* **1988**, *53*, 785–794.
2. Hattori, F.; Fukuda, K. Method for selecting myocardial cells using intracellular mitochondria labeled with fluorescent indicator. PCT Int. Appl. WO 2006022377, 2006; *Chem. Abstr.* **2006**, *144*, 270175.
3. Farkas, D. L.; Wei, M. D.; Febbroriello, P.; Carson, J. H.; Loew, L. M. Simultaneous imaging of cell and mitochondrial membrane potentials. *Biophys. J.* **1989**, *56*, 1053–1069.
4. Diaz, G.; Liu, S.; Isola, R.; Diana, A.; Falchi, A. M. Mitochondrial localization of reactive oxygen species by dihydrofluorescein probes. *Histochem. Cell Biol.* **2003**, *120*, 319–325.
5. Distelmaier, F.; Koopman, W. J.; Testa, E. R.; de Jong, A. S.; Swarts, H. G.; Mayatepek, E.; Smeitink, J. A.; Willems, P. H. Life cell quantification of mitochondrial membrane potential at the single organelle level. *Cytometry* **2008**, *73A*, 129–138.
6. Zhang, H.; Huang, H. M.; Carson, R. C.; Mahmood, J.; Thomas, H. M.; Gibson, G. E. Assessment of membrane potentials of mitochondrial populations in living cells. *Anal. Biochem.* **2001**, *298*, 170–180.
7. Galluzzi, L.; Zamzami, N.; de La Motte Rouge, T.; Lemaire, C.; Brenner, C.; Kroemer, G. Methods for the assessment of mitochondrial membrane permeabilization in apoptosis. *Apoptosis* **2007**, *12*, 803–813.
8. Gottlieb, R. A.; Granville, D. J. Analyzing mitochondrial changes during apoptosis. *Methods* **2002**, *26*, 341–347.
9. Eytan, G.; Assaraf, Y. Assay for multidrug resistance based on fluorescent dye exclusion. PCT Int. Appl. WO 9807034, 1998; *Chem. Abstr.* **1998**, *128*, 188616.

TOLUIDINE BLUE O

CAS Registry Number 92-31-9

Chemical Structure

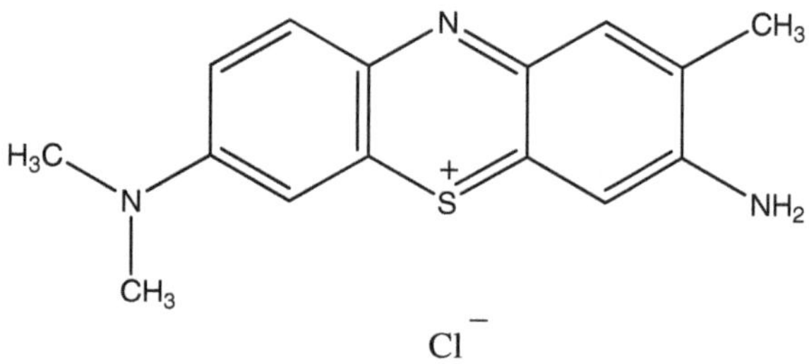

CA Index Name Phenothiazin-5-ium, 3-amino-7-(dimethylamino)-2-methyl, chloride (1:1)

Other Names Phenothiazin-5-ium, 3-amino-7-(dimethylamino)-2-methyl, chloride; 3-Amino-7-(dimethylamino)-2-methylphenazathionium chloride; 3-Amino-7-(dimethylamino)-2-methylphenothiazin-5-ium chloride; Basic Blue 17; Blutene; Blutene chloride; C.I. 52040; C.I. Basic Blue 17; Dimethyltoluthionine chloride; F Klot; Gabilin; Klot; Menodin; NSC 36758; Schultz 1041; Tolazul; Tolonium chloride; Toluidine Blue O; Toluidine Blue OO; Toluidine blue; Toluidine blue (phenothiazine dye); *o*-Toluidine blue

Merck Index Number 9520

Chemical/Dye Class Phenothiazine

Molecular Formula $C_{15}H_{16}ClN_3S$

Molecular Weight 305.83

Physical Form Dark green powder

Solubility Soluble in water, ethanol

Melting Point >200 °C

p*K*$_a$ 2.4, 11.6

Absorption (λ_{max}) 626 nm, 630 nm

Synthesis Synthetic methods[1–10]

Staining Applications Amyloid deposits;[11] cells;[12,13] gene expression;[14] glycosaminoglycans;[15] leaf cuticle;[16] Mohs' micrographic surgery;[17] neurons;[18,19] nucleic acids;[20] nuclei;[21] ocular structure;[22] oral lesions;[23–26] paraffin;[27] phenolic deposits;[28] prion protein deposits;[29] proteoglycans;[30] renal cells;[31] RNA;[32] skin lesions;[33] sperm DNA;[34] sperm chromatin;[34,35] sputum[36]

Biological Applications Bound hyaluronan assay;[37] DNA assay;[38] bioelectronic applications;[39] biofuel cells;[40,41] microbial fuel cells;[42] detecting bacteria,[43] body fluid,[44] nucleic acids,[45,46] proteins,[46] viruses;[46] identifying Gram-negative bacteria,[47] β-amyloid reducing agents;[48] photodynamic therapy;[49–56] treating eye diseases,[53] fungal infections,[54] oral lesions,[55] skin diseases,[56] nail infections,[57] neurodegenerative diseases;[58] medical devices[59]

Industrial Applications Photovoltaic cells;[60] solar cells;[60] inks;[61] toners;[62] electrorheological materials;[63] recording materials;[64] paints;[65] detergents;[66] rubber;[67] textiles;[68] analytical applications;[69] security applications[70]

Safety/Toxicity Carcinogenicity;[71,72] cytotoxicity;[73] hepatotoxicity;[74] microbial toxicity;[75] mutagenicity;[76,77] neurotoxicity;[78–80] ototoxicity;[81] photodynamic toxicity[82,83]

Certification/Approval Certified by Biological Stain Commission (BSC)

REFERENCES

1. Okolotowicz, K. Preparation of Toluidine Blue O drug for *in vivo* staining and chemotherapeutic treatment of dysplastic tissues. PCT Int. Appl. WO 2003103569, 2003; *Chem. Abstr.* **2003**, *140*, 47513.
2. Burkett, D. D. Light-stabilized *in vivo* stain composition and method of manufacture. PCT Int. Appl. WO 2003020323, 2003; *Chem. Abstr.* **2003**, *138*, 201326.
3. Burkett, D. D. Production of toluidine blue O. U.S. Patent 6194573, 2001; *Chem. Abstr.* **2001**, *134*, 194557.
4. Burkett, D. D. Toluidine blue O *in vivo* stain composition, process of manufacture, and methods of use to identify dysplastic tissue. PCT Int. Appl. WO 9925388, 1999; *Chem. Abstr.* **1999**, *130*, 349378.
5. Wang, J.; Jia, J.; Lu, M.; Lalezari, L. New method of synthesis and analysis of high-quality toluidine blue-O for medical uses. *Zhongguo Yaowu Huaxue Zazhi* **1998**, *8*, 292–293; *Chem. Abstr.* **1999**, *131*, 31915.
6. Blower, P. J.; Carter, N. J. Rapid preparation of iodine-123-labelled methylene blue and toluidine blue: potential new agents for parathyroid scintigraphy. *Nucl. Med. Commun.* **1990**, *11*, 413–420.
7. Shamolina, I. I.; Vol'f, L. A.; Abakumova, R. A.; Pal'chevskii, V. V. Preparation and study of redox fibers of phenothiazine type. *Zh. Prikl. Khim.* **1980**, *53*, 637–640; *Chem. Abstr.* **1980**, *92*, 199601.
8. Nakayama, T. Preparation of technetium-99m-labeled toluidine blue and the distribution in rats.

Kanazawa Irigaku Sosho **1974**, *95*, 19–31; *Chem. Abstr.* **1974**, *81*, 147698.

9. Slater, B. Superior stain for *Helicobacter pylori* using toluidine O. *J. Clin. Pathol.* **1990**, *43*, 961.
10. Toepfer, K. Purification of dyes by recrystallization. *Acta Histochem., Suppl.* **1973**, *13*, 291–293.
11. Richter, S.; Makovitzky, J. Topo-optical visualization reactions of carbohydrate-containing amyloid deposits in the respiratory tract. *Acta Histochem.* **2006**, *108*, 181–191.
12. Gelvan, D. J.; Goltsman, L.; Chausovsky, A. Staining methods and compositions for identifying a cell phenotype. PCT Int. Appl. WO 2007102146, 2007; *Chem. Abstr.* **2007**, *147*, 339010.
13. Burkett, D. D. Stain-directed molecular analysis for cancer prognosis and diagnosis. PCT Int. Appl. WO 2003072826, 2003; *Chem. Abstr.* **2003**, *139*, 193967.
14. Okuducu, A. F.; Janzen, V.; Hahne, J. C.; Ko, Y.; Wernert, N. Influence of histochemical stains on quantitative gene expression analysis after laser-assisted microdissection. *Int. J. Mol. Med.* **2003**, *11*, 449–453.
15. Kiraly, K.; Lapvetelainen, T.; Arokoski, J.; Torronen, K.; Modis, L.; Kiviranta, I.; Helminen, H. J. Application of selected cationic dyes for the semiquantitative estimation of glycosaminoglycans in histological sections of articular cartilage by microspectrophotometry. *Histochem. J.* **1996**, *28*, 577–590.
16. Tanaka, T.; Tanaka, H.; Machida, C.; Watanabe, M.; Machida, Y. A new method for rapid visualization of defects in leaf cuticle reveals five intrinsic patterns of surface defects in *Arabidopsis*. *Plant J.* **2004**, *37*, 139–146.
17. Todd, M. M.; Lee, J. W.; Marks, V. J. Rapid toluidine blue stain for Mohs' micrographic surgery. *Dermatol. Surg.* **2005**, *31*, 244–245.
18. Rho, O.; Miller, G. W. Laser capture microdissection to examine transporter expression in specific cell regions. *Methods Mol. Biol.* **2003**, *227*, 85–95.
19. Chelvanayagam, D. K.; Beazley, L. D. Toluidine blue-O is a Nissl bright-field counterstain for lipophilic fluorescent tracers Di-ASP, DiI and DiO. *J. Neurosci. Methods* **1997**, *72*, 49–55.
20. Inagaki, T. Staining agent for nucleic acids, and its application method. Jpn. Kokai Tokkyo Koho JP 2004163278, 2004; *Chem. Abstr.* **2004**, *141*, 3836.
21. Mello, M. L. S.; Falco, J. R. P. Critical electrolyte concentration of DNA–protein complexes in spermatozoal and somatic cell nuclei of the honey bee, *Apis mellifera*. *Insect Biochem. Mol. Biol.* **1996**, *26*, 793–795.
22. Buono, L. M. Use of injectable dyes for staining an anterior lens capsule and vitreo-retinal interface. U.S. Pat. Appl. Publ. US 2003096334, 2003; *Chem. Abstr.* **2003**, *138*, 381686.
23. Gupta, A.; Singh, M.; Ibrahim, R.; Mehrotra, R. Utility of toluidine blue staining and brush biopsy in precancerous and cancerous oral lesions. *Acta Cytol.* **2007**, *51*, 788–794.
24. Juneja, M.; Mahajan, S.; Rao, N. N.; George, T.; Boaz, K. Histochemical analysis of pathological alterations in oral lichen planus and oral lichenoid lesions. *J. Oral Sci.* **2006**, *48*, 185–193.
25. Gandolfo, S.; Pentenero, M.; Broccoletti, R.; Pagano, M.; Carrozzo, M.; Scully, C. Toluidine blue uptake in potentially malignant oral lesions *in vivo*: clinical and histological assessment. *Oral Oncol.* **2006**, *42*, 89–95.
26. Guo, Z.; Yamaguchi, K.; Sanchez-Cespedes, M.; Westra, W. H.; Koch, W. M.; Sidransky, D. Allelic losses in OraTest-directed biopsies of patients with prior upper aerodigestive tract malignancy. *Clin. Cancer Res.* **2001**, *7*, 1963–1968.
27. Kiernan, J. A. Staining paraffin sections without prior removal of the wax. *Biotech. Histochem.* **1996**, *71*, 304–310.
28. Gutmann, M. Improved staining procedures for photographic documentation of phenolic deposits in semithin sections of plant tissue. *J. Microsc.* **1995**, *179*, 277–281.
29. Sanchez, A.; Guzman, A.; Ortiz, A.; Rembao, D.; Espinosa, B.; Zenteno, E.; Guevara, J. Toluidine blue-O staining of prion protein deposits. *Histochem. Cell Biol.* **2001**, *116*, 519–524.
30. Terry, D. E.; Chopra, R. K.; Ovenden, J.; Anastassiades, T. P. Differential use of Alcian Blue and Toluidine Blue dyes for the quantification and isolation of anionic glycoconjugates from cell cultures: application to proteoglycans and a high-molecular-weight glycoprotein synthesized by articular chondrocytes. *Anal. Biochem.* **2000**, *285*, 211–219.
31. Shet, T.; Desai, S. Toluidine positivity in chromophobe renal cell carcinoma: use in intraoperative diagnosis. *Histopathology* **2005**, *47*, 534–535.
32. Meyer, W.; Zschemisch, N. H. Remarks on the usefulness of toluidine blue staining for RNA cytophotometry in plastic embedded tissues. *Cell. Mol. Biol.* **1999**, *45*, 379–382.

33. Naik, R.; Pai, M. R.; Bantwal, P. B.; Nayak, K. S.; Gandhi, A. Study of mast cells in non-neoplastic skin lesions. *Indian J. Pathol. Microbiol.* **2003**, *46*, 173–175.

34. Erenpreiss, J.; Jepson, K.; Giwercman, A.; Tsarev, I.; Erenpreisa, J.; Spano, M. Toluidine blue cytometry test for sperm DNA conformation: comparison with the flow cytometric sperm chromatin structure and TUNEL assays. *Hum. Reprod.* **2004**, *19*, 2277–2282.

35. Beletti, M. E.; Costa, L. Da F.; Viana, M. P. A computational approach to characterization of bovine sperm chromatin alterations. *Biotech. Histochem.* **2004**, *79*, 17–23.

36. Aderaye, G.; Woldeamanuel, Y.; Asrat, D.; Lebbad, M.; Beser, J.; Worku, A.; Fernandez, V.; Lindquist, L. Evaluation of Toluidine Blue O staining for the diagnosis of *Pneumocystis jiroveci* in expectorated sputum sample and bronchoalveolar lavage from HIV-infected patients in a tertiary care referral center in Ethiopia. *Infection* **2008**, *36*, 237–243.

37. Johnston, J. B. A simple, nondestructive assay for bound hyaluronan. *J. Biomed. Mater. Res.* **2000**, *53*, 188–191.

38. Passmore, L. J.; Killeen, A. A. Toluidine blue dye-binding method for measurement of genomic DNA extracted from peripheral blood leukocytes. *Mol. Diagn.* **1996**, *1*, 329–334.

39. Hassler, B. L.; Kohli, N.; Zeikus, J. G.; Lee, I.; Worden, R. M. Renewable dehydrogenase-based interfaces for bioelectronic applications. *Langmuir* **2007**, *23*, 7127–7133.

40. Minteer, S. D.; Martin, R. S.; Moore, C. M. Microfluidic biofuel cell. PCT Int. Appl. WO 2005096430, 2005; *Chem. Abstr.* **2005**, *143*, 389794.

41. Minteer, S. D.; Akers, N. L.; Moore, C. M. Enzyme immobilization for use in biofuel cells and sensors. U. S. Pat. Appl. Publ. US 2004101741, 2004; *Chem. Abstr.* **2004**, *141*, 9611.

42. Roller, S. D.; Bennetto, H. P.; Delaney, G. M.; Mason, J. R.; Stirling, J. L.; Thurston, C. F. Electron-transfer coupling in microbial fuel cells. 1. Comparison of redox-mediator reduction rates and respiratory rates of bacteria. *J. Chem. Technol. Biotechnol.* **1984**, *34*, 3–12.

43. Yazdankhah, S. P.; Gogstad, G. O.; Sorum, R. H.; Larsen, H. J. S. Diagnostic analytical method and kit containing bacteria-retaining substrate and dye for detecting bacteria. PCT Int. Appl. WO 2001049872, 2001; *Chem. Abstr.* **2001**, *135*, 89497.

44. Chaffringeon, B. Single-use device for detecting or analyzing a body fluid. PCT Int. Appl. WO 9609545, 1996; *Chem. Abstr.* **1996**, *125*, 5037.

45. Gupta, A.; Will, S. G. Light emission modifiers and their uses in nucleic acid detection, amplification and analysis. Eur. Pat. Appl. EP 1739190, 2007; *Chem. Abstr.* **2007**, *146*, 115979.

46. Althaus, J. S.; Kyonghoon, L.; Namasivayam, V.; Brahmasandra, S. N.; Handique, K. Microfluidic devices and method for electrochemical detection of nucleic acids, proteins or viruses. PCT Int. Appl. WO 2004094986, 2004; *Chem. Abstr.* **2004**, *141*, 362747.

47. Oka, A. Dyeability discriminant method for identifying Gram negative bacteria. Jpn. Kokai Tokkyo Koho JP 2004208526, 2004; *Chem. Abstr.* **2004**, *141*, 119787.

48. Cordell, B.; Higgins, L. S. Organotypic brain slice culturing methods to identify β-amyloid reducing agents. PCT Int. Appl. WO 9843094, 1998; *Chem. Abstr.* **1998**, *129*, 270601.

49. Clements, D. J.; Pearson, G. J.; Williams, J. A.; Colles, M. J. Method for photodynamic therapy and apparatus therefor. PCT Int. Appl. WO 2008110849, 2008; *Chem. Abstr.* **2008**, *149*, 363133.

50. Bevilacqua, I. M.; Nicolau, R. A.; Khouri, S.; Brugnera, A., Jr.; Teodoro, G. R.; Zangaro, R. A.; Pacheco, M. T. T. The impact of photodynamic therapy on the viability of *Streptococcus mutans* in a planktonic culture. *Photomed. Laser Surg.* **2007**, *25*, 513–518.

51. Zanin, I. C. J.; Goncalves, R. B.; Brugnera Junior, A.; Hope, C. K.; Pratten, J. Susceptibility of *Streptococcus mutans* biofilms to photodynamic therapy: an *in vitro* study. *J. Antimicrob. Chemother.* **2005**, *56*, 324–330.

52. Sharma, M.; Bansal, H.; Gupta, P. K. Virulence of *Pseudomonas aeruginosa* cells surviving photodynamic treatment with toluidine blue. *Curr. Microbiol.* **2005**, *50*, 277–280.

53. Chen, J. C. Photodynamic therapy comprising administration of a targeted photosensitizing agent, for treatment for eye disease such as diabetic retinopathy, macular degeneration, and malignant uveal melanomas. U.S. Patent 7320786, 2008; *Chem. Abstr.* **2008**, *148*, 152083.

54. Donnelly, R. F.; McCarron, P. A.; Tunney, M. M.; Woolfson, A. D. Potential of photodynamic therapy in treatment of fungal infections of the mouth. Design and characterisation of a mucoadhesive patch

containing toluidine blue O. *J. Photochem. Photobiol., B* **2007**, *86*, 59–69.

55. Kvaal, S. I.; Warloe, T. Photodynamic treatment of oral lesions. *J. Environ. Pathol., Toxicol. Oncol.* **2007**, *26*, 127–133.
56. Gilson, D. J.; Gilson, J. A.; DeBellis, J. A. Therapeutic device for delivering controlled air stream flow and/or photodynamic therapy of skin diseases. U.S. Pat. Appl. Publ. US 2008031924, 2008; *Chem. Abstr.* **2008**, *148*, 198849.
57. Loebel, N. G.; Mckenzie, F. R.; Street, C.; Andersen, R.; Scott, R. Composition containing antifungal agents and photosensitizers, therapy and device for treatment of nail infections. PCT Int. Appl. WO 2008109424, 2008; *Chem. Abstr.* **2008**, *149*, 362854.
58. Wischik, C. M.; Horsley, D.; Rickard, J. E.; Harrington, C. R. Drug screening for effectors of tau protein proteolytic processing and expression systems of controlled aggregation in treatment of neurodegenerative disease. PCT Int. Appl. WO 2002055720, 2002; *Chem. Abstr.* **2002**, *137*, 88474.
59. Ash, S. R.; Steczko, J. Antimicrobial medical devices. U.S. Pat. Appl. Publ. US 2008279907, 2008; *Chem. Abstr.* **2008**, *149*, 541798.
60. Chane, C. J. Y.; Corma, A.; Garcia, H. Multivalent metal cation-doped cerium oxide materials for photovoltaic cells and solar cells. Fr. Demande FR 2857783, 2005; *Chem. Abstr.* **2005**, *142*, 159487.
61. Coughlin, S. J.; Fraser, I. F.; Healy, T.; Niven, S. C. Pigment compositions, their production and their use in inks and coatings with improved rheology. PCT Int. Appl. WO 2003093373, 2003; *Chem. Abstr.* **2003**, *139*, 382824.
62. Yamaka, H.; Shimura, I.; Matsura, H.; Saito, H. Toner for developing electrostatic image. Jpn. Kokai Tokkyo Koho JP 10003183, 1998; *Chem. Abstr.* **1998**, *128*, 147470.
63. Weiss, K. D.; Carlson, J. D.; Duclos, T. G. Colorant-containing electrorheological materials. PCT Int. Appl. WO 9319143, 1993; *Chem. Abstr.* **1994**, *120*, 11510.
64. Maeda, Y.; Sakota, K.; Iwamoto, M. PHB (photochemical-hole-burning) recording material containing thiazine derivative. Jpn. Kokai Tokkyo Koho JP 05011402, 1993; *Chem. Abstr.* **1993**, *119*, 259650.
65. Fasano, D. M. Method for increasing the hiding power of a wet paint film. Eur. Pat. Appl. EP 549145, 1993; *Chem. Abstr.* **1993**, *119*, 252219.
66. Macdonald, J. G.; Huang, Y.; Yang, K.; Kim, J.; Wei, N. Single phase color change agents for liquid detergents for cleaning surfaces. U.S. Pat. Appl. Publ. US 2005049157, 2005; *Chem. Abstr.* **2005**, *142*, 263578.
67. Weber, K. E.; Mukamal, H. Rubber composition. Eur. Pat. Appl. EP 99710, 1984; *Chem. Abstr.* **1984**, *100*, 140656.
68. Deeds, S. M. Fabric repair coloring device and using the device to color unwanted bleached out spots on fabric. U.S. Patent 6739779, 2004; *Chem. Abstr.* **2004**, *140*, 424935.
69. Luo, H. Q.; Liu, S. P.; Liu, Z. F.; Liu, Q.; Li, N. B. Resonance Rayleigh scattering spectra for studying the interaction of heparin with some basic phenothiazine dyes and their analytical applications. *Anal. Chim. Acta* **2001**, *449*, 261–270.
70. Waning, R.; Muke, S. Method to authenticate articles and security documents. PCT Int. Appl. WO 2004087430, 2004; *Chem. Abstr.* **2004**, *141*, 327013.
71. Nicolas, P. D.; Quintero, E.; Parra, B. A. Screening the at-risk population for squamous cell carcinoma of the esophagus. *Gastroenterol. Hepatol.* **2005**, *28*, 337–346.
72. Redman, R. S.; Krasnow, S. H.; Sniffen, R. A. Evaluation of the carcinogenic potential of toluidine blue O in the hamster cheek pouch. *Oral Surg., Oral Med., Oral Pathol.* **1992**, *74*, 473–480.
73. Schuster, U.; Schmalz, G.; Thonemann, B.; Mendel, N.; Metzl, C. Cytotoxicity testing with three-dimensional cultures of transfected pulp-derived cells. *J. Endod.* **2001**, *27*, 259–265.
74. Marks, D. H.; Cooper, T.; Makovec, T.; Okerberg, C.; Lollini, L. O. Effect of polymyxin B on hemoglobin-mediated hepatotoxicity. *Military Med.* **1989**, *154*, 180–184.
75. Oros, G.; Cserhati, T.; Forgacs, E. Separation of the strength and selectivity of the microbiological effect of synthetic dyes by spectral mapping technique. *Chemosphere* **2003**, *52*, 185–193.
76. Dunipace, A. J.; Beaven, R.; Noblitt, T.; Li, Y.; Zunt, S.; Stookey, G. Mutagenic potential of toluidine blue evaluated in the Ames test. *Mutat. Res.* **1992**, *279*, 255–259.
77. Ferguson, L. R.; Baguley, B. C. Verapamil as a co-mutagen in the *Salmonella*/mammalian microsome mutagenicity test. *Mutat. Res. Lett.* **1988**, *209*, 57–62.
78. Govindaraju, S. R.; Curry, B. D.; Bain, J. L. W.; Riley, D. A. Effects of temperature on vibration-induced

damage in nerves and arteries. *Muscle Nerve* **2006**, *33*, 415–423.

79. Kristensen, B. W.; Noraberg, J.; Thiebaud, P.; Koudelka-Hep, M.; Zimmer, J. Biocompatibility of silicon-based arrays of electrodes coupled to organotypic hippocampal brain slice cultures. *Brain Res.* **2001**, *896*, 1–17.
80. Noraberg, J.; Kristensen, B. W.; Zimmer, J. Markers for neuronal degeneration in organotypic slice cultures. *Brain Res. Protoc.* **1999**, *3*, 278–290.
81. Sitaras, N.; Vrouvidou, P.; Varonos, D.; Coyas, A. Metachromasia as the key to aminoglycoside ototoxicity. Preliminary report. *Acta Oto-laryngol.* **1985**, *99*, 336–338.
82. Herkovits, J.; Perez-Coll, C. S.; Stockert, J. C.; Blazquez, A. The screening of photodynamic toxicity of dyes by means of a bioassay using amphibian embryos. *Res. J. Chem. Environ.* **2007**, *11*, 86–91.
83. Stockert, J. C.; Juarranz, A.; Villanueva, A.; Canete, M. Photodynamic damage to HeLa cell microtubules induced by thiazine dyes. *Cancer Chemother. Pharmacol.* **1996**, *39*, 167–169.

TO-PRO 1

CAS Registry Number 157199-59-2

Chemical Structure

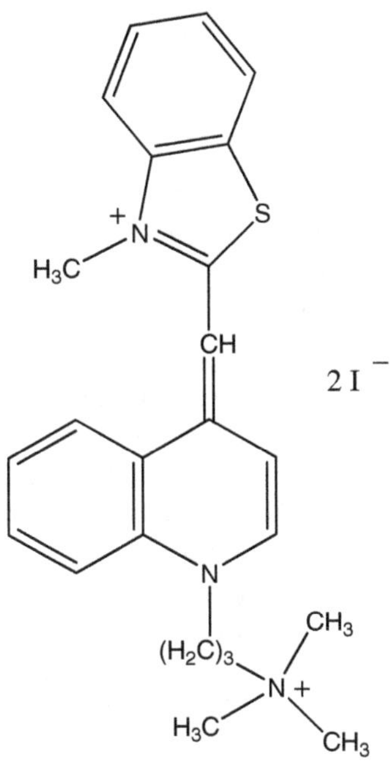

CA Index Name Quinolinium, 4-[(3-methyl-2(3*H*)-benzothiazolylidene)methyl]-1-[3-(trimethylammonio) propyl]-, iodide (1:2)

Other Names Quinolinium, 4-[(3-methyl-2(3*H*)-benzothiazolylidene)methyl]-1-[3-(trimethylammonio)propyl]-, diiodide; TO-PRO 1; TO-PRO 1 iodide

Merck Index Number Not listed

Chemical/Dye Class Cyanine

Molecular Formula $C_{24}H_{29}I_2N_3S$

Molecular Weight 645.38

Physical Form Red-brown powder

Solubility Soluble in dimethyl sulfoxide

Melting Point >250 °C

Absorption (λ_{max}) 515 nm

Emission (λ_{max}) 531 nm

Synthesis Synthetic methods[1–6]

Staining Applications Nucleic acids;[7–13] cells;[4,14,15] antibodies;[1,2] bacteria;[16] leukocytes;[17,18,27] nuclei;[27] megakaryocyte;[19] microorganisms;[20] peptides;[1,2] proteins;[1,2] reticulocytes;[21] sperms[22]

Biological Applications Nucleic acid hybridization;[23,24] nucleic acid assay;[12] nucleic acid fragment sizing;[13] nucleic acid sequencing;[11] detecting nucleic acids,[7–10] cells,[4,14,15]Human papilloma virus (HPV),[25] microstructures;[26] counting embryoblasts,[27] leukocytes;[27] monitoring cell cycle kinetics[28]

Industrial Applications Not reported

Safety/Toxicity No data available

REFERENCES

1. Hahn, K. M.; Toutchkine, A.; Muthyala, R.; Kraynov, V.; Bark, S. J.; Burton, D. R.; Chamberlain, C. Labeled peptides, proteins and antibodies and processes and intermediates useful for their preparation. U.S. Pat. Appl. Publ. US 2002055133, 2002; *Chem. Abstr*. **2002**, *136*, 366139.
2. Hahn, K. M.; Toutchkine, A.; Muthyala, R.; Kraynov, V.; Bark, S. J.; Burton, D. R.; Chamberlain, C. Labeled peptides, proteins and antibodies and processes and intermediates useful for their preparation. PCT Int. Appl. WO 2002008245, 2002; *Chem. Abstr*. **2002**, *136*, 163716.
3. Roth, B. L.; Millard, P. J.; Yue, S. T.; Wells, K. S.; Haugland, R. P. Fluorescent assay for bacterial gram reaction. U.S. Patent 5545535, 1996; *Chem. Abstr*. **1996**, *125*, 216357.
4. Millard, P. J.; Roth, B. L.; Yue, S. T.; Haugland, R. P. Fluorescent viability assay using cyclic-substituted unsymmetrical cyanine dyes. U.S. Patent 5534416, 1996; *Chem. Abstr*. **1996**, *125*, 162751.
5. Yue, S. T.; Singer, V. L.; Roth, B. L.; Mozer, T. J.; Millard, P. J.; Jones, L. J.; Jin, X.; Haugland, R. P.; Poot, M. Substituted unsymmetrical cyanine dyes with selected permeability. PCT Int. Appl. WO 9613552, 1996; *Chem. Abstr*. **1996**, *125*, 81256.
6. Yue, S. T.; Johnson, I. D.; Huang, Z.; Haugland, R. P. Unsymmetrical cyanine dyes with a cationic side chain for enhanced fluorescence on binding with nucleic acids. U.S. Patent 5321130, 1994; *Chem. Abstr*. **1994**, *121*, 129393.
7. Exner, M.; Rogers, A. Methods for identifying nucleic acids and determining melting temperature using multiple fluorophores and FRET. U.S. Pat. Appl. Publ. US 2007172836, 2007; *Chem. Abstr*. **2007**, *147*, 182489.
8. Miller, B. L.; Krauss, T. D.; Du, H.; Crnkovich, N.; Strohsahl, C. M. Use of sensor arrays containing hairpin probes for detecting nucleic acids of pathogens. PCT Int. Appl. WO 2004061127, 2004; *Chem. Abstr*. **2004**, *141*, 118285.

9. Tomita, N.; Mori, Y. Method for efficiently detecting double-stranded nucleic acid. PCT Int. Appl. WO 2002103053, 2002; *Chem. Abstr.* **2002**, *138*, 50813.
10. Baskin, D.; Brankamp, R.; Slater, M. Detection of nucleic acids in a sample by amplification in the presence of a dye-labeled precursor. U.S. Pat. Appl. Publ. US 2002137047, 2002; *Chem. Abstr.* **2002**, *137*, 227621.
11. Hoser, M. J. Methods for nucleic acid sequencing. PCT Int. Appl. WO 2004074503, 2004; *Chem. Abstr.* **2004**, *141*, 237698.
12. Kricka, L. J. Stains, labels and detection strategies for nucleic acids assays. *Ann. Clin. Biochem.* **2002**, *39*, 114–129.
13. Yan, X.; Grace, W. K.; Yoshida, T. M.; Habbersett, R. C.; Velappan, N.; Jett, J. H.; Keller, R. A.; Marrone, B. L. Characteristics of different nucleic acid staining dyes for DNA fragment sizing by flow cytometry. *Anal. Chem.* **1999**, *71*, 5470–5480.
14. Anderson, A. L.; Knutson, C. R.; Mueth, D.; Plewa, J.; Tanner, E. Methods for staining cells for identification and sorting. U.S. Pat. Appl. Publ. US 2006172315, 2006; *Chem. Abstr.* **2006**, *145*, 183714.
15. Lacombe, F.; Belloc, F.; Veriac, S.; Lefevre, D. Method of discrimination of at least two cellular populations by immunofluorescence flow cytometry and application to oncopathology. Fr. Demande FR 2895087, 2007; *Chem. Abstr.* **2007**, *147*, 93934.
16. Li, W. K. W.; Jellett, J. F.; Dickie, P. M. DNA distributions in planktonic bacteria stained with TOTO or TO-PRO. *Limnol. Oceanogr.* **1995**, *40*, 1485–1495.
17. Tsuji, T.; Mizukami, T.; Konishi, A.; Mori, Y.; Nakazawa, Y. Leukocyte classification and quantification with method comprising fluorescent dye and hemolytic agent and flow cytometer. PCT Int. Appl. WO 2004001408, 2003; *Chem. Abstr.* **2004**, *140*, 58422.
18. Sakata, T.; Mizukami, T.; Hatanaka, K. Method for classifying and counting immature leukocytes using cell hemolysis, staining and flow cytometry. Eur. Pat. Appl. EP 844481, 1998; *Chem. Abstr.* **1998**, *129*, 38386.
19. Minakami, T.; Mori, Y.; Tsuji, T.; Ikeuchi, Y. Megakaryocyte classification/counting method by double fluorescent staining and flow cytometry. Jpn. Kokai Tokkyo Koho JP 2006275985, 2006; *Chem. Abstr.* **2006**, *145*, 413679.
20. Noda, N.; Mizutani, T. Microorganism-measuring method using multiple staining. Jpn. Kokai Tokkyo Koho JP 2006340684, 2006; *Chem. Abstr.* **2006**, *146*, 77563.
21. Veriac, S. Staining solution for the flow cytometric determination of reticulocytes. Eur. Pat. Appl. EP 856735, 1998; *Chem. Abstr.* **1998**, *129*, 172758.
22. Matsumoto, T.; Okada, H.; Hamaguchi, Y. Method and reagent for counting sperm by flow cytometry. Jpn. Kokai Tokkyo Koho JP 2001242168, 2001; *Chem. Abstr.* **2001**, *135*, 207869.
23. Erikson, G. H.; Daksis, J. I. Improving the signal/noise ratio of nucleic acid hybridization assays by preincubation of primer and target with nucleic acid binding agents. U.S. Pat. Appl. Publ. US 2004180345, 2004; *Chem. Abstr.* **2004**, *141*, 255469.
24. Erikson, G. H. Method for modifying transcription and/or translation in an organism by hetero-polymeric probes and duplex, triplex or quadruplex hybridization for therapeutic, prophylactic and/or analytic uses. U.S. Pat. Appl. Publ. US 2003181412, 2003; *Chem. Abstr.* **2003**, *139*, 272000.
25. Poetter, K.; Gould, T. Human papilloma virus (HPV) detection using nucleic acid probes, microbeads, and fluorescence-activated cell sorter (FACS). PCT Int. Appl. WO 2006060872, 2006; *Chem. Abstr.* **2006**, *145*, 41223.
26. Wagner, B.; Mathis, H.; Schmidt, K.; Kalusche, G.; McCaskill, J. Single molecule detection in microstructures. *Nucleosides Nucleotides* **1997**, *16*, 635–642.
27. Heuven, B.; Wong, F. S.; Tsuji, T.; Sakata, T.; Hamaguchi, I. Method for classifying and counting erythroblasts by flow cytometry. Jpn. Kokai Tokkyo Koho JP 11326323, 1999; *Chem. Abstr.* **1999**, *132*, 1804.
28. Beisker, W.; Weller-Mewe, E. M.; Nusse, M. Fluorescence enhancement of DNA-bound TO-PRO-3 by incorporation of bromodeoxyuridine to monitor cell cycle kinetics. *Cytometry* **1999**, *37*, 221–229.

TO-PRO 3

CAS Registry Number 157199-63-8

Chemical Structure

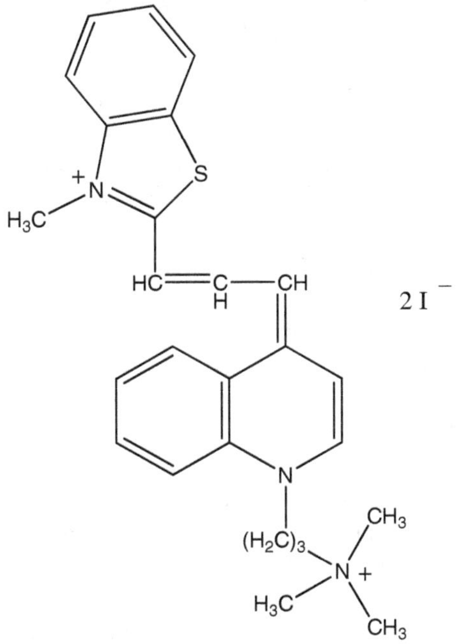

CA Index Name Quinolinium, 4-[3-(3-methyl-2(3*H*)-benzothiazolylidene)-1-propen-1-yl]-1-[3-(trimethylammonio)propyl]-, iodide (1:2)

Other Names Quinolinium, 4-[3-(3-methyl-2(3*H*)-benzothiazolylidene)-1-propenyl]-1-[3-(trimethylammonio)propyl]-, diiodide; TO-PRO 3; TO-PRO 3 iodide

Merck Index Number Not listed

Chemical/Dye Class Cyanine

Molecular Formula $C_{26}H_{31}I_2N_3S$

Molecular Weight 671.42

Physical Form Red-brown powder

Solubility Soluble in dimethyl sulfoxide

Melting Point >250 °C

Absorption (λ_{max}) 642 nm

Emission (λ_{max}) 661 nm

Synthesis Synthetic methods[1–5]

Staining Applications Nucleic acids;[5–13] cells;[4,16] bacteria;[14,15] chromatin;[17] genes;[18] leukocytes;[19] nuclei;[20] peptides;[1] proteins;[1] antibodies[1]

Biological Applications Nucleic acid hybridization;[21,22] nucleic acid assay;[9] detecting nucleic acids,[5–13] cells,[4,16] bacteria,[14,15] bromodeoxyuridine incorporation,[23] Human papilloma virus (HPV),[24] pathogens;[25] counting leukocytes[19]

Industrial Applications Microchip system and polymer waveguides[26]

Safety/Toxicity No data available

REFERENCES

1. Hahn, K. M.; Toutchkine, A.; Muthyala, R.; Kraynov, V.; Bark, S. J.; Burton, D. R.; Chamberlain, C. Labeled peptides, proteins and antibodies and processes and intermediates useful for their preparation. U.S. Pat. Appl. Publ. US 2002055133, 2002; *Chem. Abstr.* **2002**, *136*, 366139.
2. Lakowicz, J. R. Novel long lifetime luminophores with long wavelengths. PCT Int. Appl. WO 2002007779, 2002; *Chem. Abstr.* **2002**, *136*, 131225.
3. Soper, S. A.; Davidson, Y. Y.; Flanagan, J. H.; Legendre, B. L.; Owens, C.; Williams, D. C.; Hammer, R. P. Micro-DNA sequence analysis using capillary electrophoresis and near-IR fluorescence detection. *Proc. SPIE-Int. Soc. Opt. Eng.* **1996**, *2680*, 235–246.
4. Millard, P. J.; Roth, B. L.; Yue, S. T.; Haugland, R. P. Fluorescent viability assay using cyclic-substituted unsymmetrical cyanine dyes. U.S. Patent 5534416, 1996; *Chem. Abstr.* **1996**, *125*, 162751.
5. Frey, T.; Becton, D. Detection of bromodeoxyuridine incorporation by alteration of the fluorescence emission from nucleic acid binding dyes using only an argon ion laser. *Cytometry* **1994**, *17*, 310–318.
6. Exner, M.; Rogers, A. Methods for identifying nucleic acids and determining melting temperature using multiple fluorophores and FRET. U.S. Pat. Appl. Publ. US 2007172836, 2007; *Chem. Abstr.* **2007**, *147*, 182489.
7. Miller, B. L.; Krauss, T. D.; Du, H.; Crnkovich, N.; Strohsahl, C. M. Use of sensor arrays containing hairpin probes for detecting nucleic acids of pathogens. PCT Int. Appl. WO 2004061127, 2004; *Chem. Abstr.* **2004**, *141*, 118285.

8. Erikson, G. H.; Daksis, J. I.; Kandic, I.; Picard, P. Conditions for formation of three- and four-stranded nucleic acid complexes and their detection and analytical use. PCT Int. Appl. WO 2002103051, 2002; *Chem. Abstr.* **2002**, *138*, 50812.
9. Sutherland, J. W.; Patterson, D. R. Homogeneous method for assay of double-stranded nucleic acid products in nucleic acid amplification using fluorescent dyes. Eur. Pat. Appl. EP 684316, 1995; *Chem. Abstr.* **1996**, *124*, 137781.
10. Sieben, V. J.; Backhouse, C. J. Rapid on-chip postcolumn labeling and high-resolution separations of DNA. *Electrophoresis* **2005**, *26*, 4729–4742.
11. Martin, R. M.; Leonhardt, H.; Cardoso, M. C. DNA labeling in living cells. *Cytometry* **2005**, *67A*, 45–52.
12. Deka, C.; Gordon, K. M.; Gupta, R.; Horton, A. Methods and compositions for rapid staining of nucleic acids in whole cells. U.S. Patent 6271035, 2001; *Chem. Abstr.* **2001**, *135*, 149591.
13. Suzuki, T.; Fujikura, K.; Higashiyama, T.; Takata, K. DNA staining for fluorescence and laser confocal microscopy. *J. Histochem. Cytochem.* **1997**, *45*, 49–53.
14. Sakai, Y.; Kawashima, Y.; Inoue, J.; Ikeuchi, Y. Method of staining, detecting and counting bacteria, and a diluent for bacterial stain. Eur. Pat. Appl. EP 1203825, 2002; *Chem. Abstr.* **2002**, *136*, 352289.
15. Inoue, J.; Ikeuchi, Y.; Kawashima, Y. Method for staining and detecting bacteria. Jpn. Kokai Tokkyo Koho JP 2001258590, 2001; *Chem. Abstr.* **2001**, *135*, 254110.
16. Anderson, A. L.; Knutson, C. R.; Mueth, D.; Plewa, J.; Tanner, E. Methods for staining cells for identification and sorting. U.S. Pat. Appl. Publ. US 2006172315, 2006; *Chem. Abstr.* **2006**, *145*, 183714.
17. Huisman, A.; Ploeger, L. S.; Dullens, H. F. J.; Poulin, N.; Grizzle, W. E.; van Diest, P. J. Development of 3D chromatin texture analysis using confocal laser scanning microscopy. *Cell. Oncol.* **2005**, *27*, 335–345.
18. Blanco, E.; Rojas, R.; Haeger, P.; Cuevas, R.; Perez, C.; Munita, R.; Quiroz, G.; Estela Andres, M.; Forray, M. I.; Gysling, K. Intron retention as an alternative splice variant of the rat urocortin 1 gene. *Neuroscience* **2006**, *140*, 1245–1252.
19. Tsuji, T.; Mizukami, T.; Konishi, A.; Mori, Y.; Nakazawa, Y. Leukocyte classification and quantification with method comprising fluorescent dye and hemolytic agent and flow cytometer. PCT Int. Appl. WO 2004001408, 2003; *Chem. Abstr.* **2004**, *140*, 58422.
20. Bink, K.; Walch, A.; Feuchtinger, A.; Eisenmann, H.; Hutzler, P.; Hofler, H.; Werner, M. TO-PRO-3 is an optimal fluorescent dye for nuclear counterstaining in dual-color FISH on paraffin sections. *Histochem. Cell Biol.* **2001**, *115*, 293–299.
21. Erikson, G. H.; Daksis, J. I. Improving the signal/noise ratio of nucleic acid hybridization assays by preincubation of primer and target with nucleic acid binding agents. U.S. Pat. Appl. Publ. US 2004180345, 2004; *Chem. Abstr.* **2004**, *141*, 255469.
22. Erikson, G. H. Method for modifying transcription and/or translation in an organism by heteropolymeric probes and duplex, triplex or quadruplex hybridization for therapeutic, prophylactic and/or analytic uses. U.S. Pat. Appl. Publ. US 2003181412, 2003; *Chem. Abstr.* **2003**, *139*, 272000.
23. Frey, T. Detection of bromodeoxyuridine incorporation by alteration of the fluorescence emission from nucleic acid binding dyes using only an argon ion laser. *Cytometry* **1994**, *17*, 310–318.
24. Poetter, K.; Gould, T. Human papilloma virus (HPV) detection using nucleic acid probes, microbeads, and fluorescence-activated cell sorter (FACS). PCT Int. Appl. WO 2006060872, 2006; *Chem. Abstr.* **2006**, *145*, 41223.
25. Vannier, E. Methods for detection of pathogens in red blood cells. PCT Int. Appl. WO 2006031544, 2006; *Chem. Abstr.* **2006**, *144*, 307966.
26. Wang, Z.; Sekulovic, A.; Kutter, J. P.; Bang, D. D.; Wolff, A. Towards a portable microchip system with integrated thermal control and polymer waveguides for real-time PCR. *Electrophoresis* **2006**, *27*, 5051–5058.

TO-PRO 5

CAS Registry Number 177027-61-1

Chemical Structure

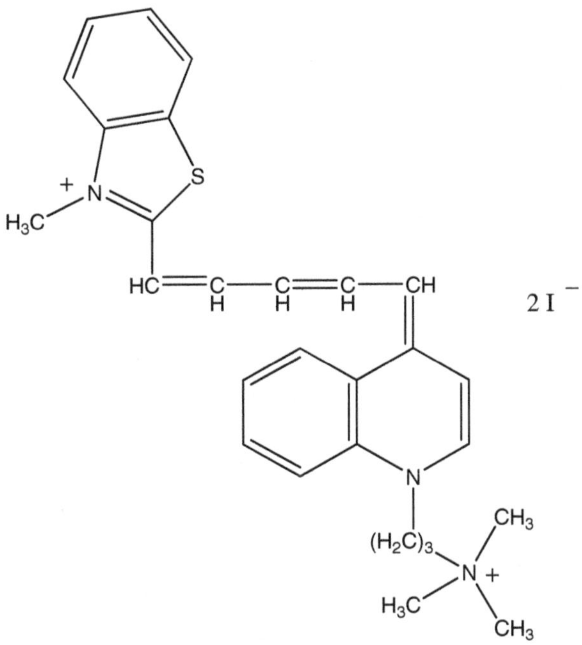

CA Index Name Quinolinium, 4-[5-(3-methyl-2(3*H*)-benzothiazolylidene)-1,3-pentadienyl]-1-[3-(trimethylammonio)propyl]-, diiodide

Other Names 3-methyl-2-[5-[1-[3-(trimethylammonio)propyl]-4(1*H*)-quinolinylidene]-1,3-pentadienyl] benzothiazolium diiodide; TO-PRO 5; TO-PRO 5 iodide

Merck Index Number Not listed

Chemical/Dye Class Cyanine

Molecular Formula $C_{28}H_{33}I_2N_3S$

Molecular Weight 697.46

Physical Form Red-brown powder

Solubility Soluble in dimethyl sulfoxide

Melting Point >250 °C

Absorption (λ_{max}) 747 nm

Emission (λ_{max}) 770 nm

Synthesis Synthetic method[1]

Staining Applications Nucleic acids;[2–4] cells[5]

Biological Applications Nucleic acid hybridization;[6,7] detecting nucleic acids,[2–4] Cells;[5] monitoring cell cycle kinetics[8]

Industrial Applications Not reported

Safety/Toxicity No data available

REFERENCES

1. Soper, S. A.; Davidson, Y. Y.; Flanagan, J. H.; Legendre, B. L.; Owens, C.; Williams, D. C.; Hammer, R. P. Micro-DNA sequence analysis using capillary electrophoresis and near-IR fluorescence detection. *Proc. SPIE-Int. Soc. Opt. Eng*. **1996**, *2680*, 235–246.
2. Exner, M.; Rogers, A. Methods for identifying nucleic acids and determining melting temperature using multiple fluorophores and FRET. U.S. Pat. Appl. Publ. US 2007172836, 2007; *Chem. Abstr*. **2007**, *147*, 182489.
3. Liu, M. S.; Chen, F. T. A. Methods and buffers for detection of nucleic acids by capillary gel electrophoresis using counter-migrating high-affinity intercalating dyes. U.S. Pat. Appl. Publ. US 2003198964, 2003; *Chem. Abstr*. **2003**, *139*, 302967.
4. Erikson, G. H.; Daksis, J. I.; Kandic, I.; Picard, P. Conditions for formation of three- and four-stranded nucleic acid complexes and their detection and analytical use. PCT Int. Appl. WO 2002103051, 2002; *Chem. Abstr*. **2002**, *138*, 50812.
5. Anderson, A. L.; Knutson, C. R.; Mueth, D.; Plewa, J.; Tanner, E. Methods for staining cells for identification and sorting. U.S. Pat. Appl. Publ. US 2006172315, 2006; *Chem. Abstr*. **2006**, *145*, 183714.
6. Erikson, G. H.; Daksis, J. I. Improving the signal/noise ratio of nucleic acid hybridization assays by preincubation of primer and target with nucleic acid binding agents. U.S. Pat. Appl. Publ. US 2004180345, 2004; *Chem. Abstr*. **2004**, *141*, 255469.
7. Erikson, G. H. Method for modifying transcription and/or translation in an organism by heteropolymeric probes and duplex, triplex or quadruplex hybridization for therapeutic, prophylactic and/or analytic uses. U.S. Pat. Appl. Publ. US 2003181412, 2003; *Chem. Abstr*. **2003**, *139*, 272000.
8. Beisker, W.; Weller-Mewe, E. M.; Nusse, M. Fluorescence enhancement of DNA-bound TO-PRO-3 by incorporation of bromodeoxyuridine to monitor cell cycle kinetics. *Cytometry* **1999**, *37*, 221–229.

TOTO 1

CAS Registry Number 143413-84-7

Chemical Structure

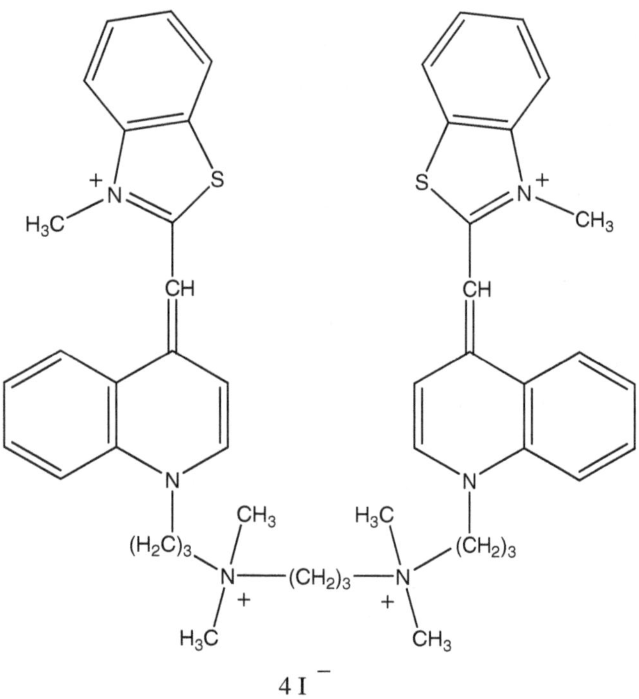

CA Index Name Quinolinium, 1,1′-[1,3-propanediylbis[(dimethyliminio)-3,1-propanediyl]]bis[4-[(3-methyl-2(3*H*)-benzothiazolylidene)methyl]-, iodide (1:4)

Other Names Quinolinium, 1,1′-[1,3-propanediylbis[(dimethyliminio)-3,1-propanediyl]]bis[4-[(3-methyl-2(3*H*)-benzothiazolylidene)methyl]-, tetraiodide; TOTO 1; TOTO 1 iodide; Thiazole Orange dimer

Merck Index Number 9308

Chemical/Dye Class Cyanine

Molecular Formula $C_{49}H_{58}I_4N_6S_2$

Molecular Weight 1302.77

Physical Form Red powder

Solubility Soluble in dimethyl sulfoxide

Melting Point >250 °C

Absorption (λ_{max}) 514 nm

Emission (λ_{max}) 533 nm

Synthesis Synthetic methods[1–6]

Staining Applications Nucleic acids;[3,5–16] cells;[17,18] bacteria;[1,20] chromatin;[24] genes;[21,22] leukocytes;[23,35] nuclei;[35] micronuclei;[24] megakaryocyte;[25] microorganisms;[26] peptides;[19] proteins;[19] antibodies;[19] sperms[27]

Biological Applications Nucleic acid hybridization;[28] nucleic acid amplification assay;[15] DNA fragment sizing;[16] detecting nucleic acids,[3,5–16] cells,[17,18] cystic fibrosis mutations,[29] genes,[21,22] human papilloma virus (HPV),[30] intact recombinant viruses,[31] oncoprotein platelet-derived growth factor,[32] PCR products,[33] stress biomarkers;[34] counting embryoblasts;[35] determining nuclease activity[36]

Industrial Applications Analyzing polymers;[37] high resolution electron-beam lithography[38]

Safety/Toxicity No data available

REFERENCES

1. Roth, B. L.; Millard, P. J.; Yue, S. T.; Wells, K. S.; Haugland, R. P. Fluorescent assay for bacterial gram reaction. U.S. Patent 5545535, 1996; *Chem. Abstr.* **1996**, *125*, 216357.
2. Yue, S. T.; Haugland, R. P. Dimers of unsymmetrical cyanine dyes containing pyridinium moieties. U.S. Patent 5410030, 1995; *Chem. Abstr.* **1995**, *123*, 259753.
3. Glazer, A. N.; Benson, S. C. Dyes designed for energy transfer as fluorescent markers for DNA. PCT Int. Appl. WO 9417397, 1994; *Chem. Abstr.* **1994**, *121*, 248036.
4. Yue, S. T.; Johnson, I. D.; Haugland, R. P. Dimers of unsymmetrical cyanine dyes having enhanced fluorescence on binding DNA or RNA. PCT Int. Appl. WO 9306482, 1993; *Chem. Abstr.* **1993**, *119*, 67272.
5. Rye, H. S.; Yue, S.; Wemmer, D. E.; Quesada, M. A.; Haugland, R. P.; Mathies, R. A.; Glazer, A. N. Stable fluorescent complexes of double-stranded DNA with bis-intercalating asymmetric cyanine dyes: properties and applications. *Nucleic Acids Res.* **1992**, *20*, 2803–2812.
6. Staerk, D.; Hamed, A. A.; Pedersen, E. B.; Jacobsen, J. P. Bisintercalation of homodimeric thiazole orange dyes in DNA: effect of modifying the linker. *Bioconjugate Chem.* **1997**, *8*, 869–877.
7. Exner, M.; Rogers, A. Methods for identifying nucleic acids and determining melting temperature using multiple fluorophores and FRET. U.S. Pat. Appl.

Publ. US 2007172836, 2007; *Chem. Abstr.* **2007**, *147*, 182489.

8. Miller, B. L.; Krauss, T. D.; Du, H.; Crnkovich, N.; Strohsahl, C. M. Use of sensor arrays containing hairpin probes for detecting nucleic acids of pathogens. PCT Int. Appl. WO 2004061127, 2004; *Chem. Abstr.* **2004**, *141*, 118285.
9. Deka, C.; Gordon, K. M.; Gupta, R.; Horton, A. Methods and compositions for rapid staining of nucleic acids in whole cells. U.S. Patent 6271035, 2001; *Chem. Abstr.* **2001**, *135*, 149591.
10. Petersen, M.; Hamed, A. A.; Pedersen, E. B.; Jacobsen, J. P. Bis-intercalation of homodimeric thiazole orange dye derivatives in DNA. *Bioconjugate Chem.* **1999**, *10*, 66–74.
11. Sailer, B. L.; Nastasi, A. J.; Valdez, J. G.; Steinkamp, J. A.; Crissman, H. A. Differential effects of deuterium oxide on the fluorescence lifetimes and intensities of dyes with different modes of binding to DNA. *J. Histochem. Cytochem.* **1997**, *45*, 165–175.
12. Hansen, L. F.; Jensen, L. K.; Jacobsen, J. P. Bis-intercalation of a homodimeric thiazole orange dye in DNA in symmetrical pyrimidine–pyrimidine–purine–purine oligonucleotides. *Nucleic Acids Res.* **1996**, *24*, 859–867.
13. Rye, H. S.; Glazer, A. N. Interaction of dimeric intercalating dyes with single-stranded DNA. *Nucleic Acids Res.* **1995**, *23*, 1215–1222.
14. Goodwin, P. M.; Johnson, M. E.; Martin, J. C.; Ambrose, W. P.; Marrone, B. L.; Jett, J. H.; Keller, R. A. Rapid sizing of individual fluorescently stained DNA fragments by flow cytometry. *Nucleic Acids Res.* **1993**, *21*, 803–806.
15. Morrison, T. Improved selective ligation and nucleic acid amplification assay. PCT Int. Appl. WO 2005059178, 2005; *Chem. Abstr.* **2005**, *143*, 72747.
16. Van Orden, A.; Keller, R. A.; Ambrose, W. P. High-throughput flow cytometric DNA fragment sizing. *Anal. Chem.* **2000**, *72*, 37–41.
17. Anderson, A. L.; Knutson, C. R.; Mueth, D.; Plewa, J.; Tanner, E. Methods for staining cells for identification and sorting. U.S. Pat. Appl. Publ. US 2006172315, 2006; *Chem. Abstr.* **2006**, *145*, 183714.
18. Lacombe, F.; Belloc, F.; Veriac, S.; Lefevre, D. Method of discrimination of at least two cellular populations by immunofluorescence flow cytometry and application to oncopathology. Fr. Demande FR 2895087, 2007; *Chem. Abstr.* **2007**, *147*, 93934.
19. Hahn, K. M.; Toutchkine, A.; Muthyala, R.; Kraynov, V.; Bark, S. J.; Burton, D. R.; Chamberlain, C. Labeled peptides, proteins and antibodies and processes and intermediates useful for their preparation. U.S. Pat. Appl. Publ. US 2002055133, 2002; *Chem. Abstr.* **2002**, *136*, 366139.
20. Li, W. K. W.; Jellett, J. F.; Dickie, P. M. DNA distributions in planktonic bacteria stained with TOTO or TO-PRO. *Limnol. Oceanogr.* **1995**, *40*, 1485–1495.
21. Golzio, M.; Teissie, J.; Rols, M. P. Direct visualization at the single-cell level of electrically mediated gene delivery. *Proc. Natl. Acad. Sci. U.S.A.* **2002**, *99*, 1292–1297.
22. Mitsuhashi, M.; Cooper, A. Detection of genes by nucleic acid hybridization using capture and reporter probes and optional nucleic acid amplification. PCT Int. Appl. WO 9402636, 1994; *Chem. Abstr.* **1994**, *120*, 237581.
23. Tsuji, T.; Mizukami, T.; Konishi, A.; Mori, Y.; Nakazawa, Y. Leukocyte classification and quantification with method comprising fluorescent dye and hemolytic agent and flow cytometer. PCT Int. Appl. WO 2004001408, 2003; *Chem. Abstr.* **2004**, *140*, 58422.
24. Dertinger, S. D.; Cairns, S. E.; Avlasevich, S. L.; Torous, D. K. Method for enumerating and differentiating between mammalian cell micronuclei by differentially staining micronuclei and the chromatin of dead, dying and viable cells. PCT Int. Appl. WO 2006007479, 2006; *Chem. Abstr.* **2006**, *144*, 124521.
25. Minakami, T.; Mori, Y.; Tsuji, T.; Ikeuchi, Y. Megakaryocyte classification/counting method by double fluorescent staining and flow cytometry. Jpn. Kokai Tokkyo Koho JP 2006275985, 2006; *Chem. Abstr.* **2006**, *145*, 413679.
26. Noda, N.; Mizutani, T. Microorganism-measuring method using multiple staining. Jpn. Kokai Tokkyo Koho JP 2006340684, 2006; *Chem. Abstr.* **2006**, *146*, 77563.
27. Matsumoto, T.; Okada, H.; Hamaguchi, Y. Method and reagent for counting sperm by flow cytometry. Jpn. Kokai Tokkyo Koho JP 2001242168, 2001; *Chem. Abstr.* **2001**, *135*, 207869.
28. Erikson, G. H.; Daksis, J. I. Improving the signal/noise ratio of nucleic acid hybridization assays by preincubation of primer and target with nucleic acid binding agents. U.S. Pat. Appl. Publ. US 2004180345, 2004; *Chem. Abstr.* **2004**, *141*, 255469.
29. Axton, R. A.; Brock, D. J. H. Use of stable dye–DNA intercalating complexes to detect cystic fibrosis mutations. *Mol. Cell. Probes* **1994**, *8*, 245–250.

30. Poetter, K.; Gould, T. Human papilloma virus (HPV) detection using nucleic acid probes, microbeads, and fluorescence-activated cell sorter (FACS). PCT Int. Appl. WO 2006060872, 2006; *Chem. Abstr.* **2006**, *145*, 41223.

31. Transfiguracion, J.; Kamen, A. Detection of intact recombinant viruses using fluorescent staining and chromatography. PCT Int. Appl. WO 2006053443, 2006; *Chem. Abstr.* **2006**, *144*, 463861.

32. Zhou, C.; Jiang, Y.; Hou, S.; Ma, B.; Fang, X.; Li, M. Detection of oncoprotein platelet-derived growth factor using a fluorescent signaling complex of an aptamer and TOTO. *Anal. Bioanal. Chem.* **2006**, *384*, 1175–1180.

33. Ohler, L. D.; Zollo, M.; Mansfield, E. S.; Rose, E. A. Use of a sensitive fluorescent intercalating dye to detect PCR products of low copy number and high molecular weight. *PCR Methods Appl.* **1993**, *3*, 115–119.

34. Cullum, M. E.; Duplessis, C. A.; Crepeau, L. J. Method for the detection of stress biomarkers including cortisol by fluorescence polarization. U. S. Pat. Appl. Publ. US 2006105397, 2006; *Chem. Abstr.* **2006**, *144*, 463810.

35. Heuven, B.; Wong, F. S.; Tsuji, T.; Sakata, T.; Hamaguchi, I. Method for classifying and counting erythroblasts by flow cytometry. Jpn. Kokai Tokkyo Koho JP 11326323, 1999; *Chem. Abstr.* **1999**, *132*, 1804.

36. Mitsuhashi, M.; Ogura, M. Determining nuclease activity with fluorescent nucleic acid dyes. U.S. Patent 5554502, 1996; *Chem. Abstr.* **1996**, *125*, 241776.

37. Chan, E. Y. Methods of analyzing polymers using ordered label strategies. U.S. Patent 6403311, 2002; *Chem. Abstr.* **2002**, *137*, 2742.

38. Bilenberg, B.; Jacobsen, S.; Schmidt, M. S.; Skjolding, L. H. D.; Shi, P.; Boggild, P.; Tegenfeldt, J. O.; Kristensen, A. High resolution 100 kV electron-beam lithography in SU-8. *Microelectron. Eng.* **2006**, *83*, 1609–1612.

TOTO 3

CAS Registry Number 166196-17-4

Chemical Structure

Solubility Soluble in dimethyl sulfoxide

Melting Point >250 °C

Absorption (λ_{max}) 642 nm

Emission (λ_{max}) 660 nm

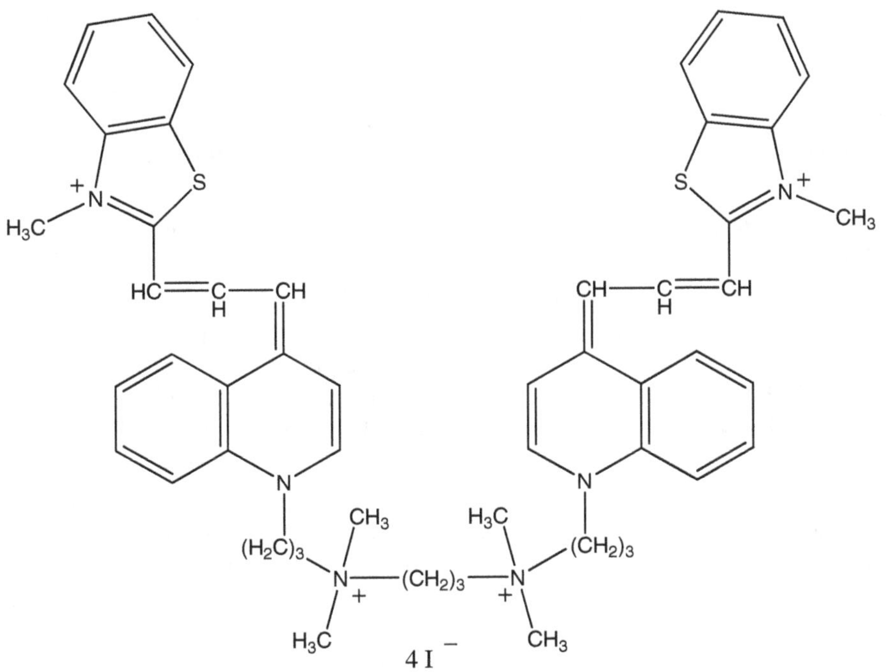

CA Index Name Quinolinium, 1,1′-[1,3-propanediylbis[(dimethyliminio)-3,1-propanediyl]]bis[4-[3-(3-methyl-2(3*H*)-benzothiazolylidene)-1-propen-1-yl]-, iodide (1:4)

Other Names Quinolinium, 1,1′-[1,3-propanediylbis[(dimethyliminio)-3,1-propanediyl]]bis[4-[3-(3-methyl-2(3*H*)-benzothiazolylidene)-1-propenyl]-, tetraiodide; TOTO 3; TOTO 3 iodide

Merck Index Number Not listed

Chemical/Dye Class Cyanine

Molecular Formula $C_{53}H_{62}I_4N_6S_2$

Molecular Weight 1354.85

Physical Form Red powder

Synthesis Synthetic methods[1–4]

Staining Applications Nucleic acids;[5–11] cells;[12] antibodies;[1,2] bacteria;[13] leukocytes;[14,21] nuclei;[21] megakaryocyte;[15] peptides;[1,2] proteins;[1,2] sperms;[16] hairs[17]

Biological Applications Nucleic acid hybridization;[18] detecting nucleic acids,[5–11] cells,[12] human papilloma virus (HPV),[19] pathogens;[20] counting embryoblasts;[21] determining nuclease activity;[22] retinal toxicity screening methods;[23] studying cellular uptake of gene transfer complexes[24]

Industrial Applications Semiconductor devices[25]

Safety/Toxicity No data available

REFERENCES

1. Hahn, K. M.; Toutchkine, A.; Muthyala, R.; Kraynov, V.; Bark, S. J.; Burton, D. R.; Chamberlain, C. Labeled peptides, proteins and antibodies and processes and intermediates useful for their preparation. U.S. Pat. Appl. Publ. US 2002055133, 2002; *Chem. Abstr.* **2002**, *136*, 366139.
2. Hahn, K. M.; Toutchkine, A.; Muthyala, R.; Kraynov, V.; Bark, Steven J.; Burton, D. R.; Chamberlain, C. Labeled peptides, proteins and antibodies and processes and intermediates useful for their preparation. PCT Int. Appl. WO 2002008245, 2002; *Chem. Abstr.* **2002**, *136*, 163716.

3. Lakowicz, J. R. Novel long lifetime luminophores with long wavelengths. PCT Int. Appl. WO 2002007779, 2002; *Chem. Abstr.* **2002**, *136*, 131225.
4. Yue, S. T.; Haugland, R. P. Dimers of unsymmetrical cyanine dyes containing pyridinium moieties. U.S. Patent 5410030, 1995; *Chem. Abstr.* **1995**, *123*, 259753.
5. Exner, M.; Rogers, A. Methods for identifying nucleic acids and determining melting temperature using multiple fluorophores and FRET. U.S. Pat. Appl. Publ. US 2007172836, 2007; *Chem. Abstr.* **2007**, *147*, 182489.
6. Martin, R. M.; Leonhardt, H.; Cardoso, M. C. DNA labeling in living cells. *Cytometry* **2005**, *67A*, 45–52.
7. Wittwer, C. T.; Dujols, V. E.; Reed, G.; Zhou, L. Genotyping by thermal denaturation analysis using loss of fluorescence of double-stranded DNA-specific dyes as a reporter to monitor denaturation. PCT Int. Appl. WO 2004038038, 2004; *Chem. Abstr.* **2004**, *140*, 386988.
8. Deka, C.; Gordon, K. M.; Gupta, R.; Horton, A. Methods and compositions for rapid staining of nucleic acids in whole cells. U.S. Patent 6271035, 2001; *Chem. Abstr.* **2001**, *135*, 149591.
9. Suzuki, T.; Fujikura, K.; Higashiyama, T.; Takata, K. DNA staining for fluorescence and laser confocal microscopy. *J. Histochem. Cytochem.* **1997**, *45*, 49–53.
10. Kim, K. H.; Min, J. H.; Lee, I. H.; Kim, A. G. Method for highly sensitive nucleic acid detection using nanopore and non-specific nucleic acid-binding agent. U.S. Pat. Appl. Publ. US 2006292605, 2006; *Chem. Abstr.* **2006**, *146*, 56513.
11. Miller, B. L.; Krauss, T. D.; Du, H.; Crnkovich, N.; Strohsahl, C. M. Use of sensor arrays containing hairpin probes for detecting nucleic acids of pathogens. PCT Int. Appl. WO 2004061127, 2004; *Chem. Abstr.* **2004**, *141*, 118285.
12. Anderson, A. L.; Knutson, C. R.; Mueth, D.; Plewa, J.; Tanner, E. Methods for staining cells for identification and sorting. U.S. Pat. Appl. Publ. US 2006172315, 2006; *Chem. Abstr.* **2006**, *145*, 183714.
13. Inoue, J.; Ikeuchi, Y.; Kawashima, Y. Method for staining and detecting bacteria. Jpn. Kokai Tokkyo Koho JP 2001258590, 2001; *Chem. Abstr.* **2001**, *135*, 254110.
14. Sakata, T.; Mizukami, T.; Hatanaka, K. Method for classifying and counting immature leukocytes using cell hemolysis, staining and flow cytometry. Eur. Pat. Appl. EP 844481, 1998; *Chem. Abstr.* **1998**, *129*, 38386.
15. Minakami, T.; Mori, Y.; Tsuji, T.; Ikeuchi, Y. Megakaryocyte classification/counting method by double fluorescent staining and flow cytometry. Jpn. Kokai Tokkyo Koho JP 2006275985, 2006; *Chem. Abstr.* **2006**, *145*, 413679.
16. Matsumoto, T.; Okada, H.; Hamaguchi, Y. Method and reagent for counting sperm by flow cytometry. Jpn. Kokai Tokkyo Koho JP 2001242168, 2001; *Chem. Abstr.* **2001**, *135*, 207869.
17. Lagrange, A. Hair dye compositions containing a polycationic direct dye. Fr. Demande FR 2848840, 2004; *Chem. Abstr.* **2004**, *141*, 76344.
18. Erikson, G. H.; Daksis, J. I. Improving the signal/noise ratio of nucleic acid hybridization assays by preincubation of primer and target with nucleic acid binding agents. U.S. Pat. Appl. Publ. US 2004180345, 2004; *Chem. Abstr.* **2004**, *141*, 255469.
19. Poetter, K.; Gould, T. Human papilloma virus (HPV) detection using nucleic acid probes, microbeads, and fluorescence-activated cell sorter (FACS). PCT Int. Appl. WO 2006060872, 2006; *Chem. Abstr.* **2006**, *145*, 41223.
20. Vannier, E. Methods for detection of pathogens in red blood cells. PCT Int. Appl. WO 2006031544, 2006; *Chem. Abstr.* **2006**, *144*, 307966.
21. Heuven, B.; Wong, F. S.; Tsuji, T.; Sakata, T.; Hamaguchi, I. Method for classifying and counting erythroblasts by flow cytometry. Jpn. Kokai Tokkyo Koho JP 11326323, 1999; *Chem. Abstr.* **1999**, *132*, 1804.
22. Mitsuhashi, M.; Ogura, M. Determining nuclease activity with fluorescent nucleic acid dyes. U.S. Patent 5554502, 1996; *Chem. Abstr.* **1996**, *125*, 241776.
23. Verdugo-Gazdik, M. E. Retinal toxicity screening methods using a fluorescently detectable integrin-specific agent. U.S. Pat. Appl. Publ. US 2005142068, 2005; *Chem. Abstr.* **2005**, *143*, 90973.
24. Ogris, M.; Wagner, E.; Steinlein, P. A versatile assay to study cellular uptake of gene transfer complexes by flow cytometry. *Biochim. Biophys. Acta* **2000**, *74*, 237–243.
25. Porta, P. A.; Summers, H. D. Vertical-cavity semiconductor devices for generation and detection of fluorescence emission on a single chip. *Appl. Phys. Lett.* **2004**, *85*, 1889–1891.

TRIPHENYL TETRAZOLIUM CHLORIDE (TTC)

CAS Registry Number 298-96-4

Chemical Structure

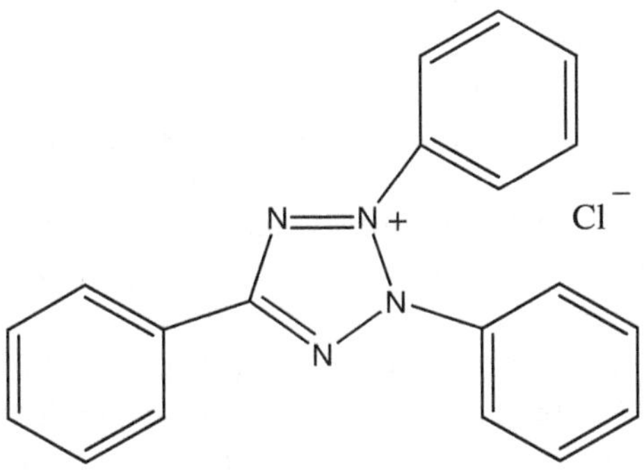

CA Index Name 2*H*-Tetrazolium, 2,3,5-triphenyl-, chloride (1:1)

Other Names 2,3,5-Triphenyl-2*H*-tetrazolium chloride; 2*H*-Tetrazolium, 2,3,5-triphenyl-, chloride; 1,3,5-Triphenyl-2*H*-tetrazolium chloride; 1,3,5-Triphenyltetrazolium chloride; 2,3,5-Triphenyltetrazolium chloride; PTB; RT; Red tetrazolium; TPTZ; TT; TTC; TTC (dye); Tetrazolium chloride; Tetrazolium red; Triphenyltetrazolium chloride; Urocheck; Uroscreen; Vitastain

Merck Index Number 9744

Chemical/Dye Class Tetrazolium salt

Molecular Formula $C_{19}H_{15}ClN_4$

Molecular Weight 334.80

Physical Form Colorless needles, turns yellow on exposure to light

Solubility Soluble in water, ethanol, acetone; insoluble in ether

Melting Point 243–253 °C (decompose)

Absorption (λ_{max}) 247 nm

Synthesis Synthetic methods[1–20]

Staining Applications Bacteria;[21] brain sections;[22] brain slices;[23] cells;[24] cerebral infarcts;[25] cortex;[26] infarcted hearts tissues;[27,28] neurons;[29] plant roots[30,31]

Biological Applications Algae viability assay;[32] microbial growth assays;[33] transketolase activity screening assays;[34] detecting bacteria,[35] γ-hydroxybutyric acid (GHB),[36] microorganisms,[37] myocardial infarction;[28] measuring dehydrogenase activity;[38] treating cancer[39]

Industrial Applications Photographic materials[40]

Safety/Toxicity Bacterial toxicity;[41] mutagenicity[42]

REFERENCES

1. Cui, X.; Vlahakis, J. Z.; Crandall, I. E.; Szarek, W. A. Anti-*Plasmodium* activity of tetrazolium salts. *Bioorg. Med. Chem.* **2008**, *16*, 1927–1947.
2. James, P. R. Tetrazolium salts as steam sterilization process indicators. PCT Int. Appl. WO 9837157, 1998; *Chem. Abstr.* **1998**, *129*, 218237.
3. Katritzky, A. R.; Belyakov, S. A.; Lam, J. N.; Durst, H. D.; Karpenko, D. V. 1-Chlorobenzotriazole-mediated ring closure of 1,3,5-triarylformazans: improved syntheses of 2,3,5-triaryl-2*H*-tetrazolium salts. *Heterocycles* **1994**, *39*, 73–80.
4. Belyakov, S. A.; Markovskii, L. N.; Silina, T. F.; Litvinov, V. K.; Merzlyak, N. I. Preparation of derivatives of 2,3,5-triaryltetrazolium chloride. U.S.S.R. SU 1549952, 1990; *Chem. Abstr.* **1990**, *113*, 97611.
5. Belyakov, S. A. Preparation of 2,3,5-triaryltetrazolium chlorides. *Zh. Org. Khim.* **1989**, *25*, 2252–2253; *Chem. Abstr.* **1990**, *112*, 216803.
6. Zhivich, A. B.; Koldobskii, G. I.; Martynova, V. P.; Ostrovskii, V. A. Preparation of triphenyltetrazolium salts. U.S.S.R. SU 1361147, 1987; *Chem. Abstr.* **1988**, *109*, 231027.
7. Zhivich, A. B.; Koldobskii, G. I.; Ostrovskii, V. A. Preparation of tetrazolium salts via interphase catalysis. *Zh. Org. Khim.* **1988**, *24*, 225–226; *Chem. Abstr.* **1988**, *109*, 170322.
8. Osipova, T. F.; Koldobskii, G. I.; Ostrovskii, V. A.; Myznikov, Yu. E. Tetrazoles. 20. Tetrazolium salts in phase-transfer catalysis. *Khim. Geterotsikl. Soedin.* **1985**, 841–845; *Chem. Abstr.* **1985**, *103*, 178201.
9. Ostrovskaya, V. M.; Lushina, O. T.; Falin, V. A. Optimization of 2,3,5-triphenyltetrazolium chloride preparation. *Zh. Prikl. Khim.* **1983**, 56, 649–653; *Chem. Abstr.* **1983**, *99*, 22435.
10. Ostrovskaya, V. M.; Lushina, O. T.; Dziomko, V. M.; Davydovskaya, Yu. A. 2*H*-Tetrazolium chloride and 2*H*-tetrazolium chloride hydrochloride. Ger. Offen. DE 2649170, 1978; *Chem. Abstr.* **1978**, *89*, 109504.
11. Vecerek, B. Tetrazolium salts. Czech. CS 147890, 1973; *Chem. Abstr.* **1973**, *79*, 53332.

12. Price, R. Metal complex formazans. I. Reaction of copper and cobalt salts with certain bi- and tridentate 1,3,5-triarylformazans. *J. Chem. Soc. A* **1971**, 3379–3385.
13. Moeckel, P.; Staerk, G. Oxidation with peroxydisulfate: preparation of 2,3,5-triphenyltetrazolium chloride. *Z. Chem.* **1967**, *7*, 107.
14. Bushmarin, N. I.; Klimova, E. I.; Kotova, V. I. Preparation of 2,3,5-triphenyltetrazolium chloride. U.S.S.R. SU 167712, 1965; *Chem. Abstr.* **1965**, *63*, 3351.
15. Moeckel, P.; Staerk, G. Substituted tetrazolium salts. German (East) DD 22410, 1961; *Chem. Abstr.* **1963**, *58*, 46792.
16. Glantz, M. D.; Fried, F. Microseparation of tetrazolium salts by paper electrophoresis. *Microchem. J.* **1959**, *3*, 214–216.
17. Kratky, J. J.; Svoboda, V. The oxidation of formazans by PbO.PbO_2. *Chem. Zvesti* **1957**, *11*, 153–156.
18. Benson, F. R.; Otten, E. A.; Schachat, R. E. Oxidation of formazans to tetrazolium chlorides with *tert*-butyl hypochlorite. *J. Am. Chem. Soc.* **1954**, *76*, 1695.
19. Hausser, K. H. Quantum balance for change of triphenyltetrazolium chloride and triphenylformazan by ultraviolet light. *Naturwissenschaften* **1949**, *36*, 314–315.
20. Mattson, A. M.; Jensen, C. O.; Dutcher, R. A. Preparation of 2,3,4-triphenyltetrazolium chloride. *J. Am. Chem. Soc.* **1948**, *70*, 1284.
21. Irtuganova, O. A.; Yushchenko, A. A.; Smirnova, N. S.; Slogotskaya, L. V.; Arkhipov, V. V. Determining drug susceptibility of *Mycobacterium tuberculosis* using a biphasic culture medium and staining by triphenyltetrazolium chloride. Russ. RU 2255977, 2005; *Chem. Abstr.* **2005**, *143*, 93610.
22. Regan, H. K.; Detwiler, T. J.; Huang, J. C.; Lynch, J. J.; Regan, C. P. An improved automated method to quantitate infarct volume in triphenyltetrazolium stained rat brain sections. *J. Pharmacol. Toxicol. Methods* **2007**, *56*, 339–343.
23. Tureyen, K.; Vemuganti, R.; Sailor, K. A.; Dempsey, R. J. Infarct volume quantification in mouse focal cerebral ischemia: a comparison of triphenyltetrazolium chloride and cresyl violet staining techniques. *J. Neurosci. Methods* **2004**, *139*, 203–207.
24. Berridge, M. V.; Herst, P. M.; Tan, A. S. Tetrazolium dyes as tools in cell biology: new insights into their cellular reduction. *Biotechnol. Annu. Rev.* **2005**, *11*, 127–152.
25. Joshi, C. N.; Jain, S. K.; Murthy, P. S. R. An optimized triphenyltetrazolium chloride method for identification of cerebral infarcts. *Brain Res. Protoc.* **2004**, *13*, 11–17.
26. Chen, S.; Feng, Z.; Zeng, S.; Luo, Q.; Li, P. Evaluation of focal cerebral ischemia development in rats by optical imaging based on the spreading depression signals. *Proc. SPIE-Int. Soc. Opt. Eng.* **2007**, *6436*, 64360A/1–64360A/10.
27. Pitts, K. R.; Stiko, A.; Buetow, B.; Lott, F.; Guo, P.; Virca, D.; Toombs, C. F. Washout of heme-containing proteins dramatically improves tetrazolium-based infarct staining. *J. Pharmacol. Toxicol. Methods* **2007**, *55*, 201–208.
28. Khalil, P. N.; Siebeck, M.; Huss, R.; Pollhammer, M.; Khalil, M. N.; Neuhof, C.; Fritz, H. Histochemical assessment of early myocardial infarction using 2,3,5-triphenyltetrazolium chloride in blood-perfused porcine hearts. *J. Pharmacol. Toxicol. Methods* **2006**, *54*, 307–312.
29. Benedek, A.; Moricz, K.; Juranyi, Z.; Gigler, G.; Levay, G.; Harsing, L. G.; Matyus, P.; Szenasi, G.; Albert, M. Use of TTC staining for the evaluation of tissue injury in the early phases of reperfusion after focal cerebral ischemia in rats. *Brain Res.* **2006**, *1116*, 159–165.
30. Sturite, I.; Henriksen, T. M.; Breland, T. A. Distinguishing between metabolically active and inactive roots by combined staining with 2,3,5-triphenyltetrazolium chloride and image colour analysis. *Plant Soil* **2005**, *271*, 75–82.
31. Duncan, D. R.; Widholm, J. M. Osmotic induced stimulation of the reduction of the viability dye 2,3,5-triphenyltetrazolium chloride by maize roots and callus cultures. *J. Plant Physiol.* **2004**, *161*, 397–403.
32. Park, S. M.; Kang, S. E.; Choi, J. S.; Cho, J. Y.; Yoon, S. J.; Ahn, D. H.; Hong, Y. K. Viability assay of coralline algae using triphenyltetrazolium chloride. *Fish. Sci.* **2006**, *72*, 912–914.
33. Storey, D.; Field, T.; Ramage, G. Microbial growth assay and assay device. PCT Int. Appl. WO 2005098023, 2005; *Chem. Abstr.* **2005**, *143*, 402159.
34. Smith, M. E. B.; Kaulmann, U.; Ward, J. M.; Hailes, H. C. A colorimetric assay for screening transketolase activity. *Bioorg. Med. Chem.* **2006**, *14*, 7062–7065.
35. Ushiyama, M. Film culture medium for detecting bacteria in water-system environment. Jpn. Kokai Tokkyo Koho JP 2007020434, 2007; *Chem. Abstr.* **2007**, *146*, 180320.
36. Parsons, S. M.; Harris, D. O.; Bravo, D. T. Methods, compositions and apparatuses for detection of

gamma-hydroxybutyric acid (GHB). U.S. Pat. Appl. Publ. US 2003175846, 2003; *Chem. Abstr.* **2003**, *139*, 256601.

37. Dinamarca, M. A.; Cereceda-Balic, F.; Fadic, X.; Seeger, M. Analysis of *s*-triazine-degrading microbial communities in soils using most-probable-number enumeration and tetrazolium-salt detection. *Int. Microbiol.* **2007**, *10*, 209–215.
38. Ghaly, A. E.; Mahmoud, N. S. Optimum conditions for measuring dehydrogenase activity of *Aspergillus niger* using TTC. *Am. J. Biochem. Biotechnol.* **2006**, *2*, 186–194.
39. Kong, Q. Method and composition for treating cancer using cellular organelle crystallizing agents. U.S. Pat. Appl. Publ. US 2006286170, 2006; *Chem. Abstr.* **2006**, *146*, 55482.
40. Hanyu, T. Heat-developable photographic material containing tetrazole derivative coloring agent. Jpn. Kokai Tokkyo Koho JP 2004054098, 2004; *Chem. Abstr.* **2004**, *140*, 189916.
41. Mazul, M. M.; Danilov, V. S. Tetrazolium salt effect on the bioluminescence of *Escherichia coli* recombinant strain. *Biotekhnologiya* **2002**, 91–96; *Chem. Abstr.* **2002**, *138*, 68140.
42. Venitt, S.; Crofton-Sleigh, C. Bacterial mutagenicity tests of phenazine methosulfate and three tetrazolium salts. *Mutat. Res.* **1979**, *68*, 107–116.

TSQ

CAS Registry Number 109628-27-5

Chemical Structure

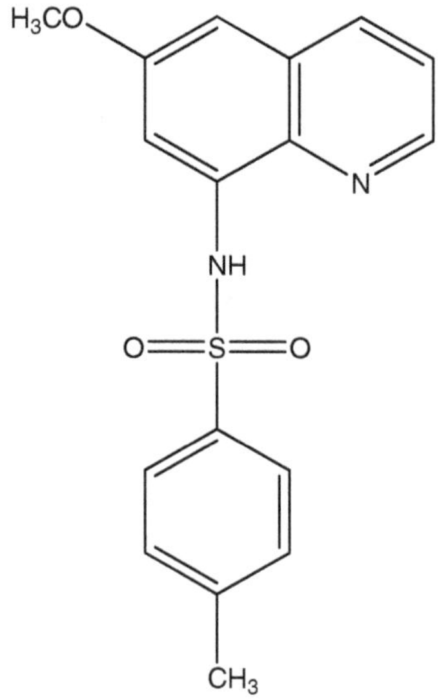

CA Index Name Benzenesulfonamide, *N*-(6-methoxy-8-quinolinyl)-4-methyl-

Other Names *p*-Toluenesulfonamide, *N*-(6-methoxy-8-quinolyl)-; NSC 120213; TSQ

Merck Index Number 9799

Chemical/Dye Class Quinoline

Molecular Formula $C_{17}H_{16}N_2O_3S$

Molecular Weight 328.39

Physical Form White crystals

Solubility Soluble in ethanol, methanol

Melting Point 133–134 °C

Boiling Point (Calcd.) 518.8 ± 60.0 °C, pressure: 760 Torr

pK_a (calcd.) 7.40 ± 0.30, most acidic, temperature: 25 °C; 3.09 ± 0.20, most basic, temperature: 25 °C

Absorption (λ_{max}) 334 nm

Emission (λ_{max}) 385 nm

Synthesis Synthetic methods[1–6]

Staining Applications Amyloid plaques;[7] pancreatic islets;[8–10,34] neurons;[7,11–16] sperms;[17] zinc ions[1,2,7–9,11–28,33]

Biological Applications Zinc indicator;[1,2,7–9,11–28,33] early diagnosis of prostate cancer;[28] treating age-related macular degeneration (AMD),[29] amyloidosis disorders,[30] autoimmune diseases,[31] herpes virus infection[32]

Industrial Applications Not reported

Safety/Toxicity Neurotoxicity;[33] pancreatic toxicity[34]

REFERENCES

1. Pearce, D. A.; Jotterand, N.; Carrico, I. S.; Imperiali, B. Derivatives of 8-hydroxy-2-methylquinoline are powerful prototypes for zinc sensors in biological systems. *J. Am. Chem. Soc.* **2001**, *123*, 5160–5161.
2. Fahrni, C. J.; O'Halloran, T. V. Aqueous coordination chemistry of quinoline-based fluorescence probes for the biological chemistry of zinc. *J. Am. Chem. Soc.* **1999**, *121*, 11448–11458.
3. Simonov, A. M. Dipolar ions formed after cleavage of a proton from the NH group. VI. *Zh. Obshch. Khim.* **1952**, *22*, 1865–1868; *Chem. Abstr.* **1953**, *47*, 41284.
4. Bachman, G. B.; Bennett, G. E.; Barker, R. S. Synthesis of substituted quinolylamines. Derivatives of 4-amino-7-chloroquinoline. *J. Org. Chem.* **1950**, *15*, 1278–1284.
5. Cerkovnikov, E.; Tomasic, P. The N-substituted derivatives of *p*-toluenesulfonamide. *Arhiv. Kem.* **1947**, *19*, 38–42; *Chem. Abstr.* **1948**, *42*, 36467.
6. Elderfield, R. C.; Kreysa, F. J.; Dunn, J. H.; Humphreys, D. D. Study of the synthesis of plasmochin by the reductive amination method with Raney nickel. *J. Am. Chem. Soc.* **1948**, *70*, 40–44.
7. Suh, S. W.; Jensen, K. B.; Jensen, M. S.; Silva, D. S.; Kesslak, P. J.; Danscher, G.; Frederickson, C. J. Histochemically-reactive zinc in amyloid plaques, angiopathy, and degenerating neurons of Alzheimer's diseased brains. *Brain Res.* **2000**, *852*, 274–278.
8. Maghasi, A. T.; Schlueter, K. T.; Halsall, H. B.; Heineman, W. R.; Rodriguez Rilo, H. L. Development and characterization of simulant pancreatic islets. *Anal. Biochem.* **2003**, *314*, 38–45.
9. Okabe, M.; Yoshida, T.; Yoshii, R.; Sawataisi, M.; Takaya, K. Zinc detection in the islet of Langerhans by SIMS. *Appl. Surf. Sci.* **2003**, *203–204*, 714–717.

10. Jindal, R. M.; Gray, D. W.; Morris, P. J. The use of TSQ as an islet-specific stain for purification of islets by fluorescence-activated sorting. *Transplantation* **1993**, *56*, 1282–1284.
11. Varea, E.; Ponsoda, X.; Molowny, A.; Danscher, G.; Lopez-Garcia, C. Imaging synaptic zinc release in living nervous tissue. *J. Neurosci. Methods* **2001**, *110*, 57–63.
12. Suh, S. W.; Listiack, K.; Bell, B.; Chen, J.; Motamedi, M.; Silva, D.; Danscher, G.; Whetsell, W.; Thompson, R.; Frederickson, C. Detection of pathological zinc accumulation in neurons: methods for autopsy, biopsy, and cultured tissue. *J. Histochem. Cytochem.* **1999**, *47*, 969–972.
13. Lee, J. M.; Zipfel, G. J.; Park, K. H.; He, Y. Y.; Hsu, C. Y.; Choi, D. W. Zinc translocation accelerates infarction after mild transient focal ischemia. *Neuroscience* **2002**, *115*, 871–878.
14. Suh, S. W.; Chen, J. W.; Motamedi, M.; Bell, B.; Listiak, K.; Pons, N. F.; Danscher, G.; Frederickson, C. J. Evidence that synaptically-released zinc contributes to neuronal injury after traumatic brain injury. *Brain Res.* **2000**, *852*, 268–273.
15. Koh, J. Y.; Suh, S. W.; Gwag, B. J.; He, Y. Y.; Hsu, C. Y.; Choi, D. W. The role of zinc in selective neuronal death after transient global cerebral ischemia. *Science* **1996**, *272*, 1013–1016.
16. Tonder, N.; Johansen, F. F.; Frederickson, C. J.; Zimmer, J.; Diemer, N. H. Possible role of zinc in the selective degeneration of dentate hilar neurons after cerebral ischemia in the adult rat. *Neurosci. Lett.* **1990**, *109*, 247–252.
17. Andrews, J. C.; Nolan, J. P.; Hammerstedt, R. H.; Bavister, B. D. Characterization of *N*-(6-methoxy-8-quinolyl)-*p*-toluenesulfonamide for the detection of zinc in living sperm cells. *Cytometry* **1995**, *21*, 153–159.
18. Quinta-Ferreira, M. E.; Matias, C. M. Tetanically released zinc inhibits hippocampal mossy fiber calcium, zinc and synaptic responses. *Brain Res.* **2005**, *1047*, 1–9.
19. Verstraeten, S. V.; Zago, M. P.; MacKenzie, G. G.; Keen, C. L.; Oteiza, P. I. Influence of zinc deficiency on cell-membrane fluidity in Jurkat, 3T3 and IMR-32 cells. *Biochem. J.* **2004**, *378*, 579–587.
20. Quinta-Ferreira, M. E.; Matias, C. M.; Arif, M.; Dionisio, J. C. Measurement of presynaptic zinc changes in hippocampal mossy fibers. *Brain Res.* **2004**, *1026*, 1–10.
21. Paski, S. C.; Covery, L.; Kummer, A.; Xu, Z. Role of metallothionein in regulating the abundance of histochemically reactive zinc in rat tissues. *Can. J. Physiol. Pharmacol.* **2003**, *81*, 815–824.
22. Suh, S. W.; Thompson, R. B.; Frederickson, C. J. Loss of vesicular zinc and appearance of perikaryal zinc after seizures induced by pilocarpine. *Neuroreport* **2001**, *12*, 1523–1525.
23. Larson, A. A.; Giovengo, S. L.; Shi, Q.; Velazquez, R. A.; Kovacs, K. J. Zinc in the extracellular area of the central nervous system is necessary for the development of kainic acid-induced persistent hyperalgesia in mice. *Pain* **2000**, *86*, 177–184.
24. Reyes, J. G.; Santander, M.; Martinez, P. L.; Arce, R.; Benos, D. J. A fluorescence method to determine picomole amounts of Zn(II) in biological systems. *Biol. Res.* **1994**, *27*, 49–56.
25. Jindal, R. M.; Taylor, R. P.; Gray, D. W. R.; Esmeraldo, R.; Morris, P. J. A new method for quantification of islets by measurement of zinc content. *Diabetes* **1992**, *41*, 1056–1062.
26. Savage, D. D.; Montano, C. Y.; Kasarskis, E. J. Quantitative histofluorescence of hippocampal mossy fiber zinc. *Brain Res.* **1989**, *496*, 257–267.
27. Frederickson, C. J.; Kasarskis, E. J.; Ringo, D.; Frederickson, R. E. A quinoline fluorescence method for visualizing and assaying the histochemically reactive zinc (bouton zinc) in the brain. *J. Neurosci. Methods* **1987**, *20*, 91–103.
28. Frederickson, C. J.; Costello, L. C.; Franklin, R. B. Zinc-based screening test and kit for early diagnosis of prostate cancer. U.S. Pat. Appl. Publ. US 2007207509, 2007; *Chem. Abstr.* **2007**, *147*, 317778.
29. Bush, A.; Masters, C. L. Method using PB-1033 and related compounds for the treatment of age-related macular degeneration (AMD). PCT Int. Appl. WO 2007118276, 2007; *Chem. Abstr.* **2007**, *147*, 480413.
30. Morgan, T. M.; Wilkins, N. F. Method and composition for treatment or prophylaxis of amyloidosis disorders. PCT Int. Appl. WO 2005049026, 2005; *Chem. Abstr.* **2005**, *143*, 20034.
31. Horrigan, S. K.; Zong, Q.; Soppet, D.; Castaneda, J.; Chen, B.; Cibotti, R.; Audoly, L. P.; Coyle, A.; Kiener, P. Compounds and methods for treating or preventing autoimmune diseases. PCT Int. Appl. WO 2008144011, 2008; *Chem. Abstr.* **2008**, *150*, 734.
32. Clements, J. B.; MacLean, A. R. Zinc finger motif sequences from herpes simplex virus protein IE63 and uses thereof in drug screening for treating herpes

virus infection. PCT Int. Appl. WO 2002004492, 2002; *Chem. Abstr*. **2002**, *136*, 112623.

33. Marin, P.; Israel, M.; Glowinski, J.; Premont, J. Routes of zinc entry in mouse cortical neurons: role in zinc-induced neurotoxicity. *Eur. J. Neurosci*. **2000**, *12*, 8–18.

34. Jindal, R. M.; Gray, D. W.; McShane, P.; Morris, P. J. Zinc-specific *N*-(6-methoxy-8-quinolyl)-*para*-toluenesulfonamide as a selective nontoxic fluorescence stain for pancreatic islets. *Biotech. Histochem*. **1993**, *68*, 196–205.

XTT

CAS Registry Number 111072-31-2

Chemical Structure

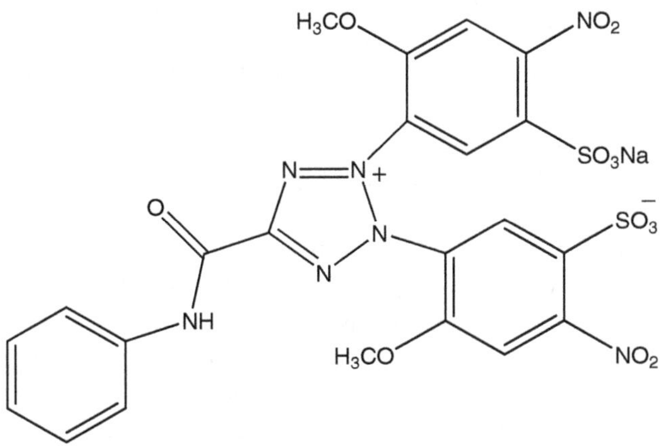

CA Index Name 2*H*-Tetrazolium, 2,3-bis(2-methoxy-4-nitro-5-sulfophenyl)-5-[(phenylamino)carbonyl]-, inner salt, sodium salt (1:1)

Other Names 2*H*-Tetrazolium, 2,3-bis(2-methoxy-4-nitro-5-sulfophenyl)-5-[(phenylamino)carbonyl]-, inner salt, monosodium salt; 2,3-Bis(2-methoxy-4-nitro-5-sulfophenyl)-2*H*-tetrazolium-5-carboxanilide inner salt; Sodium 3′-[1-(phenylaminocarbonyl)-3,4-tetrazolium]bis (4-methoxy-6-nitro)benzenesulfonic acid; XTT; XTT (tetrazolium derivative)

Merck Index Number Not listed

Chemical/Dye Class Tetrazolium salt

Molecular Formula $C_{22}H_{16}N_7NaO_{13}S_2$

Molecular Weight 673.52

Physical Form Yellow powder

Solubility Soluble in water, methanol, dimethyl sulfoxide

Melting Point 285 °C (decompose)

Absorption (λ_{max}) 286 nm

Synthesis Synthetic methods[1,2]

Staining Applications Cells[3]

Biological Applications Cell proliferation assays;[3,4] cytotoxicity assays;[5,6] ecotoxicity assays;[7] neurotoxicity assays;[8] antimicrobial susceptibility testing assays;[9,10] antifungal susceptibility testing assays;[11–15] cardiac valve viability assays;[16] cell viability assays;[17,18] bacterial viability assays;[19] fungal viability assays;[20] parasite viability assays;[21] BCG vaccines viability assays;[22] detecting DNA single strand breaks,[23] proteins;[24] generating and detecting reactive oxygen species;[25–28] treating cancer[29]

Industrial Applications Not reported

Safety/Toxicity Bacterial toxicity[30]

REFERENCES

1. Paull, K. D.; Shoemaker, R. H.; Boyd, M. R.; Parsons, J. L.; Risbood, P. A.; Barbera, W. A.; Sharma, M. N.; Baker, D. C.; Hand, E. The synthesis of XTT: a new tetrazolium reagent that is bioreducible to a water-soluble formazan. *J. Heterocycl. Chem.* **1988**, *25*, 911–914.
2. Shen, W. G.; Wang, M. L.; Gu, B. X. Synthesis of a water-soluble tetrazolium salt. *Huaxue Shiji* **2007**, *29*, 107–108; *Chem. Abstr.* **2007**, *148*, 239121.
3. Berridge, M. V.; Herst, P. M.; Tan, A. S. Tetrazolium dyes as tools in cell biology: new insights into their cellular reduction. *Biotechnol. Annu. Rev.* **2005**, *11*, 127–152.
4. Berridge, M. V.; Tan, A. S.; McCoy, K. D.; Wang, R. The biochemical and cellular basis of cell proliferation assays that use tetrazolium salts. *Biochemica* **1996**, 14–19.
5. Putnam, K. P.; Bombick, D. W.; Doolittle, D. J. Evaluation of eight *in vitro* assays for assessing the cytotoxicity of cigarette smoke condensate. *Toxicol. In Vitro* **2002**, *16*, 599–607.
6. Lamontagne, P; Maion, G; Page, M. Cytotoxicity testing using a soluble tetrazolium formazan derivative. *Cell. Pharmacol.* **1994**, *1*, 171–174.
7. Kunihiro, T.; Ichikawa, K.; Hu, H. Y.; Fujie, K. Development of ecotoxicity assay based on inhibition of respiring activity in microbial community using XTT reduction. *J. Gen. Appl. Microbiol.* **2004**, *50*, 91–96.

8. Becker, G. W.; Brems, D. N.; Chaney, M. O.; May, P. C.; Rydel, R. E.; Simmons, L. K.; Tomaselli, K. J. Antibodies to β-amyloid peptide and pharmaceutical screens for agents inhibiting neurotoxicity of β-amyloid peptide. Eur. Pat. Appl. EP 613007, 1994; *Chem. Abstr.* **1994**, *121*, 195902.
9. Brady, A. J.; Kearney, P.; Tunney, M. M. Comparative evaluation of 2,3-bis and 2-(2-methoxy-4-nitrophenyl)-3-(4-nitrophenyl)-5-(2,4-disulfophenyl)2*H*-tetrazolium, monosodium salt (WST-8) rapid colorimetric assays for antimicrobial susceptibility testing of staphylococci and ESBL-producing clinical isolates. *J. Microbiol. Methods* **2007**, *71*, 305–311.
10. Tunney, M. M.; Ramage, G.; Field, T. R.; Moriarty, T. F.; Storey, D. G. Rapid colorimetric assay for antimicrobial susceptibility testing of *Pseudomonas aeruginosa*. *Antimicrob. Agents Chemother.* **2004**, *48*, 1879–1881.
11. Antachopoulos, C.; Meletiadis, J.; Sein, T.; Roilides, E.; Walsh, T. J. Use of high inoculum for early metabolic signalling and rapid susceptibility testing of *Aspergillus* species. *J. Antimicrob. Chemother.* **2007**, *59*, 230–237.
12. Antachopoulos, C.; Meletiadis, J.; Roilides, E.; Sein, T.; Walsh, T. J. Rapid susceptibility testing of medically important zygomycetes by XTT assay. *J. Clin. Microbiol.* **2006**, *44*, 553–560.
13. Van de Sande, W. W. J.; Luijendijk, A.; Ahmed, A. O. A.; Bakker-Woudenberg, I. A. J. M.; Van Belkum, A. Testing of the *in vitro* susceptibilities of *Madurella mycetomatis* to six antifungal agents by using the Sensititre system in comparison with a viability-based 2,3-bis(2-methoxy-4-nitro-5-sulfophenyl)-5-[(phenylamino)carbonyl]-2*H*-tetrazolium hydroxide (XTT) assay and a modified NCCLS method. *Antimicrob. Agents Chemother.* **2005**, *49*, 1364–1368.
14. Hawser, S. P.; Jessup, C.; Vitullo, J.; Ghannoum, M. A. Utility of 2,3-bis(2-methoxy-4-nitro-5-sulfophenyl)-5-[(phenyl-amino)carbonyl]-2*H*-tetrazolium hydroxide (XTT) and minimum effective concentration assays in the determination of antifungal susceptibility of *Aspergillus fumigatus* to the lipopeptide class of compounds. *J. Clin. Microbiol.* **2001**, *39*, 2738–2741.
15. Tellier, R.; Krajden, M.; Grigoriew, G. A.; Campbell, I. Innovative endpoint determination system for antifungal susceptibility testing of yeasts. *Antimicrob. Agents Chemother.* **1992**, *36*, 1619–1625.
16. Lu, J. H.; Chiu, Y. T.; Sung, H. W.; Hwang, B.; Chong, C. K.; Chen, S. P.; Mao, S. J.; Yang, P. Z.; Chang, Y. XTT-colorimetric assay as a marker of viability in cryoprocessed cardiac valve. *J. Mol. Cell. Cardiol.* **1997**, *29*, 1189–1194.
17. Zeigler, F. C. Determination of cell viability using FACS. U.S. Pat. Appl. Publ. US 2006121443, 2006; *Chem. Abstr.* **2006**, *145*, 3836.
18. Knight, S. A. B.; Dancis, A. Reduction of 2,3-bis(2-methoxy-4-nitro-5-sulfophenyl)-2*H*-tetrazolium-5-carboxanilide inner salt (XTT) is dependent on CaFRE10 ferric reductase for *Candida albicans* grown in unbuffered media. *Microbiology* **2006**, *152*, 2301–2308.
19. Roslev, P.; King, G. M. Application of a tetrazolium salt with a water-soluble formazan as an indicator of viability in respiring bacteria. *Appl. Environ. Microbiol.* **1993**, *59*, 2891–2896.
20. Khot, P. D.; Suci, P. A.; Tyler, B. J. *Candida albicans* viability after exposure to amphotericin B: assessment using metabolic assays and colony forming units. *J. Microbiol. Methods* **2008**, *72*, 268–274.
21. Williams, C.; Espinosa, O. A.; Montenegro, H.; Cubilla, L.; Capson, T. L.; Ortega-Barria, E.; Romero, L. I. Hydrosoluble formazan XTT: its application to natural products drug discovery for *Leishmania*. *J. Microbiol. Methods* **2003**, *55*, 813–816.
22. Kairo, S. K.; Bedwell, J.; Tyler, P. C.; Carter, A.; Corbel, M. J. Development of a tetrazolium salt assay for rapid determination of viability of BCG vaccines. *Vaccine* **1999**, *17*, 2423–2428.
23. Nakamura, J.; Swenberg, J. A. Method of detecting DNA single strand breaks. U.S. Pat. Appl. Publ. US 2004132004, 2004; *Chem. Abstr.* **2004**, *141*, 85166.
24. Ukeda, H.; Tojo, R.; Shimamura, T.; Sawamura, M. Determination of amino group based on glutaraldehyde and XTT: its application to protein determination in food. *Food Sci. Technol. Res.* **2001**, *7*, 246–249.
25. Schopfer, P.; Heyno, E.; Drepper, F.; Krieger-Liszkay, A. Naphthoquinone-dependent generation of superoxide radicals by quinone reductase isolated from the plasma membrane of soybean. *Plant Physiol.* **2008**, *147*, 864–878.
26. Benov, L.; Fridovich, I. Is reduction of the sulfonated tetrazolium 2,3-bis (2-methoxy-4-nitro-5-sulfophenyl)-2-tetrazolium 5-carboxanilide a reliable measure of intracellular superoxide production? *Anal. Biochem.* **2002**, *310*, 186–190.
27. Ukeda, H. A method for measuring superoxide dismutase activity and reactive oxygen species. Jpn. Kokai Tokkyo Koho JP 11196896, 1999; *Chem. Abstr.* **1999**, *131*, 113119.
28. Able, A. J.; Guest, D. I.; Sutherland, M. W. Use of a new tetrazolium-based assay to study the production of superoxide radicals by tobacco cell cultures chal-

lenged with avirulent zoospores of *Phytophthora parasitica* var nicotianae. *Plant Physiol.* **1998**, *117*, 491–499.

29. Kong, Q. Method and composition for treating cancer using cellular organelle crystallizing agents. U.S. Patent 6376525, 2002; *Chem. Abstr.* **2002**, *136*, 319375.

30. Hatzinger, P. B.; Palmer, P.; Smith, R. L.; Penarrieta, C. T.; Yoshinari, T. Applicability of tetrazolium salts for the measurement of respiratory activity and viability of groundwater bacteria. *J. Microbiol. Methods* **2003**, *52*, 47–58.

YO-PRO 1

CAS Registry Number 152068-09-2

Chemical Structure

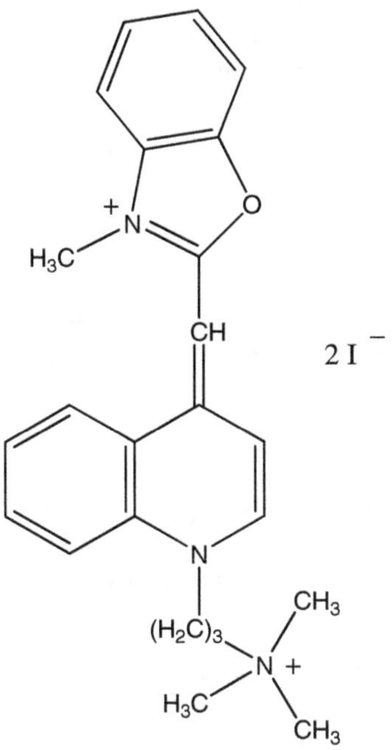

CA Index Name Quinolinium, 4-[(3-methyl-2(3*H*)-benzoxazolylidene)methyl]-1-[3-(trimethylammonio) propyl]-, iodide (1:2)

Other Names Quinolinium, 4-[(3-methyl-2(3*H*)-benzoxazolylidene)methyl]-1-[3-(trimethylammonio)propyl]-, diiodide; Oxazole yellow; YO-PRO 1; YO-PRO 1 iodide

Merck Index Number Not listed

Chemical/Dye Class Cyanine

Molecular Formula $C_{24}H_{29}I_2N_3O$

Molecular Weight 629.32

Physical Form Orange-red powder

Solubility Soluble in dimethyl sulfoxide, water

Melting Point >250 °C

Absorption (λ_{max}) 491 nm

Emission (λ_{max}) 509 nm

Synthesis Synthetic method[1]

Staining Applications Nucleic acids;[2–7] cells;[8,9] bacteria;[18] genes;[10] islets;[11] chromatin;[13] leukocytes;[22] marine prokaryotes;[12] nuclei;[22] micronuclei;[13] megakaryocyte;[14] microorganisms;[20,21] sperms;[15] viruses[23]

Biological Applications Nucleic acid hybridization;[16,17] nucleic acid sequencing;[2–7] detecting nucleic acids,[2–7] cells,[8,9] bacteria,[18] immunodeficiency virus,[19] microorganisms;[20,21] counting embryoblasts,[22] viruses;[23] apoptosis assay;[24] white blood cell functional assay[25]

Industrial Applications Not reported

Safety/Toxicity Cytotoxicity;[26] neurotoxicity;[27] vasotoxicity[28]

REFERENCES

1. Yue, S. T.; Johnson, I. D.; Huang, Z.; Haugland, R. P. Unsymmetrical cyanine dyes with a cationic side chain for enhanced fluorescence on binding with nucleic acids. U.S. Patent 5321130, 1994; *Chem. Abstr.* **1994**, *121*, 129393.
2. Exner, M.; Rogers, A. Methods for identifying nucleic acids and determining melting temperature using multiple fluorophores and FRET. U.S. Pat. Appl. Publ. US 2007172836, 2007; *Chem. Abstr.* **2007**, *147*, 182489.
3. Guillo, C.; Ferrance, J. P.; Landers, J. P. Use of a capillary electrophoresis instrument with laser-induced fluorescence detection for DNA quantitation. Comparison of YO-PRO-1 and PicoGreen assays. *J. Chromatogr., A* **2006**, *1113*, 239–243.
4. Al-Gubory, K. H. Fibered confocal fluorescence microscopy for imaging apoptotic DNA fragmentation at the single-cell level *in vivo*. *Exp. Cell Res.* **2005**, *310*, 474–481.
5. Deka, C.; Gordon, K. M.; Gupta, R.; Horton, A. Methods and compositions for rapid staining of nucleic acids in whole cells. U.S. Patent 6271035, 2001; *Chem. Abstr.* **2001**, *135*, 149591.
6. Suzuki, T.; Fujikura, K.; Higashiyama, T.; Takata, K. DNA staining for fluorescence and laser confocal microscopy. *J. Histochem. Cytochem.* **1997**, *45*, 49–53.

Handbook of Biological Dyes and Stains By R. W. Sabnis

7. Hoser, M. J. Methods for nucleic acid sequencing. PCT Int. Appl. WO 2004074503, 2004; *Chem. Abstr.* **2004**, *141*, 237698.
8. Anderson, A. L.; Knutson, C. R.; Mueth, D.; Plewa, J.; Tanner, E. Methods for staining cells for identification and sorting. U.S. Pat. Appl. Publ. US 2006172315, 2006; *Chem. Abstr.* **2006**, *145*, 183714.
9. Millard, P. J.; Roth, B. L.; Yue, S. T.; Haugland, R. P. Fluorescent viability assay using cyclic-substituted unsymmetrical cyanine dyes. U.S. Patent 5534416, 1996; *Chem. Abstr.* **1996**, *125*, 162751.
10. Stilwell, J. L.; Guan, Y.; Neve, R. M.; Gray, J. W. Systems biology in cancer research: genomics to cellulomics. *Methods Mol. Biol.* **2007**, *356*, 353–365.
11. Boffa, D. J.; Waka, J.; Thomas, D.; Suh, S.; Curran, K.; Sharma, V. K.; Besada, M.; Muthukumar, T.; Yang, H.; Suthanthiran, M.; Manova, K. Measurement of apoptosis of intact human islets by confocal optical sectioning and stereologic analysis of YO-PRO-1-stained islets. *Transplantation* **2005**, *79*, 842–845.
12. Marie, D.; Vaulot, D.; Partensky, F. Application of the novel nucleic acid dyes YOYO-1, YO-PRO-1, and PicoGreen for flow cytometric analysis of marine prokaryotes. *Appl. Environ. Microbiol.* **1996**, *62*, 1649–1655.
13. Dertinger, S. D.; Cairns, S. E.; Avlasevich, S. L.; Torous, D. K. Method for enumerating and differentiating between mammalian cell micronuclei by differentially staining micronuclei and the chromatin of dead, dying and viable cells. PCT Int. Appl. WO 2006007479, 2006; *Chem. Abstr.* **2006**, *144*, 124521.
14. Minakami, T.; Mori, Y.; Tsuji, T.; Ikeuchi, Y. Megakaryocyte classification/counting method by double fluorescent staining and flow cytometry. Jpn. Kokai Tokkyo Koho JP 2006275985, 2006; *Chem. Abstr.* **2006**, *145*, 413679.
15. Matsumoto, T.; Okada, H.; Hamaguchi, Y. Method and reagent for counting sperm by flow cytometry. Jpn. Kokai Tokkyo Koho JP 2001242168, 2001; *Chem. Abstr.* **2001**, *135*, 207869.
16. Erikson, G. H.; Daksis, J. I. Improving the signal/noise ratio of nucleic acid hybridization assays by preincubation of primer and target with nucleic acid binding agents. U.S. Pat. Appl. Publ. US 2004180345, 2004; *Chem. Abstr.* **2004**, *141*, 255469.
17. Hahn, J. H.; Park, N. Electrochemical device and methods for detection method of nucleic acid hybridization. PCT Int. Appl. WO 2003010338, 2003; *Chem. Abstr.* **2003**, *138*, 149875.
18. Stopa, P. J.; Mastromanolis, S. A. The use of blue-excitable nucleic-acid dyes for the detection of bacteria in well water using a simple field fluorometer and a flow cytometer. *J. Microbiol. Methods* **2001**, *45*, 143–153.
19. Yoshihara, N.; Suzuki, S.; Nakamura, T.; Tanaka, T. Method for detecting immunodeficiency virus. Jpn. Kokai Tokkyo Koho JP 2002085099, 2002; *Chem. Abstr.* **2002**, *136*, 259591.
20. Besson, F. I.; Hermet, J. P.; Ribault, S. Reaction medium and process for universal detection of microorganisms. Fr. Demande FR 2847589, 2004; *Chem. Abstr.* **2004**, *140*, 420343.
21. Sunamura, T.; Maruyama, A.; Kurane, R. Method for detecting and counting microorganism. Jpn. Kokai Tokkyo Koho JP 2002291499, 2002; *Chem. Abstr.* **2002**, *137*, 275332.
22. Heuven, B.; Wong, F. S.; Tsuji, T.; Sakata, T.; Hamaguchi, I. Method for classifying and counting erythroblasts by flow cytometry. Jpn. Kokai Tokkyo Koho JP 11326323, 1999; *Chem. Abstr.* **1999**, *132*, 1804.
23. Bettarel, Y.; Sime-Ngando, T.; Amblard, C.; Laveran, H. A comparison of methods for counting viruses in aquatic systems. *Appl. Environ. Microbiol.* **2000**, *66*, 2283–2289.
24. Idziorek, T.; Estaquier, J.; De Bels, F.; Ameisen, J. C. YOPRO-1 permits cytofluorometric analysis of programmed cell death (apoptosis) without interfering with cell viability. *J. Immunol. Methods* **1995**, *185*, 249–258.
25. Denlinger, L. C.; Hogan, K. J.; Bertics, P. J.; Schell, K. White blood cell functional assay. U.S. Pat. Appl. Publ. US 2004253650, 2004; *Chem. Abstr.* **2004**, *142*, 36899.
26. Reilly, T. P.; MacArthur, R. D.; Farrough, M. J.; Crane, L. R.; Woster, P. M.; Svensson, C. K. Is hydroxylamine-induced cytotoxicity a valid marker for hypersensitivity reactions to sulfamethoxazole in human immunodeficiency virus-infected individuals? *J. Pharmacol. Exp. Ther.* **1999**, *291*, 1356–1364.
27. Shimazawa, M.; Yamashima, T.; Agarwal, N.; Hara, H. Neuroprotective effects of minocycline against *in vitro* and *in vivo* retinal ganglion cell damage. *Brain Res.* **2005**, *1053*, 185–194.
28. Liao, S. D.; Puro, D. G. NAD^+-induced vasotoxicity in the pericyte-containing microvasculature of the rat retina: effect of diabetes. *Invest. Ophthalmol. Vis. Sci.* **2006**, *47*, 5032–5038.

YO-PRO 3

CAS Registry Number 157199-62-7

Chemical Structure

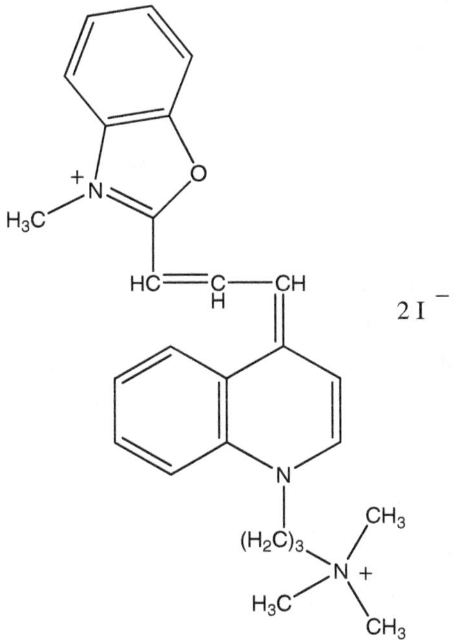

CA Index Name Quinolinium, 4-[3-(3-methyl-2(3*H*)-benzoxazolylidene)-1-propenyl]-1-[3-(trimethylammonio)propyl]-, diiodide

Other Names YO-PRO 3; YO-PRO 3 iodide

Merck Index Number Not listed

Chemical/Dye Class Cyanine

Molecular Formula $C_{26}H_{31}I_2N_3O$

Molecular Weight 655.36

Physical Form Orange-red powder

Solubility Soluble in dimethyl sulfoxide

Melting Point >250 °C

Absorption (λ_{max}) 612 nm

Emission (λ_{max}) 631 nm

Synthesis Synthetic method[1]

Staining Applications Nucleic acids;[2–4] cells[5,6]

Biological Applications Nucleic acid hybridization;[7,8] detecting nucleic acids,[2–4] cells;[5,6] monitoring cell cycle kinetics[9]

Industrial Applications Not reported

Safety/Toxicity No data available

REFERENCES

1. Yue, S. T.; Johnson, I. D.; Huang, Z.; Haugland, R. P. Unsymmetrical cyanine dyes with a cationic side chain for enhanced fluorescence on binding with nucleic acids. U.S. Patent 5321130, 1994; *Chem. Abstr.* **1994**, *121*, 129393.
2. Exner, M.; Rogers, A. Methods for identifying nucleic acids and determining melting temperature using multiple fluorophores and FRET. U.S. Pat. Appl. Publ. US 2007172836, 2007; *Chem. Abstr.* **2007**, *147*, 182489.
3. Miller, B. L.; Krauss, T. D.; Du, H.; Crnkovich, N.; Strohsahl, C. M. Use of sensor arrays containing hairpin probes for detecting nucleic acids of pathogens. PCT Int. Appl. WO 2004061127, 2004; *Chem. Abstr.* **2004**, *141*, 118285.
4. Erikson, G. H.; Daksis, J. I.; Kandic, I.; Picard, P. Conditions for formation of three- and four-stranded nucleic acid complexes and their detection and analytical use. PCT Int. Appl. WO 2002103051, 2002; *Chem. Abstr.* **2002**, *138*, 50812.
5. Anderson, A. L.; Knutson, C. R.; Mueth, D.; Plewa, J.; Tanner, E. Methods for staining cells for identification and sorting. U.S. Pat. Appl. Publ. US 2006172315, 2006; *Chem. Abstr.* **2006**, *145*, 183714.
6. Millard, P. J.; Roth, B. L.; Yue, S. T.; Haugland, R. P. Fluorescent viability assay using cyclic-substituted unsymmetrical cyanine dyes. U.S. Patent 5534416, 1996; *Chem. Abstr.* **1996**, *125*, 162751.
7. Erikson, G. H.; Daksis, J. I. Improving the signal/noise ratio of nucleic acid hybridization assays by preincubation of primer and target with nucleic acid binding agents. U.S. Pat. Appl. Publ. US 2004180345, 2004; *Chem. Abstr.* **2004**, *141*, 255469.
8. Hahn, J. H.; Park, N. Electrochemical device and methods for detection method of nucleic acid hybridization. PCT Int. Appl. WO 2003010338, 2003; *Chem. Abstr.* **2003**, *138*, 149875.
9. Beisker, W.; Weller-Mewe, E. M.; Nusse, M. Fluorescence enhancement of DNA-bound TO-PRO-3 by incorporation of bromodeoxyuridine to monitor cell cycle kinetics. *Cytometry* **1999**, *37*, 221–229.

YOYO 1

CAS Registry Number 143413-85-8

Chemical Structure

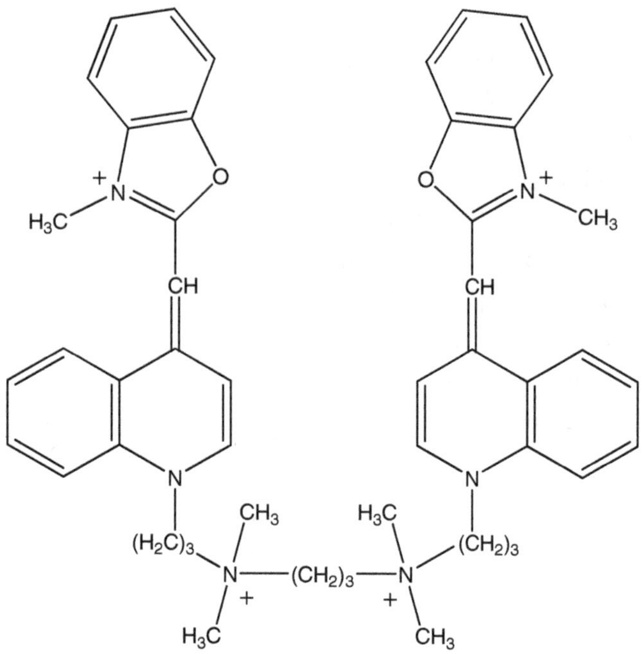

4 I$^-$

CA Index Name Quinolinium, 1,1′-[1,3-propanediylbis[(dimethyliminio)-3,1-propanediyl]]bis[4-[(3-methyl-2(3*H*)-benzoxazolylidene)methyl]-, iodide (1:4)

Other Names Quinolinium, 1,1′-[1,3-propanediylbis[(dimethyliminio)-3,1-propanediyl]]bis[4-[(3-methyl-2(3*H*)-benzoxazolylidene)methyl]-, tetraiodide; YOYO 1; YOYO 1 iodide

Merck Index Number Not listed

Chemical/Dye Class Cyanine

Molecular Formula $C_{49}H_{58}I_4N_6O_2$

Molecular Weight 1270.65

Physical Form Orange-red powder

Solubility Soluble in dimethyl sulfoxide, water

Melting Point >250 °C

Absorption (λ_{max}) 491 nm

Emission (λ_{max}) 509 nm

Synthesis Synthetic methods[1,2]

Staining Applications Nucleic acids;[3–8] cells;[9,10] antibodies;[15] genes;[11] chromatin;[12] micronuclei;[12] megakaryocyte;[13] microorganisms;[14] peptides;[15] proteins;[15] sperms[16]

Biological Applications Nucleic acid hybridization;[17,18] nucleic acid sequencing;[7] detecting nucleic acids,[3–8] cells,[9,10,23] human papilloma virus (HPV),[19] pathogens,[20] spores,[21] cancer cells,[21] stress biomarkers,[22] viruses[23,24]

Industrial Applications Not reported

Safety/Toxicity Double-strand breaks in reconstituted chromatin[25]

REFERENCES

1. Roth, B. L.; Millard, P. J.; Yue, S. T.; Wells, K. S.; Haugland, R. P. Fluorescent assay for bacterial gram reaction. U.S. Patent 5545535, 1996; *Chem. Abstr.* **1996**, *125*, 216357.
2. Rye, H. S.; Yue, S.; Wemmer, D. E.; Quesada, M. A.; Haugland, R. P.; Mathies, R. A.; Glazer, A. N. Stable fluorescent complexes of double-stranded DNA with bis-intercalating asymmetric cyanine dyes: properties and applications. *Nucleic Acids Res.* **1992**, *20*, 2803–2812.
3. Exner, M.; Rogers, A. Methods for identifying nucleic acids and determining melting temperature using multiple fluorophores and FRET. U.S. Pat. Appl. Publ. US 2007172836, 2007; *Chem. Abstr.* **2007**, *147*, 182489.
4. Miller, B. L.; Krauss, T. D.; Du, H.; Crnkovich, N.; Strohsahl, C. M. Use of sensor arrays containing hairpin probes for detecting nucleic acids of pathogens. PCT Int. Appl. WO 2004061127, 2004; *Chem. Abstr.* **2004**, *141*, 118285.
5. Kim, J. M.; Muramatsu, H.; Lee, H. Y.; Kawai, T. Near-field optical imaging of abasic sites on a single DNA molecule. *FEBS Lett.* **2003**, *555*, 611–615.
6. Zheng, J.; Yeung, E. S. Counting single DNA molecules in a capillary with radial focusing. *Aust. J. Chem.* **2003**, *56*, 149–153.
7. Williams, J. G. K.; Anderson, J. P. Field-switch single molecule DNA sequencing in a two-electrode chamber. PCT Int. Appl. WO 2005111240, 2005; *Chem. Abstr.* **2005**, *143*, 474548.

8. Nomura, Y.; Fuchigami, H.; Kii, H.; Feng, Z.; Nakamura, T.; Kinjo, M. Detection of oxidative stress-induced mitochondrial DNA damage using fluorescence correlation spectroscopy. *Anal. Biochem.* **2006**, *350*, 196–201.

9. Anderson, A. L.; Knutson, C. R.; Mueth, D.; Plewa, J.; Tanner, E. Methods for staining cells for identification and sorting. U.S. Pat. Appl. Publ. US 2006172315, 2006; *Chem. Abstr.* **2006**, *145*, 183714.

10. Lacombe, F.; Belloc, F.; Veriac, S.; Lefevre, D. Method of discrimination of at least two cellular populations by immunofluorescence flow cytometry and application to oncopathology. Fr. Demande FR 2895087, 2007; *Chem. Abstr.* **2007**, *147*, 93934.

11. Ferris, M. M.; Yoshida, T. M.; Marrone, B. L.; Keller, R. A. Fingerprinting of single viral genomes. *Anal. Biochem.* **2005**, *337*, 278–288.

12. Dertinger, S. D.; Cairns, S. E.; Avlasevich, S. L.; Torous, D. K. Method for enumerating and differentiating between mammalian cell micronuclei by differentially staining micronuclei and the chromatin of dead, dying and viable cells. PCT Int. Appl. WO 2006007479, 2006; *Chem. Abstr.* **2006**, *144*, 124521.

13. Minakami, T.; Mori, Y.; Tsuji, T.; Ikeuchi, Y. Megakaryocyte classification/counting method by double fluorescent staining and flow cytometry. Jpn. Kokai Tokkyo Koho JP 2006275985, 2006; *Chem. Abstr.* **2006**, *145*, 413679.

14. Noda, N.; Mizutani, T. Microorganism-measuring method using multiple staining. Jpn. Kokai Tokkyo Koho JP 2006340684, 2006; *Chem. Abstr.* **2006**, *146*, 77563.

15. Hahn, K. M.; Toutchkine, A.; Muthyala, R.; Kraynov, V.; Bark, S. J.; Burton, D. R.; Chamberlain, C. Labeled peptides, proteins and antibodies and processes and intermediates useful for their preparation. U.S. Pat. Appl. Publ. US 2002055133, 2002; *Chem. Abstr.* **2002**, *136*, 366139.

16. Duty, S. M.; Singh, N. P.; Ryan, L.; Chen, Z.; Lewis, C.; Huang, T.; Hauser, R. Reliability of the comet assay in cryopreserved human sperm. *Hum. Reprod.* **2002**, *17*, 1274–1280.

17. Erikson, G. H.; Daksis, J. I. Improving the signal/noise ratio of nucleic acid hybridization assays by preincubation of primer and target with nucleic acid binding agents. U.S. Pat. Appl. Publ. US 2004180345, 2004; *Chem. Abstr.* **2004**, *141*, 255469.

18. Erikson, G. H. Method for modifying transcription and/or translation in an organism by heteropolymeric probes and duplex, triplex or quadruplex hybridization for therapeutic, prophylactic and/or analytic uses. U.S. Pat. Appl. Publ. US 2003181412, 2003; *Chem. Abstr.* **2003**, *139*, 272000.

19. Poetter, K.; Gould, T. Human papilloma virus (HPV) detection using nucleic acid probes, microbeads, and fluorescence-activated cell sorter (FACS). PCT Int. Appl. WO 2006060872, 2006; *Chem. Abstr.* **2006**, *145*, 41223.

20. Vannier, E. Methods for detection of pathogens in red blood cells. PCT Int. Appl. WO 2006031544, 2006; *Chem. Abstr.* **2006**, *144*, 307966.

21. Levy, M.; Ellington, A. D.; Pai, S. Proximity ligation assays with peptide conjugate 'burrs' and aptamers for the sensitive detection of spores and cancer cells. PCT Int. Appl. WO 2007027796, 2007; *Chem. Abstr.* **2007**, *146*, 312239.

22. Cullum, M. E.; Duplessis, C. A.; Crepeau, L. J. Method for the detection of stress biomarkers including cortisol by fluorescence polarization. U. S. Pat. Appl. Publ US 2006105397, 2006; *Chem. Abstr.* **2006**, *144*, 463810.

23. Straus, D. Rapid and sensitive detection of cells and viruses. PCT Int. Appl. WO 2003073817, 2003; *Chem. Abstr.* **2003**, *139*, 210425.

24. Transfiguracion, J.; Kamen, A. Detection of intact recombinant viruses using fluorescent staining and chromatography. PCT Int. Appl. WO 2006053443, 2006; *Chem. Abstr.* **2006**, *144*, 463861.

25. Yoshikawa, Y.; Hizume, K.; Oda, Y.; Takeyasu, K.; Araki, S.; Yoshikawa, K. Protective effect of vitamin C against double-strand breaks in reconstituted chromatin visualized by single-molecule observation. *Biophys. J.* **2006**, *90*, 993–999.

YOYO 3

CAS Registry Number 156312-20-8

Chemical Structure

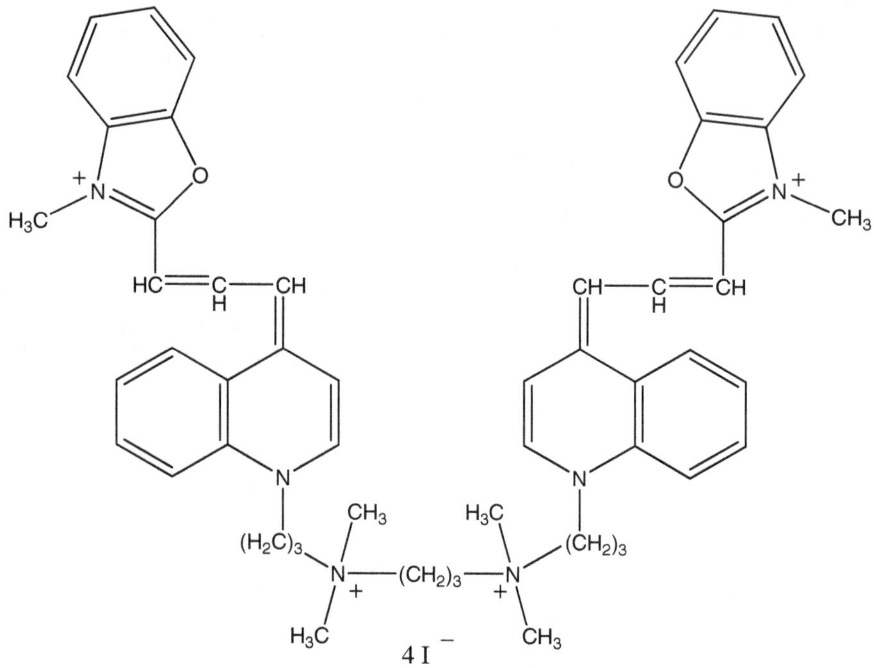

Molecular Weight 1322.73

Physical Form Orange-red powder

Solubility Soluble in dimethyl sulfoxide

Melting Point >250 °C

CA Index Name Quinolinium, 1,1′-[1,3-propanediylbis[(dimethyliminio)-3,1-propanediyl]]bis[4-[3-(3-methyl-2(3*H*)-benzoxazolylidene)-1-propen-1-yl]-, iodide (1:4)

Other Names Quinolinium, 1,1′-[1,3-propanediylbis [(dimethyliminio)-3,1-propanediyl]]bis[4-[3-(3-methyl-2 (3*H*)-benzoxazolylidene)-1-propenyl]-, tetraiodide; YO-YO 3; YOYO 3 iodide

Merck Index Number Not listed

Chemical/Dye Class Cyanine

Molecular Formula $C_{53}H_{62}I_4N_6O_2$

Absorption (λ_{max}) 612 nm

Emission (λ_{max}) 631 nm

Synthesis Synthetic method[1]

Staining Applications Nucleic acids;[2–9] cells;[10] leukocytes;[15] nuclei;[15] proteins;[9] sperms[11]

Biological Applications Nucleic acid hybridization;[12,13] detecting nucleic acids,[2–9] cells,[10] pathogens;[14] counting embryoblasts;[15] monitoring cell cycle kinetics[16]

Industrial Applications Analyzing polymers[17]

Safety/Toxicity No data available

REFERENCES

1. Millard, P. J.; Roth, B. L.; Yue, S. T.; Haugland, R. P. Fluorescent viability assay using cyclic-substituted unsymmetrical cyanine dyes. U.S. Patent 5534416, 1996; *Chem. Abstr.* **1996**, *125*, 162751.
2. Exner, M.; Rogers, A. Methods for identifying nucleic acids and determining melting temperature using multiple fluorophores and FRET. U.S. Pat. Appl. Publ. US 2007172836, 2007; *Chem. Abstr.* **2007**, *147*, 182489.
3. Wittwer, C. T.; Dujols, V. E.; Reed, G.; Zhou, L. Genotyping by thermal denaturation analysis using loss of fluorescence of double-stranded DNA-specific dyes as a reporter to monitor denaturation. PCT Int. Appl. WO 2004038038, 2004; *Chem. Abstr.* **2004**, *140*, 386988.
4. Kim, K. H.; Min, J. H.; Lee, I. H.; Kim, A. G. Method for highly sensitive nucleic acid detection using nanopore and non-specific nucleic acid-binding

agent. U.S. Pat. Appl. Publ. US 2006292605, 2006; *Chem. Abstr.* **2006**, *146*, 56513.

5. Miller, B. L.; Krauss, T. D.; Du, H.; Crnkovich, N.; Strohsahl, C. M. Use of sensor arrays containing hairpin probes for detecting nucleic acids of pathogens. PCT Int. Appl. WO 2004061127, 2004; *Chem. Abstr.* **2004**, *141*, 118285.
6. Erikson, G. H.; Daksis, J. I.; Kandic, I.; Picard, P. Conditions for formation of three- and four-stranded nucleic acid complexes and their detection and analytical use. PCT Int. Appl. WO 2002103051, 2002; *Chem. Abstr.* **2002**, *138*, 50812.
7. Li, J.; Lee, J. Y.; Yeung, E. S. Quantitative screening of single copies of human papilloma viral DNA without amplification. *Anal. Chem.* **2006**, *78*, 6490–6496.
8. Williams, J. G. K.; Anderson, J. P. Field-switch single molecule DNA sequencing in a two-electrode chamber. PCT Int. Appl. WO 2005111240, 2005; *Chem. Abstr.* **2005**, *143*, 474548.
9. Alba, F. J.; Bermudez, A.; Daban, J. R. Green-light transilluminator for the detection without photodamage of proteins and DNA labeled with different fluorescent dyes. *Electrophoresis* **2001**, *22*, 399–403.
10. Anderson, A. L.; Knutson, C. R.; Mueth, D.; Plewa, J.; Tanner, E. Methods for staining cells for identification and sorting. U.S. Pat. Appl. Publ. US 2006172315, 2006; *Chem. Abstr.* **2006**, *145*, 183714.
11. Matsumoto, T.; Okada, H.; Hamaguchi, Y. Method and reagent for counting sperm by flow cytometry. Jpn. Kokai Tokkyo Koho JP 2001242168, 2001; *Chem. Abstr.* **2001**, *135*, 207869.
12. Erikson, G. H.; Daksis, J. I. Improving the signal/noise ratio of nucleic acid hybridization assays by preincubation of primer and target with nucleic acid binding agents. U.S. Pat. Appl. Publ. US 2004180345, 2004; *Chem. Abstr.* **2004**, *141*, 255469.
13. Hahn, J. H.; Park, N. Electrochemical device and methods for detection method of nucleic acid hybridization. PCT Int. Appl. WO 2003010338, 2003; *Chem. Abstr.* **2003**, *138*, 149875.
14. Vannier, E. Methods for detection of pathogens in red blood cells. PCT Int. Appl. WO 2006031544, 2006; *Chem. Abstr.* **2006**, *144*, 307966.
15. Heuven, B.; Wong, F. S.; Tsuji, T.; Sakata, T.; Hamaguchi, I. Method for classifying and counting erythroblasts by flow cytometry. Jpn. Kokai Tokkyo Koho JP 11326323, 1999; *Chem. Abstr.* **1999**, *132*, 1804.
16. Beisker, W.; Weller-Mewe, E. M.; Nusse, M. Fluorescence enhancement of DNA-bound TO-PRO-3 by incorporation of bromodeoxyuridine to monitor cell cycle kinetics. *Cytometry* **1999**, *37*, 221–229.
17. Gilmanshin, R.; Chan, E. Y. Methods of analyzing polymers using a spatial network of fluorophores and fluorescence resonance energy transfer. U.S. Patent 6263286, 2001; *Chem. Abstr.* **2001**, *135*, 104686.

ZINQUIN

CAS Registry Number 151606-29-0

Chemical Structure

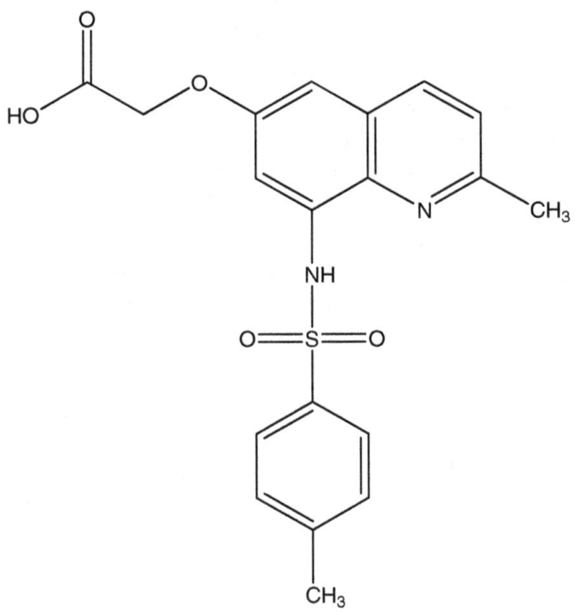

CA Index Name Acetic acid, 2-[[2-methyl-8-[[(4-methylphenyl)sulfonyl]amino]-6-quinolinyl]oxy]-

Other Names Acetic acid, [[2-methyl-8-[[(4-methylphenyl)sulfonyl]amino]-6-quinolinyl]oxy]-; Zinquin; Zinquin A; Zinquin acid

Merck Index Number 10169

Chemical/Dye Class Quinoline

Molecular Formula $C_{19}H_{18}N_2O_5S$

Molecular Weight 386.42

Physical Form Off-white crystals

Solubility Soluble in ethanol, methanol, dimethyl sulfoxide; slightly soluble in water

Melting Point 198–200 °C

Boiling Point (Calcd.) 608.0 ± 65.0 °C, pressure: 760 Torr

pK_a (Calcd.) 2.75 ± 0.40, most acidic, temperature: 25 °C; 3.88 ± 0.50, most basic, temperature: 25 °C

Absorption (λ_{max}) 368 nm

Emission (λ_{max}) 490 nm

Synthesis Synthetic methods[1–3]

Staining Applications Zinc ions[1–35]

Biological Applications Zinc indicator;[1–35] detecting apoptosis[12,13,17,21,25,32,33]

Industrial Applications Not reported

Safety/Toxicity Zinc toxicity[34]

REFERENCES

1. Kimber, M. C.; Mahadevan, I. B.; Lincoln, S. F.; Ward, A. D.; Betts, W. H. A preparative and spectroscopic study of fluorophores for zinc(II) detection. *Aust. J. Chem.* **2001**, *54*, 43–49.
2. Fahrni, C. J.; O'Halloran, T. V. Aqueous coordination chemistry of quinoline-based fluorescence probes for the biological chemistry of zinc. *J. Am. Chem. Soc.* **1999**, *121*, 11448–11458.
3. Mahadevan, I. B.; Kimber, M. C.; Lincoln, S. F.; Tiekink, E. R. T.; Ward, A. D.; Betts, W. H.; Forbes, I. J.; Zalewski, P. D. The synthesis of zinquin ester and zinquin acid, zinc(II)-specific fluorescing agents for use in the study of biological zinc(II). *Aust. J. Chem.* **1996**, *49*, 561–568.
4. Colvin, R. A.; Laskowski, M.; Fontaine, C. P. Zinquin identifies subcellular compartmentalization of zinc in cortical neurons. Relation to the trafficking of zinc and the mitochondrial compartment. *Brain Res.* **2006**, *1085*, 1–10.
5. Sarret, G.; Harada, E.; Choi, Y. E.; Isaure, M. P.; Geoffroy, N.; Fakra, S.; Marcus, M. A.; Birschwilks, M.; Clemens, S.; Manceau, A. Trichomes of tobacco excrete zinc as zinc-substituted calcium carbonate and other zinc-containing compounds. *Plant Physiol.* **2006**, *141*, 1021–1034.
6. Wilson, D.; Varigos, G.; Ackland, M. L. Apoptosis may underlie the pathology of zinc-deficient skin. *Immunol. Cell Biol.* **2006**, *84*, 28–37.
7. Zalewski, P.; Truong-Tran, A.; Lincoln, S.; Ward, D.; Shankar, A.; Coyle, P.; Jayaram, L.; Copley, A.; Grosser, D.; Murgia, C.; Lang, C.; Ruffin, R. Use of a zinc fluorophore to measure labile pools of zinc in body fluids and cell-conditioned media. *BioTechniques* **2006**, *40*, 509–520.

Handbook of Biological Dyes and Stains By R. W. Sabnis

8. Love, R.; Salazar, G.; Faundez, V. Neuronal zinc stores are modulated by non-steroidal anti-inflammatory drugs: an optical analysis in cultured hippocampal neurons. *Brain Res.* **2005**, *1061*, 1–12.
9. Huang, L.; Kirschke, C. P.; Zhang, Y.; Yu, Y. Y. The ZIP7 gene (Slc39a7) encodes a zinc transporter involved in zinc homeostasis of the Golgi apparatus. *J. Biol. Chem.* **2005**, *280*, 15456–15463.
10. Jansen, S.; Arning, J.; Dulcks, T.; Beyersmann, D. *S*-Nitroso compounds interfere with zinc probing by Zinquin. *Anal. Biochem.* **2004**, *332*, 145–152.
11. Devirgiliis, C.; Murgia, C.; Danscher, G.; Perozzi, G. Exchangeable zinc ions transiently accumulate in a vesicular compartment in the yeast *Saccharomyces cerevisiae*. *Biochem. Biophys. Res. Commun.* **2004**, *323*, 58–64.
12. Truong-Tran, A. Q.; Grosser, D.; Ruffin, R. E. Apoptosis in the normal and inflamed airway epithelium: role of zinc in epithelial protection and procaspase-3 regulation. *Biochem. Pharmacol.* **2003**, *66*, 1459–1468.
13. Kimura, E.; Aoki, S.; Kikuta, E.; Koike, T. A macrocyclic zinc(II) fluorophore as a detector of apoptosis. *Proc. Natl. Acad. Sci. U.S.A.* **2003**, *100*, 3731–3736.
14. Hendrickson, K. M.; Geue, J. P.; Wyness, O.; Lincoln, S. F.; Ward, A. D. Coordination and fluorescence of the intracellular Zn^{2+} probe [2-methyl-8-(4-toluenesulfonamido)-6-quinolyloxy]acetic acid (Zinquin A) in ternary Zn^{2+} complexes. *J. Am. Chem. Soc.* **2003**, *125*, 3889–3895.
15. Haase, H.; Beyersmann, D. Intracellular zinc distribution and transport in C6 rat glioma cells. *Biochem. Biophys. Res. Commun.* **2002**, *296*, 923–928.
16. Ranaldi, G.; Perozzi, G.; Truong-Tran, A.; Zalewski, P.; Murgia, C. Intracellular distribution of labile Zn (II) and zinc transporter expression in kidney and MDCK cells. *Am. J. Physiol.* **2002**, *283*, F1365–F1375.
17. Smith, P. J.; Wiltshire, M.; Davies, S.; Chin, S. F.; Campbell, A. K; Errington, R. J. DNA damage-induced $[Zn^{(2+)}](i)$ transients: correlation with cell cycle arrest and apoptosis in lymphoma cells. *Am. J. Physiol.* **2002**, *283*, C609–C622.
18. Michalczyk, A. A.; Allen, J.; Blomeley, R. C.; Ackland, M. L. Constitutive expression of hZnT4 zinc transporter in human breast epithelial cells. *Biochem. J.* **2002**, *364*, 105–113.
19. Colvin, R. A. pH dependence and compartmentalization of zinc transported across plasma membrane of rat cortical neurons. *Am. J. Physiol.* **2002**, *282*, C317–C329.
20. St. Croix, C. M.; Wasserloos, K. J.; Dineley, K. E.; Reynolds, I. J.; Levitan, E. S.; Pitt, B. R. Nitric oxide-induced changes in intracellular zinc homeostasis are mediated by metallothionein/thionein. *Am. J. Physiol.* **2002**, *282*, L185–L192.
21. Duffy, J. Y.; Miller, C. M.; Rutschilling, G. L.; Ridder, G. M.; Clegg, M. S.; Keen, C. L.; Daston, G. P. A decrease in intracellular zinc level precedes the detection of early indicators of apoptosis in HL-60 cells. *Apoptosis* **2001**, *6*, 161–172.
22. Snitsarev, V.; Budde, T.; Stricker, T. P.; Cox, J. M.; Krupa, D. J.; Geng, L.; Kay, A. R. Fluorescent detection of Zn^{2+}-rich vesicles with zinquin: mechanism of action in lipid environments. *Biophys. J.* **2001**, *80*, 1538–1546.
23. Qian, W. J.; Aspinwall, C. A.; Battiste, M. A.; Kennedy, R. T. Detection of secretion from single pancreatic β-cells using extracellular fluorogenic reactions and confocal fluorescence microscopy. *Anal. Chem.* **2000**, *72*, 711–717.
24. Kimber, M. C.; Mahadevan, I. B.; Lincoln, S. F.; Ward, A. D.; Tiekink, E. R. The synthesis and fluorescent properties of analogues of the zinc(II) specific fluorophore zinquin ester. *J. Org. Chem.* **2000**, *65*, 8204–8209.
25. Truong-Tran, A. Q.; Ruffin R. E.; Zalewski, P. D. Visualization of labile zinc and its role in apoptosis of primary airway epithelial cells and cell lines. *Am. J. Physiol.* **2000**, *279*, L1172–L1183.
26. Wellinghausen, N.; Martin, M.; Rink, L. Zinc inhibits interleukin-1-dependent T cell stimulation. *Eur. J. Immunol.* **1997**, *27*, 2529–2535.
27. Berendji, D.; Kolb-Bachofen, V.; Meyer, K. L.; Grapenthin, O.; Weber, H.; Wahn, V.; Kroncke, K. D. Nitric oxide mediates intracytoplasmic and intranuclear zinc release. *FEBS Lett.* **1997**, *405*, 37–41.
28. Zalewsk, P. D.; Jian, X.; Soon, L. L.; Breed, W. G.; Seamark, R. F.; Lincoln, S. F.; Ward, A. D.; Sun, F. Z. Changes in distribution of labile zinc in mouse spermatozoa during maturation in the epididymis assessed by the fluorophore Zinquin. *Reprod. Fertil., Dev.* **1996**, *8*, 1097–1105.
29. Coyle, P.; Zalewski, P. D.; Philcox, J. C.; Forbes, I. J.; Ward, A. D.; Lincoln, S. F.; Mahadevan, I.; Rofe, A. M. Measurement of zinc in hepatocytes by using a fluorescent probe, Zinquin: relationship to metallothionein and intracellular zinc. *Biochem. J.* **1994**, *303*, 781–786.

30. Kleineke, J. W.; Brand, I. A. Rapid changes in intracellular Zn^{2+} in rat hepatocytes. *J. Pharmacol. Toxicol. Methods* **1997**, *38*, 181–187.

31. Zalewski, P. D.; Millard, S. H.; Forbes, I. J.; Kapaniris, O.; Slavotinek, A.; Betts, W. H.; Ward, A. D.; Lincoln, S. F.; Mahadevan, I. Video image analysis of labile zinc in viable pancreatic islet cells using a specific fluorescent probe for zinc. *J. Histochem. Cytochem.* **1994**, *42*, 877–884.

32. Zalewski, P. D.; Forbes, I. J.; Seamark, R. F.; Borlinghaus, R.; Betts, W. H.; Lincoln, S. F.; Ward, A. D. Flux of intracellular labile zinc during apoptosis (gene-directed cell death) revealed by a specific chemical probe, Zinquin. *Chem. Biol.* **1994**, *1*, 153–161.

33. Zalewski, P. D.; Forbes, I. J.; Betts, W. H. Correlation of apoptosis with change in intracellular labile zinc (II) using zinquin [(2-methyl-8-*p*-toluenesulfonamido-6-quinolyloxy)acetic acid], a new specific fluorescent probe for zinc(II). *Biochem. J.* **1993**, *296*, 403–408.

34. Leung, K. W.; Liu, M.; Xu, X.; Seiler, M. J.; Barnstable, C. J.; Tombran-Tink, J. Expression of ZnT and ZIP zinc transporters in the human RPE and their regulation by neurotrophic factors. *Invest. Ophthalmol. Vis. Sci.* **2008**, *49*, 1221–1231.

35. Smith, P. J.; Wiltshire, M.; Furon, E.; Beattie, J. H.; Errington, R. J. Impact of overexpression of metallothionein-1 on cell cycle progression and zinc toxicity. *Am. J. Physiol.* **2008**, *295*, C1399–C1408.

ZINQUIN ESTER

CAS Registry Number 181530-09-6

Chemical Structure

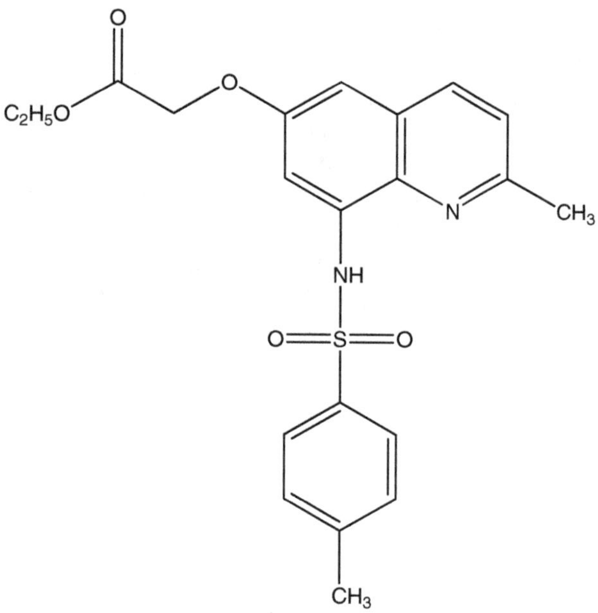

CA Index Name Acetic acid, 2-[[2-methyl-8-[[(4-methylphenyl)sulfonyl]amino]-6-quinolinyl]oxy]-, ethyl ester

Other Names Acetic acid, [[2-methyl-8-[[(4-methylphenyl)sulfonyl]amino]-6-quinolinyl]oxy]-, ethyl ester; 2-Methyl-8-[(4-methylphenyl)sulfonylamino]-6-(ethyloxycarbonylmethyloxy)quinoline; Zinquin E; Zinquin ester; Zinquin ethyl ester

Merck Index Number 10169

Chemical/Dye Class Quinoline

Molecular Formula $C_{21}H_{22}N_2O_5S$

Molecular Weight 414.47

Physical Form Colorless crystals

Solubility Soluble in ethanol, methanol, dimethyl sulfoxide; slightly soluble in water

Melting Point 111–113 °C

Boiling Point (Calcd.) 578.0 ± 60.0 °C, pressure: 760 Torr

pK_a (Calcd.) 7.38 ± 0.30, most acidic, temperature: 25 °C; 3.66 ± 0.50, most basic, temperature: 25 °C

Absorption (λ_{max}) 364 nm

Emission (λ_{max}) 385 nm

Synthesis Synthetic methods[1–3]

Staining Applications Zinc ions[1–4]

Biological Applications Zinc indicator;[1–4] detecting apoptosis[4]

Industrial Applications Not reported

Safety/Toxicity Zinc toxicity[5]

REFERENCES

1. Kimber, M. C.; Mahadevan, I. B.; Lincoln, S. F.; Ward, A. D.; Betts, W. H. A preparative and spectroscopic study of fluorophores for zinc(II) detection. *Aust. J. Chem.* **2001**, *54*, 43–49.
2. Fahrni, C. J.; O'Halloran, T. V. Aqueous coordination chemistry of quinoline-based fluorescence probes for the biological chemistry of zinc. *J. Am. Chem. Soc.* **1999**, *121*, 11448–11458.
3. Mahadevan, I. B.; Kimber, M. C.; Lincoln, S. F.; Tiekink, E. R. T.; Ward, A. D.; Betts, W. H.; Forbes, I. J.; Zalewski, P. D. The synthesis of zinquin ester and zinquin acid, zinc(II)-specific fluorescing agents for use in the study of biological zinc(II). *Aust. J. Chem.* **1996**, *49*, 561–568.
4. Kimura, E.; Aoki, S.; Kikuta, E.; Koike, T. A macrocyclic zinc(II) fluorophore as a detector of apoptosis. *Proc. Natl. Acad. Sci. U.S.A.* **2003**, *100*, 3731–3736.
5. Leung, K. W.; Liu, M.; Xu, X.; Seiler, M. J.; Barnstable, C. J.; Tombran-Tink, J. Expression of ZnT and ZIP zinc transporters in the human RPE and their regulation by neurotrophic factors. *Invest. Ophthalmol. Vis. Sci.* **2008**, *49*, 1221–1231.

APPENDIX A
INDEX OF CAS REGISTRY NUMBERS

CAS Registry Number	Dye	Page Number
4197-25-5	Sudan Black B	440
4548-53-2	Ponceau SX	375
5141-20-8	Light Green SF Yellowish	261
6359-05-3	Ethyl Eosin	193
7114-03-6	Methyl Green	300
7220-79-3	Methylene Blue Trihydrate	296
7423-31-6	Stains-All	436
8004-87-3	Methyl Violet 2B (Methyl Violet)	309
8005-03-6	Nigrosin	330
10114-58-6	Bismark Brown Y	46
10121-91-2	Dansyl Cadaverine	125
10510-54-0	Cresyl Violet Acetate	114
15391-59-0	Darrow Red	129
16423-68-0	Erythrosin	177
17372-87-1	Eosin Y	173
18472-87-2	Phloxine B	370
23491-45-4	Hoechst 33258	229
23491-52-3	Hoechst 33342	233
23555-00-2	Hoechst 34580	237
25535-16-4	Propidium Iodide	386
25956-17-6	Allura Red	16
28718-90-3	DAPI	127
28983-56-4	Aniline Blue (Methyl Blue)	24
33864-99-2	Alcian Blue 8GX	8
34373-76-7	MEQ	292
36536-22-8	DiIC$_1$(5)	156
38116-89-1	BSPT	64
41085-99-8	DiI	153
47623-98-3	DiSBAC$_2$(3)	167
50909-86-9	Coelenterazine h	100
51023-76-8	Stilbene Isothiocyanate Sulfonic Acid (SITS)	438
51811-82-6	Giemsa Stain	219
53213-81-3	DiOC$_5$(3)	160
53213-82-4	DiOC$_6$(3)	162
53213-83-5	DiOC$_7$(3)	166
53213-94-8	DiSC$_3$(5)	169
54375-47-2	Calcein Blue	78
55779-48-1	Coelenterazine	92
57576-49-5	Acridine Homodimer	3
58880-05-0	Ethidium Monoazide (EMA)	191
61389-30-8	Oxonol V	359

CAS Registry Number	Dye	Page Number
61926-22-5	Ethidium Homodimer-1 (EthD-1)	186
62669-70-9	Rhodamine 123	418
62758-13-8	Resazurin sodium salt	406
63560-89-4	DiBAC$_4$(5)	142
64724-75-0	Oxonol VI	361
67769-47-5	Lucifer Yellow CH	266
70363-83-6	DiBAC$_4$(3)	139
71231-14-6	Lucifer Yellow VS	269
71418-44-5	Monobromobimane (mBBr)	313
73630-23-6	Quin 2	400
74681-68-8	Nuclear Yellow	342
75168-11-5	Nonyl-Acridine Orange (NAO)	339
78338-22-4	Thionin	465
81029-05-2	Brilliant Cresyl Blue	60
83104-85-2	Quin 2 AM	402
83668-91-1	RH 237	408
83907-40-8	SPQ	433
85233-19-8	BAPTA	37
86701-10-2	NBD C$_6$-Ceramide	317
90134-00-2	Di-4-ANEPPS	135
90217-02-0	Cyanotolyl Tetrazolium Chloride (CTC)	120
104821-25-2	Dihydroethidium	146
107610-19-5	RH 421	412
108964-32-5	Fura 2 AM	215
109244-58-8	Dihydrorhodamine 123	151
109628-27-5	TSQ	488
111072-31-2	XTT	491
112926-02-0	Indo 1 AM	245
113694-64-7	Fura 2	213
115532-50-8	TMRM	469
115532-52-0	TMRE	467
121714-22-5	Fluo 3 AM	201
123437-16-1	Coelenterazine f	98
123437-22-9	Coelenterazine n	104
123437-25-2	Coelenterazine cp	96
123437-32-1	Coelenterazine hcp	102
123632-39-3	Fluo 3	198
124549-08-2	SBFI	428
124549-11-7	PBFI	366
124549-23-1	PBFI AM	368
126150-97-8	BAPTA AM	40

CAS Registry Number	Dye	Page Number
127274-91-3	DiD	144
129423-53-6	SBFI AM	431
130100-20-8	Mag-Fura 2 AM	278
130926-94-2	Mag-Indo 1 AM	283
132299-21-9	Mag-Indo 1	281
132319-56-3	Indo 1	243
132319-57-4	Mag-Fura 2 (Furaptra)	275
133867-53-5	BODIPY FL C_5-Ceramide	55
143413-84-7	TOTO 1	480
143413-85-8	YOYO 1	498
148504-34-1	Calcein AM	74
151606-29-0	Zinquin	503
152068-09-2	YO-PRO 1	495
153130-66-6	Calcium Green 5N	86
154324-80-8	BTC	66
154719-40-1	Calcium Green 1	81
154757-99-0	POPO 3	380
156312-20-8	YOYO 3	500
157134-53-7	Di-8-ANEPPS	137
157199-56-9	PO-PRO 1	382
157199-59-2	TO-PRO 1	475
157199-62-7	YO-PRO 3	497
157199-63-8	TO-PRO 3	477
160605-94-7	Di-2-ANEPEQ	134
161016-55-3	PO-PRO 3	384
161433-30-3	RH 414	410
162112-35-8	FM 4-64	211
162558-52-3	MQAE	316
166196-17-4	TOTO 3	483
168482-84-6	Calcein Blue AM	80
169454-13-1	BOBO 1	51
169454-15-3	POPO 1	378
169454-17-5	BOBO 3	53
170516-40-2	Calcium Green 2	84

CAS Registry Number	Dye	Page Number
170516-41-3	Magnesium Green	284
172807-13-5	RH 795	414
176767-94-5	BTC AM	68
177027-61-1	TO-PRO 5	479
180389-01-9	Ethidium Homodimer-2 (EthD-2)	189
181530-09-6	Zinquin Ester	506
192140-58-2	Fura 2FF	217
211566-66-4	Hexidium Iodide	227
214147-22-5	NBD methylhydrazine	320
217176-83-5	Dihydrorhodamine 6G	149
234075-34-4	Phen Green SK	369
254109-20-1	DAF FM	123
273221-59-3	Fluo 4	203
273221-67-3	Fluo 4 AM	205
288374-37-8	Newport Green DCF	327
296277-09-3	RedoxSensor Red CC-1	405
304014-12-8	QSY 7 Carboxylic Acid, Succinimidyl Ester	393
304014-13-9	QSY 21 Carboxylic Acid, Succinimidyl Ester	398
305801-86-9	JO-PRO 1	259
305801-87-0	JOJO 1	257
305802-06-6	LOLO 1	264
404335-95-1	FluoZin 3	209
411209-53-5	FluoZin 1	207
411209-54-6	FluoZin 2	208
522592-13-8	JC 9	256
481667-01-0	CoroNa Red	112
612502-05-3	Newport Green PDX	329
677716-65-3	RhodZin 3	421
690993-66-9	CoroNa Green	110
690993-67-0	CoroNa Green AM	111
700834-40-8	QSY 9 Carboxylic Acid, Succinimidyl Ester	396

APPENDIX B
INDEX OF BIOLOGICAL DYES/STAINS (CERTIFIED BY BIOLOGICAL STAIN COMMISSION)

Handbook of Biological Dyes and Stains By R. W. Sabnis

APPENDIX C
INDEX OF FD&C DYES (APPROVED BY FOOD & DRUG ADMINISTRATION)

Handbook of Biological Dyes and Stains By R. W. Sabnis

APPENDIX D
INDEX OF METAL INDICATORS

Handbook of Biological Dyes and Stains By R. W. Sabnis

Dye	Metal	Page Number
PBFI AM	Potassium	368
Phen Green SK	Copper; Iron	369
Quin 2	Calcium; Cadmium; Iron; Lanthanum; Zinc	400
Quin 2 AM	Calcium	402
RhodZin 3	Zinc	421
SBFI	Sodium	428
SBFI AM	Sodium	431
SPQ	Chloride; Nitrite	433
TSQ	Zinc	488
Zinquin	Zinc	503
Zinquin Ester	Zinc	506

APPENDIX E: INDEX OF NUCLEIC ACID STAINS

Handbook of Biological Dyes and Stains By R. W. Sabnis

APPENDIX F:
INDEX OF ORGANELLE PROBES

Handbook of Biological Dyes and Stains By R. W. Sabnis

APPENDIX G: INDEX OF pH INDICATORS

REFERENCE

1. Sabnis, R. W. *Handbook of Acid–Base Indicators*; CRC Press: Boca Raton, **2008**; pp 1–398.

Handbook of Biological Dyes and Stains By R. W. Sabnis

www.ingramcontent.com/pod-product-compliance
Lightning Source LLC
LaVergne TN
LVHW062136200926
848484LV00007B/1819